D0203906

$n_{\min}$ — minimum order
$n_{minQ}(F_p, F_s, K_p, K_s)$ — minimum order of minimal Q-factor design
p — normalized complex frequency
$q(k)$ — modular constant
$Q(s_i)$ — quality factor of pole/zero s_i
$Q(a, b)$ — quality factor of $H(z) = z^2/(z^2 + bz + a)$
$Q_{\min Q}(n, \xi, i)$ — quality factor of ith pole of minimal Q-factor design
$R(n, \xi, x)$ — elliptic rational function
s — complex frequency, Laplace operator (rad/s)
$\mathrm{sn}(u, k)$ — Jacobi elliptic sn function
$\mathrm{sn}^{-1}(v, k)$ — inverse Jacobi elliptic sn function
$S(n, \xi, \epsilon, i)$ — ith pole of (normalized) lowpass transfer function
S_A — attenuation-limit specification
S_G — gain-limit specification
S_K — characteristic-function-limit specification
S_M — magnitude-limit specification
S_r — magnitude-ripple specification
S_δ — magnitude-tolerance specification
$S_{minQ}(n, \xi, i)$ — ith pole of (normalized) lowpass transfer function for minimal Q-factor design
t — time (s)
x — dimensionless variable
$X(n, \xi, i)$ — ith zero of elliptic rational function
z — complex variable in the z-plane; $z = e^{j2\pi f}$ refers to the unit circle
Z — auxiliary complex variable in the z-plane
$Z_{BL}(s, f_p)$ — bilinear transformation
δ_1 — passband magnitude ripple
δ_2 — stopband magnitude ripple
δ_p — passband magnitude tolerance
δ_s — stopband magnitude tolerance
$\zeta(n, \xi, \epsilon)$ — auxiliary function
ϵ — ripple factor
ξ — selectivity factor
$\tau_{GD}(H(s), \omega)$ — group delay (s)
$\tau_{GD}(H(z), f)$ — group delay (in samples)
$\Phi(f)$ — phase response, $\Phi(f) = \arg(H(e^{j2\pi f}))$
$\Phi(\omega)$ — phase response, $\Phi(\omega) = \arg(H(j\omega))$
ω — angular frequency (rad/s), $\omega = 2\pi f$
θ — angular digital frequency, $\theta = 2\pi f$
$\lfloor x \rfloor$ — integer, $x \le \lfloor x \rfloor < x + 1$

$\underset{x_1<x<x_2}{\mathrm{FindRoot}}\{\, F(x) = G(x)\,\}$ find real x over interval $x_1 < x < x_2$ by solving $F(x) = G(x)$

$\underset{\substack{x_1<x<x_2\\ y_1<y<y_2}}{\mathrm{FindRoot}}\left\{\begin{matrix} F_1(x) = G_1(x) \\ F_2(x) = G_2(x)\end{matrix}\right\}$ find real x over interval $x_1 < x < x_2$ and real y over interval $y_1 < y < y_2$ by solving set of equations $\{F_1(x) = G_1(x),\ F_2(x) = G_2(x)\}$

We append a suffix to designate a quantity related to a specific filter type. For example, we add h to designate highpass filter; thus, A_{ph} is A_p of a highpass filter.

FILTER DESIGN FOR SIGNAL PROCESSING USING MATLAB® AND *MATHEMATICA*®

Miroslav D. Lutovac
The University of Belgrade
Belgrade, Yugoslavia

Dejan V. Tošić
The University of Belgrade
Belgrade, Yugoslavia

Brian L. Evans
The University of Texas at Austin
Austin, Texas

PERRY T. FORD LIBRARY
TRI-STATE UNIVERSITY
ANGOLA, IN 46703

PRENTICE HALL
Upper Saddle River, New Jersey 07458

Library of Congress Cataloging-in-Publication Data

Lutovac, Miroslav D.
Filter design for signal processing using Matlab and Mathematica/Miroslav D. Lutovac, Dejan V. Tosic, Brian L. Evans.
p. cm.
Includes bibliographical references and index.
ISBN: 0-201-36130-2
1. Electric filters--design and construction. 2. Matlab. 3. Mathematica (Computer file) 4. Signal processing. I. Tosic, Dejan V. II. Evans, Brian L. (Brian Lawrence). III. Title.

TK7872.F5 L88 2001
621.382'2--dc21 00-042798 CIP

Vice president and editorial director of ECS: **Marcia Horton**
Associate editor: **Alice Dworkin**
Production editor: **Irwin Zucker**
Executive managing editor: **Vince O'Brien**
Managing editor: **David A. George**
Manufacturing buyer: **Pat Brown**
Manufacturing manager: **Trudy Pisciotti**
Marketing manager: **Jenny Burger**
Copy editor: **Robert Golden**
Director of production and manufacturing, ESM: **David W. Riccardi**
Cover director: **Jayne Conte**
Editorial assistant: **Jessica Power**
Composition: **PreTEX, Inc.**

AXI3472
AXI3468
#44391540

© 2001 by Prentice Hall.
Prentice-Hall, Inc.
Upper Saddle River, New Jersey 07458

All rights reserved. No part of this book may be reproduced, in any form or by any means, without permission in writing from the publisher.

The author and publisher of this book have used their best efforts in preparing this book. These efforts include the development, research, and testing of the theories and programs to determine their effectiveness. The author and publisher make no warranty of any kind, expressed or implied, with regard to these programs or the documentation contained in this book. The author and publisher shall not be liable in any event for incidental or consequential damages in connection with, or arising out of, the furnishing, performance, or use of these programs.

MATLAB is a registered trademark of The MathWorks, Inc. *MATHEMATICA* is a registered trademark of Wolfram Research, Inc.

Printed in the United States of America

10 9 8 7 6 5 4 3 2 1

ISBN 0-201-36130-2

Prentice-Hall International (UK) Limited, *London*
Prentice-Hall of Australia Pty. Limited, *Sydney*
Prentice-Hall Canada Inc., *Toronto*
Prentice-Hall Hispanoamericana, S.A., *Mexico*
Prentice-Hall of India Private Limited, *New Delhi*
Prentice-Hall of Japan, Inc., *Tokyo*
Pearson Education Asia Pte. Ltd.
Editora Prentice-Hall do Brasil, Ltda., *Rio de Janeiro*

To my wife Anita and daughters Maja and Aleksandra
for their love, encouragement, and support
MDL

To my wife Ivana and daughter Nada
for their love, encouragement, and support
DVT

In loving memory of my dad, Harry Evans,
with love to my mom and sisters, and
with thanks to God who makes all things possible
BLE

CONTENTS

▶ 3 TRANSFORMS 64

FOREWORD

The basics of filter design theory have not changed for over half a century. Thus, irrespective of whether the prevailing or available technology favors a particular kind of filter realization; e.g., passive LCR, active RC, digital, or switched-capacitor, the initial steps from filter specifications to actual design are based on the fundamental work of such pioneers as O.J. Zobel, R.M. Foster, W. Cauer, O. Brune, S. Darlington, and many others. The resulting design theories, foremost among them insertion-loss theory, have resulted in an almost ritualistic procedure for the preliminaries of a filter design. This consists of transforming the given specifications; e.g., maximum-passband and minimum- stopband loss, transition-frequency band, impedance level, and so on, into either a transfer function in s or z (depending on whether the filter is to operate in continuous or discrete time), or into an LC filter structure. In doing so, the designer has the choice of filter type; e.g., Chebyshev, Butterworth, Elliptic, Bessel, and many more, this choice being determined by such factors as the filter order (which is generally related to the filter cost), group delay, inband ripple, band-edge selectivity, ease of tuning, and various other application-dependent requirements. Having made these decisions, and taken into account the given specifications, the filter designer then either consults a book of filter tables, or a corresponding computer program, and obtains the abovementioned transfer function or filter topology. At this point, the "preliminary design ritual" ends and the designer must either make some difficult choices in terms of available technology, or comply with the technological demands of the system to be designed; i.e., IC design, discrete-component active RC or LC design, digital signal processor (DSP), monolithic crystal, surface acoustic wave (SAW), mechanical, and so on.

This book on Advanced Filter Design by M.D. Lutovac, D.V. Tosic and B.L. Evans does away with what I have called the "preliminary design ritual," and opens up completely new vistas in basic filter design, regardless of the technology. The authors show that the conventional filter types (e.g., Butterworth, Chebyshev, Elliptic) are not unique solutions to a given set of specifications; much rather they are special cases of a *continuum of solutions*, all of which, while satisfying the specifications, permit tradeoffs to be made between a variety of optimizations that were considered inaccessible and unachievable in the past. Thus, where, for example, the poles and zeros of an elliptic filter, meeting certain gain and phase demands, were considered immutable in the past, this book can provide a new set of poles and zeros (with possibly an increased order by one or, rarely, two) which they call the "minimum Q" solution, whose poles lie not on an ellipse but on a semicircle (as with a Butterworth filter) and *whose dominant pole Q is significantly lower than that of the original elliptic filter*. Minimum pole Q, of course, implies lower sensitivity to component tolerances and, as a rule, lower thermal noise. However, this is only one of the many optimization options that this remarkable book supplies. It can provide a variety of different solutions (still meeting specs, of course, and with barely an increase in filter order). Thus, for example, it can provide a solution for minimal deviation from linear phase, or for specified group delay while maintaining minimal filter order, or, in the case of discrete-time filters, for zero-phase, or for a multiplierless elliptic IIR filter structure. These are merely examples of the unprecedented versatility in filter design that this new and unique book supplies. Since the essence of the book is to loosen the rigidity, in terms of options and optimization, that the previous "preliminary design ritual" demanded, it can be used as a new versatile preliminary design routine for any kind of subsequent filter realization—be it, for example, LC, active RC, digital or switched-capacitor. Indeed, the authors demonstrate their new and versatile design technique in conjunction with all of these filter types, and many more.

So how do the authors manage the extraordinary accomplishment of liberating themselves from the old classical preliminary design ritual, and why has this not been done long ago? The answer is simple. Beside some exceptionally elegant and creative mathematical stratagems (e.g., accurate replacement of Jacobi elliptic functions by functions comprising polynomials, square roots, and logarithms), they utilize high-power computer programs and optimization routines that were not previously available. Foremost among these are optimization routines carried out with symbolic analysis by *Mathematica*, and the advance filter design software of MATLAB. The exceptional combination of these highly advanced, modern and sophisticated design programs, together with a remarkable mathematical and algorithmic acumen, and an indepth and profound understanding of classical and modern filtering and signal-processing techniques, has produced in this book a new, and, without exaggeration, revolutionary method of filter design, that will have a pivotal effect on the way filter design is carried out in the future. This does not detract one bit from the monumental achievements of the previous pioneers in network theory and filter design; it merely reflects the fact that the modern computing tools available today, when in the hands of the right experts in mathematics, algorithmics, and network theory, can change dramatically a discipline such as filter theory and design, which was previously considered complete, definitive, and immutable. It is fortunate that the right ingredients come together in the form of these three talented authors, combined with the best that modern computer technology and algorithmics can supply (i.e. MATLAB and *Mathematica*) to produce a book and computer-software for filter design that will change the field in a profound way that could not have been anticipated but a few years ago. With the appearance of this book, filter design may never be the same again.

George S. Moschytz

Zurich

PREFACE

Analog and digital filter design is of great importance throughout engineering, applied mathematics, and computer science. Filters are the staple for designers in the controls, signal processing, and communications fields. They are commonly used in a wide variety of systems, such as chemical processing plants, instrumentation, suspension systems, modems, and digital cellular phones.

When a designer uses conventional techniques and software to design a filter, the designer receives only one possible filter that meets a set of specifications, yet an infinite number of designs may exist. This book develops alternative techniques and software to produce a comprehensive set of designs that meet the specification and represent the infinite design space. Included in the set of designs are filters that have minimal order, minimal quality factors, minimal complexity, minimal sensitivity to pole-zero locations, minimal deviation from a specified group delay, approximate linear phase, and minimized peak overshoot. For digital filters, the design space also includes filters with power-of-two coefficients. These alternative filter designs are crucial when evaluating filters for synthesis in analog circuits, digital hardware, or software.

This book overcomes the gap between filter theory and practice, and it presents new algorithms and designs developed over the last five years. The book includes ready-to-use filter design algorithms and implementations of the algorithms in both MATLAB and *Mathematica*. In order to make the material accessible to both the practitioner and the researcher, we have divided the book into two parts. Part I reviews conventional filter design techniques, presents several new ready-to-use algorithms, and discusses many case studies. The case studies present filters that cannot be designed with conventional techniques but can be designed with advanced methods. Part II discusses the theory underlying the new advanced design methods. The book also contains appendices to show examples of using advanced filter design software in MATLAB and *Mathematica*, and it includes filter design problems for the reader to solve.

In designing analog and digital IIR filters, one generally relies on canned software routines or mechanical procedures that rely on extensive tables. The primary reason for this "black box" approach is that the approximation theory that underlies filter design includes complex mathematics. Unfortunately, the conventional approach returns only one design, and it hides a wealth of alternative filter designs that are more robust when implemented in analog circuits, digital hardware, and software.

In this book, we provide advanced techniques to return multiple designs that meet the user specification. The key observations underlying our advanced filter design are as follows:

- Many designs satisfy the same user specification.
- Butterworth and Chebyshev IIR filters are special cases of elliptic IIR filters.
- Minimum-order filters may not be as efficient to implement as some higher-order filters.

Our approach is to search for a variety of design specifications that satisfies both the user specification and the limitations on the target implementation technology. Our algorithms return the following designs:

- Minimal filter order
  - Maximum stopband loss margin
  - Maximum passband loss margin
  - Minimum transition (conventional design)
  - Maximum transition
- Minimal maximum quality factor
- Minimal implementation cost
- Minimal deviation from specified group delay
- Minimal deviation from linear phase
- Elliptic IIR filters with power-of-two coefficients
- Zero-phase elliptic IIR filters
- Multiplierless elliptic half-band IIR filters
- Multiplierless Hilbert transformers
- Robust low-sensitivity sharp cutoff SC filters

For example, we can design selective elliptic IIR filters for microcontrollers and for other architectures that have fixed-point arithmetic and no hardware multiplier.

The theory underlying our advanced techniques is rooted in Jacobi elliptic functions which we use to approximate the magnitude frequency response of the filter. Jacobi elliptic functions are very complex transcendental functions. For many filter orders, however, we derive closed-form solutions to design elliptic filters that only use polynomials, square roots, and logarithms. This breakthrough allows us to derive precise relationships between the user specification, implementation constraints, and the pole-zero locations of the filter. Thus, *we have transformed the design space for IIR filters from elliptic function approximation theory into polynomial theory, which can be understood by designers with a knowledge of algebra.* In addition, final expressions are simple. Most of elliptic filters can be accurately designed ten to hundred times faster than using the classic approach.

The elliptic approximation is the most frequently used function in the design of IIR filters. In the latter part of this book, we explore many of the properties of elliptic functions such as its nesting property. These properties enable us to automate the design of filters using symbolic algebra. Transfer function poles and zeros are obtained by means of simple formulas, thereby freeing the designer from having to rely on extensive tables or canned computer programs. Symbolic design makes it possible to eliminate redundant variables, decrease the filter order, and simplify and approximate the underlying complex relations prior to the final numeric calculations.

The primary benefit of this book is convenient access to the latest advances in algorithms and software for analog and digital IIR filter design. These advanced techniques can design many types of filters that conventional techniques cannot design. A secondary benefit is a large collection of case studies for filter designs that require advanced techniques. Another benefit is a unique treatment of elliptic function filters.

The book is divided into 13 chapters.

Chapter 1 presents an overview of basic classes of continuous-time and discrete-time signals. We discuss mathematical representations of signals, and introduce the two computer environments, MATLAB and *Mathematica*, which we use to analyze and process signals.

In Chapter 2, we introduce fundamentals of linear system theory and define basic system properties. We present basic definitions and background mathematics that are used in this book. Since many readers are already familiar with this material, our aim is to be logically consistent rather than mathematically rigorous.

In Chapter 3, we review the definition and the salient properties of the most important transforms required by the filter design studied in this book. We focus on the phasor transformation, Fourier series and harmonic analysis, Fourier transformation, Laplace transformation, discrete Fourier transform, and z-transform. Step-by-step procedures for analyzing LTI systems in the transform domain are given.

Chapter 4 is intended to review the basics of classic analog filter design. Classification, salient properties and sensitivity of transfer functions are given. The most important analog filter realizations are presented. A detailed case study is given for realization of various transfer functions.

Chapter 5 reviews basic definitions of analog filter design. It introduces straightforward procedures to map the filter specification into a design space. We search this design space for the optimum solution according to given criteria. We conclude this chapter by an application example in which we design a robust selective analog filter based on commercially available integrated circuits.

In Chapter 6, we present (1) case studies of optimal analog filters that cannot be designed with classic techniques, and (2) the formal, mathematical framework that underlies their solutions. We present detailed step-by-step analog filter design algorithms.

Chapter 7 presents an extensible framework for designing analog filters that exhibit several desired behavioral properties after being realized in circuits. In the framework, we model the constrained non-linear optimization problem as a sequential quadratic programming problem. We derive the differentiable constraints and a weighted differentiable objective function for simultaneously optimizing the behavioral properties of magnitude response, phase response, and peak overshoot and the implementation property of quality factors.

Chapter 8 is intended to review the basics of classic digital IIR filter design. Classification, salient properties and sensitivity of transfer functions in the z-domain are given. The most important digital filter realizations are presented. For each realization we provide complete design equations and procedures that make the design easily applicable to a broad variety of digital filter design problems.

Chapter 9 reviews basic definitions of digital IIR filter design. It introduces straightforward procedures to map the filter specification into a design space. We search this design space for the optimum solution according to given criteria. We conclude this chapter with several important application examples in which we design low-sensitivity selective multiplierless IIR filters, power-of-two IIR filters, half-band IIR filters, 1/3-band filters, narrow-band IIR filters, Hilbert transformers, and zero-phase IIR filters. Each example design is followed by a comprehensive step-by-step procedure for computing the filter coefficients.

In Chapter 10, we present (1) case studies of optimal digital filters that cannot be designed with classic techniques, and (2) the formal, mathematical framework that underlies their solutions. We present detailed step-by-step digital filter design algorithms.

Chapter 11 presents an extensible framework for the simultaneous constrained optimization of multiple properties of digital IIR filters. The framework optimizes the pole-zero locations for behavioral properties of magnitude and phase response, and the implementation property of quality factors, subject to constraints on the same properties. We formulate the constrained nonlinear optimization problem as a sequential quadratic programming problem.

Chapter 12 introduces the basic Jacobi elliptic functions and reviews the most important relations between them. Several related theorems not found in standard textbooks are presented. Various useful approximation formulas are offered to facilitate the derivation of elliptic rational functions. A nesting property of the Jacobi elliptic functions is derived. In this chapter we present a novel approach to the design of elliptic filters in which we use exact closed-form expressions based on the nesting property.

In Chapter 13, we introduce the elliptic rational function as a natural generalization of the Chebyshev polynomial and we bypass mathematical theory of special functions required in the previous chapter. We prefer to give a reader an intuitive feel of the basic properties of the elliptic rational function. Our goal is to build the knowledge of the elliptic rational function using simple algebraic manipulations, even without mentioning the Jacobi elliptic functions.

Problems are included at the ends of chapters. The majority provide important practice with the concepts and techniques presented. Almost all the problems are suitable for solution using *Mathematica* and MATLAB. The book is supported by the following websites which contain the filter design software, *Mathematica* notebooks and MATLAB scripts:

http://galeb.etf.bg.ac.yu/~lutovac

http://iritel.iritel.bg.ac.yu/users/lutovac/www

There is also a companion website that accompanies this text:

http://www.prenhall.com/lutovac

How this book can be used in the classroom?

As the title indicates, the emphasis of this book is upon automating filter design in software (MATLAB and *Mathematica*), rather than upon studying the general filter theory.

Our overall approach to the topic has been guided by the fact that with the recent and anticipated developments in the technologies for filter design and implementation, the importance of having equal familiarity with computer-aided techniques suitable for analyzing and designing both continuous-time and discrete-time filters has increased dramatically.

We seek to leave with each reader (student, instructor, researcher, or practicing engineer) a set of software tools — *Mathematica* notebooks and MATLAB scripts — useful for solving filter design problems of practical importance.

A notable feature of the book is a detailed step-by-step exposure to the filter analysis by transform method, or in the time domain, exemplified by self-contained *Mathematica* notebooks. Students can use these notebooks to (a) automate symbolic filter analysis and design in software, (b) derive closed-form expressions for, say, transfer functions, and (c) gain insight into the relevant filter design parameters and coefficients.

This book was designed for educators who wish to integrate their curriculum with computer-based learning tools. Our goal is to provide an effective and efficient environment for students to learn the theory and problem-solving skills for contemporary filter design. To accomplish this we have used a computer-biased approach in which computer solutions and theory are viewed as mutually reinforcing rather than an either-or proposal.

We believe that students learn most effectively by solving problems following a worked-out problem as a model. Software scripts for running electronic examples (of worked-out problems) can capture the essence of a key concept, and encourage active participation in learning.

Filter analysis and design is a foundation subject for many students, due to its direct engineering applications, especially in electrical engineering. The concepts which it embodies, and the analytical techniques which it employs, are valid far outside the boundaries of electrical engineering.

The subject of filter design is an extraordinarily rich one, and a variety of approaches can be taken in structuring an introductory or advanced filter design course. This text provides a broad treatment of filter design and analysis, and contains sufficient material for a one-semester or two-semester course on the subject. Students using this book are assumed to have a basic background in calculus, complex numbers, and differential equations.

A typical one-semester introductory filter design course at a sophomore-junior level using this book could comprise the following: (a) Chapters 1–3, (b) Chapter 4, (c) a choice from Chapter 5 with the emphasis placed on specification and approximation problem, (d) Chapter 8, (e) a choice from Chapter 9 (digital specification and approximation problem). Combine the text with the MATLAB Signal Processing Toolbox and the *Mathematica* Signals and Systems Pack to illustrate the classic filter design procedures. Proceed lightly through our *Mathematica* Example Notebooks and our MATLAB filter design toolbox.

A one-semester introductory analog filter design course at a sophomore-junior level using this book could comprise the following: (a) Chapters 1–3, (b) Chapters 4, 5 and 7, (c) utilize the MATLAB Signal Processing Toolbox, and the *Mathematica* Signals and Systems Pack, to illustrate the classic analog filter design, (d) proceed through our *Mathematica* Example Notebooks, and our MATLAB analog filter design toolbox.

A one-semester introductory digital filter design course at a sophomore-junior level using this book could comprise the following: (a) Chapters 1–3, (b) Chapters 8, 9 and 11, (c) utilize the MATLAB Signal Processing Toolbox, and the *Mathematica* Signals and Systems Pack, to illustrate the classic digital filter design, (d) proceed through our *Mathematica* Example Notebooks, and our MATLAB digital filter design toolbox.

In addition to these course formats this book can be used as the basic text for a thorough, two-semester sequence on advanced filter design. The portions of the book not used in an introductory course on filter design, together with other sources, can form the basis for a senior elective course. Alternatively, for a two-semester course, we suggest coverage of the first 11 chapters, proceeding lightly through Chapters 12 and 13, and covering thoroughly Chapters 5 and 9 because they introduce the design space concept in filter design.

The book can serve as a text for a sequence of two one-semester courses on analog and digital filters for senior undergraduate or first-year graduate students. Such a course could comprise the following: (a) review of Chapters 1–3, (b) Chapters 4, 5, (c) brief discussion of analog-filter design algorithms (Chapter 6) with the emphasis placed on application rather than derivation, (d) Chapter 7 in full depth, (e) Chapters 8, 9, (f) brief discussion of digital-filter design algorithms (Chapter 10) with the emphasis placed on application rather than derivation, (g) Chapter 11 in full depth, (h) a choice from Chapters 12 and 13, depending on the course orientation desired.

The book's structure allows students who are interested in only analog filters to skip chapters on digital filters, without loss of continuity, and vice versa. It should be pointed out that not all sections in every chapter are covered in class. Also, various topics can be omitted at the discretion of the instructor. Depending upon the background the students can utilize chapters 1, 2, and 3 to review and expand their knowledge of linear system theory for continuous-time and discrete-time systems.

Advanced postgraduate courses (masters's programs and Ph.D. programs) can also be prepared from Chapters 12 and 13.

Selected topics chosen from the book chapters can be used in Electronics and Electric Circuit Theory courses, too.

We have included a collection of more that 70 *Mathematica* notebooks, numerous MATLAB scripts, and many end-of-chapter problems and exercises. This variety and quantity will hopefully provide instructors with considerable flexibility in putting together homework sets that are tailored to the specific needs of their students. Many filter realizations, both analog and digital, presented throughout the book can help lecturers organize versatile homeworks, projects and tests for the students. In addition, the filter design algorithms (Chapters 6 and 10) can be directly programmed in any language or environment such as Visual BASIC, Visual C, Maple, DERIVE, or MathCAD.

We thank Professor Ljiljana Milić for her valuable comments which have improved the book. We would like to thank Professor Marija Hribšek, Professor Antonije Djordjević and Professor Veljko Milutinović, for their encouragement throughout the period in which the book was written. The first author is grateful to general manager Siniša Davitkov for providing him with the opportunity to work on filter design. We would like to thank Professor George S. Moschytz for his encouragement and support while working on the writing this book.

The continuing encouragement, patience, technical support, and enthusiasm provided by Prentice Hall, and in particular by Alice Dworkin have been important in bringing this project to fruition.

We also want to thank the very thoughtful and careful reviewers, Professor Igor Tsukerman, The University of Akron, and Professor Michael J. Werter, The University of California at Los Angeles, for their useful comments.

Miroslav D. Lutovac is a chief scientist at the Institute for Research and Development in Telecommunications and Electronics (IRITEL) and is an Associate Professor in the School of Electrical and Computer Engineering, both of which are located at the University of Belgrade in Belgrade, Yugoslavia. His research interests include theory and implementation of active, passive, and digital networks and systems, filter approximation, symbolic analysis and synthesis of digital filters, and multiplierless digital IIR filter design. He has published over 100 papers in these fields. He received his B.Sc. (1981), M.Sc. (1985), and D.Sc. (1991) degrees in Electrical Engineering from the University of Belgrade in Belgrade, Yugoslavia. He has managed several national projects on multichip module design and voice delta coders. He teaches courses in electronics, computer-aided design, digital signal processing, and filter analysis and design.

Dejan V. Tošić is an Associate Professor in the School of Electrical and Computer Engineering at the University of Belgrade in Belgrade, Yugoslavia. His research interests include circuit theory and analysis, filter design and synthesis, neural networks, microwave circuits, and computer-aided design. He has published over 100 papers in these fields. He is currently concentrating his research efforts on creating a general framework for the symbolic analysis of linear circuits and systems, which is suitable for research, industrial, and educational applications. Using this framework, he is developing design automation tools for optimizing the design and synthesis of analog and digital filters. He received his B.Sc. (1980), M.Sc. (1986), and D.Sc. (1996) degrees in Electrical Engineering from the University of Belgrade in Belgrade, Yugoslavia. In 1992, he won the Teacher of the Year Award from the School of Electrical and Computer Engineering at the University of Belgrade. He teaches classes in circuit theory, microwave engineering, and digital image processing.

Brian L. Evans is an Associate Professor in the Department of Electrical and Computer Engineering at The University of Texas at Austin, and is the Director of the Embedded Signal Processing Laboratory, which is part of the Center for Telecommunications and Signal Processing and the Center for Vision and Image Sciences. His research interests include real-time embedded systems; signal, image and video processing systems; system-level design; electronic design automation; symbolic computation; and filter design. Dr. Evans has published over 75 refereed conference and journal papers in these fields. He developed and currently teaches Multidimensional Digital Signal Processing, Embedded Software Systems, Real-Time Digital Signal Processing Laboratory, and Linear Systems and Signals. His B.S.E.E.C.S. (1987) degree is from the Rose-Hulman Institute of Technology, and his M.S.E.E. (1988) and Ph.D.E.E. (1993) degrees are from the Georgia Institute of Technology. From 1993 to 1996, he was a postdoctoral researcher at the University of California at Berkeley with the Ptolemy Project. Ptolemy is a research project and software environment focused on design methodology for signal processing, communications, and controls systems. In addition to Ptolemy, he has played a key role in the development and release of six other computer-aided design frameworks, including the Signals and Systems Pack for Mathematica, which has been on the market since the Fall of 1995. He is an Associate Editor of the *IEEE Transactions on Image Processing*, a member of the Design and Implementation of Signal Processing System Systems Technical Committee of the IEEE Signal Processing Society, and a Senior Member of the IEEE. He is a recipient of a 1997 National Science Foundation CAREER Award.

Miroslav D. Lutovac

Dejan V. Tošić

Brian L. Evans

CHAPTER 1

SIGNALS

A *signal* is a physical quantity, or quality, which conveys information. For example, the voice of my friend is a signal which causes me to perform certain actions or react in a particular way. My friend's voice is called an *excitation*, and my action or reaction is called a *response*. The evaluation (conversion) from excitation to response is called *signal processing*.

Signals have many different origins. They may be measurements of physical phenomenon such as propagation of seismic waves, traffic noise levels, voltage variations in an electric circuit, and pressure variations in a hydraulic control system. They may be produced by a digital computer, as in text-to-speech synthesis, electronic music, computer graphics, and synthetic images.

A typical reason for signal processing is to eliminate or reduce an undesirable signal. The undesirable signal may be electrical noise in a radio signal or the power supply sinusoidal signal at a frequency of 60 Hz (or 50 Hz) in measurements of traffic noise levels. The traffic noise levels would be the desired signal for a noise detector, but the undesirable signal in a mobile speech communications system.

From a signal processing perspective, a signal must be capable of being observed. For the description and processing of signals, it is important to have a variety of representations of signals. Regardless of the origin of a signal (e.g., mechanical, electrical, acoustic, and biological), we generally convert the original signal into a form that is suitable for further processing and observation.

One fundamental representation of a signal is as a function of at least one independent variable. The variation of the signal value as a function of the independent variable is called a *waveform*. The independent variable often represents *time*. Other possibilities include spatial distance or an index of a data vector. For example, recorded data on a magnetic tape or compact disc (CD) may be a signal that is a function of time. Or, instead, it may be a function of distance (from the beginning of the tape as shown on the numeric display of the tape player) or of the CD track number and the position in the track of the digitized signal on the CD.

In this book, we define a *signal* as a function of one independent variable that contains information about the behavior or nature of a phenomenon. We assume that the independent variable is time even in cases where the independent variable is a physical quantity other than time. The rest of this chapter discusses mathematical representations of signals and introduces the two computer environments, MATLAB [1] and *Mathematica* [2], which we will use to analyze and process signals.

1.1 SIGNAL CLASSIFICATION

1.1.1 Continuous or Analog Signals

We define the *continuous signal*, $x(t)$, as a signal that exists at every instant of time, t. In the jargon of the trade, a continuous signal is often referred to as *continuous time* or *analog*. The independent variable t is a *continuous* variable. The function $x(t)$ can assume any value over a *continuous* range of numbers. For example, the functions

$$x(t) = e^{-t}$$

$$x(t) = \sin(100\pi t)$$

$$x(t) = 3 + 10t + 2|\sin(100\pi t)|$$

are continuous signals.

Signals can be represented by graphs as shown in Fig. 1.1.

Note that x is the symbol representing the signal, t is the symbol representing the independent variable, and $x(t)$ means the *value* of the signal x at time t as shown in Fig. 1.1a. Figure 1.1a plots a continuous sinusoid, $x(t) = \sin(2\pi t)$, which is a periodic signal. Figure 1.1b plots a continuous signal $x(t)$ which is zero for $t < 0$ and $t > 4$. Figure 1.1c plots a continuous signal which is periodic, as is the sinusoidal signal in Fig. 1.1a, but not sinusoidal. For the signals plotted in Figs. 1.1a–c, we can find the values for $t < 0$ as well as for $t > 3$. Figure 1.1d shows a signal that is not periodic. From the graph, we can estimate its values on the interval $-1 \leq t \leq 3$. We cannot calculate its values outside this interval without previous knowledge of the mathematical formula describing the signal.

In recent years, tremendous advances have occurred in programmable, configurable, and dedicated digital hardware as well as in digital technology. The most suitable signal to be processed by a digital computer or digital hardware is a *digital signal*, that is, a discrete-time signal whose values are represented by digits (e.g., by binary or decimal numbers). For many applications a digital signal is more convenient than an analog signal, especially when signals are transmitted and stored. For these reasons it is of interest to transmit, store, and analyze digital signals in such a way that we can restore the initial analog signal without loss or with minimum acceptable loss of information.

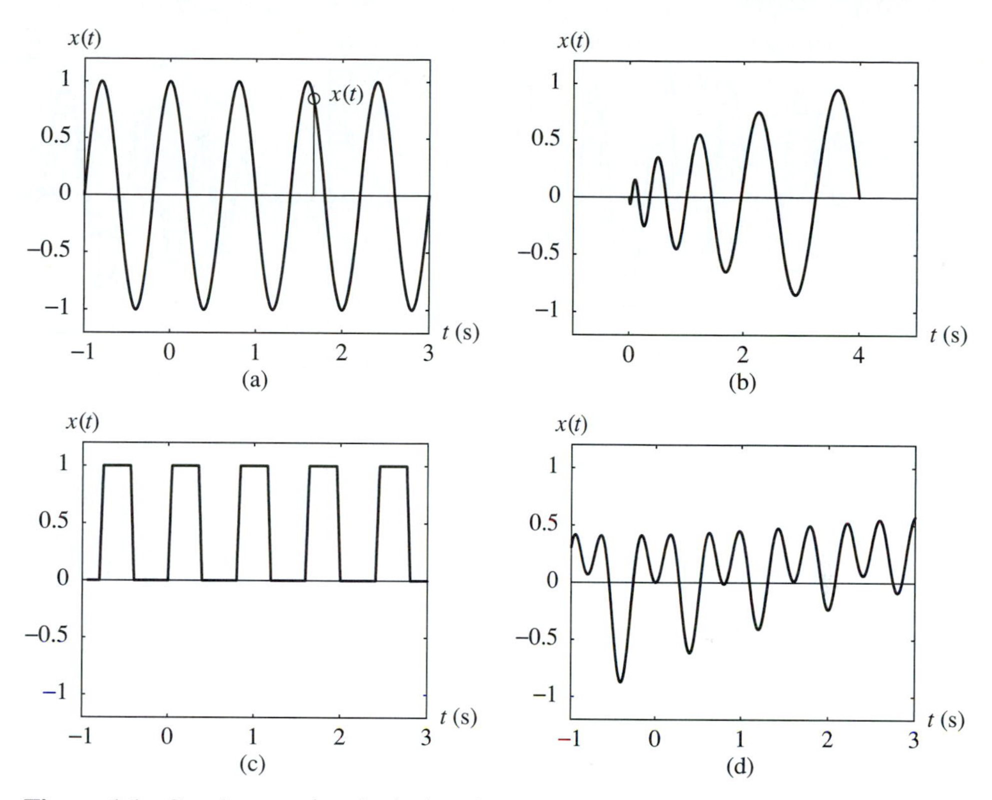

Figure 1.1 Continuous (analog) signals.

1.1.2 Discrete-Time Signals

A signal defined only for discrete values of time is called a *discrete-time signal* or simply a *discrete* signal. It may have been obtained by taking samples of an analog signal at discrete instants of time. In this book, we will only concern ourselves with uniform sampling by sampling every T units of time:

$$x(kT) = x(t)$$
$$t = 0, \pm T, \pm 2T, \pm 3T, \ldots$$

The discrete signal obtained from the continuous signal given in Fig. 1.1 is shown in Fig. 1.2.

The process of converting analog signals to digital signals takes a finite amount of time Δt at each sample, where $\Delta t \leq T$. Sampling produces a train of pulses of duration Δt, as shown in Fig. 1.3. Within the Δt time interval, the signal generally varies. If this interval is very short, as in Fig. 1.2, then the pulses are assumed to be rectangular and the pulse amplitude is uniquely defined by the instantaneous signal value. On the other hand, if Δt is long, the amplitudes of pulses are not uniquely defined and a measure for them must be specified. A typical measure is the signal value (a) at the beginning of, (b) in the middle of, (c) at the end of, or (d) the average signal value over the interval Δt.

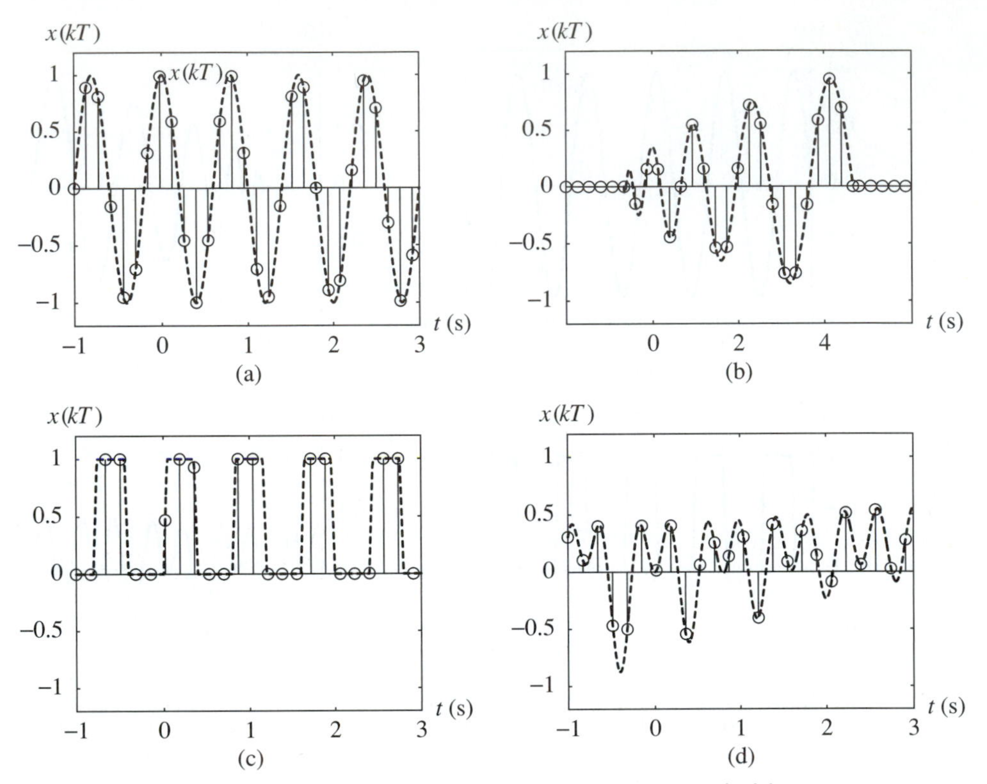

Figure 1.2 Discrete-time (discrete) signals, which are sampled but not quantized.

If the sampling process interval Δt is equal to the interval between successive samples T, and if we take the value of the signal at the beginning of Δt, the discrete-time signal looks like that in Fig. 1.4.

In this book, we will be using the MATLAB and *Mathematica* computer software environments for representing, analyzing, and manipulating signals. We used the MATLAB `plot` command to create the waveform plots in Fig. 1.1. The next three figures, Figs. 1.2, 1.3, and 1.4, were generated by the MATLAB `stem`, `bar` and `stairs` commands, respectively. We will discuss in more detail the use of MATLAB in Sections 1.5 and 1.6 and *Mathematica* in Sections 1.7 and 1.8.

All three discrete-time signals convey the same information: the values of the continuous signal $x(t)$ at regular intervals. The parameter T is called the *time step* or *sample interval.* A related quantity is the *sampling rate* or *sampling frequency* representing the number of samples per unit time

$$f_0 = \frac{1}{T}$$

When the units of T is seconds (s), the units of f_0 is Hertz (Hz).

We generally start observing a signal from an arbitrary point of time t_0. We will set $t_0 = 0$ and designate t to represent the time that has elapsed since t_0. We will uniformly sample continuous-time signals at $t = kT$, where $k = 0, \pm 1, \pm 2, \ldots$. We

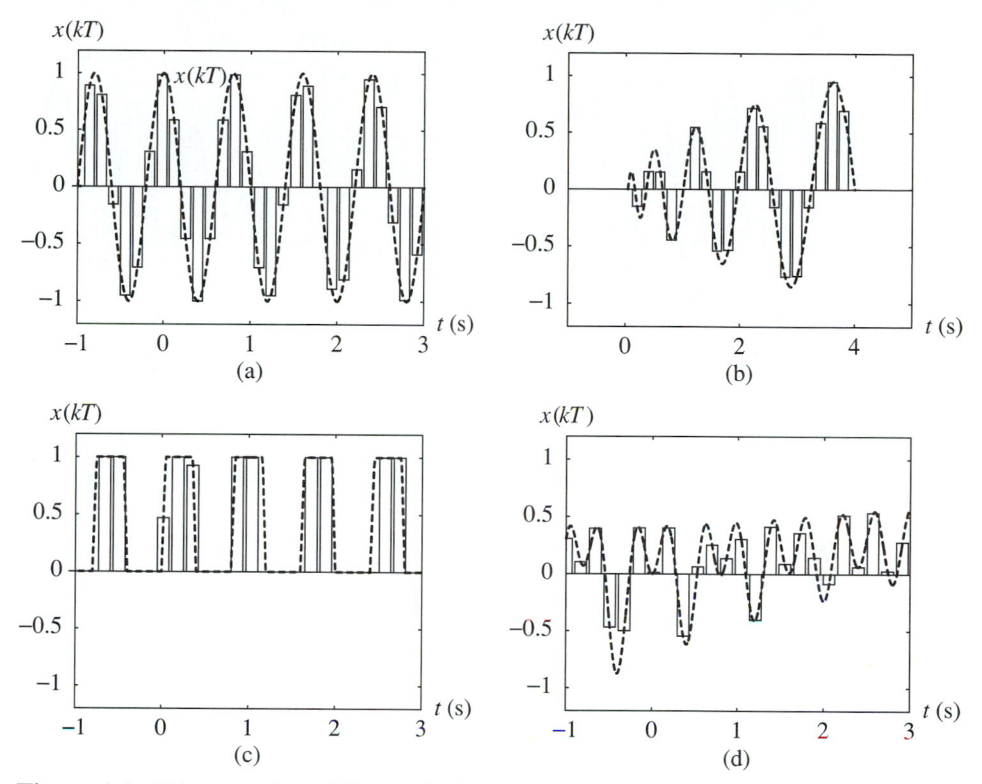

Figure 1.3 Discrete-time (discrete) signals, which are sampled but not quantized.

designate the corresponding sequence of samples $x(kT)$ by the set notation $\{x_{(k)}\}$. In order to simplify notation, we drop the curly braces and use $x(k)$ or $x_{(k)}$ or x_k rather than $x(kT)$ or $\{x_{(k)}\}$.

1.1.3 Digital Signals

The purpose of sampling a continuous signal is to transmit, store, or process a limited number of samples. It is also of interest to represent the values of samples by a limited number of digits. By using fewer digits we attain faster transmission and smaller storage requirements for the information. Thus, we utilize the quantized samples, $\hat{x}_k$, rather than the true samples of infinite accuracy, x_k,

$$\hat{x}_k = \text{quantize}(x_k) = Q[x_k]$$

When we quantize samples, we introduce an error in the representation. We define this quantization error as the difference between the quantized signal and the original signal, $e_k = \hat{x}_k - x_k$. For example, we quantize the discrete-time signals in Fig. 1.2 so that the possible values of $\hat{x}_k$ are from the set of five values $\{-1, -0.5, 0, 0.5, 1\}$:

$$\hat{x}_k \in \{-1, -0.5, 0, 0.5, 1\}$$

Figure 1.5 plots the quantized versions of these signals.

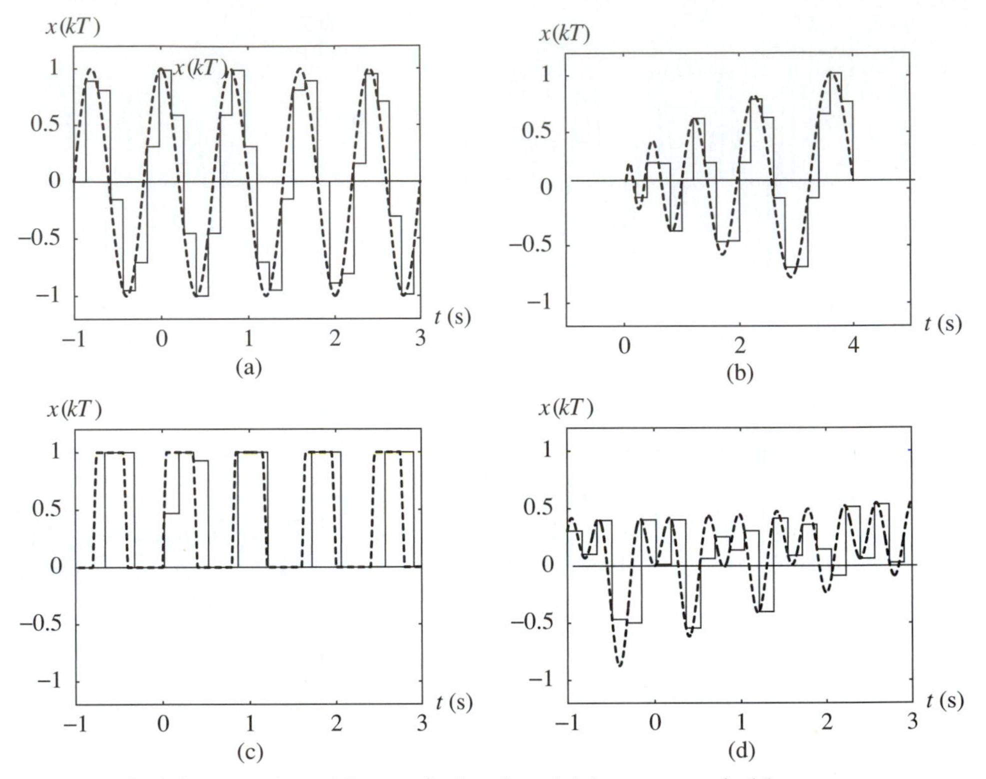

Figure 1.4 Discrete-time (discrete) signals, which are sampled but not quantized.

Each sample of $\hat{x}_k$ can be represented by its sign $\pm$ and the three magnitude values $\{0, 0.5, 1\}$ according to the following quantization strategy:

$$\begin{cases} \hat{x}_k = -1.0 & \text{for} \quad x_k < -0.6 \\ \hat{x}_k = -0.5 & \text{for} \quad -0.6 \leq x_k < -0.1 \\ \hat{x}_k = 0.0 & \text{for} \quad -0.1 \leq x_k \leq 0.1 \\ \hat{x}_k = 0.5 & \text{for} \quad 0.1 < x_k \leq 0.6 \\ \hat{x}_k = 1.0 & \text{for} \quad 0.6 < x_k \end{cases}$$

In binary notation, we can use three binary digits for representation: one for sign and two for representing the values $\{0, 0.5, 1\}$. This is known as sign-magnitude representation. Another three binary digit representation would be $110_2 = -1$, $111_2 = -0.5$, $000_2 = 0.0$, $001_2 = 0.5$, and $010_2 = 1.0$. If we extended this latter representation to include $101_2 = -1.5$, $100_2 = -2$, and $011_2 = 1.5$, then we would have a two's complement representation. The two's complement representation is the format most commonly used by programmable processors.

Although the resolution of the quantizer affects the accuracy of the data, as seen in Fig. 1.5, the error can be reduced by increasing the number of digits (possible values

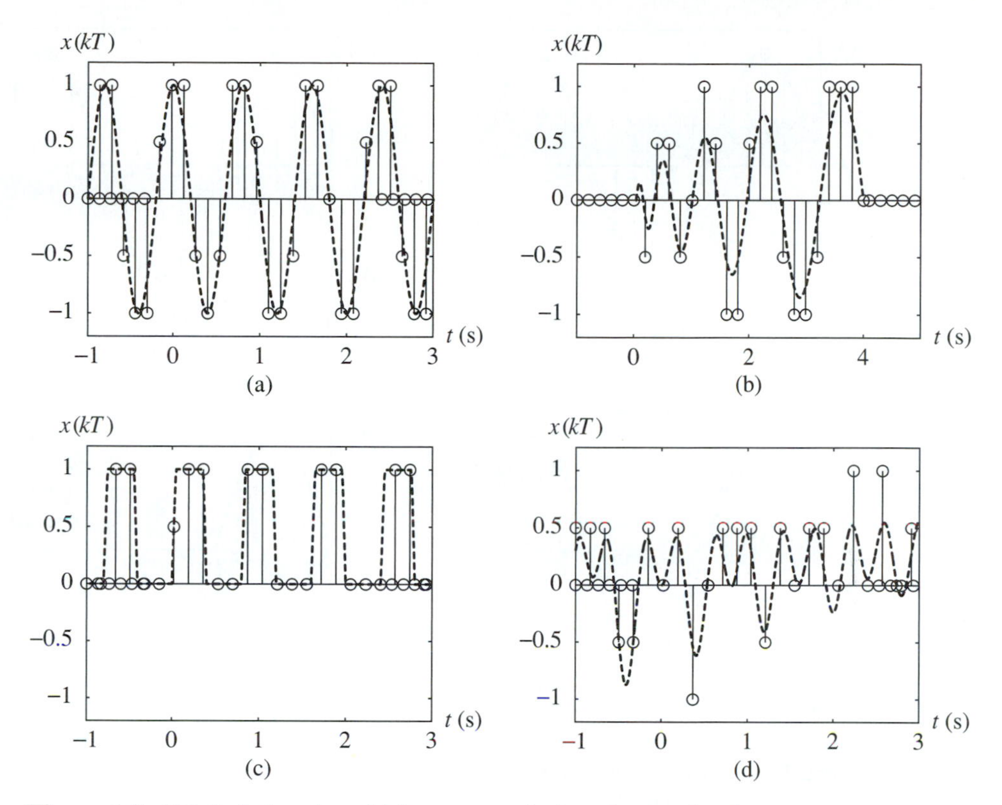

Figure 1.5 Digital signals, which are sampled and quantized.

for $\hat{x}_k$). The choice of the digits for representing the quantized signal is very important, and the quantization should be done properly: In transmitting, storing, and processing we prefer less digits, but with too small a number of digits we can lose information from the signal. The two opposing requirements must be satisfied: (1) Minimize number of digits to facilitate the signal transmission or storing, and (2) maximize number of digits to keep the quantization error as low as necessary in order to preserve the information contained in the signal.

The *quantization* is a many-to-one mapping. All the values that fall within a continuous band are mapped to the edge value of the band, as shown in Fig. 1.6. Figure 1.6 illustrates the quantization of a signal into 80 discrete levels in four different ways using the MATLAB commands `ceil`, `floor`, `round`, and `fix`. The maximum quantization error is lower than the amplitude of the undesirable signal, and the number of quantization levels is not very large. For digital signal processing, sampled values are typically mapped into an m-bit binary number. A sample of the signal shown on Fig. 1.6 can be represented by $m = 7$ bits or by 2^7 discrete values ($2^7 \geq 80$).

A *sample set* or *data sequence* is a collection of numbers which represents the quantized samples. In signal processing, we generally process finite sequences or finite portions of infinite sequences at any given time. The samples may be plotted versus the sample number k or time $t = kT$.

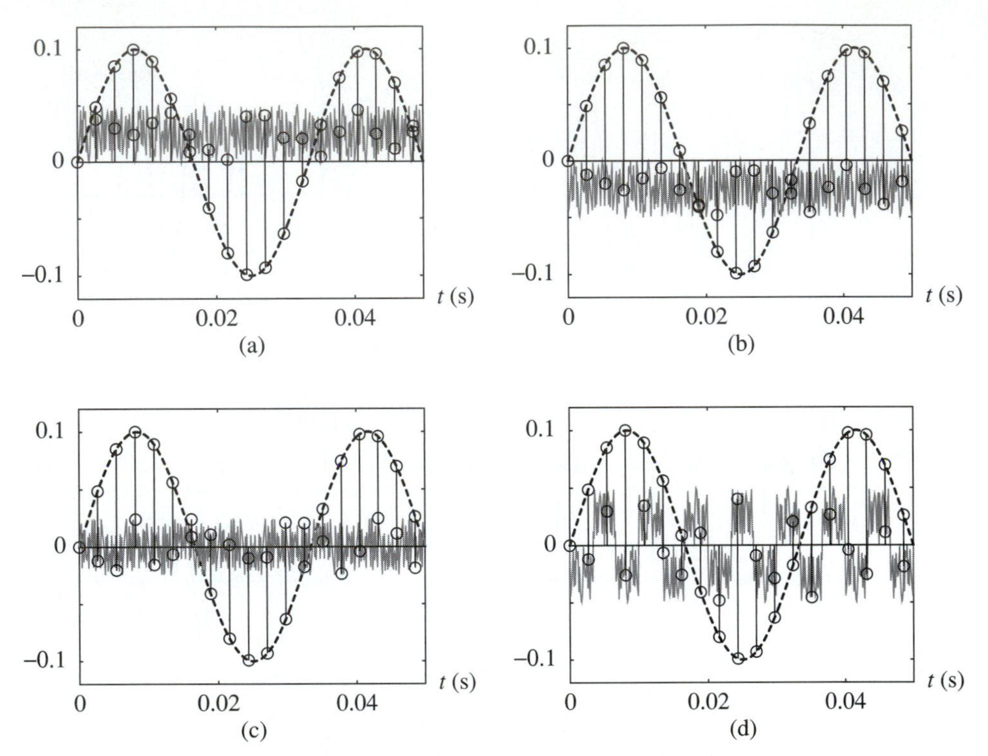

Figure 1.6 Undesirable signal $0.1\sin(60\pi t)$ and quantization error: (a) $\hat{x}(t) \geq x(t)$, (b) $\hat{x}(t) \leq x(t)$, (c) $|\hat{x}(t)| \leq |x(t)|$, (d) $|\hat{x}(t)| \geq |x(t)|$.

1.1.4 Deterministic Signals

A signal that can be described by an explicit mathematical form is *deterministic*. A deterministic signal can be *periodic* or *aperiodic*. A periodic signal consists of a basic "shape" of finite duration, T_{basic}, that is replicated infinitely as shown in Figs. 1.1a and 1.1c. Figure 1.1c plots a signal that is periodic and can be represented as an infinite sum of a constant and sinusoidal signals,

$$x(t) = X_0 + \sum_{i=1}^{+\infty} X_i \sin(i\, 2\pi f_{basic} t + \phi_i) \qquad \text{where} \qquad f_{basic} = \frac{1}{T_{basic}}$$

The sinusoidal terms $X_i \sin(i\, 2\pi f_{basic} t + \phi_i)$ are referred to as *harmonics*. An *aperiodic signal* does not have a repetitive form, such as a single rectangular pulse or the signals plotted in Figs. 1.1b and 1.1d.

1.1.5 Random (Nondeterministic) Signals

A signal that cannot be described in an explicit mathematical form is called *random*, also known as *nondeterministic* or *stochastic*. For example, the difference between a sinusoidal signal $x(t)$ and the quantized signal $\hat{x}(t)$ is a random signal as shown in Fig. 1.6. This difference we call the *quantization noise* or simply *noise*. Although we cannot explicitly describe a random signal, some of its properties can be useful in signal processing. For example, we can conclude that the *mean* (average) values of the signals in Figs. 1.6c and 1.6d are zero:

$$\mu_x = \text{mean}\, x(t) = \frac{1}{t_2 - t_1} \int_{t_1}^{t_2} x(t)\, \mathrm{d}t$$

On the other hand, the mean value is positive for Fig. 1.6a and negative for Fig. 1.6b. The *variance*

$$\sigma_x = \text{variance}\, x(t) = \frac{1}{t_2 - t_1} \int_{t_1}^{t_2} (x(t) - \mu_x)^2\, \mathrm{d}t$$

of the signal from Fig. 1.6c is lower than the variance of the signal from Fig. 1.6d. The mean and variance are important properties for random signals in the design of signal processing systems.

1.2 THE SAMPLING THEOREM

Signals occurring in nature are generally continuous. If we choose to use a digital system such as a computer to process signals, then we will need to convert the continuous signals into digital signals before processing and, possibly, convert the resulting digital signals to continuous signals. Our digital system should eliminate undesirable and interfering signals.

Prior to processing digital signals, we have to make a choice of the sampling period T and the number of quantization levels. The wrong choice of the sampling period can produce serious errors and loss of information. A key theorem, referred to as the *Sampling Theorem*, gives the guidelines to select the required T.

In order to give a reader an intuitive feel for the importance of the sampling theorem we will consider a single sinusoidal signal. The sampling theorem requires that a continuous sinusoidal signal of frequency f_a, $x_a(t) = \sin(2\pi f_a t + \phi_a)$, be sampled at a rate f_0 greater than twice f_a:

$$f_0 = \frac{1}{T} \qquad \text{such that} \qquad f_0 > 2 f_a$$

The samples of the sinusoidal signal $x_a(t)$ are

$$x_{a,k} = \sin((2\pi f_a)kT + \phi_a)$$

or

$$x_{a,k} = \sin\left(2\pi \frac{f_a}{f_0} k + \phi_a\right)$$

where ϕ_a designates the initial phase for $t = 0$, so $x_a(0) = \sin(\phi_a)$.

For the case $f_0 = 2f_a$, we have

$$x_k = \sin(\pi k + \phi_a)$$

resulting in the data sequence

$$x_k = (-1)^k \sin\phi_a, \qquad k = 0, 1, 2, \ldots$$

as shown in Fig. 1.7. The sample values x_k depend on ϕ_a and take the maximum value $(x_k)_{\max} = 1$ for $\phi_a = \frac{\pi}{2}$. For $\phi_a = 0$, as shown in Fig. 1.7a, we might conclude that there is no signal at all. For this reason, the condition for properly sampling an analog sinusoidal signal is $f_0 > 2f_a$.

In the next example we will consider two signals: (1) a signal $x_a(t)$ with $f_a > \frac{1}{2}f_0$ and (2) a signal $x_b(t) = \sin(2\pi f_b t + \phi_b)$ with $f_b < \frac{1}{2}f_0$. We assume that $f_a = \frac{1}{2}f_0 + \Delta f$ and $f_b = \frac{1}{2}f_0 - \Delta f$, where $\Delta f > 0$. The corresponding sample sequences are

$$x_{a,k} = \sin\left(2\pi\left(\frac{f_0}{2} + \Delta f\right)kT + \phi_a\right)$$

$$x_{b,k} = \sin\left(2\pi\left(\frac{f_0}{2} - \Delta f\right)kT + \phi_b\right)$$

which are shown in Fig. 1.8a, for $\Delta f = 0.1f_0$, and Fig. 1.8b, for $\Delta f = 0.25f_0$.

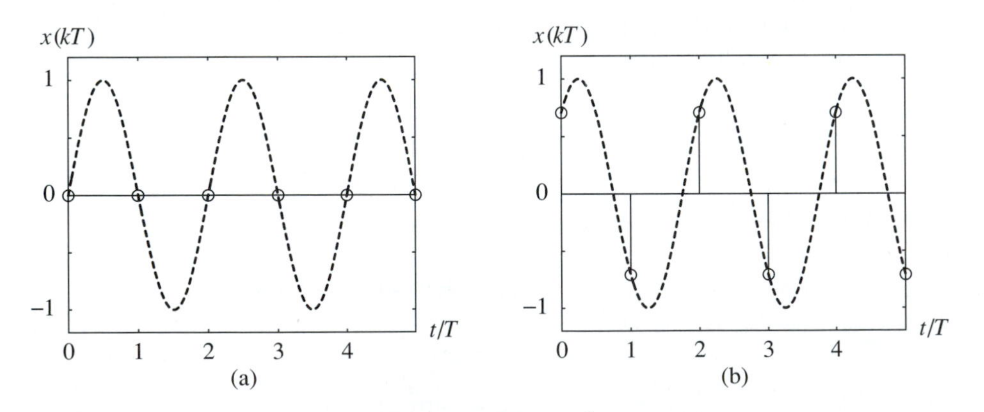

Figure 1.7 Sinusoidal signal samples for $f_a = \frac{1}{2}f_0$, and (a) $\phi_a = 0$, (b) $\phi_b = \pi/4$.

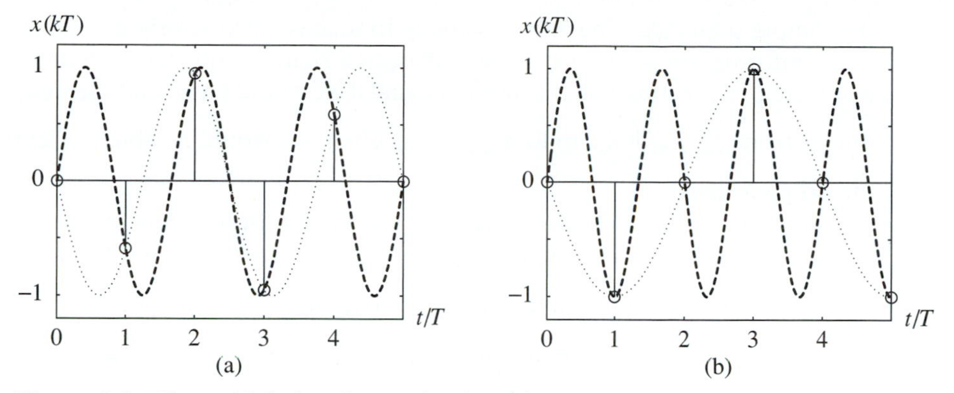

Figure 1.8 Sinusoidal signal samples for (a) $x(t) = \sin(2\pi(1/2 \pm 0.1) f_0 t)$, (b) $x(t) = \sin(2\pi(1/2 \pm 1/4) f_0 t)$.

From the samples shown in Fig. 1.8a, we cannot reconstruct the original continuous-time signal. An infinite number of continuous-time signals exist that yield the same sequence. For example, we can reconstruct the following three continuous-time signals:

$$y_a(t) = \sin\left(2\pi\left(\frac{f_0}{2} + \Delta f\right)t + \phi_a\right)$$

$$y_b(t) = \sin\left(2\pi\left(\frac{f_0}{2} - \Delta f\right)t + \phi_b\right)$$

$$y_c(t) = \frac{1}{2}\sin\left(2\pi\left(\frac{f_0}{2} + \Delta f\right)t + \phi_a\right) + \frac{1}{2}\sin\left(2\pi\left(\frac{f_0}{2} - \Delta f\right)t + \phi_b\right)$$

We use the symbol y for the reconstructed signal and use x for the sampled signal.

From Fig. 1.8b, we find that the first two sample values are $x_0 = 0$ and $x_1 = -1$. The values of the reconstructed signals are $y_a(0) = \sin(\phi_a) = x_0 = 0$, yielding $\phi_a = 0$. Next, $y_a(T) = \sin(2\pi(\Delta f + f_0/2)T) = x_1 = -1$, and we obtain $2\pi\Delta f T + \pi = 3\pi/2 \Rightarrow \Delta f = f_0/4$. Finally, $y_a(t) = \sin(1.5\pi f_0 t)$. In the same way, we find $\phi_b = 0$ and $y_b(t) = \sin(0.5\pi f_0 t)$:

$$y_a(t) = \sin(1.5\pi f_0 t)$$

$$y_b(t) = \sin(0.5\pi f_0 t)$$

$$y_c(t) = \frac{1}{2}y_a(t) + \frac{1}{2}y_b(t)$$

Since a sinusoidal signal can be represented by its amplitude, frequency, and phase, the knowledge of three adjacent samples gives us an opportunity to recover the signal from a set of three equations. In our case, we already know the amplitude to be one, so we only need two adjacent samples to recover the frequency and phase values.

In the first case, when the sampling rate (frequency) is exactly twice the sinusoidal frequency, the continuous signal cannot be reconstructed. In the second case, when the sampling rate is less than twice the sinusoidal frequency, the continuous signal cannot be uniquely reconstructed because there is an infinite number of sinusoidal signals fitting the sample sequence. The only possibility to reconstruct the original signal is to have *a priori* information about the range of expected signal frequencies. In this example, we could uniquely reconstruct the original signal if we would know that the signal frequency ranges from $f_{\min} = \frac{1}{2} f_0$, up to $f_{\max} = f_0$. Then, we would be able to identify $y_a(t)$ as the original signal.

In Fig. 1.9, the sinusoidal signal is sampled with a sampling rate satisfying the sampling theorem. The straight-line interpolation has been used to connect adjacent samples. If the signal frequency, f_a, gets closer to $\frac{1}{2} f_0$ we cannot visually identify the sine function from the simple linear interpolated diagram. Nevertheless, the continuous signal can be reconstructed exactly.

The basic results for defining relationships between continuous-time and discrete-time signals were presented by H. Nyquist, J. M. Whittaker, D. Gabor, and C. E. Shannon. At this point, we will express these results in a heuristic and intuitive way. The rigorous formulation, called the *Sampling Theorem*, will be given in Chapter 3 in Section 3.3.

First, we assume that a continuous signal consists of one or more sinusoidal signals and that the highest frequency of these sinusoidal signals is f_{max}. The continuous signal can be uniquely represented by its equally spaced samples if the sampling frequency f_0 is at least twice the highest frequency f_{max}. The original continuous signal can be reconstructed from the sample sequence by passing the sequence through a system having the property to reject sinusoidal signals of frequencies higher than f_{max}.

The minimal sampling frequency is $f_0 = 2f_{max}$, and one-half of the sampling frequency is called the *Nyquist frequency*, also called the *folding frequency*:

$$\text{Nyquist frequency} = \frac{f_0}{2} = \frac{1}{2T}$$

A sinusoidal signal with the frequency, $f_a < f_0$, which is above the Nyquist frequency, $\frac{1}{2} f_0$, is aliased by sampling into a discrete sinusoidal signal below the Nyquist frequency.

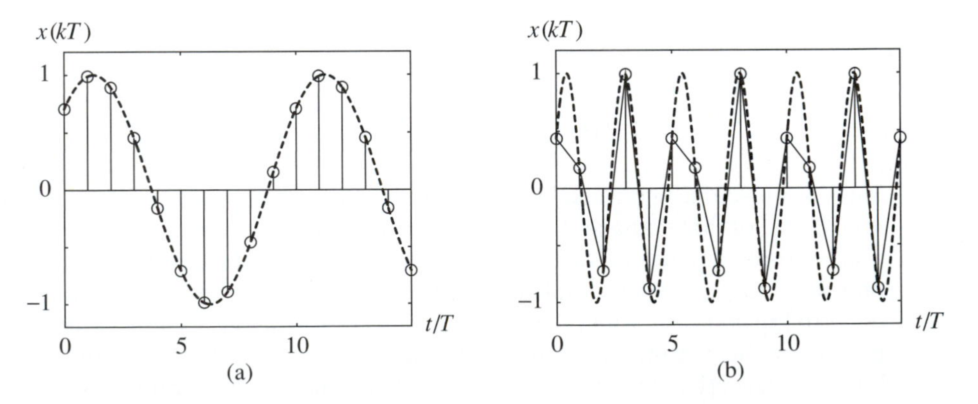

Figure 1.9 Frequency of signal is less than a half of sampling frequency.

The aliased frequency appears as if two signals existed, one at f_a and the other at $f_0 - f_a$, hence its name. The aliasing effect is eliminated by a system called a lowpass continuous *anti-aliasing* filter, which is often used before of discretizing the continuous signal. In practical applications, the actual sampling frequency is often selected to be four times the maximum expected signal frequency f_a.

1.3 BASIC CONTINUOUS-TIME SIGNALS—FUNCTIONS

In practical applications the excitation or response of a system can be represented by a combination of basic signals. In this section, all signals are assumed to be functions that depend on time t over the range $-\infty < t < +\infty$.

1.3.1 Sinusoidal Signals

A *sinusoidal* signal is defined by the sinusoidal function

$$x_s(t) = X_s \sin(2\pi f_s t + \phi_s)$$

Alternatively

$$x_s(t) = X_s \sin\left(\frac{2\pi}{T_s} t + \phi_s\right)$$

where T_s is a period of sinusoidal signal

$$T_s = \frac{1}{f_s}$$

The unit of frequency for f_s is Hertz (Hz) or cycles per second when t represents time in seconds. The period $T_s = 1/f_s$ is generally given in seconds (s). The phase shift ϕ_s is in radians (rad). The *radian* frequency $\omega_s = 2\pi f_s$ is in rad/s. The derivative or integral of a sinusoidal function is, again, a sinusoidal function.

1.3.2 Real-Valued Exponential Signals

A *real-valued exponential* signal is defined by exponential function

$$x_e(t) = X_e e^{bt}$$

where e is the Naperian constant ($e \approx 2.718$), and X_e and b are real constants. Three different signals can be derived from the exponential function:

- an increasing function of time for $b > 0$,
- a constant $x_e(t) = X_e$ for $b = 0$, and
- a decreasing function of time for $b < 0$.

In practical applications, a signal obtained as a combination of the above can be used; for example,

$$x_b(t) = X_e\left(1 - e^{bt}\right), \qquad b < 0$$

1.3.3 Unit Step Signal

A *unit step* signal, $u(t)$, is a signal that turns on from 0 to a constant (unit) value at $t = 0$:

$$u(t) = \begin{cases} 1, & t > 0 \\ 0, & t \leq 0 \end{cases}$$

The unit step signal is a continuous function with a single point of discontinuity at $t = 0$ where it jumps in amplitude from 0 to 1.

The unit step signal can be scaled in time and amplitude:

$$x_u(t) = X_u\, u(t - t_1) = \begin{cases} X_u, & t > t_1 \\ 0, & t \leq t_1 \end{cases}$$

Step signals are used to select portions of other signals. For example, $x(t)(u(t) - u(t - t_2))$ is a portion of $x(t)$ between $t = 0$ and t_2.

1.3.4 Pulse Signals

A *pulse* signal can be obtained from unit step signals as

$$x_p(t) = X_p\,(u(t - t_1) - u(t - t_2))$$

For $t_2 > t_1$ the pulse signal is constant over the range $t_1 < t \leq t_2$, $x_p(t) = X_p$, and zero otherwise. A unit pulse of width t_w can be represented by $u(t) - u(t - t_w)$.

We frequently use rectangular pulses whose area under the pulse equals 1. For this reason, we define the *pulse function* $p_\epsilon(t)$ by

$$p_\epsilon(t) = \begin{cases} 0, & t \leq 0 \\ \frac{1}{\epsilon}, & 0 < t \leq \epsilon \\ 0, & \epsilon < t \end{cases}$$

In other words, $p_\epsilon(t)$ is a pulse of height $1/\epsilon$, of width ϵ, and starting at $t = 0$. Whatever the value of the positive parameter ϵ, the area under $p_\epsilon(t)$ is 1. Note that

$$p_\epsilon(t) = \frac{u(t) - u(t - \epsilon)}{\epsilon}$$

Pulse signals are used to select portions of other signals. For example, $x(t)x_p(t)$ is a portion of $x(t)$ between $t = t_1$ and t_2, assuming $X_p = 1$.

1.3.5 Unit Ramp Signal

A *ramp* signal, $x_r(t)$, is defined as a continuous function which rises in amplitude linearly at a rate given by the constant X_r

$$x_r(t) = X_r\, t\, u(t)$$

A ramp signal shifted by t_1 is defined as

$$x_r(t - t_1) = X_r\,(t - t_1)\, u(t - t_1)$$

A *unit ramp* signal is zero for $t \le 0$ and has a unity rate of increase in amplitude, $X_r = 1$:

$$r(t) = \begin{cases} t, & t > 0 \\ 0, & t \le 0 \end{cases}$$

The unit ramp can be expressed in terms of the unit step function

$$r(t) = t\,u(t)$$

Also, the unit ramp signal results from the unit step signal by integration

$$r(t) = \int_{-\infty}^{t} u(\tau)\,\mathrm{d}\tau$$

As a consequence, the unit step signal is the derivative of the unit ramp signal

$$u(t) = \begin{cases} \dfrac{\mathrm{d}r(t)}{\mathrm{d}t}, & t \ne 0 \\ 0, & t = 0 \end{cases}$$

Notice that the derivative $\mathrm{d}r/\mathrm{d}t$ is not defined for $t = 0$ because $u(t)$ is discontinuous at $t = 0$.

A unit step signal with a linear rise from 0 to 1 can be obtained as a combination of two unit ramp signals:

$$x_{ur}(t) = \frac{r(t) - r(t - t_2)}{t_2}$$

The rise time is t_2.

1.3.6 Unit Impulse Signal

The *unit impulse* signal, $\delta(t)$, also called the *Dirac delta*, is not a function in a strict mathematical sense. We proceed intuitively [3] because a detailed exposition would require the theory of distributions. A more detailed discussion of the Dirac delta can be found in references 4–6. Our following definition of the unit impulse function, however, is satisfactory for the purposes of this book.

For our purposes, we state that

$$\delta(t) = \begin{cases} 0, & t \ne 0 \\ \text{undefined}, & t = 0 \end{cases}$$

and that the area under the unit impulse is one:

$$\int_{-\infty}^{\infty} \delta(t)\,\mathrm{d}t = 1$$

Hence, one can consider the area at the origin to be one. Intuitively, we may think of the impulse function $\delta(t)$ as the limit, as $\epsilon \to 0$, of the pulse function $p_\epsilon(t)$. From the definition of $\delta(t)$ and $u(t)$, we can formally obtain

$$u(t) = \int_{-\infty}^{t} \delta(\tau)\,\mathrm{d}\tau$$

and

$$\frac{du(t)}{dt} = \delta(t)$$

Another frequently used property of the unit impulse is the *sifting property*. Letting $x(t)$ be a continuous signal, we obtain

$$\int_{-\infty}^{\infty} x(t)\,\delta(t - t_0)\,dt = x(t_0)$$

This property is important when used in integral transforms such as the Laplace and Fourier transform, which we will cover in a subsequent chapter.

1.3.7 Causal Signals

Formally, a signal is said to be *causal* if it is zero for $t < 0$. Causal signals are readily created by multiplying any continuous signal by the unit step $u(t)$. In practical applications we often consider signals of finite duration. The instant when the signal begins is called the starting time. We usually take the starting time to be $t = 0$. In most cases, we analyze the signal from beginning at the starting time, and we consider causal signals only.

1.3.8 Combining Signals

By multiplication, addition, and subtraction of basic signals, we can generate combined signals. We have already used the multiplication by unit step. Some interesting signals that can appear in signal processing are:

- *Exponentially modulated sinusoidal* signals

$$x(t) = X_m e^{bt} \sin(2\pi f_a t + \phi_a)$$

- *Sinusoidally modulated sinusoidal* signals

$$x(t) = X_m \sin(2\pi f_b t + \phi_b) \sin(2\pi f_a t + \phi_a)$$

  which can be expressed as the sum of two sinusoidal signals

$$x(t) = \frac{1}{2} X_m(-\cos(2\pi(f_a + f_b)t + \phi_a + \phi_b) + \cos(2\pi(f_a - f_b)t + \phi_a - \phi_b))$$

- A sum of m sinusoidal signals

$$x(t) = \sum_{i=1}^{m} X_i \sin(2\pi f_i t + \phi_i)$$

  For example, if we choose $X_i = \frac{4}{\pi}\frac{1}{2i - 1}$, $f_i = (2i - 1) f_a$, $\phi_i = 0$, we obtain an approximation of a pulse train shown in Fig. 1.10, [5]. Similarly, for $X_i =$

$\frac{2}{\pi}\frac{(-1)^{i+1}}{i}$, $f_i = i f_a$, $\phi_i = 0$, we obtain an approximation of a sawtooth periodic signal shown in Fig. 1.11.

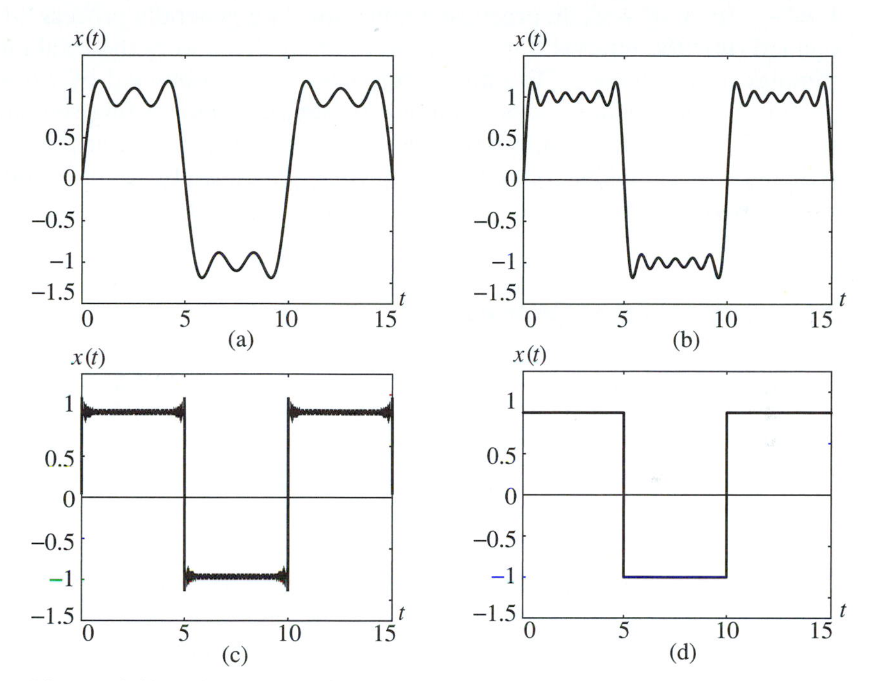

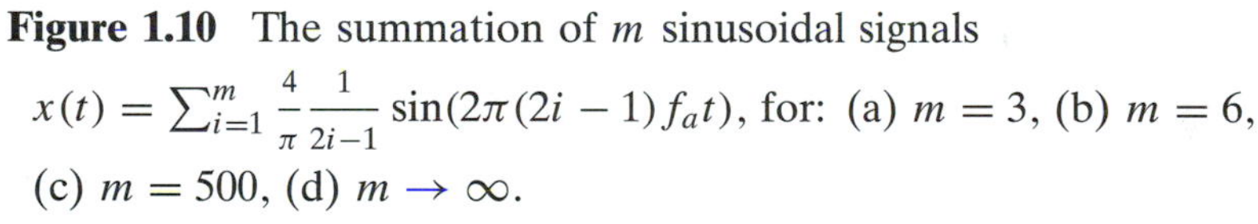

Figure 1.10 The summation of m sinusoidal signals $x(t) = \sum_{i=1}^{m} \frac{4}{\pi}\frac{1}{2i-1}\sin(2\pi(2i-1)f_a t)$, for: (a) $m = 3$, (b) $m = 6$, (c) $m = 500$, (d) $m \to \infty$.

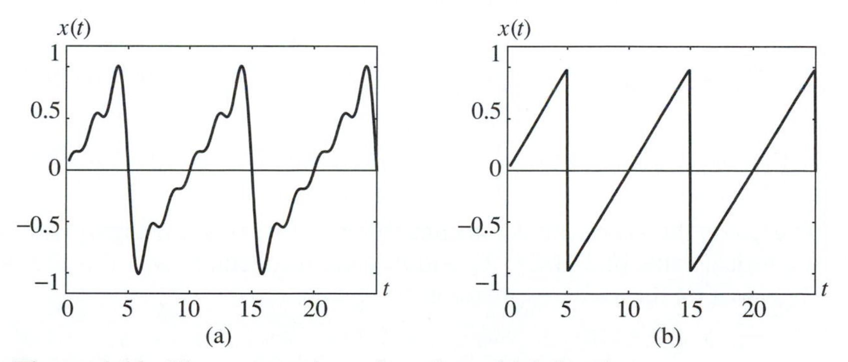

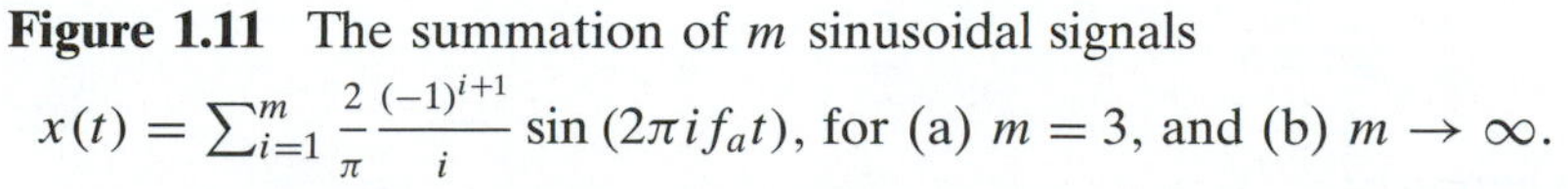

Figure 1.11 The summation of m sinusoidal signals $x(t) = \sum_{i=1}^{m} \frac{2}{\pi}\frac{(-1)^{i+1}}{i}\sin(2\pi i f_a t)$, for (a) $m = 3$, and (b) $m \to \infty$.

1.4 BASIC DISCRETE-TIME SIGNALS—SEQUENCES

A *sequence*, also known as a *data sequence* or *sample set*, is a collection of ordered samples x_k. For the general case, the sample set is defined over the entire range from $k = -\infty$ to $k = +\infty$. In practical applications, we generally process finite sequences defined over the interval $0 \leq k \leq N - 1$, where N denotes the total number of data samples in a sequence. The finite sequence is often obtained by extracting a finite portion of an existing, possibly infinite sequence. The existing sequence is often a sampled version of a continuous signal. For the purposes of analysis, we represent sequences as a combination of basic sequences which are better understood and easier to manipulate.

1.4.1 Sinusoidal Sequences

A sinusoidal sequence may be described as

$$x_{s,k} = X_s \sin\left(2\pi \frac{1}{N_s} k + \phi_s\right)$$

where X_s is the amplitude of the sinusoidal sequence (a positive real number), N_s is the period, and ϕ_s is the phase. When the sinusoidal sequence is obtained by sampling a continuous signal $x_s(t) = X_s \sin(2\pi f_s t + \phi_s)$, the resulting sample set is

$$x_{s,k} = X_s \sin\left(2\pi \frac{f_s}{f_0} k + \phi_s\right)$$

1.4.2 Real-Valued Exponential Sequences

A *real-valued exponential* sequence is defined by the exponential function

$$x_{e,k} = X_e e^{bTk}$$

where e is the exponential constant (base of natural logarithms) with an approximate numerical value of 2.71828, X_e and b are real constants, and T is the sampling period. By rewriting the above expression,

$$x_{e,k} = X_e a^k$$

where $a = e^{bT}$.

Three different sequences can be derived from the real exponential sequence:

- an increasing exponential sequence, for $a > 1$,
- a constant $x_{e,k} = X_e$, for $a = 1$,
- a decreasing exponential sequence, for $0 < a < 1$.

We combine the above sequences to form a sequence like $x_k = X_e\left(1 - a^k\right)$, $a < 1$.

1.4.3 Unit Step Sequence

A *unit step* sequence is defined as

$$u_k = \begin{cases} 1, & k \geq 0 \\ 0, & k < 0 \end{cases}$$

We can shift and scale the unit step sequence and obtain

$$X_u u_{k-k_0} = \begin{cases} X_u, & k \geq k_0 \\ 0, & k < k_0 \end{cases}$$

1.4.4 Unit Ramp Sequence

A *ramp* sequence is defined as

$$x_{r,k} = X_r\, k\, u_k$$

and the shifted ramp sequence is defined as

$$x_{r,(k-m)} = X_r\,(k - m)\, u_{k-m}$$

A *unit ramp* sequence is zero for $k < 0$:

$$r_k = \begin{cases} k, & k \geq 0 \\ 0, & k < 0 \end{cases}$$

The unit ramp sequence can be expressed in terms of the unit step sequence

$$r_k = k\, u_k$$

Also, the unit ramp sequence results from the unit step sequence by summation:

$$r_k = \sum_{m=-\infty}^{k} u_m$$

As a consequence, the unit step sequence is the difference of two adjacent samples of unit ramp sequences:

$$u_k = r_{k+1} - r_k$$

1.4.5 Unit Impulse Sequence

A *unit impulse* sequence is defined as

$$\delta_k = \begin{cases} 1, & k = 0 \\ 0, & k \neq 0 \end{cases}$$

The unit impulse sequence can be obtained as the difference of the two adjacent samples of unit step sequences:

$$\delta_k = u_k - u_{k-1}$$

Also, the unit step sequence results from the unit impulse sequence by summation:

$$u_k = \sum_{m=-\infty}^{k} \delta_m$$

The unit impulse sequence is an important signal for analysis of discrete-time systems. The impulse acts to sample a single value, x_m, of a sequence, x_k, which it multiplies:

$$x_k \delta_{k-m} = x_m \, \delta_{k-m} = x_m$$

for $-\infty < k < +\infty$. An arbitrary sequence x_k can be expressed in terms of the unit impulse sequence as

$$x_k = \sum_{m=-\infty}^{+\infty} x_m \, \delta_{k-m}$$

1.4.6 Causal Sequences

A sequence x_k that is nonzero only over a finite interval $k_{\min} < k < k_{\max}$ is called a *finite-length* sequence. A sequence containing nonzero samples for $k \geq 0$ is said to be *causal*. An *anti-causal* sequence has nonzero samples only for $k < 0$. An example of a noncausal sequence follows:

$$x_k = u_{-k-1} = \begin{cases} 1, & k < 0 \\ 0, & k \geq 0 \end{cases}$$

1.5 CONTINUOUS-TIME SIGNALS IN MATLAB

The original design of MATLAB was to provide an interactive, command-line interface to a variety of numerical analysis libraries. Arithmetic is performed using a double-precision floating-point format. MATLAB has evolved over the years from a matrix laboratory into a general-purpose programming language, an algorithm development environment through its many toolboxes, a visual programming environment through Simulink, and a graphical user interface programming environment. Throughout the book, we will draw heavily on the MATLAB Signal Processing Toolbox and the Optimization Toolbox.

For our purposes, we will focus on the matrix laboratory aspects of MATLAB. As a matrix laboratory, MATLAB primarily works with numbers, which are usually represented as sequences and stored as arrays. Therefore, we cannot use continuous signals or operate on them in an analytic manner. We can describe a continuous-time signal with sequences of numbers; that is, we sample the signal and assume that the time step between two adjacent samples is appropriate. It is expected that the signal digitalization introduces no loss of information conveyed by the signal.

Consider the signal $x(t) = X^a e^{-at} \sin(2\pi ft + \phi)$. It can be coded in MATLAB like this: `x(t) = X^a * exp(-a*t) .*sin(2*pi*f*t+phi)`. For convenience, we often use `pi` to denote π, `phi` for ϕ, `w` for $\omega = 2\pi f$, `exp(a*t)` to denote e^{at}, and * to denote multiplication.

A MATLAB script for plotting this signal follows:

```
X = 1;
a = 0.1;
f = 0.53;
phi = 0;
t = 0.:0.05:15;
x = X^a * exp(-a*t) .* sin(2*pi*f*t+phi);
plot(t,x)
ylabel('x(t)')
xlabel('t (s)')
grid
```

In this script, there are two types of multiplication: * and .*. When applied to two vectors of the same length, the * operator will give the dot product between the two vectors, whereas the .* operator will produce a new vector whose ith element is the product of the ith elements of the two vectors. Hence, the .* operator is an example of an elementwise operator. In the script, the independent variable `t` is an array, and so the expressions `exp(-a*t)` and `sin(2*pi*f*t+phi)` also create arrays because `a` and `2*pi*f` are scalars. We use the string `t (s)` as the label on the horizontal axis of the plot to denote that time is in seconds—note the space between `'t'` and `'(s)'`. No space appears when we designate a signal `x` as a function of time, `x(t)`. The plot of $x(t)$ obtained by executing the script is shown in Fig. 1.12.

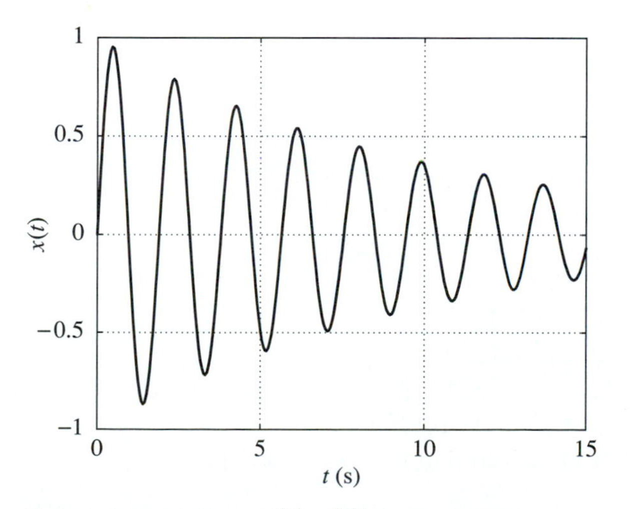

Figure 1.12 $x(t) = (1^{0.1} e^{-0.1t} \sin(2\pi 0.53t))$.

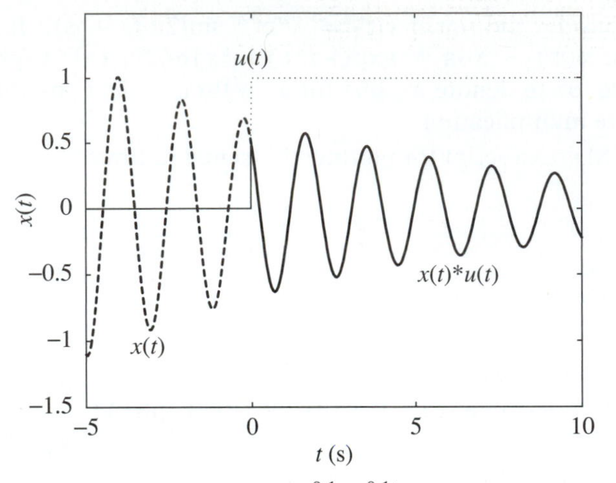

Figure 1.13 $u(t)$, $x(t) = (0.02^{0.1} e^{-0.1t} \sin(2\pi 0.53t + 3\pi/4))$ and $x(t)u(t) = (0.02^{0.1} e^{-0.1t} \sin(2\pi 0.53t + 3\pi/4))u(t)$.

In the next example, we generate a causal signal from $x(t)$. First, we calculate the unit step signal $u(t)$ by executing the MATLAB command `u=(t>0)`. For a given array `t`, this command produces the unit step sequence `u` of the same length as `t`. The logical expression `t>0` maps the ith component of `t` into 1, when `t>0` evaluates true, or into 0 when `t>0` is false. The signal is multiplied by the unit step: $x(t)\,u(t)$. Again, we have to use `.*` rather than `*` because `x` and `u` are arrays.

The signals $x(t)$, $u(t)$, and $x(t)\,u(t)$ are plotted in Fig. 1.13, and the MATLAB script that is used to create the plot is as follows:

```
X = 0.02;
a = 0.1;
f = 0.53;
phi = 3*pi/4;
t = -5.:0.05:10;
x = X^a * exp(-a*t) .* sin(2*pi*f*t+phi);
xu = (X^a * exp(-a*t) .* sin(2*pi*f*t+phi)).*(t>0);
u = (t> 0);
plot(t,x,'--',t,u,':',t,xu)
ylabel('x(t)')
xlabel('t (s)')
text(0,1.2,'u(t)')
text(-4.,-1.1,'x(t)')
text(5,-.6,'x(t)*u(t)')
axis([t(1) t(length(t)) -1.5 1.5])
```

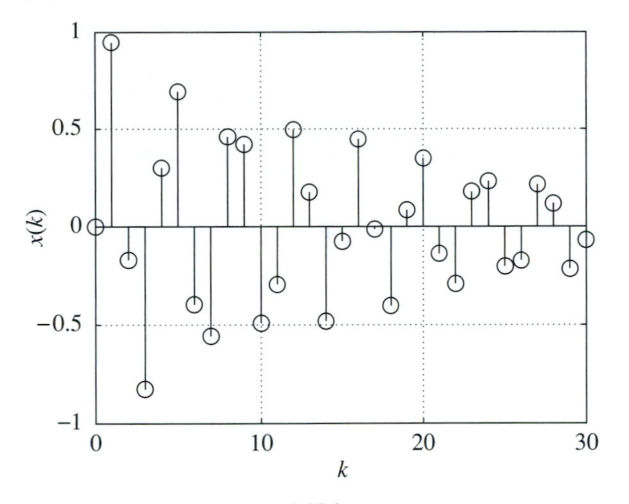

Figure 1.14 $x(k) = e^{-0.05\,k}\sin(\pi\ 0.53\ k)$.

1.6 SEQUENCES IN MATLAB

A continuous signal in MATLAB is in fact represented by ordered pairs (*time*, *value*). A digital signal is a set of numbers arranged in an array. Instead of time, we speak about index of an array element. If an analytic expression for the element of an array is known, we can generate a sequence of numbers. For example, given an index, k, we compute the array element `x(k) = X^a * exp(-a*T*k) * sin(2*pi*T*f*k+phi)` of an exponentially modulated sinusoidal sequence. The MATLAB script that is used to generate and plot the sequence follows, and the corresponding sequence is shown in Fig. 1.14:

```
X = 1;
a = 0.1;
f = 0.53;
T = 0.5;
phi = 0;
k = 0:1:30;
x = X^a * exp(-a*T*k) .* sin(2*pi*f*T*k+phi);
stem(k,x)
ylabel('x(k)')
xlabel('k')
grid
```

In the next example, we generate a causal sequence from x_k. First, we calculate the unit step sequence u_k: `u = (k>=0)`. In fact, as in the case of continuous signals, we use a logical expression so that the value of the kth component of u is 1 when the logical expression k>=0 is true, and 0 when the expression k>=0 is false.

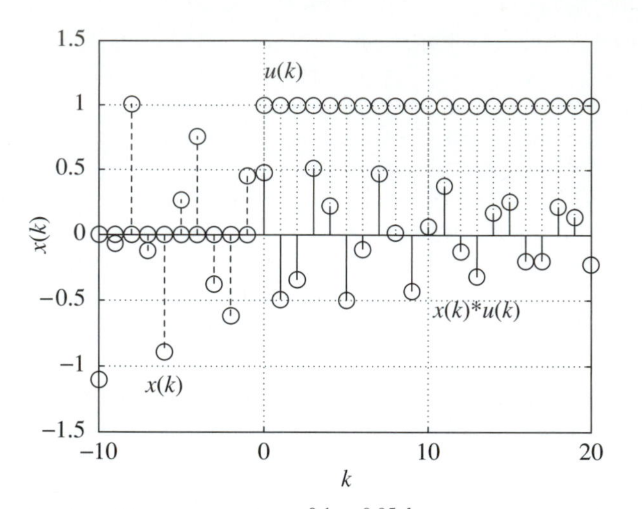

Figure 1.15 $u(k)$, $x(k) = (0.02^{0.1} e^{-0.05\ k} \sin(\pi\ 0.53\ k + 3\pi/4))$ and $x(k)\ u(k) = (0.02^{0.1} e^{-0.05\ k} \sin(\pi\ 0.53\ k + 3\pi/4))u(k)$.

The causal signal is obtained by multiplying x_k by u_k. Figure 1.15 illustrates the causal signal and its components. The MATLAB script is as follows:

```
X = 0.02;
a = 0.1;
f = 0.53;
T=.5;
phi = 3*pi/4;
k = -10:1:20;
x =  X^a * exp(-a*T*k) .* sin(2*pi*f*T*k+phi);
ux =  (X^a * exp(-a*T*k) .* sin(2*pi*f*T*k+phi)).*(k>=0);
u = (k>=0);
stem(k,x,'--')
hold on
stem(k,u,':')
stem(k,ux)
hold off
ylabel('x(k)')
xlabel('k')
text(0,1.3,'u(k)')
text(-8.,-1.12,'x(k)')
text(10,-.65,'x(k)*u(k)')
axis([k(1) k(length(k)) -1.5 1.5])
```

The primary difference between continuous and discrete-time signals in MATLAB is in the type of the corresponding independent variable. For continuous signals, it is a 'time' array, t, and can be any number; however, in the case of digital signals it is always an integer, k, and may take a positive, negative, or zero value.

1.7 CONTINUOUS-TIME SIGNALS IN MATHEMATICA

Mathematica is a system for doing mathematics by computer. Its forte is the algebra analysis and manipulation of functions and operators. One can work in an arbitrary number of dimensions and convert algebraic expressions to numbers with arbitrary precision. It has a graphical user interface which can load and store work sessions in an ASCII notebook format. When it first appeared, *Mathematica* offered a fundamentally different but complementary approach for analyzing signals and algorithms than MATLAB. *Mathematica* has a variety of *Application Packs* that extend the analysis to different fields. Throughout the book, we will draw heavily on the *Mathematica Signals and Systems Pack* [7].

Using *Mathematica*, the processing of continuous signals can be maintained in exact symbolic form. The transform of continuous signals involving integrals and derivatives are easier in symbolic form. Therefore, in symbolic processing, the signal is represented on a computer as a formula instead of a sequence of numbers. The aforementioned examples worked in MATLAB will be reconsidered here, but with the aid of *Mathematica* and the *Signals and Systems Pack* (SSP).

The *Mathematica* script that is used to create a signal is as follows:

```
X=1;
a=0.1;
f=0.53;
phi = 0;
x[t_] := X^a Exp[-a*t] Sin[2 Pi f t + phi];
Plot[ x[t], {t, 0, 14},
      AxesLabel -> {"t (s)", "x(t)"},
      PlotRange -> {-1, 1},
      GridLines -> Automatic ];
```

Two features of *Mathematica* may be readily apparent: Spaces between terms in an expression are interpreted as multiplication, and square brackets are used to delimit functions instead of parentheses. In the script, we define a function $x(t)$. In *Mathematica* notation, `x[t_]` defines a function `x` that takes any independent variable labeled `t`. For the exponential function, we can use either `Exp[-a*t]` or `E^(-a*T)`. We use `Sin[.]` to denote the sine function and use `Pi` for π. All of the built-in, protected *Mathematica* functions and constants begin with an uppercase letter.

We assign numerical values to `X`, `a`, `f`, and `phi` only for the purpose of obtaining a plot. In a symbolic algebra system such as *Mathematica*, we can work with symbols rather than with numbers. The actual plot command is `Plot[x[t],{t,0,14}]`, in which we define the independent variable to be t and its range from $t = 0$ to $t = 14$. The plot option `AxesLabel->{"t (s)","x(t)"}` adds labels to the horizontal and vertical axes, respectively. The grid lines are drawn by `GridLines->Automatic`, while the plot range is specified by `PlotRange->{-1,1}`. The resulting plot is in Fig. 1.16.

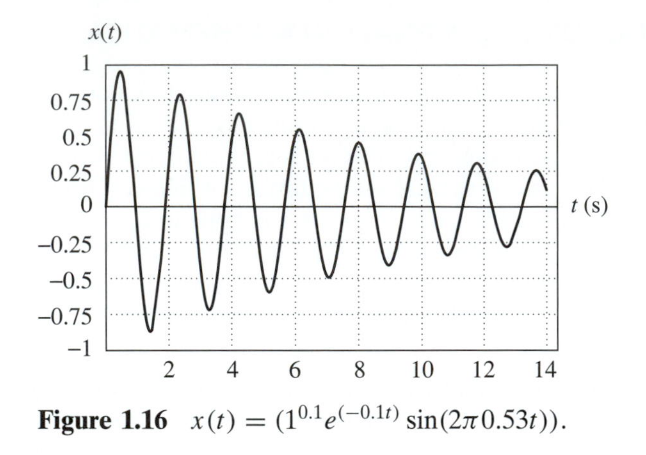

Figure 1.16 $x(t) = (1^{0.1}e^{(-0.1t)}\sin(2\pi 0.53t))$.

The second example of representing and plotting the three signals is expressed in *Mathematica* by

```
X=0.02;
a=0.1;
f=0.53;
phi = 3*Pi/4;
x[t_] :=  X^a Exp[-a*t] Sin[2*Pi*f*t+phi];
SignalPlot[{x[t]*UnitStep[t],x[t],UnitStep[t]},{t,-5,10},
     AxesLabel->{"t (s)","x(t)"},
     PlotRange->{-1.5,1.5},
     AxesOrigin->{-5,-1.5},
     PlotStyle->{GrayLevel[0],
                 Dashing[{.05}],Dashing[{.01}]}];
```

The signals are written in the *Mathematica* notation. For instance, `Sin[2 Pi f t]` designates $\sin(2\pi ft)$, and `Exp[-a*t]` represents e^{-at}. The unit step, $u(t)$, is represented by `UnitStep[t]`. Additional options define the position of the axis origin, `AxesOrigin->{-5,-1.5}`, and the type of lines, `PlotStyle->{GrayLevel[0]}`, `Dashing[{.05}]`, ... (see Fig. 1.17). We use the *Mathematica* built-in function `Plot` in Fig. 1.16 and `SignalPlot` from the SSP in Fig. 1.17. SSP contains functions for frequently used signals. For example, `ContinuousPulse[width,t]` represents a rectangular pulse with the unit height, existing for `0<t<width`. Various SSP signals are shown in Figs. 1.18–1.26. A continuous pulse signal starting at time $t = 0.4$ s and with the width of 0.2 s is plotted in Fig. 1.18.

```
SignalPlot[ContinuousPulse[.2,t-0.4],{t,0,1},
     PlotStyle->{Thickness[.01]},
     AxesLabel->{"t (s)","ContinuousPulse[.2,t-0.4]"}];
```

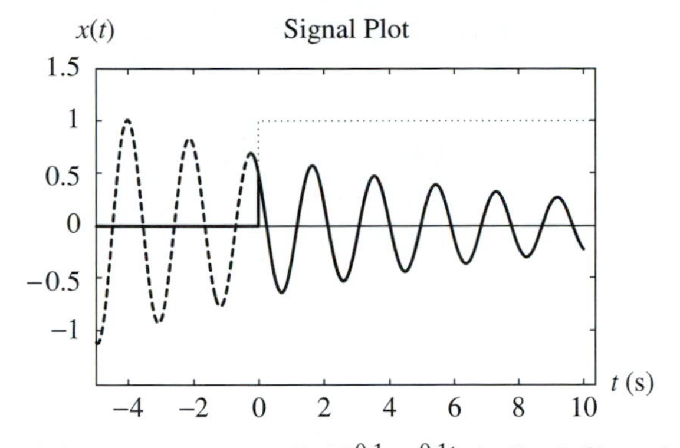

Figure 1.17 $u(t)$, $x(t) = (0.02^{0.1}e^{-0.1t}\sin(2\pi 0.53t + 3\pi/4))$ and $x(t)u(t) = (0.02^{0.1}e^{-0.1t}\sin(2\pi 0.53t + 3\pi/4))u(t)$.

The unit impulse signal can be plotted as *Dirac delta* (`DiracDelta[t-0.75]`), *continuous impulse* (`ContinuousImpulse[t-0.75]`), or as 0th-order *unit singularity function* (`UnitFunction[0][t-0.75]`, the order of the singularity is 0 and the time shift is 0.75). The height of the unit impulse represents the area under the delta function ($\int_{-\infty}^{+\infty} \delta(t)\,\mathrm{d}t = 1$). Because $\delta(t)$ has zero width and infinite amplitude at $t = 0$ but unity total area, we draw arrows at the unit height. If we multiply the unit impulse signal by 2, then the area is also 2, $\int_{-\infty}^{+\infty} 2\delta(t)\,\mathrm{d}t = 2$. We draw $2\delta(t)$ taller impulse with arrow at the doubled height. We label the arrow with a number called the *weight* of the impulse. The weight does not indicate the height of the impulse, but it indicates the area:

$$\int_{-\infty}^{+\infty} weight\delta(t)\,\mathrm{d}t = weight$$

```
SignalPlot[DiracDelta[t-0.75],{t,0,2},
     AxesLabel->{"t (s)","DiracDelta[t-0.75]"}];
```

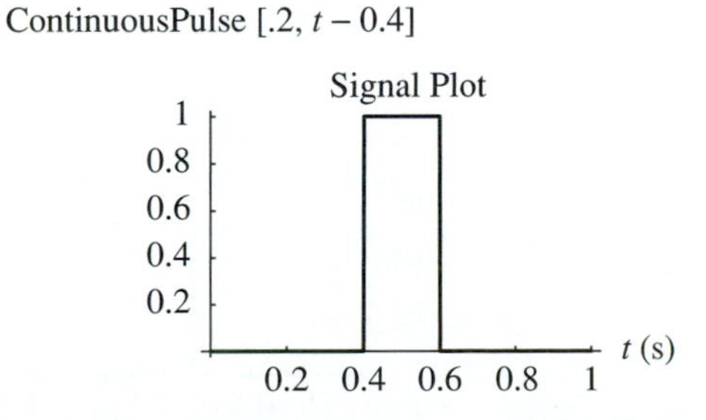

Figure 1.18 An example of a continuous unit pulse, $u(t - 0.4) - u(t - 0.6)$.

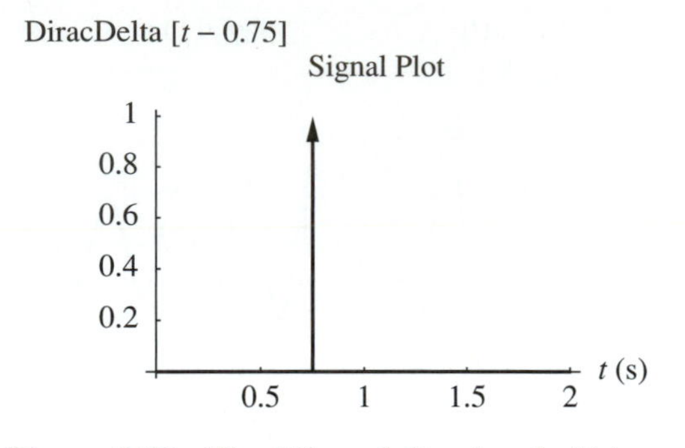

Figure 1.19 The Dirac delta signal, $\delta(t)$.

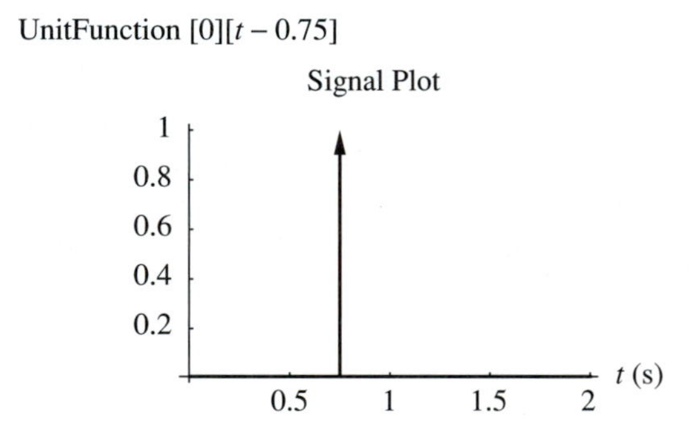

Figure 1.20 The Dirac delta signal, $\delta(t)$.

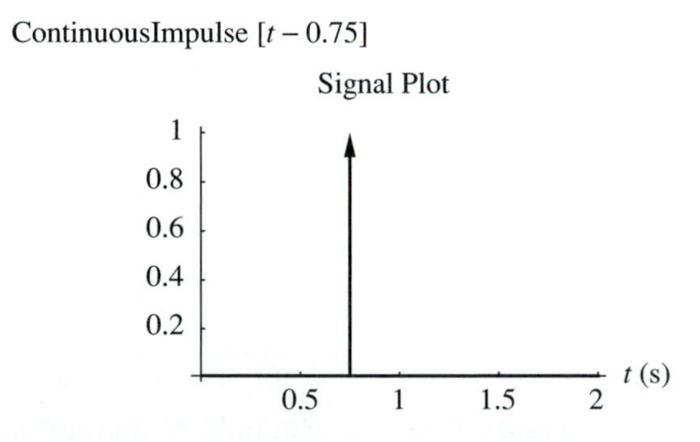

Figure 1.21 The Dirac delta signal, $\delta(t)$.

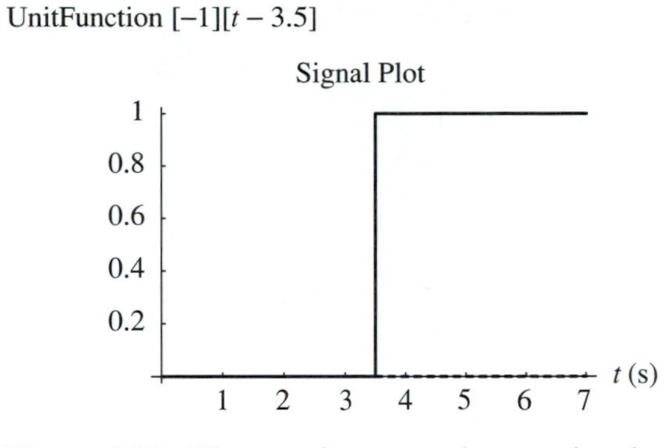

Figure 1.22 The continuous unit step signal, $u(t)$.

ContinuousStep [$t - 3.5$]

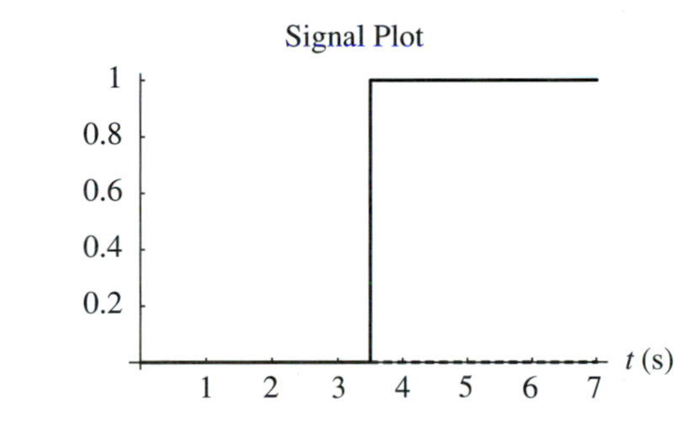

Figure 1.23 The continuous unit step signal, $u(t)$.

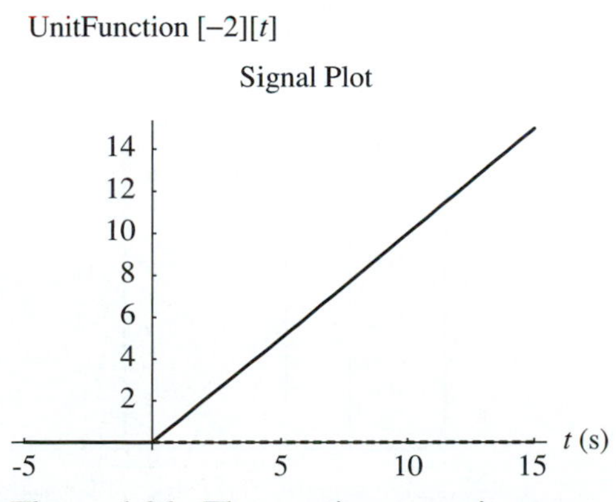

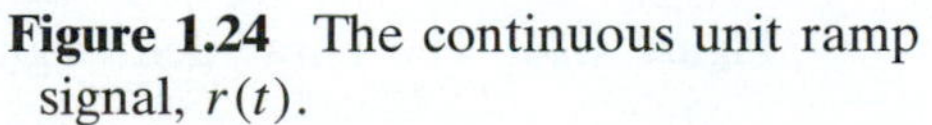

Figure 1.24 The continuous unit ramp signal, $r(t)$.

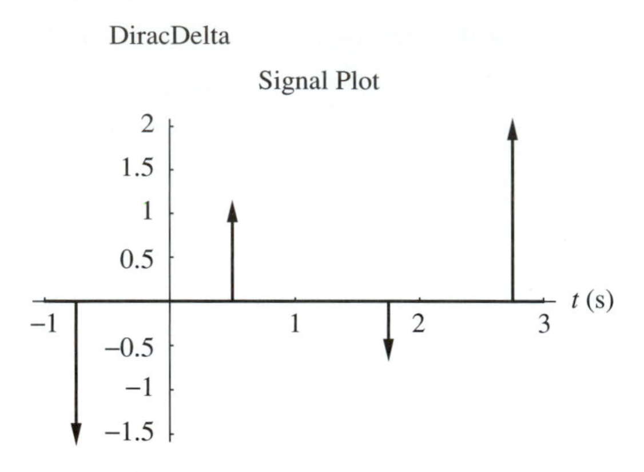

Figure 1.25 The Dirac delta signal, $\delta(t)$.

```
SignalPlot[UnitFunction[0][t-0.75],{t,0,2},PlotRange->All,
     AxesLabel->{"t (s)","UnitFunction[0][t-0.75]"}];
```

```
SignalPlot[ContinuousImpulse[t-0.75],{t,0,2},PlotRange->All,
     AxesLabel->{"t (s)","ContinuousImpulse[t-0.75]"}];
```

We draw the unit step function as `ContinuousStep[t]` or as the unit function with the order of the singularity -1, `UnitFunction[-1][t]`. When the order of the singularity of the unit function is -2, we draw the unit ramp signal.

```
SignalPlot[UnitFunction[-1][t-3.5],{t,0,7},
     PlotRange->All,PlotStyle->{Thickness[.01]},
     AxesLabel->{"t (s)","UnitFunction[-1][t-3.5]"}];
```

```
SignalPlot[ContinuousStep[t-3.5],{t,0,7},
     PlotRange->All,PlotStyle->{Thickness[.01]},
     AxesLabel->{"t (s)","ContinuousStep[t-3.5]"}];
```

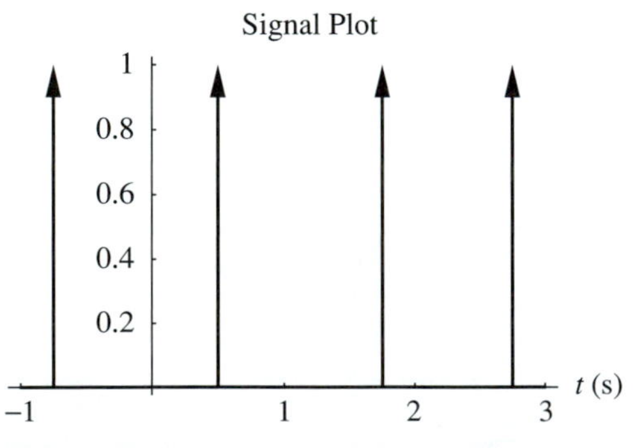

Figure 1.26 The Dirac delta signal, $\delta(t)$.

```
SignalPlot[UnitFunction[-2][t],{t,-5,15},
     PlotRange->All,PlotStyle->{Thickness[.01]},
     AxesLabel->{"t (s)","UnitFunction[-2][t]"}];
```

The weights of the impulse signal can be positive or negative. The negative impulse is with arrow toward the $-\infty$. If we represent by impulse height the value of the impulse, not the area, then the height is always zero because it makes no sense to draw taller and shorter arrows for impulse of infinite value. In SSP, we can draw impulse with always unit value by using `DiracDeltaScaling -> False`.

```
SignalPlot[{1.0 DiracDelta[t-0.5] -
            1.5 DiracDelta[t+0.75] -
            0.5 DiracDelta[t-1.75] +
            2.0 DiracDelta[t-2.75]},
           {t,-1,3},
     DiracDeltaScaling -> True,
     AxesLabel->{"t (s)","DiracDelta"}];

SignalPlot[{1.0 DiracDelta[t-0.5] -
            1.5 DiracDelta[t+0.75] -
            0.5 DiracDelta[t-1.75] +
            2.0 DiracDelta[t-2.75]},
           {t,-1,3},
     DiracDeltaScaling -> False,
     AxesLabel->{"t (s)","DiracDelta"}];
```

1.8 SEQUENCES IN MATHEMATICA

We can express a sequence in analytical form. For example, a discrete signal $x_k = e^{-0.05\,k}\sin(\pi\ 0.53\,k)$ we can describe by the *Mathematica* script as follows:

```
X=1;
a=0.1;
f=0.53;
t0=0.5;
phi = 0;
x[k_] := X^a * E^(-a t0 k) Sin[2 Pi f t0 k + phi];
DiscreteSignalPlot[x[k],{k,0,30},
     AxesLabel->{"k","x(k)"},
     PlotRange->{-1,1},
     AxesOrigin->{0,-1},
GridLines->{{0,0},{0,0}}];
```

Instead of `SignalPlot` we use `DiscreteSignalPlot`. The same example was considered in the MATLAB script. The resulting plot is in Fig. 1.27.

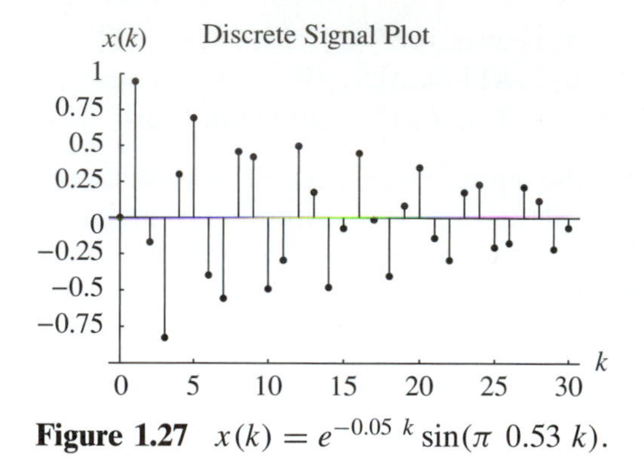

Figure 1.27 $x(k) = e^{-0.05\ k} \sin(\pi\ 0.53\ k)$.

The second example of a discrete-time signal is described by the *Mathematica* script:

```
X=0.02;
a=0.1;
t0=0.5;
f=0.53;
phi = 3*Pi/4;
x[k_] :=  X^a * E^(-a t0 k) Sin[2 Pi f t0 k + phi];
DiscreteSignalPlot[{x[k]*DiscreteStep[k],
                    x[k],DiscreteStep[k]},{k,-10,20},
     AxesLabel->{"k","x(k)"},
     PlotRange->{-1.5,1.5},
     AxesOrigin->{-10,-1.5},
     PlotStyle->{{Thickness[.01],GrayLevel[0]},
                 Dashing[{.025}],Dashing[{.01}]},
GridLines->{{-10,0},{0,0}}];
```

The resulting plot is in Fig. 1.28. A new command `DiscreteStep[k]` is used for representing discrete-time step u_k.

We use `DiscretePulse[length,k]` to draw the discrete unit pulse signal (Fig 1.29). The number of discrete unit impulses is equal to the `length` with the first impulse at k=0.

```
DiscreteSignalPlot[DiscretePulse[3,k-4],{k,0,10},
     PlotStyle->{Thickness[.01]},
     AxesLabel->{"k","DiscretePulse[3,k-4]"}];
```

A unit discrete impulse is invoked by `KroneckerDelta[k]` (Fig. 1.30). It has a unit value for k=0 and zero value everywhere else.

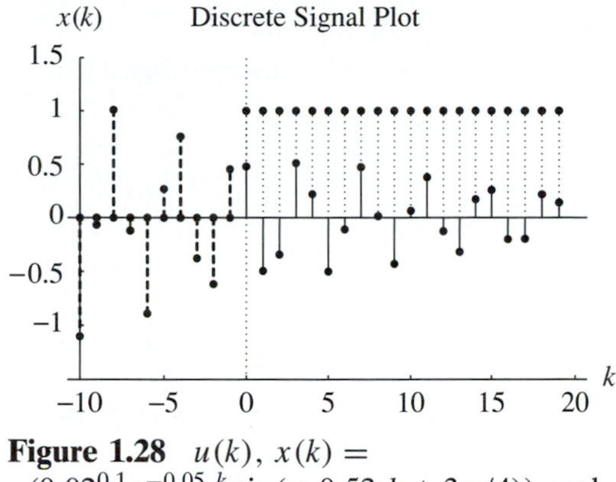

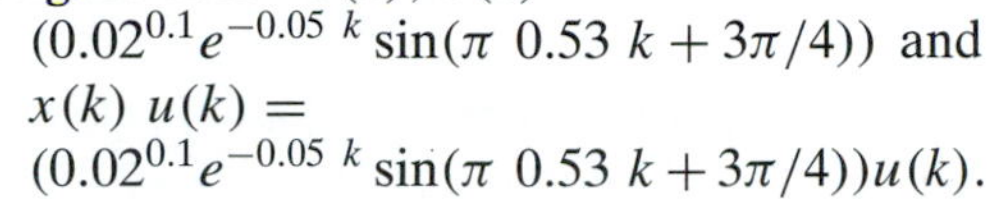

Figure 1.28 $u(k)$, $x(k) = (0.02^{0.1} e^{-0.05\ k} \sin(\pi\ 0.53\ k + 3\pi/4))$ and $x(k)\ u(k) = (0.02^{0.1} e^{-0.05\ k} \sin(\pi\ 0.53\ k + 3\pi/4))u(k)$.

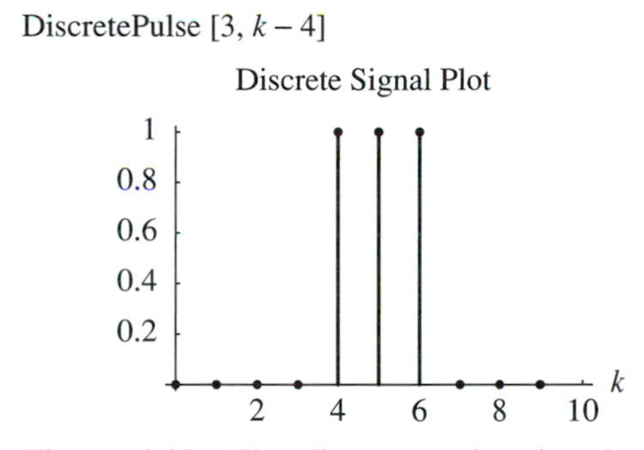

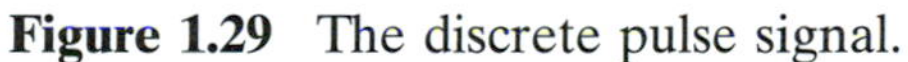

Figure 1.29 The discrete pulse signal.

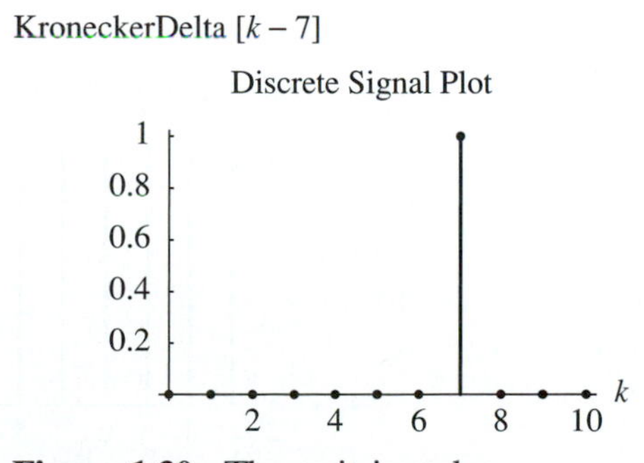

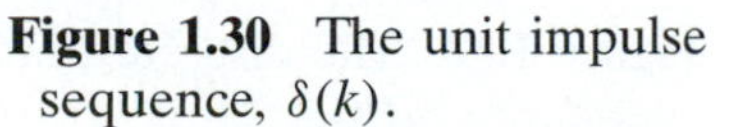

Figure 1.30 The unit impulse sequence, $\delta(k)$.

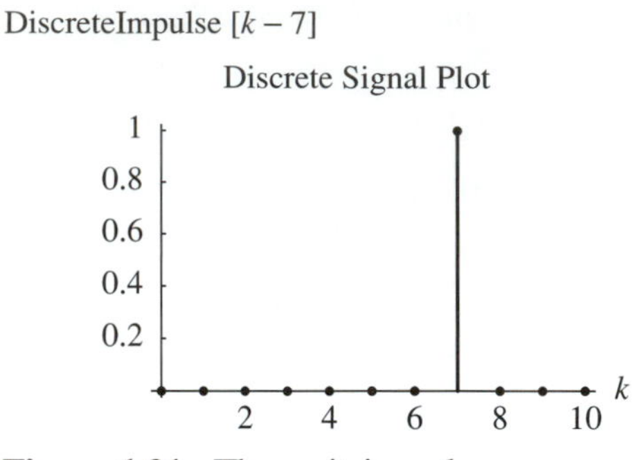

Figure 1.31 The unit impulse sequence, $\delta(k)$.

```
DiscreteSignalPlot[KroneckerDelta[k-7],{k,0,10},
     AxesLabel->{"k","KroneckerDelta[k-7]"}];
```

`KronekerDelta` is also called `DiscreteImpulse` as counterpart of `ContinuousImpulse` (Fig. 1.31).

```
DiscreteSignalPlot[DiscreteImpulse[k-7],{k,0,10},
     AxesLabel->{"k","DiscreteImpulse[k-7]"}];
```

A unit step sequence is invoked by `DiscreteStep[k]` (Fig. 1.32). It is zero for negative `k` and it has a unit value for `k=0` and `k>0`. We assume that the index of the first nonzero element in the step sequence is zero. This is opposite to standard *Mathematica* and MATLAB usage where the first index is 1.

```
DiscreteSignalPlot[DiscreteStep[k-3],{k,0,10},
     PlotRange->All,PlotStyle->{Thickness[.01]},
     AxesLabel->{"k","DiscreteStep[k-3]"}];
```

Sequences of data may be read from a file on disk or generated by a formula. We use the function `ToSampledData[function]` to create a special data structure to store the signal values (Fig. 1.33).

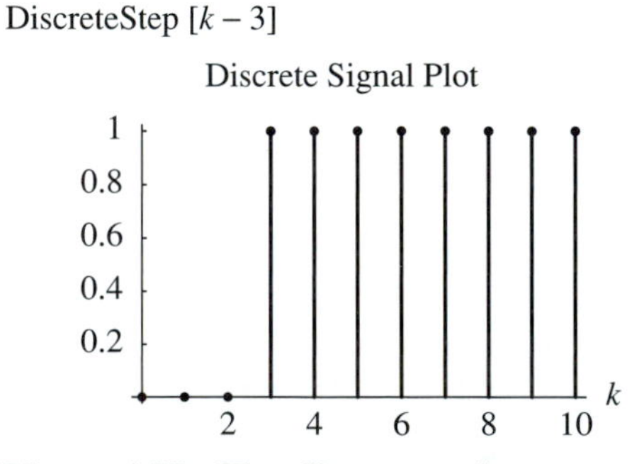

Figure 1.32 The discrete unit step sequence, $u(k)$.

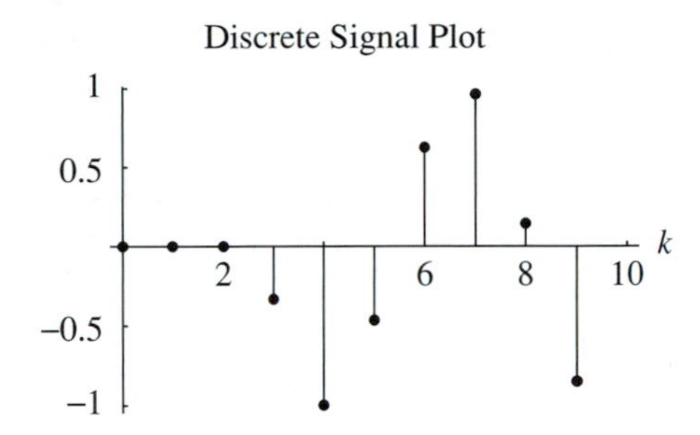

Figure 1.33 $\sin(2 \times 0.58k)u(2k - 6)$.

```
DiscreteSignalPlot[
   N[ToSampledData[Sin[0.58 k 2]*ContinuousStep[k 2 - 6],
     {k,0,10}]], {k,0,10},
     PlotRange->All,AxesLabel->{"k","ToSampledData  t -> k T"}];
```

Continuous-time and discrete-time signals represented by formulas are easily manipulated as expressions. This enables us to simplify or rearrange expressions until they take a desired form even in the case when the signal duration is infinite.

PROBLEMS

1.1 Which ones of the following signals are continuous signals?

(a) $x(t) = e^{-t}$,

(b) $x(t) = \sin(2\pi t)$,

(c) $x(t) = \sin(4\pi t + \pi/2)$,

(d) $x(t) = \frac{1}{t}\sin(2\pi t)$,

(e) $x(t) = \left(\frac{(t+1)^2}{t} - \frac{(t-1)^2}{t}\right)\cos(2\pi t)$,

(f) $x(t) = \sin(4\pi t)\ |\cos(2\pi t)| + t - 1$,

(g) $x(t) = \mathrm{F}\,(t)$, and $\mathrm{F}\,(|t|) = |t|$, $\mathrm{F}\,(-|t|) = 0$,

(h) $x(t) = \mathrm{F}\,(\sin(2\pi t + \pi/4))$, and $\mathrm{F}\,(0) = 0$, $\mathrm{F}\,(|a|) = 1$, $\mathrm{F}\,(-|a|) = -1$.

1.2 Sketch the signals given in Problem 1.1 for values of t in the range $-2 < t < 2$.

1.3 Which ones of the signals given in Problem 1.1 are periodic signals?

1.4 Determine the period of the signal

$$x(t) = \left(\frac{(t+1)^2}{t} - \frac{(t-1)^2}{t}\right)\sin(2\pi t), \quad t \neq 0$$
$$x(0) = \lim_{t\to 0} x(t)$$

1.5 Sketch the following signals:

(a) $u(t-2)$,

(b) $-\frac{1}{2}u(t-3)$,

(c) $\frac{4}{5}u(t+4)$,

(d) $u(t-2)-u(t-6)$.

1.6 Sketch the following signals:

(a) $e^{-t}u(t-2)$,

(b) $2\left(1-e^{-t}\right)\, u(t-3)$,

(c) $e^{-t}\cos(2\pi t+\frac{\pi}{4})\, u(t)$,

(d) $e^{t}\cos(\pi t)\, u(t)$.

1.7 Which ones of the following signals are digital signals for $k \in \{-4,-3,-2,-1,0,1,2,3,4\}$?

(a) $x(k)=e^{-4k}$,

(b) $x(k)=\dfrac{k(k+1)(k+2)}{3}\sin(\pi k/2+\pi/4)$,

(c) $x(k)=\dfrac{1}{3}\sin(\pi k+\pi/2)$,

(d) $x(k)=\dfrac{1}{k}\sin(\pi k/2)$,

(e) $x(k)=\left(\dfrac{(k+4)^2}{k}-\dfrac{(k-4)^2}{k}\right)\sin(\pi k/2)$,

(f) $x(k)=\sin(\pi k)\ |\cos(\pi k/2)|+5k-1$,

(g) $x(k)=\sin(\pi k/2)\ |\cos(\pi k)|+4k-1$,

(h) $x(k)=\mathrm{F}\,(k)$, and $\mathrm{F}\,(|k|)=|k|$, $\mathrm{F}\,(-\,|k|)=0$,

(i) $x(k)=\mathrm{F}\,(\sin(\pi k/2))$, and $\mathrm{F}\,(0)=0$, $\mathrm{F}\,(|k|)=1$, $\mathrm{F}\,(-\,|k|)=-1$.

1.8 Sketch the following sequences:

(a) $u(k-2)$,

(b) $-\frac{1}{2}u(k-3)$,

(c) $\frac{4}{5}u(k+4)$,

(d) $u(k-2)-u(k-6)$.

1.9 Sketch the following sequences:

(a) $e^{-k}u(k-2)$,

(b) $2\left(1-e^{-k/5}\right)\, u(k-3)$,

(c) $e^{-k}\cos(\pi k+\frac{\pi}{4})\, u(k)$,

(d) $e^{k}\cos(\frac{\pi}{4}k)\, u(k)$.

1.10 Sketch the sawtooth signal $x(t) = \frac{2}{3}t\ (u(t) - u(t-3))$. Sketch the periodic sawtooth signal $y_1(t) = \sum_{k=-\infty}^{+\infty} x(t-3k)$. Determine the period of the signals $y_1(t) = \sum_{k=-\infty}^{+\infty} x(t-3k)$, $y_2(t) = \sum_{k=-\infty}^{+\infty} x(t+4k)$, and $y_3(t) = \sum_{k=-\infty}^{+\infty} x(t-5k)$.

1.11 Sketch the triangular pulse

$$x(t) = \begin{cases} 0, & t \leq 0 \\ 4t, & 0 < t \leq 1 \\ 5 - t, & 1 < t \leq 5 \\ 0, & 5 < t \end{cases}$$

Sketch the periodic triangular signal $y_1(t) = \sum_{k=-\infty}^{+\infty} x(t+5k)$. Determine the period of the signals $y_1(t) = \sum_{k=-\infty}^{+\infty} x(t-5k)$, $y_2(t) = \sum_{k=-\infty}^{+\infty} x(t+10k)$, and $y_3(t) = \sum_{k=-\infty}^{+\infty} x(t-15k)$. Use the unit step signal to describe analytically the signal $x(t)$.

1.12 Sketch the sequence $x(k) = u(k) - u(k-3)$. Sketch the sequences $y_1(k) = \sum_{m=-\infty}^{+\infty} x(k+3m)$, $y_2(k) = \sum_{m=-\infty}^{+\infty} x(k+4m)$, $y_3(k) = \sum_{m=-\infty}^{+\infty} x(k+5m)$. Determine the period of the sequences $y_1(k)$, $y_2(k)$, and $y_3(k)$.

1.13 Sketch the sequence

$$x(k) = \begin{cases} 0, & k < 0 \\ k, & 0 \leq k < 2 \\ 2, & 2 \leq k < 4 \\ 6 - k, & 4 \leq k < 6 \\ 0, & 6 \leq k \end{cases}$$

Sketch the sequence $y_1(k) = \sum_{m=-\infty}^{+\infty} x(k-10m)$. Determine the period of the sequences $y_1(k) = \sum_{m=-\infty}^{+\infty} x(k-10m)$, $y_2(k) = \sum_{m=-\infty}^{+\infty} x(k-6m)$, and $y_3(k) = \sum_{m=-\infty}^{+\infty} x(k-4m)$. Use the unit step sequence to describe analytically the sequence $x(k)$.

1.14 Sketch the signal $x(t) = \sum_{k=1}^{3} \frac{4}{\pi} \frac{1}{2k-1} \sin(2\pi(2k-1)t + m\frac{\pi}{2})$. (a) $m = 0$; (b) $m = 1$; (c) $m = k$; (d) $m = k/5$.

MATLAB EXERCISES

1.1 Write a MATLAB script to plot the signals given in Problem 1.1 for values of t in the range $-2 < t < 2$.

1.2 Write a MATLAB script to plot the signals given in Problem 1.5 for values of t in the range $-5 < t < 15$.

1.3 Write a MATLAB script to plot the sequences given in Problem 1.7.

1.4 Write a MATLAB script to plot the sequences given in Problem 1.8 for values of k in the range $-10 \leq k \leq 20$.

1.5 Write a MATLAB script to plot the signals given in Problem 1.10 for values of t in the range $-20 < t < 20$.

1.6 Write a MATLAB script to plot the signals given in Problem 1.11 for values of t in the range $-20 \leq t \leq 40$

1.7 Write a MATLAB script to plot the sequences given in Problem 1.12 for values of k in the range $0 < k < 40$.

1.8 Write a MATLAB script to plot the sequences given in Problem 1.13 for values of k in the range $0 < k < 40$.

1.9 Define $x(t)$ as

$$x(t) = \sin(2\pi ft + \phi)$$

Write a MATLAB script to generate and plot sinusoidal sequences.

(a) Assume that $f = 0.1, \phi = \pi/3, t = \{-5, -4, \ldots, 0, \ldots, 4, 5\}$.

(b) Assume that $f = \frac{1}{2}, \phi = \pi/2, t = \{-5, -4, \ldots, 0, \ldots, 4, 5\}$.

(c) Assume that $f = \frac{1}{2}, \phi = \pi, t = \{-5, -4, \ldots, 0, \ldots, 4, 5\}$.

(d) Assume that $f = \frac{1}{2}, \phi = \pi/4, t = \{-5, -4, \ldots, 0, \ldots, 4, 5\}$.

Use the MATLAB functions `stem`, `bar`, and `stairs`.

1.10 Define $x(t)$ as

$$x(t) = A^{-t}\sin(2\pi ft + \phi), \quad t > 0$$

Write a MATLAB script to generate the quantized sequence by using the functions `ceil`, `floor`, `round`, and `fix`. Plot the sequence using the function `stem`. Assume that $f = 0.1$, $\phi = \pi/3$, $A = 1.1$, $t = 0, 1, 2, 3, \ldots, 20$. Each sample can be represented by its sign and the magnitude values

$\{-1.0, -0.9, \ldots, -0.1, 0, 0.1, \ldots, 0.9, 1.0\}$.

1.11 Define $x(t)$ as

$$x(t) = A\sin(2\pi ft + \phi)$$

Write a MATLAB script that generates the sinusoidal sequence from $x(t)$. Plot the sequence using the built-in function `stem`. Assume that $x(0) = 0$, $x(1) = -1$, $x(2) = 0$, $x(3) = 1$.

MATHEMATICA EXERCISES

1.1 Write a *Mathematica* code to plot the signals given in Problem 1.1 for values of t in the range $-2 < t < 2$.

1.2 Write a *Mathematica* code to plot the signals given in Problem 1.5 for values of t in the range $-10 < t < 20$.

1.3 Write a *Mathematica* code to plot the sequences given in Problem 1.7.

1.4 Write a *Mathematica* code to plot the sequences given in Problem 1.8 for values of k in the range $-10 \leq k \leq 20$.

1.5 Write a *Mathematica* code to plot the signals given in Problem 1.10 for values of t in the range $-10 < t < 20$.

1.6 Write a *Mathematica* code to plot the sequences given in Problem 1.12 for values of k in the range $-10 < k < 20$.

1.7 Define $x(t)$ as

$$x(t) = \sin(2\pi f t + \phi)$$

Write a *Mathematica* code to generate sinusoidal sequences from $x(t)$ and plot them using the function `ListPlot`.

(a) Assume that $f = 0.1, \phi = \pi/3, t = \{-5, -4, \ldots, 0, \ldots, 4, 5\}$;

(b) Assume that $f = \frac{1}{2}, \phi = \pi/2, t = \{-5, -4, \ldots, 0, \ldots, 4, 5\}$;

(c) Assume that $f = \frac{1}{2}, \phi = \pi, t = \{-5, -4, \ldots, 0, \ldots, 4, 5\}$;

(d) Assume that $f = \frac{1}{2}, \phi = \pi/4, t = \{-5, -4, \ldots, 0, \ldots, 4, 5\}$.

1.8 Define $x(t)$ as

$$x(t) = A^{-t} \sin(2\pi f t + \phi), \quad t > 0$$

Write a *Mathematica* code to generate a quantized sequence from $x(t)$ for $f = 0.1$, $\phi = \pi/3$, $A = 1.1$, $t = 0, 1, 2, 3, \ldots, 20$. Use the functions `Round`, `Floor`, and `Ceiling`. Assume that each sample can be represented by its sign and the magnitude value from $\{-1.0, -0.9, \ldots, -0.1, 0, 0.1, \ldots, 0.9, 1.0\}$. Plot the sequence using the function `ListPlot`.

1.9 Define $x(t)$ as

$$x(t) = A \sin(2\pi f t + \phi)$$

Write a *Mathematica* code to generate a sinusoidal sequence from $x(t)$. Assume that $x(0) = 0, x(1) = -1, x(2) = 0, x(3) = 1$. Find A, ϕ, and f. Plot the sequence using the function `ListPlot`.

CHAPTER 2

SYSTEMS

In Chapter 1, we defined a signal as a "physical quantity, or quality, which conveys information." For this book, a key mathematical representation of a signal is as a function of a single variable we call time. A *system* takes one or more signals as input, performs operations on the signals, and produces one or more signals as output. So, the input is the stimulus or excitation applied to a system from an external source, usually in order to produce a specified response. The output is the actual response obtained from a system. In an algebraic framework, we can represent a system as an operator that maps input signals onto output signals [8].

We have already talked about sampling and quantizing continuous-time signals into digital signals in Chapter 1. In practice, this conversion is accomplished by an analog-to-digital (A/D) converter. An ideal representation of an A/D conversion system is as a cascade of a lowpass filter, a sampling device, and a quantizer. The lowpass filter will ideally pass components of continuous-time signals that are below a specific frequency and reject those above the specific frequency. Lowpass filtering is necessary for guaranteeing that we can choose a sampling period T (or equivalently a sampling rate $f_0 = \frac{1}{T}$) that satisfies the Sampling Theorem.

An A/D converter is an example of a dictionary definition of a system as "a group of related parts working together, or an ordered set of ideas, methods, or ways of working" [9]. From an implementation point-of-view, a system is an arrangement of physical components connected or related in such a manner as to form and/or act as an entire unit. From a signal processing perspective, a system can be viewed as any process that results in the transformation of signals, in which systems act on signals in prescribed ways [10, 11].

In this book, we define a *system* as a mapping of N input signals onto M output signals. The mapping carries out a transformation on the input signals according to a set of rules. The rest of this chapter presents basic definitions and background mathematics that will be used in the remainder of this book. Since many readers will already be familiar with this material, we shall aim to be logically consistent rather than mathematically rigorous.

2.1 BASIC DEFINITIONS

A system is said to be a *single-variable system* if it has only one input and only one output. A system is said to be a *multivariable system* if it has more than one input or more than one output. An equation that describes the relation between the input and the output of a system is called the *input–output relationship*, also known as the external description or the input–output description, of the system. In developing this relationship, we assume that the knowledge of the internal structure of a system is unavailable to us. Instead, the only access to the system is by means of the input ports and the output ports. Under this assumption, a system may be considered as a "*black box*." We apply inputs to a black box, measure their corresponding outputs, and then try to abstract key properties of the system from these input–output pairs [12].

We represent the input–output relation of a system using the notation

$$x \to y$$

or

$$x \overset{\mathcal{R}}{\to} y$$

where $\mathcal{R}$ is an operator—a *rule*, formula, or set of formulas for transforming a signal x into a signal y. Some authors put braces around the signal designation to indicate that there are, in general, several input signals and several output signals

$$\{x\} \overset{\mathcal{R}}{\to} \{y\}$$

or explicitly specify the input and the output

$$(x_1, x_2, \ldots, x_N) \overset{\mathcal{R}}{\to} (y_1, y_2, \ldots, y_M)$$

This notation assumes that the output is excited solely and uniquely by the input. It is also legitimate to write $y = \mathcal{R}(x)$, where $\mathcal{R}$ is some operator or function that specifies uniquely the output y in terms of the input x of the system. Other authors prefer to define a system as an interconnected collection of abstract objects, defined by relations of input–output pairs rather than by operators or functions [13].

Although systems of prime concern are those required for processing a signal that is a function of the single independent variable *time*, t, it should be appreciated that the independent variable in many applications need not be time. When more than one independent variable is involved, the system is multidimensional. For example, in a video processing systems, we have two spatial variables for each frame of video and a discrete-time variable to index the frames. For now, we focus on one-dimensional systems.

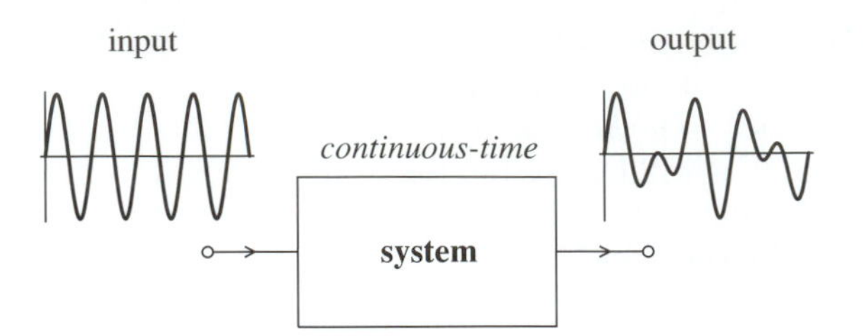

Figure 2.1 Continuous-time system.

The *time response* of a system is the output signal(s) as a function of time, following the application of a set of prescribed input signals, under specified operating conditions. In a *continuous-time system* (Fig. 2.1), the input and output signals are continuous-time. A *discrete-time system* (Fig. 2.2) has discrete-time input and output signals. A *hybrid-time system* (Fig. 2.3) is one in which both continuous-time and discrete-time signals appear at the inputs and outputs. An A/D converter is an example of a hybrid system: The input is continuous-time, but the output is discrete-time. Therefore, a continuous-time system is one in which continuous-time input signals are transformed into continuous-time output signals, and a discrete-time system is one that transforms discrete-time input signals into discrete-time output signals. A continuous-time system is symbolically represented as $x(t) \rightarrow y(t)$, and a discrete-time system is represented as $x(k) \rightarrow y(k)$ or $x_k \rightarrow y_k$, where k stands for the integer sample number.

A discrete-time system is *digital* if it operates on discrete-time signals whose amplitudes are quantized. Quantization maps each continuous amplitude level into a number. The digital system employs digital hardware either (a) explicitly in the form of the usual collection of logic circuits or (b) implicitly when the operations on the signals are executed by writing a computer program [14].

A *lumped system* is one that can be decomposed into a finite number of components, each with a finite number of inputs and outputs, and such that the values of the outputs at every time are functions of the inputs, their derivatives, and their integrals, at the same instant of time. Since the velocity of light and velocity of sound are finite, all physical objects with inputs and outputs at different places are not lumped. Nevertheless, the lumped system can provide a good approximation for a physical system.

The quantity that is responsible for activating the system to produce the output is called the *control action*. An *open-loop system* is one in which the control action is independent of the output. A *closed-loop* system is one in which the control action

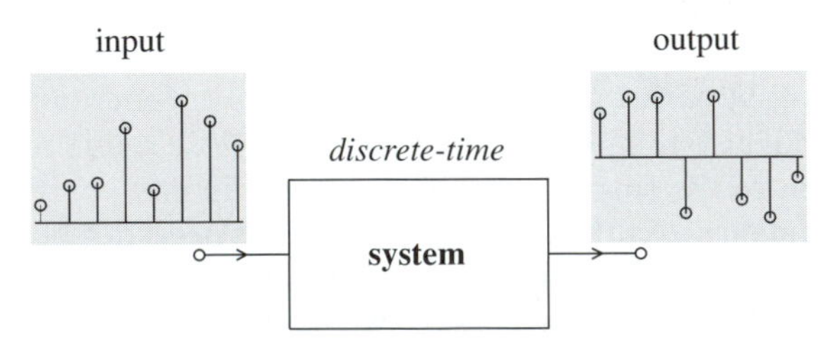

Figure 2.2 Discrete-time system.

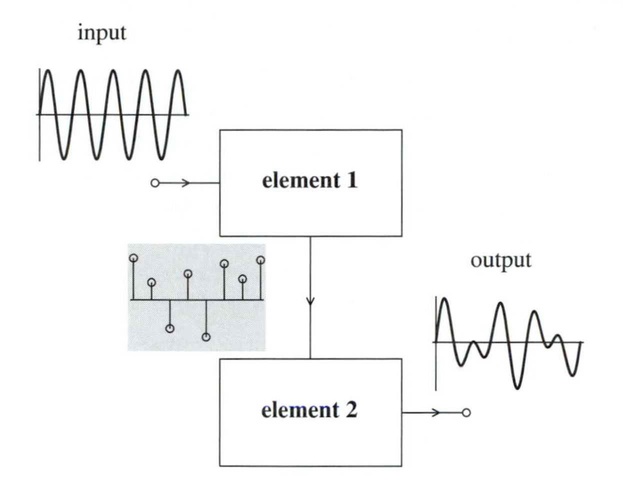

Figure 2.3 Hybrid-time system.

is dependent on the output. Closed-loop systems are commonly called feedback systems. *Feedback* is that property of a closed-loop system which permits the output to be compared with the input to the system so that the appropriate control action may be formed as some function of the output and input. A *control system* is one that commands, directs, or regulates itself or another system.

In many cases, we are presented with a specific system and are interested in characterizing it in detail to understand how it will respond to various inputs. In another context, our interest may be focused on the problem of designing systems to process signals in particular way. *Analysis* of a system is investigation of the properties and the behavior (response) of an existing system. *Design* of a system is the choice and arrangement of systems components to perform a specific task. *Design by analysis* is accomplished by modifying the characteristics of an existing system. *Design by synthesis* means that we define the form of the system directly from its specifications.

In order to analyze, design, and evaluate a system, the description of its components and their interconnections must be put into a suitable form. A mathematical or graphical representation of a system is called the *model.* The model resembles the system in its salient features but is easier to study. Some authors refer to models of physical systems as systems. Therefore, a physical system is a device or a collection of devices existing in the real world, and a system is a model of a physical system [12].

A *mathematical model* is a set of mathematical relations representing the system. These relations may assume many forms, including linear equations, nonlinear equations, integral equations, differential equations, and difference equations. The solution of these equations represents the system's behavior. Often, this solution is difficult, if not impossible, to find, and certain simplifying assumptions must be made in the mathematical description. Frequently, these approximations and simplifications lead to systems describable by linear ordinary differential equations or difference equations.

2.2 BLOCK DIAGRAMS

A *block diagram* is a shorthand, pictorial representation of the cause and effect relationship between the input and output of a system. It provides a convenient and useful method for characterizing the functional relationships among the various components of a system. Block diagrams are representations of either (a) the schematic diagram of a physical system or (b) the set of mathematical equations characterizing its parts. The simplest form of the block diagram is the single *block*, with one input and one output (Fig. 2.4). The interior of the rectangle representing the block usually contains (a) the name of the component, (b) a description of the component, or (c) the symbol for the mathematical operation to be performed on the input to yield the output. The *arrows* represent the direction of unilateral information or signal flow. The standard symbols used to represent various types of blocks are shown in Fig. 2.5.

The operations of addition and subtraction are represented by a circle, called a *summing point*, with the appropriate plus or minus sign associated with the arrows entering the circle (Fig. 2.6). The output is the algebraic sum of the inputs. Any number of inputs may enter a summing point. Some authors put the plus sign "+" or the Greek letter "Σ" in the circle.

In order to employ the same signal or variable as an input to more than one block or summing point, a *takeoff point* is used, as shown in Fig. 2.7. This permits the signal to proceed unaltered along several different paths to several destinations. A block with N inputs and M outputs is represented by a rectangle as shown in Fig. 2.8. The blocks representing the various components of a system are connected in a fashion which characterizes their functional relationship within the system. The arrows connecting one block with another represent the direction of flow of signals or information. In general, a block diagram consists of a specific configuration of four types of elements: blocks, summing points, takeoff points, and arrows representing unidirectional signal flow.

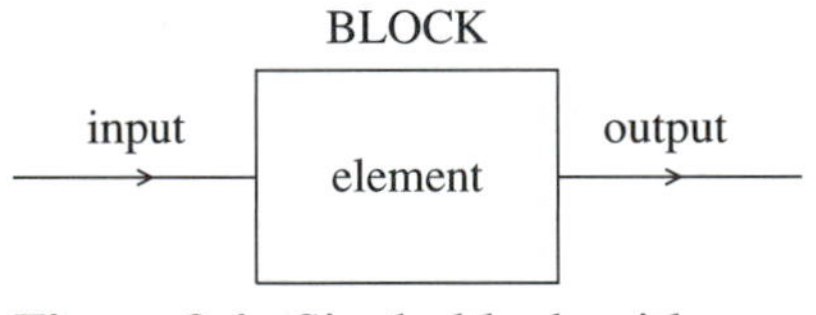

Figure 2.4 Single block with one input and one output.

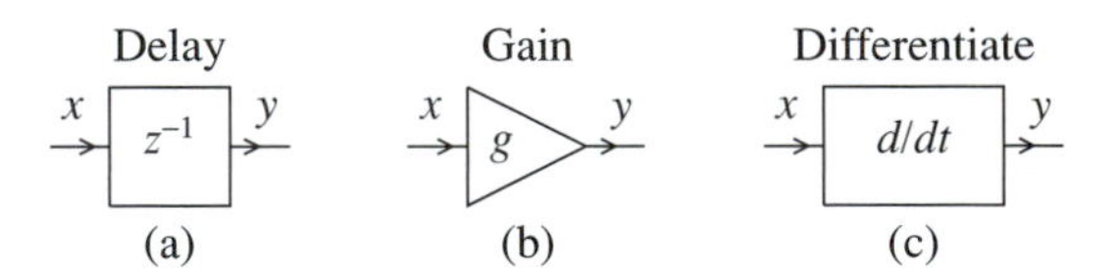

Figure 2.5 Various blocks: (a) Delay, $y(t) = x(t - T)$, some authors put the amount of delay T in the box, (b) Gain, multiplier by constant, or amplifier, $y = gx$, (c) Differentiate with respect to time, $y = \mathrm{d}x/\mathrm{d}t$.

Basically, two blocks may be connected in cascade (Fig. 2.9), in parallel (Fig. 2.10), or in feedback (Fig. 2.11). We combine cascade, parallel, and feedback interconnections to obtain more complicated interconnections. Next, we can use these interconnections to construct new systems or models out of existing ones.

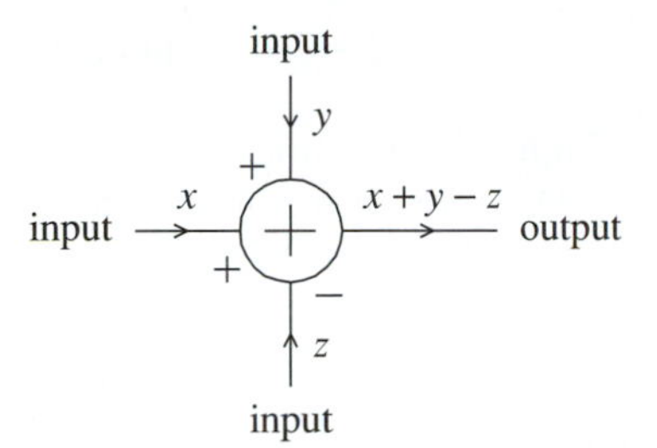

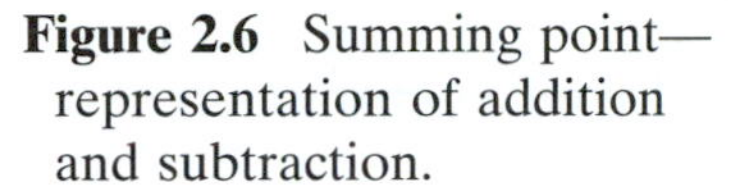
Figure 2.6 Summing point—representation of addition and subtraction.

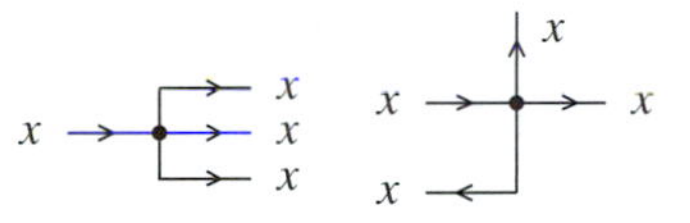

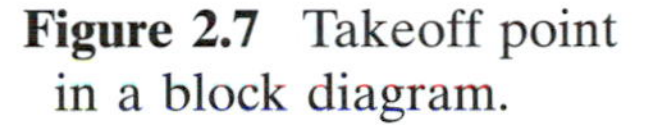
Figure 2.7 Takeoff point in a block diagram.

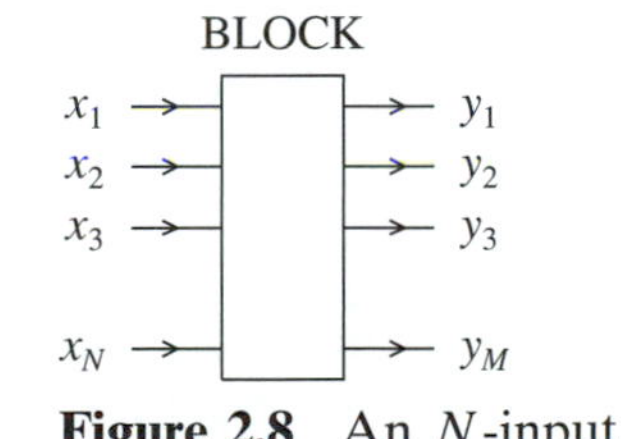

Figure 2.8 An N-input, M-output block.

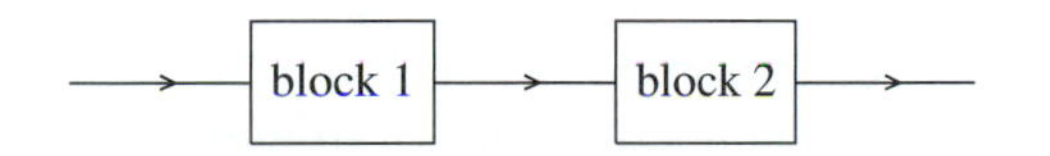

Figure 2.9 Blocks connected in cascade.

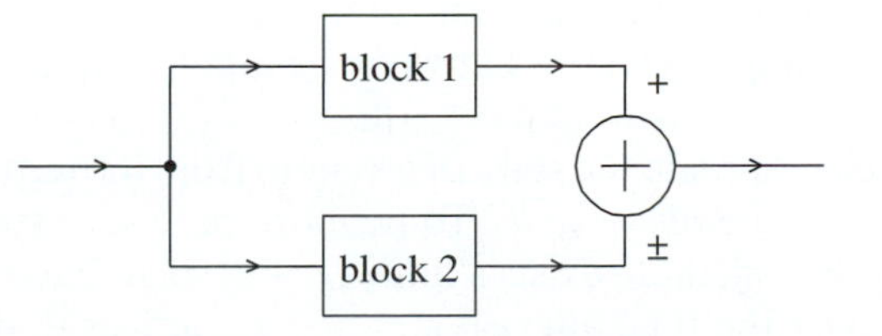

Figure 2.10 Blocks connected in parallel.

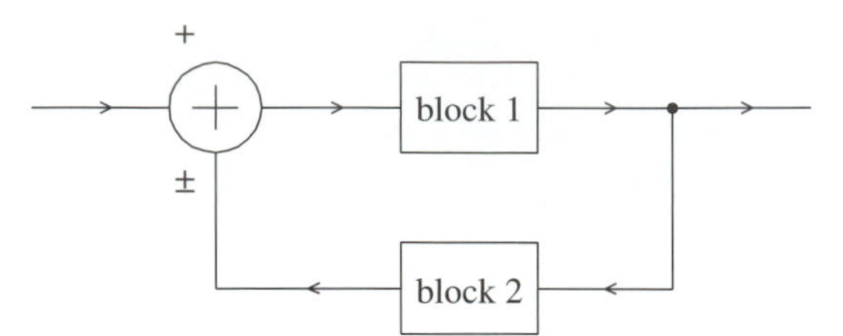

Figure 2.11 Blocks connected in feedback. The plus sign refers to a *positive* feedback, and the minus sign refers to a *negative* feedback.

2.3 SYSTEM PROPERTIES

2.3.1 State and Relaxed Systems

For most systems of interest, the output at time t_0 depends not only on the input applied at t_0, but also on the input applied before and after t_0. If an input $x(t)$, $t \geq t_0$, is applied to a system, unless we know the input applied before t_0, the output $y(t)$, $t \geq t_0$, is generally not uniquely determinable. For different inputs applied before t_0, we will obtain different output $y(t)$, $t \geq t_0$, although the same input $x(t)$, $t \geq t_0$, is applied. Hence, in developing the input–output description, before an input is applied, the system must be assumed to be *relaxed* or *at rest*, and that **the output is excited solely and uniquely by the input applied**. This definition is general and it is valid for an arbitrary system, continuous-time, discrete-time, and so on. If the concept of energy is applicable to a system, the system is said to be relaxed at time t_0 if no energy is stored in the system at that instant. A system is said to be relaxed at time t_0 if the output $y(t)$, $t \geq t_0$, is solely and uniquely excited by the input $x(t)$ defined for $t \geq t_0$.

We shall assume that every system is relaxed at time $-\infty$. We shall call a system that is initially relaxed at $-\infty$ an *initially relaxed system*, or a *relaxed system*, for short. In this section, whenever we talk about the input–output pairs of a system, we mean only those input–output pairs derived or formulated under relaxed system assumptions.

The input–output description of a system is applicable only when the system is initially relaxed. If a system is not initially relaxed, say at time t_0, then it depends also on the set of initial conditions at t_0, which we call the *state*. The state is the information that, together with the input, determines **uniquely** the output. The state of a system at time t_0 is the minimum amount of information needed about the system at time t_0, in addition to the system diagram (or system equations) and the system inputs between t_0 and $t_1 > t_0$, so that all outputs at any time t_1 can be determined. The set of equations that describes the unique relations between the input, output, and state is called a *dynamical equation*.

If we can estimate the state of a system from its output, in a finite time interval, the system is said to be *observable*. To be more precise, a system is observable at t_0, if for any state at time t_0, there exists a finite $t_1 > t_0$ such that the knowledge of the input and the output over the time interval $t_0 \leq t \leq t_1$ suffices to determine the state. If we can steer the state of a system from the input, in a finite time interval, the system is referred to as *controllable*. We can find inputs capable of moving any state of a controllable system to any other state in a finite time. Furthermore, there is no constraint imposed on the input—its magnitude can be as large as desired.

2.3.2 Causality and Realizable Systems

A system in which time is the independent variable is called *causal* if the output depends only on the present and past values of the input. If $y(t)$ is the output, then $y(t)$ depends only on the input $x(\tau)$ for values of $\tau \leq t$. Such a system is often referred to as being *nonanticipative*, because the system output does not anticipate future values of the input.

Causal systems are important, but they are not the only systems that are of practical significance. Causality is not of fundamental importance in

- applications, such as image processing, in which the independent variable is not time;
- processing recorded data for which time is the independent variable, as often happens with speech; and
- many applications, such as in stock market analysis and demographic studies, when we want to determine a slowly varying trend in data, and when we average data over an interval to smooth out the fluctuations and keep only the trend.

Finally, noncausal systems may arise in the course of analysis, when decomposing or recombining causal systems. Causal systems are sometimes called *physically realizable* systems.

2.3.3 Stability

The stability of a system is determined by its response to inputs or disturbances. Intuitively, a stable system is one that will remain at rest unless excited by an external source and will return to rest if all excitations are removed. The definition of a stable system can be based upon the response of the system to *bounded inputs*. Bounded inputs have magnitudes that are less than some finite value for all time. A relaxed system is said to be bounded-input bounded-output (BIBO) stable if every bounded input produces a bounded output. The stability that is defined in terms of the input–output description is applicable only to relaxed systems. This stability is referred to as the *input–output stability*. A zero-input system is said to be *asymptotically stable* if the response approaches zero asymptotically as time t approaches infinity. If the response remains bounded for $t > t_0$, we speak about the *bounded stability*. There are several methods (criteria) for determining system stability: Routh, Hurwitz, Liénard–Chipart, Nyquist, Lyapunov, and so on.

Consideration of the stability of a system provides valuable information about its behavior and is an important issue in system design. Often, it is desirable to determine a range of values of a particular system parameter for which the system is stable. The concept of stability is extremely important, because almost every workable system is designed to be stable. If a system is not stable, it is usually of no use in practice.

2.3.4 Time Invariance

A relaxed system is *time-invariant*, also known as fixed or stationary, if a time shift in the input signal causes a time shift in the output signal. Therefore, if an input $x(t)$ produces an output $y(t)$, then an input $x(t - t_0)$ produces an output $y(t - t_0)$. In other words, no matter at what time an input is applied to a time-invariant relaxed system, the waveform of the output is the same. We can also say that a (relaxed) system is time-invariant if delaying the input by t_0 seconds merely delays the response by t_0 seconds. A relaxed discrete-time system is said to be time-invariant if a shifted input $x(k - k_0)$ produces a shifted output $y(k - k_0)$. In the case of discrete-time digital systems, we often use the term *shift-invariant* instead of time-invariant. The characteristics and parameters of a time-invariant system do not change with time. A relaxed system that is not time-invariant is said to be *time-varying*.

2.3.5 Linearity and Superposition

Consider a relaxed system in which there is one independent variable t. A *linear system* is a system which has the property that if:

(a) an input $x_1(t)$ produces an output $y_1(t)$ and

(b) an input $x_2(t)$ produces an output $y_2(t)$, then

(c) an input $c_1x_1(t) + c_2x_2(t)$ produces an output $c_1y_1(t) + c_2y_2(t)$ for all pairs of inputs $x_1(t)$ and $x_2(t)$ and all pairs of constants c_1 and c_2.

Otherwise the (relaxed) system is *nonlinear*. This is the concept of zero-state linearity. The concept of linearity can be represented by the principle of superposition.

Theorem 2.1 Principle of Superposition: The response $y(t)$ of a linear system due to several inputs $x_1(t), x_2(t), \ldots, x_N(t)$ acting simultaneously is equal to the sum of the responses of each input acting alone, that is, if $y_i(t)$ is the response due to the input $x_i(t)$, then $y(t) = \sum_{i=1}^{N} y_i(t)$.

The principle of superposition follows directly from the definition of linearity. Any system which satisfies the principle of superposition is linear.

In the engineering literature, linearity is often defined in terms of the following two properties:

1. Property of *additivity*. If an input $x_1(t)$ produces an output $y_1(t)$, and an input $x_2(t)$ produces an output $y_2(t)$, then an input $x_1(t) + x_2(t)$ produces an output $y_1(t) + y_2(t)$ for all pairs of inputs $x_1(t)$ and $x_2(t)$.
2. Property of *homogeneity*. If an input $x(t)$ produces an output $y(t)$, then an input $a\,x(t)$ produces an output $a\,y(t)$ for any input $x(t)$ and any constant a.

The property of homogeneity does not imply the property of additivity and vice versa.

Linear systems can often be represented by linear differential equations or difference equations. Also, any continuous-time system is linear if its input–output relationship can be described by the ordinary linear differential equation

$$\sum_{i=0}^{n} a_i(t)\frac{\mathrm{d}^i y}{\mathrm{d}t^i} = \sum_{k=0}^{m} b_k(t)\frac{\mathrm{d}^k x}{\mathrm{d}t^k} \tag{2.1}$$

where $y = y(t)$ is the system output and $x = x(t)$ is the system input. The coefficients $a_i(t)$ and $b_k(t)$ depend only upon the independent variable t (e.g., time).

Initially relaxed linear systems possess the following property: Zero input yields zero output. An *incrementally linear system* is one that responds linearly to changes in the input; that is, the difference in the response to any two inputs is a linear function of the difference between the two inputs. Many of the characteristics of such systems can be analyzed using the techniques developed for linear systems. In reality, no physical system can be described exactly by a linear differential or difference equation. Many systems, however, can be represented over a limited operating range, or approximated by such equations.

2.3.6 Memoryless Systems

A system is said to be *memoryless*, also known as zero-memory or instantaneous, if its output for each value of the independent variable is dependent only on the input at that same time. The simplest memoryless system is the *identity system*, whose output is identical to its input. The input–output relationship for the continuous-time identity system is $y(t) = x(t)$, and the corresponding relationship in discrete time is $y(k) = x(k)$. All memoryless systems are causal.

2.3.7 Sensitivity

A first step in the analysis or design of a system is the generation of models for the various elements in the system. The system characteristics are fixed when a finite number of constant parameters have been chosen. The values given to these parameters are called the *nominal values*, and the corresponding characteristics are called the *nominal characteristics*. The accuracy of the model depends on how closely these nominal parameter values approximate the actual parameter values, and how much these parameters deviate from the nominal values during the course of system operation.

The *sensitivity* of a system can be defined as a measure of the amount by which a system characteristic differs from its nominal value when one of its parameters differs from the number chosen as its nominal value. Consider a characteristic C that depends on a parameter p; that is, the mathematical model of the characteristic is $C = C(p)$.

Usually, p is a real or complex quantity representing some identifiable parameter of the system. The sensitivity of $C(p)$ with respect to the parameter p is defined by

$$S_p^{C(p)} = \frac{\mathrm{d}C(p)/C(p)}{\mathrm{d}p/p} = \frac{\mathrm{d}C(p)}{\mathrm{d}p}\frac{p}{C(p)} = \frac{\mathrm{d}\ln C(p)}{\mathrm{d}\ln p} \tag{2.2}$$

where p is regarded as a variable; $S_p^{C(p)}$ is called the *single-parameter relative sensitivity*. The relative change of C due to the variation of N parameters, $p_1, p_2, \ldots, p_N$, is given by

$$\frac{\Delta C}{C} = \sum_{i=1}^{N} S_{p_i}^{C} \frac{\Delta p_i}{p_i} \tag{2.3}$$

If we define the *variation* of a function C due to a relative change in a parameter p by

$$V_p^C = S_p^C \frac{\mathrm{d}p}{p} \tag{2.4}$$

then the *worst-case* variation follows as

$$W^C = \sum_{i=1}^{N} |V_{p_i}^C| \tag{2.5}$$

Another useful figure-of-merit is the so-called *Schoeffler criterion*, or the sum-of-squares sensitivity:

$$S^C = \sum_{i=1}^{N} |V_{p_i}^C|^2 \tag{2.6}$$

In some cases it is useful to define the *semirelative sensitivity*

$$H_p^C = C(p)\, S_p^C \tag{2.7}$$

In general, the sensitivity is a complex number [15].

2.3.8 Optimal and Adaptive Systems

The basic goal of system design is meeting performance specifications. *Performance specifications* are the constraints put on mathematical functions describing system characteristics. They may be stated in any number of ways. The desired (or prescribed) system characteristics specify important properties of the system: speed of response, stability, system accuracy, allowable error, sensitivity, and so on. Satisfactory performance is determined by the application and the characteristics of the particular system.

The system measure of performance is called *performance index*. In more general designs, it is not specified. Instead, the system components, their interconnections, and the parameter values are chosen so that the performance index is maximized or minimized. The systems designed in this way are called *optimal systems*.

In many problems, the performance index is a measure or function of the error between the actual and ideal (or desired) responses. It is formulated in terms of the design parameters chosen, subject to existing physical constraints, to optimize the performance index. The measures of system performance are essentially *criteria of optimality*, and systems are designed taking them into account.

In some systems, certain parameters are either not constant or vary in an unknown manner. It may be desirable to design for the capability of continuously measuring them and changing the system so that the system performance criteria are always satisfied. Systems designed with these objectives are called *adaptive systems*.

Adaptive systems are an important issue in channel equalization, which is the recovery of a signal distorted in transmission through a communication channel with a nonflat magnitude or a nonlinear phase response. When the channel response is unknown, the process of signal recovery is called *blind equalization*.

2.3.9 Invertibility

A system is said to be *invertible* if distinct inputs lead to distinct outputs. By observing its output we can determine its input. An *inverse system* of a given system is one which, when cascaded with the given system, yields an output equal to the input: that is, if $x(t)$ is an input to the cascaded interconnection, its output will be $y(t) = x(t)$.

2.4 LINEAR TIME-INVARIANT SYSTEMS

When a system is both linear and time-invariant, it is called a *linear time-invariant* (LTI) system, and it is amenable to analysis using many techniques. A lumped continuous-time system is linear time-invariant if its input–output relationship can be described by the ordinary linear constant coefficient differential equation

$$\sum_{i=0}^{n} a_i \frac{\mathrm{d}^i y}{\mathrm{d}t^i} = \sum_{k=0}^{m} b_k \frac{\mathrm{d}^k x}{\mathrm{d}t^k} \tag{2.8}$$

where $y = y(t)$ is the system output, and $x = x(t)$ is the system input. The coefficients a_i and b_k do not depend on the independent variable t (time). If the equation describes a physical system, then generally $m \leq n$, and n is called the *order* of the system.

A lumped discrete-time system is linear time-invariant if its input–output relationship can be described by means of a linear constant coefficients difference equation

$$\sum_{i=0}^{n} a_i\, y(\nu - i) = \sum_{k=0}^{m} b_k\, x(\nu - k) \tag{2.9}$$

where the notation $y(\nu)$ means that the value of the sequence y at the νth sample. Some authors prefer to write (2.9) in the form

$$y_{(\nu)} + \sum_{i=1}^{n} a_i\, y_{(\nu-i)} = \sum_{k=0}^{m} b_k\, x_{(\nu-k)} \tag{2.10}$$

The parameter n is called the order of the difference equation. Discrete-time systems described by (2.10) are known as *recursive systems*, because the output depends on the previous values of the output as well as on the input. Another class of discrete systems is described by

$$y_{(\nu)} = \sum_{k=0}^{m} b_k \, x_{(\nu-k)} \tag{2.11}$$

and is known as *nonrecursive*, because the previous values of the output do not appear.

Differential and difference equations have broad application in the description of physical laws and are useful for relating rates of change of variables and other parameters of a system or its components. In actuality, LTI systems do not exist. All physical systems are nonlinear to some extent. Fortunately, a large percentage of systems can be represented by LTI models over a limited operating range. Many systems always operate within this linear range. Other systems exceed the limits of linear operation, but may be approximated by LTI systems.

2.4.1 The Differential Operator

Differential equations can be presented more compactly by introducing a *differential operator*

$$\mathrm{D} = \frac{\mathrm{d}}{\mathrm{d}t}$$

an ith-*order differential operator*

$$\mathrm{D}^i = \frac{\mathrm{d}^i}{\mathrm{d}t^i}$$

and more generally a *polynomial differential operator* (PDO)

$$A(\mathrm{D}) = \sum_{i=0}^{n} a_i \mathrm{D}^i$$

The differential equation (2.8) can be written as

$$A(\mathrm{D})\, y(t) = B(\mathrm{D})\, x(t)$$

where $A(\mathrm{D})$ maps the function $y(t)$ into the expression $\sum_{i=0}^{n} a_i \mathrm{D}^i y(t)$, and $B(\mathrm{D})\, x(t)$ represents the expression $\sum_{k=0}^{m} b_k \mathrm{D}^k x(t)$.

The polynomial in complex variable s

$$A(s) = \sum_{i=0}^{n} a_i s^i = a_n s^n + a_{n-1} s^{n-1} + \cdots + a_1 s + a_0$$

is called the *characteristic polynomial*. The equation $A(s) = 0$ is called the *characteristic equation*. From the definition of polynomial differential operators it follows:

$$A(\mathrm{D})\ (K\ y(t)) = (K\ A(\mathrm{D}))\ y(t)$$

$$A(\mathrm{D})\ (y_1(t) \pm y_2(t)) = A(\mathrm{D})\, y_1(t) \pm A(\mathrm{D})\, y_2(t)$$

$$A_1(\mathrm{D})\ (A_2(\mathrm{D})\, y(t)) = (A_1(\mathrm{D})\, A_2(\mathrm{D}))\ y(t) = A_2(\mathrm{D})\ (A_1(\mathrm{D})\, y(t))$$

where K is a constant, $y(t)$, $y_1(t)$, and $y_2(t)$ are functions of t, and $A(\mathrm{D})$, $A_1(\mathrm{D})$, and $A_2(\mathrm{D})$ are polynomial differential operators. Addition, $A_1(\mathrm{D}) + A_2(\mathrm{D})$, subtraction, $A_1(\mathrm{D}) - A_2(\mathrm{D})$, and multiplication, $A_1(\mathrm{D})\, A_2(\mathrm{D})$, are performed following the rules for the corresponding operations defined for algebraic polynomials in complex variables. Division $A_1(\mathrm{D})/A_2(\mathrm{D})$ is not defined.

2.4.2 Response of Continuous-Time LTI Systems

Polynomial differential operators are important in analysis of continuous-time linear systems. If a system is specified by its block diagram, we can use polynomial differential operators to simplify derivation of the system input–output description. To specify the problem completely so that the unique solution $y(t)$ can be obtained for a known time function $x(t)$ and the coefficients a_i and b_k, we must specify (1) the interval of time over which a solution is desired and (2) a set of n initial conditions for $y(t)$ and its first $n-1$ derivatives. In many cases the time interval is defined by $0 \le t < +\infty$ and the set of conditions is

$$y(0),\ \left.\frac{\mathrm{d}y}{\mathrm{d}t}\right|_{t=0},\ \ldots,\ \left.\frac{\mathrm{d}^{n-1}y}{\mathrm{d}t^{n-1}}\right|_{t=0}$$

The response of a continuous-time LTI system—that is, the solution $y(t)$—can be divided into two parts, namely, a free response and a forced response. The sum of two responses constitutes the total response.

The *free response* of an LTI system, $y_0(t)$, is the solution of the differential equation when the input $x(t)$ is identically zero. The free response depends only on the n initial conditions; some authors call it the *zero-input response*. The *forced response* of an LTI system, $y_x(t)$, is the solution of the differential equation when all initial conditions are identically zero; some authors call it the *zero-state response*. The forced response depends only on the input $x(t)$. The *total response* of an LTI system is the sum of the free response and the forced response, $y(t) = y_0(t) + y_x(t)$. The total response can be viewed, also, as the sum of the steady-state response and transient response. These terms are often used for specifying system performance. The *steady-state response* of an LTI system is that part of the total response which does not approach zero as time approaches infinity. The *transient response* is that part of the total response which approaches zero as time goes to infinity.

The general procedure for analyzing a system is as follows:

1. Determine the equations for each system component.
2. Choose a model for representing the system (e.g., block diagram).
3. Formulate the system model by appropriately connected the components.
4. Determine the system characteristics.

The direct solution of the system differential equation may be employed to find the total response or its parts, as well as the steady-state response and the transient response.

In the study of systems and differential equations which describe them, a particular family of functions (signals), the *singularity functions*, is used. These functions represent idealizations of physical phenomena and allow us to obtain simple representations of signals and systems. Assuming that signals described by singularity functions excite a system, we define the following special responses:

- *Unit step response* is the output of an LTI system to a unit step input when all initial conditions are zero.
- *Unit ramp response* is the output of an LTI system to a unit ramp input when all initial conditions are zero.
- *Unit impulse response* is the output of an LTI system to a unit impulse input when all initial conditions are zero.

If we know the unit impulse response, $y(t) = y_\delta(t)$, of an initially relaxed causal CTLTI system, then the forced response, $y_x(t)$, of the system, when the input is a known time function $x(t)$, can be written in terms of the *convolution integral*:

$$y_x(t) = \int_{-\infty}^{t} y_\delta(t - \tau)\, x(\tau)\, d\tau$$

This formula directly follows from the principle of superposition and the sifting property of the unit impulse function.

2.4.3 Transient System Specifications

In many cases the desired, or prescribed, system characteristics are specified in terms of time response. Time-domain specifications are customarily defined in terms of unit step function. Typical transient performance specifications are:

1. *Overshoot*—the maximum difference between the transient and steady-state response for a unit step function input (Fig. 2.12).
2. *Delay time*—the time required for the response to a unit step function input to reach 50% of its final value (Fig. 2.13).
3. *Rise time*—the time required for the response to a unit step function input to rise from 10% to 90% of its final value (Fig. 2.13).
4. *Settling time*—the time required for the response to a unit step function input to reach and remain within a specified percentage (frequently 2% or 5%) of its final value (Fig. 2.12).

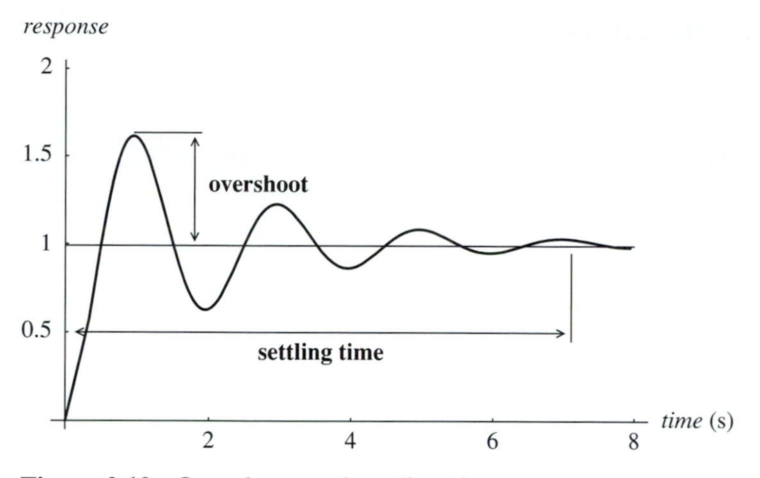

Figure 2.12 Overshoot and settling time.

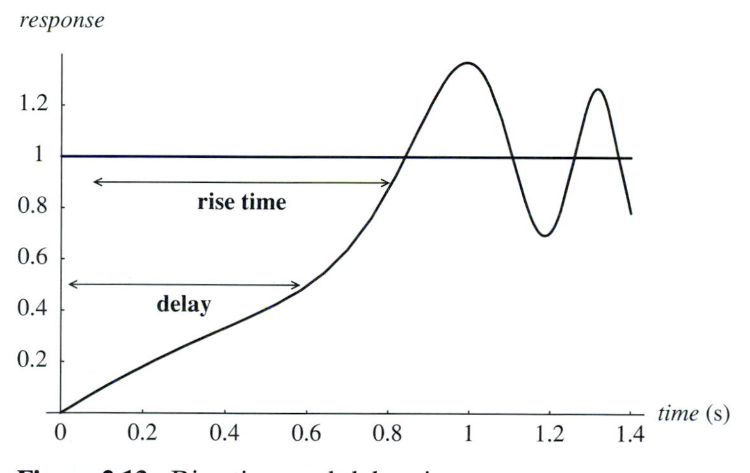

Figure 2.13 Rise time and delay time.

2.4.4 The Shifting Operator for Difference Equations

Difference equations can be presented more compactly by introducing a *shifting operator*

$$\mathrm{Q}x(\nu) = x(\nu + 1)$$

an i*th-order shifting operator*

$$\mathrm{Q}^i x(\nu) = x(\nu + i)$$

and more generally a polynomial shifting operator

$$A(\mathrm{Q}) = \sum_{i=0}^{n} a_i \mathrm{Q}^i$$

The difference equation (2.9) can be written as

$$A(\mathrm{Q})\, y(\nu) = B(\mathrm{Q})\, x(\nu)$$

where $A(\mathrm{Q})$ maps the sequence $y(\nu)$ into the expression $\sum_{i=0}^{n} a_i \mathrm{Q}^{-i} y(\nu)$, and $B(\mathrm{Q})x(\nu)$ represents the expression $\sum_{k=0}^{m} b_k \mathrm{Q}^{-k} x(\nu)$.

The polynomial in complex quantity z^{-1}

$$A(z) = \sum_{i=0}^{n} a_i z^{-i} = a_0 + a_1 z^{-1} + \cdots + a_{n-1} z^{-n+1} + a_n z^{-n}$$

is called the *characteristic polynomial.* The equation $A(z) = 0$ is called the *characteristic equation.*

From the definition of polynomial shifting operators, it follows that

$$A(\mathrm{Q})\ (K\ y(\nu)) = (K\ A(\mathrm{Q}))\ y(\nu)$$

$$A(\mathrm{Q})\ (y_1(\nu) \pm y_2(\nu)) = A(\mathrm{Q})\, y_1(\nu) \pm A(\mathrm{Q})\, y_2(\nu)$$

$$A_1(\mathrm{Q})\ (A_2(\mathrm{Q})\, y(\nu)) = (A_1(\mathrm{Q})\, A_2(\mathrm{Q}))\ y(\nu) = A_2(\mathrm{Q})\ (A_1(\mathrm{Q})\, y(\nu))$$

where K is a constant, $y(\nu)$, $y_1(\nu)$, and $y_2(\nu)$ are sequences of ν, and $A(\mathrm{Q})$, $A_1(\mathrm{Q})$, and $A_2(\mathrm{Q})$ are polynomial shifting operators. Addition, $A_1(\mathrm{Q}) + A_2(\mathrm{Q})$, subtraction, $A_1(\mathrm{Q}) - A_2(\mathrm{Q})$, and multiplication, $A_1(\mathrm{Q})A_2(\mathrm{Q})$, are performed following the rules for the corresponding operations defined for algebraic polynomials in complex variables. Division, $A_1(\mathrm{Q})/A_2(\mathrm{Q})$, is not defined.

2.4.5 Discrete-Time LTI Systems

Polynomial shifting operators are useful in analysis of discrete-time linear systems. If a system is specified by its block diagram, we can use polynomial shifting operators to derive the system input–output relationship in a simpler way. To specify the problem completely so that a unique solution $y(\nu)$ can be obtained, for a known sequence $x(\nu)$ and the coefficients a_i and b_k, we must specify (1) the range over which a solution is desired and (2) a set of n initial conditions for $y(\nu)$. In many cases the range is defined by $0 \leq \nu < +\infty$ and the set of conditions is

$$y(-1),\ y(-2),\ \ldots,\ y(-n)$$

The response of a discrete-time LTI system—that is, the solution $y(\nu)$—can be divided into two parts, namely, a free response and a forced response. The sum of two responses constitutes the total response. The *free response* (or *initial condition response*), $y_0(\nu)$, is the solution of the difference equation when the input $x(\nu)$ is identically zero. The free response depends only on the n initial conditions. Some authors call it the *zero-input response*. The *forced response*, $y_x(\nu)$, is the solution of the difference equation when all initial conditions are identically zero. Some authors call it the *zero-state response*. The forced response depends only on the input $x(\nu)$. The *total response* is the sum of the free response and the forced response, $y(\nu) = y_0(\nu) + y_x(\nu)$. The total response can be viewed, also, as the sum of the steady-state response and transient response. These terms are often used for specifying system performance. The *steady-state response* is that part of the total response which does not approach zero as ν approaches infinity. The *transient response* is that part of the total response which approaches zero as ν goes to infinity.

In the study of discrete-time systems and difference equations which describe them, two special sequences (signals) are used that allow us to obtain simple representations of signals and systems. Assuming that signals described by these sequences excite a system, we define the following special responses:

- *Unit step response* of a discrete-time LTI system is the output of the system to a unit step input with all initial conditions set to zero.
- *Unit impulse response* of a discrete-time LTI system is the output of the system to a unit impulse input with all initial conditions set to zero.

If we know the unit impulse response, $y(\nu) = y_\delta(\nu)$, of an initially relaxed causal discrete-time LTI system, the forced response, $y_x(\nu)$, of the system, then when the input is a known sequence $x(\nu)$, can be written in terms of the *convolution sum*:

$$y_x(\nu) = \sum_{i=-\infty}^{\nu} y_\delta(\nu - i)\, x(i)$$

This formula directly follows from the principle of superposition and the sifting property of the unit impulse sequence.

2.4.6 Properties of LTI Systems

In the preceding sections we showed that the response of LTI systems can be expressed in terms of the unit impulse responses. It implies that the characteristics of an LTI system are completely determined by its impulse response. In the following paragraphs we revisit several important system properties and define them in terms of the impulse responses of LTI systems:

- Memoryless: A continuous-time LTI system is memoryless if $y_\delta(t) = K\delta(t)$, and a discrete-time LTI system is memoryless if $y_\delta(\nu) = K\delta(\nu)$, where K represents a constant. If $K = 1$, the systems become **identity** systems.
- Causal: A continuous-time LTI system is causal if $y_\delta(t) = 0|_{t<0}$, and a discrete-time LTI system is causal if $y_\delta(\nu) = 0|_{\nu<0}$.

- Stability: A sufficient condition for BIBO stability of continuous-time LTI systems is

$$\int_{-\infty}^{\infty} |y_\delta(t)| \mathrm{d}t < \infty$$

which means that the impulse response is absolutely integrable. The corresponding condition for discrete-time LTI systems is

$$\sum_{\nu=-\infty}^{\infty} |y_\delta(\nu)| < \infty$$

which means that the impulse response is absolutely summable.

The **unit step response**, $s(t)$ or $s(\nu)$, can be computed from the unit impulse response by evaluating $s(t) = \int_{-\infty}^{t} y_\delta(\tau)\,\mathrm{d}\tau$, or $s(\nu) = \sum_{i=-\infty}^{\nu} y_\delta(i)$. The above-listed properties can be examined and verified by using the convolution integral and convolution sum as a model of an LTI system.

PROBLEMS

2.1 Construct block diagrams for each of the following equations:

(a) $y = Ax + B$

(b) $y = K_0 + \sum_{i=1}^{n} K_i x_i$

where $A = 100$, $B = 1$, $K_0 = 0.1$, $n = 3$, $K_1 = -2$, $K_2 = \pi$, $K_3 = \sqrt{2}$, x and x_i denote the inputs, and y is the output. Assume that the following basic blocks are available: gain (multiplication by a constant), adder (summing/subtracting point), takeoff point, generator (excitation).

2.2 Consider the following equations in which x_1, x_2, x, represent the inputs and y represents the output of a system:

(a) $a_2 \frac{\mathrm{d}^2}{\mathrm{d}t^2} y + a_1 \frac{\mathrm{d}}{\mathrm{d}t} y + a_0 y = B \frac{\mathrm{d}}{\mathrm{d}t} x_1 + C x_2$

(b) $A \frac{\mathrm{d}}{\mathrm{d}t} y + B \int y\,\mathrm{d}t + Cy = Kx$

(c) $y = A \frac{\mathrm{d}}{\mathrm{d}t} x + B \int_0^t x\,\mathrm{d}t + Cx$

where $a_2 = 1$, $a_1 = 1$, $a_0 = 4$, $A = -1$, $B = 3$, $C = \frac{1}{2}$, $K = \pi^2$, and t designates time. Draw a block diagram for each equation. Assume that the following basic blocks are available: gain (multiplication by a constant), adder (summing/subtracting point), takeoff point, generator (excitation), differentiator (with respect to time t), integrator (with respect to time t).

2.3 Classify the following systems, described by the input-output relation, according to whether they are time-variable or time-invariant:

(a) $y = 100x_1 - 10^{-6}x_2$

(b) $y = \frac{1}{1200} \frac{\mathrm{d}}{\mathrm{d}t} x$

(c) $12t \int y \, dt = \pi x$

(d) $\frac{d^2}{dt^2} y + (\sin(2\pi t))y = 0$

2.4 A system is described by the convolution integral

$$y(t) = \int_{-\infty}^{t} h(t - \tau) x(\tau) \, d\tau$$

where x is the input and y is the output. Show that the system is linear.

2.5 Use the *Principle of Superposition* to find the output y of the two-input single-output system

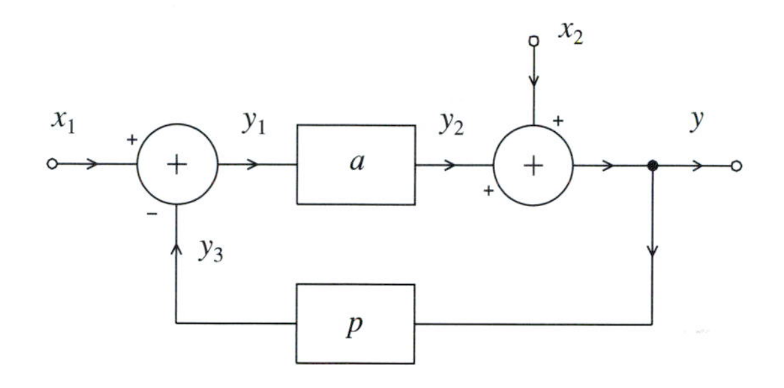

A two-input single-output system.

2.6 Are either of the following systems, described by the input–output relation, causal?

(a) $y(t) = 120 x(t - \frac{1}{125})$

(b) $y(t) = 10^{-3} t^2 x(t + 0.01)$

2.7 A system is characterized by the following equation

$$y(t) = \sqrt{x(t)^2}$$

where $x(t)$ is the system input and $y(t)$ is the system output. Is this system a linear system?

2.8 The input x and the output y of a system are related by

$$y = A \exp(-kx)$$

where $A = -1000$ and $k = 0.01$. Is this system invertible? If yes, find the inverse system to the given system?

2.9 Consider systems described by the following equations

(1) $y = 2x - \sqrt{3}$

(2) $\frac{d}{dt} y = x$

(3) $y = x(t - 0.1)$

(4) $y(\nu) = 5x(\nu - 1) - 7x(\nu - 2)$, ν is integer.

Classify the systems according to whether they are lumped, linear

(a) continuous-time, time-invariant systems

(b) discrete-time, shift-invariant systems.

2.10 Find the output y of a system described by the differential equation

$$a_1 \frac{dy}{dt} + a_0 y = b_1 \frac{dx}{dt} + b_0 x, \quad x = u(t)$$

with an initial condition $y(0-) = K$. Assume that $a_1 = 1$, $a_0 = 2$, $b_0 = -1$, $K = 5$, and

(a) $b_1 = 0$

(b) $b_1 = 4$.

2.11 Identify the transient response and the steady-state response if the output of a system is

(a) $y(t) = (2 - 2\exp(-0.12t))\, u(t)$

(b) $y(n) = (2^{-n} + 0.1 \sin(n))\, u(n)$.

2.12 Find the transient and steady-state response of a system described by the differential equation

$$a_1 \frac{dy}{dt} + a_0 y = b_1 \frac{dx}{dt} + b_0 x$$

with an initial condition $y(0-) = K$. Assume that $a_1 = 2$, $a_0 = \frac{1}{2}$, $b_1 = 1$, $b_0 = -1$, $K = 3$, and

(a) $x = 2\sin(8\pi t + \frac{\pi}{4})\, u(t)$

(b) $x = \sqrt{2}\cos(\pi t - \frac{\pi}{3})\, u(t)$

2.13 Show that the unit step response $y_u(t)$, of a causal CTLTI system, is related to the unit impulse response $y_\delta(t)$ by the equation

$$y_u(t) = \int_0^t y_\delta(\tau)\, d\tau$$

2.14 Show that the unit ramp response $y_r(t)$, of a causal CTLTI system, is related to the unit step response $y_u(t)$ by the equation

$$y_r(t) = \int_0^t y_u(\tau)\, d\tau$$

2.15 The unit impulse response of a continuous-time system is given by

$$y_\delta(t) = (1 - \exp(-t))\, u(t)$$

Find the system output, $y_x(t)$, if the system is excited by $x(t) = \sin(t)\, u(t)$. Assume that the system is initially relaxed.

2.16 We know that the unit impulse response of an initially relaxed discrete-time system is

$$y_\delta(\nu) = \delta(\nu) - \delta(\nu - 1)$$

Determine the system output, $y_x(\nu)$, for the input $x(\nu) = u(\nu)$.

2.17 Two CTLTI systems are connected in cascade. The unit impulse responses of the systems are $h_1(t) = 2(1 - \exp(-0.2t))\, u(t)$ and $h_2(t) = \cos(\pi t)\, u(t)$. Find the overall unit impulse response of the cascaded systems.

2.18 Two DTLTI systems are connected in parallel. The unit impulse responses of the systems are $h_1(\nu) = 4^{-\nu} u(\nu)$ and $h_2(\nu) = 0.4\cos(2\pi\nu/3)$. Find the overall unit impulse response of the systems.

2.19 Show that the response y of a relaxed causal DTLTI system, excited by a causal input sequence x, is

$$y(n) = \sum_{\nu=0}^{n} x(\nu)\, h(n-\nu) = \sum_{\nu=0}^{n} h(\nu)\, x(n-\nu)$$

where h is the unit impulse response of the system.

2.20 The response of a relaxed LTI system, $y_x(t)$, to the unit step input, $x = u(t)$, is

$$y_x(t) = \exp(-0.1t)\,(0.3\sin(2t) - \cos(2t))\,u(t) + u(t)$$

Determine

(a) the peak overshoot,

(b) the delay time,

(c) the rise time, and

(d) the settling time.

2.21 Unit impulse responses of two discrete-time LTI systems are given by

$$h_1(\nu) = \begin{cases} 0, \; \nu < -1 \\ 2^{-\nu}, \; -1 \le \nu \le 1 \\ 0, \; \nu > 1 \end{cases}$$

$h_2(\nu) = h_1(\nu - 2)$

Which one of the two systems is causal?

2.22 Find the response of the system described in Problem 2.20 if the system is excited by $x(t) = t\,u(t)$.

2.23 The impulse responses of several systems are given below. For each case determine if the impulse response represents a stable or an unstable system.

(a) $y_\delta(t) = (-2t + 2)\exp(0.7t)u(t)$

(b) $y_\delta(t) = (10\cos(t) + \sqrt{5}\sin(t))\exp(t)u(t)$

(c) $y_\delta(t) = (t + 2)\sin(\pi t)\exp(-0.6t)u(t)$

(d) $y_\delta(t) = \cos(100\pi t)\sin(\pi t)u(t)$.

2.24 The unit step signal is applied to the input of a system and the output is of the form

$$y_u(t) = 2t\sin(4\pi t)\,u(t)$$

Is the system stable or unstable?

2.25 The gain of the system from Problem 2.5 is defined as

$$G = \frac{y}{x_1}, \quad x_2 = 0$$

Find the sensitivity of the gain with respect to the parameter p at the nominal value $p = \frac{1}{2}$, $a = 10$. Compute the gain variation if p changes $\pm 10\%$ about the nominal value.

2.26 The unit impulse response of a DTLTI system is $h(n) = 0.7^n\, u(n)$. Find the response $y(n)$ if the signal $x(n) = u(n) - u(n-4)$ is applied to the input of the system. Assume that the system is initially relaxed.

2.27 A DTLTI system is described by the equation

$$y(n) = x(n) - ax(n-1)$$

Find the response $y(n)$ to the input $x(n) = (\sin(\pi n/3) + \sin(\pi n/21))\, u(n)$, for

(a) $a = 0.1$

(b) $a = 0.9$

(c) $a = 10$.

Assume that the system is initially relaxed.

MATLAB EXERCISES

2.1 Write a MATLAB script that computes and plots the output $y(t)$, $0 < t < 5$, from Problem 2.4 for $h(t) = \exp(-t)\, u(t)$, and sinusoidal input $x(t) = \sin(\omega t)\, u(t)$, if the angular frequency takes a value (a) $\omega = 0.5$, (b) $\omega = 1$, (c) $\omega = 5$, (d) $\omega = 10^3$, (e) $\omega = 10^6$.

2.2 Use MATLAB to find the response from Problem 2.10 and Problem 2.12.

2.3 Use MATLAB to compute the response from Problem 2.15 for $0 \le t \le 5$.

2.4 Write a MATLAB script that computes (a) the peak overshoot, (b) the delay time, (c) the rise time, and (d) the settling time, from Problem 2.20.

2.5 Write a MATLAB script that computes and plots the response from Problem 2.26 for $0 \le n \le 100$.

2.6 A DTLTI system is described by the equation

$$y(n) = \frac{1}{M+1} \sum_{\nu=0}^{M} x(n-\nu)$$

Write a MATLAB script that computes and plots the response $y(n)$, $0 \le n \le 10$, to the input $x(n) = 0.6^{-n}\, u(n)$, for

(a) $M = 3$

(b) $M = 7$.

Assume that the system is initially relaxed.

2.7 Automate the computation of Problem 2.27 in MATLAB. Plot the response and the input on the same diagram.

MATHEMATICA EXERCISES

2.1 Analyze the system from Problem 2.5 symbolically in *Mathematica*.

(a) Write a notebook that sets the analysis equations and derives the output.

(b) Find the gain and the sensitivity from Problem 2.25.

(c) Determine the unit impulse response for the input x_1.

(d) Determine the unit step response for the input x_2.

2.2 Use *Mathematica* to find the response of each system from Problem 2.23 if the following signals are applied at the system input:

(a) $x(t) = -4u(t)$

(b) $x(t) = 0.6\exp(0.25t)u(t)$

(c) $x(t) = \sqrt{2}\cos(\pi t)u(t)$

(d) $x(t) = 2\sin(2\pi(t-5))u(t-5)$.

Assume that the systems are initially relaxed.

2.3 Automate response computation from Problem 2.10 and Problem 2.12 in *Mathematica*.

2.4 Write a *Mathematica* code that computes (a) the peak overshoot, (b) the delay time, (c) the rise time, and (d) the settling time, from Problem 2.20.

2.5 Can you find the inverse system from Problem 2.8 in *Mathematica*?

2.6 Use *Mathematica* to compute the response from Problem 2.15.

2.7 Write a *Mathematica* code that computes and plots the response from Problem 2.26.

CHAPTER 3

TRANSFORMS

The term *transform* refers to a mathematical operation that takes a given function, called the *original* and returns a new function, referred to as the *image*. The transformation is often done by means of an integral or summation formula. Commonly used transforms are named after Laplace and Fourier. Transforms are frequently used to change a complicated problem into a simpler one. The simpler problem is then solved in the image domain; next, by using the inverse transform we obtain the solution in the original domain. A standard example is the use of the Laplace transform to solve a differential equation, or the z transform to solve a difference equation.

The theory of transforms has two principal aspects:

- examining the nature of the signals or sequences and
- solving LTI systems by transforming differential or difference equations into algebraic equations.

The concept of transforms is based upon suitable mapping of

- functions representing signals or sequences into new (complex) functions, which we call the *images*, and
- the set of differential and difference equations, describing systems under analysis, into algebraic equations in a new (complex) variable.

We investigate signals and systems by investigating the images or complex equations in new (complex) variables.

We seek a suitable transformation $\mathcal{T}$,and the inverse transformation $\mathcal{T}^{-1}$, such that $\mathcal{T}(x) = X$ and $\mathcal{T}^{-1}(X) = x$, which has to satisfy the four important properties:

1. **uniqueness**, $\mathcal{T}(x_1) = \mathcal{T}(x_2) \Leftrightarrow x_1 = x_2$;
2. **homogeneity**, $\mathcal{T}(Kx) = K\mathcal{T}(x)$, $\mathcal{T}^{-1}(KX) = K\mathcal{T}^{-1}(X)$, K is a constant;
3. **additivity**, $\mathcal{T}(x_1 + x_2) = \mathcal{T}(x_1) + \mathcal{T}(x_2)$, $\mathcal{T}^{-1}(X_1 + X_2) = \mathcal{T}^{-1}(X_1) + \mathcal{T}^{-1}(X_2)$;
4. **differentiating** and **differencing**, the operation of differentiating or differencing maps into the algebraic operation of multiplication.

The subsequent sections review the definition and the salient properties of the most important transforms required by the advanced filter design studied in this book.

For an in-depth study of transforms readers can consult excellent books such as references 5–7, 10, 11, 14, and 16–18.

3.1 PHASOR TRANSFORMATION

Sinusoidal waveforms play an important role in science and engineering. If we know the response of a linear time-invariant (LTI) system to any sinusoidal signal, we know, in principle, its response to any signal [3]. In this chapter we shall develop a method for calculating the response of LTI systems to sinusoidal inputs, which is based on the idea of representing a sinusoid of a given frequency by a complex quantity.

3.1.1 The Representation of a Sinusoid by a Phasor

A *sinusoidal signal* (also called a *sinusoid*) of angular frequency ω is any function of time t defined on $(-\infty < t < +\infty)$ and of the form

$$x(t) = X_{\mathrm{m}} \cos(\omega t + \xi) \tag{3.1}$$

where the real constants X_{m}, ω, and ξ are called the *amplitude*, the *angular frequency*, and the *phase* of the sinusoid, respectively. The amplitude is taken to be positive, $X_{\mathrm{m}} > 0$. The angular frequency is measured in rad/s and is given by

$$\omega = 2\pi f$$

where f is the *frequency* of the sinusoid measured in Hz.

The algebraic sum of any number of sinusoids of the same angular frequency, say ω, and of any number of their derivatives of any order is also a sinusoid of the same angular frequency ω.

Any sinusoid can be expressed in terms of the exponential function of complex argument

$$x(t) = X_{\mathrm{m}} \cos(\omega t + \xi) = \frac{1}{2} X_{\mathrm{m}}\, \mathrm{e}^{j(\omega t+\xi)} + \frac{1}{2} X_{\mathrm{m}}\, \mathrm{e}^{-j(\omega t+\xi)}$$

or, equivalently,

$$x(t) = \frac{1}{2} X_{\mathrm{m}}\, \mathrm{e}^{j\xi}\, \mathrm{e}^{j\omega t} + \frac{1}{2} X_{\mathrm{m}}\, \mathrm{e}^{-j\xi}\, \mathrm{e}^{-j\omega t}$$

and we may write

$$x(t) = \frac{1}{2}\left(X_{\mathrm{m}}\,\mathrm{e}^{j\xi}\right)\left(\mathrm{e}^{j\omega t}\right) + \frac{1}{2}\left(\left(X_{\mathrm{m}}\,\mathrm{e}^{j\xi}\right)\left(\mathrm{e}^{j\omega t}\right)\right)^*$$

where j designates the imaginary unit, $j = \sqrt{-1}$, and the asterisk denotes complex conjugate values.

A sinusoid with angular frequency ω is completely specified by its amplitude X_{m} and its phase ξ, leading to the idea of representing the sinusoid by the complex quantity

$$X = X_{\mathrm{m}}\,\mathrm{e}^{j\xi} \tag{3.2}$$

Given the complex number X and the angular frequency ω, we recover the sinusoid (3.1) as follows:

$$x(t) = \mathrm{Re}(X\,\mathrm{e}^{j\omega t}) \tag{3.3}$$

The complex quantity X is called the *phasor* representing the sinusoid (3.1).

The mapping of a sinusoid $x(t)$ into a complex quantity X can be viewed as a transformation that we shall call the *phasor transformation*. Formally, we can establish this transformation by an integral formula

$$X = \mathcal{P}(x(t)) = \frac{\omega}{\pi}\int_0^{2\pi/\omega} x(t)\,\mathrm{e}^{-j\omega t}\,\mathrm{d}t \tag{3.4}$$

and we can establish the *inverse phasor transformation* by

$$x(t) = \mathcal{P}^{-1}(X) = \mathrm{Re}(X\,\mathrm{e}^{j\omega t}) \tag{3.5}$$

We shall use the following notation to explicitly designate the angular frequency, ω, to which the phasor, X, has been associated:

$$X^{(\omega)} = \mathcal{P}_\omega(x(t)) = \frac{\omega}{\pi}\int_0^{2\pi/\omega} x(t)\,\mathrm{e}^{-j\omega t}\,\mathrm{d}t \tag{3.6}$$

and

$$x(t) = \mathcal{P}_\omega^{-1}(X) = \mathrm{Re}(X\,\mathrm{e}^{j\omega t}) \tag{3.7}$$

When performing calculations with phasors we assume that the angular frequency is known and that it is the same for all phasors; that is, the operators $\mathcal{P}$ and $\mathcal{P}^{-1}$ assume the notion of ω.

Unless otherwise stated, the symbol t will always represent time.

3.1.2 Properties of the Phasor Transform

The phasor transform is unique for all t:

$$x_1(t) = x_2(t) \Leftrightarrow X_1 = X_2 \tag{3.8}$$

where $X_1 = \mathcal{P}(x_1(t))$, $X_2 = \mathcal{P}(x_2(t))$, and $x_1(t)$, $x_2(t)$ are sinusoids of the form (3.1).

The operators $\mathcal{P}$ and $\mathcal{P}^{-1}$ are *homogeneous*:

$$\begin{aligned} \mathcal{P}(Kx(t)) &= K\mathcal{P}(x(t)) \\ \mathcal{P}^{-1}(KX) &= K\mathcal{P}^{-1}(X) \end{aligned} \tag{3.9}$$

for any real constant K, with $X = \mathcal{P}(x(t))$.

Also, the two operators are *additive*:

$$\begin{aligned} \mathcal{P}(x_1(t) + x_2(t)) &= \mathcal{P}(x_1(t)) + \mathcal{P}(x_2(t)) \\ \mathcal{P}^{-1}(X_1 + X_2) &= \mathcal{P}^{-1}(X_1) + \mathcal{P}^{-1}(X_2) \end{aligned} \tag{3.10}$$

The phasor transform, $\mathcal{P}$, maps the operation of differentiating, D, into the multiplication by $j\omega$:

$$\begin{aligned} \mathcal{P}(\mathrm{D}x(t)) &= j\omega\mathcal{P}(x(t)) = j\omega X \\ \mathcal{P}^{-1}(j\omega X) &= \mathrm{D}\mathcal{P}^{-1}(X) = \mathrm{D}x(t) \end{aligned} \tag{3.11}$$

where $X = \mathcal{P}(x(t))$, and D denotes differentiation with respect to time, $\mathrm{D}x(t) = \mathrm{d}x(t)/\mathrm{d}t$.

3.1.3 Application of the Phasor Transform

The phasor representation of sinusoids is used mainly in the computation of the sinusoidal particular solution of ordinary linear differential equations with real constant coefficients when the forcing function is sinusoid; this computation is referred to as the *phasor method*.

Consider a lumped LTI system with a single input $x(t)$ and a single output $y(t)$, described by the following differential equation:

$$\begin{aligned} &a_n\mathrm{D}^n y + a_{n-1}\mathrm{D}^{n-1} y + \cdots + a_1\mathrm{D}y + a_0 y \\ &= b_m\mathrm{D}^m x + b_{m-1}\mathrm{D}^{m-1} x + \cdots + b_1\mathrm{D}x + b_0 x \end{aligned} \tag{3.12}$$

that is,

$$\sum_{i=0}^{n} a_i\mathrm{D}^i y = \sum_{k=0}^{m} b_k\mathrm{D}^k x \tag{3.13}$$

or, by introducing the polynomial differential operators,

$$A(\mathrm{D})\, y = B(\mathrm{D})\, x \tag{3.14}$$

where $a_0, a_1, \ldots, a_n$ and $b_0, b_1, \ldots, b_m$ are real numbers. If the input is a sinusoid given by (3.1), then we assume that a **sinusoidal** particular solution of Eq. (3.12) is of the form

$$y_\mathrm{p}(t) = Y_\mathrm{m}\cos(\omega t + \eta) \tag{3.15}$$

We apply the phasor transform (3.4) to both sides of Eq. (3.13) to obtain

$$\mathcal{P}\left(\sum_{i=0}^{n} a_i\mathrm{D}^i y_\mathrm{p}\right) = \mathcal{P}\left(\sum_{k=0}^{m} b_k\mathrm{D}^k x\right)$$

From the additivity (3.10) and homogeneity (3.9) property this becomes

$$\sum_{i=0}^{n} a_i \mathcal{P}(\mathrm{D}^i y_\mathrm{p}) = \sum_{k=0}^{m} b_k \mathcal{P}(\mathrm{D}^k x)$$

and from the differentiating property (3.11), applied repeatedly, we get

$$\sum_{i=0}^{n} a_i \, (j\omega)^i Y = \sum_{k=0}^{m} b_k \, (j\omega)^k X \tag{3.16}$$

with $Y = \mathcal{P}(y_\mathrm{p}(t))$, or, equivalently,

$$\left(\sum_{i=0}^{n} a_i \, (j\omega)^i\right) Y = \left(\sum_{k=0}^{m} b_k \, (j\omega)^k\right) X$$

By using the compact notation similar to Eq. (3.14) we may write

$$A(j\omega)\, Y = B(j\omega) X \tag{3.17}$$

Thus, if

$$A(j\omega) \neq 0 \tag{3.18}$$

we compute the phasor of the sinusoidal particular solution

$$Y = \frac{B(j\omega)}{A(j\omega)} X \tag{3.19}$$

where

$$A(j\omega) = \sum_{i=0}^{n} a_i \, (j\omega)^i \tag{3.20}$$

$$B(j\omega) = \sum_{k=0}^{m} b_k \, (j\omega)^k \tag{3.21}$$

Equation (3.16) can be obtained directly from Eq. (3.13) by replacing the ith derivatives of $y(t)$ with $(j\omega)^i Y$, for $i = 0$ to n, and by replacing the kth derivatives of $x(t)$ with $(j\omega)^k X$, for $k = 0$ to m. Also, we can write Eq. (3.17) directly from Eq. (3.14) by replacing the differential operator D with $j\omega$, $x(t)$ with X, and $y(t)$ with Y. Note that polynomial differential operators $A(\mathrm{D})$ and $B(\mathrm{D})$ are mapped into algebraic polynomials in complex variable $j\omega$ [Eqs. (3.20) and (3.21)].

The sinusoidal particular solution is found by the inverse phasor transform (3.5)

$$y_\mathrm{p}(t) = \mathcal{P}^{-1}(Y) = \mathcal{P}^{-1}(Y_\mathrm{m}\, \mathrm{e}^{j\eta}) = Y_\mathrm{m} \cos(\omega t + \eta) \tag{3.22}$$

where

$$Y_\mathrm{m} = \left| \frac{B(j\omega)}{A(j\omega)} X \right| = \left| \frac{B(j\omega)}{A(j\omega)} \right| X_\mathrm{m} \tag{3.23}$$

and

$$\eta = \arg\left(\frac{B(j\omega)}{A(j\omega)} X\right) = \arg\left(\frac{B(j\omega)}{A(j\omega)} \mathrm{e}^{j\xi}\right) \tag{3.24}$$

Notice that the condition (3.18) must be satisfied for the value of ω under consideration. The sinusoidal particular solution (3.22) does not exist if this condition is not met—that is, if ω is such that $A(j\omega) = 0$.

The complex quantity $j\omega$ is often referred to as the *complex frequency of the excitation*. We shall frequently designate the complex frequency by s.

3.1.4 Sinusoidal Steady-State Response

Let us consider an LTI system characterized by Eq. (3.12) and driven by a single causal sinusoidal excitation:

$$x(t) = X_{\mathrm{m}} \cos(\omega t + \xi)\, u(t) = x_{\mathrm{s}}(t) u(t) \tag{3.25}$$

where $u(t)$ stands for the unit step function, and $x_{\mathrm{s}}(t)$ represents a sinusoid. We shall assume that the initial conditions are defined at time $t = t_0^- = 0^-$,

$$t_0^- = \lim_{\substack{t \to t_0 \\ t < t_0}} t$$

and the complete response of the system will be analyzed for $t \geq t_0$. Our goal is to examine what will response become as time approaches infinity.

The *complete response* of a system is the response of the system to both an input and the initial conditions. The complete response, $y(t)$, is the sum of the zero-input response, $y_0(t)$, and the zero-state response, $y_x(t)$:

$$y(t) = y_0(t) + y_x(t)$$

First, let us examine the *zero-input response* that is defined as a response of a system with no applied input. We have to solve the homogeneous differential equation

$$A(\mathrm{D})\, y_0 = 0 \tag{3.26}$$

If the corresponding characteristic equation

$$A(s) = \sum_{i=0}^{n} a_i s^i = 0 \tag{3.27}$$

has n distinct roots $s_1, s_2, \ldots, s_n$, any solution of (3.26), for $t \geq t_0$, may be written in the form

$$y_0(t) = \sum_{i=1}^{n} K_i\, \mathrm{e}^{s_i t} \tag{3.28}$$

where the constants $K_1, K_2, \ldots, K_n$ are appropriately chosen so that the solution satisfies the prescribed initial conditions

$$\begin{aligned} y_0(t_0^-) &= y(t_0^-) \\ \mathrm{D}y_0(t_0^-) &= \mathrm{D}y(t_0^-) \\ &\vdots \\ \mathrm{D}^{n-1}y_0(t_0^-) &= \mathrm{D}^{n-1}y(t_0^-) \end{aligned} \tag{3.29}$$

Here, we use the notation

$$\begin{aligned} y_0(t_0^-) &= \lim_{\substack{t\to t_0 \\ t<t_0}} y_0(t) \\ \mathrm{D}y_0(t_0^-) &= \lim_{\substack{t\to t_0 \\ t<t_0}} \frac{\mathrm{d}y_0(t)}{\mathrm{d}t} \\ &\vdots \\ \mathrm{D}^k y_0(t_0^-) &= \lim_{\substack{t\to t_0 \\ t<t_0}} \frac{\mathrm{d}^k y_0(t)}{\mathrm{d}t^k} \end{aligned}$$

If the characteristic equation (3.27) has ν distinct roots $s_1, s_2, \ldots, s_\nu$, where $\nu < n$, and if the multiplicities of these roots are $k_1, k_2, \ldots, k_\nu$ respectively, then

$$k_1 + k_2 + \cdots + k_\nu = n$$

and any solution of (3.26) can be written as

$$y_0(t) = \sum_{i=1}^{\nu} p_i(t)\, \mathrm{e}^{s_i t} \tag{3.30}$$

where $p_1(t), p_2(t), \ldots, p_\nu(t)$ are polynomials in the variable t of degree $k_1 - 1, k_2 - 1, \ldots, k_\nu - 1$, respectively. The coefficients of these polynomials must be chosen so that the solution satisfies the initial conditions (3.29).

The roots of the characteristic equation (3.27), $s_1, s_2, \ldots$, are referred to as the *natural frequencies* of the system. If the coefficients of the characteristic equations are real, the natural frequencies are real or occur in complex conjugate pairs. We shall be concerned with systems that are characterized by equations with real coefficients.

Equations (3.28) and (3.30) indicate that irrespective of the initial state and provided that all natural frequencies are in the open left-half plane, the zero-input response tends to **zero** as $t \to \infty$. The *open left-half plane* consists of the left half of the complex plane with the imaginary axis excluded. In other words, the open left-half plane includes all points with negative real parts.

When an LTI system has all its natural frequencies in the open left-half plane, we say that the system is *asymptotically stable*. Any zero-input response of an asymptotically stable system vanishes (approaches zero) as $t \to \infty$. Real systems, especially filters, are designed to be stable. If one or more natural frequencies are in the open right-half plane, we say that the system is *unstable*. For most initial states, the zero-input response of an unstable system becomes infinite as $t \to \infty$. Unstable systems are not within the scope of this book.

If the system has only one complex conjugate pair of purely imaginary simple natural frequencies, say $j\omega_0$ and $-j\omega_0$, and all other natural frequencies are in the open left-half plane, the zero-input response contains an oscillatory part at the angular frequency ω_0, and it becomes sinusoidal as $t \to \infty$. This response is called the *zero-input sinusoidal steady-state* response.

If the system has several imaginary natural frequencies that are simple, $\pm j\omega_{01}$, $\pm j\omega_{02}, \ldots$, and all other natural frequencies are in the open left-half plane, the zero-input response will tend to the sum of sinusoids with angular frequencies $\omega_{01}, \omega_{02}, \ldots$, as $t \to \infty$. We shall call this response the *zero-input steady-state* response; it is periodic but it is not sinusoidal.

The *zero-state response* of an LTI system, $y_x(t)$, is the response of the system to an input applied at some time t_0 subject to the condition that the system be in zero state just prior to the application of the input (that is, at time t_0^-). In calculating zero-state responses, our primary interest is the behavior of the response for $t \geq t_0$. For this reason we adopt the convention that the input and the zero-state response are taken to be identically zero for $t < t_0$. Unless otherwise stated, we assume that $t_0 = 0$. We have to solve the equation

$$\sum_{i=0}^{n} a_i \mathrm{D}^i y_x = \sum_{k=0}^{m} b_k \mathrm{D}^k x \tag{3.31}$$

and we shall assume that $n \geq m$. By definition, the initial conditions are zero and

$$\begin{aligned} y_x(t_0^-) &= 0 \\ \mathrm{D}y_x(t_0^-) &= 0 \\ &\vdots \\ \mathrm{D}^{n-1}y_x(t_0^-) &= 0 \end{aligned} \tag{3.32}$$

The causal input $x(t)$ is the product of an elementary function, $x_s(t)$, and the unit step function $u(t)$. *Elementary functions* are polynomials, exponential functions, trigonometric functions, and so on, and combination of these functions; elementary functions do not contain step functions and impulse functions.

Under the condition $n \geq m$ the response $y_x(t)$ must be of the form

$$y_x(t) = y_{xs}(t)u(t) \tag{3.33}$$

where $y_{xs}(t)$ represents an elementary function and Eq. (3.31) becomes

$$\sum_{i=0}^{n} a_i \mathrm{D}^i (y_{xs}(t)u(t)) = \sum_{k=0}^{m} b_k \mathrm{D}^k (x_s(t)u(t)) \tag{3.34}$$

Next, we apply rules for differentiating the product of an elementary function and the step function

$$\begin{aligned}
\mathrm{D}(x_s(t)u(t)) &= (\mathrm{D}x_s(t))\,u(t) + \big(x_s(0^+)\big)\mathrm{D}u(t) \\
\mathrm{D}^2(x_s(t)u(t)) &= \big(\mathrm{D}^2x_s(t)\big)u(t) + \big(\mathrm{D}x_s(0^+)\big)\mathrm{D}u(t) + \big(x_s(0^+)\big)\mathrm{D}^2u(t) \\
\mathrm{D}^k(x_s(t)u(t)) &= \big(\mathrm{D}^kx_s(t)\big)\,u(t) + \sum_{q=1}^{k}\big(\mathrm{D}^{k-q}x_s(0^+)\big)\mathrm{D}^qu(t) \qquad (3.35)
\end{aligned}$$

where

$$\begin{aligned}
x_s(0^+) &= \lim_{\substack{t\to 0\\ t>0}} x_s(t) \\
\mathrm{D}^kx_s(t) &= \frac{\mathrm{d}^kx_s(t)}{\mathrm{d}t^k} \\
\mathrm{D}^0x_s(t) &= x_s(t) \\
\mathrm{D}^kx_s(0^+) &= \lim_{\substack{t\to 0\\ t>0}} \frac{\mathrm{d}^kx_s(t)}{\mathrm{d}t^k}
\end{aligned}$$

Equation (3.34) transforms into

$$\begin{aligned}
&\sum_{i=0}^{n} a_i\left(\big(\mathrm{D}^iy_{xs}(t)\big)\,u(t) + \sum_{q=1}^{i}\big(\mathrm{D}^{i-q}y_{xs}(0^+)\big)\,\mathrm{D}^qu(t)\right) \\
&\quad = \sum_{k=0}^{m} b_k\left(\big(\mathrm{D}^kx_s(t)\big)\,u(t) + \sum_{q=1}^{k}\big(\mathrm{D}^{k-q}x_s(0^+)\big)\,\mathrm{D}^qu(t)\right) \\
&\sum_{i=0}^{n} a_i\,\big(\mathrm{D}^iy_{xs}(t)\big)\,u(t) + \sum_{i=1}^{n} a_i\sum_{q=1}^{i}\big(\mathrm{D}^{i-q}y_{xs}(0^+)\big)\,\mathrm{D}^qu(t) \\
&\quad = \sum_{k=0}^{m} b_k\,\big(\mathrm{D}^kx_s(t)\big)\,u(t) + \sum_{k=1}^{m} b_k\sum_{q=1}^{k}\big(\mathrm{D}^{k-q}x_s(0^+)\big)\,\mathrm{D}^qu(t)
\end{aligned}$$

which yields

$$\sum_{i=0}^{n} a_i\mathrm{D}^iy_{xs}(t) = \sum_{k=0}^{m} b_k\mathrm{D}^kx_s(t) \qquad (3.36)$$

or

$$A(\mathrm{D})\,y_{xs} = B(\mathrm{D})\,x_s \qquad (3.37)$$

and

$$\sum_{i=1}^{n} a_i \sum_{q=1}^{i} \left(\mathrm{D}^{i-q} y_{xs}(0^+)\right) \mathrm{D}^q u(t) = \sum_{k=1}^{m} b_k \sum_{q=1}^{k} \left(\mathrm{D}^{k-q} x_s(0^+)\right) \mathrm{D}^q u(t) \tag{3.38}$$

Equating the like terms in (3.38), we formulate a system of linear equations from which we determine the initial conditions $y_{xs}(0^+)$, $\mathrm{D}y_{xs}(0^+)$, $\mathrm{D}^2 y_{xs}(0^+)$, ..., $\mathrm{D}^{n-1} y_{xs}(0^+)$. We use these initial conditions to solve Eq. (3.36)

$$\begin{aligned}
\mathrm{D}u(t) &\leftrightarrow \sum_{i=1}^{n} a_i \mathrm{D}^{i-1} y_{xs}(0^+) = \sum_{k=1}^{m} b_k \mathrm{D}^{k-1} x_s(0^+) \\
\mathrm{D}^2 u(t) &\leftrightarrow \sum_{i=2}^{n} a_i \mathrm{D}^{i-2} y_{xs}(0^+) = \sum_{k=2}^{m} b_k \mathrm{D}^{k-2} x_s(0^+) \\
&\vdots
\end{aligned} \tag{3.39}$$

If $n > m$ the response $y_x(t)$ is continuous at the instant $t = t_0 = 0$ and

$$y_x(0^+) = y_{xs}(0^+) = y_x(0^-) = 0 \tag{3.40}$$

We say that the initial conditions are *consistent*. For $n = m$, Eq. (3.40) does not hold, the response $y_x(t)$ has a jump at t_0, and the initial conditions are *inconsistent*. The set of equations (3.39) must be solved to find the required initial conditions at $t = t_0^+ = 0^+$.

The general solution of Eq. (3.36) is the sum of two terms:

$$y_{xs}(t) = y_{\mathrm{xsh}}(t) + y_{\mathrm{xsp}}(t) \tag{3.41}$$

where $y_{\mathrm{xsh}}(t)$ is the solution of the homogeneous equation

$$A(\mathrm{D})\, y_{\mathrm{xsh}} = 0 \tag{3.42}$$

and $y_{\mathrm{xsp}}(t)$ is a particular solution.

The term $y_{\mathrm{xsh}}(t)$ is of the form (3.30):

$$y_{\mathrm{xsh}}(t) = \sum_{i=1}^{\nu} P_i(t)\, \mathrm{e}^{s_i t}$$

where $P_1(t)$, $P_2(t)$, ..., $P_\nu(t)$ are polynomials in the variable t of degree $k_1 - 1$, $k_2 - 1$, ..., $k_\nu - 1$, respectively; k_1, k_2, ..., k_ν are multiplicities of the natural frequencies.

The particular solution is sinusoidal

$$y_{\mathrm{xsp}}(t) = K_{\mathrm{c}} \cos(\omega t) + K_{\mathrm{s}} \sin(\omega t) = Y_{\mathrm{m}} \cos(\omega t + \eta) \tag{3.43}$$

if the complex frequency $j\omega$ is not a root of the characteristic equation; K_{c} and K_{s} are real constants. If $j\omega$ is a natural frequency of order λ, then

$$y_{\mathrm{xsp}}(t) = p_{\mathrm{c}}(t) \cos(\omega t) + p_{\mathrm{s}}(t) \sin(\omega t) \tag{3.44}$$

where $p_{\mathrm{c}}(t)$ and $p_{\mathrm{s}}(t)$ are polynomials in t of degrees λ.

The complete response is

$$y(t) = y_0(t) + y_{\mathrm{xsh}}(t)u(t) + y_{\mathrm{xsp}}(t)u(t), \qquad t \geq t_0 \tag{3.45}$$

or generally

$$y(t) = \sum_{i=1}^{\nu} p_i(t)\,\mathrm{e}^{s_i t} + \left(\sum_{i=1}^{\nu} P_i(t)\,\mathrm{e}^{s_i t}\right)u(t) + (p_{\mathrm{c}}(t)\cos(\omega t) + p_{\mathrm{s}}(t)\sin(\omega t))u(t)$$

and the above analysis allows us to state the following facts [3]:

- Irrespective of the initial state and provided that all the natural frequencies are in the open left-half plane, the complete response of an LTI system driven by a sinusoidal input will become sinusoidal as $t \to \infty$. This sinusoidal response is called the *sinusoidal steady-state response.*
- The sinusoidal steady-state response always has the same frequency as the input.
- The sinusoidal steady-state response can be obtained efficiently by the phasor method.

The systems that we analyze in this book have natural frequencies in the open left-half plane; therefore, the sinusoidal steady-state response exists for these systems at any angular frequency of the sinusoidal input.

3.1.5 Nonsinusoidal Steady-State Response

Practical systems are often driven by several excitations. A system variable, the response $y(t)$, can be found from a differential equation of the form

$$A(\mathrm{D})\,y = \sum_{g=1}^{N} B_g(\mathrm{D})x_g \tag{3.46}$$

The inputs to the system, $x_g(t)$, are supposed to be sinusoidal

$$x_g(t) = X_{\mathrm{m}g}\cos(\omega_g t + \xi_g)\,u(t) = x_{\mathrm{s}g}(t)\,u(t) \tag{3.47}$$

with arbitrary and different angular frequencies $\omega_1, \omega_2, \ldots, \omega_N$. The degree of $A(\mathrm{D})$ is assumed not to be less than the degree of $B_g(\mathrm{D})$.

We focus on lumped LTI systems with natural frequencies in the open left-half plane.

According to the principal of superposition the complete response, $y(t)$, is the sum of the zero-input response, $y_0(t)$, and the zero-state responses that would exist if each input sinusoid were acting alone on the system:

$$y(t) = y_0(t) + \sum_{g=1}^{N} y_{xg}(t) = y_0(t) + \sum_{g=1}^{N} y_{\mathrm{xh}g}(t) + \sum_{g=1}^{N} y_{\mathrm{xp}g}(t), \qquad t \geq 0 \tag{3.48}$$

where $y_{\mathrm{xh}g}(t)$ is the solution of the homogeneous equation, and $y_{\mathrm{xp}g}(t)$ is the particular solution of the form

$$y_{\mathrm{xp}g}(t) = Y_{\mathrm{m}g}\cos(\omega_g t + \eta_g)u(t) \tag{3.49}$$

for $g = 1, 2, \ldots, N$.

Whatever the initial conditions may be, as $t \to \infty$, the complete response $y(t)$ becomes arbitrarily close to the value of its particular solution given by

$$y_{\mathrm{p}}(t) = \sum_{g=1}^{N} Y_{\mathrm{m}g} \cos(\omega_g t + \eta_g) u(t) \tag{3.50}$$

that is,

$$y(t) \to \sum_{g=1}^{N} Y_{\mathrm{m}g} \cos(\omega_g t + \eta_g), \qquad t \to \infty$$

The response $y(t)$, as $t \to \infty$, is called the *steady state* (not sinusoidal steady state).

The steady state resulting from several input sinusoids is the sum of the sinusoidal steady states that would exist if each input sinusoid were acting alone on the system.

We can use the phasor method to obtain $y_{\mathrm{xp}g}(t)$. First, we find the phasors of the input sinusoids (3.47):

$$X^{(\omega_g)} = \mathcal{P}_{\omega_g}(x_g(t)) = X_{\mathrm{m}g}\, \mathrm{e}^{j\xi_g}, \qquad g = 1, 2, \ldots, N \tag{3.51}$$

Next, we compute the output phasors as

$$Y^{(\omega_g)} = \frac{B_g(j\omega_g)}{A(j\omega_g)} X^{(\omega_g)}, \qquad g = 1, 2, \ldots, N \tag{3.52}$$

Finally, we determine the particular solution (i.e., steady state) by taking the inverse phasor transform of the output phasors:

$$y_{\mathrm{p}}(t) = u(t) \sum_{g=1}^{N} \mathcal{P}_{\omega_g}^{-1}(Y^{(\omega_g)}) \tag{3.53}$$

As we expect, the superposition holds in the steady state.

The *constant steady state* can be readily obtained as a special (trivial) case when

$$\omega_g = 0, \qquad \xi_g = 0$$
$$g = 1, 2, \ldots, N$$

that is, when all inputs are constant (time-invariant) signals.

3.1.6 Transfer Function and Frequency Response

The principal advantage of the phasor method is efficient computation of the sinusoidal steady-state response by solving an algebraic equation rather than a differential equation. Equation (3.19) shows that a linear relation exists between the output phasor and the input phasor. For a single-input lumped LTI system in sinusoidal steady state we can represent the input, output, and other system variables with phasors. Our target is to study complex functions that relate phasors in an LTI system.

We define a *transfer function* (also called the *frequency response*), for a single-input lumped LTI system in the sinusoidal steady state, to be the ratio of the output phasor to the input phasor

$$H(j\omega) = \frac{Y}{X} = \frac{B(j\omega)}{A(j\omega)} \tag{3.54}$$

The transfer function depends on system parameters and the angular frequency ω. Generally, $H(j\omega)$ is a rational function of the complex frequency $j\omega$, and it contains all the needed information concerning the sinusoidal steady-state response.

The magnitude of the transfer function

$$M(\omega) = |H(j\omega)| = \left|\frac{B(j\omega)}{A(j\omega)}\right| \tag{3.55}$$

and its phase

$$\Phi(\omega) = \arg(H(j\omega)) = \arg\left(\frac{B(j\omega)}{A(j\omega)}\right) \tag{3.56}$$

are called the *frequency response*; $M(\omega)$ is the *magnitude response*, and $\Phi(\omega)$ is the *phase response*. The curves of $M(\omega)$ and $\Phi(\omega)$ versus ω or $\log(\omega)$ are called the *frequency characteristics* for a specified input and output. The amplitude response is often expressed in *decibels* (dB) as follows:

$$M_{\mathrm{dB}}(\omega) = 20\log_{10}(M(\omega)) \tag{3.57}$$

The quantity $M_{\mathrm{dB}}(\omega)$ is usually called the *gain*. The negative of the gain is called the *attenuation* or *loss*. The plot of the attenuation versus ω is the *attenuation characteristic* or *loss characteristic*. Given the phase response $\Phi(\omega)$, the *group delay* is defined as

$$\tau(\omega) = -\frac{\mathrm{d}\Phi(\omega)}{\mathrm{d}\omega} \tag{3.58}$$

and the corresponding plot against ω is known as *delay characteristic*. The frequency characteristics can also be plotted versus f or $\log(f)$.

Given the transfer function $H(j\omega) = M(\omega)\,\mathrm{e}^{j\Phi(\omega)}$ and the sinusoidal input $x(t) = X_{\mathrm{m}}\cos(\omega t + \xi)$, the sinusoidal steady-state response is $y(t) = M(\omega)X_{\mathrm{m}}\cos(\omega t + \xi + \Phi(\omega))$. In other words, the amplitude of the output is obtained by taking the product of the transfer function magnitude and the amplitude of the input; the phase of the output is obtained by adding the phase of the transfer function and the phase of the input.

If we denote the complex frequency $j\omega$ by the symbol s, we can formally write

$$H(j\omega) = H(s) = \frac{B(s)}{A(s)}, \qquad s = j\omega \tag{3.59}$$

The roots of the transfer function denominator $A(s)$ are called the *poles* of the transfer function. In fact, the poles are the natural frequencies of the system. We compute the poles from the equation

$$A(s) = \text{denominator}(H(s)) = 0$$

and designate these poles by $s_{\mathrm{p}1}, s_{\mathrm{p}2}, s_{\mathrm{p}3}, \ldots, s_{\mathrm{p}n}$.

In a similar way, the roots of the transfer function numerator $B(s)$ are called the *zeros* of the transfer function. We obtain zeros from the equation

$$B(s) = \text{numerator}(H(s)) = 0$$

and designate these zeros by $s_{z1}, s_{z2}, s_{z3}, \ldots, s_{zm}$.

The transfer function can be written as a function of $s = j\omega$ in the factored form as

$$H(s) = H_0 \frac{\prod_{k=1}^{m}(s - s_{zk})}{\prod_{i=1}^{n}(s - s_{pi})}, \qquad s = j\omega \tag{3.60}$$

The form (3.60) is known as the *pole-zero representation* of the transfer function. The real constant H_0 is called the *scale factor*.

A transfer function with no zeros is called the *all-pole transfer function* and is of the form

$$H(s) = H_0 \frac{1}{\prod_{i=1}^{n}(s - s_{pi})}, \qquad s = j\omega \tag{3.61}$$

Since the polynomials $A(s)$ and $B(s)$ have real coefficients, zeros and poles must be real or occur in complex conjugate pairs.

For each complex pole $s_{pi} = \sigma_{pi} + j\omega_{pi}$ we define the *pole magnitude*

$$p_i = |s_{pi}| = \sqrt{\sigma_{pi}^2 + \omega_{pi}^2} \tag{3.62}$$

and the pole *quality factor* (also called the Q-factor)

$$Q_{pi} = \frac{|s_{pi}|}{-2\sigma_{pi}} = \frac{p_i}{-2\sigma_{pi}} \tag{3.63}$$

For the systems that we study, the poles are in the open left-half plane and the pole Q-factor is a positive number. The minimal value of Q-factor is $1/2$ and occurs for poles with vanishingly small imaginary parts. Purely imaginary poles have infinite quality factors. In the same way we define the magnitude and the Q-factor of the transfer-function zeros.

The simplest transfer function is the first-order transfer function

$$H(s) = \frac{b_1 s + b_0}{a_1 s + a_0}, \qquad s = j\omega, \quad a_1 \neq 0 \tag{3.64}$$

Systems described by (3.64) we call *first-order sections*.

Transfer functions of the form

$$H(s) = \frac{b_2 s^2 + b_1 s + b_0}{a_2 s^2 + a_1 s + a_0}, \qquad s = j\omega, \quad a_2 \neq 0 \tag{3.65}$$

are called *second-order transfer functions*, and they play an important role in analysis and design of LTI systems. A system characterized by (3.65) we call a *biquadratic section* or *biquad*.

Any transfer function (3.54) can be expressed as a product of first-order (3.64) and second-order (3.65) transfer functions, which implies that any LTI system can be resolved into first-order sections and biquads.

Partial Transfer Function and Transfer-Function Matrix. Consider a multiple-input multiple-output system with N sinusoidal inputs, $x_1, x_2, \ldots, x_k, \ldots, x_N$, with equal angular frequencies $\omega_1 = \omega_2 = \cdots = \omega_N = \omega$, and assume that the system has L outputs $y_1, y_2, \ldots, y_i, \ldots, y_L$. We define the *partial transfer function* between the ith output and the kth input to be the ratio of the output phasor Y_i to the input phasor X_k, with the other inputs being identically zero:

$$H_{ik}(j\omega) = \left.\frac{Y_i}{X_k}\right|_{x_1=\cdots=x_{k-1}=x_{k+1}=\cdots=x_N=0} \tag{3.66}$$

The input–output description of the system can be extended to phasors and expressed in a matrix form

$$\begin{bmatrix} Y_1 \\ Y_2 \\ \vdots \\ Y_i \\ \vdots \\ Y_L \end{bmatrix} = \begin{bmatrix} H_{11}(j\omega) & H_{12}(j\omega) & \cdots & H_{1k}(j\omega) & \cdots & H_{1N}(j\omega) \\ H_{21}(j\omega) & H_{22}(j\omega) & \cdots & H_{2k}(j\omega) & \cdots & H_{2N}(j\omega) \\ \vdots & \vdots & \ddots & \vdots & & \vdots \\ H_{i1}(j\omega) & H_{i2}(j\omega) & \cdots & H_{ik}(j\omega) & \cdots & H_{iN}(j\omega) \\ \vdots & \vdots & & \vdots & \ddots & \vdots \\ H_{L1}(j\omega) & H_{L2}(j\omega) & \cdots & H_{Lk}(j\omega) & \cdots & H_{LN}(j\omega) \end{bmatrix} \begin{bmatrix} X_1 \\ X_2 \\ \vdots \\ X_k \\ \vdots \\ X_N \end{bmatrix}$$

where the matrix

$$\mathbf{H}(j\omega) = \begin{bmatrix} H_{11}(j\omega) & H_{12}(j\omega) & \cdots & H_{1k}(j\omega) & \cdots & H_{1N}(j\omega) \\ H_{21}(j\omega) & H_{22}(j\omega) & \cdots & H_{2k}(j\omega) & \cdots & H_{2N}(j\omega) \\ \vdots & \vdots & \ddots & \vdots & & \vdots \\ H_{i1}(j\omega) & H_{i2}(j\omega) & \cdots & H_{ik}(j\omega) & \cdots & H_{iN}(j\omega) \\ \vdots & \vdots & & \vdots & \ddots & \vdots \\ H_{L1}(j\omega) & H_{L2}(j\omega) & \cdots & H_{Lk}(j\omega) & \cdots & H_{LN}(j\omega) \end{bmatrix}$$

is called the *transfer-function matrix* of the system. Whenever a transfer-function matrix $\mathbf{H}(j\omega)$ is used, the inputs are assumed to have equal angular frequencies ω. The principle of superposition holds for phasors if all inputs have the same angular frequency ω.

3.1.7 Application Example of Phasor Method

The computational efficiency of the phasor method is best demonstrated by illustrative application examples. We shall solve a problem that is important in analysis and design of LTI systems.

Solving a Second-Order System

Problem 3.1.1

A system is characterized by the equation

$$a_2 \mathrm{D}^2 y + a_1 \mathrm{D} y + a_0 y = b_2 \mathrm{D}^2 x + b_1 \mathrm{D} x + b_0 x$$

with real coefficients such that

$$a_2 > 0$$

$$a_1 > 0$$

$$a_0 > 0$$

$$a_1^2 < a_2 a_0$$

$$b_2 \neq 0$$

The input is a causal sinusoid:

$$x(t) = X_\mathrm{m} \cos(\omega t) u(t)$$

The initial conditions are given at $t = t_0 = 0$:

$$y(0^-) = U$$

$$\mathrm{D}y(0^-) = V$$

Find the complete response for $t \geq 0$, and check whether the system can reach the sinusoidal steady state (numerical values: $a_2 = 1$, $a_1 = 2$, $a_0 = 5$, $b_2 = 3$, $b_1 = 4$, $b_0 = 6$, $U = 2$, $V = 6$, $\omega = 1$, $X_\mathrm{m} = 1$).

Solution

The characteristic equation is

$$A(s) = a_2 s^2 + a_1 s + a_0 = 0$$

and the natural frequencies are

$$s_1 = \sigma_1 + j\omega_1 = \frac{-a_1 - \sqrt{a_1^2 - 4a_2 a_0}}{2a_2} = \frac{-a_1}{2a_2} + j\frac{-\sqrt{4a_2 a_0 - a_1^2}}{2a_2} = -1 - j2$$

$$s_2 = \sigma_2 + j\omega_2 = \sigma_1 - j\omega_1 = s_1^* = -1 + j2$$

with

$$\sigma_1 = \frac{-\,a_1}{2a_2} = -1 < 0, \quad \omega_1 = \frac{-\,\sqrt{4a_2 a_0 - a_1^2}}{2a_2} = -2$$

The natural frequencies are different, are simple, occur in a complex conjugate pair, and are in the open left-half complex plane.

The zero-input response, $y_0(t)$, is the solution of the equation

$$a_2 \mathrm{D}^2 y_0 + a_1 \mathrm{D} y_0 + a_0 y_0 = 0$$

and it is of the form

$$y_0(t) = K_1 \, \mathrm{e}^{s_1 t} + K_2 \, \mathrm{e}^{s_2 t}$$

The real constants K_1 and K_2 are chosen to meet the initial conditions:

$$y_0(0) = y(0^-) = U$$

$$\mathrm{D} y_0(0) = \mathrm{D} y(0^-) = V$$

It follows that

$$K_1 + K_2 = U$$

$$s_1 K_1 + s_2 K_2 = V$$

$$K_1 = \frac{s_2 U - V}{s_2 - s_1} = \frac{U}{2} + j \frac{\sigma_1 U - V}{2\omega_1} = 1 + j2$$

$$K_2 = \frac{s_1 U - V}{s_1 - s_2} = \frac{U}{2} - j \frac{\sigma_1 U - V}{2\omega_1} = K_1^* = 1 - j2$$

$$y_0(t) = \mathrm{e}^{\sigma_1 t} \left(U \cos(\omega_1 t) + \frac{V - \sigma_1 U}{\omega_1} \sin(\omega t) \right), \qquad t \geq 0$$

$$y_0(t) = \mathrm{e}^{-t} \left(2\cos(2t) + 4\sin(2t) \right), \qquad t \geq 0$$

The zero-state response $y_x(t)$ is of the form

$$y_x(t) = y_{\mathrm{xs}}(t) \, u(t)$$

The differential equation that characterizes the system becomes

$$a_2 \mathrm{D}^2 (y_{\mathrm{xs}}(t) u(t)) + a_1 \mathrm{D} (y_{\mathrm{xs}}(t) u(t)) + a_0 (y_{\mathrm{xs}}(t) u(t))$$
$$= b_2 \mathrm{D}^2 (x_{\mathrm{s}}(t) u(t)) + b_1 \mathrm{D} (x_{\mathrm{s}}(t) u(t)) + b_0 (x_{\mathrm{s}}(t) u(t))$$

where $x_{\mathrm{s}}(t) = X_{\mathrm{m}} \cos(\omega t)$. It expands to

$$a_2 \left(\mathrm{D}^2 y_{\mathrm{xs}}(t) \right) u(t) + a_2 \left(\mathrm{D} y_{\mathrm{xs}}(0^+) \right) \mathrm{D} u(t) + a_2 \left(y_{\mathrm{xs}}(0^+) \right) \mathrm{D}^2 u(t)$$
$$+ a_1 (\mathrm{D} y_{\mathrm{xs}}(t)) \, u(t) + a_1 \left(y_{\mathrm{xs}}(0^+) \right) \mathrm{D} u(t)$$
$$+ a_0 y_{\mathrm{xs}}(t) \, u(t)$$
$$= b_2 \left(\mathrm{D}^2 x_{\mathrm{s}}(t) \right) u(t) + b_2 \left(\mathrm{D} x_{\mathrm{s}}(0^+) \right) \mathrm{D} u(t) + b_2 \left(x_{\mathrm{s}}(0^+) \right) \mathrm{D}^2 u(t)$$
$$+ b_1 (\mathrm{D} x_{\mathrm{s}}(t)) \, u(t) + b_1 \left(x_{\mathrm{s}}(0^+) \right) \mathrm{D} u(t)$$
$$+ b_0 x_{\mathrm{s}}(t) u(t)$$

with

$$x_{\mathrm{s}}(0^+) = X_{\mathrm{m}} \cos(\omega t)|_{t \to 0+} = X_{\mathrm{m}} = 1$$

$$\mathrm{D} x_{\mathrm{s}}(0^+) = -\omega X_{\mathrm{m}} \sin(\omega t)|_{t \to 0+} = 0$$

Equating the like terms we obtain the differential equation in $y_{xs}(t)$

$$a_2 D^2 y_{xs} + a_1 D y_{xs} + a_0 y_{xs} = b_2 D^2 x_s + b_1 D x_s + b_0 x_s$$

and the set of linear algebraic equations for the corresponding initial conditions that $y_{xs}(t)$ must meet

$$a_2 D y_{xs}(0^+) + a_1 y_{xs}(0^+) = b_2 \cdot 0 + b_1 X_m$$

$$a_2 y_{xs}(0^+) = b_2 X_m$$

The required initial conditions are

$$y_{xs}(0^+) = \frac{b_2}{a_2} X_m = 3$$

$$D y_{xs}(0^+) = \frac{a_2 b_1 - a_1 b_2}{a_2^2} X_m = -2$$

and are inconsistent, i.e. $y_x(t)$ is discontinuous at $t = 0$.

The function $y_{xs}(t)$ is of the form

$$y_{xs}(t) = y_{xsh}(t) + y_{xsp}(t)$$

$y_{xsh}(t)$ is a solution of the homogeneous differential equation

$$a_2 D^2 y_{xsh} + a_1 D y_{xsh} + a_0 y_{xsh} = 0$$

and it takes the form

$$y_{xsh}(t) = C_1 e^{s_1 t} + C_2 e^{s_2 t}$$

The natural frequencies s_1 and s_2 are in the open left-half plane, thus, the particular solution $y_{xsp}(t)$ is sinusoidal

$$y_{xsp}(t) = Y_m \cos(\omega t + \eta)$$

and can be computed by the phasor method

$$y_{xsp}(t) = \mathcal{P}^{-1}\left(\frac{b_2(j\omega)^2 + b_1(j\omega) + b_0}{a_2(j\omega)^2 + a_1(j\omega) + a_0} X_m e^{j0}\right)$$

$$Y_m = \left|\frac{b_2(j\omega)^2 + b_1(j\omega) + b_0}{a_2(j\omega)^2 + a_1(j\omega) + a_0} X_m\right| = \frac{2}{5}\sqrt{5}$$

$$\eta = \arg\left(\frac{b_2(j\omega)^2 + b_1(j\omega) + b_0}{a_2(j\omega)^2 + a_1(j\omega) + a_0} e^{j0}\right) = -\operatorname{atan}\left(\frac{1}{2}\right)$$

Finally, we have

$$y_{xs}(t) = C_1 e^{s_1 t} + C_2 e^{s_2 t} + Y_m \cos(\omega t + \eta)$$

Constants C_1 and C_2 are chosen to satisfy the initial conditions $y_{xs}(0^+)$ and $D y_{xs}(0^+)$:

$$y_{xs}(0^+) = C_1 + C_2 + Y_m \cos(\eta)$$

$$D y_{xs}(0^+) = s_1 C_1 + s_2 C_2 - \omega Y_m \sin(\eta)$$

that is, the set of linear algebraic equations

$$\begin{aligned}\frac{b_2}{a_2}X_\mathrm{m} &= C_1 + C_2 + Y_\mathrm{m}\cos(\eta)\\ \frac{a_2b_1 - a_1b_2}{a_2^2}X_\mathrm{m} &= s_1C_1 + s_2C_2 - \omega Y_\mathrm{m}\sin(\eta)\end{aligned}$$

uniquely determines $C_1 = \frac{11}{10} - j\frac{1}{20}$ and $C_2 = \frac{11}{10} + j\frac{1}{20}$.

The complete response $y(t)$, for $t \geq 0$, is as follows:

$$y(t) = \underbrace{K_1\,\mathrm{e}^{s_1t} + K_2\,\mathrm{e}^{s_2t}}_{y_0} + \underbrace{\left(C_1\,\mathrm{e}^{s_1t} + C_2\,\mathrm{e}^{s_2t}\right)u(t)}_{y_{\mathrm{xsh}}} + \underbrace{Y_\mathrm{m}\cos(\omega t + \eta)\,u(t)}_{y_{\mathrm{xsp}}}$$

where K_1, K_2, C_1, C_2, Y_m, and η are known constants.

$$\begin{aligned}y(t) &= \mathrm{e}^{-t}(2\cos(2t) + 4\sin(2t))\\ &\quad + \mathrm{e}^{-t}(\frac{11}{5}\cos(2t) - \frac{1}{10}\sin(2t))\,u(t)\\ &\quad + \frac{2}{5}\sqrt{5}\cos(t - \mathrm{atan}(\frac{1}{2}))\,u(t)\\ t &\geq 0\end{aligned}$$

As $t \to \infty$ the exponential terms tend to zero because

$$\begin{aligned}\mathrm{Re}(s_1) &= \mathrm{Re}(s_2) < 0\\ \lim_{t\to\infty}\mathrm{e}^{s_1t} &= \lim_{t\to\infty}\left(\mathrm{e}^{\sigma_1t}(\cos(\omega_1t) + j\sin(\omega_1t))\right) = 0, \qquad \sigma_1 < 0\\ \lim_{t\to\infty}\mathrm{e}^{s_2t} &= \lim_{t\to\infty}\mathrm{e}^{s_1^*t} = \left(\lim_{t\to\infty}\mathrm{e}^{s_1t}\right)^* = 0\end{aligned}$$

Therefore, as time tends to infinity the complete response approaches the particular solution.

The sinusoidal steady-state response is $Y_\mathrm{m}\cos(\omega t + \eta)$ for $t \gg t_0 = 0$.

3.2 FOURIER SERIES AND HARMONIC ANALYSIS

The theory of phasors and the "$j\omega$" method was entirely founded on the assumption of sinusoidal variation of signals. Nonsinusoidal periodic signals play an important part in electronics, telecommunications, control engineering, power engineering, and signal processing in general. The theory of nonsinusoidal periodic signals is based upon resolving them into sinusoidal components. Next, according to the principle of superposition, we can find the steady-state response of an LTI system to an arbitrary periodic input by applying the phasor method to each harmonic component.

3.2.1 The Fourier Series

Consider a real signal $x(t)$ of real variable t that satisfies the following conditions:

- $x(t+T) = x(t)$; that is, the signal is periodic having a period T.
- $x(t)$ is defined in the interval $\tau < t < \tau + T$.
- $x(t)$ and $\mathrm{d}x(t)/\mathrm{d}t$ are sectionally continuous in $\tau < t < \tau + T$.

A signal is called *sectionally continuous* or piecewise continuous in an interval if the interval can be subdivided into finite number of intervals in each of which the signal is continuous and has finite right- and left-hand limits. Then, at every point of continuity the periodic signal $x(t)$ can be represented by a series of the form

$$x(t) = C_0 + \sum_{n=1}^{+\infty} (A_n \cos(n\omega_1 t) + B_n \sin(n\omega_1 t)) \tag{3.67}$$

where

$$\omega_1 = \frac{2\pi}{T} \tag{3.68}$$

$$C_0 = \frac{1}{T} \int_{\tau}^{\tau+T} x(t)\,\mathrm{d}t \tag{3.69}$$

$$A_n = \frac{2}{T} \int_{\tau}^{\tau+T} x(t)\cos(n\omega_1 t)\,\mathrm{d}t \tag{3.70}$$

$$B_n = \frac{2}{T} \int_{\tau}^{\tau+T} x(t)\sin(n\omega_1 t)\,\mathrm{d}t \tag{3.71}$$

At a point of discontinuity, say t_0, the series converges to the mean value

$$\frac{x(t_0^+) + x(t_0^-)}{2} \tag{3.72}$$

The series (3.67) with coefficients (3.68)–(3.71) is called the *Fourier series* of the signal $x(t)$, or the *trigonometric Fourier series* of $x(t)$. The coefficients C_0, A_n, and B_n are called the *Fourier coefficients*. The frequency ω_1 is the *fundamental angular frequency*.

The above conditions are known as the *Dirichlet's conditions* and are sufficient, but not necessary, for convergence of Fourier series; these conditions are satisfied by virtually all signals arising in physical and engineering problems.

The kth partial sum of the Fourier series

$$S_k(t) = C_0 + \sum_{n=1}^{k} (A_n \cos(n\omega_1 t) + B_n \sin(n\omega_1 t)) \tag{3.73}$$

is often referred to as the *truncated Fourier series* of $x(t)$.

Computation of the Fourier coefficients can be simplified if $x(t)$ has symmetry. An even signal, $x(-t) = x(t)$, has no sine terms, $B_n = 0$; an odd signal, $x(-t) = -x(t)$, has no cosine terms, $A_n = 0$, and $C_0 = 0$.

The process of resolving a signal into its Fourier series is called *spectral analysis* or *harmonic analysis*.

3.2.2 Complex Form of the Fourier Series

In complex notation, the Fourier series can be written in a more compact representation as follows:

$$x(t) = \sum_{n=-\infty}^{+\infty} C_n \, e^{jn\omega_1 t} \tag{3.74}$$

where

$$C_n = \frac{1}{T} \int_{\tau}^{\tau+T} x(t) \, e^{-jn\omega_1 t} \, dt \tag{3.75}$$

The series (3.74) is the *complex Fourier series* of $x(t)$, also called the *exponential Fourier series*.

Obviously

$$C_{-n} = C_n^* \tag{3.76}$$

and

$$C_n = \frac{A_n - jB_n}{2} \tag{3.77}$$

for $n \neq 0$. (C^* denotes the conjugate of C.)

The complex Fourier coefficients C_n contain all the information about the signal $x(t)$.

3.2.3 Parseval's Identity

Parseval's identity or *Parseval's theorem* states that

$$\frac{1}{T} \int_{\tau}^{\tau+T} |x(t)|^2 \, dt = C_0^2 + \frac{1}{2} \sum_{n=1}^{+\infty} (A_n^2 + B_n^2) = \sum_{n=-\infty}^{+\infty} |C_n|^2 \tag{3.78}$$

where A_n and B_n are given by (3.68)–(3.71), and C_n is given by (3.75). As a consequence,

$$\lim_{n \to \infty} A_n = 0 \tag{3.79}$$

$$\lim_{n \to \infty} B_n = 0 \tag{3.80}$$

which is known as the *Riemann's theorem*.

If the signal $x(t)$ represents an electric current or voltage across a resistor, the expression (3.78) is proportional to the resistor average power over a period.

3.2.4 Harmonics of Periodic Signals

The Fourier series (3.67) can be written as

$$x(t) = \sum_{n=0}^{+\infty} x^{(n)}(t) = X^{(0)} + \sum_{n=1}^{+\infty} X_{\mathrm{m}}^{(n)} \cos(n\omega_1 t + \xi^{(n)}) \tag{3.81}$$

where

$$x^{(0)}(t) = X^{(0)} = C_0 \tag{3.82}$$

$$x^{(n)}(t) = X_{\mathrm{m}}^{(n)} \cos(n\omega_1 t + \xi^{(n)}) \tag{3.83}$$

$$X_{\mathrm{m}}^{(n)} = \sqrt{A_n^2 + B_n^2} \tag{3.84}$$

$$\xi^{(n)} = \arg(A_n - jB_n) \tag{3.85}$$

The term $x^{(1)}(t)$ is the *fundamental component* or the first harmonic, $x^{(2)}(t)$ is the second harmonic, and so on. If the signal $x(t)$ represents an electrical quantity, the constant term $X^{(0)}$ is called the *dc component*, and the higher-order harmonics are known as *ac components*; ω_1 is the *fundamental angular frequency*, and $n\omega_1$ is the nth *harmonic angular frequency*.

According to Parseval's identity, the root-mean-square (rms) value of a signal expressed as a Fourier series is the square root of the sum of squares of the rms values of the separate components:

$$X_{\mathrm{rms}} = \sqrt{\frac{1}{T}\int_{\tau}^{\tau+T} |x(t)|^2 \,\mathrm{d}t} = \sqrt{\left(X^{(0)}\right)^2 + \frac{1}{2}\sum_{n=1}^{+\infty}\left(X_{\mathrm{m}}^{(n)}\right)^2} \tag{3.86}$$

The quantities $\left(X^{(0)}\right)^2$ and $\frac{1}{2}\left(X_{\mathrm{m}}^{(n)}\right)^2$ show how the power in a periodic signal $x(t)$ is distributed in the frequency domain.

3.2.5 Gibbs Phenomenon

When a sudden change of amplitude occurs in a signal and the attempt is made to represent it by a finite number of terms in a Fourier series, the overshoot at the corners (at the points of abrupt change) is always found. As the number of terms is increased, the overshoot is still found; this is called the *Gibbs phenomenon*.

The best way to illustrate the Gibbs phenomenon is to analyze a unit periodic square pulse train defined as

$$x(t) = \begin{cases} 1, & 0 \le t < \dfrac{T}{2} \\ -1, & \dfrac{T}{2} \le t < T \end{cases} \tag{3.87}$$

$$x(t + kT) = x(t)$$

$$k = \pm 1, \pm 2, \pm 3, \ldots$$

which may be represented by the series

$$x(t) = \sum_{n=1,3,5,\ldots,\infty} \frac{4}{\pi n} \sin\left(2\pi n \frac{t}{T}\right)$$

If we approximate $x(t)$ with a finite sum

$$x_N(t) = \sum_{n=1,3,5,\ldots,N} \frac{4}{\pi n} \sin\left(2\pi n \frac{t}{T}\right)$$

the sum will exhibit an overshoot at points very near the corners $t = k\frac{T}{2}$, and the overshoot will not vanish as N increases—it will have a value of approximately 1.12. [14]

3.2.6 Amplitude and Phase Spectrum of Periodic Signals

A graphical representation of a periodic signal $x(t)$, produced by drawing a series of vertical lines at intervals on a horizontal axis, where the intervals represent an increase of n and the vertical lines are proportional to $X_{\mathrm{m}}^{(n)}$, is called the *amplitude spectrum* of the periodic signal. Often, angular frequencies $n\omega_1$ may replace n as the abscissa.

Similarly, the phase angles $\xi^{(n)}$ plotted against n or $n\omega_1$ are called the *phase spectrum* of the periodic signal. Notice that $\xi^{(0)} \equiv 0$.

The two sequences of numbers

$$\{X^{(0)}, X_{\mathrm{m}}^{(1)}, X_{\mathrm{m}}^{(2)}, \ldots, X_{\mathrm{m}}^{(n)}, \ldots\}$$

$$\{\xi^{(0)}, \xi^{(1)}, \xi^{(2)}, \ldots, \xi^{(n)}, \ldots\}$$

along with the period T, contain all the information embodied in the Fourier series.

The plot of quantities

$$\left(X^{(0)}\right)^2, \frac{1}{2}\left(X_{\mathrm{m}}^{(1)}\right)^2, \frac{1}{2}\left(X_{\mathrm{m}}^{(2)}\right)^2, \ldots, \frac{1}{2}\left(X_{\mathrm{m}}^{(n)}\right)^2, \ldots \tag{3.88}$$

versus n or $n\omega_1$ is called the *power spectrum*.

The spectra defined above are frequently referred to as the *natural spectra* and are defined for non-negative n.

In considering representations of $x(t)$ in terms of the coefficients C_n the integer n is arbitrary (positive, negative, or zero) and we can define the amplitude spectrum as the plot of $|C_n|$ against n or $n\omega_1$; this spectrum is known as the *mathematical amplitude spectrum*. Similarly, the *mathematical power spectrum* is the plot of $|C_n|^2$ against n or $n\omega_1$ $(n = 0, \pm 1, \pm 2, \pm 3, \ldots)$.

The spectra of a periodic signal are *discrete spectra* or *line spectra*.

Sequence of Short Square Pulses. Consider a train of square pulses which occurs at regular intervals T. The amplitude of the pulses is A, and each pulse is of duration τ:

$$x(t) = \begin{cases} A, & 0 < t \leq \tau \\ 0, & \tau < t \leq T \end{cases}$$
$$x(t + kT) = x(t) \tag{3.89}$$
$$k = \pm 1, \pm 2, \pm 3, \ldots$$

The Fourier series for the pulse train $x(t)$ is

$$x(t) = C_0 + \sum_{n=1}^{+\infty} (A_n \cos(n\omega_1 t) + B_n \sin(n\omega_1 t))$$

with

$$C_0 = A\frac{\tau}{T}, \qquad A_n = A\frac{\sin(n\omega_1\tau)}{n\pi}, \qquad B_n = 2A\frac{\sin^2(n\omega_1\tau/2)}{n\pi}, \qquad \omega_1 = \frac{2\pi}{T}$$

or, in the complex form,

$$x(t) = \sum_{n=-\infty}^{+\infty} C_n\, e^{jn\omega_1 t}, \qquad C_n = \frac{A_n - jB_n}{2}$$

A limiting case of great theoretical significance is that of very short pulses ($\tau \to 0$), when pulses are of infinite magnitude and infinitesimal duration, but the product of these quantities is finite, $A\tau \equiv 1$, or $A = 1/\tau$:

$$c(t) = \lim_{\tau\to 0} x(t) = \sum_{k=-\infty}^{+\infty} \delta(t - kT) = \frac{1}{T} + \frac{2}{T}\sum_{n=1}^{+\infty}\cos(n\omega_1 t) = \frac{1}{T}\sum_{n=-\infty}^{+\infty} e^{jn\omega_1 t} \tag{3.90}$$

The signal $c(t)$ is a train of equidistant Dirac's delta pulses and is frequently called the *comb signal*.

3.2.7 Steady-State Response of a System to a Nonsinusoidal Periodic Stimulus

Consider a single-input single-output lumped LTI system excited by a nonsinusoidal periodic stimulus $x(t)$ resolved into its harmonics (3.81). Again, we focus on the systems with natural frequencies in the open left-half complex plane.

We can find the phasors for each input harmonic component (3.83):

$$X^{(n\omega_1)} = \mathcal{P}_{n\omega_1}(x^{(n)}(t)) = X_{\mathrm{m}}^{(n)}\, e^{j\xi^{(n)}} \tag{3.91}$$

According to the principle of superposition, the steady state resulting from the input harmonics is the sum of the sinusoidal steady states that would exist if each input harmonic were acting alone on the system.

We may use the phasor method to obtain the sinusoidal steady state to a single input harmonic $x^{(n)}(t)$. If the transfer function of the system is known, say $H(j\omega)$, the output harmonic phasors are

$$Y^{(n\omega_1)} = H(jn\omega_1)X^{(n\omega_1)} \tag{3.92}$$

Finally, we determine the steady-state response by taking the inverse phasor transform of the output harmonic phasors

$$y(t) = \sum_{n=0}^{\infty} \mathcal{P}_{n\omega_1}^{-1}(Y^{(n\omega_1)}) \tag{3.93}$$

In other words, the steady state resulting from a nonsinusoidal periodic input is the sum of the sinusoidal steady states that would exist if each input harmonic were acting alone on the system.

3.3 FOURIER TRANSFORM

The Fourier transform occupies so central a place in analysis of signals and systems as to demand at least an introductory treatment. A major reason for this is that the most convenient way of measuring and specifying the performance of a system or signal is based on frequency. If we apply continuous sinusoidal stimuli over a wide range of frequencies and then measure the relation between response and stimuli both in magnitude and phase, the Fourier transform would then make it possible in principle to deduce the response to any other form of stimulus. Therefore, the Fourier transform can be thought of as the ultimate generalization of the phasor transform.

3.3.1 Definition of the Fourier Transform

Consider a nonperiodic signal $x(t)$. The *Fourier transform* of $x(t)$ is

$$X(j\omega) = \mathcal{F}(x(t)) = \int_{-\infty}^{+\infty} x(t)\, \mathrm{e}^{-j\omega t}\, \mathrm{d}t \tag{3.94}$$

and the *inverse Fourier transform* of $X(j\omega)$ is

$$x(t) = \mathcal{F}^{-1}(X(j\omega)) = \frac{1}{2\pi}\int_{-\infty}^{+\infty} X(j\omega)\, \mathrm{e}^{j\omega t}\, \mathrm{d}\omega \tag{3.95}$$

The two functions $x(t)$ and $X(j\omega)$ form a *Fourier transform pair* which is designated by $x(t) \leftrightarrow X(j\omega)$. Some authors drop the imaginary unit, j, and denote the Fourier transform of $x(t)$ by $X(\omega)$. The variable ω is called a *continuous frequency variable.*

The Fourier transform exists if

$$\lim_{T\to\infty} \int_{-T}^{T} |x(t)|\, \mathrm{d}t < \infty \tag{3.96}$$

exists. This is a sufficient, but not necessary, condition. The more general version of the existence conditions requires $x(t)$ to have a finite number of maxima and minima, as well as a finite number of discontinuities over the entire range $-\infty < t < \infty$; $x(t)$ may become infinite at some isolated points provided that (3.96) is finite.

The inverse Fourier transform (3.95) gives the value of $x(t)$ in terms of $X(j\omega)$, at any point t where $x(t)$ is continuous. However, at a point of discontinuity, say t_0, (3.95) gives the arithmetic mean $\frac{1}{2}(x(t_0^-) + x(t_0^+))$.

The process through which signals of a real variable t are associated with corresponding complex functions of a new complex variable $j\omega$ is known as a *Fourier transformation*. The complex function $X(j\omega)$ is called the *image* of $x(t)$, and $x(t)$ is known as the *original* of $X(j\omega)$. The process of going back from $X(j\omega)$ to $x(t)$ is referred to as an *inverse Fourier transformation*.

The plots of $|X(j\omega)|$ and $\arg X(j\omega)$ versus ω are called the *amplitude spectrum* and *phase spectrum* of $x(t)$, respectively.

Fourier transform (3.94) can be viewed as the harmonic content per unit interval of frequency $f = \frac{1}{2\pi}\omega$. Thus, it is convenient to call $x(t)$ the signal possessing the spectrum $X(j\omega)$.

Notice that apart from the factor 2π and the sign of the exponents, Eqs. (3.94) and (3.95) are identical in form.

3.3.2 Properties of the Fourier Transform

We summarize the salient properties of the Fourier transform subject always to the proviso that the transform exists.

The Fourier transform is *unique* (except for a finite number of isolated points of discontinuity) for all t:

$$x_1(t) = x_2(t) \Leftrightarrow X_1(j\omega) = X_2(j\omega) \tag{3.97}$$

where $X_1(j\omega) = \mathcal{F}(x_1(t))$, $X_2(j\omega) = \mathcal{F}(x_2(t))$, and $x_1(t)$, $x_2(t)$ are nonperiodic signals. For physically generated signals and for the signals dealt with in this book, the Fourier transform is unique for all t.

The operators $\mathcal{F}$ and $\mathcal{F}^{-1}$ are *homogeneous*

$$\begin{aligned} \mathcal{F}(Kx(t)) &= K\mathcal{F}(x(t)) \\ \mathcal{F}^{-1}(KX(j\omega)) &= K\mathcal{F}^{-1}(X(j\omega)) \end{aligned} \tag{3.98}$$

for any constant K, with $X(j\omega) = \mathcal{F}(x(t))$.

Also, the two operators are *additive*:

$$\begin{aligned} \mathcal{F}(x_1(t) + x_2(t)) &= \mathcal{F}(x_1(t)) + \mathcal{F}(x_2(t)) \\ \mathcal{F}^{-1}(X_1(j\omega) + X_2(j\omega)) &= \mathcal{F}^{-1}(X_1(j\omega)) + \mathcal{F}^{-1}(X_2(j\omega)) \end{aligned} \tag{3.99}$$

The Fourier transform, $\mathcal{F}$, maps the operation of differentiating, D, into the multiplication by $j\omega$:

$$\begin{aligned} \mathcal{F}(\mathrm{D}x(t)) &= j\omega\mathcal{F}(x(t)) = j\omega X(j\omega) \\ \mathcal{F}^{-1}(j\omega X(j\omega)) &= \mathrm{D}\mathcal{F}^{-1}(X(j\omega)) = \mathrm{D}x(t) \end{aligned} \tag{3.100}$$

where $X(j\omega) = \mathcal{F}(x(t))$, and D denotes differentiation with respect to time, $\mathrm{D}x(t) = \mathrm{d}x(t)/\mathrm{d}t$.

A set of properties (theorems) can be derived from the near-symmetry of the direct and inverse Fourier transformations:

Symmetry

$$\mathcal{F}(X(jt)) = 2\pi x(-\omega) \tag{3.101}$$

$$\mathcal{F}^{-1}(x(\omega)) = \frac{1}{2\pi} X(-jt) \tag{3.102}$$

Time shifting

$$\mathcal{F}(x(t - T)) = X(j\omega)\,\mathrm{e}^{-j\omega T} \tag{3.103}$$

Frequency shifting

$$\mathcal{F}^{-1}(X(j(\omega - \Omega))) = x(t)\mathrm{e}^{j\Omega t} \tag{3.104}$$

Time convolution

$$\mathcal{F}\int_{-\infty}^{\infty} x(\tau)h(t - \tau)\,\mathrm{d}\tau = X(j\omega)H(j\omega) \tag{3.105}$$

Frequency convolution

$$\mathcal{F}^{-1}\frac{1}{2\pi}\int_{-\infty}^{\infty} X(jv)Y(j(\omega - v))\,\mathrm{d}v = x(t)y(t) \tag{3.106}$$

where $X(j\omega) = \mathcal{F}(x(t))$, $H(j\omega) = \mathcal{F}(h(t))$, $Y(j\omega) = \mathcal{F}(y(t))$, $T = const$, $\Omega = const$.

Time scaling

$$\mathcal{F}(x(at)) = \frac{1}{|a|} X\left(\frac{j\omega}{a}\right) \tag{3.107}$$

Frequency scaling

$$\mathcal{F}^{-1}(X(ja\omega)) = \frac{1}{|a|} x\left(\frac{t}{a}\right) \tag{3.108}$$

for any real $a \neq 0$.

3.3.3 Convolution

The *convolution* of two signals $x(t)$ and $h(t)$ is denoted by $x(t) * h(t)$ and defined as

$$x(t) * h(t) = \int_{-\infty}^{\infty} x(\tau)h(t - \tau)\,\mathrm{d}\tau = \int_{-\infty}^{\infty} x(t - \tau)h(\tau)\,\mathrm{d}\tau \tag{3.109}$$

The *convolution in the frequency domain* is defined by

$$X(j\omega) * Y(j\omega) = \int_{-\infty}^{\infty} X(jv)Y(j(\omega - v))\,\mathrm{d}v = \int_{-\infty}^{\infty} X(j(\omega - v))Y(jv)\,\mathrm{d}v \quad (3.110)$$

where $X(j\omega)$ and $Y(j\omega)$ are the Fourier transforms of $x(t)$ and $y(t)$, respectively.

Dirac delta impulse, $\delta(t)$, is the *identity element* in the convolution operation

$$x(t) * \delta(t) = x(t) \quad (3.111)$$

Also, $x(t) * h(t) = h(t) * x(t)$, and it can be shown that

$$x(t) * \delta(t - T) = x(t - T) \quad (3.112)$$

for any real time shift T.

The property (3.112) can be useful in representing periodic signals. If the signal $x(t)$ is *time-limited*—that is, it is zero outside a time interval, say $x(t) = 0$ for $|t| > \frac{T}{2}$—then the periodic continuation of $x(t)$, with period T, can be represented as a convolution of $x(t)$ and the comb signal $c(t) = \sum_{k=-\infty}^{+\infty} \delta(t - kT)$:

$$x_{\text{per}}(t) = x(t) * c(t) \quad (3.113)$$

where $x_{\text{per}}(t)$ denotes the periodic signal, and $x(t) = x_{\text{per}}(t)$ for $|t| < \frac{T}{2}$.

Modulation Property. The frequency shifting property can be employed to obtain the Fourier transform of a signal of the form

$$x_{\text{am}}(t) = x(t)\cos(\Omega t) \quad (3.114)$$

in which $x(t)$ is called the *modulating signal*. The term $\cos(\Omega t)$ is said to be *modulated in amplitude*, and is called the *carrier*.

The Fourier transform of $x_{\text{am}}(t)$ can be found by expressing the $\cos(\Omega t)$ in terms of the exponential function

$$\begin{aligned} X_{\text{am}}(j\omega) &= \mathcal{F}(x_{\text{am}}(t)) = \mathcal{F}(x(t)(\mathrm{e}^{j\Omega t} + \mathrm{e}^{-j\Omega t})/2) \\ &= \frac{1}{2}\mathcal{F}(x(t)\,\mathrm{e}^{j\Omega t}) + \frac{1}{2}\mathcal{F}(x(t)\mathrm{e}^{-j\Omega t}) \end{aligned}$$

After employing the frequency shifting property, we find

$$X_{\text{am}}(j\omega) = \frac{1}{2}X(j(\omega - \Omega)) + \frac{1}{2}X(j(\omega + \Omega)) \quad (3.115)$$

where $X(j\omega) = \mathcal{F}(x(t))$ is called the *base-band spectrum*, and the two terms are said to be the *side-bands*.

3.3.4 Parseval's Theorem and Energy Spectral Density

Parseval's formula (theorem) states that

$$\int_{-\infty}^{\infty} |x(t)|^2 \,\mathrm{d}t = \frac{1}{2\pi} \int_{-\infty}^{\infty} |X(j\omega)|^2 \,\mathrm{d}\omega \tag{3.116}$$

If $x(t)$ represents an electric voltage or current, then the left-hand integral in (3.116) represents the total energy that would be delivered to a 1-Ω resistor.

The quantity

$$E(\omega) = |X(j\omega)|^2 \tag{3.117}$$

represents the energy per unit bandwidth of frequency (not angular frequency) and is called the *energy spectral density*. A plot of $E(\omega)$ versus ω is known as the *energy spectrum* of $x(t)$. The *total energy* of the signal is

$$W = \int_{-\infty}^{\infty} |x(t)|^2 \,\mathrm{d}t = \frac{1}{2\pi} \int_{-\infty}^{+\infty} E(\omega) \,\mathrm{d}\omega \tag{3.118}$$

Notice that for a periodic signal the total signal power was obtained, in terms of the Fourier series coefficients, as the sum of the power contents of all the discrete frequency components. For a nonperiodic signal the total signal energy is obtained, in terms of the amplitude spectrum, as the integral of the energy contents of all the continuous frequency components.

Assume that the real signal $x(t)$ has finite energy. The *autocorrelation* function of $x(t)$ is defined as

$$\rho_{xx}(\tau) = \int_{-\infty}^{\infty} x(t)x(t+\tau) \,\mathrm{d}t \tag{3.119}$$

It can be shown that the autocorrelation function, $\rho_{xx}(\tau)$, and the energy spectral density, $E(\omega)$, constitute a Fourier transform pair

$$\rho_{xx}(\tau) \leftrightarrow E(\omega) \tag{3.120}$$

that is,

$$\mathcal{F}(\rho_{xx}(\tau)) = E(\omega)$$

or

$$\rho_{xx}(\tau) = \mathcal{F}^{-1}(E(\omega))$$

which is known as the *Wiener–Kintchine theorem*.

3.3.5 Properties of the Fourier Transform of Real Signals

Assume that the signal $x(t)$ is a real function of time—that is, of the real variable t—and that its Fourier transform $X(j\omega)$ may be expressed as either

$$X(j\omega) = M_x(\omega)\,\mathrm{e}^{j\Phi_x(\omega)}$$

or

$$X(j\omega) = X_{\mathrm{re}}(\omega) + jX_{\mathrm{im}}(\omega)$$

where $M_x(\omega) = |X(j\omega)|$, $\Phi_x(\omega) = \arg(X(j\omega))$, $X_{\text{re}}(\omega) = \text{Re}(X(j\omega))$, and $X_{\text{im}}(\omega) = \text{Im}(X(j\omega))$.

From the definition of the Fourier transform (3.94) the real and imaginary parts of $X(j\omega)$ become, respectively,

$$X_{\text{re}}(\omega) = \int_{-\infty}^{\infty} x(t)\cos(\omega t)\,\mathrm{d}t \tag{3.121}$$

$$X_{\text{im}}(\omega) = -\int_{-\infty}^{\infty} x(t)\sin(\omega t)\,\mathrm{d}t \tag{3.122}$$

It can be shown (after employing the definition and/or the properties of the Fourier transform) that the following holds:

- The real part of $X(j\omega)$ is an **even** function of ω, $X_{\text{re}}(-\omega) = X_{\text{re}}(\omega)$.
- The imaginary part of $X(j\omega)$ is an **odd** function of ω, $X_{\text{im}}(-\omega) = -X_{\text{im}}(\omega)$.
- Changing the sign of ω in $X(j\omega)$ is equivalent to taking the complex conjugate of $X(j\omega)$, $X(-j\omega) = X^*(j\omega)$ (the asterisk denotes the complex conjugate).
- The amplitude spectrum is an **even** function of ω, $M_x(-\omega) = M_x(\omega)$.
- The phase spectrum is an **odd** function of ω, $\Phi_x(-\omega) = -\Phi_x(\omega)$.
- The energy spectral density is an **even** function of ω, $E(-\omega) = E(\omega)$.
- If $x(t)$ is an **even** function of t, $X(j\omega)$ is **real**, $X(j\omega) = X_{\text{re}}(\omega)$.
- If $x(t)$ is an **odd** function of t, $X(j\omega)$ is **imaginary**, $X(j\omega) = jX_{\text{im}}(\omega)$.
- Changing the sign of t in $x(t)$ is equivalent to changing the sign of ω in $X(j\omega)$, $\mathcal{F}(x(-t)) = X(-j\omega)$, implying that $x(-t)$ and $X(-j\omega)$ constitute a Fourier transformation pair, $x(-t) \leftrightarrow X(-j\omega)$, or $x(-t) \leftrightarrow X^*(j\omega)$.

The real signal $x(t)$ can be decomposed into the sum of its even part

$$\text{Ev}(x(t)) = \frac{x(t) + x(-t)}{2} \tag{3.123}$$

and its odd part

$$\text{Od}(x(t)) = \frac{x(t) - x(-t)}{2} \tag{3.124}$$

that is,

$$x(t) = \text{Ev}(x(t)) + \text{Od}(x(t))$$

The even and the odd parts of $x(t)$ form Fourier pairs with the real and the imaginary parts of the Fourier transform of $x(t)$:

$$\text{Ev}(x(t)) \leftrightarrow X_{\text{re}}(\omega)$$

$$\text{Od}(x(t)) \leftrightarrow jX_{\text{im}}(\omega)$$

A number of useful Fourier transform pairs for some real signals follows

$$\begin{array}{ccc}
x(t) & \leftrightarrow & X(j\omega) \\
\delta(t) & \leftrightarrow & 1 \\
1 & \leftrightarrow & 2\pi\delta(\omega) \\
\cos(\Omega t) & \leftrightarrow & \pi\delta(\omega+\Omega)+\pi\delta(\omega-\Omega) \\
\sin(\Omega t) & \leftrightarrow & j\pi\delta(\omega+\Omega)-\mathrm{j}\pi\delta(\omega-\Omega) \\
\mathrm{e}^{j\Omega t} & \leftrightarrow & 2\pi\delta(\omega-\Omega) \\
p_T(t) = \begin{cases} 1, & |t| \le T \\ 0, & |t| > T \end{cases} & \leftrightarrow & \frac{2}{\omega}\sin(\omega T) \\
\frac{1}{\pi t}\sin(\Omega t) & \leftrightarrow & p_\Omega(\omega) = \begin{cases} 1, & |\omega| \le \Omega \\ 0, & |\omega| > \Omega \end{cases} \\
u(t) & \leftrightarrow & \frac{1}{j\omega}+\pi\delta(\omega) \\
\mathrm{e}^{-at}\,u(t) & \leftrightarrow & \frac{1}{a+j\omega}
\end{array} \tag{3.125}$$

The pairs are generated by employing the definition of the transform or its properties. A thorough treatment of the Fourier transform of special signals, like Dirac delta $\delta(t)$, unity function $x(t) = 1$, and unit step $u(t)$ (i.e., generalized functions or distributions), is not given here; it can be found in references 5 and 14.

3.3.6 Causal Signals and the Hilbert Transform

Time-varying signals that can be generated by physically realizable signal sources are causal and do not exist prior to $t = t_0$. By choosing the convenient time origin we can always adjust $t_0 = 0$. Therefore, we define a *causal signal* as a signal which exists only for positive values of time and is equal to zero otherwise:

$$x(t) = 0, \qquad t < 0 \tag{3.126}$$

An important property of a real causal signal is that it can be uniquely determined from either the real part or the imaginary part of its Fourier transform:

$$x(t) = \frac{2}{\pi}\int_0^\infty X_{\mathrm{re}}(\omega)\cos(\omega t)\,\mathrm{d}t \tag{3.127}$$

$$x(t) = -\frac{2}{\pi}\int_0^\infty X_{\mathrm{im}}(\omega)\sin(\omega t)\,\mathrm{d}t \tag{3.128}$$

Also, we can write

$$x(t) \leftrightarrow 2X_{\mathrm{re}}(\omega)$$
$$x(t) \leftrightarrow j2X_{\mathrm{im}}(\omega)$$

or equivalently

$$\mathcal{F}(x(t)) = 2X_{\text{re}}(\omega)$$

$$\mathcal{F}(x(t)) = j2X_{\text{im}}(\omega)$$

The real and imaginary part of a real causal signal are related by integrals of the form

$$X_{\text{re}}(\omega) = \frac{1}{\pi}\int_{-\infty}^{\infty} \frac{X_{\text{im}}(v)}{\omega - v}\,\mathrm{d}v \tag{3.129}$$

$$X_{\text{im}}(\omega) = -\frac{1}{\pi}\int_{-\infty}^{\infty} \frac{X_{\text{re}}(v)}{\omega - v}\,\mathrm{d}v \tag{3.130}$$

The expressions (3.129) and (3.130) are known as the *Hilbert transform pair.*

3.3.7 Application of the Fourier Transform

The Fourier transformation can be used in the computation of the solution of ordinary linear differential equations with real constant coefficients when the forcing function is nonperiodic and when all initial conditions are zero; this computation is referred to as the *Fourier transform method.*

Consider a lumped LTI system with a single input $x(t)$ and a single output $y(t)$, described by the following differential equation:

$$\begin{aligned} a_n \mathrm{D}^n y &+ a_{n-1}\mathrm{D}^{n-1} y + \cdots + a_1 \mathrm{D} y + a_0 y \\ &= b_m \mathrm{D}^m x + b_{m-1}\mathrm{D}^{m-1} x + \cdots + b_1 \mathrm{D} x + b_0 x \end{aligned} \tag{3.131}$$

that is,

$$\sum_{i=0}^{n} a_i \mathrm{D}^i y = \sum_{k=0}^{m} b_k \mathrm{D}^k x \tag{3.132}$$

or, by introducing the polynomial differential operators,

$$A(\mathrm{D})\, y = B(\mathrm{D})\, x \tag{3.133}$$

where $a_0, a_1, \ldots, a_n$ and $b_0, b_1, \ldots, b_m$ are real numbers.

We apply the Fourier transform (3.94) to both sides of Eq. (3.132) to obtain

$$\mathcal{F}\left(\sum_{i=0}^{n} a_i \mathrm{D}^i y\right) = \mathcal{F}\left(\sum_{k=0}^{m} b_k \mathrm{D}^k x\right)$$

From the additivity (3.99) and homogeneity (3.98) property this becomes

$$\sum_{i=0}^{n} a_i \mathcal{F}\left(\mathrm{D}^i y\right) = \sum_{k=0}^{m} b_k \mathcal{F}\left(\mathrm{D}^k x\right)$$

and from the differentiating property (3.100), applied repeatedly, we get

$$\sum_{i=0}^{n} a_i \, (j\omega)^i Y(j\omega) = \sum_{k=0}^{m} b_k \, (j\omega)^k X(j\omega) \tag{3.134}$$

with $Y(j\omega) = \mathcal{F}(y(t))$, or, equivalently,

$$\left(\sum_{i=0}^{n} a_i \, (j\omega)^i \right) Y(j\omega) = \left(\sum_{k=0}^{m} b_k \, (j\omega)^k \right) X(j\omega)$$

By using the compact notation similar to Eq. (3.133) we may write

$$A(j\omega) Y(j\omega) = B(j\omega) X(j\omega) \tag{3.135}$$

Thus, if

$$A(j\omega) \neq 0 \tag{3.136}$$

we compute the Fourier transform of the system output

$$Y(j\omega) = \frac{B(j\omega)}{A(j\omega)} X(j\omega) \tag{3.137}$$

where

$$A(j\omega) = \sum_{i=0}^{n} a_i \, (j\omega)^i \tag{3.138}$$

$$B(j\omega) = \sum_{k=0}^{m} b_k \, (j\omega)^k \tag{3.139}$$

Equation (3.134) can be obtained directly from Eq. (3.132) by replacing the ith derivatives of $y(t)$ with $(j\omega)^i Y(j\omega)$, for $i = 0$ to n, and by replacing the kth derivatives of $x(t)$ with $(j\omega)^k X(j\omega)$, for $k = 0$ to m. Also, we can write Eq. (3.135) directly from Eq. (3.133) by replacing the differential operator D with $j\omega$, $x(t)$ with $X(j\omega)$, and $y(t)$ with $Y(j\omega)$. Note that polynomial differential operators $A(\mathrm{D})$ and $B(\mathrm{D})$ are mapped into algebraic polynomials in complex variable $j\omega$ [Eqs. (3.138) and (3.139)].

The solution is found by the inverse Fourier transform (3.95)

$$y(t) = \mathcal{F}^{-1}(Y(j\omega)) = \mathcal{F}^{-1}(H(j\omega) X(j\omega)) \tag{3.140}$$

where

$$H(j\omega) = \frac{B(j\omega)}{A(j\omega)} \tag{3.141}$$

is the transfer function of the system. The response $y(t)$ is the zero-state response, $y(t) = y_x(t)$, and is excited solely by the input signal $x(t)$.

If the excitation is the Dirac delta, $x(t) = \delta(t)$, whose Fourier transform is unity, $X(j\omega) = \mathcal{F}(\delta(t)) = 1$, then the response, $y_\delta(t)$, can be obtained from (3.140) as

$$y(t) = y_\delta(t) = \mathcal{F}^{-1}(H(j\omega) X(j\omega)) = \mathcal{F}^{-1}(H(j\omega)) \tag{3.142}$$

It follows that the impulse response and the transfer function constitute Fourier transform pair. It is customarily to designate the impulse response by $h(t)$, so

$$h(t) \leftrightarrow H(j\omega) \tag{3.143}$$

Knowledge of the transfer function allows us to find the zero-state response due to any type of excitation. For the given transfer function $H(j\omega)$ and the input $x(t)$, we can find the zero-state response from

$$y_x(t) = \int_{-\infty}^{\infty} x(\tau)h(t-\tau)\,\mathrm{d}\tau, \qquad h(t) = \mathcal{F}^{-1}(H(j\omega)) \tag{3.144}$$

Generally, it is more efficient to compute the Fourier transform of the input signal, multiply it by the transfer function, and take the inverse Fourier transform to find the zero-state response, that is,

$$y_x(t) = \mathcal{F}^{-1}(H(j\omega)\mathcal{F}(x(t))) \tag{3.145}$$

The impulse response $h(t)$ is a real signal if the system under consideration is real. According to the properties of the Fourier transform of real signals, it follows that the magnitude response, $M(\omega) = |H(j\omega)|$, is an even function of ω, while the phase response, $\Phi(\omega) = \arg(H(j\omega))$, is an odd function of ω:

$$M(-\omega) = M(\omega)$$

$$\Phi(-\omega) = -\Phi(\omega)$$

If we can compute (or measure) the transfer function for $\omega \geq 0$, we can find it for $\omega < 0$, too.

A system is said to be *causal* if its impulse response is a causal signal:

$$h(t) = 0, \qquad t < 0 \tag{3.146}$$

If we excite a relaxed causal system by a causal input signal

$$x(t) = 0, \qquad t < 0$$

the zero-state response will be causal:

$$y_x(t) = 0, \qquad t < 0$$

In addition, the real part and the imaginary part of the transfer function of a causal system are not independent, because they form a Hilbert transform pair:

$$H_{\mathrm{re}}(\omega) = \frac{1}{\pi}\int_{-\infty}^{\infty} \frac{H_{\mathrm{im}}(v)}{\omega - v}\,\mathrm{d}v \tag{3.147}$$

$$H_{\mathrm{im}}(\omega) = -\frac{1}{\pi}\int_{-\infty}^{\infty} \frac{H_{\mathrm{re}}(v)}{\omega - v}\,\mathrm{d}v \tag{3.148}$$

where $H_{\mathrm{re}}(\omega) = \mathrm{Re}(H(j\omega))$ and $H_{\mathrm{im}}(\omega) = \mathrm{Im}(H(j\omega))$.

3.3.8 Fourier Transform of Sampled Signals and the Sampling Theorem

The infinite train of Dirac delta impulses, which we call the comb signal, can be expressed as

$$c(t) = \sum_{k=-\infty}^{+\infty} \delta(t-kT) = \frac{1}{2\pi}\Omega \sum_{n=-\infty}^{+\infty} \mathrm{e}^{jn\Omega t}, \qquad \Omega = \frac{2\pi}{T} \tag{3.149}$$

and is useful in describing the important operation of sampling a continuous signal.

The Fourier transform of the comb signal can be found as

$$C(j\omega) = \mathcal{F}(c(t)) = \frac{1}{2\pi}\Omega \sum_{n=-\infty}^{+\infty} \mathcal{F}(\mathrm{e}^{jn\Omega t}) = \frac{1}{2\pi}\Omega \sum_{n=-\infty}^{+\infty} 2\pi\delta(\omega - n\Omega)$$

that is,

$$C(j\omega) = \Omega \sum_{n=-\infty}^{+\infty} \delta(\omega - n\Omega)$$

or

$$\sum_{k=-\infty}^{+\infty} \delta(t-kT) \leftrightarrow \Omega \sum_{n=-\infty}^{+\infty} \delta(\omega - n\Omega)$$
$$\Omega T = 2\pi \tag{3.150}$$

In other words, the Fourier transform of an infinite train of equidistant Dirac impulses is another infinite train of equidistant Dirac impulses.

A *sampled signal*, denoted as $x_{\mathrm{samp}}(t)$, can be generated by multiplying a continuous-time signal $x(t)$ by the comb signal $c(t)$:

$$x_{\mathrm{samp}}(t) = x(t)c(t)$$

The comb signal acts as an ideal impulse sampler and transforms $x(t)$ into a sequence of Dirac impulses, and each impulse is of strength $x(kT)$.

The sampled signal can be expressed in the form

$$x_{\mathrm{samp}}(t) = \sum_{k=-\infty}^{+\infty} x(kT)\delta(t-kT) = \frac{1}{2\pi}\Omega \sum_{n=-\infty}^{+\infty} x(t)\,\mathrm{e}^{jn\Omega t}, \qquad \Omega = \frac{2\pi}{T}$$

and its Fourier transform is

$$X_{\mathrm{samp}}(j\omega) = \mathcal{F}(x_{\mathrm{samp}}(t)) = \sum_{k=-\infty}^{+\infty} x(kT)\mathrm{e}^{-jk\omega T} = \frac{1}{2\pi}\Omega \sum_{n=-\infty}^{+\infty} X(j(\omega - n\Omega)) \tag{3.151}$$

The spectrum of the sampled signal, $X_{\mathrm{samp}}(j\omega)$, can be viewed as a periodic continuation of the base-band spectrum $X(j\omega)$.

The equation (3.151) can be rewritten, for $\omega = 0$, as *Poisson's summation formula* (theorem):

$$\sum_{n=-\infty}^{\infty} x(nT) = \frac{1}{T}\sum_{n=-\infty}^{\infty} X(jn\Omega) \tag{3.152}$$

with $\Omega = \frac{2\pi}{T}$. If the signal is causal, $x(t) = 0$ for $t < 0$, then

$$x(0^+) + \sum_{n=1}^{\infty} x(nT) = \frac{x(0^+)}{2} + \frac{1}{T} \sum_{n=-\infty}^{\infty} X(jn\Omega) \tag{3.153}$$

where $x(0^+) = \lim_{\substack{t \to 0 \\ t > 0}} x(t)$.

For many practical signals the spectrum may be zero outside a certain frequency band

$$X(j\omega) = 0, \qquad |\omega| > \Omega_x \tag{3.154}$$

Such signals are called *band-limited signals.*

The *sampling theorem* states that a band-limited signal $x(t)$ for which

$$X(j\omega) = 0, \qquad |\omega| \geq \frac{1}{2}\Omega \tag{3.155}$$

can be uniquely determined from its samples $x(kT)$, where $\Omega T = 2\pi$. The frequency at which we take samples is called the *sampling frequency*, $f_{\text{samp}} = \frac{1}{T} = \frac{\Omega}{2\pi}$, and must be greater than or equal to twice the maximum frequency in the spectrum of $x(t)$:

$$f_{\text{samp}} \geq 2F_x = \frac{\Omega_x}{\pi} \tag{3.156}$$

The sampled signal can be converted back into the original continuous-time signal if we pass the sampled signal through a system characterized by the transfer function

$$\begin{aligned} H(j\omega) &= T, \qquad |\omega| < \frac{\Omega}{2} \\ H(j\omega) &= 0, \qquad |\omega| \geq \frac{\Omega}{2} \end{aligned} \tag{3.157}$$

Such a system is called an *ideal lowpass filter.* If the sampling frequency is lower than required by (3.156), the output of the ideal lowpass filter will at best yield a distorted version of $x(t)$; this effect is called *aliasing* or *frequency folding.*

3.3.9 Fourier Transform of Periodic Signals and Power Spectral Density

Consider a real periodic signal $x_{\text{per}}(t)$ with period T. According to (3.113) the periodic signal can be expressed as a convolution

$$x_{\text{per}}(t) = x(t) * c(t) \tag{3.158}$$

of a nonperiodic time-limited signal $x(t)$, which coincides with $x_{\text{per}}(t)$ over one period $|t| < \frac{T}{2}$,

$$x(t) = \begin{cases} x_{\text{per}}(t), & -\frac{T}{2} < t < \frac{T}{2} \\ 0, & |t| \geq \frac{T}{2} \end{cases} \tag{3.159}$$

and the comb signal (3.149)

$$c(t) = \sum_{k=-\infty}^{+\infty} \delta(t - kT) \tag{3.160}$$

The Fourier transform of the periodic signal $x_{\text{per}}(t)$ can be found by using the time-convolution property (3.105) and the Fourier transform of the comb function (3.150):

$$X_{\text{per}}(j\omega) = \mathcal{F}(x_{\text{per}}(t)) = \mathcal{F}(x(t) * c(t)) = X(j\omega)C(j\omega)$$

that is,

$$X_{\text{per}}(j\omega) = X(j\omega)\Omega \sum_{n=-\infty}^{+\infty} \delta(\omega - n\Omega) = \Omega \sum_{n=-\infty}^{+\infty} X(jn\Omega)\delta(\omega - n\Omega)$$

with $\Omega = \frac{2\pi}{T}$.

The signal $x(t)$ is time-limited, so its Fourier transform becomes

$$X(j\omega) = \mathcal{F}(x(t)) = \int_{-T/2}^{T/2} x(t)e^{-j\omega t}\,dt \tag{3.161}$$

We may write

$$X_{\text{per}}(j\omega) = \sum_{n=-\infty}^{+\infty} \frac{X(jn\Omega)}{T} 2\pi\delta(\omega - n\Omega) \tag{3.162}$$

and

$$x_{\text{per}}(t) = \mathcal{F}^{-1}(X_{\text{per}}(j\omega)) = \sum_{n=-\infty}^{+\infty} \frac{X(jn\Omega)}{T} \mathcal{F}^{-1}(2\pi\delta(\omega - n\Omega)) \tag{3.163}$$

or

$$x_{\text{per}}(t) = \sum_{n=-\infty}^{+\infty} \frac{X(jn\Omega)}{T} e^{jn\Omega t} \tag{3.164}$$

Obviously, the expression $\frac{X(jn\Omega)}{T}$ is the coefficient C_n, given by (3.75), in the complex form of the Fourier series (3.74) of the periodic signal $x_{\text{per}}(t)$:

$$C_n = \frac{X(jn\Omega)}{T} = \frac{1}{T} \int_{-T/2}^{T/2} x(t)e^{-jn\Omega t}\,dt \tag{3.165}$$

with the fundamental angular frequency $\omega_1 = \Omega$.

We conclude from (3.162) that the Fourier transform of a periodic function is an infinite train of equidistant Dirac delta impulses at the harmonic frequencies; the strength of impulses is proportional to the complex coefficients of the Fourier series. Therefore, periodic signals have discrete Fourier transform spectra.

We define the *power spectral density* of a periodic signal as

$$P_{\mathrm{per}}(\omega) = 2\pi \sum_{n=-\infty}^{+\infty} |C_n|^2 \delta(\omega - n\Omega) \tag{3.166}$$

The plot of $P_{\mathrm{per}}(\omega)$ against ω is called the *power spectrum* of a periodic signal.

The *autocorrelation function of periodic signals* is defined by

$$\rho_{\mathrm{per}}(\tau) = \frac{1}{T} \int_{-T/2}^{T/2} x_{\mathrm{per}}(t) x_{\mathrm{per}}(t + \tau)\, \mathrm{d}t \tag{3.167}$$

and can be expressed as

$$\rho_{\mathrm{per}}(\tau) = \sum_{n=-\infty}^{+\infty} |C_n|^2\, \mathrm{e}^{jn\Omega\tau} \tag{3.168}$$

The quantity $\rho_{\mathrm{per}}(0)$ is equal to the *average power* of the periodic signal

$$P_{\mathrm{av}} = \frac{1}{T} \int_{-T/2}^{T/2} x_{\mathrm{per}}^2(t)\, \mathrm{d}t \tag{3.169}$$

computed over one period.

The power spectral density and the autocorrelation function of a periodic signal form a Fourier transform pair

$$\rho_{\mathrm{per}}(\tau) \leftrightarrow P_{\mathrm{per}}(\omega) \tag{3.170}$$

which is the Wiener–Kintchine theorem for periodic signals.

3.3.10 Fourier Transform and the Phasor Method

Consider a relaxed, single-input single-output, lumped LTI causal system characterized by its transfer function

$$H(j\omega) = M(\omega)\mathrm{e}^{j\Phi(\omega)}$$

and assume that the excitation to the system is sinusoidal:

$$x(t) = X_{\mathrm{m}} \cos(\omega_x t + \xi)$$

for all t.

The zero-state response of the system can be obtained by (3.145)

$$y_x(t) = \mathcal{F}^{-1}(H(j\omega)\mathcal{F}(x(t)))$$

from which it follows

$$\begin{aligned} y_x(t) &= \mathcal{F}^{-1}(H(j\omega)\mathcal{F}(X_\mathrm{m}\cos(\omega_x t + \xi))) \\ &= X_\mathrm{m}\mathcal{F}^{-1}(H(j\omega)\mathcal{F}(\mathrm{e}^{j\omega_x t + j\xi} + \mathrm{e}^{-j\omega_x t - j\xi})/2) \\ &= \frac{1}{2}X_\mathrm{m}\mathcal{F}^{-1}(H(j\omega)\mathrm{e}^{j\xi}\,\mathcal{F}(\mathrm{e}^{j\omega_x t}) + H(j\omega)\mathrm{e}^{-j\xi}\,\mathcal{F}(\mathrm{e}^{-j\omega_x t})) \end{aligned}$$

By using the Fourier pair $\mathcal{F}(\mathrm{e}^{j\omega_x t}) = 2\pi\delta(\omega - \omega_x)$ we obtain

$$y_x(t) = \frac{1}{2}X_\mathrm{m}\mathcal{F}^{-1}(H(j\omega)\mathrm{e}^{j\xi}\,2\pi\delta(\omega - \omega_x) + H(j\omega)\mathrm{e}^{-j\xi}\,2\pi\delta(\omega + \omega_x))$$

Since $f(t)\delta(t - T) = f(T)\delta(t - T)$, we find

$$\begin{aligned} y_x(t) &= \frac{1}{2}X_\mathrm{m}\mathcal{F}^{-1}(H(j\omega_x)\mathrm{e}^{j\xi}\,2\pi\delta(\omega - \omega_x) + H(-j\omega_x)\mathrm{e}^{-j\xi}\,2\pi\delta(\omega + \omega_x)) \\ &= \frac{1}{2}X_\mathrm{m}H(j\omega_x)\mathrm{e}^{j\xi}\,\mathrm{e}^{j\omega_x t} + \frac{1}{2}X_\mathrm{m}H(-j\omega_x)\mathrm{e}^{-j\xi}\mathrm{e}^{-j\omega_x t} \end{aligned}$$

The magnitude response is an even function in ω, $M(-\omega) = M(\omega)$, and the phase response is an odd function in ω, $\Phi(-\omega) = -\Phi(\omega)$, so

$$\begin{aligned} y_x(t) &= \frac{1}{2}X_\mathrm{m}M(\omega_x)\mathrm{e}^{j(\omega_x t + \xi + \Phi(\omega_x))} + \frac{1}{2}X_\mathrm{m}M(\omega_x)\mathrm{e}^{-j(\omega_x t + \xi + \Phi(\omega_x))} \\ &= \mathrm{Re}(X_\mathrm{m}M(\omega_x)\mathrm{e}^{j(\omega_x t + \xi + \Phi(\omega_x))}) \end{aligned}$$

Finally,

$$y_x(t) = X_\mathrm{m}M(\omega_x)\cos(\omega_x t + \xi + \Phi(\omega_x)) \tag{3.171}$$

under the condition that $M(\omega_x)$ and $\Phi(\omega_x)$ exist at ω_x.

The response $y_x(t)$ is the steady-state sinusoidal response and is exactly the same as the result obtained by the phasor method:

$$y_x(t) = \mathcal{P}_{\omega_x}^{-1}(H(j\omega_x)X^{(\omega_x)}), \quad X^{(\omega_x)} = X_\mathrm{m}\,\mathrm{e}^{j\xi}$$

Therefore, the Fourier transform can be thought of as the ultimate generalization of the phasor transform.

3.4 LAPLACE TRANSFORM

The Laplace transform is perhaps the mathematical signature of the electrical engineer, having a long history of application to problems of electrical engineering.

The term *transform* refers to a mathematical operation that takes a given function and returns a new function. The transformation is often done by means of an integral formula. Commonly used transforms are named after Laplace and Fourier. Transforms are frequently used to change a complicated problem into a simpler one. The simpler problem is then solved, usually using the inverse transform. A standard example is the use of the Laplace transform to solve a differential equation.

The Laplace transform is particularly useful as an analytical tool in the analysis, characterization, and study of linear time-invariant (LTI) systems. It plays a particularly important role in analyzing causal systems specified by linear constant-coefficient differential equations with nonzero initial conditions (i.e., which are not initially at rest).

3.4.1 Definition of the Laplace Transform

The *unilateral Laplace transform* $X(s)$ of a signal $x(t)$ is defined as

$$X(s) = \mathcal{L}(x(t)) = \int_0^{\infty} x(t)\mathrm{e}^{-st}\,\mathrm{d}t. \tag{3.172}$$

The range of values s for which the integral in Eq. (3.172) converges is referred to as the *region of convergence* (ROC) of the Laplace transform. The complex variable s is in general of the form $s = \sigma + j\omega$, with σ and ω the real and imaginary parts, respectively.

The *inverse Laplace transform* is

$$x(t) = \mathcal{L}^{-1}(X(s)) = \frac{1}{2\pi j}\int_{\sigma-j\infty}^{\sigma+j\infty} X(s)\mathrm{e}^{st}\,\mathrm{d}s. \tag{3.173}$$

The contour of integration is a straight line in the complex plane, parallel to the $j\omega$-axis and determined by any value of σ so that $X(\sigma + j\omega)$ converges. For the class of rational transforms, the inverse Laplace transform can be determined without direct evaluation of the integral in Eq. (3.173) by utilizing the partial fraction expansion. [3] Basically, the procedure consists of expanding the rational algebraic expression into a linear combination of lower-order terms of the same type.

The two functions $x(t)$ and $X(s)$ form a *Laplace transform pair* which is designated by $x(t) \leftrightarrow X(s)$. The variable s is called a *complex frequency variable.*

The process through which signals of a real variable t are associated with corresponding complex functions of a new complex variable s is known as a *Laplace transformation*. The complex function $X(s)$ is called the *image* of $x(t)$, and $x(t)$ is known as the *original* of $X(s)$. The process of going back from $X(s)$ to $x(t)$ is referred to as an *inverse Laplace transformation*.

3.4.2 Properties of the Laplace Transform

We summarize the salient properties of the Laplace transform that is always subject to the proviso that the transform exists.

Uniqueness. The Laplace transform is *unique* (except for a finite number of isolated points of discontinuity) for all t:

$$x_1(t) = x_2(t) \Leftrightarrow X_1(s) = X_2(s) \tag{3.174}$$

where $X_1(s) = \mathcal{L}(x_1(t))$, $X_2(s) = \mathcal{L}(x_2(t))$. For physically generated signals and for the signals dealt with in this book, the Laplace transform is unique for all t.

Homogeneity. The operators $\mathcal{L}$ and $\mathcal{L}^{-1}$ are *homogeneous*:

$$\begin{aligned} \mathcal{L}(Kx(t)) &= K\mathcal{L}(x(t)) \\ \mathcal{L}^{-1}(KX(s)) &= K\mathcal{L}^{-1}(X(s)) \end{aligned} \tag{3.175}$$

for any constant K, with $X(s) = \mathcal{L}(x(t))$.

Additivity. Also, the two operators are *additive*:

$$\begin{aligned} \mathcal{L}(x_1(t) + x_2(t)) &= \mathcal{L}(x_1(t)) + \mathcal{L}(x_2(t)) \\ \mathcal{L}^{-1}(X_1(s) + X_2(s)) &= \mathcal{L}^{-1}(X_1(s)) + \mathcal{L}^{-1}(X_2(s)) \end{aligned} \tag{3.176}$$

Differentiation. The Laplace transform, $\mathcal{L}$, maps the operation of differentiating, D, into the multiplication by s:

$$\mathcal{L}(\mathrm{D}x(t)) = s\mathcal{L}(x(t)) - x(0^-) = sX(s) - x(0^-) \tag{3.177}$$

where $X(s) = \mathcal{L}(x(t))$, and D denotes differentiation with respect to time, $\mathrm{D}x(t) = \mathrm{d}x(t)/\mathrm{d}t$.

Convolution. The *convolution* of two causal signals $x(t)$ and $h(t)$ is denoted by $x(t) * h(t)$ and defined as

$$x(t) * h(t) = \int_0^\infty x(\tau)h(t-\tau)\,\mathrm{d}\tau = \int_0^\infty x(t-\tau)h(\tau)\,\mathrm{d}\tau \tag{3.178}$$

Dirac delta impulse, $\delta(t)$, is the *identity element* in the convolution operation

$$x(t) * \delta(t) = x(t) \tag{3.179}$$

Also, $x(t) * h(t) = h(t) * x(t)$, and it can be shown that

$$x(t) * \delta(t - T) = x(t - T) \tag{3.180}$$

for any real time shift T.

The convolution property of the Laplace transform is

$$\mathcal{L}\int_0^\infty x(\tau)h(t-\tau)\,\mathrm{d}\tau = X(s)H(s) \tag{3.181}$$

The Laplace transform is particularly useful as an analytical tool in the analysis, characterization, and study of causal linear time-invariant (LTI) systems excited by causal signals. Its role for this class of systems stems directly from the convolution property of the transform, from which it follows that the Laplace transform of the input and output of an LTI system are related through multiplication by the Laplace transform of the system impulse response (assuming zero initial conditions). Thus, $Y(s) = H(s)X(s)$, where $X(s)$, $Y(s)$, and $H(s)$ are the Laplace transforms of the system input, output, and impulse response, respectively.

3.4.3 The Inverse Laplace Transform of Rational Functions

We shall be concerned with *proper* rational functions $X(s)$ with real coefficients. Specifically, suppose that the denominator of $X(s)$ has distinct roots $p_1, p_2, \ldots, p_r$ with *multiplicities* $m_1, m_2, \ldots, m_r$. In this case the inverse Laplace transform of $X(s)$ is of the form

$$x(t) = \mathcal{L}^{-1}(X(s)) = \sum_{i=1}^{r} \sum_{k=1}^{m_i} A_{ik} \frac{t^{k-1}}{(k-1)!} \mathrm{e}^{p_i t} \tag{3.182}$$

where the A_{ik} are computed from the equation

$$A_{ik} = \frac{1}{(m_i - k)!} \lim_{s \to p_i} \left(\frac{\mathrm{d}^{m_i - k}}{\mathrm{d}s^{m_i - k}} \left((s - p_i)^{m_i} X(s)\right) \right) \tag{3.183}$$

We use the factorial notation $n!$ for the product $n(n-1)(n-2)\ldots 2 \cdot 1$; the quantity $0!$ is defined to be equal to 1.

Standard Laplace Transform Pairs. A number of useful Laplace transform pairs for some real causal signals $x(t)$ follows:

$$\begin{array}{lcl}
x(t) & \leftrightarrow & X(s) \\
\delta(t) & \leftrightarrow & 1 \\
u(t) & \leftrightarrow & \dfrac{1}{s} \\
\mathrm{e}^{-at}\, u(t) & \leftrightarrow & \dfrac{1}{s+a} \\
t^n\, u(t) & \leftrightarrow & \dfrac{n!}{s^{n+1}} \\
\sin(\omega t)\, u(t) & \leftrightarrow & \dfrac{\omega}{s^2+\omega^2} \\
\cos(\omega t)\, u(t) & \leftrightarrow & \dfrac{s}{s^2+\omega^2} \\
\mathrm{e}^{-\alpha t} \sin(\omega t)\, u(t) & \leftrightarrow & \dfrac{\omega}{(s+\alpha)^2+\omega^2} \\
\mathrm{e}^{-\alpha t} \cos(\omega t)\, u(t) & \leftrightarrow & \dfrac{(s+\alpha)}{(s+\alpha)^2+\omega^2}
\end{array} \tag{3.184}$$

The pairs are generated by employing the definition of the transform or its properties.

3.4.4 Transfer Function of Continuous-Time Systems

Consider a relaxed, single-input, single-output, continuous-time LTI system described by means of a linear constant-coefficients differential equation:

$$\sum_{m=0}^{M} a_m \mathrm{D}^m y(t) = \sum_{l=0}^{L} b_l \mathrm{D}^l x(t) \tag{3.185}$$

Assume that the system is excited by an input causal signal $x(t)$, and observe the output signal $y(t)$. By applying the Laplace transform to both sides of the differential equation, and after employing the linearity property and the differentiating property, we obtain an equation relating the Laplace transforms of the two signals:

$$\left(\sum_{m=0}^{M} a_m s^m\right) Y(s) = \left(\sum_{l=0}^{L} b_l s^l\right) X(s)$$

The system transforms the input signal by multiplying the Laplace transform of the input signal with the factor

$$H(s) = \frac{\sum_{l=0}^{L} b_l s^l}{\sum_{m=0}^{M} a_m s^m} \tag{3.186}$$

The function $H(s)$ is a rational function in s:

$$H(s) = \frac{B(s)}{A(s)}, \qquad B(s) = \sum_{l=0}^{L} b_l s^l, \qquad A(s) = \sum_{m=0}^{M} a_m s^m$$

where $A(s)$ and $B(s)$ are polynomials in s.

The function $H(s)$ characterizes the continuous-time system and is called the *system function* or the *transfer function* of the system.

The roots of the transfer function denominator $A(s)$ are called the *poles* of the transfer function. We compute the poles from the equation

$$A(s) = \text{denominator}(H(s)) = 0$$

and designate by $s_{\mathrm{p}1}, s_{\mathrm{p}2}, s_{\mathrm{p}3}, \ldots$.

In a similar way, the roots of the transfer function numerator $B(s)$ are called the *zeros* of the transfer function. We obtain the zeros from the equation

$$B(s) = \text{numerator}(H(s)) = 0$$

and designate these zeros by $s_{\mathrm{z}1}, s_{\mathrm{z}2}, s_{\mathrm{z}3}, \ldots$.

The transfer function can be written in the factored form as

$$H(s) = H_0 \frac{\prod_k (s - s_{zk})}{\prod_i (s - s_{pi})} \tag{3.187}$$

The form (3.187) is known as the *pole-zero representation* of the transfer function. The real constant H_0 is called the *scale factor*.

Since the polynomials $A(s)$ and $B(s)$ have real coefficients, zeros and poles must be real or occur in complex conjugate pairs.

The simplest transfer function is the first-order transfer function

$$H(s) = \frac{b_1 s + b_0}{a_1 s + a_0}, \qquad a_1 \neq 0 \tag{3.188}$$

Systems described by (3.188) we call *first-order sections*.

Transfer functions of the form

$$H(s) = \frac{b_2 s^2 + b_1 s + b_0}{a_2 s^2 + a_1 s + a_0}, \qquad a_2 \neq 0 \tag{3.189}$$

are called *second-order transfer functions*, and they play an important role in analysis and design of continuous-time LTI systems. A system characterized by (3.189) we call a *biquadratic section* or *biquad*.

Any transfer function (3.186) can be expressed as a product of first-order (3.188) and second-order (3.189) transfer functions, which implies that any LTI system can be resolved into first-order sections and biquads.

For $s = j\omega$, $H(s)$ is the frequency response of the LTI system. Many properties of LTI systems can be closely associated with the characteristics of the system function in the s-plane, and in particular with the pole locations.

Partial Transfer Function and Transfer-Function Matrix. Consider a relaxed multiple-input multiple-output system with N inputs, $x_1, x_2, \ldots, x_k, \ldots, x_N$, and assume that the system has L outputs $y_1, y_2, \ldots, y_i, \ldots, y_L$. We define the *partial transfer function* between the ith output and the kth input to be the ratio of the output Laplace transform $Y_i(s)$ to the input Laplace transform $X_k(s)$, with the other inputs being identically zero:

$$H_{ik}(s) = \left. \frac{Y_i(s)}{X_k(s)} \right|_{x_1 = \cdots = x_{k-1} = x_{k+1} = \cdots = x_N = 0} \tag{3.190}$$

The input–output description of the system can be expressed in a matrix form:

$$\begin{bmatrix} Y_1(s) \\ Y_2(s) \\ \vdots \\ Y_i(s) \\ \vdots \\ Y_L(s) \end{bmatrix} = \begin{bmatrix} H_{11}(s) & H_{12}(s) & \cdots & H_{1k}(s) & \cdots & H_{1N}(s) \\ H_{21}(s) & H_{22}(s) & \cdots & H_{2k}(s) & \cdots & H_{2N}(s) \\ \vdots & \vdots & \ddots & \vdots & & \vdots \\ H_{i1}(s) & H_{i2}(s) & \cdots & H_{ik}(s) & \cdots & H_{iN}(s) \\ \vdots & \vdots & & \vdots & \ddots & \vdots \\ H_{L1}(s) & H_{L2}(s) & \cdots & H_{Lk}(s) & \cdots & H_{LN}(s) \end{bmatrix} \begin{bmatrix} X_1(s) \\ X_2(s) \\ \vdots \\ X_k(s) \\ \vdots \\ X_N(s) \end{bmatrix}$$

where the matrix

$$\mathbf{H}(s) = \begin{bmatrix} H_{11}(s) & H_{12}(s) & \cdots & H_{1k}(s) & \cdots & H_{1N}(s) \\ H_{21}(s) & H_{22}(s) & \cdots & H_{2k}(s) & \cdots & H_{2N}(s) \\ \vdots & \vdots & \ddots & \vdots & & \vdots \\ H_{i1}(s) & H_{i2}(s) & \cdots & H_{ik}(s) & \cdots & H_{iN}(s) \\ \vdots & \vdots & & \vdots & \ddots & \vdots \\ H_{L1}(s) & H_{L2}(s) & \cdots & H_{Lk}(s) & \cdots & H_{LN}(s) \end{bmatrix}$$

is called the *transfer-function matrix* of the system.

3.5 DISCRETE FOURIER TRANSFORM

Discrete Fourier transform (DFT) is one of the most important tools for signal processing techniques. Originally, it has been developed for calculating the Fourier coefficients and the Fourier transform on a digital computer.

3.5.1 Definition of the Discrete Fourier Transform

Consider a finite-length sequence

$$\{x_{(n)}\}_N = \{x_{(0)}, x_{(1)}, x_{(2)}, \ldots, x_{(n)}, \ldots, x_{(N-1)}\} \tag{3.191}$$

where the curly brackets are used to denote a sequence, and $x_{(n)}$ is the nth member of the sequence $(n = 0, 1, 2, \ldots, N - 1)$.

The *discrete Fourier transform* (DFT) is defined as another sequence of the same length:

$$\{X_{(k)}\}_N = \{X_{(0)}, X_{(1)}, X_{(2)}, \ldots, X_{(k)}, \ldots, X_{(N-1)}\} \tag{3.192}$$

with

$$X_{(k)} = \sum_{n=0}^{N-1} x_{(n)} w^{-kn} \tag{3.193}$$

and

$$w = \mathrm{e}^{j\frac{2\pi}{N}} \tag{3.194}$$

Note that (the asterisk denotes the complex conjugate)

$$w^* = w^{-1}$$

and

$$w^{l+mN} = w^l$$

for arbitrary integers l and m.

The *inverse discrete Fourier transform* (IDFT) is defined as

$$x_{(n)} = \frac{1}{N} \sum_{k=0}^{N-1} X_{(k)} w^{kn} \tag{3.195}$$

The two sequences, $\{x_{(n)}\}_N$ and $\{X_{(k)}\}_N$, are said to form a *discrete Fourier transform pair*, which is symbolized by

$$\{x_{(n)}\}_N \leftrightarrow \{X_{(k)}\}_N \tag{3.196}$$

or

$$\begin{aligned} \mathcal{D}\{x_{(n)}\}_N &= \{X_{(k)}\}_N \\ \mathcal{D}^{-1}\{X_{(k)}\}_N &= \{x_{(n)}\}_N \end{aligned} \tag{3.197}$$

Generally, the members of these sequences are complex numbers, and the discrete Fourier transform establishes a one-to-one correspondence between the two sequences.

The finite-length sequence $\{x_{(n)}\}_N$ can be periodically extended to form a periodic sequence of numbers, $\{p_{(n)}\}$, with period N:

$$\begin{aligned} p_{(n)} &= x_{(n)}, \quad (n = 0, 1, 2, \ldots, N-1) \\ p_{(n+mN)} &= p_{(n)}, \quad (m = \pm 1, \pm 2, \ldots) \end{aligned} \tag{3.198}$$

The discrete Fourier transform of the periodic sequence $\{p_{(n)}\}$ is defined as another periodic sequence, $\{P_{(k)}\}$, with the same period N:

$$\begin{aligned} P_{(k)} &= X_{(k)} \qquad (k = 0, 1, 2, \ldots, N-1) \\ P_{(k+mN)} &= P_{(k)} \qquad (m = \pm 1, \pm 2, \ldots) \end{aligned} \tag{3.199}$$

Often, the sequence $\{x_{(n)}\}_N$ is generated by sampling a continuous-time signal $x(t)$. We observe the signal over an interval of length T, and we take N equidistant samples at the points $t_0, t_0 + \Delta t, t_0 + 2\Delta t, \ldots, t_0 + (N-1)\Delta t$; frequently, $t_0 = 0$. The members of the sequence $\{x_{(n)}\}_N$ are

$$\begin{aligned} x_{(n)} &= x(t_0 + n\Delta t) \\ n &= 0, 1, 2, \ldots, N-1 \end{aligned}$$

The number of samples per unit interval of time defines the *sampling rate*

$$f_{\text{samp}} = \frac{1}{\Delta t}$$

also called the *sampling frequency*.

We associate a sequence of frequencies

$$\begin{aligned}\{f_{(k)}\}_N &= \{0, \Delta f, 2\Delta f, \ldots, k\Delta f, \ldots, (N-1)\Delta f\} \\ \Delta f &= \frac{1}{N} f_{\text{samp}}\end{aligned} \tag{3.200}$$

to the DFT sequence $\{X_{(k)}\}_N$; that is, the frequency

$$f_{(k)} = k\Delta f$$

corresponds to $X_{(k)}$. Sometimes, it is more convenient to introduce a sequence of angular frequencies:

$$\begin{aligned}\{\omega_{(k)}\}_N &= \{0, \Delta\omega, 2\Delta\omega, \ldots, k\Delta\omega, \ldots, (N-1)\Delta\omega\} \\ \Delta\omega &= 2\pi\Delta f = 2\pi\frac{1}{N} f_{\text{samp}}\end{aligned} \tag{3.201}$$

Definition equations of DFT and IDFT can be written in matrix form:

$$\begin{bmatrix} X_{(0)} \\ X_{(1)} \\ X_{(2)} \\ \vdots \\ X_{(N-1)} \end{bmatrix} = \begin{bmatrix} 1 & 1 & 1 & \cdots & 1 \\ 1 & w^{-1} & w^{-2} & \cdots & w^{-(N-1)} \\ 1 & w^{-2} & w^{-4} & \cdots & w^{-2(N-1)} \\ \vdots & \vdots & \vdots & \ddots & \vdots \\ 1 & w^{-(N-1)} & w^{-2(N-1)} & \cdots & w^{-(N-1)^2} \end{bmatrix} \begin{bmatrix} x_{(0)} \\ x_{(1)} \\ x_{(2)} \\ \vdots \\ x_{(N-1)} \end{bmatrix}$$

or

$$\mathbf{X} = \mathbf{W}\,\mathbf{x}$$

and

$$\begin{bmatrix} x_{(0)} \\ x_{(1)} \\ x_{(2)} \\ \vdots \\ x_{(N-1)} \end{bmatrix} = \frac{1}{N}\begin{bmatrix} 1 & 1 & 1 & \cdots & 1 \\ 1 & w^{1} & w^{2} & \cdots & w^{(N-1)} \\ 1 & w^{2} & w^{4} & \cdots & w^{2(N-1)} \\ \vdots & \vdots & \vdots & \ddots & \vdots \\ 1 & w^{(N-1)} & w^{2(N-1)} & \cdots & w^{(N-1)^2} \end{bmatrix} \begin{bmatrix} X_{(0)} \\ X_{(1)} \\ X_{(2)} \\ \vdots \\ X_{(N-1)} \end{bmatrix}$$

or

$$\mathbf{x} = \frac{1}{N}\hat{\mathbf{W}}\,\mathbf{X}$$

where the matrix $\hat{\mathbf{W}}$ is obtained from $\mathbf{W}$ by changing w^{-k} to w^{k}.

Any efficient computational algorithms for the calculation of DFT and IDFT, which speeds up the evaluation of (3.193) and (3.195), is called a *fast Fourier transform* (FFT).

3.5.2 Properties of the Discrete Fourier Transform

Consider sequences $\{x_{(n)}\}$, $\{h_{(n)}\}$, and $\{y_{(n)}\}$ with transforms $\{X_{(k)}\}$, $\{H_{(k)}\}$, and $\{Y_{(k)}\}$, that is, $\{x_{(n)}\} \leftrightarrow \{X_{(k)}\}$, $\{h_{(n)}\} \leftrightarrow \{H_{(k)}\}$, and $\{y_{(n)}\} \leftrightarrow \{Y_{(k)}\}$. Unless otherwise stated, we assume that the sequences are either periodic or nonperiodic finite-length sequences.

We review several important properties relevant for the theory of discrete-time signals and systems.

Uniqueness. The discrete Fourier transform is *unique*:

$$\{x_{(n)}\} = \{y_{(n)}\} \Leftrightarrow \{X_{(k)}\} = \{Y_{(k)}\} \tag{3.202}$$

Homogeneity. The operators $\mathcal{D}$ and $\mathcal{D}^{-1}$ are *homogeneous*:

$$\begin{aligned} \mathcal{D}(K\{x_{(n)}\}) &= K\{X_{(k)}\} \\ \mathcal{D}^{-1}(K\{X_{(k)}\}) &= K\{x_{(n)}\} \end{aligned} \tag{3.203}$$

for any real or complex constant K.

Remark 3.1. The notation $K\{x_{(n)}\}$ symbolizes multiplication of a sequence by a constant; it means that each member of the sequence $\{x_{(n)}\}$ is multiplied by the constant K.

Additivity. The operators $\mathcal{D}$ and $\mathcal{D}^{-1}$ are *additive*:

$$\begin{aligned} \mathcal{D}(\{x_{(n)}\} + \{y_{(n)}\}) &= \{X_{(k)}\} + \{Y_{(k)}\} \\ \mathcal{D}^{-1}(\{X_{(k)}\} + \{Y_{(k)}\}) &= \{x_{(n)}\} + \{y_{(n)}\} \end{aligned} \tag{3.204}$$

The two sequences can be periodic with the same period N, or they can be nonperiodic finite-length sequences of the same length N.

Remark 3.2. The notation $\{x_{(n)}\}+\{y_{(n)}\}$ symbolizes addition of two sequences element by element.

Linearity. The homogeneity and additivity properties can be combined into the *linearity* property as follows:

$$\begin{aligned} \mathcal{D}(A\{x_{(n)}\} + B\{y_{(n)}\}) &= A\{X_{(k)}\} + B\{Y_{(k)}\} \\ \mathcal{D}^{-1}(A\{X_{(k)}\} + B\{Y_{(k)}\}) &= A\{x_{(n)}\} + B\{y_{(n)}\} \end{aligned} \tag{3.205}$$

for arbitrary real or complex constants A and B.

Shifting. The operation of shifting a periodic sequence maps into the operation of multiplication its transform by a constant:

$$\begin{aligned} \mathcal{D}\{x_{(n-\mu)}\} &= \{w^{-\mu k} X_{(k)}\} \\ \mathcal{D}^{-1}\{w^{-\mu k} X_{(k)}\} &= \{x_{(n-\mu)}\} \end{aligned} \tag{3.206}$$

for an arbitrary integer μ, and

$$w = e^{j\frac{2\pi}{N}}$$

with N being the period of the sequence.

Cyclic Convolution. *Cyclic convolution* of two periodic sequences with period N is defined as

$$\{y_{(n)}\} = \{x_{(n)}\} * \{h_{(n)}\}$$
$$y_{(n)} = \sum_{\mu=0}^{N-1} x_{(\mu)} h_{(n-\mu)} \tag{3.207}$$

and is a periodic sequence with period N. The cyclic convolution is a commutative operation.

DFT of the cyclic convolution of two periodic sequences is a periodic sequence whose members are products of the corresponding members of the individual DFTs:

$$\mathcal{D}\{x_{(n)}\} * \{h_{(n)}\} = \mathcal{D}\{h_{(n)}\} * \{x_{(n)}\} = \{X_{(k)} H_{(k)}\}$$
$$\mathcal{D}^{-1}\{X_{(k)} H_{(k)}\} = \{x_{(n)}\} * \{h_{(n)}\} = \{h_{(n)}\} * \{x_{(n)}\}$$

Symmetry. For any periodic sequence we have

$$\{x^*_{(-n)}\} \leftrightarrow \{X^*_{(k)}\}$$
$$\{x_{(-n)}\} \leftrightarrow \{X_{(-k)}\} \tag{3.208}$$

If a periodic sequence is real, it follows that

$$X_{(-k)} = X^*_{(k)}$$

showing that we effectively need to know only $X_{(k)}$ for $k = 0, 1, 2, \ldots, \frac{N}{2}$.

Parseval's Identity and Power Spectrum. For any periodic sequence with period N we have

$$\sum_{n=0}^{N-1} |x_{(n)}|^2 = \frac{1}{N} \sum_{k=0}^{N-1} |X_{(k)}|^2 \tag{3.209}$$

which is known as the *Parseval's identity*.

The sequence of numbers $\{\frac{1}{N}|X_{(k)}|^2\}$ is called the *power spectral density*, and the plot of $\frac{1}{N}|X_{(k)}|^2$ against the integer k is the *power spectrum* of the sequence ($k = 0, 1, 2, \ldots, N-1$).

Cyclic Correlation. *Cyclic correlation* of two periodic sequences with period N is defined as

$$\{R_{xy(n)}\} = \{x_{(n)}\} \star \{y_{(n)}\}$$
$$R_{xy(n)} = \frac{1}{N} \sum_{\mu=0}^{N-1} x_{(\mu)} y_{(n+\mu)} \tag{3.210}$$

and is a periodic sequence with period N.

DFT of the cyclic correlation is

$$\mathcal{D}\{x_{(n)}\} \star \{y_{(n)}\} = \left\{\frac{1}{N} X^*_{(k)} Y_{(k)}\right\}$$

Cyclic autocorrelation of a periodic sequence is $\{R_{xx(n)}\}$, and it forms a DFT pair with the power spectral density

$$\{R_{xx(n)}\} \leftrightarrow \left\{\frac{1}{N} |X_{(k)}|^2\right\} \tag{3.211}$$

3.5.3 Computation of the Fourier Series Coefficients by DFT

Consider a real continuous-time periodic signal $x(t)$, with period T, that can be represented by the Fourier series

$$x(t) = \sum_{k=-\infty}^{+\infty} C_k \, e^{jk\omega_1 t} \tag{3.212}$$

with the coefficients

$$C_k = \frac{1}{T} \int_0^T x(t) \, e^{-jk\omega_1 t} \, dt \tag{3.213}$$

and the fundamental angular frequency

$$\omega_1 = \frac{2\pi}{T} \tag{3.214}$$

Also, we know that for real signals

$$C_{-k} = C_k^*$$

The integral in (3.213) which determines the Fourier coefficients C_k can be evaluated by using numerical techniques; therefore, the integral must be approximated by a summation:

$$C_k = \frac{1}{T} \sum_{n=0}^{N-1} \int_{n\Delta t}^{(n+1)\Delta t} x(t) e^{-jk\omega_1 t} \, dt$$

$$C_k \approx \frac{1}{T} \sum_{n=0}^{N-1} x(n\Delta t) e^{-jk\omega_1 n\Delta t} \, \Delta t$$

where we choose the number of subintervals N and compute the time step

$$\Delta t = \frac{T}{N}$$

After some simplification we may write

$$C_k \approx \frac{1}{N} \sum_{n=0}^{N-1} x_{(n)} w^{-kn} = \frac{1}{N} X_{(k)} \tag{3.215}$$

where

$$x_{(n)} = x(n\Delta t), \qquad w = e^{j\frac{2\pi}{N}}, \quad \{X_{(k)}\} = \mathcal{D}\{x_{(n)}\}$$

The summation in (3.215) gives only N different coefficients because w is periodic with period N. It means that the approximation (3.215) will be accurate only if we can accurately represent the signal $x(t)$ by a truncated series

$$x(t) = \sum_{k=-K}^{K} C_k \, e^{jk\omega_1 t}$$

which implies that either we assume $C_k = 0$, for $|k| > K$, or we neglect the Fourier coefficients for $|k| > K$.

According to the sampling theorem the maximum spectral frequency, $f_{\max} = \frac{1}{2\pi} K\omega_1 = K\frac{1}{T}$, and the sampling frequency, $f_{\text{samp}} = \frac{1}{\Delta t} = N\frac{1}{T}$, must satisfy the condition $f_{\max} < 2 f_{\text{samp}}$, which yields

$$K < 2N$$

We conclude that the Fourier series coefficients, C_k, of a real periodic function, $x(t)$, can be computed by DFT if we may assume that

$$C_k = \begin{cases} \frac{1}{N} X_{(k)}, & |k| < \frac{N}{2} \\ 0, & |k| \geq \frac{N}{2} \end{cases}$$

and if the signal is uniformly sampled over one period. In other words, we expect that the signal has negligible spectral components at frequencies higher than $\frac{N}{2T}$.

3.5.4 Computation of the Fourier Integral by DFT

Consider a real nonperiodic signal $x(t)$ that can be represented by the Fourier transform

$$x(t) = \frac{1}{2\pi} \int_{-\infty}^{+\infty} X(j\omega) e^{j\omega t} \, d\omega$$

where

$$X(j\omega) = \int_{-\infty}^{+\infty} x(t) e^{-j\omega t} \, dt \tag{3.216}$$

For real signals

$$X(-j\omega) = X^*(j\omega)$$

Without lack of generality, assume that the origin of the time axis has been adjusted so that the integral (3.216) can be approximated by one with finite limits

$$X(j\omega) \approx \int_0^T x(t)e^{-j\omega t}\,dt$$

and evaluated by the summation

$$X(j\omega) \approx \sum_{n=0}^{N-1} x_{(n)} e^{-j\omega n\Delta t}\,\Delta t$$

where

$$\Delta t = \frac{T}{N}, \qquad x_{(n)} = x(n\Delta t)$$

The number of samples, N, is chosen to ensure accurate numerical integration.

Our goal is to compute the Fourier transform of $x(t)$, in terms of its samples, by means of DFT. We sample the signal at the rate $f_{\text{samp}} = \frac{1}{\Delta t}$, and we assume that the transform $X(j\omega)$ can be neglected for $|\omega| \geq \omega_{\max} = \frac{1}{2}2\pi f_{\text{samp}}$; next, we want the signal samples to be accurately represented by the finite-limits integral

$$x(n\Delta t) = \frac{1}{2\pi}\int_{-\omega_{\max}}^{+\omega_{\max}} X(j\omega)e^{j\omega n\Delta t}\,d\omega$$

DFT can be used to determine only N values of $X(j\omega)$ from N values (samples) of $x(t)$. Therefore, we compute

$$X(jk\Delta\omega) = \Delta t \sum_{n=0}^{N-1} x_{(n)} e^{-jk\Delta\omega n\Delta t}$$

with

$$\Delta\omega = \frac{1}{N}2\pi\frac{1}{\Delta t}$$

because DFT and IDFT must be periodic with period N. We obtain

$$X(jk\Delta\omega) = \Delta t \sum_{n=0}^{N-1} x_{(n)} w^{-kn} = \frac{1}{f_{\text{samp}}} X_{(k)}$$

where

$$w = e^{j\frac{2\pi}{N}}, \qquad \{X_{(k)}\} = \mathcal{D}\{x_{(n)}\}$$

We conclude that the Fourier transform, $X(j\omega)$, of a real nonperiodic function, $x(t)$, can be computed by DFT if we may assume that

$$X(jk\Delta\omega) = \begin{cases} \dfrac{1}{f_{\text{samp}}} X_{(k)}, & |k| < \dfrac{N}{2} \\ 0, & |k| \geq \dfrac{N}{2} \end{cases}$$

and if the signal is uniformly sampled at the rate $f_{\text{samp}} = \frac{1}{\Delta t} = \frac{T}{N}$ over a time interval of duration T. In other words, we expect that the signal has negligible spectral components at frequencies higher than $\frac{N}{2T}$.

3.5.5 Frequency Response of Discrete-Time Systems

Consider a relaxed, single-input, single-output, discrete-time LTI system described by means of a linear constant-coefficients difference equation

$$\sum_{m=0}^{M} a_m y_{(n-m)} = \sum_{l=0}^{L} b_l x_{(n-l)}$$

(We drop the curly braces around $y_{(n-m)}$ and $x_{(n-l)}$ for the case of simplicity.)

Assume that the system is excited by an input sequence $\{x_{(n)}\}$ of length N, and observe the output sequence $\{y_{(n)}\}$ of the same length N. By applying the N-point DFT to both sides of the difference equation, and after employing the linearity property and the shifting property, we obtain an equation relating the discrete Fourier transforms of the two sequences

$$\left(\sum_{m=0}^{M} a_m w^{-mk}\right) Y_{(k)} = \left(\sum_{l=0}^{L} b_l w^{-lk}\right) X_{(k)}$$

where $w = e^{(j2\pi/N)}$, and $k = 0, 1, 2, \ldots, N$. The system transforms the input sequence by multiplying each member of the input DFT with the factor

$$H_{(k)} = \frac{\sum_{l=0}^{L} b_l w^{-lk}}{\sum_{m=0}^{M} a_m w^{-mk}}$$

The sequence $\{H_{(k)}\}$ can be generated from a rational function of complex variable

$$H(z) = \frac{N(z)}{D(z)}, \qquad N(z) = \sum_{l=0}^{L} b_l z^{-l}, \qquad D(z) = \sum_{m=0}^{M} a_m z^{-m}$$

if we make the substitution

$$z = e^{j\theta}$$

with

$$\theta = 2\pi \frac{k}{N}$$

The function $H(z)$ characterizes the discrete-time system and is called the *system function* or the *transfer function* of the system. The quantity θ is known as the *digital angular frequency*. In this book the quantity $\frac{\theta}{2\pi}$ is referred to as the *digital frequency*. The complex function $H(e^{j\theta})$ in terms of real variable θ is called the *frequency response* of the system.

3.6 THE z TRANSFORM

Linear time-invariant discrete-time systems, characterized by linear constant-coefficient difference equations, are efficiently analyzed by using the z transform that transforms the difference equations into algebraic equations which are easier to manipulate.

3.6.1 Definition of the z Transform

The *two-sided z transform* (or bilateral z transform) of a sequence $\{x_{(n)}\}$ is defined as

$$X_b(z) = Z_b\{x_{(n)}\} = \sum_{n=-\infty}^{\infty} x_{(n)} z^{-n} \tag{3.217}$$

for all z for which $X_b(z)$ converges.

The *one-sided z transform* (or unilateral z transform) of a sequence $\{x_{(n)}\}$ is defined as

$$X(z) = Z\{x_{(n)}\} = \sum_{n=0}^{\infty} x_{(n)} z^{-n} \tag{3.218}$$

for all z for which $X(z)$ converges.

We shall be concerned with z transforms whose singularities are poles. Mostly, we shall consider causal sequences whose members are zero for $n < 0$. Unless otherwise stated, in this book the term *z transform* refers to the one-sided z transform.

The region of convergence (ROC) for $X(z)$ is an annular ring in the z plane

$$R_{\min} < |z| < R_{\max}$$

and (unless obvious) must be included in the specification of $X(z)$ in order for the z transform to be complete.

Sequence $\{x_{(n)}\}$ is said to be the *inverse z transform* of $X(z)$ and can be uniquely determined by

$$x_{(n)} = \frac{1}{2\pi j} \oint_{\Gamma} X(z) z^{n-1}\, dz \tag{3.219}$$

where Γ is a contour in the counterclockwise sense enclosing all the singularities of $X(z)$.

The sequence $\{x_{(n)}\}$ and the complex function $X(z)$ are said to form a *z transform pair*, which is symbolized by

$$\{x_{(n)}\} \leftrightarrow X(z) \tag{3.220}$$

or

$$\begin{aligned} \mathcal{Z}\{x_{(n)}\} &= X(z) \\ \mathcal{Z}^{-1}X(z) &= \{x_{(n)}\} \end{aligned} \tag{3.221}$$

3.6.2 Properties of the z Transform

Consider sequences $\{x_{(n)}\}$, $\{h_{(n)}\}$, and $\{y_{(n)}\}$ with transforms $X(z)$, $H(z)$, and $Y(z)$, that is, $\{x_{(n)}\} \leftrightarrow X(z)$, $\{h_{(n)}\} \leftrightarrow H(z)$, and $\{y_{(n)}\} \leftrightarrow Y(z)$. We review several important properties relevant for the theory of discrete-time signals and systems.

Uniqueness. The z transform is *unique*:

$$\{x_{(n)}\} = \{y_{(n)}\} \Leftrightarrow X(z) = Y(z) \tag{3.222}$$

Homogeneity. The operators $\mathcal{Z}$ and $\mathcal{Z}^{-1}$ are *homogeneous*:

$$\begin{aligned} \mathcal{Z}(K\{x_{(n)}\}) &= KX(z) \\ \mathcal{Z}^{-1}(KX(z)) &= K\{x_{(n)}\} \end{aligned} \tag{3.223}$$

for any real or complex constant K.

Remark 3.3. The notation $K\{x_{(n)}\}$ symbolizes multiplication of a sequence by a constant; it means that each member of the sequence $\{x_{(n)}\}$ is multiplied by the constant K.

Additivity. The operators $\mathcal{Z}$ and $\mathcal{Z}^{-1}$ are *additive*:

$$\begin{aligned} \mathcal{Z}(\{x_{(n)}\} + \{y_{(n)}\}) &= X(z) + Y(z) \\ \mathcal{Z}^{-1}(X(z) + Y(z)) &= \{x_{(n)}\} + \{y_{(n)}\} \end{aligned} \tag{3.224}$$

Remark 3.4. The notation $\{x_{(n)}\}+\{y_{(n)}\}$ symbolizes addition of two sequences element by element.

Linearity. The homogeneity and additivity properties can be combined into the *linearity* property as follows:

$$\begin{aligned} \mathcal{Z}(A\{x_{(n)}\} + B\{y_{(n)}\}) &= AX(z) + BY(z) \\ \mathcal{Z}^{-1}(AX(z) + BY(z)) &= A\{x_{(n)}\} + B\{y_{(n)}\} \end{aligned} \tag{3.225}$$

for arbitrary real or complex constants A and B.

Shifting. The operation of *shifting* (*translation*) a causal sequence maps into the operation of multiplication of its z transform:

$$\begin{aligned} Z\{x_{(n-\mu)}\} &= z^{-\mu}X(z) \\ Z^{-1}z^{-\mu}X(z) &= \{x_{(n-\mu)}\} \\ Z\{x_{(n+\mu)}\} &= z^{\mu}X(z) - z^{\mu}\sum_{k=0}^{\mu-1} x_{(k)}z^{-k} \end{aligned} \tag{3.226}$$

for an arbitrary positive integer μ. Since a negative shift (delay) of $\mu = 1$ causes $X(z)$ to be multiplied by z^{-1}, z^{-1} is referred to as the *unit delay operator*.

Convolution. *Convolution* of two causal sequences is defined as

$$\begin{aligned} \{y_{(n)}\} &= \{x_{(n)}\} * \{h_{(n)}\} \\ y_{(n)} &= \sum_{\mu=0}^{\infty} x_{(\mu)}h_{(n-\mu)} \end{aligned} \tag{3.227}$$

The convolution is a commutative operation.

The z transform of the convolution of two causal sequences is a product of the z transforms of the sequences

$$\begin{aligned} Z\{x_{(n)}\} * \{h_{(n)}\} &= Z\{h_{(n)}\} * \{x_{(n)}\} = X(z)H(z) \\ Z^{-1}(X(z)H(z)) &= \{x_{(n)}\} * \{h_{(n)}\} = \{h_{(n)}\} * \{x_{(n)}\} \end{aligned}$$

Scaling. Multiplying each member of a sequence by w^{-n} maps into the z transform of the scaled argument:

$$\begin{aligned} Z(w^{-n}\{x_{(n)}\}) &= X(wz) \\ Z^{-1}X(wz) &= w^{-n}\{x_{(n)}\} \end{aligned} \tag{3.228}$$

Differentiation.

$$\begin{aligned} Z(n\{x_{(n)}\}) &= -z\frac{\mathrm{d}X(z)}{\mathrm{d}z} \\ Z^{-1}z\frac{\mathrm{d}X(z)}{\mathrm{d}z} &= -n\{x_{(n)}\} \end{aligned} \tag{3.229}$$

Standard z Transform Pairs. Some useful z transform pairs for some real causal sequences $\{x_{(n)}\}$ are as follows:

$$
\begin{aligned}
x_{(n)} &\leftrightarrow X(z) \\
\delta_{(n)} &\leftrightarrow 1 \\
u_{(n)} &\leftrightarrow \frac{z}{z-1} = \frac{1}{1-z^{-1}} \\
a^n u_{(n)} &\leftrightarrow \frac{z}{z-a} = \frac{1}{1-az^{-1}} \\
n\, u_{(n)} &\leftrightarrow \frac{z}{(z-1)^2} = \frac{z^{-1}}{(1-z^{-1})^2} \\
\sin(\theta n)\, u_{(n)} &\leftrightarrow \frac{z\sin(\theta)}{z^2 - 2z\cos(\theta) + 1} \\
\cos(\theta n)\, u_{(n)} &\leftrightarrow \frac{z(z-\cos(\theta))}{z^2 - 2z\cos(\theta) + 1} \\
\mathrm{e}^{-\alpha n}\sin(\theta n)\, u_{(n)} &\leftrightarrow \frac{z\,\mathrm{e}^{-\alpha}\sin(\theta)}{z^2 - 2z\,\mathrm{e}^{-\alpha}\cos(\theta) + \mathrm{e}^{-2\alpha}} \\
\mathrm{e}^{-\alpha n}\cos(\theta n)\, u_{(n)} &\leftrightarrow \frac{z(z-\mathrm{e}^{-\alpha}\cos(\theta))}{z^2 - 2z\,\mathrm{e}^{-\alpha}\cos(\theta) + \mathrm{e}^{-2\alpha}}
\end{aligned}
\tag{3.230}
$$

The pairs are generated by employing the definition of the transform or its properties.

3.6.3 Transfer Function of Discrete-Time Systems

Consider a relaxed, single-input, single-output, discrete-time LTI system described by means of a linear constant-coefficients difference equation

$$\sum_{m=0}^{M} a_m y_{(n-m)} = \sum_{l=0}^{L} b_l x_{(n-l)} \tag{3.231}$$

(We drop the curly braces around $y_{(n-m)}$ and $x_{(n-l)}$ for the case of simplicity.)

Assume that the system is excited by an input causal sequence $\{x_{(n)}\}$, and observe the output sequence $\{y_{(n)}\}$. By applying the z transform to both sides of the difference equation, and after employing the linearity property and the shifting property, we obtain an equation relating the z transforms of the two sequences:

$$\left(\sum_{m=0}^{M} a_m z^{-m}\right) Y(z) = \left(\sum_{l=0}^{L} b_l z^{-l}\right) X(z)$$

The system transforms the input sequence by multiplying the z transform of the input sequence with the factor

$$H(z) = \frac{\sum_{l=0}^{L} b_l z^{-l}}{\sum_{m=0}^{M} a_m z^{-m}} \tag{3.232}$$

The function $H(z)$ is a rational function in z^{-1}:

$$H(z) = \frac{N(z)}{D(z)}, \quad N(z) = \sum_{l=0}^{L} b_l z^{-l}, \quad D(z) = \sum_{m=0}^{M} a_m z^{-m}$$

which can, equivalently, be expressed as a rational function in z:

$$H(z) = \frac{B(z)}{A(z)}$$

where $A(z)$ and $B(z)$ are polynomials in z.

The function $H(z)$ characterizes the discrete-time system and is called the *system function* or the *transfer function* of the system. For causal systems the degree of the numerator polynomial, $B(z)$, is equal to or less than that of the denominator polynomial, $A(z)$.

The roots of the transfer function denominator $A(z)$ are called the *poles* of the transfer function. We compute the poles from the equation

$$A(z) = \text{denominator}(H(z)) = 0$$

and designate these poles by $z_{\text{p}1}, z_{\text{p}2}, z_{\text{p}3}, \ldots$.

In a similar way, the roots of the transfer function numerator $B(z)$ are called the *zeros* of the transfer function. We obtain the zeros from the equation

$$B(z) = \text{numerator}(H(z)) = 0$$

and designate these zeros by $z_{\text{z}1}, z_{\text{z}2}, z_{\text{z}3}, \ldots$.

The transfer function can be written in the factored form as

$$H(z) = H_0 \frac{\prod_k (z - z_{\text{z}k})}{\prod_i (z - z_{\text{p}i})} \tag{3.233}$$

The form (3.233) is known as the *pole-zero representation* of the transfer function. The real constant H_0 is called the *scale factor*.

Since the polynomials $A(z)$ and $B(z)$ have real coefficients, zeros and poles must be real or occur in complex-conjugate pairs.

The simplest transfer function is the first-order transfer function

$$H(z) = \frac{b_1 z + b_0}{a_1 z + a_0}, \quad a_1 \neq 0 \tag{3.234}$$

Systems described by (3.234) are called *first-order sections*.

Transfer functions of the form

$$H(z) = \frac{b_2 z^2 + b_1 z + b_0}{a_2 z^2 + a_1 z + a_0}, \quad a_2 \neq 0 \tag{3.235}$$

are called *second-order transfer functions*, and they play an important role in analysis and design of discrete-time LTI systems. A system characterized by (3.235) is called a *biquadratic section* or *biquad*.

Any transfer function (3.232) can be expressed as a product of first-order (3.234) and second-order (3.235) transfer functions, which implies that any discrete-time LTI system can be resolved into first-order sections and biquads.

Partial Transfer Function and Transfer-Function Matrix. Consider a relaxed multiple-input multiple-output system with N inputs, $x_1, x_2, \ldots, x_k, \ldots, x_N$, and assume that the system has L outputs $y_1, y_2, \ldots, y_i, \ldots, y_L$. We define the *partial transfer function* between the ith output and the kth input to be the ratio of the output z transform $Y_i(z)$ to the input z transform $X_k(z)$, with the other inputs being identically zero:

$$H_{ik}(z) = \left.\frac{Y_i(z)}{X_k(z)}\right|_{x_1=\cdots=x_{k-1}=x_{k+1}=\cdots=x_N=0} \tag{3.236}$$

The input–output description of the system can be expressed in a matrix form

$$\begin{bmatrix} Y_1(z) \\ Y_2(z) \\ \vdots \\ Y_i(z) \\ \vdots \\ Y_L(z) \end{bmatrix} = \begin{bmatrix} H_{11}(z) & H_{12}(z) & \cdots & H_{1k}(z) & \cdots & H_{1N}(z) \\ H_{21}(z) & H_{22}(z) & \cdots & H_{2k}(z) & \cdots & H_{2N}(z) \\ \vdots & \vdots & \ddots & \vdots & & \vdots \\ H_{i1}(z) & H_{i2}(z) & \cdots & H_{ik}(z) & \cdots & H_{iN}(z) \\ \vdots & \vdots & & \vdots & \ddots & \vdots \\ H_{L1}(z) & H_{L2}(z) & \cdots & H_{Lk}(z) & \cdots & H_{LN}(z) \end{bmatrix} \begin{bmatrix} X_1(z) \\ X_2(z) \\ \vdots \\ X_k(z) \\ \vdots \\ X_N(z) \end{bmatrix}$$

where the matrix

$$\mathbf{H}(z) = \begin{bmatrix} H_{11}(z) & H_{12}(z) & \cdots & H_{1k}(z) & \cdots & H_{1N}(z) \\ H_{21}(z) & H_{22}(z) & \cdots & H_{2k}(z) & \cdots & H_{2N}(z) \\ \vdots & \vdots & \ddots & \vdots & & \vdots \\ H_{i1}(z) & H_{i2}(z) & \cdots & H_{ik}(z) & \cdots & H_{iN}(z) \\ \vdots & \vdots & & \vdots & \ddots & \vdots \\ H_{L1}(z) & H_{L2}(z) & \cdots & H_{Lk}(z) & \cdots & H_{LN}(z) \end{bmatrix}$$

is called the *transfer-function matrix* of the system.

3.7 ANALYSIS OF LTI SYSTEMS BY TRANSFORM METHOD

Linear time-invariant (LTI) systems, characterized by linear constant-coefficient differential or difference equations, are efficiently analyzed by using the Laplace transform or the z transform. The two transforms map the differential or difference equations into algebraic equations which are easier to manipulate. This section illustrates step-by-step procedures for analyzing LTI systems in the transform domain. For the given block diagram of a system the required equations are formulated and mapped by a suitable transform into a system of algebraic equations. The set of algebraic equations is solved to find the system response in the transform domain. Next, by the inverse transform, the system response is computed as a continuous-time function or sequence [19–22]

3.7.1 Continuous-Time LTI Systems

Consider a continuous-time linear time-invariant (CTLTI) system specified by its block diagram, as shown in Fig. 3.1, and assume zero initial conditions—that is, that the system is at rest. Our target is to find the response of the system—that is, the signals at all nodes—for an excitation applied at node 1. To make the analysis simple we proceed as follows:

1. Label all nodes by consecutive integer numbers starting with 1.
2. Assume that the signals at the nodes are y_1, y_2, and so on.
3. Assume that the excitation is a known function of time, $x(t)$.
4. Write the equations characterizing each block of the diagram. Number of equations equals the number of blocks and equals the number of nodes.
5. Apply the Laplace transform to both sides of each equation.
6. Solve the set of algebraic equations, obtained as a result of the Laplace transform, to find the transforms of the signals at the nodes, that is, compute $Y_1(s)$, $Y_2(s)$, in terms of $X(s)$ and system parameters.
7. Find the required transfer functions by dividing the transform at a node, $Y_i(s)$, by the transform of the excitation, $X(s)$; assume zero initial conditions.
8. Find the inverse Laplace transform of $Y_1(s)$, $Y_2(s)$, ..., to obtain the response $y_1(t)$, $y_2(t)$, ..., of the system.

We analyze the system depicted in Fig. 3.1 as follows:

1. There are six nodes that we label 1, 2, 3, 4, 5, and 6.
2. The signals at the nodes are designated by y_1, y_2, y_3, y_4, y_5, and y_6.
3. The excitation is a known given function of time, $x(t)$.

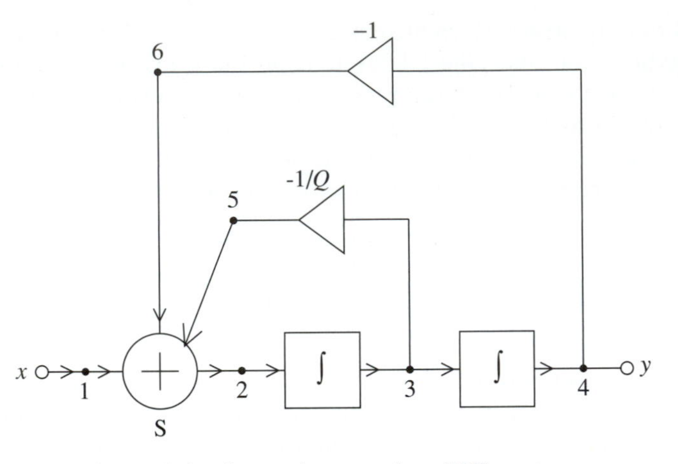

Figure 3.1 A continuous-time LTI system.

4. The equations characterizing each block of the diagram are

$$
\begin{aligned}
y_1 &= x(t) \\
y_2 &= y_1 + y_5 + y_6 \\
y_3 &= \omega \int_0^t y_2(\tau)\, d\tau \\
y_4 &= \omega \int_0^t y_3(\tau)\, d\tau \\
y_5 &= -\frac{1}{Q} y_3 \\
y_6 &= -y_4
\end{aligned}
\tag{3.237}
$$

The initial conditions are assumed to be zero. Notice that we have six blocks (excitation, adder, two integrators, two amplifiers), six nodes, and six equations.

5. We apply the Laplace transform to both sides of each equation in (3.237). We employ the four properties of the Laplace transform (uniqueness, homogeneity, additivity, differentiating) and obtain

$$
\begin{aligned}
Y_1(s) &= X(s) \\
Y_2(s) &= Y_1(s) + Y_5(s) + Y_6(s) \\
Y_3(s) &= \omega \frac{Y_2(s)}{s} \\
Y_4(s) &= \omega \frac{Y_3(s)}{s} \\
Y_5(s) &= -\frac{1}{Q} Y_3(s) \\
Y_6(s) &= -Y_4(s)
\end{aligned}
\tag{3.238}
$$

6. Solution of the set of algebraic equations (3.238) yields the complex response $Y_1(s), \ldots, Y_6(s)$ in terms of the complex excitation $X(s)$ and the system parameters Q and ω:

$$Y_1(s) = X(s)$$

$$Y_2(s) = \frac{s^2}{s^2 + \frac{\omega}{Q}s + \omega^2} X(s)$$

$$Y_3(s) = \frac{\omega s}{s^2 + \frac{\omega}{Q}s + \omega^2} X(s)$$

$$Y_4(s) = \frac{\omega^2}{s^2 + \frac{\omega}{Q}s + \omega^2} X(s)$$

$$\ldots$$

7. Assume that we want to compute the transfer functions $H_i(s) = Y_i(s)/X(s)$, $(i = 2, 3, 4)$. We find

$$H_2(s) = \frac{Y_2(s)}{X(s)} = \frac{s^2}{s^2 + \frac{\omega}{Q}s + \omega^2}$$

$$H_3(s) = \frac{Y_3(s)}{X(s)} = \frac{\omega s}{s^2 + \frac{\omega}{Q}s + \omega^2}$$

$$H_4(s) = \frac{Y_4(s)}{X(s)} = \frac{\omega^2}{s^2 + \frac{\omega}{Q}s + \omega^2}$$

8. If required, we can find the inverse Laplace transform of $Y_1(s)$, $Y_2(s)$, ... to obtain the response $y_1(t)$, $y_2(t)$, ... for the given excitation; for example, the step response of the system at node 3, for $Q = \frac{3}{2}$, $\omega = 1$, is

$$y_3(t) = \mathcal{L}^{-1}(H_3(s)X(s)) = \mathcal{L}^{-1}\frac{H_3(s)}{s} = \frac{3}{2\sqrt{2}} \mathrm{e}^{\frac{-1}{3}t} \sin\left(\frac{2\sqrt{2}}{3}t\right) u(t)$$

Notice that we have assumed the zero initial conditions. Nonzero initial conditions would change the equations

$$y_3 = y_3(0^-) + \omega \int_0^t y_2(\tau)\,d\tau$$
$$y_4 = y_4(0^-) + \omega \int_0^t y_3(\tau)\,d\tau$$

and the algebraic equation

$$Y_3(s) = \frac{y_3(0^-)}{s} + \omega \frac{Y_2(s)}{s}$$
$$Y_4(s) = \frac{y_4(0^-)}{s} + \omega \frac{Y_3(s)}{s}$$

By definition, when computing a transfer function, the initial conditions must be taken to be zero.

3.7.2 Discrete-Time LTI Systems

Consider a discrete-time linear time-invariant (DTLTI) system specified by its block diagram, as shown in Fig. 3.2, and assume zero initial conditions—that is, that the system is at rest. Our target is to find the response of the system—that is, the sequences at all nodes—for an excitation applied at node 1. To make the analysis simple we proceed as follows:

1. Label all nodes by consecutive integer numbers starting with 1.
2. Assume that the sequences at the nodes are $\{y_{1(n)}\}$, $\{y_{2(n)}\}$, and so on.
3. Assume that the excitation is a known sequence, $\{x_{(n)}\}$.
4. Write the equations characterizing each block of the diagram. Number of equations equals the number of blocks and equals the number of nodes.
5. Apply the z transform to both sides of each equation.
6. Solve the set of algebraic equations, obtained as a result of the z transform, to find the transforms of the sequences at the nodes—that is, compute $Y_1(z)$, $Y_2(z)$, ..., in terms of $X(z)$ and system parameters.
7. Find the required transfer functions by dividing the transform at a node, $Y_i(z)$, by the transform of the excitation, $X(z)$; assume zero initial conditions.
8. Find the inverse z transform of $Y_1(z)$, $Y_2(z)$, ... to obtain the response $\{y_{1(n)}\}$, $\{y_{2(n)}\}$, ... of the system.

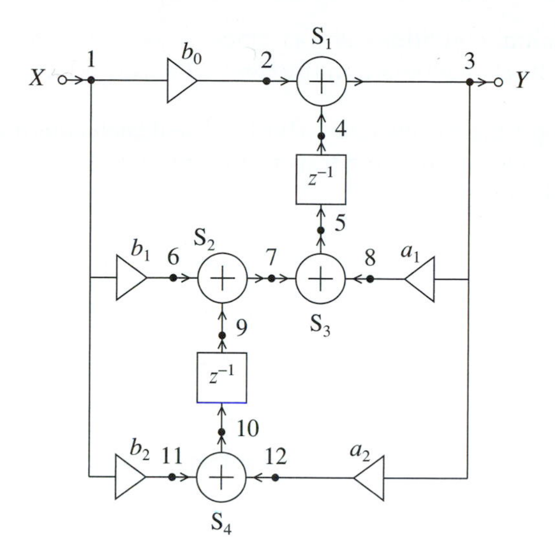

Figure 3.2 Discrete-time LTI system.

We analyze the system depicted in Fig. 3.2 as follows:

1. There are 12 nodes that we label 1, 2, 3,. . ., 12.
2. We designate the signals at the nodes by $y_1(n)$, $y_2(n)$, $y_3(n)$, . . ., $y_{12}(n)$.
3. The excitation is a known given sequence, $x(n)$.
4. The equations characterizing each block of the diagram are

$$
\begin{aligned}
y_1(n) &= x(n) \\
y_2(n) &= b_0 y_1(n) + x_{b0}(n) \\
y_3(n) &= y_2(n) + y_4(n) \\
y_4(n) &= y_5(n-1) \\
y_5(n) &= y_7(n) + y_8(n) \\
y_6(n) &= b_1 y_1(n) + x_{b1}(n) \\
y_7(n) &= y_6(n) + y_9(n) \\
y_8(n) &= a_1 y_3(n) + x_{a1}(n) \\
y_9(n) &= y_{10}(n-1) \\
y_{10}(n) &= y_{11}(n) + y_{12}(n) \\
y_{11}(n) &= b_2 y_1(n) + x_{b2}(n) \\
y_{12}(n) &= a_2 y_3(n) + x_{a2}(n)
\end{aligned}
\tag{3.239}
$$

The initial conditions are assumed to be zero. Notice that we have 12 blocks (excitation, 4 adders, 5 multipliers, 2 delays), 12 nodes, and 12 equations.

5. We apply the z transform to both sides of each equation in (3.239). We employ the four properties of the z transform (uniqueness, homogeneity, additivity, shifting) and obtain

$$
\begin{aligned}
Y_1(z) &= X(z) \\
Y_2(z) &= b_0 Y_1(z) + X_{b0}(z) \\
Y_3(z) &= Y_2(z) + Y_4(z) \\
Y_4(z) &= z^{-1} Y_5(z) \\
Y_5(z) &= Y_7(z) + Y_8(z) \\
Y_6(z) &= b_1 Y_1(z) + X_{b1}(z) \\
Y_7(z) &= Y_6(z) + Y_9(z) \\
Y_8(z) &= a_1 Y_3(z) + X_{a1}(z) \\
Y_9(z) &= z^{-1} Y_{10}(z) \\
Y_{10}(z) &= Y_{11}(z) + Y_{12}(z) \\
Y_{11}(z) &= b_2 Y_1(z) + X_{b2}(z) \\
Y_{12}(z) &= a_2 Y_3(z) + X_{a2}(z)
\end{aligned}
\tag{3.240}
$$

6. Solution of the set of algebraic equations (3.240) yields the complex response $Y_1(z), \ldots, Y_{12}(z)$ in terms of the complex excitation $X(z)$, noise sources due to rounding the output of multipliers, $X_{a1}(z)$, $X_{a2}(z)$, $X_{b0}(z)$, $X_{b1}(z)$, $X_{b2}(z)$, and the multiplier coefficients a_1, a_2, b_0, b_1, b_2.

$$
\begin{aligned}
Y_1(z) &= X(z) \\
&\vdots \\
Y_3(z) &= \cdots \\
&\vdots \\
Y_{12}(z) &= \cdots
\end{aligned}
$$

Desired response, say y_3, in the z domain in terms of all excitations is

$$
Y_3(z) = \frac{(b_0 + b_1 z^{-1} + b_2 z^{-2}) X(z) + z^{-1} X_{a1}(z) + z^{-2} X_{a2}(z) + X_{b0}(z) + z^{-1} X_{b1}(z) + z^{-2} X_{b2}(z)}{1 - a_1 z^{-1} - a_2 z^{-2}}
$$

or, equivalently,

$$\begin{aligned} Y_3(z) = & \frac{b_0 + b_1 z^{-1} + b_2 z^{-2}}{1 - a_1 z^{-1} - a_2 z^{-2}} X(z) \\ & + \frac{z^{-1}}{1 - a_1 z^{-1} - a_2 z^{-2}} X_{a1}(z) \\ & + \frac{z^{-2}}{1 - a_1 z^{-1} - a_2 z^{-2}} X_{a2}(z) \\ & + \frac{1}{1 - a_1 z^{-1} - a_2 z^{-2}} X_{b0}(z) \\ & + \frac{z^{-1}}{1 - a_1 z^{-1} - a_2 z^{-2}} X_{b1}(z) \\ & + \frac{z^{-2}}{1 - a_1 z^{-1} - a_2 z^{-2}} X_{b2}(z) \end{aligned}$$

7. Assume that we want to compute the transfer function of the system and the noise transfer functions for all multipliers. The transfer function of the system is defined by

$$H_3(z) = \left.\frac{Y_3(z)}{X(z)}\right|_{X_{a1}(z)=0, X_{a2}(z)=0, X_{b0}(z)=0, X_{b1}(z)=0, X_{b2}(z)=0}$$

and the noise transfer functions for each multiplier are defined by

$$\begin{aligned} H_{3a1}(z) &= \left.\frac{Y_3(z)}{X_{a1}(z)}\right|_{X(z)=0, X_{a2}(z)=0, X_{b0}(z)=0, X_{b1}(z)=0, X_{b2}(z)=0} \\ H_{3a2}(z) &= \left.\frac{Y_3(z)}{X_{a2}(z)}\right|_{X(z)=0, X_{a1}(z)=0, X_{b0}(z)=0, X_{b1}(z)=0, X_{b2}(z)=0} \\ H_{3b0}(z) &= \left.\frac{Y_3(z)}{X_{b0}(z)}\right|_{X(z)=0, X_{a1}(z)=0, X_{a2}(z)=0, X_{b1}(z)=0, X_{b2}(z)=0} \\ H_{3b1}(z) &= \left.\frac{Y_3(z)}{X_{b1}(z)}\right|_{X(z)=0, X_{a1}(z)=0, X_{a2}(z)=0, X_{b0}(z)=0, X_{b2}(z)=0} \\ H_{3b2}(z) &= \left.\frac{Y_3(z)}{X_{b2}(z)}\right|_{X(z)=0, X_{a1}(z)=0, X_{a2}(z)=0, X_{b0}(z)=0, X_{b1}(z)=0} \end{aligned}$$

We find the transfer function as

$$H_3(z) = \frac{b_0 + b_1 z^{-1} + b_2 z^{-2}}{1 - a_1 z^{-1} - a_2 z^{-2}}$$

and find the noise transfer functions as

$$H_{3a1}(z) = \frac{z^{-1}}{1 - a_1 z^{-1} - a_2 z^{-2}}$$

$$H_{3a2}(z) = \frac{z^{-2}}{1 - a_1 z^{-1} - a_2 z^{-2}}$$

$$H_{3b0}(z) = \frac{1}{1 - a_1 z^{-1} - a_2 z^{-2}}$$

$$H_{3b1}(z) = \frac{z^{-1}}{1 - a_1 z^{-1} - a_2 z^{-2}}$$

$$H_{3b2}(z) = \frac{z^{-2}}{1 - a_1 z^{-1} - a_2 z^{-2}}$$

All transfer functions have the same denominator $1 - a_1 z^{-1} - a_2 z^{-2}$.

8. If required, we can find the inverse z transform of $Y_1(z)$, $Y_2(z)$, ... to obtain the response $y_1(n)$, $y_2(n)$, ... for the given excitation; for example, the impulse response of the system at node 3, for $a_1 = 0$, $a_2 = \frac{1}{2}$, $b_0 = 1$, $b_1 = 2$, and $b_2 = 1$, is

$$\begin{aligned} y_3(n) &= Z^{-1}(H_3(z)X(z)) = Z^{-1}(H_3(z)) \\ &= 2^{-n/2}\left(-\cos(n\frac{\pi}{2}) + \sqrt{8}\sin(n\frac{\pi}{2})\right)u(n) + 2\delta(n) \end{aligned}$$

and is shown in Fig. 3.3.

Multipliers of discrete-time systems are single-input single-output blocks defined by the equation $y(n) = ax(n)$, where a is the multiplier coefficient, $y(n)$ is the multiplier output, and $x(n)$ is the multiplier input. When quantization to B bits is performed after each multiplication, an error occurs; this error is considered as a noise sequence superimposed on the signal at the output of the multiplier. Therefore, the multiplier with product quantization at the output is modeled by the equation

$$y(n) = ax(n) + x_a(n)$$

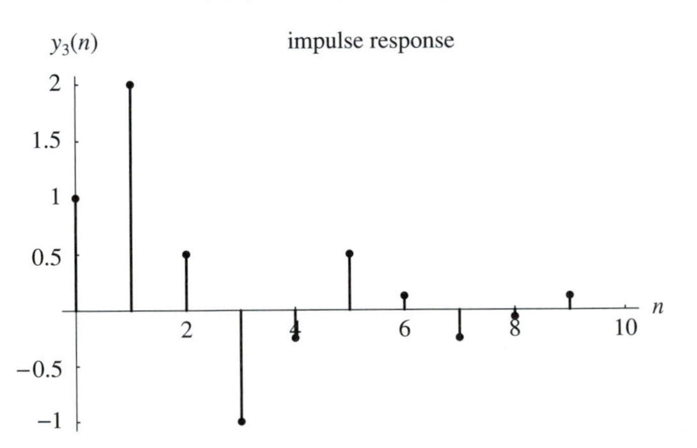

Figure 3.3 Impulse response of the discrete-time LTI system.

or in the z domain, taking the z transform of the left and right side of the above equation:

$$Y(z) = aX(z) + X_a(z)$$

The term $x_a(n)$ is the noise sequence due to product quantization superimposed on the signal at the output of the multiplier. Noise sequences are treated as inputs and are used for derivation of noise transfer functions. The noise transfer function $H_a(z)$, of a multiplier with the coefficient a, is defined as a ratio of the output z transform $Y(z)$ and the noise sequence z transform $X_a(z)$, assuming that all other inputs are set to zero.

3.7.3 Analog LTI Circuits

Consider a linear time-invariant analog circuit specified by its schematic, as shown in Fig. 3.4, and assume zero initial conditions—that is, that no energy exists in the circuit capacitors and inductors. Our target is to find the response of the circuit—that is, the currents and the voltages of the circuit branches. To make the analysis simple we proceed as follows:

1. Label all nodes by consecutive integer numbers starting from 0. The zero node is the ground node.
2. Assume that the node voltages are v_1, v_2, and so on. The voltage of the ground is zero, $v_0 = 0$.
3. Assume that the excitation (the voltage of an independent voltage source, or the current of an independent current source) is a known function of time, $x(t)$.
4. Write the Kirchhoff-current-law (KCL) equations for each node except node 0. Express the branch currents in terms of the node voltages. If the current of a branch cannot be expressed in terms of the node voltages, write the equation characterizing that branch, in terms of the node voltages and the current of that branch.
5. Apply the Laplace transform to both sides of each equation.

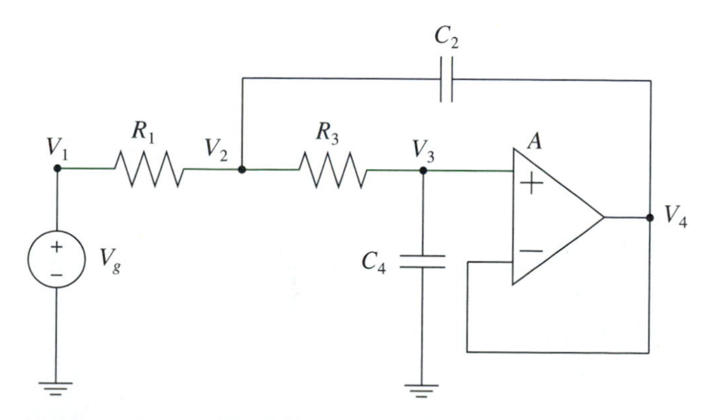

Figure 3.4 Analog LTI circuit.

6. Solve the set of algebraic equations, obtained as a result of the Laplace transform, to find the transforms of the circuit variables (that we call the *complex response*) in terms of the circuit parameters and the transform of the excitation.

7. Find the required transfer functions (also called *network functions*) by dividing the transform of a response by the transform of the excitation; assume zero initial conditions.

8. Find the inverse Laplace transform of the complex response to obtain the time-domain response of the circuit.

We analyze the analog LTI circuit depicted in Fig. 3.4 as follows:

1. There are five nodes that we label 0, 1, 2, 3, and 4.

2. We designate the node voltages by v_0, v_1, v_2, v_3, and v_4.

3. The excitation is a known given function of time, $v_g(t)$.

4. The circuit equations are

$$
\begin{aligned}
v_1 &= v_g(t) \\
\frac{v_2 - v_1}{R_1} + \frac{v_2 - v_3}{R_3} + C_2 \mathrm{D}(v_2 - v_4) &= 0 \\
\frac{v_3 - v_2}{R_3} + C_4 \mathrm{D} v_3 &= 0 \\
v_3 &= v_4
\end{aligned}
\tag{3.241}
$$

The initial conditions are assumed to be zero. Notice that we have four equations (number of nodes minus one).

5. We apply the Laplace transform to both sides of each equation in (3.241). We employ the four properties of the Laplace transform (uniqueness, homogeneity, additivity, differentiating) and obtain

$$
\begin{aligned}
V_1(s) &= V_g(s) \\
\frac{V_2(s) - V_1(s)}{R_1} + \frac{V_2(s) - V_3(s)}{R_3} + C_2 s(V_2(s) - V_4(s)) &= 0 \\
\frac{V_3(s) - V_2(s)}{R_3} + C_4 s V_3(s) &= 0 \\
V_3(s) &= V_4(s)
\end{aligned}
\tag{3.242}
$$

6. Solution of the set of algebraic equations (3.242) yields the complex response $V_1(s), \ldots, V_4(s)$ in terms of the complex excitation $V_g(s)$ and the circuit parameters R_1, R_3, C_2, and C_4

$$
\begin{aligned}
V_1(s) &= V_g(s) \\
V_2(s) &= \frac{1 + C_4 R_3 s}{C_2 C_4 R_1 R_3 s^2 + C_4(R_1 + R_3)s + 1} V_g(s) \\
V_3(s) &= \frac{1}{C_2 C_4 R_1 R_3 s^2 + C_4(R_1 + R_3)s + 1} V_g(s) \\
V_4(s) &= \frac{1}{C_2 C_4 R_1 R_3 s^2 + C_4(R_1 + R_3)s + 1} V_g(s)
\end{aligned}
$$

7. Assume that we want to compute the transfer functions $H_i(s) = V_i(s)/V_g(s)$ $(i = 1, 2, 3, 4)$. We find

$$
\begin{aligned}
H_1(s) &= \frac{V_1(s)}{V_g(s)} = 1 \\
H_2(s) &= \frac{V_2(s)}{V_g(s)} = \frac{1 + C_4 R_3 s}{C_2 C_4 R_1 R_3 s^2 + C_4(R_1 + R_3)s + 1} \\
H_3(s) &= \frac{V_3(s)}{V_g(s)} = \frac{1}{C_2 C_4 R_1 R_3 s^2 + C_4(R_1 + R_3)s + 1} \\
H_4(s) &= \frac{V_4(s)}{V_g(s)} = \frac{1}{C_2 C_4 R_1 R_3 s^2 + C_4(R_1 + R_3)s + 1}
\end{aligned}
$$

8. If required, we can find the inverse Laplace transform of $V_1(s)$, $V_2(s)$, ... to obtain the response $v_1(t)$, $v_2(t)$, ... for the given excitation; for example, the step response of the system at node 4, for $C_2 = 4C$, $C_4 = C$, and $R_3 = R_1 = R$, is

$$
\begin{aligned}
v_4(t) &= \mathcal{L}^{-1}(H_4(s)V_g(s)) = \mathcal{L}^{-1}\frac{H_4(s)}{s} \\
&= \left(1 - e^{-t/(4CR)}\left(\cos(\frac{\sqrt{3}}{4CR}t) - \frac{1}{\sqrt{3}}\sin(\frac{\sqrt{3}}{4CR}t)\right)\right)u(t)
\end{aligned}
$$

Figure 3.5 shows the step response for $C = 10$ nF and $R = 10$ kΩ.

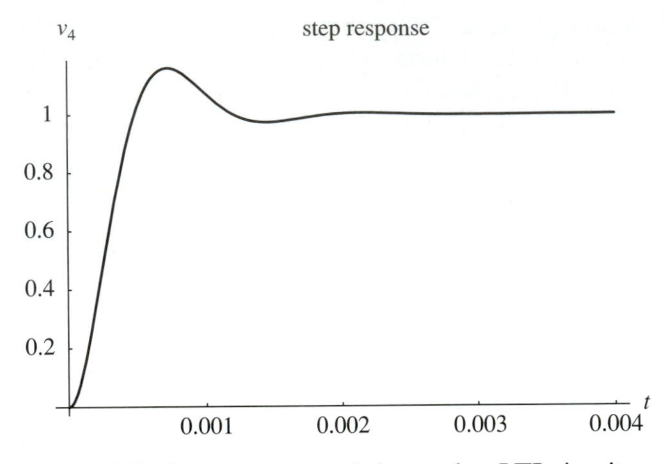

Figure 3.5 Step response of the analog LTI circuit.

PROBLEMS

3.1 Find the phasor representing the sinusoid

(a) $x_1(t) = 2\sin(\pi t)$,

(b) $x_2(t) = \sqrt{2}\sin(\frac{1}{2}\pi t + \frac{\pi}{6})$,

(c) $x_3(t) = 2\sqrt{2}\cos(3\pi t)$,

(d) $x_4(t) = 0.8\cos(0.25\pi t - \frac{\pi}{4})$,

(e) $x_5(t) = \frac{1}{2}\sin(\pi t/2) + \frac{1}{3}\cos(\pi t/2)$;

use the integral formula that establishes the phasor transformation.

3.2 Can you find the phasor for the following signal: $x(t) = \sqrt{2}\sin(125t) + 2\cos(250t)$? Explain the answer.

3.3 What signal quantities do we have to know when we perform the inverse phasor transformation? Find sinusoids corresponding to the phasors

(a) $X_1 = j\sqrt{2}$ at $f_1 = 1$ kHz,

(b) $X_2 = 2$ at $f_2 = 500$ Hz,

(c) $X_3 = 3\exp(j\pi/6)$ at $f_3 = 2$ kHz.

3.4 Use phasor transformation to find a sinusoidal signal, $x_s(t)$, equivalent to the signal

$$x(t) = \sin(2\pi t) + \frac{1}{2}\sin\left(2\pi t - \frac{\pi}{2}\right) + \frac{1}{3}\sin\left(2\pi t - \frac{\pi}{3}\right) + \frac{1}{4}\sin\left(2\pi t - \frac{\pi}{4}\right)$$

3.5 A system is defined by the input–output relation

$$\mathrm{D}^2 y(t) + a_1 \mathrm{D}y(t) + a_0 y(t) = b_0 x(t)$$

and excited by a sinusoidal signal $x(t) = X_m\cos(\omega t + \xi)$. Determine the steady-state response. Assume $a_1 = 2$, $a_0 = 4$, $b_0 = 1$, $X_m = \sqrt{2}$, $\omega = 2$, $\xi = 0$.

3.6 Compute and plot the steady-state response of a system specified by

$$\mathrm{D}^2 y(t) + \frac{\omega}{Q}\mathrm{D}y(t) + \omega^2 y(t) = \frac{\omega}{Q}x(t)$$

and excited by $x(t) = X_m \cos(\omega t) + X_m \cos(2\omega t)$, for $\omega = 2,\ Q = 4,\ X_m = 1$.

3.7 Find the transfer function (frequency response) of a system described by

$$a_2\mathrm{D}^2 y(t) + a_1\mathrm{D}y(t) + a_0 y(t) = b_2\mathrm{D}^2 x(t) + b_1\mathrm{D}x(t) + b_0 x(t)$$

in the sinusoidal steady state. Assume $a_2 = 1$, $a_1 = 0.5$, $a_0 = 1$, $b_2 = 4$, $b_1 = 0$, $b_0 = 4$.

3.8 Consider the system from Problem 3.6.

(a) Find the frequency response.

(b) Plot the magnitude response in decibels, the phase response, and the group delay.

(c) Compute the poles and zeros of the system. For each complex pole find the pole magnitude and the pole quality factor.

3.9 The frequency response of a system is given by

$$H(j\omega) = \frac{9 - \omega^2}{2(4 + 5j\omega - 2\omega^2 - j\omega^3)}$$

(a) Plot the magnitude response in linear and log scale.

(b) Resolve the system into first-order sections and biquads.

(c) Find the frequency at which the magnitude response is 3 dB below the maximum.

3.10 Consider the system in Fig. 3.1 and assume identical integrators, characterized by $y_{int}(t) = \int x_{int}(t)\,\mathrm{d}t$, where $x_{int}(t)$ is integrator input and $y_{int}(t)$ is integrator output. The system input is at node 1 and the output is at node 4.

(a) Find the frequency response of the system.

(b) Compute phasors and steady-state signals at all nodes if the system is excited by $x(t) = \sin(\pi t)$. Assume $Q = 1$.

(c) Can this system reach a steady state for an arbitrary sinusoidal input?

3.11 Circuit in Fig. 3.4 is in steady state.

(a) Find the frequency response of the circuit (assume that the output is at node 4).

(b) Derive the input impedance seen by the generator.

(c) Compute phasors and steady-state voltages at all nodes if the circuit is excited by $v_g(t) = \cos(5000\pi t)$. Assume $R_1 = R_3 = 10\,\mathrm{k\Omega}$, $C_2 = 40\,\mathrm{nF}$, $C_4 = 10\,\mathrm{nF}$.

(d) Can this circuit reach a steady state for an arbitrary sinusoidal input?

(e) Use phasor method to find the steady-state output for $v_g(t) = 1 + \sin(5000\pi t)$.

3.12 Derive the Fourier series for a sawtooth signal which rises linearly from 0 to A over the range $0 \le t \le T$, then drops instantaneously to 0 and repeats the cycle. Find the amplitude spectrum of the sawtooth signal. Assume $A = 5$, $T = 2$.

3.13 Use the results of Problem 3.12 to deduce the amplitude spectrum of the signal $x(t) = At$ for $|t| < \frac{1}{2}T$, $x(t + kT) = x(t)$ for integer k.

3.14 Derive the Fourier series for a unit periodic square pulse train defined as

$$x(t) = \begin{cases} 1, & 0 \le t < \frac{T}{2} \\ -1, & \frac{T}{2} \le t < T \end{cases}$$

$x(t + kT) = x(t)$ for integer k. Assume $T = 2$.

Approximate $x(t)$ with a finite sum of the first five harmonics. Plot the approximation and observe an overshoot at points very near the corners $t = k\frac{T}{2}$.

3.15 Consider a train of square pulses

$$x(t) = \begin{cases} A, & 0 < t \le \tau \\ 0, & \tau < t \le T \end{cases}$$

$x(t + kT) = x(t)$ for integer k. Derive the Fourier series in the complex form (the exponential Fourier series). Assume $A = 1/\tau$ and find the complex Fourier coefficients for very short pulses when τ tends to zero.

3.16 A periodic signal is given by

$$x(t) = A\frac{t^2}{T^2}, \qquad 0 \le t < T$$

where $x(t + kT) = x(t)$ for integer k. Derive the trigonometric Fourier series and the exponential Fourier series. Observe the relation between the Fourier coefficients. Assume $A = 1, T = 1$.

3.17 Employ the definition of the Fourier transform to compute the transform of the signal

$$x(t) = A\cos(\pi\frac{t}{T})^2, \qquad |t| < \frac{T}{2}$$

$$x(t) = 0, \qquad |t| \ge \frac{T}{2}$$

Assume $A = 1, T = 1$.

3.18 Find the Laplace transform of the signals

(a) $x(t) = A\cos(\pi\frac{t}{T})^2\, u(t)$

(b) $x(t) = A\exp(-\alpha t)\cos(\omega t + \xi)u(t)$.

Assume $A = 2, T = 1, \alpha = 0.1, \omega = 100\pi, \xi = 0.2\pi$.

3.19 The system from Problem 3.5 is excited by $x(t) = tu(t)$. Find the zero-state response. Use the Laplace transform.

3.20 Find the zero-state response of the system from Problem 3.6 for the causal input $x(t) = (X_m\cos(\omega t) + X_m\cos(2\omega t))\,u(t)$. Assume $X_m = \sqrt{2}, \omega = 2$.

3.21 Derive the transfer function in the Laplace domain for the system specified in Problem 3.7. What is the value of the impulse response of the system at $t \to +\infty$?

3.22 The Laplace transform of a signal is given by

$$X(s) = A\frac{s + a}{s(s + b)}, \qquad a \ne b$$

Does this signal vanish as $t \to +\infty$? Assume $A = 4, a = 2, b = 3$.

3.23 Use the Laplace transform to find the output of the system specified in Problem 3.10 for the input $x(t) = \cos(\pi t)\,u(t)$, and the initial conditions $y_3(0^-) = K_3 = 1.25$ and $y_4(0^-) = K_4 = -0.75$. Try to solve this problem in the time domain. Compute the impulse response of the system.

3.24 Find the step response of the circuit specified in Problem 3.11. Use the Laplace transform. Compute the zero-input response if the voltage across C_2 has a nonzero value $(v_2 - v_4)_{t=0-} = K = 0.025$.

3.25 Consider the periodic signal

$$x(t) = 4\sin(2\pi t) + \sin(200\pi t)$$

Sample this signal uniformly at n points over the interval $0 \leq t < 1$, and find the DFT for (a) $n = 8$, (b) $n = 32$, (c) $n = 64$, (d) $n = 128$, (e) $n = 256$. Plot and compare the signal spectrum for each n.

3.26 Find the transfer function of the digital filter shown in Fig. 3.2 and its poles and zeros.

3.27 Compute the zero-input response of the digital filter in Fig. 3.2, if the initial conditions are $y_5(-1) = 0.2$, $y_{10}(-1) = 0.1$.

MATLAB EXERCISES

3.1 Write a MATLAB script that finds and plots sinusoids corresponding to the phasors in Problem 3.3.

3.2 Write a MATLAB script that performs the computation required in Problem 3.4. Plot $x(t)$ and $x_e(t)$ on the same diagram, for $0 \leq t \leq 2$, to validate the equivalence.

3.3 Find and plot the steady-state response from Problem 3.5 in MATLAB.

3.4 Write a MATLAB script that finds and plots the steady-state response from Problem 3.6.

3.5 Plot the magnitude response and the phase response from Problem 3.7 for $a_2 = 1$, $a_1 = 2$, $a_0 = 5$, $b_2 = 3$, $b_1 = 4$, $b_0 = 6$, by using the MATLAB function `freqs`.

3.6 Write a MATLAB script that carries out the computation required in Problem 3.8.

3.7 Can you automate the analysis from Problem 3.9 in MATLAB? If yes, write the corresponding code.

3.8 Write a MATLAB script for computation of the Fourier coefficients for signals from Problems 3.12, 3.13, 3.14, 3.15, 3.16.

(a) Write your own code from the scratch (without the `fft` function).

(b) Use the MATLAB function `fft`.

3.9 Find the Fourier transform from Problem 3.17 in MATLAB. Use the MATLAB function `fft`.

MATHEMATICA EXERCISES

3.1 Write the corresponding code in *Mathematica* for symbolic computation of the phasor transform in Problem 3.1.

3.2 Find and plot sinusoids from Problem 3.3 in *Mathematica*.

3.3 Write a *Mathematica* notebook that performs the computation required in Problem 3.4. Plot $x(t)$ and $x_e(t)$ on the same diagram, for $0 \leq t \leq 2$, to validate the equivalence.

3.4 Write a *Mathematica* code that finds the steady-state response from Problem 3.5.

3.5 Find symbolically the steady-state response from Problem 3.6 in *Mathematica*.

3.6 Automate the manipulation from Problem 3.7 in *Mathematica*, and find the transfer function symbolically. Plot the magnitude response and the phase response for $a_2 = 1$, $a_1 = 2$, $a_0 = 5$, $b_2 = 3$, $b_1 = 4$, $b_0 = 6$, $0 \leq \omega \leq 8$.

3.7 Analyze the system from Problem 3.8 symbolically in *Mathematica*. Find the required quantities symbolically.

3.8 Use *Mathematica* for symbolic analysis of the system from Problem 3.9. Try to find the required quantities symbolically.

3.9 Write a *Mathematica* notebook to automate the analysis required in Problem 3.10.

3.10 Consider the circuit from Problem 3.11.

- **(a)** Carry out the general symbolic analysis in *Mathematica*. Formulate the circuit equations in terms of phasors, find the frequency response, and compute the phasors for all node voltages.
- **(b)** Plot the magnitude response, phase response, and group delay.
- **(c)** Find the steady-state node voltages.

3.11 Write a *Mathematica* code to derive the Fourier series. Next, find the Fourier series for signals from Problems 3.12, 3.13, 3.14, 3.15, and 3.16.

3.12 Use *Mathematica* to derive the Fourier transform in Problem 3.17.

- **(a)** Write your own code.
- **(b)** Use the *Mathematica* standard packages.

3.13 Automate the Laplace domain manipulation from Problems 3.18, 3.19, 3.20, 3.21, 3.22, 3.23, and 3.24 in *Mathematica*.

- **(a)** Write your own code.
- **(b)** Use the *Mathematica* standard packages.

CHAPTER 4

CLASSIC ANALOG FILTER DESIGN

This chapter is intended to review the basics of classical analog filter design. Classification, salient properties, and sensitivity of transfer functions are given. The most important analog filter realizations are presented, including operational amplifier (op amp) active *RC*, switched-capacitor (SC), passive *RLC*, operational transconductance amplifier (OTA), and current-conveyor (CC) realizations. For each realization we provide complete design equations and procedures that make the design easily applicable to a broad variety of analog filter design problems. A detailed case study is given for the realization of various transfer functions.

4.1 INTRODUCTION TO ANALOG FILTERS

Analog filters are frequency-selective electrical circuits that are used to amplify or attenuate a single sinusoidal signal component or a portion of the signal frequency spectrum. The range of frequencies in which the sinusoidal signals are amplified or passed without considerable attenuation is called the *passband*. The frequency range in which the sinusoidal signals are significantly attenuated is called the *stopband*. The required minimum and maximum of the attenuation or amplification, along with the corresponding edge frequencies of the passbands and stopbands, are called the *specification*.

An *analog filter design* is a process in which we construct an electrical circuit that meets the given specification. The design starts with the specification and it consists of four basic steps: *approximation*, *realization*, *study of imperfections*, and *implementation* [16, XIII.65]. There is an infinite number of circuits that meet the specification; therefore, the filter design is by no means unique.

The example to follow details on the four design steps. A typical analog filter design is shown in Fig. 4.1.

Assume that we want to design a filter that passes sinusoidal signals over the frequency range $0 \leq f \leq F_p$, but attenuates the signals for $f \geq F_s > F_p$. This type of filter is called the *analog lowpass filter*. The frequency range $0 \leq f \leq F_p$ is the passband, and the range $F_s \leq f$ is the stopband; F_p is the *passband edge frequency* and F_s is the *stopband edge frequency*. We require that the attenuation in passband must not exceed A_p dB, and that the attenuation in stopband should be no less than A_s dB. Obviously, A_p is the *maximum passband attenuation*, and A_s is the *minimum stopband attenuation*. The four quantities that specify our lowpass filter requirements can be written in the form of the list $S = \{F_p, F_s, A_p, A_s\}$, which we simply call the *lowpass specification*.

In the first design step, the *approximation step*, we construct the filter transfer function $H(s)$, which is a rational function in the complex frequency s. The attenuation,

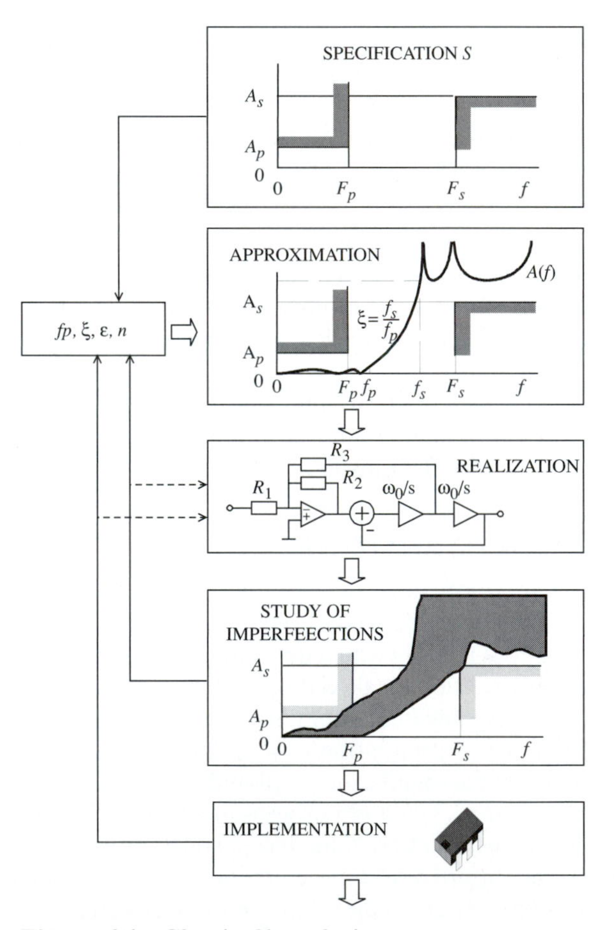

Figure 4.1 Classic filter design.

$A(f) = -20\log_{10}|H(j2\pi f)|$, must satisfy the specification S, that is

$$0 \le A(f) \le A_p, \qquad 0 \le f \le F_p \tag{4.1}$$

$$A_s \le A(f), \qquad F_s \le f \tag{4.2}$$

The function $A(f)$ is called the *attenuation approximation function*, or simply the *attenuation*.

Suppose that we choose the elliptic transfer function, known as the most selective function, which depends on four parameters: the order n, the actual passband edge frequency f_p, the actual stopband edge frequency f_s, and the ripple factor ϵ. It is convenient to define the ratio $\xi = f_s/f_p$, which is called the *selectivity factor*. We adjust the transfer-function parameters f_p, ξ, ϵ, and n to meet the specification; usually, we set $f_p = F_p$ and $\xi = F_s/F_p$. The maximal passband attenuation is controlled by the passband ripple factor ϵ, and we traditionally choose $\epsilon = \sqrt{10^{A_p/10} - 1}$. Generally, we prefer the minimum order n of the transfer function.

The *realization step* of an analog filter is the process of converting the transfer function into an electrical circuit; this circuit is sometimes called the *realization*. The designer is interested in realizations which are economical, simple, cheap, with small noise and distortion, and with high dynamic range and which are not seriously affected by small changes in the element values (tolerances, temperature and humidity variations, aging drift). In Fig. 4.1 a realization of a second-order universal SC filter is presented. Numerical values of the resistors are calculated from known $H(s)$.

In practice, the filter is implemented with nonideal elements and the designer must accomplish the *study of imperfections* which includes tolerance analysis and study of parasitics. If the specification can be satisfied only with high-precision expensive components, then the designer has to choose another transfer function $H(s)$ and reevaluate the realization or approximation step.

In the *implementation step* a device called the product prototype, also called the *implementation*, is constructed and tested. The cost of the mass production depends on the type of components, packaging, methods of manufacturing, testing, and tuning. The best implementation is a device with no need for tuning. If the requirements are not met, then the realization step (new circuit) or the approximation step (new f_p, ξ, ϵ, or n) must be redone.

The classical analog filters can be classified according to the frequency range they pass or reject:

- *Lowpass filter* passes sinusoidal signals over the range $0 \le f \le F_p$, but attenuates the signals for $f \ge F_s > F_p$.
- *Highpass filter* passes sinusoidal signals for $f \ge F_p$, but attenuates the signals over the range $0 \le f \le F_s < F_p$.
- *Bandpass filter* passes sinusoidal signals over the range $F_{p1} \le f \le F_{p2}$, but attenuates the signals for $0 \le f \le F_{s1} < F_{p1}$ and $f \ge F_{s2} > F_{p2}$.
- *Bandreject* or *bandstop filter* passes sinusoidal signals for $0 \le f \le F_{p1} < F_{s1}$ and $f \ge F_{p2} > F_{s2}$, but attenuates the signals over the range $F_{s1} \le f \le F_{s2}$.
- *Allpass filter* or *phase equalizer* passes sinusoidal signals without attenuation, and it shapes the phase response.

- *Lowpass-notch filter* rejects sinusoidal signals at frequencies $f \approx f_z$, but it passes signals at high frequencies $(f \gg f_z)$ with some attenuation.
- *Highpass-notch filter* rejects sinusoidal signals at frequencies $f \approx f_z$, but it passes signals at low frequencies $(f \ll f_z)$ with some attenuation.
- *Amplitude equalizer* or *bump filter* slightly amplifies or attenuates signals over a range of frequencies.

We mention here several, among many, excellent books on classical analog filter theory, analysis, and design: 15, 16, 23–34.

4.2 BASIC FILTER TRANSFER FUNCTIONS

The filter transfer function is a rational function in the complex frequency s and can be written in the form

$$H(s) = K\frac{(s - s_{z1})(s - s_{z2})\cdots(s - s_{zm})}{\left(s - s_{p1}\right)\left(s - s_{p2}\right)\cdots\left(s - s_{pn}\right)} \tag{4.3}$$

where K is a real constant, $s_{z1}, \ldots, s_{zm}$ are zeros, and $s_{p1}, \ldots, s_{pn}$ are poles of the transfer function. Poles and zeros can be real or complex. Complex poles or zeros, s_i, occur in complex-conjugate pairs:

$$\begin{aligned} s_i &= \mathrm{Re}(s_i) + j\mathrm{Im}(s_i) \\ s_{i+1} &= \mathrm{Re}(s_i) - j\mathrm{Im}(s_i) \end{aligned} \tag{4.4}$$

The corresponding factors of the transfer function can be expressed as

$$(s - s_i)(s - s_{i+1}) = s^2 - (s_i + s_{i+1})s + s_i s_{i+1} \tag{4.5}$$

or can be presented by a second-order polynomial with real coefficients $a_i = -(s_i + s_{i+1})$ and $\omega_i^2 = s_i s_{i+1}$:

$$(s - s_i)(s - s_{i+1}) = s^2 + a_i s + \omega_i^2, \quad a_i = -2\ \mathrm{Re}(s_i), \quad \omega_i = |s_i| \tag{4.6}$$

or equivalently

$$(s - s_i)(s - s_{i+1}) = s^2 + \frac{\omega_i}{Q_i}s + \omega_i^2 \tag{4.7}$$

In this book we prefer the form (4.7) because it is more significant from a practical point of view. The angular frequency ω_i is the *magnitude* of s_i. The quantity Q_i is called the *Q-factor* of s_i.

The Q-factor is used to characterize the ratio of the magnitude and the real part of a complex-conjugate pole or zero pair

$$Q_i = -\frac{|s_i|}{2\ \mathrm{Re}(s_i)} \tag{4.8}$$

Pole Q-factors of stable systems are always positive because $\mathrm{Re}(s_i) < 0$. Formally, when the imaginary part vanishes the Q-factor reaches its minimal value $1/2$.

A complex pole-zero pair can be represented by

$$H_i(s) = \frac{s^2 + \dfrac{\omega_{zi}}{Q_{zi}}s + \omega_{zi}^2}{s^2 + \dfrac{\omega_{pi}}{Q_{pi}}s + \omega_{pi}^2} \tag{4.9}$$

and is an example of the *second-order transfer function*, or *biquad*, for short.

To reduce the sensitivity of a transfer function with respect to deviations of the element values and to simplify the tuning process of the filter, it is preferable to realize the filter by a cascade of the first-order and second-order filter sections [30]. The cascade approach consists of realizing each of the biquads by an appropriate circuit and connecting these circuits in cascade. The overall transfer function, $H(s)$, could be expressed as a product of the first-order and second-order functions $H_i(s)$

$$H(s) = \prod_{i=1}^{N} H_i(s) \tag{4.10}$$

The advantage of the cascade approach is that the realization of a higher-order transfer function is reduced to the much simpler design of first-order and second-order filters. The individual low-order filters are isolated so that any change in one filter does not affect any other filter in the cascade. This property is useful for adjusting and testing the filter at the time of manufacture.

We characterize each second-order filter section by the frequencies of the magnitude response extremes and the 3-dB frequencies. The *3-dB frequencies*, denoted by $\omega_{3\mathrm{dB}} = 2\pi f_{3\mathrm{dB}}$, are frequencies at which the magnitude response $|H(j\omega)|$ is $\sqrt{2}$ times smaller than the magnitude response at some reference frequency $\omega_r = 2\pi f_r$ (3 dB only approximately corresponds to $\sqrt{2}$, strictly speaking it is $10^{3/20}$)

$$\frac{|H(j\omega_{3\mathrm{dB}})|}{|H(j\omega_r)|} = \frac{1}{\sqrt{2}} \tag{4.11}$$

For example, for lowpass filters, the reference frequency is $\omega_r = 0$.

In filter design, we prefer to use the *normalized transfer function*, $H_n(s)$, defined by

$$H_n(s) = \frac{H(s)}{\max\limits_{\omega} |H(j\omega)|} \tag{4.12}$$

The transfer function can be obtained by scaling the normalized transfer function by a constant

$$H(s) = kH_n(s) \tag{4.13}$$

Quite generally, the normalization constant k can take any real value.

4.2.1 Second-Order Transfer Functions

In this section we analyze the properties of the basic second-order transfer functions. We examine the magnitude response $|H(j\omega)|$ for real positive angular frequencies ω. The angular frequencies of the magnitude response local extrema are designated by ω_e.

Lowpass Transfer Function. The second-order *lowpass transfer function* is defined as

$$H_{LP}(s) = \frac{\omega_p^2}{s^2 + \frac{\omega_p}{Q_p}s + \omega_p^2} \tag{4.14}$$

At high frequencies, $f \gg \omega_p/(2\pi)$, the magnitude response $|H_{LP}(j2\pi f)|$ decrease as f^2 and, thus, high-frequency sinusoidal signals are rejected.

The key properties of the lowpass transfer function are summarized below:

$$\begin{aligned}
&H_{LP}(0) = 1, && s = 0, \quad \omega = 0 \\
&H_{LP}(s) \to 0, && s \to +\infty, \quad \omega \to +\infty \\
&|H_{LP}(j\omega_p)| = Q_p, && s = j\omega_p, \quad \omega = \omega_p
\end{aligned} \tag{4.15}$$

$$|H_{LP}(j\omega_e)| = \frac{Q_p}{\sqrt{1 - \frac{1}{4Q_p^2}}}, \qquad \omega_e = \omega_p\sqrt{1 - \frac{1}{2Q_p^2}}$$

$$|H_{LP}(j\omega_{3\text{dB}})| = \frac{1}{\sqrt{2}}, \qquad \omega_{3\text{dB}} = \omega_p\sqrt{1 - \frac{1}{2Q_p^2} + \sqrt{1 - \left(1 - \frac{1}{2Q_p^2}\right)^2}} \tag{4.16}$$

The maximal value of the magnitude response is approximately equal to Q_p for $Q_p \gg 1$ as shown in Fig. 4.2:

$$\max_{\omega} |H_{LP}(j\omega)| = |H_{LP}(j\omega_e)| \approx Q_p, \quad \omega_e \approx \omega_p, \quad Q_p \gg 1 \tag{4.17}$$

This fact is very important for active filters. Suppose that the output voltage of operational amplifier must be from the range ±1 V. The amplitude of the sinusoidal signal of frequency ω_e, at the input of the lowpass second-order filter, must be smaller than

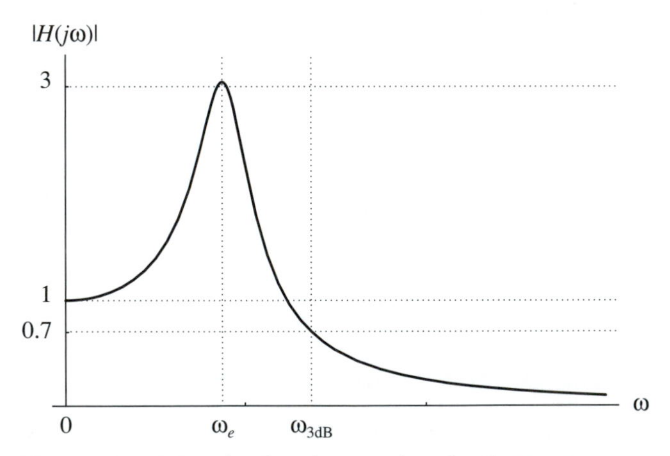

Figure 4.2 Magnitude of second-order lowpass transfer function: $Q_p = 3$ and $\omega_p = 0.9$.

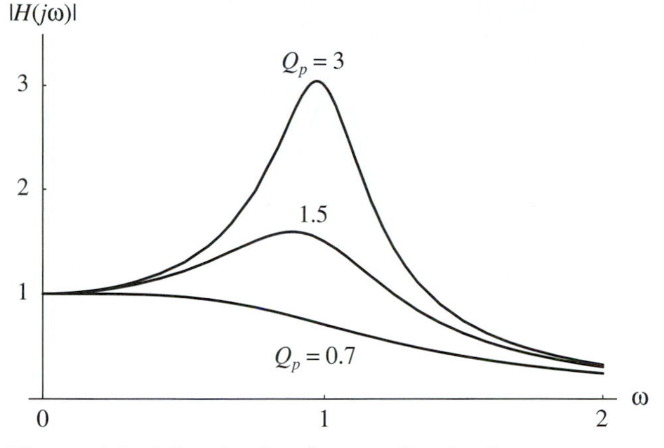

Figure 4.3 Magnitude of second-order lowpass transfer functions: $Q_p = 3$, 1.5, 0.7 and $\omega_p = 0.9$.

$1/Q_p$, so that, after filtering, the amplitude of the output signal remains within the prescribed range ± 1 V.

For $Q_p \leq 1/\sqrt{2}$ we find $\omega_e = 0$ as shown in Fig. 4.3.

The reference frequency for $\omega_{3\text{dB}}$, for lowpass filters, is $\omega_r = 0$.

Highpass Transfer Function. The second-order *highpass transfer function* is defined as

$$H_{HP}(s) = \frac{s^2}{s^2 + \dfrac{\omega_p}{Q_p}s + \omega_p^2} \tag{4.18}$$

The sinusoidal signals at very low frequencies, $f \ll \omega_p/(2\pi)$, are rejected while the high-frequency sinusoidal signals, $f \geq \omega_p/(2\pi)$, pass without attenuation.

The key properties of the highpass transfer function are summarized below:

$$\begin{aligned} H_{HP}(0) &= 0, & s &= 0, \quad \omega = 0 \\ H_{HP}(s) &\to 1, & s &\to +\infty, \quad \omega \to +\infty \\ |H_{HP}(j\omega_p)| &= Q_p, & s &= j\omega_p, \quad \omega = \omega_p \end{aligned} \tag{4.19}$$

$$\begin{aligned} |H_{HP}(j\omega_e)| &= \frac{Q_p}{\sqrt{1 - \dfrac{1}{4Q_p^2}}}, & \omega_e &= \frac{\omega_p}{\sqrt{1 - \dfrac{1}{2Q_p^2}}} \\ |H_{HP}(j\omega_{3\text{dB}})| &= \frac{1}{\sqrt{2}}, & \omega_{3\text{dB}} &= \frac{\omega_p}{\sqrt{1 - \dfrac{1}{2Q_p^2} + \sqrt{1 - \left(1 - \dfrac{1}{2Q_p^2}\right)^2}}} \end{aligned} \tag{4.20}$$

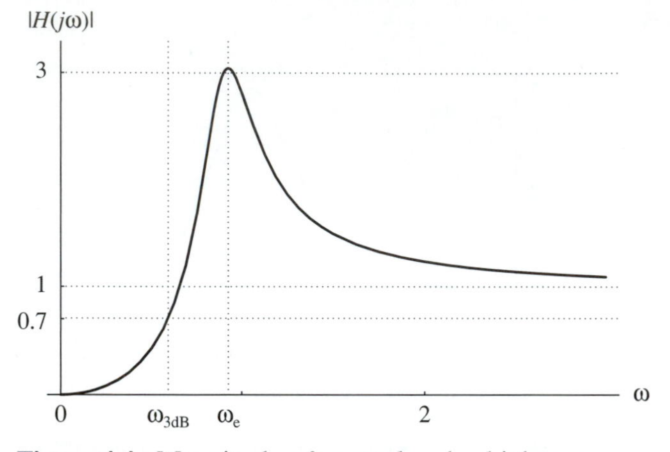

Figure 4.4 Magnitude of second-order highpass transfer function: $Q_p = 3$, and $\omega_p = 0.9$.

As in the case of lowpass filters, the maximal value of the magnitude response is approximately equal to Q_p, as shown in Fig. 4.4. For $Q_p \leq 1/\sqrt{2}$ we find that ω_e is at infinity, as shown in Fig. 4.5.

The reference frequency for $\omega_{3\text{dB}}$, for highpass filters, is at infinity $\omega_r \to +\infty$.

Bandpass Transfer Function. The second-order *bandpass transfer function* is defined as

$$H_{BP}(s) = \frac{\dfrac{\omega_p}{Q_p} s}{s^2 + \dfrac{\omega_p}{Q_p} s + \omega_p^2} \tag{4.21}$$

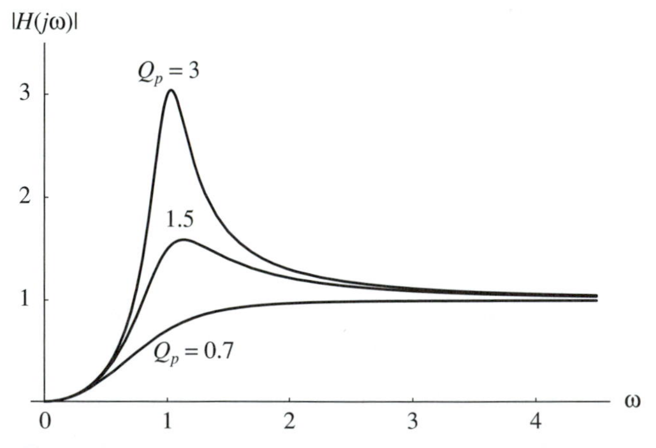

Figure 4.5 Magnitude of second-order highpass transfer functions: $Q_p = 3$, 1.5, 0.7 and $\omega_p = 0.9$.

The key properties of the bandpass transfer function are summarized below:

$$\begin{aligned}
H_{BP}(0) &= 0, && s = 0, \quad \omega = 0 \\
H_{BP}(s) &\to 0, && s \to +\infty, \quad \omega \to +\infty \\
|H_{BP}(j\omega_p)| &= 1, && s = j\omega_p, \quad \omega = \omega_p \\
|H_{BP}(j\omega_e)| &= 1, && \omega_e = \omega_p
\end{aligned} \tag{4.22}$$

The angular frequency at which the magnitude response reaches its maximum, ω_e, is equal to the pole magnitude, ω_p, and is sometimes called the *resonant angular frequency*, or the *central angular frequency*.

$$|H_{BP}(j\omega_{low,3\text{dB}})| = \frac{1}{\sqrt{2}}, \qquad \omega_{low,3\text{dB}} = \omega_p \sqrt{1 + \frac{1}{2Q_p^2} - \frac{1}{Q_p}\sqrt{1 + \left(\frac{1}{2Q_p}\right)^2}}$$

$$|H_{BP}(j\omega_{high,3\text{dB}})| = \frac{1}{\sqrt{2}}, \qquad \omega_{high,3\text{dB}} = \omega_p \sqrt{1 + \frac{1}{2Q_p^2} + \frac{1}{Q_p}\sqrt{1 + \left(\frac{1}{2Q_p}\right)^2}} \tag{4.23}$$

Second-order bandpass filters pass sinusoidal signals from the band of frequencies $\omega_{low,3\text{dB}}/(2\pi) < f < \omega_{high,3\text{dB}}/(2\pi)$ with insignificant attenuation, but reject sinusoidal signals whose frequencies are on either side of this band. The *3-dB bandwidth* of a bandpass filter is defined as

$$\omega_{BW} = \omega_{high,3\text{dB}} - \omega_{low,3\text{dB}} \tag{4.24}$$

and the *relative bandwidth* is

$$\text{relative bandwidth} = \frac{\omega_{BW}}{\omega_p}$$

The maximum of the magnitude response is 1. The 3-dB bandwidth is affected by Q_p (Fig. 4.6). Higher Q-factors produce narrower bandwidths (Fig. 4.7).

The reference frequency for $\omega_{3\text{dB}}$, for bandpass filters, is $\omega_r = \omega_p$.

Bandreject Transfer Function. The second-order *bandreject transfer function* is defined as

$$H_{BR}(s) = \frac{s^2 + \omega_p^2}{s^2 + \dfrac{\omega_p}{Q_p}s + \omega_p^2} \tag{4.25}$$

The key properties of the bandreject transfer function are summarized below:

$$\begin{aligned}
H_{BR}(0) &= 1, && s = 0, \quad \omega = 0 \\
H_{BR}(s) &= 1, && s \to +\infty, \quad \omega \to +\infty \\
|H_{BR}(j\omega_p)| &= 0, && s = j\omega_p, \quad \omega = \omega_p
\end{aligned} \tag{4.26}$$

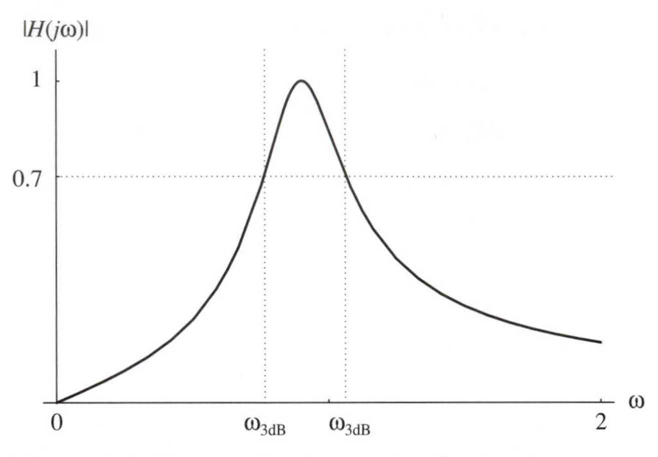

Figure 4.6 Magnitude of second-order bandpass transfer function: $Q_p = 3$ and $\omega_p = 0.9$.

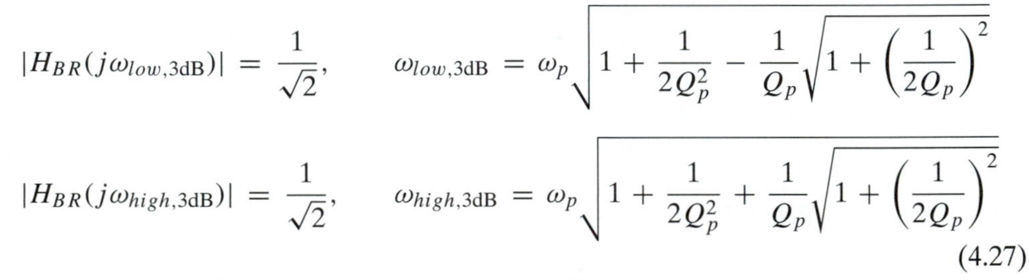

$$|H_{BR}(j\omega_{low,3\text{dB}})| = \frac{1}{\sqrt{2}}, \qquad \omega_{low,3\text{dB}} = \omega_p \sqrt{1 + \frac{1}{2Q_p^2} - \frac{1}{Q_p}\sqrt{1 + \left(\frac{1}{2Q_p}\right)^2}}$$

$$|H_{BR}(j\omega_{high,3\text{dB}})| = \frac{1}{\sqrt{2}}, \qquad \omega_{high,3\text{dB}} = \omega_p \sqrt{1 + \frac{1}{2Q_p^2} + \frac{1}{Q_p}\sqrt{1 + \left(\frac{1}{2Q_p}\right)^2}} \tag{4.27}$$

Bandreject filters reject sinusoidal signals from the band of frequencies $\omega_{low,3\text{dB}}/(2\pi) < f < \omega_{high,3\text{dB}}/(2\pi)$, but they pass signals whose frequencies are on either side of this band. The *3-dB bandwidth* of a bandreject filter is defined as

$$\omega_{BW} = \omega_{high,3\text{dB}} - \omega_{low,3\text{dB}} \tag{4.28}$$

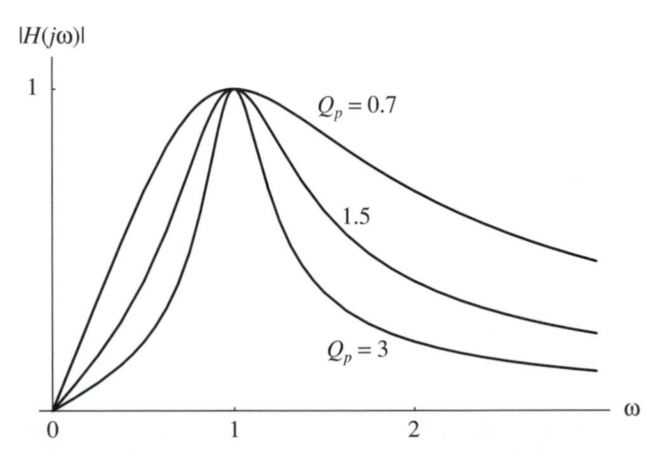

Figure 4.7 Magnitude of second-order bandpass transfer functions: $Q_p = 3$, 1.5, 0.7 and $\omega_p = 0.9$.

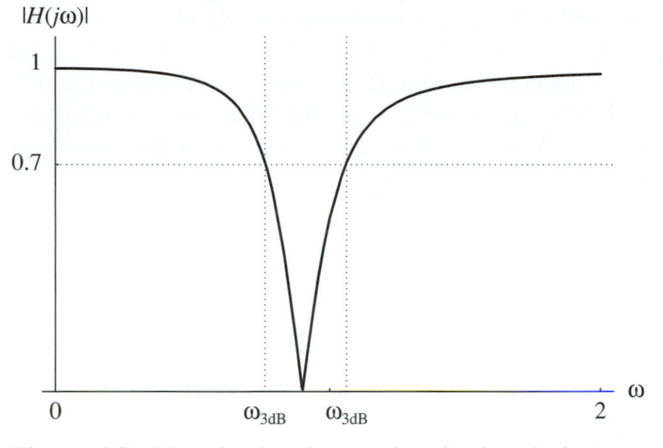

Figure 4.8 Magnitude of second-order bandreject transfer function: $Q_p = 3$ and $\omega_p = 0.9$.

The maximum of the magnitude response is 1 (Fig. 4.8), and the magnitude response reaches its minimum at $\omega = \omega_p$. The 3-dB bandwidth is affected by Q_p (Fig. 4.9); higher Q-factors produce narrower bandwidths.

The reference frequency for $\omega_{3\text{dB}}$, for bandreject filters, is $\omega_r = 0$.

Lowpass-Notch and Highpass-Notch Transfer Function. The second-order *lowpass-notch* and *highpass-notch transfer functions* are defined as

$$\begin{aligned} H_{LPN}(s) &= \frac{s^2 + \omega_z^2}{s^2 + \dfrac{\omega_p}{Q_p}s + \omega_p^2}, \quad \omega_z^2 > \omega_p^2 \\ H_{HPN}(s) &= \frac{s^2 + \omega_z^2}{s^2 + \dfrac{\omega_p}{Q_p}s + \omega_p^2}, \quad \omega_z^2 < \omega_p^2 \end{aligned} \tag{4.29}$$

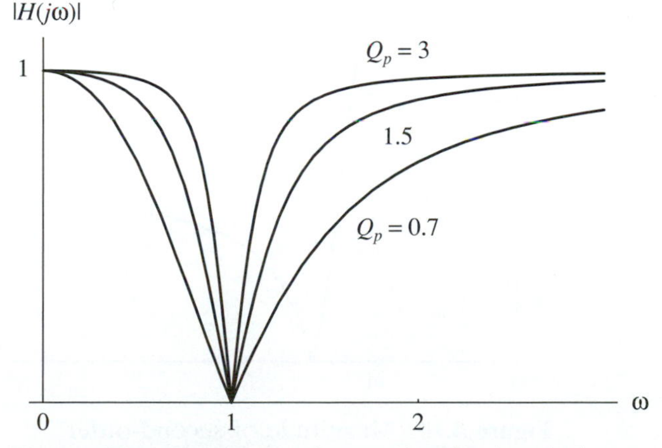

Figure 4.9 Magnitude of second-order bandreject transfer functions: $Q_p = 3$, 1.5, 0.7 and $\omega_p = 0.9$.

Lowpass-notch filters reject sinusoidal signals whose frequencies are $f \approx \omega_z/(2\pi)$, but they pass signals at high frequencies ($f \gg \omega_z/(2\pi)$) with some attenuation. Highpass-notch filters reject sinusoidal signals whose frequencies are $f \approx \omega_z/(2\pi)$, but they pass signals at low frequencies ($f \ll \omega_z/(2\pi)$) with some attenuation.

The key properties of the lowpass-notch and highpass-notch transfer function are summarized below:

$$\begin{aligned}
H_{LPN}(0) &= H_{HPN}(0) = \frac{\omega_z^2}{\omega_p^2}, && s = 0, \quad \omega = 0 \\
H_{LPN}(s) &= H_{HPN}(s) = 1, && s \to +\infty, \quad \omega \to +\infty \\
|H_{LPN}(j\omega_p)| &= |H_{HPN}(j\omega_p)| = \left| Q_p \left(-1 + \frac{\omega_z^2}{\omega_p^2} \right) \right|, && s = j\omega_p, \quad \omega = \omega_p \\
H_{LPN}(j\omega_z) &= H_{HPN}(j\omega_z) = 0, && s = j\omega_z, \quad \omega = \omega_z
\end{aligned} \tag{4.30}$$

The maximum of the magnitude response occurs at ω_e and is given below:

$$|H_{LPN}(j\omega_e)| = |H_{HPN}(j\omega_e)| = \frac{2Q_p\omega_z^2}{\omega_p^2} \sqrt{\frac{Q_p^2 + (1 - 2Q_p^2)\frac{\omega_p^2}{\omega_z^2} + Q_p^2 \frac{\omega_p^4}{\omega_z^4}}{4Q_p^2 - 1}} \tag{4.31}$$

$$\omega_e = \omega_p \sqrt{\frac{2Q_p^2\omega_p^2 + \omega_z^2 - 2Q_p^2\omega_z^2}{-\omega_p^2 + 2Q_p^2\omega_p^2 - 2Q_p^2\omega_z^2}}$$

The reference frequency for $\omega_{3\mathrm{dB}}$, for notch filters, is at infinity $\omega_r \to +\infty$, as shown in Fig. 4.10.

The maximum of the magnitude response in terms of ω_p/ω_z is plotted in Fig. 4.11; notice that it reaches the minimum for $\omega_p = \omega_z$.

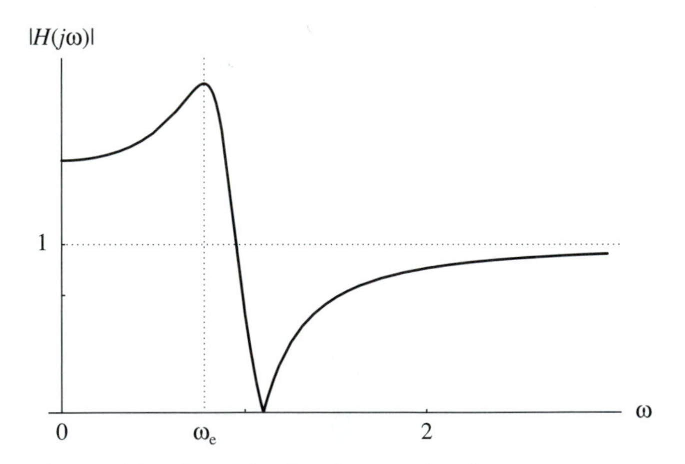

Figure 4.10 Magnitude of second-order lowpass-notch transfer function, $Q_p = 3$, $\omega_p = 0.9$, and $\omega_z = 1.1$.

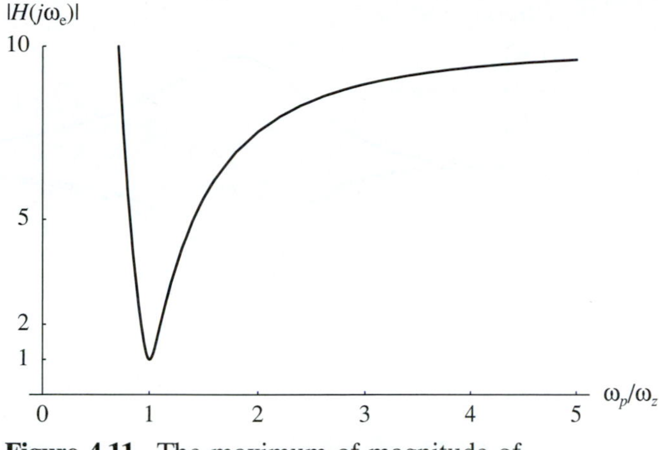

Figure 4.11 The maximum of magnitude of second-order notch transfer function, $Q_p = 10$.

Allpass Transfer Function. The second-order *allpass transfer function* is defined as

$$H_{AP}(s) = \frac{s^2 - \dfrac{\omega_p}{Q_p}s + \omega_p^2}{s^2 + \dfrac{\omega_p}{Q_p}s + \omega_p^2} \tag{4.32}$$

The magnitude of the allpass transfer function is constant for all ω:

$$|H_{AP}(j\omega)| = 1 \tag{4.33}$$

Thus far, we have discussed the magnitude response only. However, for allpass transfer functions the phase response is of key interest. Allpass filters can be used to modify the phase responses of filters without changing their magnitude responses. Usually, they are used to satisfy magnitude and phase specifications simultaneously.

The phase response of second-order allpass filters is shown in Fig. 4.12. If the Q-factor is smaller, then the phase response of an allpass filter is more linear.

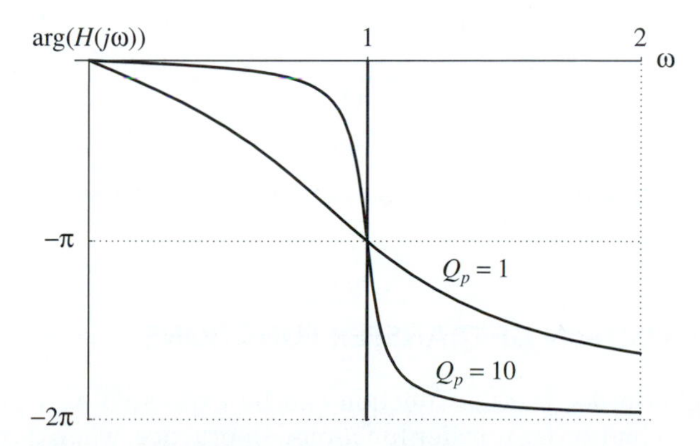

Figure 4.12 Phase of second-order allpass transfer functions, $Q_p = 10$, 1, and $\omega_p = 1$.

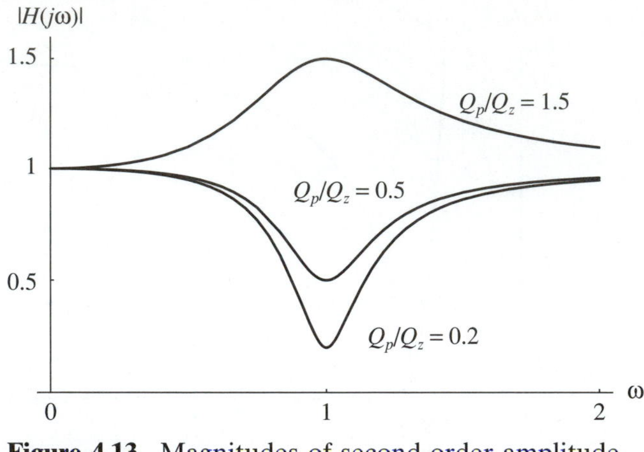

Figure 4.13 Magnitudes of second-order amplitude equalizer transfer function, $\omega_p = 1$.

The zeros of an allpass transfer function are mirror images of the transfer function poles, in the complex plane, with respect to the imaginary axis.

Amplitude Equalizer Transfer Function. *Amplitude equalizer* or *bump filter* slightly amplifies (boosts) or attenuates signals over a range of frequencies and has the transfer function

$$H_{AE}(s) = \frac{s^2 + \dfrac{\omega_p}{Q_z}s + \omega_p^2}{s^2 + \dfrac{\omega_p}{Q_p}s + \omega_p^2} \tag{4.34}$$

The key properties of the amplitude equalizer transfer function are summarized below:

$$\begin{aligned} H_{AE}(0) &= 1, & s &= 0, \quad \omega = 0 \\ H_{AE}(s) &= 1, & s &\to +\infty, \quad \omega \to +\infty \\ H_{AE}(j\omega_p) &= \frac{Q_p}{Q_z}, & s &= j\omega_p, \quad \omega = \omega_p \end{aligned} \tag{4.35}$$

The magnitude response of the amplitude equalizer is shown in Fig. 4.13.

4.3 DECOMPOSITION OF TRANSFER FUNCTIONS

Any higher-order transfer function can be expressed as a product of the first-order and the second-order transfer functions. In practice, when designing a filter, we prefer transfer functions with complex-conjugate poles; however, if the transfer-function order is odd, we prefer transfer functions with only one real pole:

$$H_{even}(s) = \prod_i \frac{b_{2,i}s^2 + b_{1,i}s + b_{0,i}}{s^2 + \frac{\omega_{pi}}{Q_{pi}}s + \omega_{pi}^2} \tag{4.36}$$

$$H_{odd}(s) = \frac{b_1 s + b_0}{s + \omega_p} \prod_i \frac{b_{2,i}s^2 + b_{1,i}s + b_{0,i}}{s^2 + \frac{\omega_{pi}}{Q_{pi}}s + \omega_{pi}^2} \tag{4.37}$$

The first-order and the second-order transfer functions are realized separately. Next, the transfer function is realized by cascading these low-order filter sections. The cascade design is attractive for various reasons:

- The filter design is straightforward.
- The case study for selecting the appropriate low-order filter realizations is facilitated.
- The effort for testing and tuning the filter is reduced.

4.4 POLE-ZERO PAIRING

The *pole-zero pairing* involves (a) a decomposition of the numerator and the denominator of the transfer function into products of constant terms, and the first-order and the second-order functions in s and, then, (b) constructing the first-order and the second-order rational functions in s by pairing the numerator and denominator terms.

The pole-zero pairing is not unique and depends on implementation, as well as on the selected transfer function. A detailed analysis of this topic can be found in reference 29.

The most frequently used procedure is pairing the poles with higher Q-factors with zeros that are as close as possible to the poles. This pairing can achieve the maximal dynamic range of the second-order sections.

The most important criteria that are used in practice, as criteria for pole-zero pairing, are as follows:

1. *Maximal dynamic range*: A figure-of-merit of the dynamic range of the second-order filter section can be determined as a ratio of (a) the maximum magnitude response computed over the whole frequency range to (b) the minimum magnitude response in the passband. This criterion is of importance when large signals are processed.
2. *Maximal signal-to-noise ratio*: The maximal signal-to-noise ratio that does not drive the filter out of the linear mode of operation, assuming that the noise is generated within the active device.
3. *Minimal inband losses*: The minimum magnitude response of a filter section, computed over the passband, defines the maximum inband loss. Our goal is to find pole-zero pairing that keeps this loss as low as possible.
4. *Minimal sensitivity*: The overall sensitivity of the filter can be reduced by appropriate pole-zero pairing.

The choice of the scaling factors of the first-order and the second-order sections is called the *gain distribution*. There exist procedures for optimizing the gain distribution for maximal dynamic range or for minimal sensitivity [29].

4.5 OPTIMUM CASCADING SEQUENCE

After a higher-order transfer function is decomposed into, say m, first-order and second-order functions, with appropriate scaling factors, the designer has to choose the *optimal cascading sequence*. The number of possible cascade realizations is $m!$, where m is, also, the number of cascaded filter sections.

For example, a sixth-order transfer function can be decomposed into the product $H(s) = H_1(s)H_2(s)H_3(s)$, and realized as a cascade of three second-order sections. The number of possible realizations is $m! = 3! = 6$ as shown in Fig. 4.14.

Frequently, for the maximal dynamic range, the optimal sequence of the second-order filter sections is the sequence in which the preceding section has lower Q-factor than the following section.

In some cases, some other criteria may be more important in choosing the optimal cascading sequence. For example, if the input signal contains an undesired signal with very large amplitude, this signal has to be filtered out by the very first section (the first section should have a transfer function zero at the frequency of the undesired signal). Thus, we avoid possible distortion of the desired signal at the remaining sections.

Special care must be taken when we filter pulse signals that contain very high frequency components. These signals may pass through the filter (e.g., through the passive components) without attenuation because active devices do not have the expected or assumed characteristics at higher frequencies (e.g., operational amplifiers can have no gain at very high frequencies). Therefore, the first filter section must be carefully chosen and, sometimes, only passive components (such as resistors and capacitors) should be used to ensure required attenuation at very high frequencies.

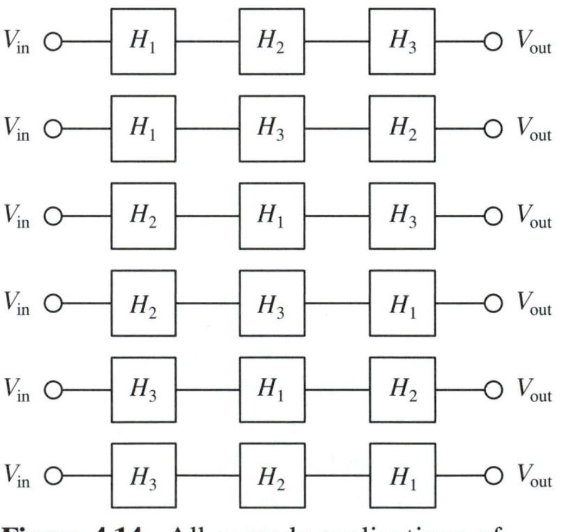

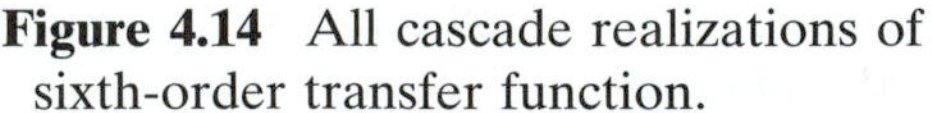

Figure 4.14 All cascade realizations of sixth-order transfer function.

4.6 SENSITIVITY

4.6.1 Basic Definitions

Performances of manufactured filters cannot be guaranteed to correspond exactly to the designed filter performances. The differences are due to component imperfections such as the following:

- The element value is not an exact number, it is always within prescribed tolerances; for the nominal element value x_{nom} and tolerance x_t (%) the actual value is from the range $x_{nom} - x_t/100 < x < x_{nom} + x_t/100$.
- Ambient variations (temperature, humidity) and chemical changes due to aging change the element value $x_{nom} - \Delta x_1 < x < x_{nom} + \Delta x_2$, usually, $\Delta x_1 \neq \Delta x_2$.

Many circuit realizations can be used to satisfy the prescribed filter transfer function. If we assume ideal circuit elements (perfect components), various realizations will exhibit the same performances. In practice, real components are imperfect and their values deviate from the nominal values; therefore, performances of the manufactured filter differ from the performances of the filter with perfect components.

Before the filter is manufactured, the effects of manufacturing tolerances and component imperfections have to be analyzed. Such an analysis allows the designer to predict variations of the filter performances and to predict the production yield. The *yield* is the ratio of (a) the number of manufactured filters satisfying the specification to (b) the total number of manufactured filters. Obviously, a high yield is desirable for profitability.

The simplest way to predict the yield is to use the concept of network sensitivity, assuming that the component changes are small.

The *single-parameter relative sensitivity* of a function $F = F(x_1, \ldots, x_i, \ldots, x_n)$, due to a change in quantity x_i, is

$$S_{x_i}^F = \frac{x_i}{F}\frac{\partial F}{\partial x_i} \tag{4.38}$$

and, for a small change in x_i, the relative variation of the function F is expected to be

$$\frac{\Delta F}{F} = S_{x_i}^F \frac{\Delta x_i}{x_i} \tag{4.39}$$

where $\frac{\Delta x_i}{x_i}$ is called the *variability* of x.

For simpler notation we use $S(F, x_i)$ instead of $S_{x_i}^F$:

$$S(F, x_i) = S_{x_i}^F \tag{4.40}$$

The expected *relative variation* in function $F = F(x_1, \ldots, x_i, \ldots, x_n)$, due to the changes in all quantities $x_1, \ldots, x_i, \ldots, x_n$, is given by

$$\frac{\Delta F}{F} = \sum_{i=1}^{n} S_{x_i}^{F} \frac{\Delta x_i}{x_i} \tag{4.41}$$

The function $F(x_1, \ldots, x_i, \ldots, x_n)$ can be a transfer function, the magnitude of a transfer function, the filter attenuation, the group delay, or any other function derived from the transfer function, such as a pole of the transfer function, the Q-factor of a pole, or a transfer function coefficient.

If $S_{x_i}^{F} = 1$, then a 1% change in the quantity x_i will cause a 1% change in the function F. When a sensitivity is zero, $S_{x_i}^{F} = 0$, then any change in x_i will not affect the function F.

In practice, the upper limit of the relative variation $\dfrac{\Delta F}{F}$ is estimated by three methods:

- The *worst-case* method gives the absolute relative variation

$$\left.\frac{\Delta F}{F}\right|_{\text{worst case}} = \sum_{i=1}^{n} \left| S_{x_i}^{F} \frac{\Delta x_i}{x_i} \right| \tag{4.42}$$

- The *Schoeffler criterion* method is based on the fomula

$$\left.\frac{\Delta F}{F}\right|_{\text{Schoeffler}} = \sqrt{\sum_{i=1}^{n} \left| S_{x_i}^{F} \frac{\Delta x_i}{x_i} \right|^2} \tag{4.43}$$

- The *Monte Carlo* method relies on extensive simulation of the filter realization (circuit) with randomly chosen element values. The performances are interpreted statistically.

The worst-case variation is a rather pessimistic figure-of-merit, assuming that all changes are at their extreme. The *Schoeffler* variation is more realistic, and it is closer to the variation obtained by statistical computations. The most preferable method is the *Monte Carlo* method. It evaluates filter performances for random combination of element values. We can use different statistical distributions of element values to model component imperfections for the *Monte Carlo* method. All three sensitivity analyses are useful for quantifying the probability of successful implementation, increasing the production yield, and minimizing the cost of filter manufacturing.

A list of sensitivities of elementary functions is presented below:

$F(x)$	$S_x^{F(x)}$	$F(y(x), v(x))$	$S_x^{F(y(x),v(x))}$
x	1	$y(x)$	$S_{y(x)}^{F(y(x))} = S_y^{F(y)}$
cx	1	$y(x) \cdot v(x)$	$S_x^{y(x)} + S_x^{v(x)}$
c^x	$x \ \ln c$	$\lvert y(x)\rvert e^{jv(x)}$	$\mathrm{Re}\ S_x^{\lvert y(x)\rvert e^{jv(x)}} + j\dfrac{\mathrm{Im}\ S_x^{\lvert y(x)\rvert e^{jv(x)}}}{v(x)}$
$c + x$	$\dfrac{x}{c + x}$	$y(x) + v(x)$	$\dfrac{y(x)S_x^{y(x)} + v(x)S_x^{v(x)}}{y(x) + v(x)}$
$\dfrac{1}{x}$	-1	$\dfrac{y(x)}{v(x)}$	$S_x^{y(x)} - S_x^{v(x)}$
c	0	$\lvert y(x)\rvert$	$\mathrm{Re}\ S_x^{y(x)}$
$x - c$	$\dfrac{x}{x - c}$	$y(x) - v(x)$	$\dfrac{y(x)S_x^{y(x)} - v(x)S_x^{v(x)}}{y(x) - v(x)}$
$\ln(x)$	$\dfrac{1}{\ln x}$	$e^{y(x)}$	$y(x)S_x^{y(x)}$
$\sqrt{x}$	$\frac{1}{2}$	$\ln(y(x))$	$\dfrac{1}{\ln(y(x))}S_x^{y(x)}$
$v^2(x)$	$2S_x^{v(x)}$	$y(v(x))$	$S_v^{y(v)} S_x^{v(x)}$

(4.44)

c is a constant, $y(x)$ and $v(x)$ are functions of x.

The *semirelative sensitivity* is defined as

$$\mathcal{S}_{x_i}^F = x_i \frac{\partial F}{\partial x_i} = F\, S_{x_i}^F \tag{4.45}$$

4.6.2 Sensitivity of Second-Order Transfer Function

Let us consider a second-order transfer function of the form

$$H(s) = \frac{1}{s^2 + \dfrac{\omega_p}{Q_p}s + \omega_p^2} \tag{4.46}$$

Our goal is to find the sensitivities of the magnitude response, $M(\omega) = |H(j\omega)|$, with respect to the pole magnitude, ω_p, and the pole Q-factor, Q_p.

The squared magnitude response is

$$M^2(\omega) = \frac{1}{\left(-\omega^2 + \omega_p^2\right)^2 + \left(\dfrac{\omega_p}{Q_p}\right)^2 \omega^2} \tag{4.47}$$

The sensitivity of the magnitude response to the pole magnitude is

$$S_{\omega_p}^{M(\omega)}(\omega) = -\frac{2\left(1-\dfrac{\omega^2}{\omega_p^2}\right)+\dfrac{\omega^2}{\omega_p^2 Q_p^2}}{\left(1-\dfrac{\omega^2}{\omega_p^2}\right)^2+\dfrac{\omega^2}{\omega_p^2 Q_p^2}} \tag{4.48}$$

and the sensitivity to the pole Q-factor is

$$S_{Q_p}^{M(\omega)}(\omega) = \frac{\dfrac{\omega^2}{\omega_p^2 Q_p^2}}{\left(1-\dfrac{\omega^2}{\omega_p^2}\right)^2+\dfrac{\omega^2}{\omega_p^2 Q_p^2}} \tag{4.49}$$

To gain a better insight into the order of magnitudes of $S_{\omega_p}^{M(\omega)}$ and $S_{Q_p}^{M(\omega)}$, we plot them against the normalized frequency ω/ω_p in Figs. 4.15 and 4.16.

The extremes of $S_{\omega_p}^{M(\omega)}(\omega)$ are

$$\begin{aligned} S_{\omega_p}^{M(\omega)}(\omega)\Big|_{\min} &\approx -Q_p \qquad \text{for} \quad \frac{\omega}{\omega_p} \approx 1-\frac{1}{2Q_p} \quad \text{and } Q_p \gg 1 \\ S_{\omega_p}^{M(\omega)}(\omega)\Big|_{\max} &\approx Q_p \qquad \text{for} \quad \frac{\omega}{\omega_p} \approx 1+\frac{1}{2Q_p} \quad \text{and } Q_p \gg 1 \end{aligned} \tag{4.50}$$

while the extreme of $S_{Q_p}^{M(\omega)}$ is

$$S_{Q_p}^{M(\omega)}(\omega)\Big|_{\max} = 1 \tag{4.51}$$

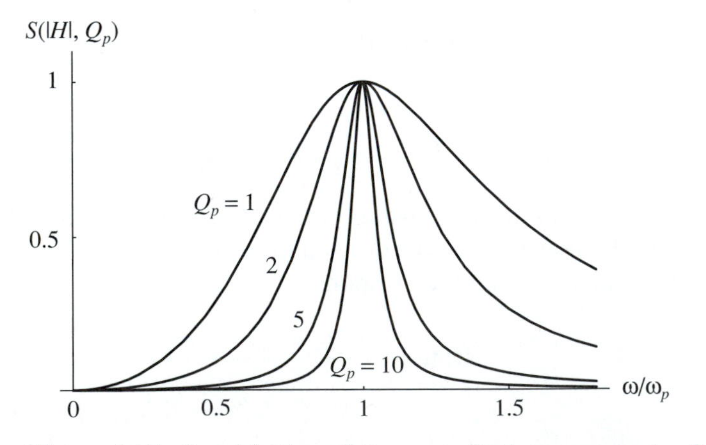

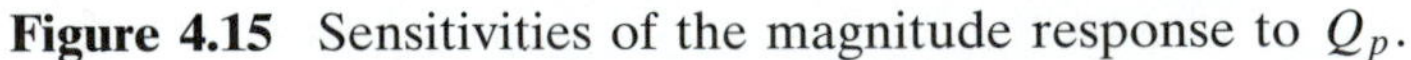
Figure 4.15 Sensitivities of the magnitude response to Q_p.

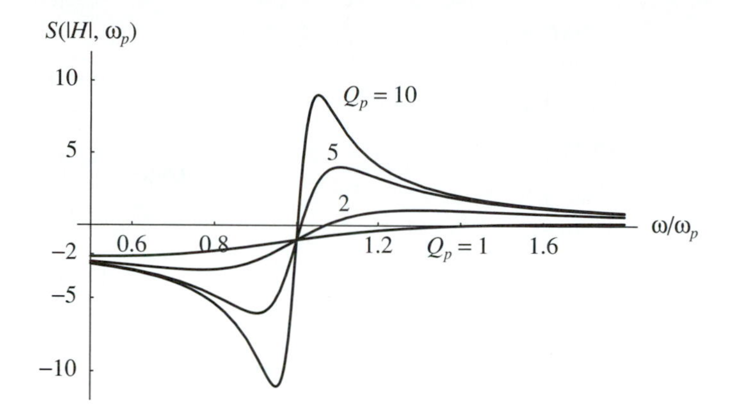

Figure 4.16 Sensitivities of the magnitude response to ω_p.

and is independent of Q_p. In the frequency range $0 \leq \omega \leq \omega_p$ the maximal value of the absolute sensitivity $\left|S_{\omega_p}^{M(\omega)}(\omega)\right|$ is Q_p times larger than $\left|S_{Q_p}^{M(\omega)}(\omega)\right|$:

$$\left|S_{\omega_p}^{M(\omega)}(\omega)\right|_{\max} \approx Q_p \left|S_{Q_p}^{M(\omega)}(\omega)\right|_{\max} \tag{4.52}$$

Summarizing the above analysis, we conclude that we can obtain the filter performance less sensitive to component deviations by decreasing Q_p and, also, by keeping the frequency $\omega_p\left(1 \pm 1/Q_p\right)$ away from the filter passband. Also, we have to pay more attention to the pole-magnitude sensitivity to element values than to the pole Q-factor sensitivity to component values.

The upper limit of the relative variation of the magnitude response can be expressed in the form

$$\left.\frac{\Delta M(\omega)}{M(\omega)}\right|_{\text{worst case}} = \sum_i \left|S_{\omega_p}^{M(\omega)} S_{x_i}^{\omega_p} \frac{\Delta x_i}{x_i}\right| + \sum_i \left|S_{Q_p}^{M(\omega)} S_{x_i}^{Q_p} \frac{\Delta x_i}{x_i}\right| \tag{4.53}$$

where x_i represents an element value of the chosen realization. The magnitude-response sensitivity is affected by

- the sensitivities $S_{\omega_p}^{M(\omega)}$, $S_{Q_p}^{M(\omega)}$ that depend on the transfer function—that is, on the approximation step;
- the sensitivities $S_{x_i}^{\omega_p}$, $S_{x_i}^{Q_p}$, determined by the chosen realization (circuit)—that is, by the realization step;
- the component tolerances $\dfrac{\Delta x_i}{x_i}$, imposed by the available technologies (the implementation step).

It can be shown that the gain deviation in dB, $\Delta G(\omega)$, is proportional to the magnitude response sensitivity:

$$\Delta G(\omega) \approx S_{x_i}^{G(\omega)} \frac{\Delta x_i}{x_i} = \frac{20}{\ln 10} S_{x_i}^{M(\omega)} \frac{\Delta x_i}{x_i} \approx 8.686\, S_{x_i}^{M(\omega)} \frac{\Delta x_i}{x_i} \tag{4.54}$$

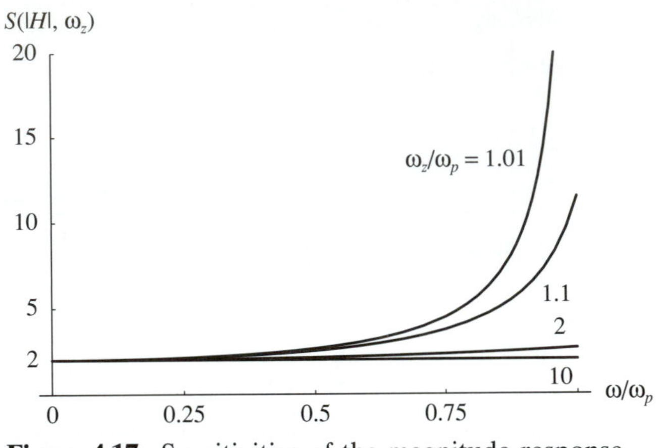

Figure 4.17 Sensitivities of the magnitude response to ω_z, for $\omega_z > \omega_p$.

where $G(\omega) = 20\log_{10} M(\omega)$, and x_i is an element value. Obviously, when optimizing a filter design it is irrelevant whether we optimize the gain sensitivity $S_{x_i}^{G(\omega)}$ or the magnitude response sensitivity $S_{x_i}^{M(\omega)}$.

Let us consider a second-order notch transfer function of the form

$$H(s) = \frac{s^2 + \omega_z^2}{s^2 + \dfrac{\omega_p}{Q_p}s + \omega_p^2} \tag{4.55}$$

The sensitivity of the magnitude response to the magnitude of the transfer function zero is

$$S_{\omega_z}^{M(\omega)}(\omega) = \frac{2}{1 - \dfrac{\omega^2}{\omega_p^2}\dfrac{\omega_p^2}{\omega_z^2}} \tag{4.56}$$

We plot the sensitivity $S_{\omega_z}^{M(\omega)}$ against the normalized frequency ω/ω_p in Fig. 4.17.

4.6.3 Sensitivity to Passive Components

Analog filter realizations consist of passive circuit elements, resistors, capacitors, and inductors. We define the *summed sensitivity* of a function $F(x_1, \ldots, x_n)$ to all elements of the same type $x_1, \ldots, x_n$ as

$$S_x^F = \sum_{i=1}^{n} S_{x_i}^F \tag{4.57}$$

where $S_{x_i}^F$ is defined by Eq. (4.38).

We mention without proof an important property that holds for all active RC filters with n_R resistors and n_C capacitors [28]:

$$\sum_{i=1}^{n_R} S_{R_i}^{Q_p} = \sum_{k=1}^{n_C} S_{C_k}^{Q_p} = 0 \tag{4.58}$$

$$\sum_{i=1}^{n_R} S_{R_i}^{\omega_p} = \sum_{k=1}^{n_C} S_{C_k}^{\omega_p} = -1 \tag{4.59}$$

where the summation is over all the resistors (capacitors).

In practice, the relative changes of elements of the same type can be $\frac{\Delta R_i}{R_i} = \frac{\Delta R}{R}$ and $\frac{\Delta C_k}{C_k} = \frac{\Delta C}{C}$. The relative variation of the magnitude response of an RC filter with $n = n_R + n_C$ passive elements becomes

$$\frac{\Delta M(\omega)}{M(\omega)} = S_{\omega_p}^{M(\omega)} \sum_{i=1}^{n} S_{x_i}^{\omega_p} \frac{\Delta x_i}{x_i} + S_{Q_p}^{M(\omega)} \sum_{i=1}^{n} S_{x_i}^{Q_p} \frac{\Delta x_i}{x_i} \tag{4.60}$$

$$\begin{aligned}\frac{\Delta M(\omega)}{M(\omega)} &= S_{\omega_p}^{M(\omega)} \left(\frac{\Delta R}{R} \sum_{i=1}^{n_R} S_{R_i}^{\omega_p} + \frac{\Delta C}{C} \sum_{k=1}^{n_C} S_{C_k}^{\omega_p} \right) \\ &+ S_{Q_p}^{M(\omega)} \left(\frac{\Delta R}{R} \sum_{i=1}^{n_R} S_{R_i}^{Q_p} + \frac{\Delta C}{C} \sum_{k=1}^{n_C} S_{C_k}^{Q_p} \right)\end{aligned} \tag{4.61}$$

According to Eq. (4.58) the second term is zero, and

$$\frac{\Delta M(\omega)}{M(\omega)} = S_{\omega_p}^{M(\omega)} \left(\frac{\Delta R}{R} \sum_{i=1}^{n_R} S_{R_i}^{\omega_p} + \frac{\Delta C}{C} \sum_{k=1}^{n_C} S_{C_k}^{\omega_p} \right) \tag{4.62}$$

For $\frac{\Delta R}{R} = -\frac{\Delta C}{C} = \frac{\Delta x}{x}$

$$\frac{\Delta M(\omega)}{M(\omega)} = S_{\omega_p}^{M(\omega)} \frac{\Delta x}{x} \left(\sum_{i=1}^{n_R} S_{R_i}^{\omega_p} - \sum_{k=1}^{n_C} S_{C_k}^{\omega_p} \right) \tag{4.63}$$

According to Eq. (4.59) the expression in parentheses is zero, and

$$\frac{\Delta M(\omega)}{M(\omega)} = 0 \tag{4.64}$$

which shows that the magnitude response of active RC filters can be made insensitive to the variation of passive components if $\frac{\Delta R_i}{R_i} = \frac{\Delta R}{R}$, $\frac{\Delta C_k}{C_k} = \frac{\Delta C}{C}$, and $\frac{\Delta R}{R} = -\frac{\Delta C}{C}$ hold.

The element values x, such as R_i or C_k, are spread about their nominal values x_0:

$$x = x_0(1 + \gamma_x) \tag{4.65}$$

where γ_x is a random number whose range is the tolerance x_t:

$$-\frac{x_t}{100} < \gamma_x < +\frac{x_t}{100} \tag{4.66}$$

The random number γ_x typically has a Gaussian distribution characterized by mean $\mu(\gamma_x)$ and standard deviation $\sigma(\gamma_x)$. In that case the tolerance corresponds to $3\sigma(\gamma_x)$—that is, $x_t = 300\sigma(\gamma_x)\%$.

The element value at temperature $T = T_0 + \Delta T$ is given by

$$x = x_0(1 + \alpha_x \Delta T) \tag{4.67}$$

where x_0 is the value at room temperature T_0 and α_x is the temperature coefficient of the component x. If a resistor has a temperature coefficient of 100 ppm/K (parts per million per kelvin), that is $\alpha_x = 100 \cdot 10^{-6}$ 1/K.

In a similar manner we can describe the effects of aging

$$x = x_0(1 + \beta_x \sqrt{t}) \tag{4.68}$$

where β_x is in ppm/yr. (parts per million per year) and t is time.

For small deviation from the nominal value the element value can be described as

$$x = x_0(1 + \gamma_x + \alpha_x \Delta T + \beta_x \sqrt{t}) \tag{4.69}$$

The relative deviation (variability) is

$$\frac{\Delta x}{x_0} = \gamma_x + \alpha_x \Delta T + \beta_x \sqrt{t} \tag{4.70}$$

the mean deviation is

$$\mu(\frac{\Delta x}{x_0}) \approx \mu(\gamma_x) + \mu(\alpha_x)\Delta T + \mu(\beta_x)\sqrt{t} \tag{4.71}$$

and the standard deviation per unit change in element value x is

$$\sigma\left(\frac{\Delta x}{x_0}\right) \approx \sqrt{(\sigma(\gamma_x))^2 + (\sigma(\alpha_x)\Delta T)^2 + (\sigma(\beta_x))^2 t} \tag{4.72}$$

4.6.4 Gain-Sensitivity Product (GSP)

Practical operational amplifiers (op amp) have finite and frequency-dependent gain that must be taken into account when designing active *RC* filters. Typically, the voltage gain, A, of an op amp can be approximated by

$$A(s) = A_0 \frac{1}{1 + \dfrac{s}{\omega_{3\text{dB}}}} = \frac{\omega_G}{s + \omega_{3\text{dB}}} \tag{4.73}$$

where A_0 is the DC gain ($A_0 \gg 1$), $\omega_{3\text{dB}}$ is the angular frequency at which $|A(j\omega_{3\text{dB}})| = A_0/\sqrt{2}$, and $\omega_G = A_0\omega_{3\text{dB}}$ is the gain–bandwidth product ($\omega_G \gg \omega_{3\text{dB}}$). Notice that at $s = j\omega_G$ we have

$$|A(j\omega_G)| \approx 1 \tag{4.74}$$

At DC and low frequencies we have

$$A(0) = A_0 = \frac{\omega_G}{\omega_{3dB}} \gg 1, \qquad \omega_G \gg \omega_{3dB} \tag{4.75}$$

At high frequencies we have

$$A(s) \approx A_0 \frac{\omega_{3dB}}{s} = \frac{\omega_G}{s}, \qquad \omega \gg \omega_{3dB} \tag{4.76}$$

In practice, we choose an op amp with a very large gain $|A(j\omega)|$ in the filter passband.

The finite op amp gain can produce considerable degradation of filter performances, so we have to study the effects of the op amp gain on the filter transfer function.

In order to quantify op amp imperfections, Moschytz [29] defines the *gain–sensitivity product* (GSP), Γ_A^F of a function F with respect to gain A:

$$\Gamma_A^F = A \; S_A^F \tag{4.77}$$

The main reason for using Γ_A^F instead of S_A^F is that S_A^F tends to be 0 for infinite values of A, and we cannot investigate the influence of A on filter performances. On the other hand, Γ_A^F is usually nonzero when $A \to \infty$, and the relative deviation of the function F is

$$\frac{\Delta F}{F} = \Gamma_A^F \frac{\Delta A}{A^2} \tag{4.78}$$

In this book we use the simplified figure-of-merit, I_A, recommended by reference 29, for the worst-case deviation of the magnitude response:

$$I_A = Q_p \Gamma_A^{\omega_p} + \frac{1}{2} \Gamma_A^{Q_p} \tag{4.79}$$

If a more complicated amplifier model than the single-pole model Eq. (4.73) is considered, then more complex relations for GSP and I_A have to be used as reported in reference 29. Fortunately, I_A is a good measure of performance for second-order active *RC* filters of practical importance.

For many second-order single op amp filters we have $\Gamma_A^{\omega_p} = 0$, and we can minimize the magnitude response sensitivity to A by optimizing $\Gamma_A^{Q_p}$ only.

4.7 ANALOG FILTER REALIZATIONS

After having accomplished the approximation step, the filter transfer function is known, and the designer must choose a realization—that is, an electric circuit. Analog filters can be classified on the basis of their constituent components as

- Passive *RLC* filters
- Operational amplifier *RC* filters (op amp active *RC* filters)
- Switched-capacitor (SC) filters
- Operational transconductance amplifier (OTA) filters

- Current-conveyor (CC) filters
- Microwave filters
- Electromechanical filters
- Crystal filters

Passive RLC filters consist of passive macrocomponents: resistors, capacitors, and inductors (coils and transformers). They do not require a power supply. The main drawbacks of passive filters are as follows: (a) They often exhibit a significant passband loss, and (b) the inductors cannot be miniaturized as required by modern applications. These filters are practical at frequencies up to a few hundred MHz. Passive *RLC* realizations are important in deriving realizations of some active filters, such as active *RC* filters, OTA filters, and current-conveyor filters.

Active RC filters, as well as *SC filters*, can be reduced in size and weight, especially when implemented as integrated circuits. Manufacturing of these filters can be automated with high production yield. They are made out of resistors, capacitors, switches and operational amplifiers. The disadvantages of active *RC* and SC filters are as follows: (a) They require a power supply, (b) the output signal can be distorted if the input signal is too large, and (c) an extra noise is generated in active devices. These filters operate over the frequency rage from 0.1 Hz up to 500 kHz.

Microwave filters are used for realization of filters above 300 MHz. They consist of microwave passive components, such as transmission line sections, coupled transmission lines, resonators, and cavities. They do not require a power supply.

Crystal filters are made of piezoelectric resonators. In filter applications, the Q-factor of a crystal resonator is very large, between 10^4 and 10^5, and the useful frequency range is from 10 kHz to 200 MHz. The relative bandwidth of a bandpass crystal filter using quartz resonators is very small, being less than 10^{-4}.

Electromechanical filters are made of mechanical resonators. The electrical signals are converted into mechanical vibrations and, after filtering, the resultant mechanical vibration is converted back to an electrical signal. The method of transforming electrical energy to mechanical energy and vice versa makes use of the magnetostrictive effect. Various mechanical solids can be made to vibrate in any one of many different modes such as longitudinal, extensional, torsional, and flexural. The Q-factor of the mechanical resonators can be 10^4. They can be used up to 200 MHz. The mechanical and piezoelectric resonators are based on the fact that at lower frequencies the exchange of the potential and kinetic energies in acoustical waves occurs much more efficiently than does a corresponding energy exchange in electromagnetic devices.

In this book we will focus on the realizations of passive *RLC* filters, op amp active *RC* filters, SC filters, OTA filters, and current-conveyor filters.

4.8 OP AMP ACTIVE *RC* FILTERS

Suppose that the filter transfer function has been determined and decomposed into second-order transfer functions. The filter can be realized as a cascade of second-order sections. Thus, the designer has to choose a realization for each biquadratic section. We adopt the classification of biquadratic realizations suggested by Moschytz and Horn [15 pp. 38–84].

The classification is based on the pole Q-factor:

- Low Q-factor realizations ($Q \leq 2$) exhibit, in general, no problems with tolerances, and there is no need for tuning. Therefore, the minimal number of passive components and operational amplifiers (one op amp per second-order section) has been chosen.
- Medium Q-factor realizations ($2 < Q \leq 20$) were selected on the basis of minimum gain-sensitivity product, simple tuning, minimal number of passive components, and only one operational amplifier.
- High Q-factor realizations ($20 < Q$) require two operational amplifiers; the sensitivities of these realizations are lower than the sensitivity of single-amplifier biquads.

We also present a general-purpose biquad with three amplifiers.

4.8.1 Low-Q-Factor Biquadratic Realizations

The most important feature of the presented low-Q-factor biquads is $\Gamma_A^{\omega_p} = 0$, which means that the pole magnitude is insensitive to the finite op amp gain.

Lowpass Low-Q-Factor Realization. The lowpass low-Q-factor realization is shown in Fig. 4.18 [15, pp. 38–39; 35].

The transfer function of the biquad from Fig. 4.18 is

$$H_{LP}(s) = \frac{V_4}{V_1} = K \frac{\omega_p^2}{s^2 + \dfrac{\omega_p}{Q_p} s + \omega_p^2} \tag{4.80}$$

The constant K is called the *gain constant* of the lowpass biquad and

$$K \leq 1 \tag{4.81}$$

The resistor and capacitor values are functions of the transfer function parameters K, Q_p, ω_p, the capacitance C_x, and an auxiliary quantity P. (P is resistnance ratio $P = R_3/R_1$). We set the value of C_x and P and compute the element values as

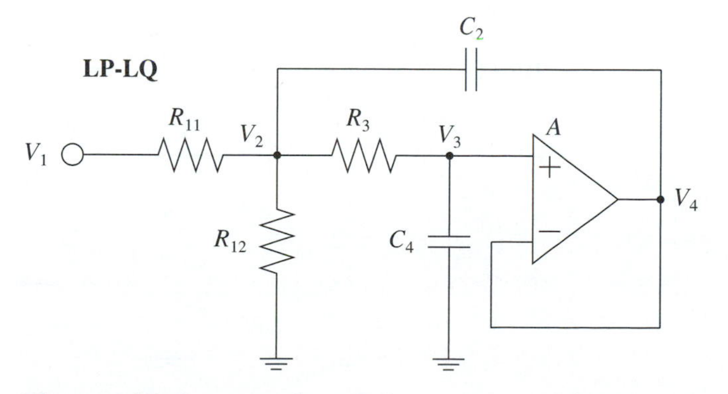

Figure 4.18 Lowpass low-Q-factor op amp biquad.

$$\boxed{\begin{aligned} R_1 &= \frac{1}{Q_p \omega_p C_x (1+P)} \\ R_{11} &= R_{11}(K, Q_p, \omega_p, C_x, P) = \frac{R_1}{K} \\ R_{12} &= R_{12}(K, Q_p, \omega_p, C_x, P) = \frac{R_1}{1-K} \\ C_2 &= C_2(K, Q_p, \omega_p, C_x, P) = Q_p^2 C_x \frac{(1+P)^2}{P} \\ R_3 &= R_3(K, Q_p, \omega_p, C_x, P) = P R_1 \\ C_4 &= C_4(K, Q_p, \omega_p, C_x, P) = C_x \end{aligned}} \tag{4.82}$$

Notice that P is the ratio of two resistances, $P = R_3/R_1$; these resistances affect the pole magnitude, $\omega_p = 1/(R_1 R_3 C_2 C_4)$. In practice, we prefer to choose P from the range $0.1 < P < 10$.

The intermediate quantity, R_1, is indented in (4.82).

If we can choose $K = 1$, then R_{12} becomes infinite and the resistor R_{12} degenerates to the open circuit (disappears from the schematic).

The quantity P can be used for optimizing an element value, or the element-value spread ratio. For example, we can vary P until C_2 gets a value from a prescribed set of values.

Highpass Low-Q-Factor Realization. The highpass low-Q-factor realization is shown in Fig. 4.19 [15, pp. 44–45; 35].

The transfer function of the biquad from Fig. 4.19 is

$$H_{HP}(s) = \frac{V_4}{V_1} = K \frac{s^2}{s^2 + \frac{\omega_p}{Q_p} s + \omega_p^2} \tag{4.83}$$

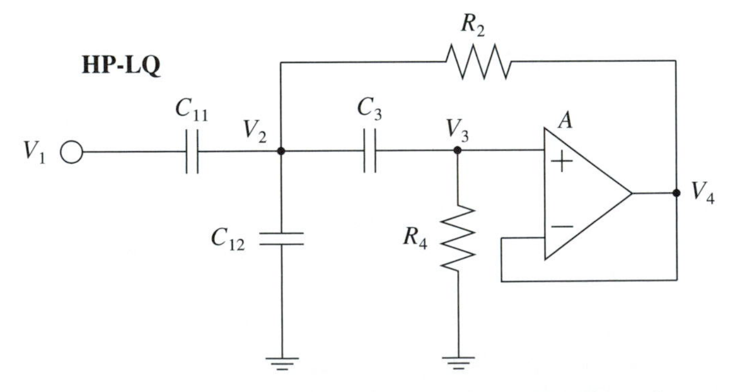

Figure 4.19 Highpass low-Q-factor op amp biquad.

The constant K is the gain constant of the highpass biquad and

$$K \leq 1 \tag{4.84}$$

The resistor and capacitor values are functions of the transfer function parameters K, Q_p, and ω_p, the capacitance C_x, and an auxiliary quantity P. (P is the resistance ratio $P = R_4/R_2$). We set the value of C_x and P and compute the element values as

$$\boxed{\begin{aligned} C_1 &= C_x \\ C_{11} &= C_{11}(K, Q_p, \omega_p, C_x, P) = KC_x \\ C_{12} &= C_{12}(K, Q_p, \omega_p, C_x, P) = C_x - C_{11} \\ C_3 &= C_3(K, Q_p, \omega_p, C_x, P) = C_x \frac{P - 2Q_p^2 - \sqrt{P^2 - 4PQ_p^2}}{2Q_p^2} \\ R_2 &= R_2(K, Q_p, \omega_p, C_x, P) = \frac{1}{Q_p\omega_p(C_1 + C_3)} \\ R_4 &= R_4(K, Q_p, \omega_p, C_x, P) = PR_2 \end{aligned}} \tag{4.85}$$

A good choice for P is $0.1 < P < 10$.

Bandpass Low-Q-Factor Realization. The bandpass low-Q-factor realization is shown in Fig. 4.20 [15, pp. 40–41; 36].

The transfer function of the biquad from Fig. 4.20 is

$$H_{BP}(s) = \frac{V_4}{V_1} = K \frac{\dfrac{\omega_p}{Q_p} s}{s^2 + \dfrac{\omega_p}{Q_p} s + \omega_p^2} \tag{4.86}$$

The constant K is the gain constant of the bandpass biquad.

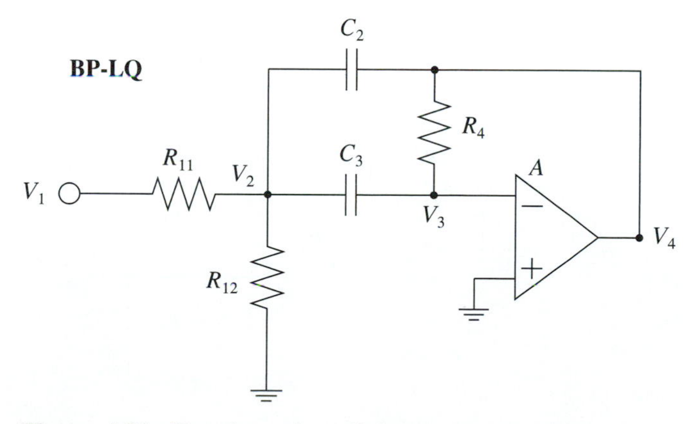

Figure 4.20 Bandpass low-Q-factor op amp biquad.

The resistor and capacitor values are functions of the transfer function parameters K, Q_p, and ω_p, the capacitance C_x, and an auxiliary quantity P. (P is resistance ratio $P = R_4/R_1$). We set the value of C_x and P and compute the element values as

$$
\boxed{
\begin{aligned}
C_2 &= C_2(K, Q_p, \omega_p, C_x, P) = C_x \\
C_3 &= C_3(K, Q_p, \omega_p, C_x, P) = C_x \frac{P - 2Q_p^2 - \sqrt{P^2 - 4PQ_p^2}}{2Q_p^2} \\
R_1 &= \frac{1}{Q_p\omega_p(C_2 + C_3)} \\
R_4 &= R_4(K, Q_p, \omega_p, C_x, P) = PR_1 \\
R_{11} &= R_{11}(K, Q_p, \omega_p, C_x, P) = \frac{C_3R_4}{K(C_2 + C_3)} \\
R_{12} &= R_{12}(K, Q_p, \omega_p, C_x, P) = \frac{C_3R_1R_4}{C_3R_4 - R_1K(C_2 + C_3)}
\end{aligned}
}
\tag{4.87}
$$

Notice that P is the ratio of two resistances, $P = R_4/R_1$; these resistances affect the pole magnitude, $\omega_p = 1/(R_1R_4C_2C_3)$, where $R_1 = 1/(1/R_{11} + 1/R_{12})$. Obviously,

$$
P \geq 4Q_p^2 \tag{4.88}
$$

For $P = 4Q_p^2$ the capacitors have the same value $C_2 = C_3$.

For $P = 4Q_p^2$, and $K = 2Q_p^2$, the resistance R_{12} becomes infinite; that is, R_{12} disappears from the schematic.

Bandreject Low-Q-Factor Realization. The bandreject low-Q-factor realization is shown in Fig. 4.21 [15, pp. 50–51; 37].

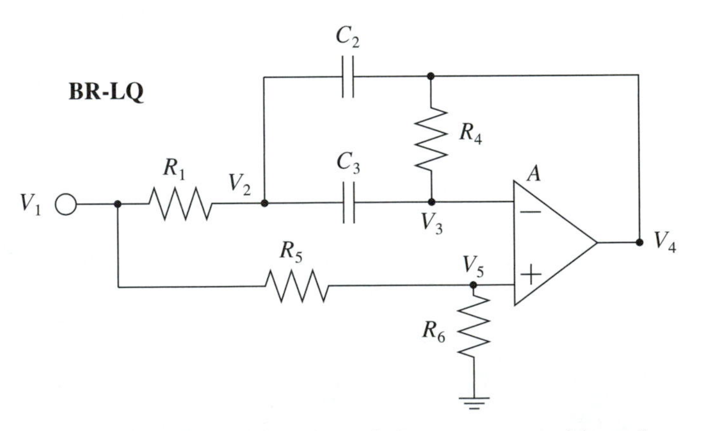

Figure 4.21 Bandreject low-Q-factor op amp biquad.

The transfer function of the biquad from Fig. 4.21 is

$$H_{BR}(s) = \frac{V_4}{V_1} = K \frac{s^2 + \omega_p^2}{s^2 + \dfrac{\omega_p}{Q_p}s + \omega_p^2} \tag{4.89}$$

The constant K is the gain constant of the bandreject biquad, and

$$K < 1 \tag{4.90}$$

The resistor and capacitor values are functions of the transfer function parameters K, Q_p, and ω_p, the capacitance C_x, and the resistance R_x. We set the value of C_x and R_x and compute the element values as

$$\begin{aligned}
C_2 &= C_2(K, Q_p, \omega_p, C_x, R_x) = C_x \\
P &= \frac{4}{\left(\dfrac{1}{K} - 1\right)\left(2 - \left(\dfrac{1}{K} - 1\right) Q_p^2\right)} \\
C_3 &= C_3(K, Q_p, \omega_p, C_x, P, R_x) = C_x \frac{P - 2Q_p^2 - \sqrt{P^2 - 4PQ_p^2}}{2Q_p^2} \\
R_1 &= R_1(K, Q_p, \omega_p, C_x, P, R_x) = \frac{1}{Q_p\omega_p(C_2 + C_3)} \\
R_4 &= R_4(K, Q_p, \omega_p, C_x, P, R_x) = PR_1 \\
R_5 &= R_5(K, Q_p, \omega_p, C_x, P, R_x) = R_x\left(\frac{1}{K} - 1\right) \\
R_6 &= R_6(K, Q_p, \omega_p, C_x, P, R_x) = R_x
\end{aligned} \tag{4.91}$$

It should be noticed that K and Q_p must satisfy

$$Q_p < \frac{1}{\sqrt{\dfrac{1}{K} - 1}} \tag{4.92}$$

and

$$\frac{1}{1 + \dfrac{1}{Q_p^2}} < K < 1 \tag{4.93}$$

to ensure positive element values.

If we choose

$$K = \frac{1}{1 + \dfrac{1}{2Q_p^2}} \tag{4.94}$$

we obtain

$$P \geq 4Q_p^2 \tag{4.95}$$

and the capacitor values become equal, $C_2 = C_3 = C_x$.

Notice that P is the ratio of two resistances, $P = R_4/R_1$; these resistances affect the pole magnitude, $\omega_p = 1/(R_1 R_4 C_2 C_3)$.

Allpass Low-Q-Factor Realization. The allpass low-Q-factor realization is shown in Fig. 4.22 [15, pp. 48–49; 37].

The transfer function of the biquad from Fig. 4.22 is

$$H_{AP}(s) = \frac{V_4}{V_1} = K\frac{s^2 - \dfrac{\omega_p}{Q_p}s + \omega_p^2}{s^2 + \dfrac{\omega_p}{Q_p}s + \omega_p^2} \tag{4.96}$$

The constant K is the gain constant of the allpass biquad.

The resistor and capacitor values are functions of the transfer function parameters K, Q_p, and ω_p, the capacitance C_x, and the resistance R_x. We set the value of C_x and

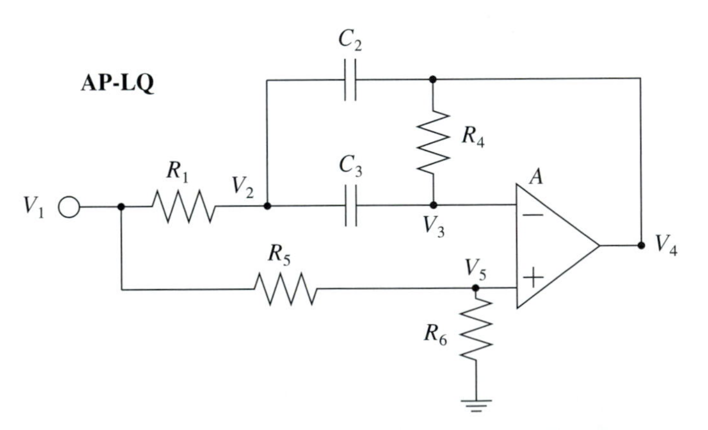

Figure 4.22 Allpass low-Q-factor op amp biquad.

R_x and compute the element values as

$$
\begin{aligned}
C_2 &= C_2(K, Q_p, \omega_p, C_x, R_x) = C_x \\
P &= \frac{4}{\left(\frac{1}{K} - 1\right)\left(2 - \left(\frac{1}{K} - 1\right) Q_p^2\right)} \\
C_3 &= C_3(K, Q_p, \omega_p, C_x, P, R_x) = C_x \frac{P - 2Q_p^2 - \sqrt{P^2 - 4PQ_p^2}}{2Q_p^2} \\
R_1 &= R_1(K, Q_p, \omega_p, C_x, P, R_x) = \frac{1}{Q_p \omega_p (C_2 + C_3)} \\
R_4 &= R_4(K, Q_p, \omega_p, C_x, P, R_x) = P R_1 \\
R_5 &= R_5(K, Q_p, \omega_p, C_x, P, R_x) = R_x \left(\frac{1}{K} - 1\right) \\
R_6 &= R_6(K, Q_p, \omega_p, C_x, P, R_x) = R_x
\end{aligned}
\tag{4.97}
$$

It should be noticed that K must satisfy

$$
\frac{1}{1 + \frac{2}{Q_p^2}} < K < 1 \tag{4.98}
$$

to ensure positive element values.

If we choose

$$
K = \frac{1}{1 + \frac{1}{Q_p^2}} \tag{4.99}
$$

the capacitor values become equal, $C_2 = C_3 = C_x$.

4.8.2 Medium-Q-Factor Biquadratic Realizations

The most important feature of the presented medium-Q-factor biquads, except for notch biquads, is $\Gamma_A^{\omega_p} = 0$, which means that the pole magnitude is insensitive to the finite op amp gain.

Lowpass Medium-Q-Factor Realization. The lowpass medium-Q-factor realization is shown in Fig. 4.23[15, pp. 52–53; 35].

The transfer function of the biquad from Fig. 4.23 is

$$
H_{LP}(s) = \frac{V_4}{V_1} = K \frac{\omega_p^2}{s^2 + \frac{\omega_p}{Q_p} s + \omega_p^2} \tag{4.100}
$$

The constant K is the gain constant of the lowpass biquad.

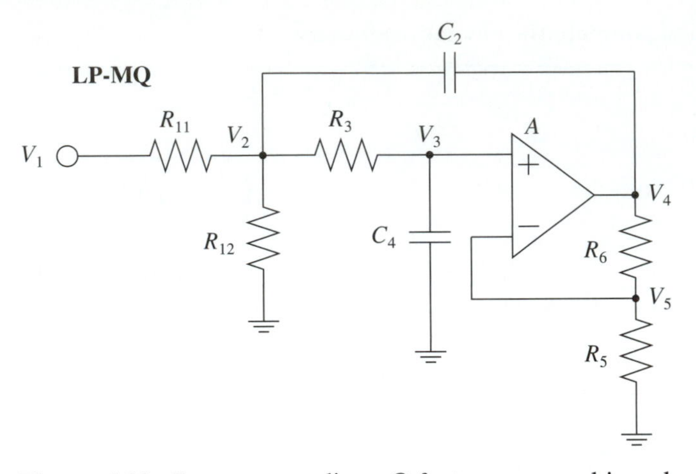

Figure 4.23 Lowpass medium-Q-factor op amp biquad.

The resistor and capacitor values are functions of the transfer function parameters K, Q_p, and ω_p, the capacitances C_{2x} and C_{4x}, the resistance R_x, and an auxiliary quantity P. (P is resistance ratio $P = R_3/R_1$). We set the value of C_{2x}, C_{4x}, R_x, and P and compute the element values as

$$
\begin{aligned}
C_2 &= C_2(K, Q_p, \omega_p, C_{2x}, C_{4x}, P, R_x) = C_{2x} \\
C_4 &= C_4(K, Q_p, \omega_p, C_{2x}, C_{4x}, P, R_x) = C_{4x} \\
R_1 &= \frac{1}{\omega_p \sqrt{C_{2x} C_{4x} P}} \\
R_3 &= R_3(K, Q_p, \omega_p, C_{2x}, C_{4x}, P, R_x) = P R_1 \\
R_5 &= R_5(K, Q_p, \omega_p, C_{2x}, C_{4x}, P, R_x) = R_x \\
R_6 &= R_6(K, Q_p, \omega_p, C_{2x}, C_{4x}, P, R_x) = R_x \left(\frac{C_4(1+P)}{C_2} - \frac{\sqrt{P \frac{C_4}{C_2}}}{Q_p} \right) \\
K_0 &= 1 + \frac{R_6}{R_x} \\
R_{11} &= R_{11}(K, Q_p, \omega_p, C_{2x}, C_{4x}, P, R_x) = \frac{R_1 K_0}{K} \\
R_{12} &= R_{12}(K, Q_p, \omega_p, C_{2x}, C_{4x}, P, R_x) = \frac{R_1 K_0}{K_0 - K}
\end{aligned}
\tag{4.101}
$$

The gain constant K can be larger than 1.

Highpass Medium-Q-Factor Realization. The highpass medium-Q-factor realization is shown in Fig. 4.24 [15, pp. 58–59; 35].

The transfer function of the biquad from Fig. 4.24 is

$$H_{HP}(s) = \frac{V_4}{V_1} = K\frac{s^2}{s^2 + \dfrac{\omega_p}{Q_p}s + \omega_p^2} \tag{4.102}$$

The constant K is the gain constant of the highpass biquad.

The resistor and capacitor values are functions of the transfer function parameters K, Q_p, and ω_p, the capacitances C_{1x} and C_{3x}, the resistance R_x, and an auxiliary quantity P. (P is resistance ratio $P = R_4/R_2$). We set the value of C_{1x}, C_{3x}, R_x, and P and compute the element values as

$$\boxed{\begin{aligned}
C_1 &= C_{1x} \\
C_3 &= C_3(K, Q_p, \omega_p, C_{1x}, C_{3x}, P, R_x) = C_{3x} \\
R_2 &= R_2(K, Q_p, \omega_p, C_{1x}, C_{3x}, P, R_x) = \frac{1}{\omega_p\sqrt{C_{1x}C_{3x}P}} \\
R_4 &= R_4(K, Q_p, \omega_p, C_{1x}, C_{3x}, P, R_x) = PR_2 \\
R_5 &= R_5(K, Q_p, \omega_p, C_{1x}, C_{3x}, P, R_x) = R_x \\
R_6 &= R_6(K, Q_p, \omega_p, C_{1x}, C_{3x}, P, R_x) = R_x\left(\frac{1 + \dfrac{C_1}{C_3}}{P} - \frac{\sqrt{\dfrac{C_1}{PC_3}}}{Q_p}\right) \\
K_0 &= 1 + \frac{R_6}{R_x} \\
C_{11} &= C_{11}(K, Q_p, \omega_p, C_{1x}, C_{3x}, P, R_x) = \frac{C_1K}{K_0} \\
C_{12} &= C_{12}(K, Q_p, \omega_p, C_{1x}, C_{3x}, P, R_x) = C_1 - C_{11}
\end{aligned}} \tag{4.103}$$

The gain constant K can be larger than 1.

Bandpass Medium-Q-Factor Realization. The bandpass medium-Q-factor realization is shown in Fig. 4.25 [15, pp. 54–55; 38].

The transfer function of the biquad from Fig. 4.25 is

$$H_{BP}(s) = \frac{V_4}{V_1} = K\frac{\dfrac{\omega_p}{Q_p}s}{s^2 + \dfrac{\omega_p}{Q_p}s + \omega_p^2} \tag{4.104}$$

The constant K is the gain constant of the bandpass biquad.

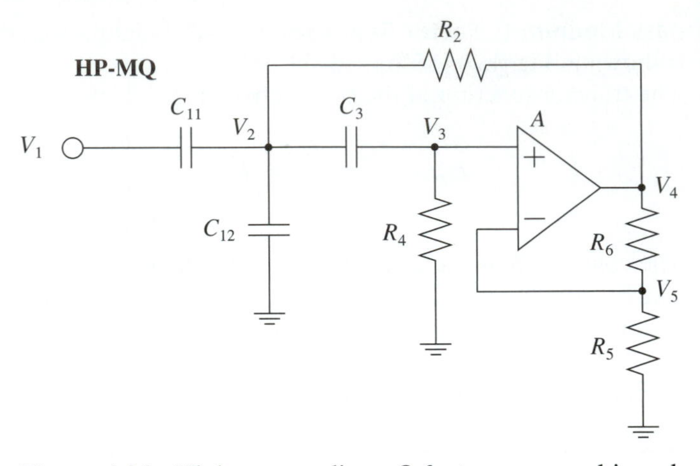

Figure 4.24 Highpass medium-Q-factor op amp biquad.

The resistor and capacitor values are functions of the transfer function parameters K, Q_p, and ω_p, the capacitances C_{2x} and C_{3x}, the resistance R_x, and an auxiliary quantity P. (P is resistance ratio $P = R_4/R_1$). We set the value of C_{2x}, C_{3x}, R_x, and P and compute the element values as

$$
\boxed{
\begin{aligned}
C_2 &= C_2(K, Q_p, \omega_p, C_{2x}, C_{3x}, P, R_x) = C_{2x} \\
C_3 &= C_3(K, Q_p, \omega_p, C_{2x}, C_{3x}, P, R_x) = C_{3x} \\
R_1 &= \frac{1}{\omega_p \sqrt{C_{2x} C_{3x} P}} \\
R_4 &= R_4(K, Q_p, \omega_p, C_{2x}, C_{3x}, P, R_x) = P R_1 \\
R_6 &= R_6(K, Q_p, \omega_p, C_{2x}, C_{3x}, P, R_x) = R_x \\
R_5 &= R_5(K, Q_p, \omega_p, C_{2x}, C_{3x}, P, R_x) = R_x \left(\frac{1 + \dfrac{C_2}{C_3}}{P} - \frac{\sqrt{\dfrac{C_2}{P C_3}}}{Q_p} \right) \\
K_0 &= Q_p \left(1 + \frac{R_5}{R_x} \right) \sqrt{\frac{P C_3}{C_2}} \\
R_{11} &= R_{11}(K, Q_p, \omega_p, C_{2x}, C_{3x}, P, R_x) = \frac{R_1 K_0}{K} \\
R_{12} &= R_{12}(K, Q_p, \omega_p, C_{2x}, C_{3x}, P, R_x) = \frac{R_1 K_0}{K_0 - K}
\end{aligned}
}
\quad (4.105)
$$

The gain constant K can be larger than 1.

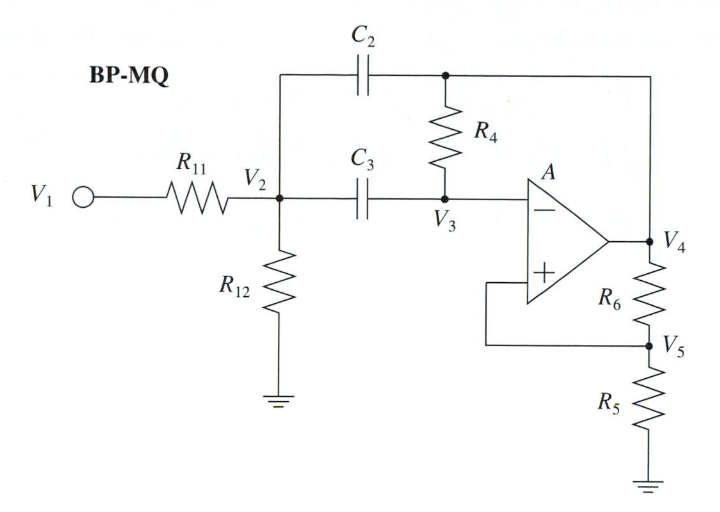

Figure 4.25 Bandpass medium-Q-factor op amp biquad.

Bandreject Medium-Q-Factor Realization. The bandreject medium-Q-factor realization is shown in Fig. 4.26 [15, pp. 62–63; 29].

The transfer function of the biquad from Fig. 4.26 is

$$H_{BR}(s) = \frac{V_4}{V_1} = \frac{s^2 + \omega_p^2}{s^2 + \dfrac{\omega_p}{Q_p}s + \omega_p^2} \tag{4.106}$$

The resistor and capacitor values are functions of the transfer function parameters Q_p and ω_p, the capacitances C_{2x} and C_{3x}, the resistance R_x, and an auxiliary quantity

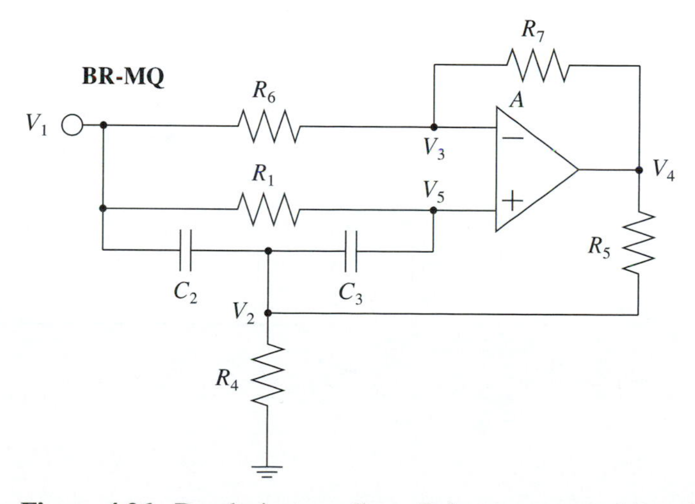

Figure 4.26 Bandreject medium-Q-factor op amp biquad.

P. We set the value of C_{2x}, C_{3x}, R_x, and P and compute the element values as

$$
\begin{aligned}
C_2 &= C_2(Q_p, \omega_p, C_{2x}, C_{3x}, P, R_x) = C_{2x} \\
C_3 &= C_3(Q_p, \omega_p, C_{2x}, C_{3x}, P, R_x) = C_{3x} \\
R_1 &= R_1(Q_p, \omega_p, C_{2x}, C_{3x}, P, R_x) = \frac{1}{\omega_p \sqrt{C_{2x} C_{3x} P}} \\
R_p &= P R_1 \\
R_6 &= R_6(Q_p, \omega_p, C_{2x}, C_{3x}, P, R_x) = R_x \\
R_7 &= R_7(Q_p, \omega_p, C_{2x}, C_{3x}, P, R_x) = R_x P\left(1 + \frac{C_2}{C_3}\right) \\
a &= 1 - \frac{\sqrt{\dfrac{P C_2}{C_3}}}{Q_p\left(1 + \dfrac{R_7}{R_x}\right)} \\
R_5 &= R_5(Q_p, \omega_p, C_{2x}, C_{3x}, P, R_x) = \frac{R_p}{a} \\
R_4 &= R_4(Q_p, \omega_p, C_{2x}, C_{3x}, P, R_x) = \frac{R_p}{1 - a}
\end{aligned}
\tag{4.107}
$$

Notice that the gain constant is 1.

Allpass Medium-Q-Factor Realization. The allpass medium-Q-factor realization is shown in Fig. 4.27 [15, pp. 60–61; 29].

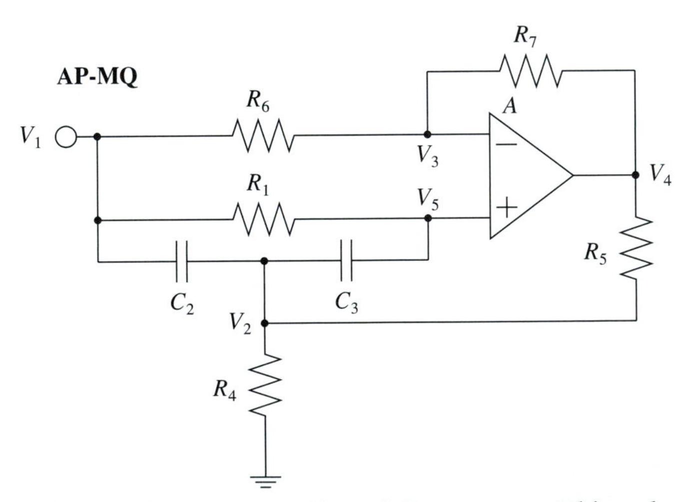

Figure 4.27 Allpass medium-Q-factor op amp biquad.

The transfer function of the biquad from Fig. 4.27 is

$$H_{AP}(s) = \frac{V_4}{V_1} = \frac{s^2 - \dfrac{\omega_p}{Q_p}s + \omega_p^2}{s^2 + \dfrac{\omega_p}{Q_p}s + \omega_p^2} \tag{4.108}$$

The resistor and capacitor values are functions of the transfer function parameters Q_p and ω_p, the capacitances C_{2x} and C_{3x}, the resistance R_x, and an auxiliary quantity P. We set the value of C_{2x}, C_{3x}, R_x, and P and compute the element values as

$$\begin{aligned}
C_2 &= C_2(Q_p, \omega_p, C_{2x}, C_{3x}, P, R_x) = C_{2x} \\
C_3 &= C_3(Q_p, \omega_p, C_{2x}, C_{3x}, P, R_x) = C_{3x} \\
R_1 &= R_1(Q_p, \omega_p, C_{2x}, C_{3x}, P, R_x) = \frac{1}{\omega_p\sqrt{C_{2x}C_{3x}P}} \\
R_p &= PR_1 \\
R_6 &= R_6(Q_p, \omega_p, C_{2x}, C_{3x}, P, R_x) = R_x \\
R_7 &= R_7(Q_p, \omega_p, C_{2x}, C_{3x}, P, R_x) = R_x\left(P\left(1 + \frac{C_2}{C_3}\right) + \frac{\sqrt{P\dfrac{C_2}{C_3}}}{Q_p}\right) \\
a &= 1 - \frac{2\sqrt{\dfrac{PC_2}{C_3}}}{Q_p\left(1 + \dfrac{R_7}{R_x}\right)} \\
R_5 &= R_5(Q_p, \omega_p, C_{2x}, C_{3x}, P, R_x) = \frac{R_p}{a} \\
R_4 &= R_4(Q_p, \omega_p, C_{2x}, C_{3x}, P, R_x) = \frac{R_p}{1 - a}
\end{aligned} \tag{4.109}$$

Notice that the gain constant is 1.

Lowpass Notch Medium-Q-Factor Realization. The lowpass notch medium-Q-factor realization is shown in Fig. 4.28 [15, pp. 64–65; 39].

The transfer function of the biquad from Fig. 4.28 is

$$H_{LPN}(s) = \frac{V_4}{V_1} = K\frac{s^2 + \omega_z^2}{s^2 + \dfrac{\omega_p}{Q_p}s + \omega_p^2}, \qquad \omega_z > \omega_p \tag{4.110}$$

The constant K is the gain constant of the notch biquad.

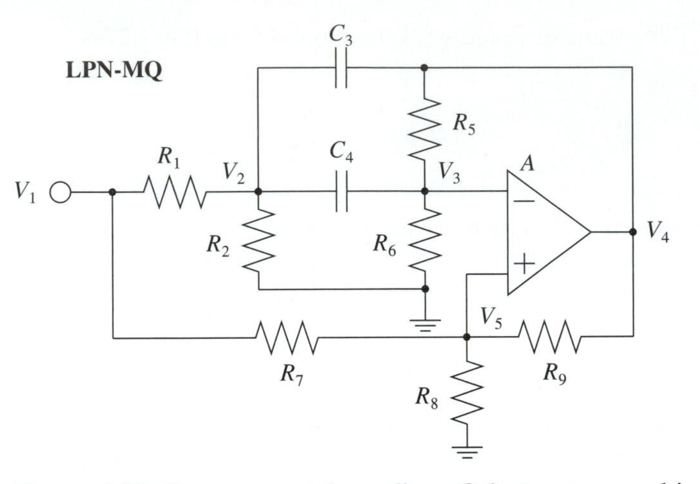

Figure 4.28 Lowpass notch medium-Q-factor op amp biquad.

The resistor and capacitor values are functions of the transfer function parameters K, Q_p, ω_p, and ω_z, the capacitances C_{3x} and C_{4x}, the resistance R_x, and an auxiliary quantity P. We set the value of C_{3x}, C_{4x}, R_x, and P and compute the element values as

$$
\begin{aligned}
C_3 &= C_3(K, Q_p, \omega_p, \omega_z, P, C_{3x}, C_{4x}, R_x) = C_{3x} \\
C_4 &= C_4(K, Q_p, \omega_p, \omega_z, P, C_{3x}, C_{4x}, R_x) = C_{4x} \\
G &= \frac{C_3\omega_p}{2PQ_p}\left(\sqrt{1 + 4Q_p^2 P\left(1 + \frac{C_4}{C_3}\right)} - 1\right) \\
K_0 &= \frac{1 + P}{1 + \left(1 + \frac{C_4}{C_3}\right)\omega_z^2 \frac{C_3^2}{G^2}} \\
R_1 &= R_1(K, Q_p, \omega_p, \omega_z, P, C_{3x}, C_{4x}, R_x) = \frac{K_0}{KG} \\
R_2 &= R_2(K, Q_p, \omega_p, \omega_z, P, C_{3x}, C_{4x}, R_x) = \frac{K_0}{G(K_0 - K)} \\
R_6 &= R_6(K, Q_p, \omega_p, \omega_z, P, C_{3x}, C_{4x}, R_x) = \frac{G(1 + P)}{C_3C_4(\omega_z^2 - \omega_p^2)} \\
R_5 &= R_5(K, Q_p, \omega_p, \omega_z, P, C_{3x}, C_{4x}, R_x) = \frac{1}{\frac{C_3C_4\omega_p^2}{G} + \frac{P}{R_6}} \\
R_7 &= R_7(K, Q_p, \omega_p, \omega_z, P, C_{3x}, C_{4x}, R_x) = \frac{PR_x}{K} \\
R_8 &= R_8(K, Q_p, \omega_p, \omega_z, P, C_{3x}, C_{4x}, R_x) = \frac{PR_x}{1 - K} \\
R_9 &= R_9(K, Q_p, \omega_p, \omega_z, P, C_{3x}, C_{4x}, R_x) = R_x
\end{aligned}
\tag{4.111}
$$

The gain constant K can be greater than 1.

Highpass Notch Medium-Q Factor Realization. The highpass notch medium-Q-factor realization is shown in Fig. 4.29 [15, pp. 64–65; 39].

The transfer function of the biquad from Fig. 4.29 is

$$H_{HPN}(s) = \frac{V_4}{V_1} = K\frac{s^2 + \omega_z^2}{s^2 + \dfrac{\omega_p}{Q_p}s + \omega_p^2}, \qquad \omega_z < \omega_p \tag{4.112}$$

The constant K is the gain constant of the notch biquad.

The resistor and capacitor values are functions of the transfer function parameters K, Q_p, ω_p, and ω_z, the capacitances C_{3x}, and C_{4x}, the resistance R_x, and an auxiliary quantity P. We set the value of C_{3x}, C_{4x}, R_x, and P and compute the element values as

$$\boxed{\begin{aligned}
C_3 &= C_3(K, Q_p, \omega_p, \omega_z, P, C_{3x}, C_{4x}, R_x) = C_{3x}\\
C_4 &= C_4(K, Q_p, \omega_p, \omega_z, P, C_{3x}, C_{4x}, R_x) = C_{4x}\\
G &= \frac{C_3\omega_p}{2PQ_p}\left(\sqrt{1 + 4Q_p^2P\left(1 + \frac{C_4}{C_3}\right)} - 1\right)\\
K_0 &= \frac{1 + P}{1 + \left(1 + \dfrac{C_4}{C_3}\right)\omega_z^2\dfrac{C_3^2}{G^2}}\\
R_1 &= R_1(K, Q_p, \omega_p, \omega_z, P, C_{3x}, C_{4x}, R_x) = \frac{K_0}{KG}\\
R_2 &= R_2(K, Q_p, \omega_p, \omega_z, P, C_{3x}, C_{4x}, R_x) = \frac{K_0}{G(K_0 - K)}\\
R_6 &= R_6(K, Q_p, \omega_p, \omega_z, P, C_{3x}, C_{4x}, R_x) = \frac{G(1 + P)(1 - \frac{1}{K})}{C_3C_4(\omega_z^2 - \omega_p^2)}\\
R_5 &= R_5(K, Q_p, \omega_p, \omega_z, P, C_{3x}, C_{4x}, R_x) = \frac{1}{\dfrac{C_3C_4\omega_p^2}{G} + \dfrac{P}{R_6}}\\
R_7 &= R_7(K, Q_p, \omega_p, \omega_z, P, C_{3x}, C_{4x}, R_x) = \frac{PR_x}{K}\\
R_8 &= R_8(K, Q_p, \omega_p, \omega_z, P, C_{3x}, C_{4x}, R_x) = \frac{PR_x}{1 - K}\\
R_9 &= R_9(K, Q_p, \omega_p, \omega_z, P, C_{3x}, C_{4x}, R_x) = R_x
\end{aligned}} \tag{4.113}$$

The gain constant K can be greater than 1.

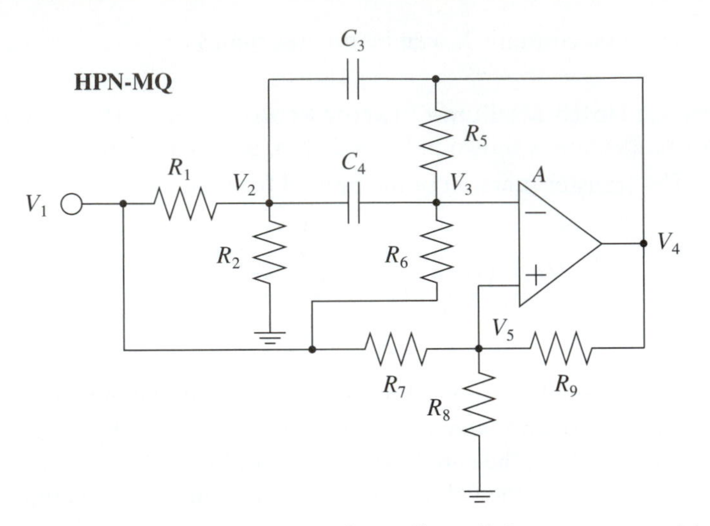

Figure 4.29 Highpass notch medium-Q-factor op amp biquad.

4.8.3 High-Q-Factor Biquadratic Realizations

High-Q-factor realizations require two operational amplifiers. The sensitivities of these realizations are lower than the sensitivity of single-amplifier biquads. The gain constant is equal to 1 for all presented realizations.

Lowpass High-Q-Factor Realization. The lowpass high-Q-factor realization is shown in Fig. 4.30 [15, pp. 68–69; 40].

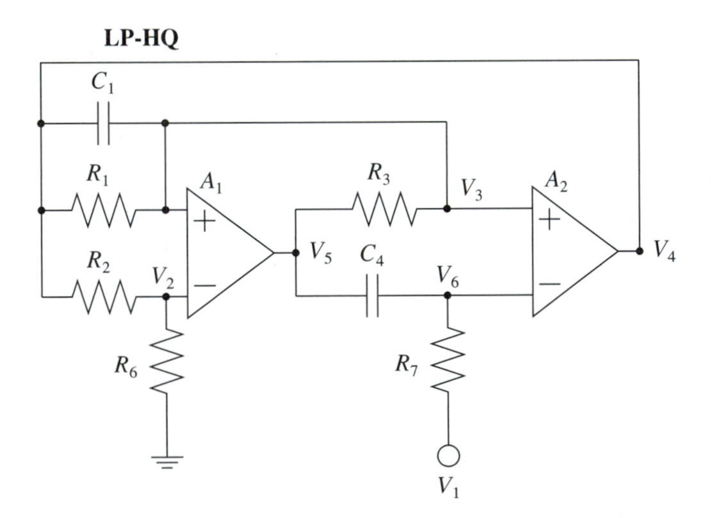

Figure 4.30 Lowpass high-Q-factor op amp biquad.

The transfer function of the biquad from Fig. 4.30 is

$$H_{LP}(s) = \frac{V_5}{V_1} = \frac{\omega_p^2}{s^2 + \dfrac{\omega_p}{Q_p}s + \omega_p^2} \tag{4.114}$$

The resistor and capacitor values are functions of the transfer function parameters Q_p and ω_p, the capacitance C_x, and the resistance R_x. We set the value of C_x, and R_x and compute the element values as

$$\boxed{\begin{aligned} C_1 &= C_1(Q_p, \omega_p, C_x, R_x) = C_x \\ C_4 &= C_4(Q_p, \omega_p, C_x, R_x) = C_x \\ R_0 &= \frac{1}{\omega_p C_x} \\ R_2 &= R_2(Q_p, \omega_p, C_x, R_x) = R_x \\ R_3 &= R_3(Q_p, \omega_p, C_x, R_x) = R_x \\ R_6 &= R_6(Q_p, \omega_p, C_x, R_x) = R_x \\ R_1 &= R_1(Q_p, \omega_p, C_x, R_x) = Q_p R_0 \\ R_7 &= R_7(Q_p, \omega_p, C_x, R_x) = \frac{R_0^2}{R_x} \end{aligned}} \tag{4.115}$$

Highpass High-Q-Factor Realization. The highpass high-Q-factor realization is shown in Fig. 4.31 [15, pp. 72–73; 40, 41].

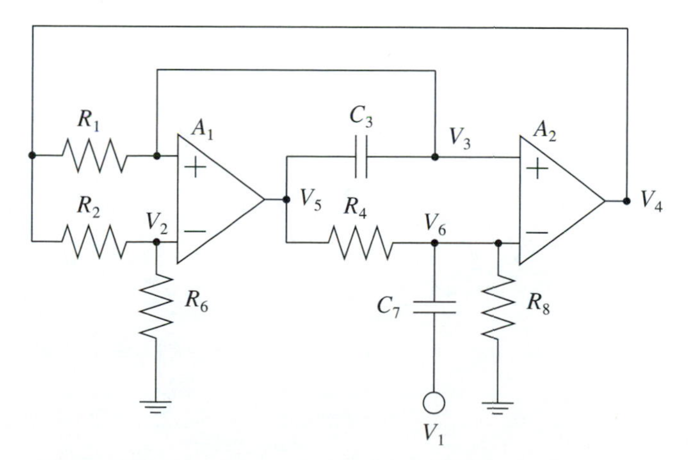

Figure 4.31 Highpass high-Q-factor op amp biquad.

The transfer function of the biquad from Fig. 4.31 is

$$H_{HP}(s) = \frac{V_4}{V_1} = \frac{s^2}{s^2 + \dfrac{\omega_p}{Q_p}s + \omega_p^2} \tag{4.116}$$

The resistor and capacitor values are functions of the transfer function parameters Q_p and ω_p, the capacitance C_x, and the resistance R_x. We set the value of C_x and R_x and compute the element values as

$$\boxed{\begin{aligned} C_3 &= C_3(Q_p, \omega_p, C_x, R_x) = C_x \\ C_7 &= C_7(Q_p, \omega_p, C_x, R_x) = C_x \\ R_0 &= \frac{1}{\omega_p C_x} \\ R_1 &= R_1(Q_p, \omega_p, C_x, R_x) = R_x \\ R_2 &= R_2(Q_p, \omega_p, C_x, R_x) = R_x \\ R_6 &= R_6(Q_p, \omega_p, C_x, R_x) = R_x \\ R_8 &= R_8(Q_p, \omega_p, C_x, R_x) = Q_p R_0 \\ R_4 &= R_4(Q_p, \omega_p, C_x, R_x) = \frac{R_0^2}{R_x} \end{aligned}} \tag{4.117}$$

Bandpass High-Q-Factor Realization. The bandpass high-Q-factor realization is shown in Fig. 4.32 [15, pp. 70–71; 40, 41].

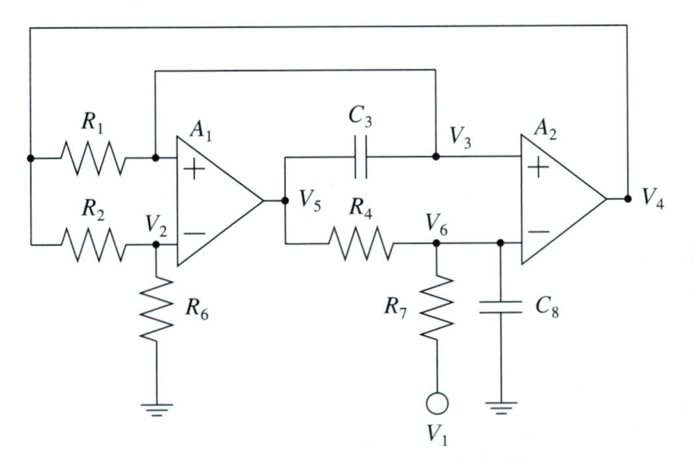

Figure 4.32 Bandpass high-Q-factor op amp biquad.

The transfer function of the biquad from Fig. 4.32 is

$$H_{BP}(s) = \frac{V_4}{V_1} = \frac{\dfrac{\omega_p}{Q_p} s}{s^2 + \dfrac{\omega_p}{Q_p} s + \omega_p^2} \tag{4.118}$$

The resistor and capacitor values are functions of the transfer function parameters Q_p and ω_p, the capacitance C_x, and the resistance R_x. We set the value of C_x and R_x and compute the element values as

$$\boxed{\begin{aligned} C_3 &= C_3(Q_p, \omega_p, C_x, R_x) = C_x \\ C_8 &= C_8(Q_p, \omega_p, C_x, R_x) = C_x \\ R_0 &= \frac{1}{\omega_p C_x} \\ R_1 &= R_1(Q_p, \omega_p, C_x, R_x) = R_x \\ R_2 &= R_2(Q_p, \omega_p, C_x, R_x) = R_x \\ R_6 &= R_6(Q_p, \omega_p, C_x, R_x) = R_x \\ R_7 &= R_7(Q_p, \omega_p, C_x, R_x) = Q_p R_0 \\ R_4 &= R_4(Q_p, \omega_p, C_x, R_x) = \frac{R_0^2}{R_x} \end{aligned}} \tag{4.119}$$

Bandreject High-Q-Factor Realization. The bandreject high-Q-factor realization is shown in Fig. 4.33 [15, pp. 76–77; 41].

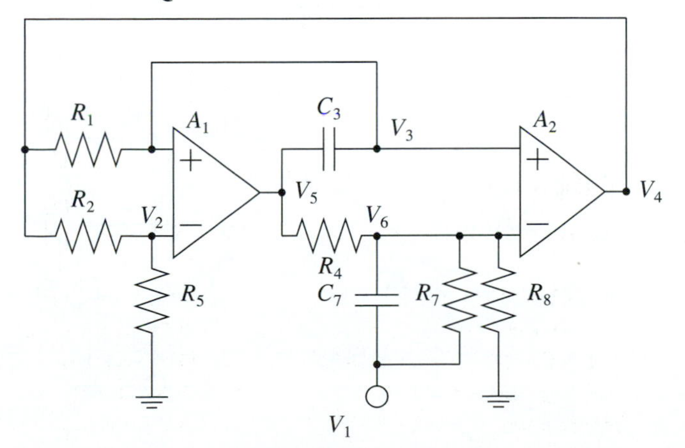

Figure 4.33 Bandreject high-Q-factor op amp biquad.

The transfer function of the biquad from Fig. 4.33 is

$$H_{BR}(s) = \frac{V_4}{V_1} = \frac{s^2 + \omega_p^2}{s^2 + \dfrac{\omega_p}{Q_p}s + \omega_p^2} \tag{4.120}$$

The resistor and capacitor values are functions of the transfer function parameters Q_p and ω_p, the capacitance C_x, and the resistance R_x. We set the value of C_x and R_x and compute the element values as

$$\boxed{\begin{aligned}
C_3 &= C_3(Q_p, \omega_p, C_x, R_x) = C_x \\
C_7 &= C_7(Q_p, \omega_p, C_x, R_x) = C_x \\
R_0 &= \frac{1}{\omega_p C_x} \\
R_1 &= R_1(Q_p, \omega_p, C_x, R_x) = R_x \\
R_2 &= R_2(Q_p, \omega_p, C_x, R_x) = R_x \\
R_5 &= R_5(Q_p, \omega_p, C_x, R_x) = R_x \\
R_7 &= R_7(Q_p, \omega_p, C_x, R_x) = 2Q_p R_0 \\
R_8 &= R_8(Q_p, \omega_p, C_x, R_x) = 2Q_p R_0 \\
R_4 &= R_4(Q_p, \omega_p, C_x, R_x) = \frac{R_0^2}{R_x}
\end{aligned}} \tag{4.121}$$

Allpass High-Q-Factor Realization. The allpass high-Q-factor realization is known in Fig. 4.34 [15, pp. 74–75; 40, 41].

AP-HQ

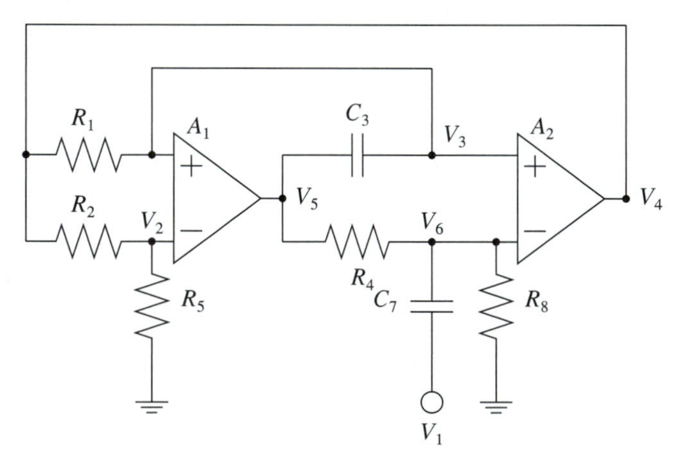

Figure 4.34 Allpass high-Q-factor op amp biquad.

The transfer function of the biquad from Fig. 4.34 is

$$H_{AP}(s) = \frac{V_4}{V_1} = \frac{s^2 - \dfrac{\omega_p}{Q_p}s + \omega_p^2}{s^2 + \dfrac{\omega_p}{Q_p}s + \omega_p^2} \tag{4.122}$$

The resistor and capacitor values are functions of the transfer function parameters Q_p and ω_p, the capacitance C_x, and the resistance R_x. We set the value of C_x and R_x and compute the element values as

$$\boxed{\begin{aligned} C_3 &= C_3(Q_p, \omega_p, C_x, R_x) = C_x \\ C_7 &= C_7(Q_p, \omega_p, C_x, R_x) = C_x \\ R_0 &= \frac{1}{\omega_p C_x} \\ R_1 &= R_1(Q_p, \omega_p, C_x, R_x) = R_x \\ R_2 &= R_2(Q_p, \omega_p, C_x, R_x) = R_x \\ R_5 &= R_5(Q_p, \omega_p, C_x, R_x) = R_x \\ R_8 &= R_8(Q_p, \omega_p, C_x, R_x) = Q_p R_0 \\ R_4 &= R_4(Q_p, \omega_p, C_x, R_x) = \frac{R_0^2}{R_x} \end{aligned}} \tag{4.123}$$

Lowpass Notch High-Q-Factor Realization. The lowpass notch high-Q-factor realization is shown in Fig. 4.35 [15, pp. 78–79; 41].

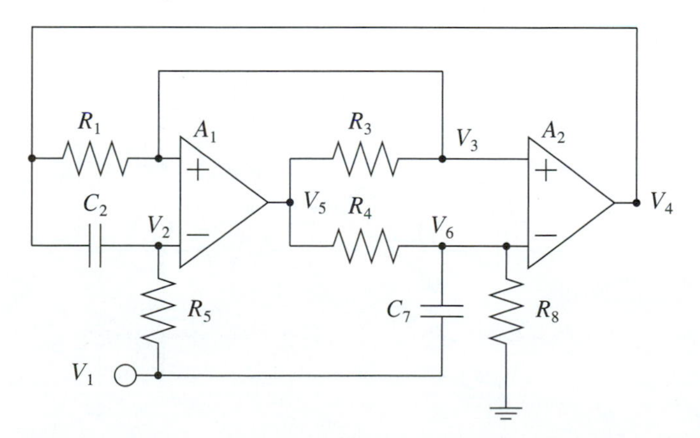

Figure 4.35 Lowpass notch high-Q-factor op amp biquad.

The transfer function of the biquad from Fig. 4.35 is

$$H_{LPN}(s) = \frac{V_5}{V_1} = \frac{s^2 + \omega_z^2}{s^2 + \dfrac{\omega_p}{Q_p}s + \omega_p^2}, \qquad \omega_z > \omega_p \tag{4.124}$$

The resistor and capacitor values are functions of the transfer function parameters Q_p, ω_p, and ω_z, the capacitance C_x, and the resistance R_x. We set the value of C_x and R_x and compute the element values as

$$\boxed{\begin{aligned}
C_2 &= C_2(Q_p, \omega_p, \omega_z, C_x, R_x) = C_x \\
C_7 &= C_7(Q_p, \omega_p, \omega_z, C_x, R_x) = C_x \\
R_0 &= \frac{1}{\omega_p C_x} \\
R_1 &= R_1(Q_p, \omega_p, \omega_z, C_x, R_x) = R_x \\
R_3 &= R_3(Q_p, \omega_p, \omega_z, C_x, R_x) = R_x \\
R_8 &= R_8(Q_p, \omega_p, \omega_z, C_x, R_x) = Q_p R_0 \\
R_4 &= R_4(Q_p, \omega_p, \omega_z, C_x, R_x) = R_8\left(\frac{\omega_z^2}{\omega_p^2} - 1\right) \\
R_5 &= R_5(Q_p, \omega_p, \omega_z, C_x, R_x) = \frac{R_0^2}{R_4}
\end{aligned}} \tag{4.125}$$

Highpass Notch High-Q-Factor Realization. The highpass notch high-Q-factor realization is shown in Fig. 4.36 [15, pp. 78–79; 41].

HPN-HQ

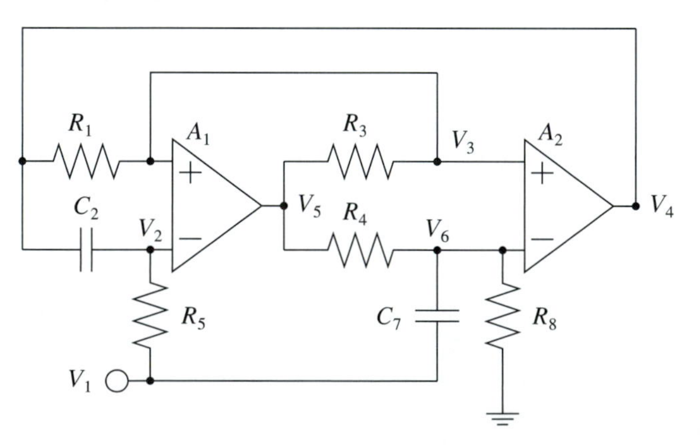

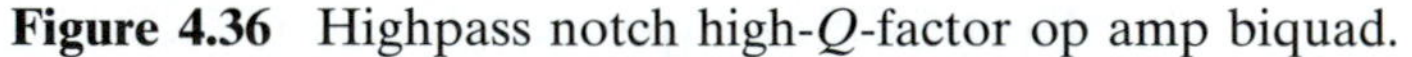
Figure 4.36 Highpass notch high-Q-factor op amp biquad.

The transfer function of the biquad from Fig. 4.36 is

$$H_{HPN}(s) = \frac{V_4}{V_1} = \frac{s^2 + \omega_z^2}{s^2 + \dfrac{\omega_p}{Q_p}s + \omega_p^2} \qquad \omega_z < \omega_p \tag{4.126}$$

The resistor and capacitor values are functions of the transfer function parameters Q_p, ω_p, and ω_z, the capacitance C_x, and the resistance R_x. We set the value of C_x and R_x and compute the element values as

$$\boxed{\begin{aligned}
C_2 &= C_2(Q_p, \omega_p, \omega_z, C_x, R_x) = C_x \\
C_7 &= C_7(Q_p, \omega_p, \omega_z, C_x, R_x) = C_x \\
R_0 &= \frac{1}{\omega_p C_x} \\
R_1 &= R_1(Q_p, \omega_p, \omega_z, C_x, R_x) = R_x \\
R_3 &= R_3(Q_p, \omega_p, \omega_z, C_x, R_x) = R_x \\
R_8 &= R_8(Q_p, \omega_p, \omega_z, C_x, R_x) = Q_p R_0 \\
R_4 &= R_4(Q_p, \omega_p, \omega_z, C_x, R_x) = R_8\left(1 - \frac{\omega_z^2}{\omega_p^2}\right) \\
R_5 &= R_5(Q_p, \omega_p, \omega_z, C_x, R_x) = \frac{R_0^2}{R_4}
\end{aligned}} \tag{4.127}$$

General-Purpose Realization. The general-purpose realization, sometimes called the *KHN filter* (Kerwin–Huelsman–Newcomb), is shown in Fig. 4.37 [15, pp. 80–81; 42].

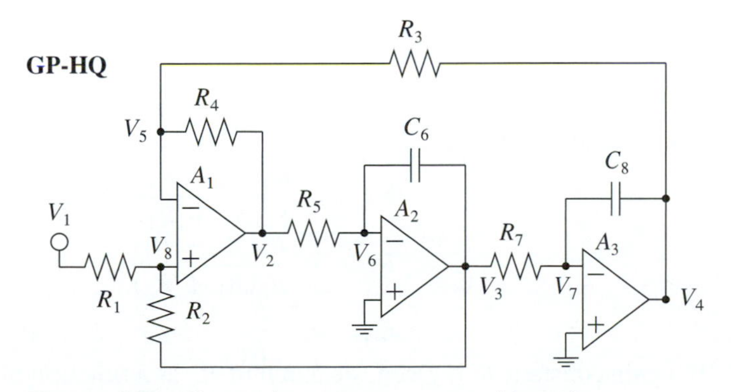

Figure 4.37 General-purpose op amp biquad.

The realizable transfer functions of the biquad from Fig. 4.37 are

$$\begin{aligned} H_{LP}(s) &= \frac{V_4}{V_1} = K_{LP}\frac{\omega_p^2}{s^2 + \frac{\omega_p}{Q_p}s + \omega_p^2} \\ H_{HP}(s) &= \frac{V_2}{V_1} = K_{HP}\frac{s^2}{s^2 + \frac{\omega_p}{Q_p}s + \omega_p^2} \\ H_{BP}(s) &= \frac{V_3}{V_1} = K_{BP}\frac{\frac{\omega_p}{Q_p}s}{s^2 + \frac{\omega_p}{Q_p}s + \omega_p^2} \end{aligned} \tag{4.128}$$

The resistor and capacitor values are functions of the transfer function parameters Q_p and ω_p, the capacitance C_x, and the resistance R_x. We set the value of C_x and R_x and compute the element values as

$$\boxed{\begin{aligned} C_6 &= C_6(Q_p, \omega_p, C_x, R_x) = C_x \\ C_8 &= C_8(Q_p, \omega_p, C_x, R_x) = C_x \\ R_0 &= \frac{1}{\omega_p C_x} \\ R_1 &= R_1(Q_p, \omega_p, C_x, R_x) = R_x \\ R_3 &= R_3(Q_p, \omega_p, C_x, R_x) = R_x \\ R_5 &= R_5(Q_p, \omega_p, C_x, R_x) = R_x \\ R_7 &= R_7(Q_p, \omega_p, C_x, R_x) = R_x \\ R_4 &= R_4(Q_p, \omega_p, C_x, R_x) = \frac{R_x^3}{R_0^2} \\ R_2 &= R_2(Q_p, \omega_p, C_x, R_x) = R_x\left(\frac{Q_p\left(1 + \frac{R_4}{R_x}\right)}{\sqrt{\frac{R_4}{R_x}}} - 1\right) \end{aligned}} \tag{4.129}$$

The gain constants are

$$\begin{aligned} K_{LP} &= \frac{R_2(R_3 + R_4)}{R_4(R_1 + R_2)} \\ K_{HP} &= \frac{R_2(R_3 + R_4)}{R_3(R_1 + R_2)} \\ K_{BP} &= -\frac{R_2}{R_1} \end{aligned}$$

If a gain constant K is given, we can find R_x as a function of Q_p, ω_p, C_x and that gain constant K. Next, we proceed according to Eq. (4.129).

4.9 SWITCHED-CAPACITOR (SC) FILTERS

Switched-capacitor (SC) filters implemented as integrated circuits offer high accuracy, relatively low price, straightforward design, and small number of external components. The benefits of SC filters are as follows:

- no attenuation in passband;
- possible gain in passband;
- realization of all basic break transfer-function types (lowpass, highpass, bandpass, bandreject, allpass, notch and bump) with one universal circuit;
- fully inductorless implementation;
- no external capacitors;
- small number of external resistors, or no external resistors in fully integrated implementation;
- high-input impedance;
- low-output impedance;
- small size and weight;
- easy tuning;
- low-frequency operation (as low as 0.1 Hz)
- simple design equations;
- short time-to-market interval.

The drawbacks of SC filters are as follows:

- noise associated with active devices;
- limited dynamic range to about 80 dB;
- small amount of clock-frequency signal feedthrough appears at the output;
- high-frequency operation limited to approximately 200 kHz.

The operation of SC filters is based on the fact that resistance can be simulated by using switches and capacitors. By definition, the resistance of a resistor is the ratio of the voltage across the resistor to the current through the resistor. When a grounded capacitor C is repeatedly switched at the clock frequency f_{CLK}, between a constant voltage source V_s and the ground, the average current that flows into the capacitor is equal to the capacitor's charge $q = V_s C$, multiplied by the clock frequency f_{CLK}:

$$I_{\text{average}} = V_s C f_{CLK} \tag{4.130}$$

We define the equivalent resistance of the switched capacitor as a ratio of the source voltage to the average current. It follows that the resistance is inversely proportional to the product of the capacitance, C, and the clock frequency, f_{CLK}:

$$R = \frac{V_s}{I_{\text{average}}} = \frac{1}{C f_{CLK}} \tag{4.131}$$

Typically, the capacitor tolerances of integrated SC filters can be more than 30%, which is too large for filter applications. However, for a unity capacitance ratio, $r_c = C_1/C_2 \approx 1$, a 0.1% accuracy of r_c can be achieved. In fact, for filter applications, a high

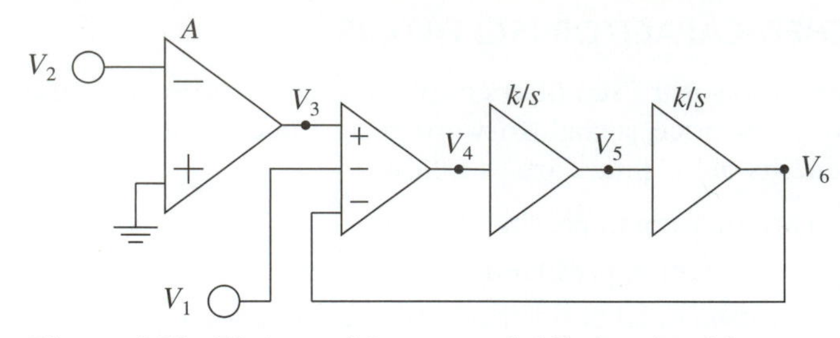

Figure 4.38 Universal integrated SC circuit with internal feedback.

accuracy of pole magnitudes should be obtained. The pole magnitude of an SC filter, ω_p, is proportional to the capacitance ratio $r_c = C_1/C_2$ and the clock frequency f_{CLK}:

$$\omega_p = \frac{1}{R_1C_2} = \frac{1}{\dfrac{1}{C_1 f_{CLK}}C_2} = \frac{C_1}{C_2}f_{CLK} = r_c f_{CLK} \tag{4.132}$$

where R_1 is the equivalent resistance of the switched capacitor C_1. Therefore, SC integrated filters can accurately realize poles because they have a very good matching of integrated capacitors, r_c, and an accurate clock frequency, f_{CLK}.

A universal integrated SC filter is based on the state-variable structure. This structure consists of four active components: an uncommitted operational amplifier, a unique three-input summing stage, and two integrators. The transfer function of the integrators is $\frac{k}{s}$. One of the inputs of the summing stage can be connected through an internal switch to the output of the second integrator (Fig. 4.38) or to the ground (Fig. 4.39).

The integration constant k is proportional to the clock frequency f_{CLK}:

$$k = \frac{2\pi}{P}f_{CLK}, \qquad P = 50 \quad \text{or} \quad P = 100 \tag{4.133}$$

where P can be set by an external control signal.

The grounded mode of operation (Fig. 4.39) has a double pole at infinity. Finite poles can be realized by feeding the signal from the output of the second integrator

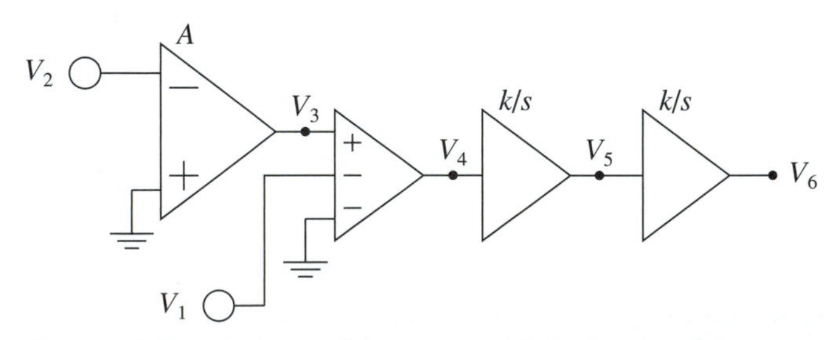

Figure 4.39 Universal integrated SC circuit with grounded summing stage input.

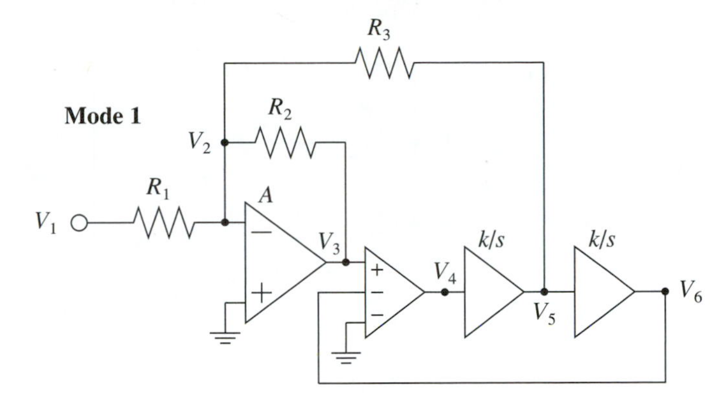

Figure 4.40 Mode 1 SC biquad.

(through a resistor) to the operational amplifier. In this case, the operational amplifier acts as a summing amplifier.

The feedback mode of operation (Fig. 4.38) realizes finite poles, $\omega_p = \frac{2\pi}{P} f_{CLK}$. Pole magnitudes different from $\frac{2\pi}{P} f_{CLK}$ can be realized by additional feedback from the output of the second integrator (through a resistor) to the operational amplifier.

The universal integrated SC circuit, along with additional resistors, operates in several modes that are classified according to the type of realized transfer function or feedback connection (Figs. 4.40–4.55) [43]–[45].

Modes of operation 1, 2, 3, 4, and 5 are used for realizations of second-order transfer functions Eqs. (134)–(140). Modes of operation 6 and 7 are used for realizations of first-order transfer functions, Eqs. (134) and (141).

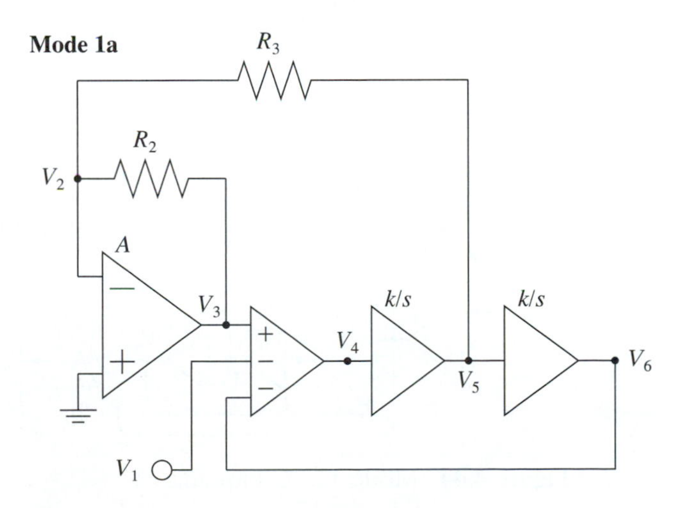

Figure 4.41 Mode 1a SC biquad.

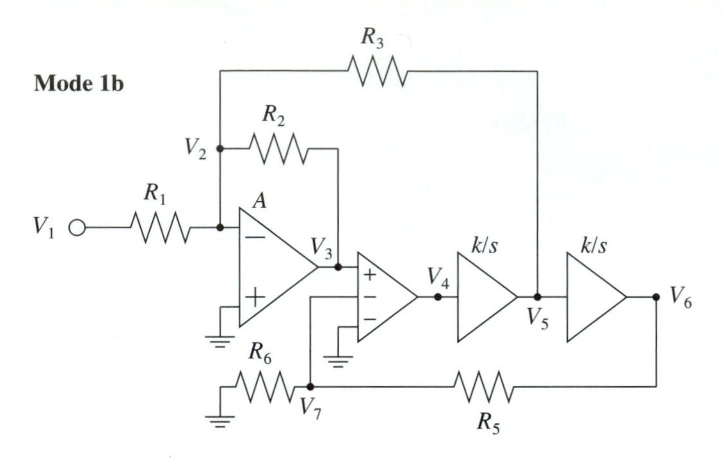

Figure 4.42 Mode 1b SC biquad.

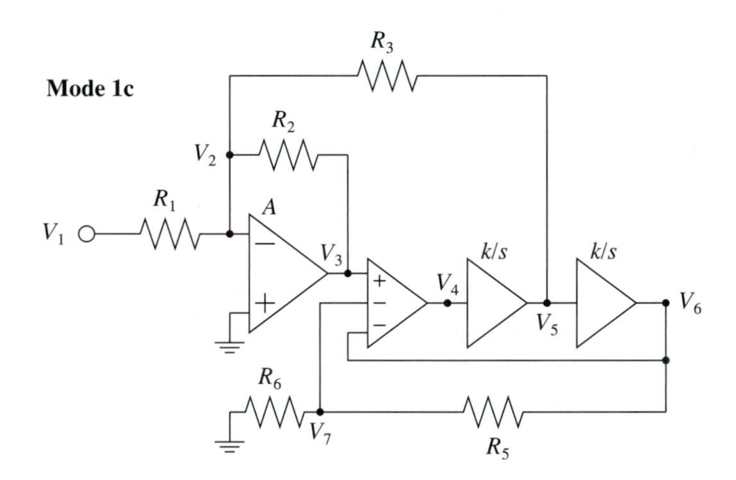

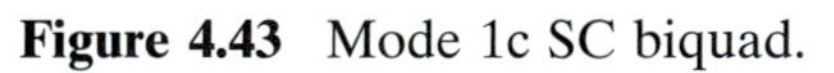
Figure 4.43 Mode 1c SC biquad.

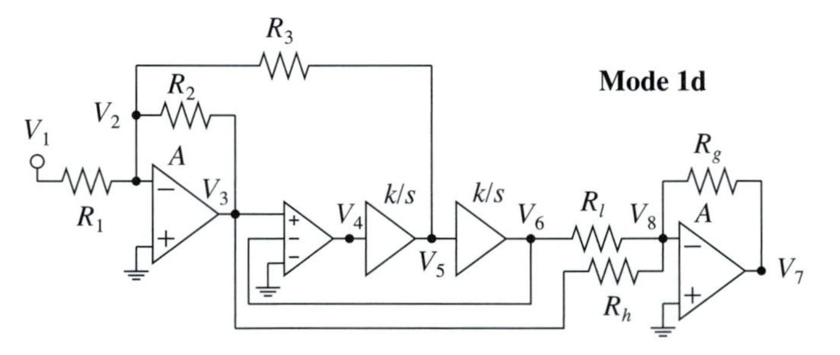

Figure 4.44 Mode 1d SC biquad.

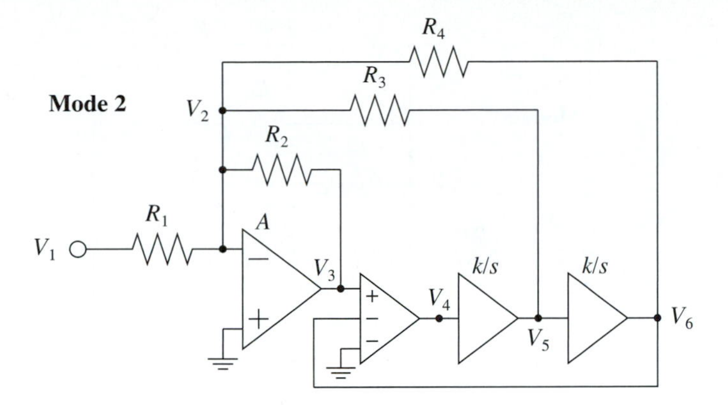

Figure 4.45 Mode 2 SC biquad.

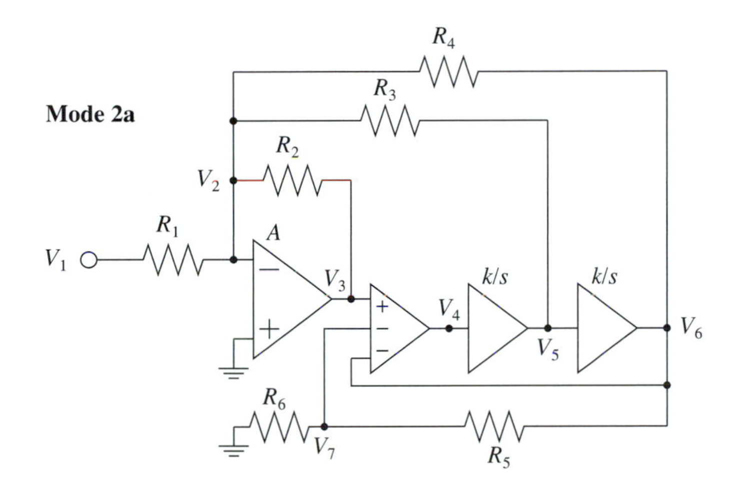

Figure 4.46 Mode 2a SC biquad.

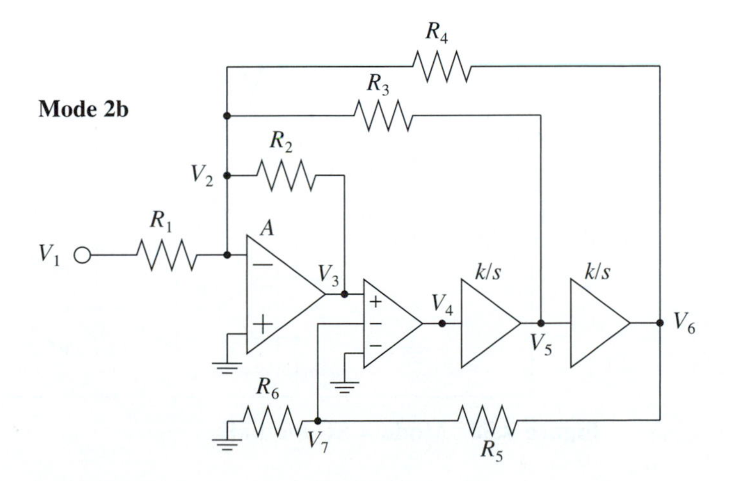

Figure 4.47 Mode 2b SC biquad.

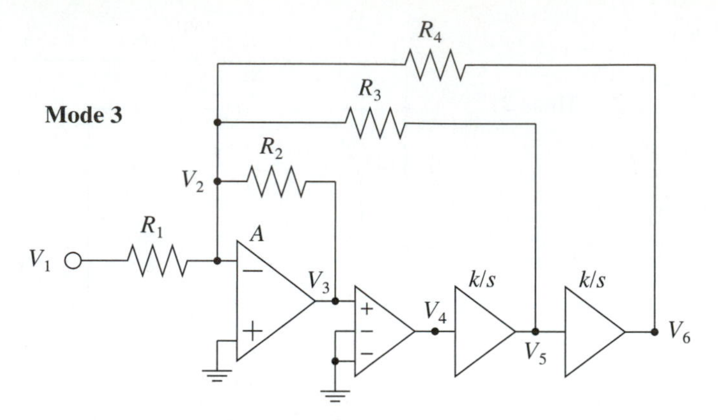

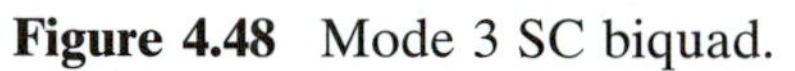
Figure 4.48 Mode 3 SC biquad.

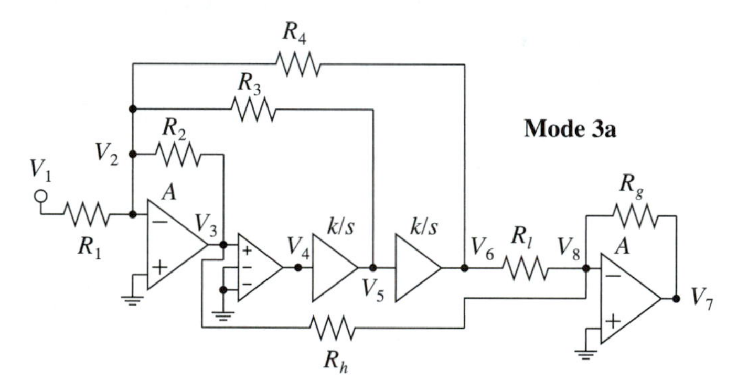

Figure 4.49 Mode 3a SC biquad.

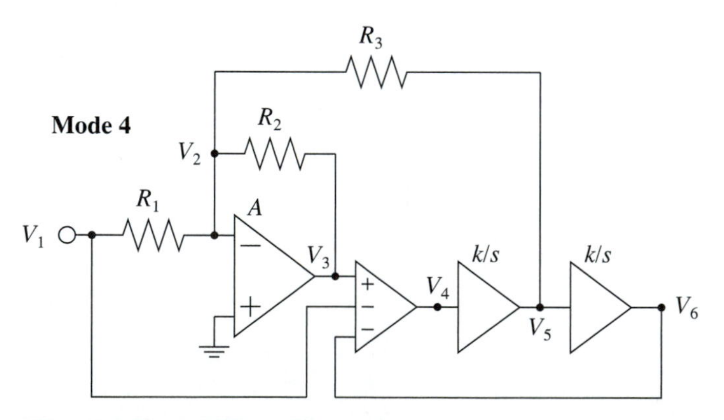

Figure 4.50 Mode 4 SC biquad.

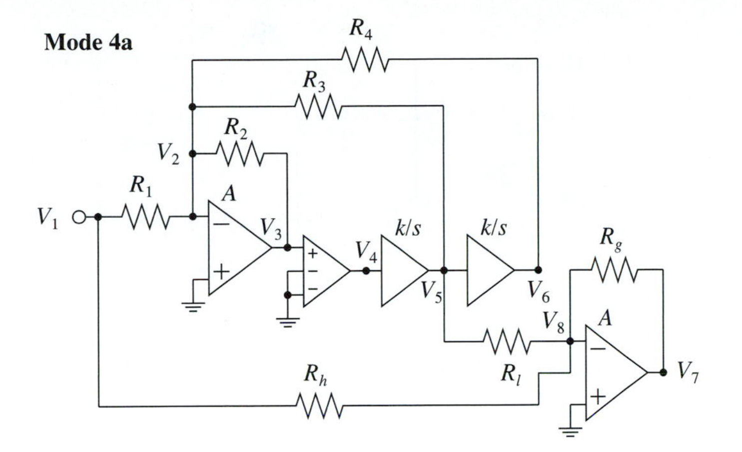

Figure 4.51 Mode 4a SC biquad.

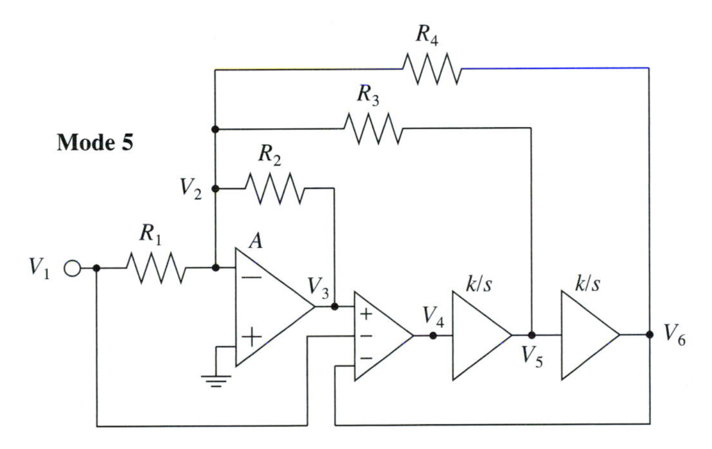

Figure 4.52 Mode 5 SC biquad.

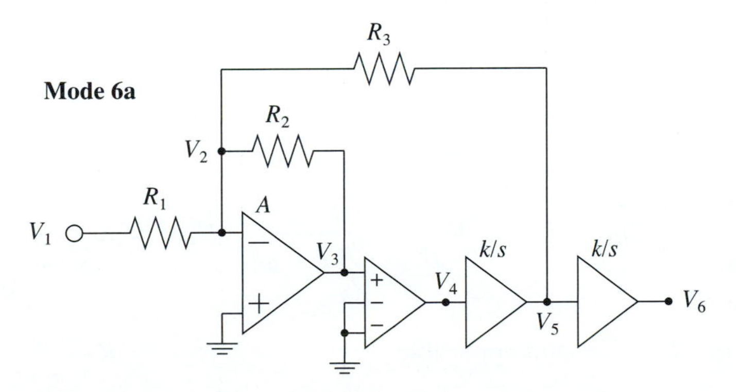

Figure 4.53 Mode 6a SC first-order realization.

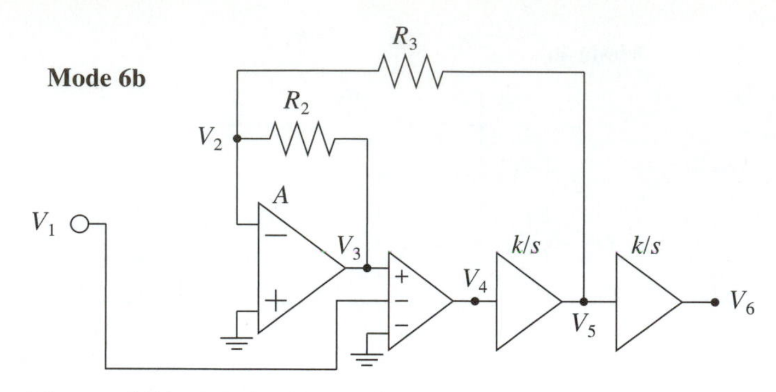

Figure 4.54 Mode 6b SC first-order realization.

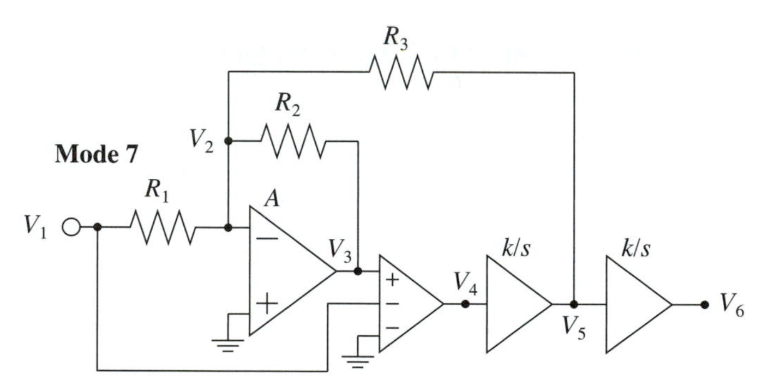

Figure 4.55 Mode 7 SC first-order realization.

The output signal of an SC filter section can be taken at the output of the operational amplifier, V_3, at the output of the first integrator, V_5, and at the output of the second integrator, V_6. The output signal of the three-input summing stage is not available. Modes of operation 6 and 7 do not make use of the output signal V_6. The signal V_1 is the input.

In all modes of operation, the ratio V_6/V_1 is the lowpass transfer function, and V_5/V_1 is the bandpass or lowpass transfer function,

Mode	$\dfrac{V_6}{V_1}$	$\dfrac{V_5}{V_1}$
1, 1a, 1b, 1c, 1d, 2, 2a, 2b, 3, 3a, 4, 4a, 5	$H_{LP} = K\dfrac{\omega_p^2}{s^2 + \dfrac{\omega_p}{Q_p}s + \omega_p^2}$	$H_{BP} = K\dfrac{\dfrac{\omega_p}{Q_p}s}{s^2 + \dfrac{\omega_p}{Q_p}s + \omega_p^2}$
6a, 6b, 7	Not applicable	$H_{LP} = K\dfrac{\omega_p}{s + \omega_p}$

(4.134)

The pole magnitude of an SC section can be $\omega_p = \frac{2\pi}{P} f_{CLK}$ (modes 1, 1a, 1d, 4), $\omega_p > \frac{2\pi}{P} f_{CLK}$ (modes 1c, 2, 2a, 4, 5), or $\omega_p < \frac{2\pi}{P} f_{CLK}$ (mode 1b), or it can be set to an arbitrary value (modes 2b, 3, 3a, 4a).

Most modes of operation are used without any additional operational amplifiers (modes 1, 1a, 1b, 1c, 2, 2a, 2b, 3, 4, 5), while some modes (modes 1d, 3a, 4a) require an additional operational amplifier. In practice, this additional amplifier is often the operation amplifier of the next biquad in the cascade. Thus, the whole filter can be realized with no external operational amplifier.

The output of the operational amplifier can be used for realizing the highpass, bandreject, and allpass transfer functions:

Mode	$H_{HP} = \dfrac{V_3}{V_1} = K \dfrac{s^2}{s^2 + \dfrac{\omega_p}{Q_p} s + \omega_p^2}$
3	$\omega_p = \dfrac{2\pi f_{CLK}}{P} \sqrt{\dfrac{R_2}{R_4}}$

(4.135)

Mode	$H_{BR} = \dfrac{V_3}{V_1} = K \dfrac{s^2 + \omega_p^2}{s^2 + \dfrac{\omega_p}{Q_p} s + \omega_p^2}$
1	$\omega_p = \dfrac{2\pi f_{CLK}}{P}$
1b	$\omega_p = \dfrac{2\pi f_{CLK}}{P} \sqrt{\dfrac{R_6}{R_5 + R_6}}$
1c	$\omega_p = \dfrac{2\pi f_{CLK}}{P} \sqrt{1 + \dfrac{R_6}{R_5 + R_6}}$

(4.136)

Mode	$H_{AP} = \dfrac{V_3}{V_1} = K \dfrac{s^2 - \dfrac{\omega_p}{Q_p} s + \omega_p^2}{s^2 + \dfrac{\omega_p}{Q_p} s + \omega_p^2}$
4	$\omega_p = \dfrac{2\pi f_{CLK}}{P}$
4a	$\omega_p = \dfrac{2\pi f_{CLK}}{P} \sqrt{\dfrac{R_2}{R_4}}$

(4.137)

$$\begin{array}{ll} \text{Mode} & H_{BP} = \dfrac{V_3}{V_1} = K \dfrac{\dfrac{\omega_p}{Q_p} s}{s^2 + \dfrac{\omega_p}{Q_p} s + \omega_p^2} \\ \hline \text{1a} & \omega_p = \dfrac{2\pi f_{CLK}}{P} \end{array} \tag{4.138}$$

The highpass notch transfer function can be obtained without an additional amplifier:

$$\begin{array}{lll} \text{Mode} & H_{HPN} = \dfrac{V_3}{V_1} = K \dfrac{s^2 + \omega_z^2}{s^2 + \dfrac{\omega_p}{Q_p} s + \omega_p^2} & \dfrac{R_2}{R_4} > \dfrac{R_h}{R_l} \\ \hline 2 & \omega_p = \dfrac{2\pi f_{CLK}}{P}\sqrt{1 + \dfrac{R_2}{R_4}} & \omega_z = \dfrac{2\pi f_{CLK}}{P} \\ \text{2a} & \omega_p = \dfrac{2\pi f_{CLK}}{P}\sqrt{\dfrac{R_2}{R_4} + \dfrac{R_5 + 2R_6}{R_5 + R_6}} & \omega_z = \dfrac{2\pi f_{CLK}}{P}\sqrt{\dfrac{R_5 + 2R_6}{R_5 + R_6}} \\ \text{2b} & \omega_p = \dfrac{2\pi f_{CLK}}{P}\sqrt{\dfrac{R_2}{R_4} + \dfrac{R_6}{R_5 + R_6}} & \omega_z = \dfrac{2\pi f_{CLK}}{P}\sqrt{\dfrac{R_6}{R_5 + R_6}} \\ \text{3a} & \omega_p = \dfrac{2\pi f_{CLK}}{P}\sqrt{\dfrac{R_2}{R_4}} & \omega_z = \dfrac{2\pi f_{CLK}}{P}\sqrt{\dfrac{R_h}{R_l}} \end{array} \tag{4.139}$$

while the lowpass notch transfer function requires an additional op amp:

$$\begin{array}{lll} \text{Mode} & H_{LPN} = \dfrac{V_3}{V_1} = K \dfrac{s^2 + \omega_z^2}{s^2 + \dfrac{\omega_p}{Q_p} s + \omega_p^2} & \dfrac{R_2}{R_4} < \dfrac{R_h}{R_l} \\ \hline \text{1d} & \omega_p = \dfrac{2\pi f_{CLK}}{P} & \omega_z = \dfrac{2\pi f_{CLK}}{P}\sqrt{1 + \dfrac{R_h}{R_l}} \\ \text{3a} & \omega_p = \dfrac{2\pi f_{CLK}}{P}\sqrt{\dfrac{R_2}{R_4}} & \omega_z = \dfrac{2\pi f_{CLK}}{P}\sqrt{\dfrac{R_h}{R_l}} \end{array} \tag{4.140}$$

The highpass, lowpass, and allpass first-order transfer functions can be obtained using the output of the operational amplifier:

Mode	Transfer function	Pole magnitude
6a	$H_{HP} = \dfrac{V_3}{V_1} = K\dfrac{s}{s + \omega_p}$	$\omega_p = \dfrac{2\pi f_{CLK}}{P}\dfrac{R_2}{R_3}$
6b	$H_{LP} = \dfrac{V_3}{V_1} = K\dfrac{\omega_p}{s + \omega_p}$	$\omega_p = \dfrac{2\pi f_{CLK}}{P}\dfrac{R_2}{R_3}$
7	$H_{AP} = \dfrac{V_3}{V_1} = K\dfrac{s - \omega_p}{s + \omega_p}$	$\omega_p = \dfrac{2\pi f_{CLK}}{P}\dfrac{R_2}{R_3}$

(4.141)

Realizations in mode 6 and 7 do not use the second integrator.

4.9.1 Mode 1 SC Realization

The main feature of the mode 1 operation and its derivatives (modes 1a, 1b, 1c, 1d) is that they do not use the feedback from the second integrator to the operational amplifier. Thus, the gain-sensitivity product of the pole magnitude to the gain of the operational amplifier is zero. This property enables implementation of SC filters at higher frequencies.

Mode 1 requires only three resistors, while mode 1a can be realized with only two external resistors.

4.9.2 Mode 2 SC Realization

The main characteristic of the mode 2 operation and its derivatives (2a, 2b) is that they can realize the highpass notch transfer function without an additional amplifier. The magnitude of the transfer-function zero, ω_z, can be tuned independently of the pole magnitude, ω_p.

Mode 2 requires only four external resistors.

4.9.3 Mode 3 SC Realization

The mode 3 operation is intended for highpass biquads. The pole magnitude, ω_p, can be adjusted to an arbitrary value.

Mode 3a requires an additional operational amplifier. This mode is frequently used for realization of lowpass and highpass notch transfer functions. The magnitude of the transfer-function zero, ω_z, can be adjusted to be higher or lower than the pole magnitude, ω_p.

The notch transfer functions are realized by summing the highpass, V_3, and the lowpass, V_6, outputs using an external op amp and three external resistors.

Mode 3a can realize lowpass notch and highpass notch transfer functions

$$
\begin{aligned}
H_{LPN} &= K \frac{s^2 + \omega_z^2}{s^2 + \frac{\omega_p}{Q_p} s + \omega_p^2}, \qquad \omega_z > \omega_p \\
H_{HPN} &= K \frac{s^2 + \omega_z^2}{s^2 + \frac{\omega_p}{Q_p} s + \omega_p^2}, \qquad \omega_z < \omega_p
\end{aligned}
\tag{4.142}
$$

The resistor values are functions of the transfer-function parameters K, Q_p, ω_p, and ω_z, the clock frequency f_{CLK}, the parameter P, and the resistances R_{1x}, R_{2x}, and R_{hx}. We set the value of R_{1x}, R_{2x}, R_{hx}, and P and compute the element values as

$$
\boxed{
\begin{aligned}
R_1 &= R_1(K, Q_p, \omega_p, \omega_z, f_{CLK}, P, R_{1x}, R_{2x}, R_{hx}) &&= R_{1x} \\
R_2 &= R_2(K, Q_p, \omega_p, \omega_z, f_{CLK}, P, R_{1x}, R_{2x}, R_{hx}) &&= R_{2x} \\
R_4 &= R_4(K, Q_p, \omega_p, \omega_z, f_{CLK}, P, R_{1x}, R_{2x}, R_{hx}) &&= R_{2x} \left(\frac{2\pi f_{CLK}}{P \omega_p} \right)^2 \\
R_3 &= R_3(K, Q_p, \omega_p, \omega_z, f_{CLK}, P, R_{1x}, R_{2x}, R_{hx}) &&= Q_p \sqrt{R_{2x} R_4} \\
R_h &= R_h(K, Q_p, \omega_p, \omega_z, f_{CLK}, P, R_{1x}, R_{2x}, R_{hx}) &&= R_{hx} \\
R_l &= R_l(K, Q_p, \omega_p, \omega_z, f_{CLK}, P, R_{1x}, R_{2x}, R_{hx}) &&= R_{hx} \left(\frac{2\pi f_{CLK}}{P \omega_z} \right)^2 \\
R_g &= R_g(K, Q_p, \omega_p, \omega_z, f_{CLK}, P, R_{1x}, R_{2x}, R_{hx}) &&= \frac{K R_{hx} R_{1x}}{R_{2x}}
\end{aligned}
}
\tag{4.143}
$$

The operational amplifier of the next biquad in the cascade can be used instead of the external op amp. In order to avoid the external op amp for the last biquad in the cascade, the highest zero magnitude of the lowpass notch biquad can be moved to infinity, transforming the last biquad into a lowpass biquad.

4.9.4 Mode 4 SC Realization

Modes 4 and 4a are intended for allpass biquads. The pole magnitude of the mode 4 biquad is $\omega_p = \frac{2\pi}{P} f_{CLK}$, while the pole magnitude for mode 4a can be set to an arbitrary value.

The mode 4a operation requires an additional amplifier.

4.9.5 Mode 5 SC Realization

The mode 5 operation can realize second-order transfer functions of the form

$$
H_{CZ} = K \frac{s^2 + \frac{\omega_z}{Q_z} s + \omega_z^2}{s^2 + \frac{\omega_p}{Q_p} s + \omega_p^2}, \qquad \omega_z < \omega_p \tag{4.144}
$$

where

$$\begin{aligned}\omega_p &= \frac{2\pi f_{CLK}}{P}\sqrt{1+\frac{R_2}{R_4}}\\ \omega_z &= \frac{2\pi f_{CLK}}{P}\sqrt{1-\frac{R_1}{R_4}}\\ Q_p &= \frac{R_3}{R_4}\sqrt{1+\frac{R_2}{R_4}}\\ Q_z &= \frac{R_3}{R_1}\sqrt{1-\frac{R_1}{R_4}}\end{aligned} \tag{4.145}$$

Obviously, the zero Q-factor, Q_z, can be different from the pole Q-factor, Q_p, and the zero magnitude, ω_z, is lower than the pole magnitude, ω_p.

4.9.6 Modes 6 and 7 SC Realizations

By using only the first integrator, modes 6 and 7 realize first-order transfer functions that appear in odd-order filters. The output of the operational amplifier is used for realizing highpass (mode 6a), lowpass (mode 6b), and allpass (mode 7) transfer functions.

4.9.7 Low-Sensitive Lowpass Notch Realization

The extreme of the magnitude response sensitivity of a biquad to the Q -factor, $S_{Q_p}^{M(\omega)}$, is

$$S_{Q_p}^{M(\omega)}(\omega)\Big|_{\max} = 1 \tag{4.146}$$

In the frequency range $0 \le \omega \le \omega_p$ the maximal value of $\left|S_{\omega_p}^{M(\omega)}(\omega)\right|$ is Q_p times larger than the maximal value of $\left|S_{Q_p}^{M(\omega)}(\omega)\right|$:

$$\left|S_{\omega_p}^{M(\omega)}(\omega)\right|_{\max} \approx Q_p \left|S_{Q_p}^{M(\omega)}(\omega)\right|_{\max} \tag{4.147}$$

The upper limit of the magnitude response relative variation of SC filters can be approximated by

$$\frac{\Delta M(\omega)}{M(\omega)}\bigg|_{\text{worst case}} = \sum_i \left|S_{\omega_p}^{M(\omega)} S_{x_i}^{\omega_p}\frac{\Delta x_i}{x_i}\right| + \sum_i \left|S_{Q_p}^{M(\omega)} S_{x_i}^{Q_p}\frac{\Delta x_i}{x_i}\right| \tag{4.148}$$

where x_i is a resistance or the clock frequency. In practice, the relative variation of the clock frequency is several times smaller than the relative variation of the resistance of the resistors, and it can be neglected in Eq. (4.148). We can assume that all relative variations are the same $\frac{\Delta x_i}{x_i} = \cdots = \frac{\Delta x_j}{x_j}$. Also, we can substitute the approximate maximal values from Eqs. (4.146) and (4.147) into Eq. (4.148):

$$\left.\frac{\Delta M(\omega)}{M(\omega)}\right|_{\text{worst case}} = \left(\sum_i Q_p \left|S_{x_i}^{\omega_p}\right| + \sum_i \left|S_{x_i}^{Q_p}\right|\right)\frac{\Delta x_i}{x_i} \tag{4.149}$$

For most SC filters, $S_{x_i}^{Q_p}$ is 1 or lower. Therefore, for high Q-factor filters, the dominant term can be $Q_p \left|S_{x_i}^{\omega_p}\right|$.

In almost all modes of operation, except modes 1 and 4, $S_{x_i}^{\omega_p} = \frac{1}{2}$. In modes 1 and 4, $S_{x_i}^{\omega_p} = 0$. Mode 1 is obviously superior to other modes with respect to the sensitivity. Nevertheless, this mode is used more in theory than in practice because different clock frequencies must be provided for biquads with different pole magnitudes. So, we could benefit from mode 1 if we could design an SC filter with constant pole magnitudes. In that case we could use only one clock frequency and achieve a very robust design exhibiting the very low passive sensitivity.

We propose an efficient solution to the mode 1 SC filter design based on special elliptic transfer functions which have poles on a circle centered at the origin of the complex s-plane [47].

The lowpass elliptic transfer function can be decomposed into a product of second-order notch transfer functions:

$$H_{LPN}(s) = K\frac{s^2 + \omega_z^2}{s^2 + \dfrac{\omega_p}{Q_p}s + \omega_p^2}, \qquad \omega_z > \omega_p \tag{4.150}$$

The pole magnitudes of all sections are equal to the geometric mean of the passband edge and stopband edge frequency:

$$\omega_p = 2\pi\sqrt{f_{pass} f_{stop}}$$

Therefore, the maximal magnitude response deviation occurs at the frequency $\sqrt{f_{pass} f_{stop}}$, which is in the transition band. Practically, the passband variation is insensitive to the changes of external resistor values.

We propose mode 1d [46] for realization of the notch transfer function (4.150). The transfer function of the mode 1d biquad is

$$H_{LPN}(s) = \frac{R_g R_2}{R_1 R_h}\frac{s^2 + \omega_p^2\left(1 + \dfrac{R_h}{R_l}\right)}{s^2 + \dfrac{R_2}{R_3}\omega_p s + \omega_p^2} \tag{4.151}$$

The pole Q-factor depends on two resistors

$$Q_p = \frac{R_3}{R_2} \tag{4.152}$$

and the notch frequency is given by

$$\omega_z = \frac{2\pi f_{CLK}}{P}\sqrt{1 + \frac{R_h}{R_l}} \tag{4.153}$$

The magnitude response of notch filters is sensitive not only to ω_p and Q_p but also to ω_z:

$$\left.\frac{\Delta M(\omega)}{M(\omega)}\right|_{\text{worst case}} = \sum_i \left| S_{\omega_z}^{M(\omega)} S_{x_i}^{\omega_z} \frac{\Delta x_i}{x_i} \right| + \sum_i \left| S_{\omega_p}^{M(\omega)} S_{x_i}^{\omega_p} \frac{\Delta x_i}{x_i} \right| + \sum_i \left| S_{Q_p}^{M(\omega)} S_{x_i}^{Q_p} \frac{\Delta x_i}{x_i} \right| \tag{4.154}$$

It has been shown that $S_{\omega_z}^{M(\omega)}$ tends to infinity if ω_z approaches ω_p. In mode 1d the sensitivity $S_{x_i}^{\omega_z}$ is zero for $\omega_z = \omega_p$. The maximal value of the product $\left| S_{\omega_z}^{M(\omega)} S_{x_i}^{\omega_z} \right|$ is $\frac{1}{2}$. Therefore, the zeros of the magnitude response are practically insensitive to small changes of the resistor values.

The resistor values are functions of the transfer-function parameters K, Q_p, ω_p, and ω_z, the clock frequency f_{CLK}, the parameter P, and the resistances R_{1x}, R_{2x}, and R_{hx}. We set the value of R_{1x}, R_{2x}, R_{hx}, and P and compute the element values as

$$\begin{aligned}
R_1 &= R_1(K, Q_p, \omega_z, f_{CLK}, P, R_{1x}, R_{2x}, R_{hx}) &&= R_{1x} \\
R_2 &= R_2(K, Q_p, \omega_z, f_{CLK}, P, R_{1x}, R_{2x}, R_{hx}) &&= R_{2x} \\
R_3 &= R_3(K, Q_p, \omega_z, f_{CLK}, P, R_{1x}, R_{2x}, R_{hx}) &&= Q_p R_{2x} \\
R_h &= R_h(K, Q_p, \omega_z, f_{CLK}, P, R_{1x}, R_{2x}, R_{hx}) &&= R_{hx} \\
R_l &= R_l(K, Q_p, \omega_z, f_{CLK}, P, R_{1x}, R_{2x}, R_{hx}) &&= \frac{R_h}{\left(\frac{\omega_z P}{2\pi f_{CLK}}\right)^2 - 1} \\
R_g &= R_g(K, Q_p, \omega_z, f_{CLK}, P, R_{1x}, R_{2x}, R_{hx}) &&= \frac{K R_{hx} R_{1x}}{R_{2x}}
\end{aligned} \tag{4.155}$$

The pole magnitudes of all cascaded biquads are identical, and we are allowed to use single clock-frequency. This way we minimize the influence of resistor values on the pole magnitudes.

4.9.8 Programmable Lowpass/Highpass SC Filters

Generally, our target is to design a programmable cost-effective filter with reduced complexity. A simple modification of mode 3a of a universal SC filter accomplishes this task. A double switch, controlled by an external signal, is introduced to enable the SC biquad to operate as a lowpass or highpass filter.

The second-order lowpass and highpass notch transfer functions can be obtained using mode 3a:

Lowpass filter

$$H_{LP}(s) = k_l \frac{s^2 + \omega_{lz}^2}{s^2 + \dfrac{\omega_{lp}}{Q_{lp}}s + \omega_{lp}^2}$$

$$Q_{lp} = \frac{R_3}{\sqrt{R_2 R_4}}$$

$$\omega_{lp} = \frac{2\pi f_{CLK}}{100}\sqrt{\frac{R_2}{R_4}}$$

$$\omega_{lz} = \frac{2\pi f_{CLK}}{100}\sqrt{\frac{R_h}{R_l}}$$

$$k_l = \frac{R_g R_2}{R_h R_1}$$

Highpass filter

$$H_{HP}(s) = k_h \frac{s^2 + \omega_{hz}^2}{s^2 + \dfrac{\omega_{hp}}{Q_{hp}}s + \omega_{hp}^2}$$

$$Q_{hp} = \frac{R_3'}{\sqrt{R_2' R_4'}}$$

$$\omega_{hp} = \frac{2\pi f_{CLK}}{100}\sqrt{\frac{R_2'}{R_4'}} \tag{4.156}$$

$$\omega_{hz} = \frac{2\pi f_{CLK}}{100}\sqrt{\frac{R_h'}{R_l'}}$$

$$k_h = \frac{R_g' R_2'}{R_h' R_1'}$$

where R_1, R_2, R_3, R_4, R_g, R_h, and R_l are external resistors of the lowpass biquad and R_1', R_2', R_3', R_4', R_g', R_h', and R_l' are external resistors of the highpass biquad.

If we substitute the complex frequency s in the lowpass notch transfer function with $4\pi^2 F_p F_s \frac{1}{s}$, which we symbolically designate by $s \to 4\pi^2 F_p F_s / s$, we obtain the transfer function of a highpass notch filter:

$$H_{HP}(s) = H_{LP}\left(\frac{4\pi^2 F_p F_s}{s}\right) = k_l \frac{\left(4\pi^2 F_p F_s\right)^2 + \omega_{lz}^2 s^2}{\left(4\pi^2 F_p F_s\right)^2 + 4\pi^2 F_p F_s \dfrac{\omega_{lp}}{Q_{lp}} s + \omega_{lp}^2 s^2} \tag{4.157}$$

or

$$H_{HP}(s) = k_l \frac{\omega_{lz}^2}{\omega_{lp}^2} \frac{s^2 + \dfrac{\left(4\pi^2 F_p F_s\right)^2}{\omega_{lz}^2}}{s^2 + \dfrac{4\pi^2 F_p F_s}{\omega_{lp} Q_{lp}} s + \dfrac{\left(4\pi^2 F_p F_s\right)^2}{\omega_{lp}^2}} \tag{4.158}$$

It is important to notice that $|H_{LP}(j2\pi F_p)| = |H_{HP}(j2\pi F_s)|$, and $|H_{LP}(j2\pi F_s)| = |H_{HP}(j2\pi F_p)|$; this means that the passband edge frequency, F_p, of the lowpass filter is equal to the stopband edge frequency of the highpass filter; the stopband edge frequency, F_s, of the lowpass filter is equal to the passband edge frequency of the highpass filter.

By equating the highpass transfer function from Eqs. (4.156) to (4.158), we find the parameters of the highpass transfer functions k_h, Q_{hp}, ω_{hp}, and ω_{hz}:

$$\omega_{hz} = \frac{4\pi^2 F_p F_s}{\omega_{lz}} \tag{4.159}$$

$$\omega_{hp} = \frac{4\pi^2 F_p F_s}{\omega_{lp}} \tag{4.160}$$

$$Q_{hp} = Q_{lp} \tag{4.161}$$

$$k_h = k_l \frac{\omega_{n,i}^2}{\omega_{0,i}^2} \tag{4.162}$$

and

$$\left(\frac{2\pi f_{CLK}}{100}\right)^2 \sqrt{\frac{R_2'}{R_4'}} = 4\pi^2 F_p F_s \sqrt{\frac{R_4}{R_2}} \tag{4.163}$$

$$\left(\frac{2\pi f_{CLK}}{100}\right)^2 \sqrt{\frac{R_h'}{R_l'}} = 4\pi^2 F_p F_s \sqrt{\frac{R_l}{R_h}} \tag{4.164}$$

If we choose the clock frequency f_{CLK} to meet the condition

$$\left(\frac{f_{CLK}}{100}\right)^2 = F_p F_s \tag{4.165}$$

then the relations between resistors are

$$\frac{R_2'}{R_4'} = \frac{R_4}{R_2} \tag{4.166}$$

$$\frac{R_h'}{R_l'} = \frac{R_l}{R_h} \tag{4.167}$$

Our goal is to use the same set of external resistors to implement both types of filters. We meet this requirement by choosing

$$\begin{aligned} R_2' &= R_4 \\ R_4' &= R_2 \\ R_h' &= R_l \\ R_l' &= R_h \end{aligned} \tag{4.168}$$

According to the above analyses we propose a modification to the mode 3a operation which transforms a lowpass filter to a highpass filter: We have to swap the resistors according to Eq. (4.168).

The modification of the mode 3a operation, which implements a programmable lowpass/highpass filter, is shown in Fig. 4.56. The control signal is applied at input LP/HP, and it controls the double switch which connects the resistors R_2 and R_l or R_4

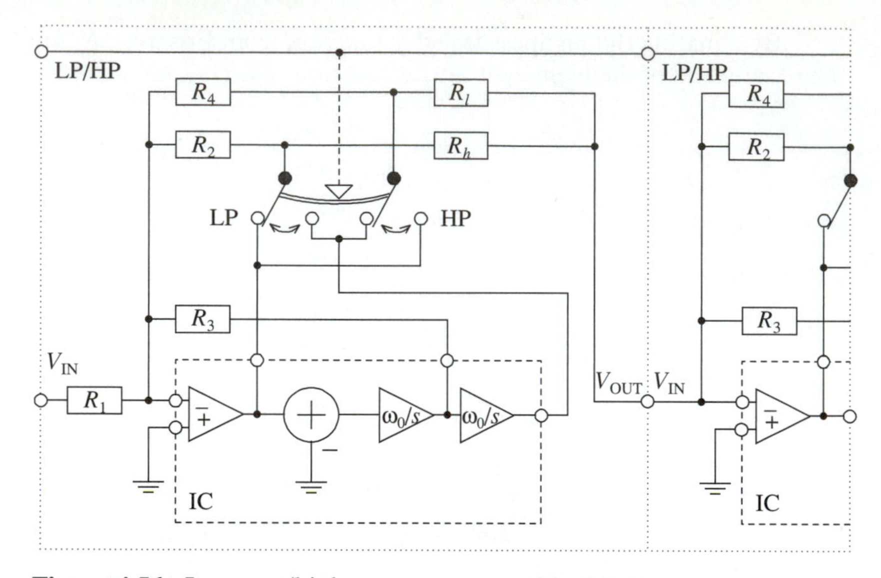

Figure 4.56 Lowpass/highpass programmable SC filter.

and R_h to the output of the summing amplifier, or to the output of the second integrator of the integrated second-order universal SC block.

A simplified second-order section realized with the second-order integrated SC block and resistors R_2, R_3, R_4, R_h and R_l from Fig. 4.56, is shown in Fig. 4.57.

The transfer function of the ith biquad, $T_i(s)$, is defined as the ratio of the output current to the input current:

$$T_i(s) = -\frac{R_2}{R_h} \frac{s^2 + \left(\frac{2\pi f_{CLK}}{100}\right)^2 \frac{R_h}{R_l}}{s^2 + \left(\frac{2\pi f_{CLK}}{100}\right) \frac{R_2}{R_3} s + \left(\frac{2\pi f_{CLK}}{100}\right)^2 \frac{R_2}{R_4}} \tag{4.169}$$

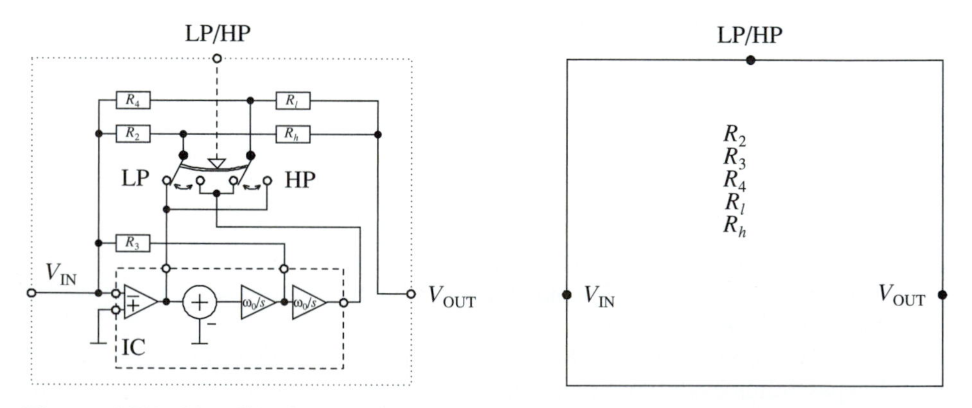

Figure 4.57 Simplified second-order programmable SC filter section.

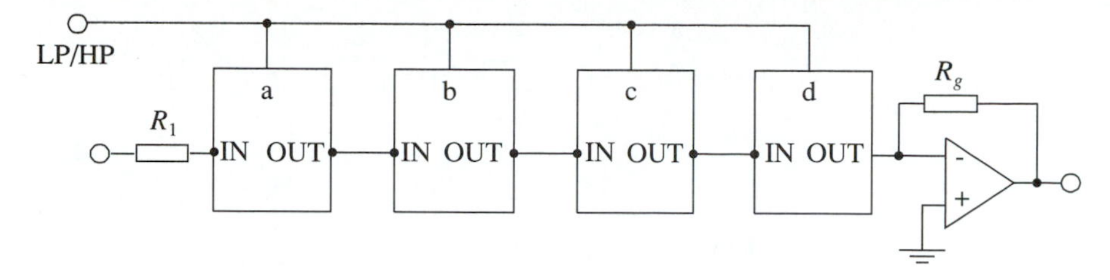

Figure 4.58 Cascade realization of the programmable SC filter.

The cascade realization of the programmable filter is shown in Fig. 4.58. The overall filter transfer function, $H(s)$, is the ratio of the output voltage to the input voltage:

$$H(s) = -\frac{R_g}{R_1} \prod_i T_i(s) \tag{4.170}$$

By changing the external control signal, applied at the LP/HP port, all biquads are simultaneously set to operate as lowpass or highpass filter sections.

4.10 PASSIVE *RLC* FILTERS

In this section we consider realizations of passive RLC filters. We present two basic realizations known as

1. *Singly terminated ladder realization*
2. *Doubly terminated ladder realization*

The doubly terminated RLC filters can be designed to have the lowest sensitivity. These filters consist of capacitors and inductors, and they are terminated at both ends by resistors. The low sensitivity is based on the property that at the frequencies of the magnitude response maxima the generator delivers the maximum power to the filter and to the resistor at the other end which we call the load. Small changes of the capacitors and inductors can slightly change the frequencies of the magnitude response maxima, but these changes do not affect the maximum power delivered to the load. Therefore, the sensitivity must be zero at those frequencies. The frequencies of the magnitude response maxima are in the passband. In the case of elliptic-type filters, as well as in the case of Chebyshev-type filters, those frequencies are distributed over the passband, and thus the sensitivity can not be too large in the passband.

The doubly terminated ladder realization is shown in Fig. 4.59. For $R_g = 0$ this realization becomes the singly terminated ladder realization. The resistor R_g is the output resistance of the voltage generator while R_o is the resistance of the purely resistive load. The transfer function is the ratio of the load voltage and the generator voltage:

$$H(s) = \frac{V_o}{V_g} \tag{4.171}$$

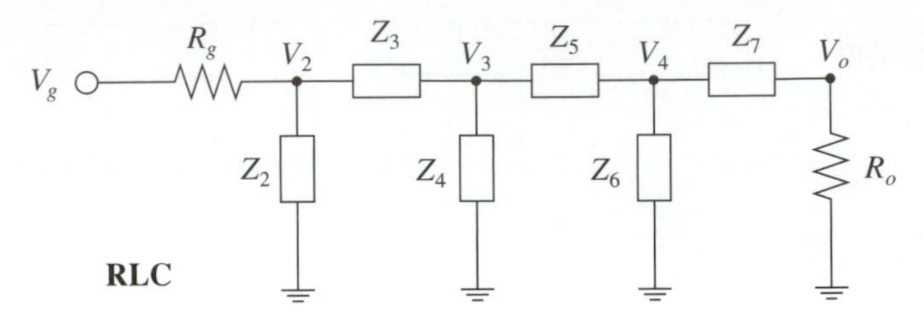

Figure 4.59 Ladder realization.

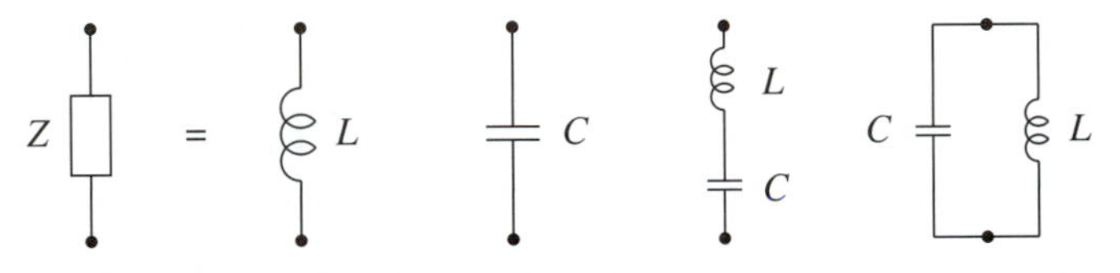

Figure 4.60 Types of ladder branches.

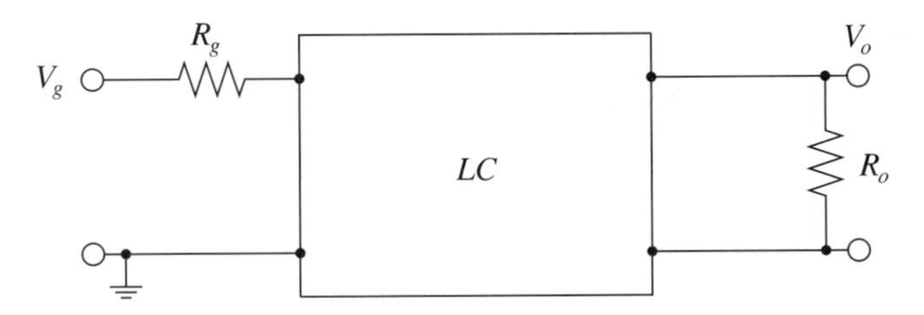

Figure 4.61 Lossless *LC* filter.

A branch of the ladder, whose impedance is Z_i, can be an inductor, a capacitor, the series connection of an inductor and a capacitor, or the shunt connection of an inductor and a capacitor (Fig. 4.60). The first parallel branch can be an open circuit (Z_2 can be omitted from the ladder). The last series branch can be a short circuit ($Z_7 = 0$). The number of ladder branches can be arbitrary.

Any filter that is implemented with capacitors and inductors, only, is called the lossless LC filter, because all filter components are (lossless) capacitors and inductors, except the internal resistor of the voltage source (generator), R_g, and the resistive load, R_o. A lossless LC filter can be represented by a two-port network, the LC network, as shown in Fig. 4.61.

4.10.1 Singly Terminated Ladder Realization

The transfer function of the singly terminated ladder realization, shown in Fig. 4.61, is obtained for $R_g = 0$:

$$H(s) = -\frac{y_{21} R_o}{1 + y_{22} R_o} \tag{4.172}$$

where y_{21} is the ratio of the current through R_o, for $R_o = 0$, to V_g:

$$y_{21} = -\frac{I_{Ro}}{V_g}, \qquad R_o = 0 \tag{4.173}$$

and y_{22} is the ratio of the current through Z_7 to the voltage across R_o, for $V_g = 0$ and $R_o \neq 0$:

$$y_{22} = \frac{I_{Z_7}}{V_{Ro}}, \qquad V_g = 0 \tag{4.174}$$

It can be shown that y_{21} and y_{22} are odd rational functions in complex frequency s (i.e., the ratio of an odd to an even polynomial, or the ratio of an even to an odd polynomial). Therefore, the transfer function of the singly terminated ladder realization is

$$H(s) = \frac{N_{\text{even}} + N_{\text{odd}}}{D_{\text{even}} + D_{\text{odd}}}, \qquad N_{\text{even}} = 0 \quad \text{or} \quad N_{\text{odd}} = 0 \tag{4.175}$$

where N_{even} is the even part and N_{odd} is the odd part of the numerator, while D_{even} and D_{odd} are even and odd part of the denominator, respectively. The transfer function can be rewritten as

$$\begin{aligned} H(s) &= \frac{\dfrac{N_{\text{odd}}}{D_{\text{even}}}}{1 + \dfrac{D_{\text{odd}}}{D_{\text{even}}}}, \quad N_{\text{even}} = 0 \\ H(s) &= \frac{\dfrac{N_{\text{even}}}{D_{\text{odd}}}}{1 + \dfrac{D_{\text{even}}}{D_{\text{odd}}}}, \quad N_{\text{odd}} = 0 \end{aligned} \tag{4.176}$$

yielding

$$y_{21}R_o = \frac{N_{\text{odd}}}{D_{\text{even}}} \quad \text{and} \quad y_{22}R_o = \frac{D_{\text{odd}}}{D_{\text{even}}} \tag{4.177}$$

or

$$y_{21}R_o = \frac{N_{\text{even}}}{D_{\text{odd}}} \quad \text{and} \quad y_{22}R_o = \frac{D_{\text{even}}}{D_{\text{odd}}} \tag{4.178}$$

The admittance y_{22} can be obtained from the transfer function, and it can be realized by the classic Foster or Cauer synthesis procedures. After y_{22} has been realized, we have to determine the generator port, by taking into account the zeros of y_{21}.

The algorithmic details for the singly terminated ladder realization are illustrated by the following examples.

Singly Terminated Ladder Realization with Zeros at the Origin. Let us realize the transfer function

$$H(s) = \frac{s}{s^4 + 3s^3 + 3s^2 + 3s + 1}$$

assuming $R_o = 1\,\Omega$.

First, we find the even and odd parts of the numerator and denominator:

$$N_{\text{odd}} = s$$
$$N_{\text{even}} = 0$$
$$D_{\text{odd}} = 3s^3 + 3s$$
$$D_{\text{even}} = s^4 + 3s^2 + 1$$

Next, we compute y_{21} and y_{22} from

$$y_{21} = \frac{N_{\text{odd}}}{R_o D_{\text{even}}} = \frac{s}{s^4 + 3s^2 + 1}$$
$$y_{22} = \frac{D_{\text{odd}}}{R_o D_{\text{even}}} = \frac{3s^3 + 3s}{s^4 + 3s^2 + 1}$$

Since the order of the y_{22} denominator is larger than the order of the y_{22} numerator, we proceed with the impedance Z_{22}:

$$Z_{22} = \frac{1}{y_{22}} = \frac{s^4 + 3s^2 + 1}{3s^3 + 3s}$$

The impedance Z_{22} is an improper rational function, and we have to extract its polynomial part (by dividing the numerator by the denominator):

$$Z_{22} = \frac{s}{3} + \frac{2s^2 + 1}{3s^3 + 3s}$$

The first term, $\frac{s}{3}$, corresponds to the first serial branch, it is of the form $L_1 s$, and it identifies the inductance of the series inductor L_1:

$$Z_{22} = L_1 s + Z_2, \qquad L_1 = \frac{1}{3}, \qquad Z_2 = \frac{2s^2 + 1}{3s^3 + 3s}$$

The order of the Z_2 denominator is larger than the order of the Z_2 numerator, and we proceed with the admittance

$$Y_2 = \frac{1}{Z_2} = \frac{3s^3 + 3s}{2s^2 + 1}$$

which can be written as (after extracting the polynomial part)

$$Y_2 = \frac{3}{2}s + \frac{\frac{3}{2}s}{2s^2 + 1}$$

The first term, $\frac{3}{2}s$, corresponds to the first parallel branch, it is of the form $C_1 s$, and it identifies the capacitance of the parallel capacitor C_1:

$$Y_2 = C_1 s + Y_3, \qquad C_1 = \frac{3}{2}, \qquad Y_3 = \frac{\frac{3}{2}s}{2s^2 + 1}$$

The order of the Y_3 denominator is larger than the order of the Y_3 numerator, and we proceed with the impedance

$$Z_3 = \frac{1}{Y_3} = \frac{2s^2 + 1}{\frac{3}{2}s}$$

which simplifies to

$$Z_3 = \frac{4}{3}s + \frac{1}{\frac{3}{2}s}$$

and identifies the series connection of an inductor and a capacitor:

$$Z_3 = L_2 s + \frac{1}{C_2 s}, \qquad L_2 = \frac{4}{3}, \qquad C_2 = \frac{3}{2}$$

The LC-ladder realization of y_{22} is shown in Fig. 4.62, and the element values refer to

$$L_1 = \frac{1}{3}\,\text{H}, \qquad C_1 = \frac{3}{2}\,\text{F}, \qquad L_2 = \frac{4}{3}\,\text{H}, \qquad C_2 = \frac{3}{2}\,\text{F}$$

Finally, we have to determine the terminals at which the voltage generator should be connected. The generator can be placed in one of the ladder branches.

The transfer function $H(s)$ has one real zero, $s_z = 0$, and it implies that the generator should be inserted in series with a capacitor. We examine the transfer function for large s, $s \to \infty$:

- If we place the generator in series with C_1, the transfer function $H_1(s) = V_o/V_g$ asymptotically tends to $\frac{1}{s}$, which disagrees with $H(s)$.
- If we place the generator in series with C_2, the transfer function $H_2(s) = V_o/V_g$ asymptotically tends to $\frac{1}{s^3}$, which is in agreement with $H(s)$. This realization is shown in Fig. 4.62.

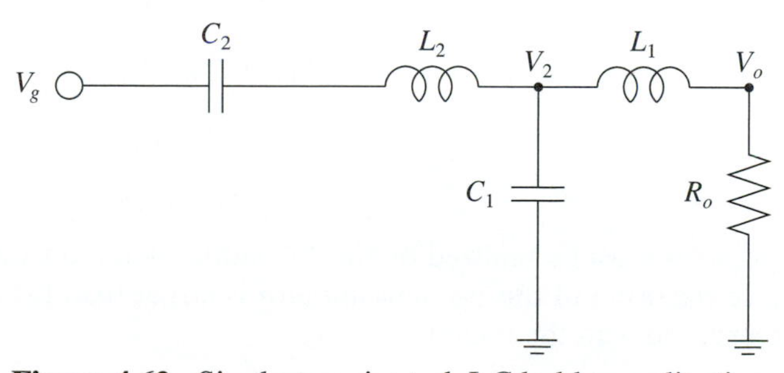

Figure 4.62 Singly terminated LC-ladder realization with zeros at the origin.

Let us derive the transfer function of the circuit from Fig. 4.62 to validate the realization

$$H_2(s) = \frac{\frac{3}{2}s}{s^4 + 3s^3 + 3s^2 + 3s + 1} = \frac{3}{2}H(s)$$

This transfer function has the desired poles and zeros, but the gain constant is different; that is, it is $\frac{3}{2}$ instead of 1. Therefore, the above procedure can realize the desired transfer function within a constant multiplier.

This procedure is applicable only to transfer functions with zeros at the origin.

Singly Terminated Ladder Realization with Complex Zeros. Consider the transfer function with complex zeros:

$$H(s) = \frac{(s^2 + 3.476896154)(s^2 + 8.227391422)}{55.3858\,(s + 0.60913)\left(s^2 + 0.263147s + 1.166357185\right)\left(s^2 + 0.85422659s + 0.7269594794\right)}$$

assuming $R_o = 1\,\Omega$.

First, we find the even and odd parts of the numerator and denominator:

$$N_{\text{odd}} = 0$$

$$N_{\text{even}} = 0.516482303994966 + 0.2113228946047543s^2 + 0.01805516937554391s^4$$

$$D_{\text{odd}} = 1.571315794907603s + 2.79872960375543s^3 + s^5$$

$$D_{\text{even}} = 0.5164779231828084 + 2.477831112275122s^2 + 1.72650359s^4$$

Next, we compute y_{21} and y_{22}:

$$y_{21} = \frac{0.516482303994966 + 0.2113228946047543s^2 + 0.01805516937554391s^4}{1.571315794907603s + 2.79872960375543s^3 + s^5}$$

$$y_{22} = \frac{0.5164779231828084 + 2.477831112275122s^2 + 1.72650359s^4}{1.571315794907603s + 2.79872960375543s^3 + s^5}$$

The transfer function zeros coincide with the zeros of y_{21}:

$$s_1 = +j2.868342974959585$$

$$s_2 = -j2.868342974959585$$

$$s_3 = +j1.864643706985332$$

$$s_4 = -j1.864643706985332$$

or

$$s_1^2 = s_2^2 = -8.2273914225$$

$$s_3^2 = s_4^2 = -3.476896154$$

and these zeros must be realized by the LC ladder derived from y_{22}.

Since the order of the y_{22} denominator is larger than the order of the y_{22} numerator, we proceed with the impedance Z_{22}:

$$Z_{22} = \frac{1}{y_{22}} = \frac{1.5713157949076s + 2.79872960375543s^3 + s^5}{0.5164779231828 + 2.477831112275s^2 + 1.72650359s^4}$$

The impedance Z_{22} is an improper rational function in s, and it can be expressed as a sum of a linear term $L_1 s$ and an impedance Z_2; our goal is to find L_1 such that Z_2 has at least one complex pair of zeros that coincide with the transfer-function zeros (say, the zeros s_1 and s_2):

$$Z_{22} = L_1 s + Z_2, \qquad Z_2(s_1) = Z_2(s_2) = 0$$

The inductance L_1 is computed from

$$L_1 = \left.\frac{Z_{22}}{s}\right|_{s=s_1} = \frac{1.5713157949 + 2.7987296s_1^2 + s_1^4}{0.516477923 + 2.47783s_1^2 + 1.7265s_1^4} = 0.47666285$$

and the impedance Z_2 is found as

$$Z_2 = Z_{22} - L_1 s = s\frac{0.76752227 + 0.93694538s^2 + 0.102542432s^4}{0.2991467415 + 1.43517286997s^2 + s^4}$$

or, equivalently,

$$Z_2 = s\frac{(s^2 + 8.2273914225)(0.093288654 + 0.102542432s^2)}{0.2991467415 + 1.43517286997s^2 + s^4}$$

We proceed with the admittance Y_2

$$Y_2 = \frac{1}{Z_2} = \frac{0.2991467415 + 1.43517286997s^2 + s^4}{s(s^2 + 8.2273914225)(0.093288654 + 0.102542432s^2)}$$

and extract a biquadratic term with the pole pair s_1, s_2, which corresponds to a simple LC connection:

$$Y_2 = \frac{1}{sL_2 + \frac{1}{sC_2}} + Y_3 = \frac{\frac{s}{L_2}}{s^2 + \frac{1}{C_2L_2}} + Y_3 = \frac{\frac{s}{L_2}}{s^2+s_1s_2} + Y_3$$

We find L_2

$$L_2 = Z_2\left.\frac{s}{s^2+8.2273914225}\right|_{s=s_1} = s_1^2\frac{(0.093288654 + 0.102542432s_1^2)}{0.2991467415 + 1.43517286997s_1^2 + s_1^4}$$
$$L_2 = 0.1098864$$

and C_2

$$C_2 = -\frac{1}{L_2s_1^2} = \frac{1}{8.2273914225L_2} = 1.1060985$$

The impedance Y_3 can be calculated from

$$Y_3 = Y_2 - \frac{\frac{s}{0.1098864}}{s^2+8.2273914225}$$
$$= \frac{0.2991467415 + 1.43517286997s^2 + s^4}{s(s^2 + 8.2273914225)(0.093288654 + 0.102542432s^2)} - \frac{\frac{s}{0.1098864}}{s^2+8.2273914225}$$

yielding

$$Y_3 = \frac{0.06683258(0.5440438232 + s^2)}{s(0.093288654 + 0.102542432s^2)} = \frac{0.65175537(0.5440438232 + s^2)}{s(0.909756595 + s^2)}$$

Since the order of the Y_3 denominator is larger than the order of the Y_3 numerator, we proceed with the impedance Z_3:

$$Z_3 = \frac{1}{Y_3} = \frac{s(0.909756595 + s^2)}{0.65175537(0.5440438232 + s^2)}$$

The impedance Z_3 is an improper rational function in s, and it can be expressed as a sum of a linear term $L_3 s$ and an impedance Z_4; our goal is to find L_3 such that Z_4 has at least one complex pair of zeros that coincide with the transfer-function zeros (excluding the zeros s_1 and s_2, but using s_3 and s_4):

$$Z_3 = L_3 s + Z_4$$

We obtain

$$L_3 = \left.\frac{Z_3}{s}\right|_{s=s_3} = \frac{0.909756595 + s_3^2}{0.65175537(0.5440438232 + s_3^2)} = 1.342996$$

and

$$Z_4 = Z_3 - L_3 s = s\frac{0.665207337596 + 0.1913221759s^2}{0.54404382321885 + s^2}$$

$$Z_4 = s\frac{0.1913221759(3.476896154 + s^2)}{0.54404382321885 + s^2}$$

Again, we proceed with

$$Y_4 = \frac{1}{Z_4} = \frac{0.54404382321885 + s^2}{0.1913221759s(3.476896154 + s^2)}$$

and extract a biquadratic term with the pole pair s_3, s_4, which corresponds to a simple LC connection:

$$Y_4 = \frac{1}{sL_4 + \frac{1}{sC_4}} + Y_5 = \frac{\frac{s}{L_4}}{s^2 + \frac{1}{C_4 L_4}} + Y_5 = \frac{\frac{s}{L_4}}{s^2 + s_3 s_4} + Y_5$$

We find

$$L_4 = \left. Z_4 \frac{s}{s^2 + 3.476896154}\right|_{s=s_3} = s_3^2 \frac{0.1913221759}{0.54404382321885 + s_3^2} = 0.226812421$$

and

$$C_4 = -\frac{1}{L_4 s_3^2} = \frac{1}{3.476896154 L_4} = 1.2680648$$

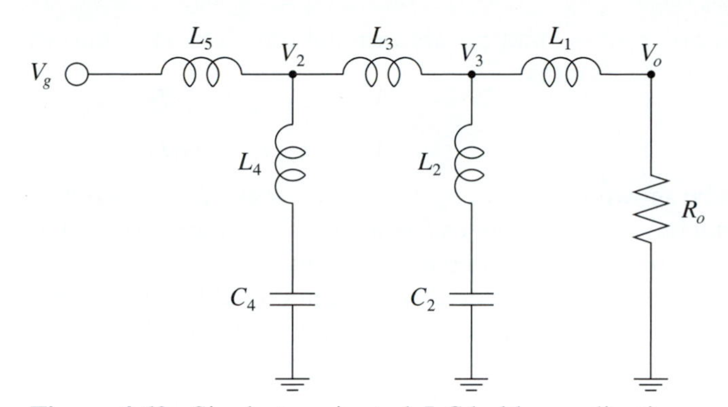

Figure 4.63 Singly terminated LC-ladder realization with complex zeros.

The impedance Y_5 can be calculated from

$$Y_5 = Y_4 - \frac{\frac{s}{L_4}}{s^2 - s_3^2} = \frac{0.54404382321885 + s^2}{0.1913221759s(3.476896154 + s^2)} - \frac{\frac{s}{0.226812421}}{s^2+3.47689615}$$
$$= \frac{0.156474(3.476896154 + s^2)}{0.1913221759s(3.476896154 + s^2)} = \frac{1}{1.22270919s} = \frac{1}{L_5 s}$$

and identified as an admittance of an inductor

$$L_5 = 1.22270919$$

The realization is shown in Fig. 4.63; the element values are

$$L_1 = 0.47666\,\mathrm{H}$$
$$L_2 = 0.10989\,\mathrm{H}$$
$$C_2 = 1.1061\,\mathrm{F}$$
$$L_3 = 1.343\,\mathrm{H}$$
$$L_4 = 0.22681\,\mathrm{H}$$
$$C_4 = 1.268\,\mathrm{F}$$
$$L_5 = 1.2227\,\mathrm{H}$$

The above procedure is known as the *zero shifting technique*.

4.10.2 Doubly Terminated Ladder Networks

The transfer function of the doubly terminated ladder realization, shown in Fig. 4.61, is obtained for $R_g \neq 0$:

$$H(s) = \frac{V_o}{V_g} = \frac{R_o z_{21}}{R_g R_o + R_o z_{11} - z_{12}z_{21} + R_g z_{22} + z_{11}z_{22}} \tag{4.179}$$

where z_{11}, z_{12}, z_{21}, and z_{22} are rational functions in s, known as z-parameters of the LC network, such that

$$\begin{aligned} V_1 &= z_{11}I_1 + z_{12}I_2 \\ V_2 &= z_{21}I_1 + z_{22}I_2 \end{aligned} \tag{4.180}$$

It can be shown that $z_{12} = z_{21}$ holds for passive LC networks; also, z_{11} and z_{22} are odd rational functions in complex frequency s (i.e., the ratio of an odd and even polynomial, or the ratio of an even and odd polynomial).

In order to find the doubly terminated ladder realization whose transfer function is $H(s)$, the first step is to find z_{11} and z_{22} from $H(s)$. Then, we proceed with the realization of z_{11} or z_{22}. The transfer function zeros coincide with the zeros of z_{21} [Eq. (4.179)], so we do not have to find z_{21}.

The basic idea used in deriving z_{11} and z_{22} is that the average power entering the LC network must be equal to the average power leaving the LC network, which is delivered to the load R_o. For this purpose we define auxiliary rational functions $K(s)$ and $\mathcal{K}(\omega^2)$ such that

$$K(s)K(-s)|_{s^2=-\omega^2} = \mathcal{K}(\omega^2) = \mathcal{K}(-s^2) = \frac{M_{\max}^2}{M^2(\omega)} - 1 \tag{4.181}$$

where $M(\omega) = |H(j\omega)|$ is the magnitude response, and $M_{\max} = \max_{\omega} M(\omega)$ is the maximum value of the magnitude response. $\mathcal{K}(\omega^2)$ is an even rational function in ω, $\mathcal{K}(-s^2)$ is an even rational function in s with real coefficients, and

$$\mathcal{K}(\omega^2) \geq 0 \tag{4.182}$$

The complex zeros of $\mathcal{K}(-s^2)$ appear in quadruples, that is, $s_1 = \sigma_1 + j\omega_1$, $s_2 = \sigma_1 - j\omega_1$, $s_3 = -\sigma_1 + j\omega_1$, and $s_4 = -\sigma_1 - j\omega_1$; σ_1, ω_1 are real and positive. Since $K(s)$ is a rational function with real coefficients, we must choose complex-conjugate pairs of zeros from the left, or right, complex s-plane; that is, we can choose s_1 and s_2, or s_3 and s_4. If the zeros are purely imaginary, it is irrelevant which pair we choose.

The real zeros of $\mathcal{K}(-s^2)$ appear in pairs—that is, $s_1 = \sigma_1$ and $s_2 = -\sigma_1$—and we select only one zero for $K(s)$.

The normalized transfer function $H_n(s)$, used in finding the doubly terminated ladder realization,

$$H_n(s) = \frac{H(s)}{M_{\max}} \tag{4.183}$$

is expressed as

$$H_n(s) = \frac{P(s)}{D_{\text{even}} + D_{\text{odd}}} = \frac{P_{\text{even}} + P_{\text{odd}}}{D_{\text{even}} + D_{\text{odd}}} \tag{4.184}$$

where D_{even} is an even polynomial in s, D_{odd} is an odd polynomial in s, and $P(s)$ is determined by the transfer function zeros. P_{even} is an even polynomial in s, and P_{odd} is an odd polynomial in s.

From (4.181) we find

$$K(s) = \frac{N_{\text{even}} + N_{\text{odd}}}{P(s)} = \frac{N_{\text{even}} + N_{\text{odd}}}{P_{\text{even}} + P_{\text{odd}}} \tag{4.185}$$

where N_{even} is an even polynomial in s, and N_{odd} is an odd polynomial in s. We adopt that the poles of $K(s)$ are the transfer function zeros.

It can be shown [24] that

$$z_{11} = R_g \frac{D_{\text{even}} - N_{\text{even}}}{D_{\text{odd}} + N_{\text{odd}}}, \qquad z_{22} = R_o \frac{D_{\text{even}} + N_{\text{even}}}{D_{\text{odd}} + N_{\text{odd}}} \tag{4.186}$$

or

$$z_{11} = R_g \frac{D_{\text{odd}} - N_{\text{odd}}}{D_{\text{even}} + N_{\text{even}}}, \qquad z_{22} = R_o \frac{D_{\text{odd}} + N_{\text{odd}}}{D_{\text{even}} + N_{\text{even}}} \tag{4.187}$$

The impedance z_{11} can be realized by the classical Foster or Cauer synthesis procedures.

The algorithmic details for the doubly terminated ladder realization are illustrated by the following example.

Doubly Terminated Ladder Realization with Complex Zeros. Consider the transfer function with complex zeros:

$$H(s) = \frac{(s^2 + 3.476896154)(s^2 + 8.227391422)}{55.3858\,(s + 0.60913)\left(s^2 + 0.263147s + 1.166357185\right)\left(s^2 + 0.85422659s + 0.7269594794\right)}$$

assuming $R_g = R_o = 1\,\Omega$.

The magnitude response maximum is

$$M_{\max} = \max_{\omega} |H(j\omega)| = 1.00000848$$

The normalized transfer function is given by

$$H_n(s) = \frac{H(s)}{1.00000848}$$

and expressed as

$$H_n(s) = \frac{P_{\text{even}} + P_{\text{odd}}}{D_{\text{even}} + D_{\text{odd}}} \tag{4.188}$$

with

$$\begin{aligned}
P_{\text{odd}} &= 0 \\
P_{\text{even}} &= 28.6058 + 11.7043s^2 + s^4 \\
D_{\text{odd}} &= 87.0293s + 155.011s^3 + 55.3863s^5 \\
D_{\text{even}} &= 28.6058 + 137.238s^2 + 95.6246s^4
\end{aligned}$$

The auxiliary function $\mathcal{K}(-s^2)$ is computed from

$$\mathcal{K}(-s^2) = \frac{D_{\text{even}}^2 - D_{\text{odd}}^2}{P_{\text{even}}^2 - P_{\text{odd}}^2} - 1 = \frac{D_{\text{even}}^2 - D_{\text{odd}}^2 - P_{\text{even}}^2 + P_{\text{odd}}^2}{P_{\text{even}}^2 - P_{\text{odd}}^2} = \frac{N_{\text{even}}^2 - N_{\text{odd}}^2}{P_{\text{even}}^2 - P_{\text{odd}}^2} \tag{4.189}$$

The 10 zeros of $\mathcal{K}(-s^2)$, obtained from the algebraic equation in s,

$$D_{\text{even}}^2 - D_{\text{odd}}^2 - P_{\text{even}}^2 + P_{\text{odd}}^2 = 0$$

are

$$
\begin{aligned}
s_{k1} &= 0 & s_{k6} &= 0 \\
s_{k2} &= -0.00102181 + 0.95906j & s_{k7} &= 0.00102181 + 0.95906j \\
s_{k3} &= -0.00102181 - 0.95906j & s_{k8} &= 0.00102181 - 0.95906j \\
s_{k4} &= -0.00284197 + 0.623457j & s_{k9} &= 0.00284197 + 0.623457j \\
s_{k5} &= -0.00284197 - 0.623457j & s_{k10} &= 0.00284197 - 0.623457j
\end{aligned}
$$

The five zeros of $K(s)$ can be summarized as

$$
\begin{aligned}
s_{k1} &= 0 \\
s_{k2} &= \alpha_1 0.00102181 + 0.95906j \\
s_{k3} &= \alpha_1 0.00102181 - 0.95906j \\
s_{k4} &= \alpha_2 0.00284197 + 0.623457j \\
s_{k5} &= \alpha_2 0.00284197 - 0.623457j
\end{aligned}
\qquad
\begin{aligned}
\alpha_1 &\in \{1, -1\} \\
\alpha_2 &\in \{1, -1\}
\end{aligned}
$$

where the real part can be positive for $\alpha_i = 1$ or negative for $\alpha_i = -1$.

Let us select $\alpha_1 = 1$ and $\alpha_2 = 1$; we find $K(s)$ and the even and odd parts of the numerator of $K(s)$ and the denominator of $H_n(s)$:

$$
\begin{aligned}
N_{\text{odd}} &= 0.357531s + 1.30852s^3 + s^5 \\
N_{\text{even}} &= 0.00602244s^2 + 0.00772756s^4 \\
D_{\text{odd}} &= 1.57132s + 2.79873s^3 + s^5 \\
D_{\text{even}} &= 0.516478 + 2.47783s^2 + 1.7265s^4
\end{aligned}
$$

New D_{odd} and D_{even} are obtained after normalizing the numerator of $H_n(s)$ in such a way that the coefficient of the highest term, s^5, is unity.

Using Eq. (4.186), we find z_{11}:

$$
z_{11} = R_g \frac{D_{\text{even}} - N_{\text{even}}}{D_{\text{odd}} + N_{\text{odd}}} = \frac{0.516478 + 2.47181s^2 + 1.71878s^4}{1.92885s + 4.10724s^3 + 2s^5}
$$

The transfer function zeros are

$$
\begin{aligned}
s_1 &= +j2.868342974959585 \\
s_2 &= -j2.868342974959585 \\
s_3 &= +j1.864643706985332 \\
s_4 &= -j1.864643706985332
\end{aligned}
$$

or

$$
\begin{aligned}
s_1^2 &= s_2^2 = -8.2273914225 \\
s_3^2 &= s_4^2 = -3.476896154
\end{aligned}
$$

and they must be realized by the LC ladder derived from z_{11}.

Since the order of the z_{11} denominator is larger than the order of the z_{11} numerator, we proceed with the admittance Y_{11}:

$$Y_{11} = \frac{1}{z_{11}} = \frac{1.92885s + 4.10724s^3 + 2s^5}{0.516478 + 2.47181s^2 + 1.71878s^4}$$

The admittance Y_{11} is an improper rational function in s, and it can be expressed as a sum of a linear term $C_1 s$ and an admittance Y_2; our goal is to find C_1 such that Y_2 has at least one complex pair of zeros that coincide with the transfer-function zeros (say, the zeros s_1 and s_2):

$$Y_{11} = C_1 s + Y_2, \qquad Y_2(s_1) = Y_2(s_2) = 0$$

The capacitance C_1 is computed from

$$C_1 = \left.\frac{Y_{11}}{s}\right|_{s=s_1} = \frac{1.92885 + 4.10724s^2 + 2s^4}{0.516478 + 2.47181s^2 + 1.71878\,s^4} = 1.07245$$

and the admittance Y_2 is found as

$$Y_2 = Y_{11} - C_1 s = s\frac{1.37495 + 1.45636s^2 + 0.156701s^4}{0.516478 + 2.47181s^2 + 1.71878s^4}$$

or, equivalently,

$$Y_2 = s\frac{0.0911699\,(8.22739 + s^2)\,(1.06648 + s^2)}{(0.253704 + s^2)\,(1.18442 + s^2)}$$

We proceed with the impedance Z_2,

$$Z_2 = \frac{1}{Y_2} = \frac{(0.253704 + s^2)\ (1.18442 + s^2)}{0.0911699s\,(8.22739 + s^2)\,(1.06648 + s^2)}$$

and extract a biquadratic term with the pole pair s_1, s_2, which corresponds to a simple LC connection

$$Z_2 = \frac{1}{sC_2 + \dfrac{1}{sL_2}} + Z_3 = \frac{\dfrac{s}{C_2}}{s^2 + \dfrac{1}{L_2C_2}} + Z_3 = \frac{\dfrac{s}{C_2}}{s^2+s_1s_2} + Z_3$$

We find C_2,

$$C_2 = \left.Y_2\frac{s}{s^2+8.2273914225}\right|_{s=s_1} = s_1^2\frac{0.0911699\ (1.06648 + s_1^2)}{(0.253704 + s_1^2)\ \ (1.18442 + s_1^2)}$$
$$C_2 = 0.0956459$$

and L_2,

$$L_2 = -\frac{1}{C_2 s_1^2} = \frac{1}{8.2273914225C_2} = 1.27078$$

The impedance Z_3 can be calculated from

$$Z_3 = Z_2 - \frac{\frac{s}{0.0956459}}{s^2+8.2273914225}$$

yielding

$$Z_3 = \frac{0.513303\,(0.78045 + s^2)}{s(1.06648 + s^2)}$$

Since the order of the Z_3 denominator is larger than the order of the Z_3 numerator, we proceed with the admittance Y_3:

$$Y_3 = \frac{1}{Z_3} = \frac{s(1.06648 + s^2)}{0.513303\,(0.78045 + s^2)}$$

The admittance Y_3 is an improper rational function in s, and it can be expressed as a sum of a linear term $C_3 s$ and an admittance Y_4; our goal is to find C_3 such that Y_4 has at least one complex pair of zeros that coincide with the transfer-function zeros (excluding the zeros s_1 and s_2, but using s_3 and s_4):

$$Y_3 = C_3 s + Y_4$$

We obtain

$$C_3 = \left.\frac{Y_3}{s}\right|_{s=s_3} = \frac{1.06648 + s_3^2}{0.513303\,(0.78045 + s_3^2)} = 1.74151$$

and

$$Y_4 = Y_3 - C_3 s = s\frac{0.718527 + 0.206658\,s^2}{0.78045 + s^2} = s\frac{0.206658\,(3.4769 + s^2)}{0.78045 + s^2}$$

Again, we proceed with

$$Z_4 = \frac{1}{Y_4} = \frac{0.78045 + s^2}{0.206658s\,(3.4769 + s^2)}$$

and extract a biquadratic term with the pole pair s_3, s_4, which corresponds to a simple LC connection:

$$Z_4 = \frac{1}{sC_4 + \frac{1}{sL_4}} + Z_5 = \frac{\frac{s}{C_4}}{s^2 + \frac{1}{L_4 C_4}} + Z_5 = \frac{\frac{s}{C_4}}{s^2+s_3 s_4} + Z_5$$

We find

$$C_4 = \left.Y_4\frac{s}{s^2 + 3.476896154}\right|_{s=s_3} = s_3^2\frac{0.206658}{0.78045 + s_3^2} = 0.266472$$

and

$$L_4 = -\frac{1}{C_4 s_3^2} = \frac{1}{3.476896154 C_4} = 1.07934$$

The impedance Z_5 can be calculated from

$$Z_5 = Z_4 - \frac{\frac{s}{C_4}}{s^2 - s_3^2} = \frac{1.08618}{s} = \frac{1}{C_5 s}$$

and identified as an impedance of a capacitor:

$$C_5 = 0.920658$$

The same procedure can be repeated for all possible combinations of α_1 and α_2. The realization is shown in Fig. 4.64, and the element values for various α_1 and α_2 follow:

α_1	α_2	C_1 (F)	C_2 (F)	C_3 (F)	C_4 (F)	C_5 (F)	L_1 (H)	L_2 (H)
1	1	1.0724	0.095646	1.7415	0.26647	0.92066	1.2708	1.0793
−1	1	1.0695	0.095821	1.7413	0.26590	0.92382	1.2685	1.0817
1	−1	1.0649	0.095526	1.7417	0.26687	0.92796	1.2724	1.0777
−1	−1	1.0620	0.095700	1.7414	0.26629	0.93117	1.2701	1.0801
0	0	1.0672	0.095675	1.7415	0.26639	0.92594	1.2704	1.0797
1%	→	1.07	0.096	1.74	0.27	0.93	1.27	1.08

Approximate element values, which are within 1% error, are given in the last row. The approximate element values are important for implementation because commercial components are available within some tolerances.

Formally, we can examine the case $\alpha_1 = \alpha_2 = 0$, which implies purely imaginary zeros. This is legitimate if the real parts of the zeros are negligibly small, or when we know that the function $K(s)$ has purely imaginary zeros (which is exactly the case when we design elliptic-type filters).

Let us derive the transfer function of the circuit from Fig. 4.64 to validate the realization:

$$H_{\text{val}}(s) = \frac{(3.4769+s^2)(8.22739+s^2)}{110.773\ (0.60913+s)(1.16636+0.263147s+s^2)\ (0.726959+0.854227s+s^2)}$$

$$H_{\text{val}}(s) = \frac{1}{2}H(s)$$

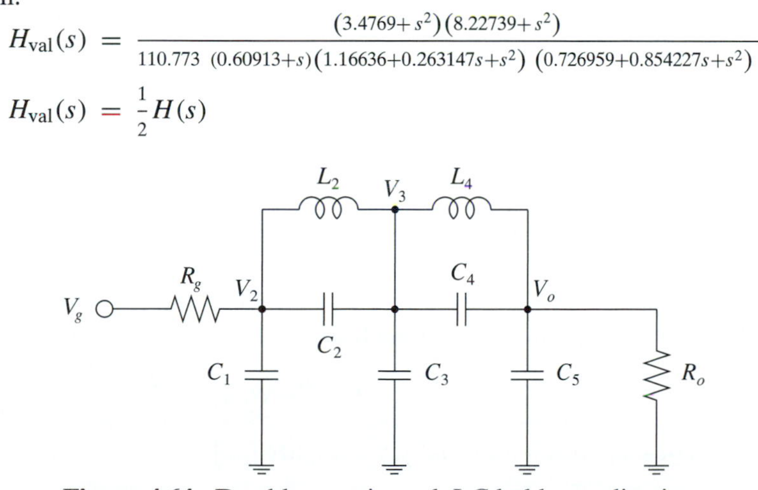

Figure 4.64 Doubly terminated LC-ladder realization with complex zeros.

The transfer function $H_{\text{val}}(s)$ has the desired poles and zeros, but the gain constant is different; that is, it is $\frac{1}{2}$ instead of 1. Therefore, the above procedure can realize the desired transfer function within a constant multiplier.

Instead of using z_{11},

$$z_{11} = R_g \frac{D_{\text{even}} - N_{\text{even}}}{D_{\text{odd}} + N_{\text{odd}}}$$

we could use z_{22},

$$z_{22} = R_o \frac{D_{\text{even}} + N_{\text{even}}}{D_{\text{odd}} + N_{\text{odd}}}$$

to realize the required transfer function $H(s)$. Obviously, the two circuits should be the same. If this is not the case we have to change R_o to provide same element values (capacitances and inductances) in both realizations [24].

4.10.3 Quality Factor of Inductors and Capacitors

For some practical applications a critical quantity might be the quality factor of inductors and capacitors. The quality factor of an inductor, Q_L, and a capacitor, Q_C, is defined as [30]

$$Q_L = \frac{2\pi f L}{R}$$

$$Q_C = \frac{2\pi f C}{G}$$

where L is the inductance in H, C is the capacitance in F, R is the inductor loss resistance in Ω, and G is the capacitor loss conductance in S.

An LC filter with $R = 0$ and $G = 0$ is called the lossless filter. $R > 0$ and $G > 0$ cause dissipation or loss. A transfer-function pole, s_p, of an LC filter implemented with lossy elements ($R > 0$, $G > 0$) can be approximately computed from the transfer-function pole, s_{p0}, of the lossless LC filter from

$$s_p \approx s_{p0} + \frac{1}{2Q_L} + \frac{1}{2Q_C}$$

assuming that the Q-factors of all inductors are Q_L and that the Q-factors of all capacitors are Q_C [30].

In general, $Q_C \gg Q_L$ and we can neglect the capacitor losses. Q_L must be sufficiently larger than the maximal transfer-function pole Q-factor, $Q_{\max}$:

$$Q_L \gg Q_{\max}$$

For practical applications, usually, we require [30]

$$Q_L \geq 3Q_{\max}$$

Lossy components increase the passband attenuation by an amount proportional to the group delay of the lossless filter. Therefore, we prefer the transfer function with lower pole Q-factors and smaller group delays. This fact should be seriously taken into account in the approximation step of the filter design.

4.11 OPERATIONAL TRANSCONDUCTANCE AMPLIFIER (OTA) FILTERS

In this section we consider realizations of continuous-time active RC filters with operational transconductance amplifiers (OTA). There are two types of these filters:

- *OTA-C filters* (integrated filters implemented with OTAs and poly-silicon capacitors)
- *OTA-R-C filters* (integrated filters implemented with OTAs, poly-silicon capacitors, and linear full CMOS resistors)

The actual trend in integrated circuits technology is to incorporate in a chip as many digital and analog blocks as possible. The mixed-mode (analog and digital) integrated circuits may require high-quality analog filters before analog-to-digital signal conversion, as well as after the digital-to-analog signal conversion. In order to use very small silicon area and to manufacture low-cost integrated circuits, it is preferable to use the same technologies for digital and analog circuits. Available technologies for integrated circuits are complementary-metal-oxide-semiconductor (CMOS), bipolar, bipolar-CMOS (BICMOS), gallium-arsenide.

In many communication applications, the frequency range of filters is from a few kHz up to several GHz. The conventional operational amplifier cannot be used as an amplifier at very high frequencies. High-frequency and high-selectivity continuous-time filters demand new active devices acting as amplifiers. The useful frequency range of the operational amplifier is lower than the useful frequency range of OTA.

OTA is basically a voltage-to-current transducer. OTA converts the input voltage $(V_+ - V_-)$ into the output current I_o:

$$I_o = g_m(V_+ - V_-)$$

where g_m is the OTA *transconductance* with units of ampere/volt or siemens, abbreviated S [16 pp. 2472–2490]. Typical values for g_m are tens to hundreds of μS in CMOS technology, and up to mS in bipolar technology.

The most commonly used circuit symbol and a simplified small-signal equivalent circuit for OTA is shown in Fig. 4.65.

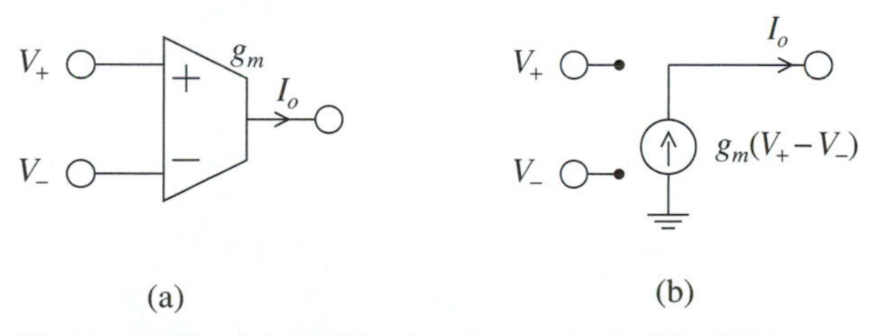

Figure 4.65 (a) OTA circuit symbol, (b) OTA small-signal equivalent cicuit.

Most of the fully integrated OTA filters use OTAs and capacitors as the main components. They are sensitive to parasitic capacitances. If the parasitic capacitances and temperature variations are considered, typically, the tolerances of the ratio $r_{cg} = C/g_m$ can be greater than 30%. However, for a unity capacitor ratio, $r_c = C_1/C_2 \approx 1$, a 0.1% tolerance of r_c can be achieved. The tolerance of the unity transconductance ratio $r_g = g_{m1}/g_{m2} \approx 1$ is smaller than 0.5%. For a larger transconductance ratio $r_g \gg 1$, the tolerance can be 2%.

Another limitation of OTAs can be a small input voltage swing required to maintain linearity.

All of these technological imperfections must be taken into consideration in filter design.

The parasitic effects and the large parameter tolerances are the disadvantages of OTA filters. Therefore, it is necessary to include an on-chip tuning system to compensate for the technological imperfections. We can achieve a better performance of OTA filters by adjusting the transfer function parameters, so that the transfer-function pole magnitude can be successfully tuned by the on-chip tuning system. The OTA filter is often implemented on the same chip with the self-tuning system.

OTA filters can be designed using the doubly terminated *RLC* realization or as a cascade connection of biquads. It has been shown that the dynamic range of the ladder filters is superior to the dynamic range of the cascaded biquad filters. Also, the ladder filters are less sensitive to the tolerances than the cascaded biquads. The main disadvantage of a higher-order selective ladder filter is its complex and less precise tuning procedure. For high-frequency applications and sharp specifications, realizations with cascaded biquads are more attractive because they can be precisely tuned.

4.11.1 Biquadratic OTA-C Realizations

In this chapter we present filter realizations with OTAs and capacitors, only. We consider several types of OTA biquads, including very simple realizations with the minimal number of elements (OTAs and capacitors).

Simple OTA-C Biquads. A simple universal OTA biquad realized with only four elements is shown in Fig. 4.66. This biquad can be used as a lowpass, highpass, bandpass, or bandreject second-order filter section. The required transfer function is realized by connecting the terminals V_a, V_b, and V_c to the ground or to the input voltage source V_g. The output voltage is V_3.

The realizable transfer functions of the biquad from Fig. 4.66 are

$$\frac{V_3}{V_g} = \frac{s^2\dfrac{V_c}{V_g} + s\dfrac{g_{m2}}{C_2}\dfrac{V_b}{V_g} + \dfrac{g_{m1}g_{m2}}{C_1C_2}\dfrac{V_a}{V_g}}{s^2 + \dfrac{g_{m2}}{C_2}s + \dfrac{g_{m1}g_{m2}}{C_1C_2}}$$

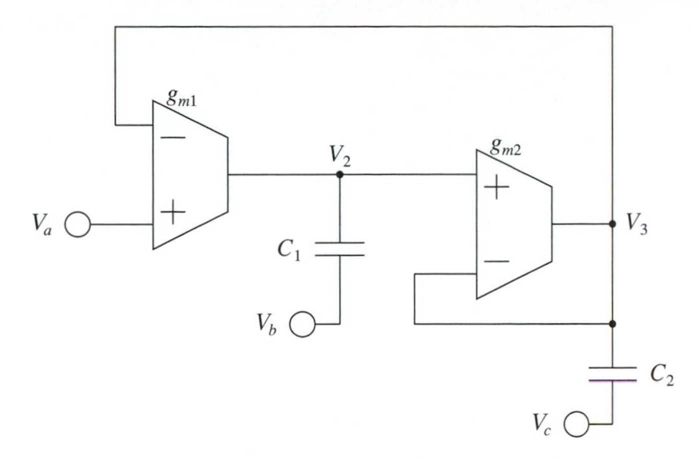

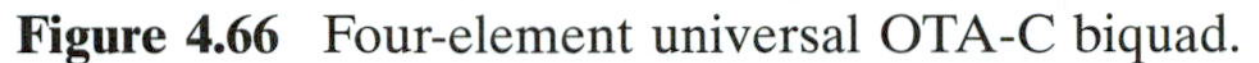

Figure 4.66 Four-element universal OTA-C biquad.

$$
\begin{aligned}
H_{LP}(s) &= \frac{\dfrac{g_{m1}g_{m2}}{C_1C_2}}{s^2 + \dfrac{g_{m2}}{C_2}s + \dfrac{g_{m1}g_{m2}}{C_1C_2}}, \quad V_a = V_g,\ V_b = 0,\ V_c = 0 \\
H_{HP}(s) &= \frac{s^2}{s^2 + \dfrac{g_{m2}}{C_2}s + \dfrac{g_{m1}g_{m2}}{C_1C_2}}, \quad V_a = 0,\ V_b = 0,\ V_c = V_g \\
H_{BP}(s) &= \frac{s\dfrac{g_{m2}}{C_2}}{s^2 + \dfrac{g_{m2}}{C_2}s + \dfrac{g_{m1}g_{m2}}{C_1C_2}}, \quad V_a = 0,\ V_b = V_g,\ V_c = 0 \\
H_{BR}(s) &= \frac{s^2 + \dfrac{g_{m1}g_{m2}}{C_1C_2}}{s^2 + \dfrac{g_{m2}}{C_2}s + \dfrac{g_{m1}g_{m2}}{C_1C_2}}, \quad V_a = V_g,\ V_b = 0,\ V_c = V_g
\end{aligned}
\tag{4.190}
$$

The pole magnitude, ω_p, and the pole Q-factor, Q_p, are given by

$$
\begin{aligned}
\omega_p &= \sqrt{\frac{g_{m1}g_{m2}}{C_1C_2}} \\
Q_p &= \sqrt{\frac{g_{m1}C_2}{g_{m2}C_1}}
\end{aligned}
\tag{4.191}
$$

The Q-factor is determined by the capacitance ratio, $\frac{C_2}{C_1}$, and the transconductance ratio, $\frac{g_{m1}}{g_{m2}}$, which can be accurately maintained in monolithic design. The most sensitive parameter, ω_p, is a function of the transconductance-capacitance ratio, $\frac{g_m}{C}$, which is difficult to manufacture accurately.

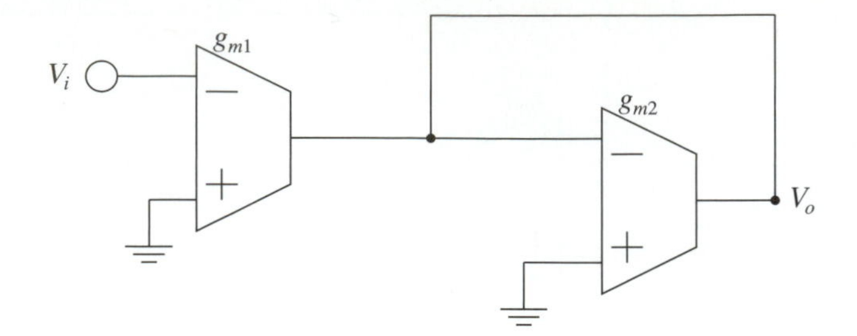

Figure 4.67 Inverting all-OTA amplifier.

For higher-order filters obtained by cascading these biquads, a buffer (voltage follower) is required to prevent interstage loading. Only lowpass filter sections can be cascaded without using buffers because the output V_3 of one biquad can be directly connected to the input V_a of the next biquad.

Typical inverting and noninverting all-OTA amplifiers are shown in Figs. 4.67 and 4.68. The gain of the inverting amplifier is

$$\frac{V_o}{V_i} = -\frac{g_{m1}}{g_{m2}}$$

and the gain of the noninverting amplifier is

$$\frac{V_o}{V_i} = \frac{g_{m1}}{g_{m2}}$$

For $g_{m1} = g_{m2}$ the gain is -1 or 1.

Simple OTA-C Lowpass Notch Biquad. The universal biquad from Fig. 4.66 can be modified to become a second-order lowpass notch filter section as shown in Fig. 4.69. The capacitor C_2 in Fig. 4.66 has been split into two capacitors C_2 and C_3. The transfer function is found to be

$$H(s) = \frac{V_3}{V_1} = \frac{C_2}{C_2 + C_3} \frac{s^2 + \dfrac{g_{m1}g_{m2}}{C_1C_2}}{s^2 + \dfrac{g_{m2}}{C_2 + C_3}s + \dfrac{g_{m1}g_{m2}}{C_1(C_2 + C_3)}} \tag{4.192}$$

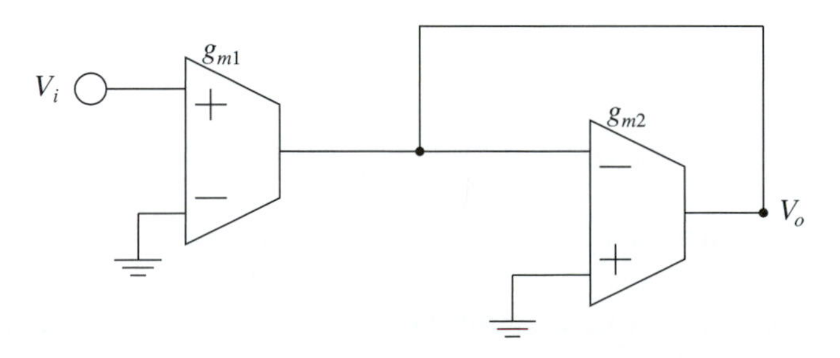

Figure 4.68 Noninverting all-OTA amplifier.

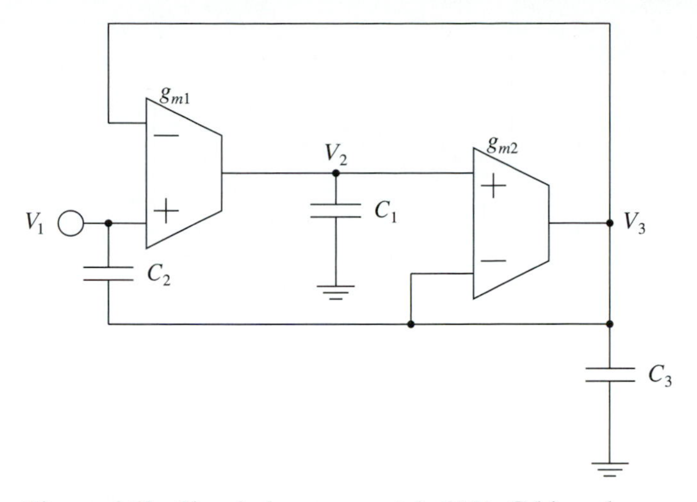

Figure 4.69 Simple lowpass notch OTA-C biquad.

The pole magnitude, ω_p, the pole Q-factor, Q_p, and the magnitude of the transfer function zero, ω_z, are

$$\omega_p = \sqrt{\frac{g_{m1}g_{m2}}{C_1(C_2 + C_3)}}$$
$$Q_p = \sqrt{\frac{g_{m1}(C_2 + C_3)}{g_{m2}C_1}} \tag{4.193}$$
$$\omega_z = \sqrt{\frac{g_{m1}g_{m2}}{C_1C_2}}$$

Obviously,

$$\frac{\omega_p}{\omega_z} = \sqrt{\frac{C_2}{C_2 + C_3}} < 1 \tag{4.194}$$

which implies that we can realize only the lowpass notch transfer function with this biquad.

Notice that the gain constant $\frac{C_2}{C_2+C_3}$ is always smaller than 1.

The disadvantages of this realization are as follows: (a) It requires three capacitors instead of two, and (b) buffers between biquads are required in cascade realizations of higher-order filters. Therefore, at least seven components (four OTAs and three capacitors) are required for a lowpass notch biquad.

General OTA-C Biquads. A universal OTA-C second-order filter section that does not require any additional buffers is shown in Fig. 4.70.

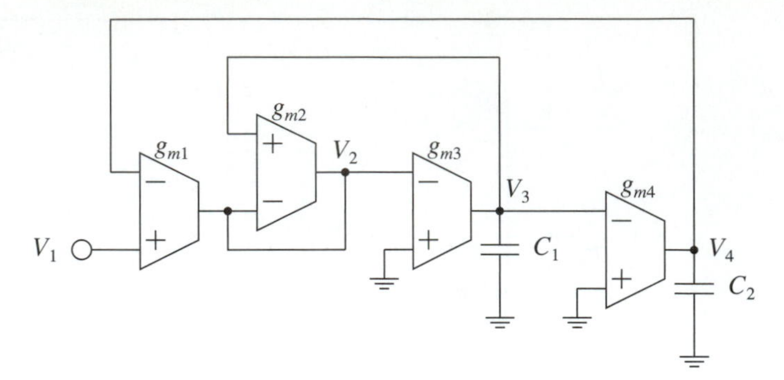

Figure 4.70 Four-OTA general biquad.

This biquad can realize the lowpass $H_{LP}(s)$, highpass $H_{HP}(s)$, and bandpass $H_{BP}(s)$ transfer function:

$$
\begin{aligned}
H_{LP}(s) &= \frac{V_4}{V_1} = \frac{\omega_p^2}{s^2 + \dfrac{\omega_p}{Q_p}s + \omega_p^2} \\
H_{BP}(s) &= \frac{V_3}{V_1} = \frac{\dfrac{\omega_p}{Q_p}s}{s^2 + \dfrac{\omega_p}{Q_p}s + \omega_p^2} \\
H_{HP}(s) &= \frac{V_2}{V_1} = \frac{s^2}{s^2 + \dfrac{\omega_p}{Q_p}s + \omega_p^2} \\
\omega_p &= \sqrt{\frac{g_{m1}g_{m3}}{C_1C_2}\frac{g_{m4}}{g_{m2}}} \\
Q_p &= \sqrt{\frac{C_1g_{m1}}{C_2g_{m3}}\frac{g_{m4}}{g_{m2}}}
\end{aligned}
\tag{4.195}
$$

The design equations for this biquad are simple: We can assign arbitrary values to g_{m1}, g_{m2}, C_1, and C_2, and then we can calculate the other transconductances g_{m3} and g_{m4} from

$$
\begin{aligned}
g_{m3} &= \frac{\omega_p}{Q_p}C_1 \\
g_{m4} &= \omega_p Q_p C_2 \frac{g_{m2}}{g_{m1}}
\end{aligned}
\tag{4.196}
$$

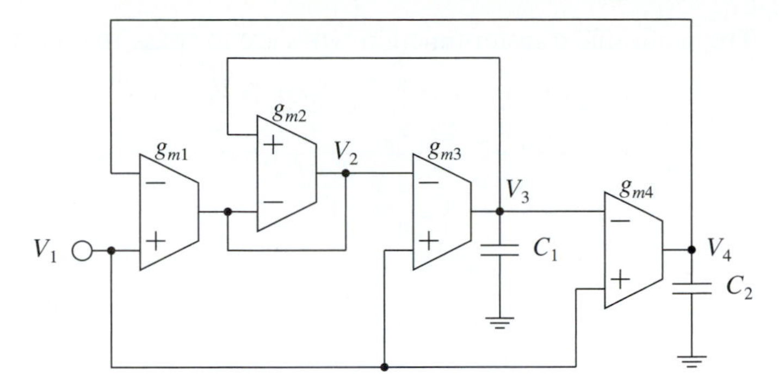

Figure 4.71 Four-OTA notch biquad.

For $g_{m1} = g_{m2}$ and $C_1 = C_2 = C$, the Q-factor can be tuned by the ratio $\frac{g_{m4}}{g_{m3}}$, and the pole magnitude can be tuned by the product $g_{m3}g_{m4}$:

$$\omega_p = \sqrt{\frac{g_{m3}g_{m4}}{C^2}}$$

$$Q_p = \sqrt{\frac{g_{m4}}{g_{m3}}}$$

which means that we have to tune only two biquad parameters, g_{m3} and g_{m4}.

A modification of the realization shown in Fig. 4.70, performed by connecting the $+$ inputs of the third and fourth OTA to the input V_1, is presented in Fig. 4.71. Such a realization can be used as a second-order notch filter section.

Another versatile OTA-C biquad is shown in Fig. 4.72.

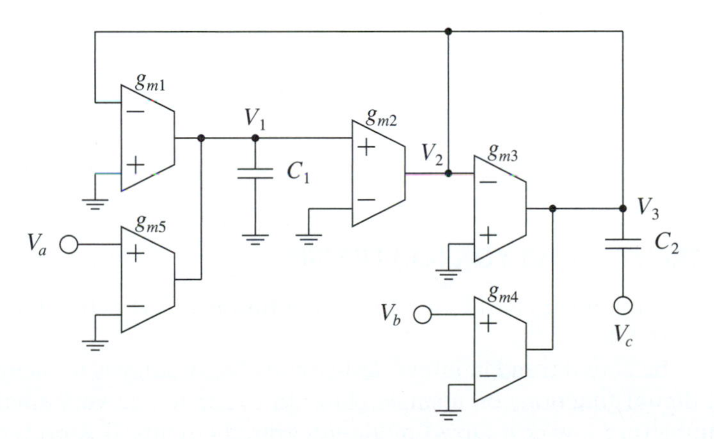

Figure 4.72 Five-OTA universal biquad.

The realizable transfer-function types are lowpass, highpass, and bandpass:

$$\frac{V_3}{V_g} = \frac{s^2\dfrac{V_c}{V_g} + s\dfrac{g_{m4}}{C_2}\dfrac{V_b}{V_g} + \dfrac{g_{m2}g_{m5}}{C_1C_2}\dfrac{V_a}{V_g}}{s^2 + \dfrac{g_{m3}}{C_2}s + \dfrac{g_{m1}g_{m2}}{C_1C_2}}$$

$$H_{LP}(s) = \frac{\dfrac{g_{m2}g_{m5}}{C_1C_2}}{s^2 + \dfrac{g_{m3}}{C_2}s + \dfrac{g_{m1}g_{m2}}{C_1C_2}}, \quad V_a = V_g, \ V_b = 0, \ V_c = 0$$

$$H_{HP}(s) = \frac{s^2}{s^2 + \dfrac{g_{m3}}{C_2}s + \dfrac{g_{m1}g_{m2}}{C_1C_2}}, \quad V_a = 0, \ V_b = 0, \ V_c = V_g \tag{4.197}$$

$$H_{BP}(s) = \frac{s\dfrac{g_{m4}}{C_2}}{s^2 + \dfrac{g_{m3}}{C_2}s + \dfrac{g_{m1}g_{m2}}{C_1C_2}}, \quad V_a = 0, \ V_b = V_g, \ V_c = 0$$

The bandreject, lowpass-notch, and highpass-notch transfer functions can be realized for $V_b = 0$ and $V_a = V_c = V_g$, and by proper selection of the ratio of transconductances g_{m1} and g_{m5} we obtain

$$H_{BR}(s) = \frac{s^2 + \dfrac{g_{m2}g_{m5}}{C_1C_2}}{s^2 + \dfrac{g_{m3}}{C_2}s + \dfrac{g_{m1}g_{m2}}{C_1C_2}}, \quad g_{m1} = g_{m5}$$

$$H_{LPN}(s) = \frac{s^2 + \dfrac{g_{m2}g_{m5}}{C_1C_2}}{s^2 + \dfrac{g_{m3}}{C_2}s + \dfrac{g_{m1}g_{m2}}{C_1C_2}}, \quad g_{m1} < g_{m5} \tag{4.198}$$

$$H_{HPN}(s) = \frac{s^2 + \dfrac{g_{m2}g_{m5}}{C_1C_2}}{s^2 + \dfrac{g_{m3}}{C_2}s + \dfrac{g_{m1}g_{m2}}{C_1C_2}}, \quad g_{m1} > g_{m5}$$

The design equations can be readily derived from the above transfer functions.

4.12 CURRENT-CONVEYOR (CC) FILTERS

In this section we consider design of continuous-time active filters using current conveyors (CC).

The actual trend in integrated-circuits technology is to incorporate mixed analog and digital functions on a single chip. In order to use very small silicon area and to manufacture low cost mixed-mode integrated circuits, it is preferable to use the same technologies for digital and analog circuits. Generally, the technology has been optimized for digital circuits. Therefore, the mixed-mode (analog and digital) integrated circuits require an advanced level of analog continuous-time filter design.

Analog continuous-time active filters consist of active amplifiers and passive components like capacitors and resistors. The basic element of an amplifier is a transistor. The small-signal model of a transistor is a voltage-controlled current output device. Usually, we assemble transistors into voltage-oriented circuits because we tend to think in voltage terms rather than in current terms. This "voltage" approach reduces the useful frequency range 10 or 100 times the useful frequency range of the "current" approach.

In this book we consider only the second generation of current conveyors, denoted as CCII, which is more useful than the first generation, CCI, for filter design.

A great deal of work has been reported on the design of CC filters. The frequency range of the CC filters can be up to GHz. The useful frequency range of current conveyors is limited only by the transistor gain–bandwidth product.

The most commonly used circuit symbol for CCII is shown in Fig. 4.73. The ideal CCII is described by three equations:

$$\begin{aligned} I_y &= 0 \\ V_x &= V_y \\ I_z &= aI_x \end{aligned} \tag{4.199}$$

The input terminal Y exhibits an infinite input impedance, and the current that flows into that terminal is identically zero. The voltage applied to terminal Y appears at terminal X. The current flowing into terminal X is conveyed to the output terminal Z. The real constant a is the *current gain* of CCII. Traditionally, the CCII operation has largely been associated with the unity current gain. For $a = 1$ the device is called the positive current conveyor and denoted by CCII+. For $a = -1$ the device is called the negative current conveyor and denoted by CCII–. However, a nonunity current gain is just as easily implemented [48, p. 115].

Most CC filters are based on the following:

- Simulating inductors of a passive doubly terminated RLC filter with CCs, capacitors, and resistors
- Converting an op amp active RC filter into a CC active RC equivalent filter

The inductor simulation is attractive because the corresponding CC active filter retains low passive-component sensitivities. The conversion of op amp active RC circuits into CC active RC circuits retains exactly the same component sensitivities, and we can use

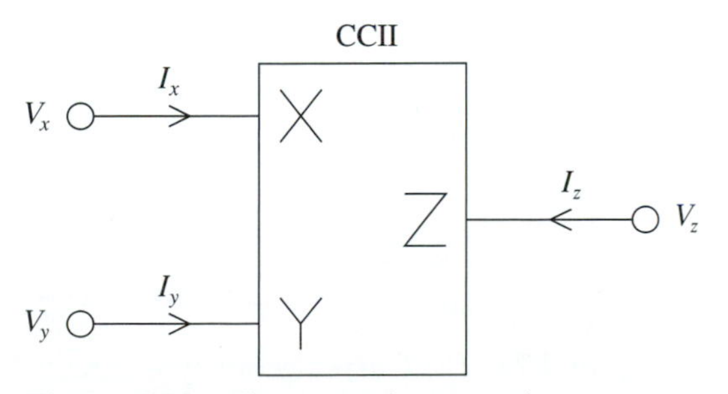

Figure 4.73 The second generation current-conveyor (CCII) circuit symbol.

the well-known theory of op amp active RC filters without having to reinvent the new low-sensitive CC circuits.

4.12.1 Current-Conveyor Filters Based on Passive RLC Filters

The current conveyor is an active device that is ideally suited for integrated-circuits technology. Thus, the filter design with CCs should rely on inductorless realizations because inductors cannot be successfully integrated. A straightforward way to obtain a CC filter is to start from a passive RLC realization and simulate the inductors by CCs, capacitors and resistors.

A grounded inductor can be simulated with the one-port active RC network shown in Fig. 4.74a. The input impedance of the one-port is

$$Z_{in1} = \frac{V_1}{I_1} = \frac{CR_1R_2}{-\,a_1a_2}s \tag{4.200}$$

The input impedance of the one-port network containing single inductor (Fig. 4.74b), is

$$Z_{in2} = \frac{V_2}{I_2} = Ls \tag{4.201}$$

To make the two one-ports equivalent, $Z_{in1} = Z_{in2}$ must hold, or

$$L = \frac{CR_1R_2}{-\,a_1a_2} > 0 \tag{4.202}$$

We conclude that one CC must be CCII+, and the other must be CCII–. The simplest case is the unity gain conveyor, $a_1 = 1$ and $a_2 = -1$, which yields $L = CR_1R_2$.

Consider a doubly terminated passive RLC realization of the fifth-order filter displayed in Fig. 4.75. If we replace the grounded inductors with the one-port active

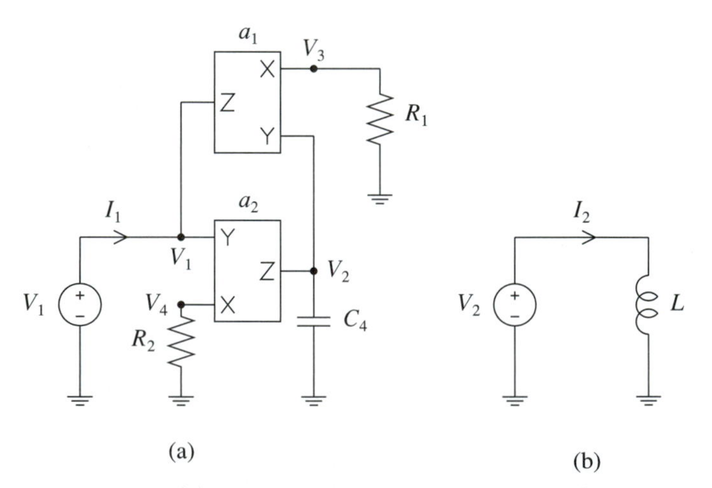

Figure 4.74 (a) Current conveyor circuit that simulates a grounded inductor (b) Simple circuit with the grounded inductor.

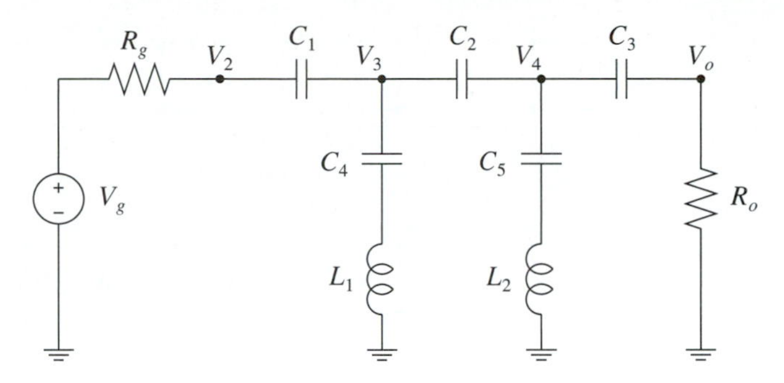

Figure 4.75 Fifth-order highpass filter.

RC network from Fig. 4.74, we obtain a CC active *RC* realization. The transfer function of the CC realization is

$$H(s) = H_1(s)H_2(s)$$

$$H_1(s) = \frac{1+C_4Ls^2}{C_1+C_4+C_1C_4Rs+C_1C_4Ls^2}$$

$$H_2(s) = \frac{C_1^2C_2R(1+C_4Ls^2)s}{C_1+2C_2+C_4+2C_1C_2Rs+C_1C_4Rs+C_1C_4Ls^2+2C_2C_4Ls^2+2C_1C_2C_4LRs^3}$$

$$R_g = R_o = R, \qquad C_3 = C_1, \qquad C_5 = C_4, \qquad L_2 = L_1 = L$$

It is important to notice that the CC filters derived from doubly terminated ladder realizations inherit low sensitivities to passive components.

4.12.2 Current-Conveyor Filters Derived from Op Amp Active *RC* Filters

Well-established theory and practice of monolithic op amp active *RC* filters can serve as a basis for obtaining high performance current-conveyor filters. In this section we demonstrate how op amp active *RC* realizations can be directly converted to CC realizations with the same transfer function and component sensitivities.

The conversion is based on the reciprocal behavior of two-port networks. *Two-port reciprocity* means that a voltage, say V, applied across the input port of a linear two-port network produces a short-circuited output current that is identical to the short-circuited input port current that would result if V were to be removed from the input port and impressed instead across the output port [16, p. 544].

In other words, a two-port network is considered reciprocal when the same transfer function results as the excitation and the response are interchanged (Fig. 4.76a):

$$\frac{V_o}{V_g} = \frac{I_o}{I_g} \tag{4.203}$$

where V_g is the voltage of the voltage source, I_g is the current of the current source, V_o is the voltage of the open-circuited port, and I_o is the current through the short-circuited port.

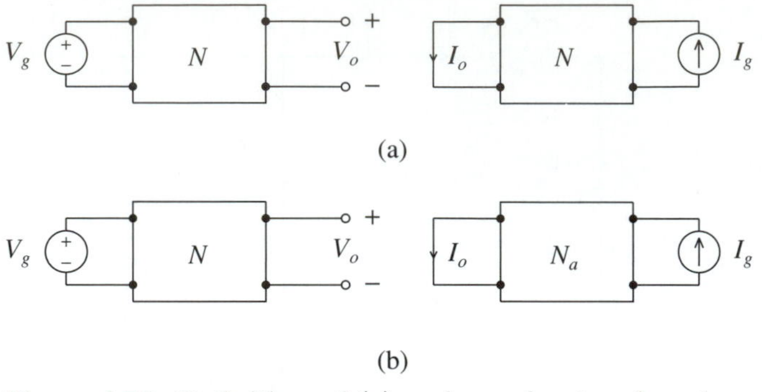

Figure 4.76 Definition of (a) reciprocal network and (b) interreciprocal networks, $\frac{V_o}{V_g} = \frac{I_o}{I_g}$.

Two different two-port networks (Fig. 4.76b), satisfying the Eq. (4.203) are called *interreciprocal networks*. An interreciprocal network N_a to a given network N is called an *adjoint network*. The adjoint network is obtained from the network N by using the following conversion rules (Fig. 4.77):

- Passive R and C elements in N_a are the same as those in N.
- The ideal input voltage source of N is converted to a short circuit, and the current through it becomes the output of N_a.
- The port of N at which the output voltage is taken becomes a current source in N_a.

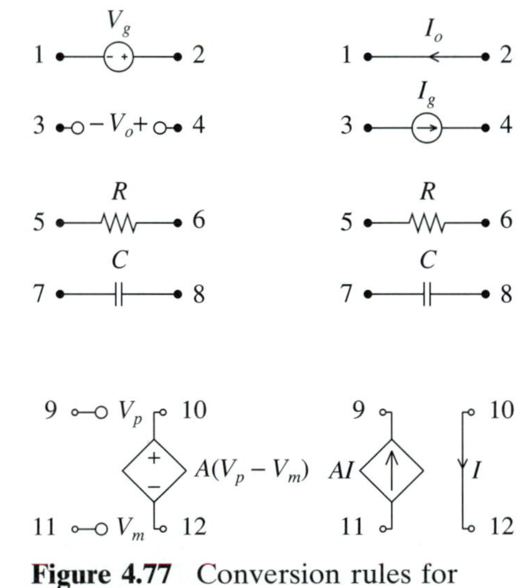

Figure 4.77 Conversion rules for deriving adjoint network.

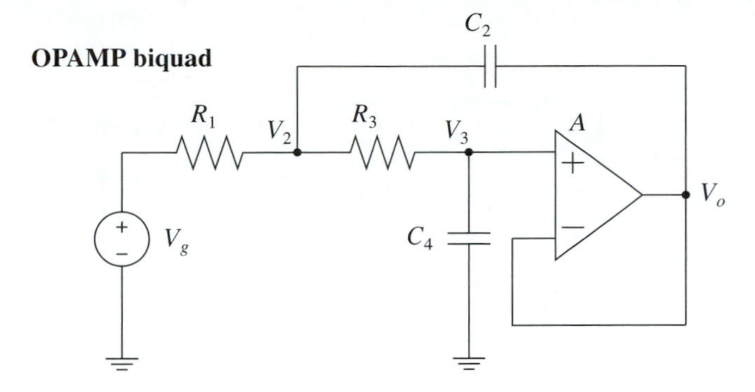

Figure 4.78 Sallen–Key lowpass op amp biquad.

- A voltage-controlled voltage source is converted to a current-controlled current source: The controlled voltage source, AV, is converted to a short circuit through which flows the controlling current I, while the controlling voltage V is converted to a controlled current source, AI.

For example, the Sallen–Key lowpass low-Q-factor op amp realization (Fig. 4.78) can be directly converted to a CC realization shown in Fig. 4.79. The current conveyor with the grounded Y terminal acts as a current amplifier. The transfer function of both circuits is the lowpass transfer function:

$$H(s) = \frac{\dfrac{1}{C_2C_4R_1R_3}}{s^2 + \dfrac{R_1 + R_3}{C_2R_1R_3}s + \dfrac{1}{C_2C_4R_1R_3}}$$

A fourth-order highpass filter realized as a cascade of two biquads is converted to the corresponding current-conveyor filter as shown in Figs. 4.80 and 4.81.

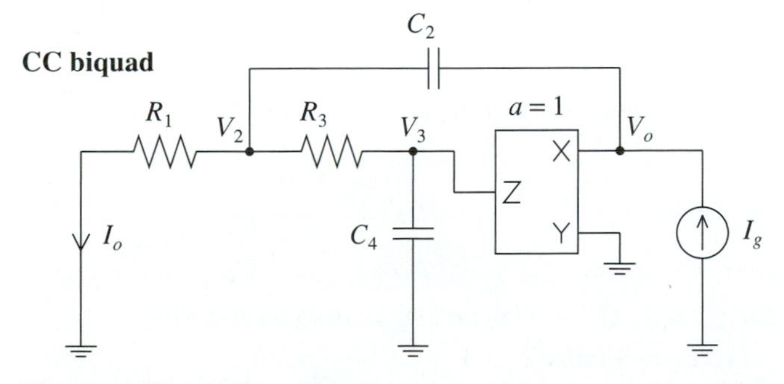

Figure 4.79 Sallen–Key lowpass current-conveyor biquad.

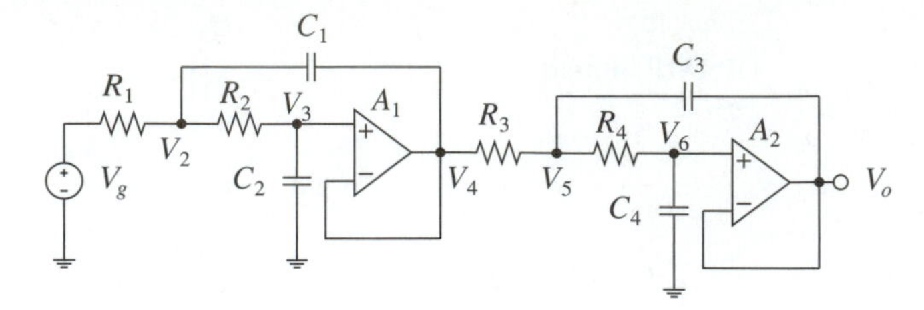

Figure 4.80 Cascaded op amp biquads.

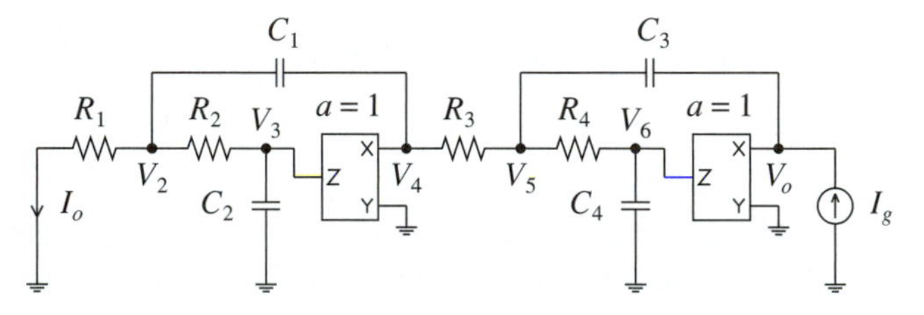

Figure 4.81 Cascaded current-conveyor biquads.

PROBLEMS

4.1 Find the transfer function coefficients of the second-order lowpass ($c = 1$, $a = b = 0$), highpass ($a = 1$, $b = c = 0$), and bandpass ($b = Q_p$, $a = c = 0$) filter, and find the maximum of the magnitude response if it occurs at $f_e = 10$ kHz

$$H(s) = \frac{as^2 + b\dfrac{\omega_p}{Q_p}s + c\omega_p^2}{s^2 + \dfrac{\omega_p}{Q_p}s + \omega_p^2}$$

Assume $Q_p = 2$.

4.2 Sketch the magnitude response of the second-order lowpass, highpass, and bandpass filter with the same $Q_p = 100$ and the maximum occurring at $f_e = 10$ kHz. What do you conclude about relationship between the Q-factor and the magnitude response?

4.3 Sketch the magnitude response of the second-order lowpass, highpass, and bandreject ($a = c = Q_p$, $b = 0$) filter with the same $Q_p = 10$ and $\omega_p = 10^4$ rad/s. Determine the 3-dB frequencies.

4.4 Sketch the magnitude response of the second-order filter

$$H(s) = \frac{s^2 + \dfrac{\omega_z}{Q_z}s + \omega_z^2}{s^2 + \dfrac{\omega_p}{Q_p}s + \omega_p^2}$$

for $Q_p = 2$, $Q_z = 1000$, and $\omega_p = 10$ rad/s and with

(a) $\omega_z = 3$ rad/s,

(b) $\omega_z = 6$ rad/s,

(c) $\omega_z = 9$ rad/s,

(d) $\omega_z = 10$ rad/s,

(e) $\omega_z = 10.5$ rad/s,

(f) $\omega_z = 12$ rad/s,

(g) $\omega_z = 15$ rad/s.

4.5 Sketch the magnitude response of the second-order filter

$$H(s) = \frac{s^2 + \dfrac{\omega_z}{Q_z}s + \omega_z^2}{s^2 + \dfrac{\omega_p}{Q_p}s + \omega_p^2}$$

for $Q_p = 3$, $\omega_z = 11$ rad/s, and $\omega_p = 10$ rad/s and with

(a) $Q_z = 1$,

(b) $Q_z = 3$,

(c) $Q_z = 10$,

(d) $Q_z = 100$,

(e) $Q_z = 1000$.

4.6 Decompose the fourth-order transfer function

$$H_4(s) = \frac{\left(\dfrac{\omega_p}{Q_p}s\right)^2}{\left(s^2 + \dfrac{\omega_p}{Q_p}s + \omega_p^2\right)^2}$$

into second-order transfer functions:

(a) into a product of a lowpass and a highpass transfer function and

(b) into a product of two bandpass transfer functions.

For $Q_p = 10$ and $\omega_p = 10$ rad/s, find the maximal values of the magnitude responses of each second-order transfer function. Suggest a decomposition suitable for implementation and give reasons for your choice.

4.7 Decompose the second-order transfer function

$$H_2(s) = \frac{\omega_p^2}{s^2 + \dfrac{\omega_p}{0.5 + q_p}s + \omega_p^2}$$

into a product of two first-order transfer functions for $\omega_p = 1$ rad/s and $q_p = -0.25$. Can you find q_p that yields the minimum pole Q-factor of $H_2(s)$.

4.8 Consider the lowpass low-Q-factor op amp biquad. Find the transfer function parameters K, Q_p, and ω_p

$$H(s) = K\frac{\omega_p^2}{s^2 + \dfrac{\omega_p}{Q_p}s + \omega_p^2}$$

in terms of element values. Assume $R_{11} = 200\ \text{k}\Omega$, $R_{12} = 100\ \text{k}\Omega$, $R_3 = 50\ \text{k}\Omega$, $C_2 = 100$ nF, $C_4 = 10$ nF. Compute the passive sensitivities of ω_p, K, and Q_p.

4.9 Consider the biquad from Problem 4.8. Assume that a parasitic capacitor, $C = 100$ pF, exists between each node and ground. Derive the transfer function and calculate the magnitude response deviation in dB with respect to the ideal case ($C = 0$).

4.10 Consider the biquad from Problem 4.8 which is driven by a voltage source with the serial internal resistance $R_s = 10$ kΩ. Find the transfer function and calculate the magnitude response deviation in dB with respect to the ideal case ($R_s = 0$).

4.11 Consider the biquad from Problem 4.8 with finite and frequency dependent op amp gain A that can be approximated by

$$A = 10^5 \frac{2\pi 10^2}{s + 2\pi 10^2} \frac{2\pi 10^4}{s + 2\pi 10^4} \frac{\pi 10^6}{s + \pi 10^6}$$

Find the transfer function and calculate the magnitude response deviation in dB with respect to the ideal case ($A \to +\infty$).

4.12 Consider the biquad from Problem 4.8 with the resistor and capacitor tolerances $\pm 2\%$, the resistor temperature coefficient 1000 ppm/K, and the capacitor temperature coefficient -150 ppm/K. The filter is required to work over the temperature range from $T_1 = 250$ K to $T_2 = 350$ K. Calculate the maximum magnitude response deviation in dB with respect to the ideal case (zero tolerances and temperature independent components). Assume the room temperature $T_0 = 293$ K.

MATLAB EXERCISES

4.1 Write a MATLAB script to plot the magnitude responses of the second-order lowpass, highpass, and bandpass transfer functions given in Problem 4.1 for $0 \le f \le 30$ kHz. Verify your results with the function `freqs`.

4.2 Write a MATLAB script to plot the magnitude responses of the second-order lowpass, highpass, and bandpass transfer functions given in Problem 4.2 for $0 \le f \le 30$ kHz. Verify your results with the function `freqs`.

4.3 Write a MATLAB script to plot the magnitude responses of the second-order transfer functions given in Problem 4.3 for $0 \le f \le 3$ kHz. Verify your results with the function `freqs`.

4.4 Write a MATLAB script to plot the magnitude responses of the second-order transfer functions given in Problem 4.4 for $0 \le \omega \le 20$ rad/s. Which ones of the transfer functions are lowpass-notch transfer functions?

4.5 Write a MATLAB script to plot the magnitude responses of the second-order transfer functions given in Problem 4.5 for $0 \le \omega \le 20$ rad/s. Which ones of the transfer functions are lowpass-notch transfer functions?

4.6 Write a MATLAB script to plot the magnitude responses of the transfer functions given in Problem 4.6 for $0 \le \omega \le 20$ rad/s. Which one of the transfer functions has the largest maximum of the magnitude responses?

4.7 Write a MATLAB script to plot the magnitude response of the second-order transfer function given in Problem 4.7 for $0 \le f \le 1$ Hz. Plot the transfer function poles for $q_p \in \{-1, -0.5, 0, 0.5, 1, 1.5, 2\}$. In which case is the transfer function causal?

4.8 Write a MATLAB script to plot the magnitude response of the transfer function given in Problem 4.8. Write a MATLAB program to calculate K, Q_p, and ω_p in terms of R_{11}, R_{12}, R_3, C_2, and C_4. If C_2 changes by $+10\%$, while C_4 changes by -10%, calculate the magnitude response deviation in dB.

4.9 Write MATLAB programs to (1) draw the circuit schematic, (2) compute the transfer function, poles, zeros, and Q-factors in terms of element values, (3) find the element values in terms of design parameters, (4) verify the filter realization, and plot the frequency response of the following second-order filters:

- **(a)** lowpass low-Q-factor op amp active RC second-order filter,
- **(b)** highpass low-Q-factor op amp active RC filter
- **(c)** bandpass low-Q-factor op amp active RC filter
- **(d)** bandreject low-Q-factor op amp active RC filter
- **(e)** allpass low-Q-factor op amp active RC filter
- **(f)** lowpass medium-Q-factor op amp active RC filter
- **(g)** highpass medium-Q-factor op amp active RC filter
- **(h)** bandpass medium-Q-factor op amp active RC filter
- **(i)** bandreject medium-Q-factor op amp active RC filter
- **(j)** allpass medium-Q-factor op amp active RC filter
- **(k)** lowpass notch medium-Q-factor op amp active RC filter
- **(l)** highpass notch medium-Q-factor op amp active RC filter
- **(m)** lowpass high-Q-factor op amp active RC filter
- **(n)** highpass high-Q-factor op amp active RC filter
- **(o)** bandpass high-Q-factor op amp active RC filter
- **(p)** bandreject high-Q-factor op amp active RC filter
- **(q)** allpass high-Q-factor op amp active RC filter
- **(r)** lowpass-notch high-Q-factor op amp active RC filter
- **(s)** highpass-notch high-Q-factor op amp active RC filter
- **(t)** general-purpose op amp active RC filter
- **(u)** OTA-C general biquad

MATHEMATICA EXERCISES

4.1 Write a *Mathematica* code to plot the magnitude responses of the second-order lowpass, highpass, and bandpass transfer functions given in Problem 4.1 for $0 \le f \le 30$ kHz. Determine the 3-dB frequencies.

4.2 Write a *Mathematica* code to plot the magnitude responses of the second-order lowpass, highpass, and bandpass transfer functions given in Problem 4.2 for $0 \le f \le 30$ kHz. Determine the 3-dB frequencies.

4.3 Write a *Mathematica* code to plot the magnitude responses of the second-order transfer functions given in Problem 4.3 for $0 \le f \le 3$ kHz.

4.4 Write a *Mathematica* code to plot the magnitude responses of the second-order transfer functions given in Problem 4.4 for $0 \le \omega \le 20$ rad/s. Which ones of the transfer functions are lowpass-notch transfer functions?

4.5 Write a *Mathematica* code to plot the magnitude responses of the second-order transfer functions given in Problem 4.5 for $0 \le \omega \le 20$ rad/s. Which ones of the transfer functions are lowpass-notch transfer functions?

4.6 Write a *Mathematica* code to plot the magnitude responses of the transfer functions given in Problem 4.6 for $0 \le \omega \le 20$ rad/s. Which one of the transfer function has the largest maximum of the magnitude responses?

4.7 Write a *Mathematica* code to plot the magnitude response of the second-order transfer functions given in Problem 4.7 for $0 \le f \le 1$ Hz. Plot the poles for $q_p \in \{-1, -0.5, 0, 0.5, 1, 1.5, 2\}$. In which case is the transfer function causal?

4.8 Write a *Mathematica* code to plot the magnitude response of the transfer function given in Problem 4.8. Write a *Mathematica* code to calculate K, Q_p, and ω_p in terms of R_{11}, R_{12}, R_3, C_2, and C_4. If C_2 changes by $+10\%$ while C_4 changes by -10%, calculate the magnitude response deviation in dB.

4.9 Write *Mathematica* programs to (1) draw the circuit schematic and formulate circuit equations from the schematic, (2) compute the transfer function, poles, zeros, and Q-factors in terms of element values, (3) find sensitivity functions and the gain-sensitivity product, (4) find the element values in terms of design parameters, (5) verify the filter realization, and plot the frequency response of the following second-order filters:

- **(a)** lowpass low-Q-factor op amp active RC second-order filter,
- **(b)** highpass low-Q-factor op amp active RC filter
- **(c)** bandpass low-Q-factor op amp active RC filter
- **(d)** bandreject low-Q-factor op amp active RC filter
- **(e)** allpass low-Q-factor op amp active RC filter
- **(f)** lowpass medium-Q-factor op amp active RC filter
- **(g)** highpass medium-Q-factor op amp active RC filter
- **(h)** bandpass medium-Q-factor op amp active RC filter
- **(i)** bandreject medium-Q-factor op amp active RC filter
- **(j)** allpass medium-Q-factor op amp active RC filter
- **(k)** lowpass notch medium-Q-factor op amp active RC filter
- **(l)** highpass notch medium-Q-factor op amp active RC filter
- **(m)** lowpass high-Q-factor op amp active RC filter
- **(n)** highpass high-Q-factor op amp active RC filter
- **(o)** bandpass high-Q-factor op amp active RC filter
- **(p)** bandreject high-Q-factor op amp active RC filter
- **(q)** allpass high-Q-factor op amp active RC filter
- **(r)** lowpass-notch high-Q-factor op amp active RC filter
- **(s)** highpass-notch high-Q-factor op amp active RC filter
- **(t)** general-purpose op amp active RC filter
- **(u)** OTA-C general biquad

CHAPTER 5

ADVANCED ANALOG FILTER DESIGN CASE STUDIES

This chapter reviews basic definitions of analog filter design. It introduces straightforward procedures to map the filter specification into a design space—that is, a set of ranges for parameters that we use in the filter design. We search this design space for the optimum solution according to given criteria, such as minimal Q-factor.

The principal drawback of the classical analog filter design is in returning only one solution, which can be unacceptable for many practical implementations. We propose an advanced approach to the filter design by using a mixture of symbolic and numeric computation and discrete nonlinear optimization. This approach should provide reduced filter complexity for a desired performance, or better performances for the required complexity.

We conclude this chapter by an application example in which we design a robust selective analog switched-capacitor (SC) filter based on commercially available integrated SC circuits.

5.1 BASIC DEFINITIONS

A *filter* is a system that can be used to modify or reshape the frequency spectrum of a signal according to some prescribed requirements.

An *electrical filter* may be used to amplify or attenuate a range of frequency components (sinusoidal signals) or to reject or isolate one specific frequency component. The applications are numerous: to eliminate signal contamination such as noise in communication systems, to separate relevant from irrelevant frequency components, to bandlimit signals before sampling, to convert sampled signals into continuous-time signal, to improve quality of audio equipment, in time-division to frequency-division multiplex systems, in speech synthesis, in the equalization of transmission lines and cables, in the design of artificial cochleas [16] in audio, video, speech, voiceband modems, control, instrumentation, radio signaling and radar, high definition television, radio modems, seismic modeling, financial modeling, and weather modeling.

Generally, the purpose of most filters is to separate the desired signals from undesired signals or noise. Often, the descriptions of the signals and noise are given in terms of their frequency content or the energy of the signals in the frequency bands. For this reason, the filter specifications are usually given in the frequency domain as magnitude response or by gain or attenuation.

The range of frequencies in which the sinusoidal signals are rejected is called a *stopband*. The range of frequencies in which the sinusoidal signals pass with tolerated distortion is called a *passband*. A region between the passband and stopband, where neither desired nor undesired signals exist or the spectra of the desired and undesired signals are overlapped, can be defined as a *transition region*.

In this chapter we will consider a filter with single passband referred to as *lowpass* filter. All other types of filters (*highpass*, *bandpass*, and *bandstop*) can be easily obtained by simple transformation from the lowpass filter.

Once the filter requirements are known, the filter specification can be established; for example, we specify the passband and stopband edge frequencies and tolerances. Next, we proceed with the analog filter design.

The *design* is a set of processes that starts with the specification and ends with the implementation of an analog (product) filter prototype. It comprises four general steps, as follows:

- Approximation
- Realization
- Study of imperfections
- Implementation

The *approximation* step is the process of generating a transfer function that satisfies the desired specification.

The *realization* step is the process of converting the transfer function of the filter into an electric circuit.

The *study of imperfections* investigates the effects of element imperfections, which determine the highest tolerance that can be tolerated without violating the specification of the filter throughout its working life.

The *implementation* step is constructing the product prototype of the filter in hardware. Decisions to be made involve (a) the type of components and packaging and (b) the methods to be used for the manufacture, testing, and tuning of the filter, and so on.

Usually, those four design steps are considered separately, although they are not independent of each other. The main goal is to find the most economical solution in short time. Which filter is better depends on the hardware used for the implementation. Many different constrains have to be fulfilled. The component tolerances and parasitic effects have significant influence on fulfilling the specification. In this case, classical approaches are not adequate for optimizing both the behavior (performance) and implementation (complexity and cost).

We propose an advanced approach to the analog filter design by using a mixture of symbolic and numeric computation and discrete nonlinear optimization. Opposite to the conventional approaches, which return only one design and hide a wealth of alternative filter designs, the advanced design techniques (that we introduce) find a comprehensive set of optimal designs to represent the infinite solution space.

5.2 SPECIFICATION OF AN ANALOG FILTER

Usually, the filter specification is given, but in many cases the designer has to establish the specification by himself. This is the most important prerequisite for the filter design. Namely, if the specification is too restrictive (e.g., very low passband and stopband tolerances, narrow transition region), the filter may not be feasible.

Special care must be taken in determining the passband and stopband tolerances. For example, if the noise at the output of an amplifier is −60 dB, it will not be reasonable to require 80-dB attenuation in stopband. The generated noise in the filter will be much higher than the attenuated undesired signal.

When down-sampling is performed in a digital filter, the lower or higher half of the spectra has to be rejected in order to prevent aliasing effects. In many cases just 20- or 30-dB attenuation in stopband will be sufficient.

The proper selection of the specification must be done according to the nature of signals (i.e., frequency bands and the corresponding levels of the desired and undesired signals or noise) and the available hardware (element tolerances, parasitic effects, etc.).

In this chapter we assume that the specification is given. Next, we examine the feasibility of the practical filter design. Finally, if the filter is not feasible, we propose the minimum changes in the given specification to make the filter design possible.

In practice, there are several ways in presenting the analog filter specification. Usually, designers of analog filters prefer attenuation or gain expressed in dB, while magnitude tolerances are more convenient for the designers of digital filters.

To provide a unified and consistent design, we adopt one form of presenting the specification.

Let us consider a *linear time-invariant system* (LTI), with the input sine signal $x_{in}(t)$

$$x_{in}(t) = X_m \sin(\omega t + \zeta) \tag{5.1}$$

and the steady-state output signal $y_{out}(t)$,

$$y_{out}(t) = Y_m \sin(\omega t + \eta) \tag{5.2}$$

as shown in Fig. 5.1. The ratio $M(\omega) = Y_m / X_m$ describes the change in the amplitude of the input sine signal; we call $M(\omega)$ the *magnitude response*. With $\Phi(\omega) = \eta - \zeta$ we designate the change in phase of the input sine signal; we call $\Phi(\omega)$ the *phase response*; both quantities are defined at the angular frequency in rad/s, $\omega = 2\pi f$, from the frequency range of interest. Some authors prefer to express $M(\omega)$ and $\Phi(\omega)$ in terms of frequency f in Hz. The *frequency response* of the system is defined as $M(\omega)e^{j\Phi(\omega)}$.

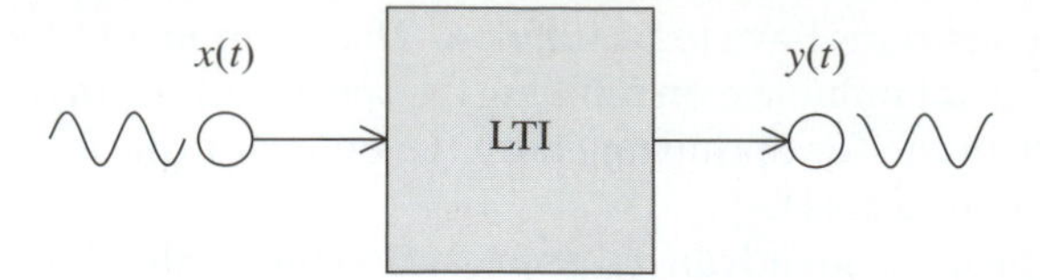

Figure 5.1 Linear time-invariant system.

In other words, $M(\omega)e^{j\Phi(\omega)}$ shows how the input signal is transferred through the system at the specific angular frequency ω rad/s. From the frequency response we can derive the *transfer function*. The transfer function of an LTI system, $H(s)$, is a rational function in the complex variable s; for $s = j\omega$ the transfer function becomes the frequency response $H(j\omega) = M(\omega)e^{j\Phi(\omega)}$.

Several functions are derived from the magnitude response, $M(\omega) = |H(j\omega)|$, and are frequently used in practice. The reciprocal of the squared magnitude is called the *loss function*:

$$L_F(\omega) = \frac{1}{M^2(\omega)} \tag{5.3}$$

We call the function $\sqrt{L_F(\omega) - 1}$ the *characteristic function*:

$$K(\omega) = \sqrt{\frac{1}{M^2(\omega)} - 1} \tag{5.4}$$

Attenuation (in dB) or *loss characteristic* is defined by

$$A(\omega) = 20 \log_{10} \frac{1}{M(\omega)} \tag{5.5}$$

Gain (in dB) is the negative of the attenuation:

$$G(\omega) = -A(\omega) = 20 \log_{10} M(\omega) \tag{5.6}$$

We assume that the maximal value of the magnitude response is 1:

$$\max_{\omega} M(\omega) = \max_{\omega} |H(j\omega)| = 1$$

If this is not the case, the required attenuation or gain can be easily compensated for by multiplying $H(s)$ by a constant.

The magnitude response $M(\omega)$, of an analog lowpass filter, against frequency $f = \frac{1}{2\pi}\omega$, is plotted in Fig. 5.2. Theoretically, upper limit frequency does not exist;

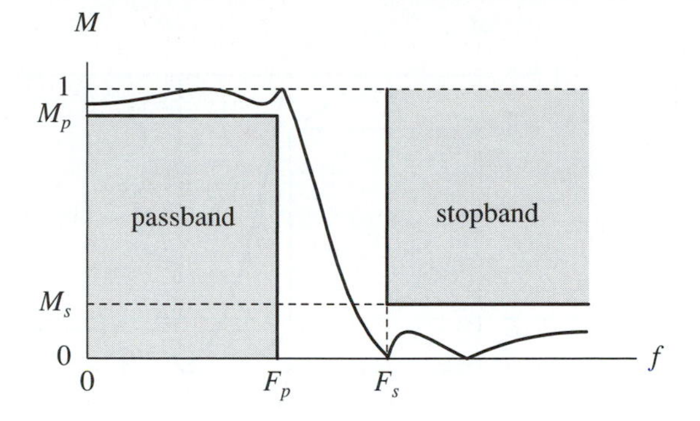

Figure 5.2 Magnitude-limit specification.

however, we can assume that the practical upper frequency is up to 100 times higher than the maximal frequency of interest.

The filter specification can be expressed in several ways:

1. The magnitude limits (Fig. 5.2) define the minimum magnitude in passband, M_p, and the maximum magnitude in stopband, M_s.
2. The magnitude tolerances (Fig. 5.3), specify the maximum magnitude decrease in passband, $\delta_p = 1 - M_p$, and the maximum magnitude in stopband, $\delta_s = M_s$.
3. The magnitude ripple tolerances (Fig. 5.4) describe the maximum magnitude variation, in passband, δ_1, and in stopband, δ_2.
4. The attenuation limits in dB (Fig. 5.5) specify the maximum attenuation in passband, A_p, and the minimum attenuation in stopband, A_s.
5. The gain limits in dB (Fig. 5.6) specify the minimum gain in passband, $G_p = -A_p$, and the maximum gain in stopband, $G_s = -A_s$.

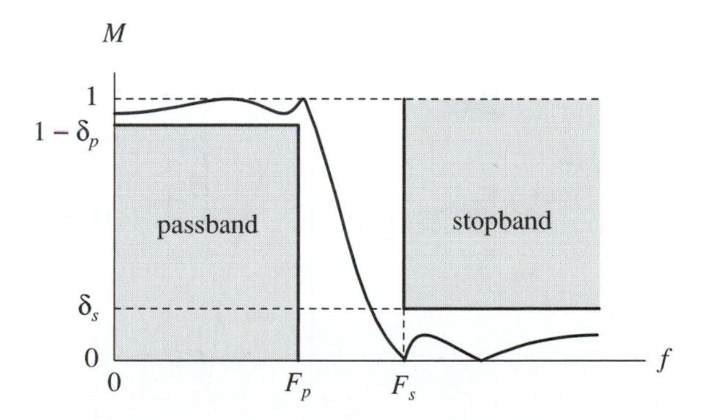

Figure 5.3 Magnitude-tolerance specification.

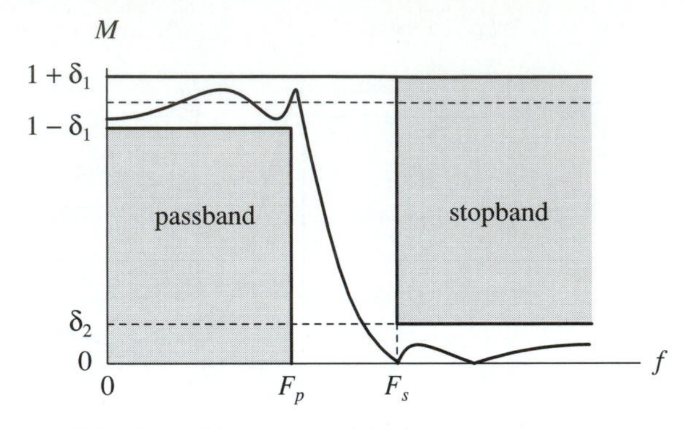

Figure 5.4 Magnitude-ripple specification.

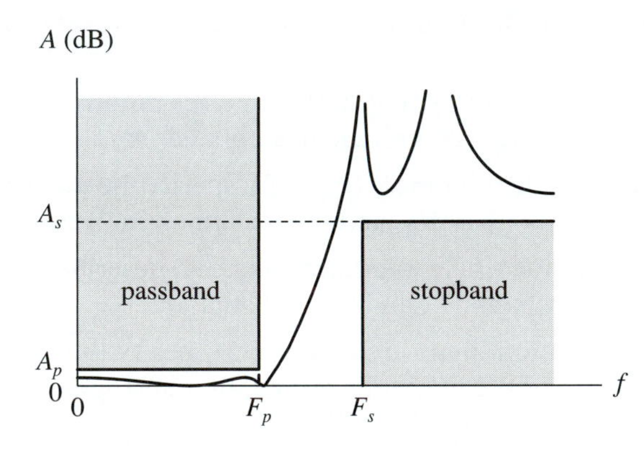

Figure 5.5 Attenuation-limit specification.

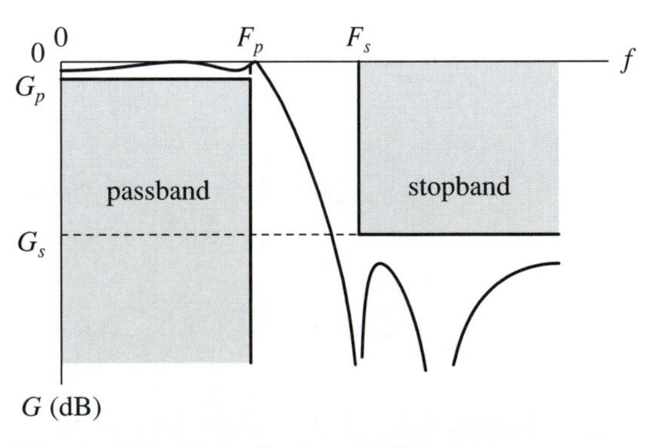

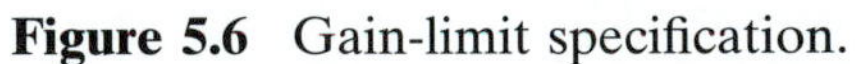

Figure 5.6 Gain-limit specification.

Relations between the specification quantities are summarized in Eqs. (5.7)–(5.12):

$$
\begin{array}{ccccccc}
\delta_p & = & 1 - M_p & = & \dfrac{2\delta_1}{1+\delta_1} & = & \dfrac{-1+\sqrt{1+K_p^2}}{\sqrt{1+K_p^2}} \\
1-\delta_p & = & M_p & = & \dfrac{1-\delta_1}{1+\delta_1} & = & \dfrac{1}{\sqrt{1+K_p^2}} \\
\dfrac{\delta_p}{2-\delta_p} & = & \dfrac{1-M_p}{1+M_p} & = & \delta_1 & = & \dfrac{-1+\sqrt{1+K_p^2}}{1+\sqrt{1+K_p^2}} \\
\dfrac{\sqrt{\delta_p(2-\delta_p)}}{1-\delta_p} & = & \dfrac{\sqrt{1-M_p^2}}{M_p} & = & \dfrac{2\sqrt{\delta_1}}{1-\delta_1} & = & K_p \\
-20\log_{10}(1-\delta_p) & = & -20\log_{10} M_p & = & 20\log_{10}\dfrac{1+\delta_1}{1-\delta_1} & = & 10\log_{10}(1+K_p^2) \\
20\log_{10}(1-\delta_p) & = & 20\log_{10} M_p & = & 20\log_{10}\dfrac{1-\delta_1}{1+\delta_1} & = & -10\log_{10}(1+K_p^2)
\end{array}
\tag{5.7}
$$

$$
\begin{array}{ccccc}
\dfrac{-1+\sqrt{1+K_p^2}}{\sqrt{1+K_p^2}} & = & 1-10^{-A_p/20} & = & 1-10^{G_p/20} \\
\dfrac{1}{\sqrt{1+K_p^2}} & = & 10^{-A_p/20} & = & 10^{G_p/20} \\
\dfrac{-1+\sqrt{1+K_p^2}}{1+\sqrt{1+K_p^2}} & = & \dfrac{1-10^{-A_p/20}}{1+10^{-A_p/20}} & = & \dfrac{1-10^{G_p/20}}{1+10^{G_p/20}} \\
K_p & = & \dfrac{\sqrt{1-10^{-A_p/10}}}{10^{-A_p/20}} & = & \dfrac{\sqrt{1-10^{G_p/10}}}{10^{G_p/20}} \\
10\log_{10}(1+K_p^2) & = & A_p & = & -G_p \\
-10\log_{10}(1+K_p^2) & = & -A_p & = & G_p
\end{array}
\tag{5.8}
$$

$$
\begin{array}{ccccccc}
\delta_s & = & M_s & = & \dfrac{\delta_2}{1+\delta_1} & = & \dfrac{1}{\sqrt{1+K_s^2}} \\
\dfrac{2\delta_s}{2-\delta_p} & = & \dfrac{2M_s}{1+M_p} & = & \delta_2 & = & 2\dfrac{\sqrt{\dfrac{1+K_p^2}{1+K_s^2}}}{1+\sqrt{1+K_p^2}} \\
\dfrac{\sqrt{1-\delta_s^2}}{\delta_s} & = & \dfrac{\sqrt{1-M_s^2}}{M_s} & = & \dfrac{\sqrt{(1+\delta_1)^2-\delta_2^2}}{\delta_2} & = & K_s \\
-20\log_{10}(\delta_s) & = & -20\log_{10} M_s & = & -20\log_{10}\dfrac{\delta_2}{1+\delta_1} & = & 10\log_{10}(1+K_s^2) \\
20\log_{10}(\delta_s) & = & 20\log_{10} M_s & = & 20\log_{10}\dfrac{\delta_2}{1+\delta_1} & = & -10\log_{10}(1+K_s^2)
\end{array}
\tag{5.9}
$$

$$
\begin{array}{ccccc}
\dfrac{1}{\sqrt{1+K_s^2}} & = & 10^{-A_s/20} & = & 10^{G_s/20} \\
2\dfrac{\sqrt{\dfrac{1+K_p^2}{1+K_s^2}}}{1+\sqrt{1+K_p^2}} & = & 2\dfrac{10^{-A_s/20}}{1+10^{-A_p/20}} & = & 2\dfrac{10^{G_p/20}}{1+10^{G_p/20}} \\
K_s & = & \dfrac{\sqrt{1-10^{-A_s/10}}}{10^{-A_s/20}} & = & \dfrac{\sqrt{1-10^{G_s/10}}}{10^{G_s/20}} \\
10\log_{10}(1+K_s^2) & = & A_s & = & -G_s \\
-10\log_{10}(1+K_s^2) & = & -A_s & = & G_s
\end{array}
\tag{5.10}
$$

$$
\begin{array}{llll}
\delta_p = 1-\dfrac{\sqrt{2}}{2} & M_p = \dfrac{\sqrt{2}}{2} & & K_p = 1 \\
\delta_s \approx 0.1 & M_s \approx 0.1 & & K_s = 10 \\
\delta_s \approx 0.01 & M_s \approx 0.01 & & K_s = 10^2 \\
\delta_s \approx 0.0001 & M_s \approx 0.0001 & & K_s = 10^4 \\
\delta_s \approx 0.00001 & M_s \approx 0.00001 & & K_s = 10^5 \\
\delta_p \approx 0.005 & M_p \approx 0.995 & \delta_1 \approx 0.0099 & K_p = 1/10 \\
\delta_p \approx 0.00005 & M_p \approx 0.99995 & \delta_1 \approx 0.00001 & K_p = 1/100
\end{array}
\tag{5.11}
$$

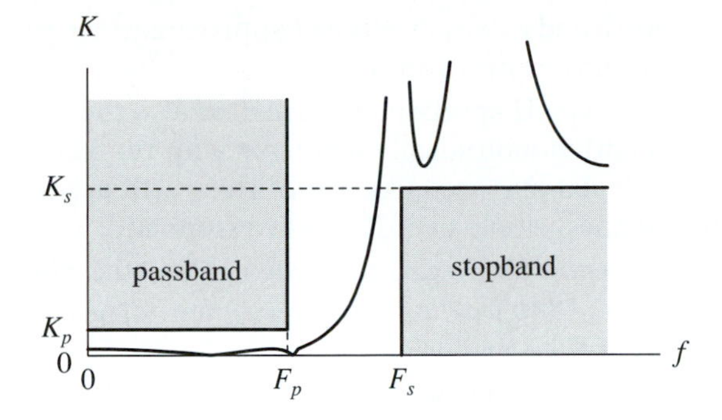

Figure 5.7 Characteristic-function-limit specification.

$$
\begin{array}{lll}
K_p = 1 & A_p \approx 3 & G_p \approx -3 \\
K_s = 10 & A_s \approx 20 & G_s \approx -20 \\
K_s = 10^2 & A_s \approx 40 & G_s \approx -40 \\
K_s = 10^4 & A_s \approx 80 & G_s \approx -80 \\
K_s = 10^5 & A_s \approx 100 & G_s \approx -100 \\
K_p = 1/10 & A_p \approx 0.043 & G_p \approx -0.043 \\
K_p = 1/100 & A_p \approx 0.0004 & G_p \approx -0.0004
\end{array}
\tag{5.12}
$$

The underlining idea of the design that we propose is to map any specification into a new one, expressed in terms of the characteristic function limits (Fig. 5.7) specifying the maximum value in passband, K_p, and the minimum value in stopband, K_s. This provides a unified start for the subsequent design steps.

5.3 APPROXIMATION PROBLEM

In this section we consider the approximation step of the analog filter design.

Let us consider the specification shown in Fig. 5.7. The first step is to generate the characteristic function $K(\omega)$. A function that is a candidate for $K(\omega)$ is known as the *approximation function*, also called the *approximation*. From the approximation the transfer function that meets the specification can be derived. Besides several classical approximations there are numerous closed-form expressions and numerical procedures for generating approximation functions.

The *Butterworth* approximation is smooth and monotonically increases with respect to frequency. It is maximally flat at $\omega = 0$.

The *Chebyshev* approximation, sometimes called the *Chebyshev type I* approximation, gives the smallest ripple over the entire passband. In the stopband, this approximation monotonically increases with respect to frequency.

The Butterworth and Chebyshev type I approximations yield *allpole* transfer functions; they have no transfer function zeros.

The Chebyshev type II approximation, called also the *inverse Chebyshev* approximation, is smooth and monotonically increases with respect to frequency in the passband. It is maximally flat at $\omega = 0$ like the Butterworth approximation. This approximation gives the smallest ripple over the entire stopband.

The *elliptic function approximation*, also called the *elliptic approximation*, the *Cauer approximation* or the *Darlington approximation*, gives the smallest ripple over the entire passband and stopband.

The *Bessel* approximation yields an allpole transfer function like Butterworth and Chebyshev type I. Its magnitude response is smooth and monotonically decreases with respect to frequency. The main characteristics of this approximation are (1) maximally flat group delay at $\omega = 0$ and (2) low step-response overshoot.

Other types of approximations exist, and they exhibit good properties of group delay or time domain [33]. There is a class of approximations that combines the properties of the classical ones; we call this class of approximation functions the *transitional approximations* [49].

Which approximation should the designer choose?

One approach is to analyze all known approximations and to use all known procedures for calculating the magnitude response. The examination of all known approximations has a very high computational cost. Such an approach can be time-consuming, too.

In this chapter we focus on only one approximation, the elliptic approximation. Next, we find the design space—that is, the range of design parameters that satisfy the specification—and keep the design parameters as symbols.

5.4 DESIGN SPACE

In this section we define the design space. First, we map the specification into a standard form. Next, we identify the design parameters. Finally, we calculate the limits of the design parameters.

In the previous section we have shown several ways of presenting required specifications. Any analog lowpass filter can be specified by a set of four quantities as follows:

$$S_\delta = \{F_p, F_s, \delta_p, \delta_s\} \tag{5.13}$$

$$S_M = \{F_p, F_s, M_p, M_s\} \tag{5.14}$$

$$S_r = \{F_p, F_s, \delta_1, \delta_2\} \tag{5.15}$$

$$S_K = \{F_p, F_s, K_p, K_s\} \tag{5.16}$$

$$S_A = \{F_p, F_s, A_p, A_s\} \tag{5.17}$$

$$S_G = \{F_p, F_s, G_p, G_s\} \tag{5.18}$$

and relations between them have been summarized in Eqs. (5.7)–(5.12). The symbol F_p designates the passband edge frequency in Hz, and F_s stands for the stopband edge frequency in Hz.

It is more convenient to transform a given specification S into the specification S_K because it provides a clearer relationship between the design parameters and the specification. Since we have to find a characteristic function $K(\omega)$ in the approximation step, S_K is the most suitable way of presenting the specification.

An infinite number of characteristic functions that fit S exists. We consider the *elliptic approximation*, because it fulfills the requirements with the minimal transfer function order. The minimal order can often lead to the most economical solution, such as the minimal number of components. Also, it will be shown that some other classic approximations are special cases of the elliptic approximation.

The *prototype elliptic approximation*, K_e, is an nth-order rational function in the real variable x:

$$K_e(x) = \epsilon|R(n, \xi, x)| \tag{5.19}$$

where R, referred to as the *elliptic rational function*, satisfies the conditions

$$0 \le |R(n, \xi, x)| \le 1, \qquad |x| \le 1 \tag{5.20}$$

$$L(n, \xi) \le |R(n, \xi, x)|, \qquad |x| \ge \xi \tag{5.21}$$

and L is the *discrimination factor*—that is, the minimal value of the magnitude of R for $|x| \ge \xi$—and can be calculated as

$$L(n, \xi) = |R(n, \xi, \xi)| \tag{5.22}$$

The normalized transition band $1 < x < \xi$ is defined by

$$1 < |R(n, \xi, x)| < L(n, \xi), \qquad 1 < |x| < \xi \tag{5.23}$$

The parameter ξ is called the *selectivity factor*.

The parameter ϵ determines the maximal variation of K_e in the normalized passband $0 \le x \le 1$:

$$0 \le K_e(x) \le \epsilon, \qquad |x| \le 1 \tag{5.24}$$

and is called the *ripple factor*.

The *elliptic approximation*, $K(\omega)$, is a rational function in angular frequency ω rad/s (Figs. 5.8 and 5.9):

$$K(\omega) = K_e(x), \qquad x = \frac{\omega}{2\pi f_p} \tag{5.25}$$

where f_p represents a design parameter that we call the *actual passband edge*. Traditionally, it has been set to $f_p = F_p$.

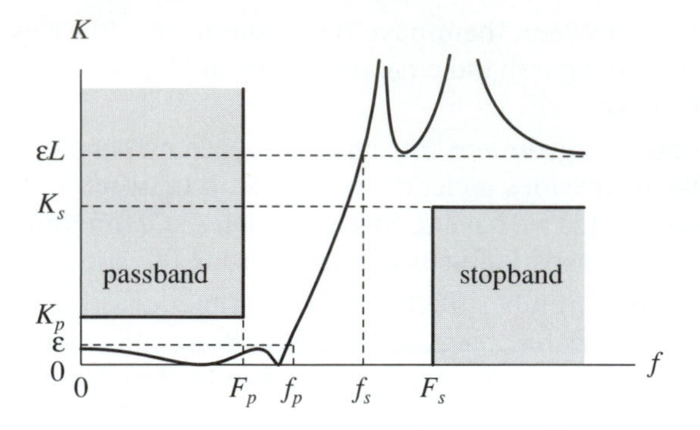

Figure 5.8 Elliptic approximation.

The four quantities n, ξ, ϵ, and f_p are collectively referred to as *design parameters* and can be expressed as a list of the form

$$D = \{n, \xi, \epsilon, f_p\} \tag{5.26}$$

Quantities ξ, ϵ, and f_p take a value over a continuous range of numbers. The order n can take a value from a discrete range of numbers, and is referred to as the *filter order* or *transfer function order*.

It is known [50] that the elliptic approximation provides the minimal order, $n_{min} = n_{ellip}$, for a given specification. The maximal order, from the practical viewpoint, can be assumed to be $n_{max} = 2n_{min}$.

The selectivity factor, ξ, falls within the limits which are found by solving the equations [51]

$$R(n, \xi, \xi) = \frac{K_s}{K_p} \quad \Rightarrow \quad \xi_{min} = \xi_{min}(n) \tag{5.27}$$

$$R\left(n, \xi, \frac{F_s}{F_p}\right) = \frac{K_s}{K_p}, \qquad \xi > \frac{F_s}{F_p} \quad \Rightarrow \quad \xi_{max} = \xi_{max}(n) \tag{5.28}$$

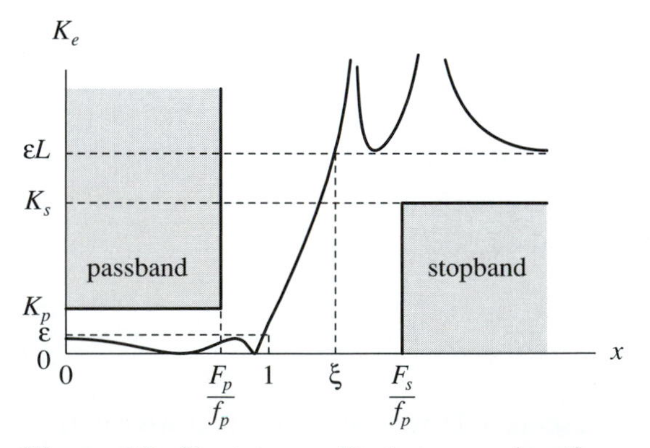

Figure 5.9 Prototype elliptic approximation.

It follows from Eqs. (5.19) and (5.20) that the maximal value of ϵ must be equal to or less than K_p:

$$\epsilon \leq K_p \tag{5.29}$$

The ripple factor quantifies the output signal amplitude, Y_m, with respect to the input signal amplitude, X_m. When $K(\omega) = 0$, both amplitudes have the same value; that is,

$$Y_m = X_m \qquad \text{for } K(\omega) = 0 \tag{5.30}$$

With $K(\omega) = \epsilon$ the amplitudes are different:

$$Y_m = \frac{X_m}{\sqrt{1 + \epsilon^2}} \qquad \text{for } K(\omega) = \epsilon \tag{5.31}$$

From the previous equations it follows that the smaller the value of ϵ, the smaller the difference between the input and output amplitude.

What is the lower limit of ϵ?

The ripple factor, for $x > \xi$, must meet another condition $K_e(x) \geq K_s$ (Fig. 5.9); thus

$$\epsilon L(n, \xi) \geq K_s \tag{5.32}$$

Therefore, the maximal and minimal values of ϵ has to be determined:

$$\epsilon_{min} \leq \epsilon \leq \epsilon_{max} \tag{5.33}$$

From Eq. (5.29) we find the upper bound

$$\epsilon_{max} = K_p \tag{5.34}$$

From Eq. (5.32) we determine the lower bound:

$$\epsilon_{min} = \frac{K_s}{L(n, \xi)} = \epsilon_{min}(n, \xi) \tag{5.35}$$

The maximal value of ϵ directly follows from specification, while the minimal value of ϵ depends on the order n and the selectivity factor ξ.

The actual passband edge, f_p, can take a value from the interval

$$f_{p,min} = \frac{F_s}{\xi_{max}} \leq f_p \leq f_{p,max} = \frac{F_s}{\xi_{min}} \tag{5.36}$$

Obviously, $f_{p,min} = f_{p,min}(n)$, and $f_{p,max} = f_{p,max}(n)$.

The set of all quadruples $D = \{n, \xi, \epsilon, f_p\}$ satisfying the constraints $\{n_{min} \leq n \leq n_{max}, \xi_{min} \leq \xi \leq \xi_{max}, \epsilon_{min} \leq \epsilon \leq \epsilon_{max}, f_{p,min} \leq f_p \leq f_{p,max}\}$ is called the *design space*:

$$D_S = \{D_{S,n}\}_{n=n_{min}, n_{min}+1, \ldots, n_{max}} \tag{5.37}$$

$$D_{S,n} = \left\{ \begin{array}{ccc} & n & \\ \xi_{min}(n) & \leq \xi \leq & \xi_{max}(n) \\ \epsilon_{min}(n, \xi) & \leq \epsilon \leq & K_p \\ f_{p,min}(n) & \leq f_p \leq & f_{p,max}(n) \end{array} \right\} \tag{5.38}$$

The order n is an integer; it takes only the discrete numeric values, so, it is more convenient to express the design space, D_S, as a list of subspaces, $D_{S,n}$:

$$\left\{\begin{array}{l} \left\{\begin{array}{ccc} & n = n_{min} & \\ \xi_{min}(n) & \leq \xi \leq & \xi_{max}(n) \\ \epsilon_{min}(n,\xi) & \leq \epsilon \leq & K_p \\ f_{p,min}(n) & \leq f_p \leq & f_{p,max}(n) \end{array}\right\} \\ \left\{\begin{array}{ccc} & n = n_{min}+1 & \\ \xi_{min}(n) & \leq \xi \leq & \xi_{max}(n) \\ \epsilon_{min}(n,\xi) & \leq \epsilon \leq & K_p \\ f_{p,min}(n) & \leq f_p \leq & f_{p,max}(n) \end{array}\right\} \\ \cdots \\ \left\{\begin{array}{ccc} & n = n_{max} & \\ \xi_{min}(n) & \leq \xi \leq & \xi_{max}(n) \\ \epsilon_{min}(n,\xi) & \leq \epsilon \leq & K_p \\ f_{p,min}(n) & \leq f_p \leq & f_{p,max}(n) \end{array}\right\} \end{array}\right. \tag{5.39}$$

where

$$\begin{array}{ccccc} 0 & < & \epsilon_{min}(n+1) & < & \epsilon_{min}(n) \\ 1 & < & \xi_{min}(n+1) & < & \xi_{min}(n) \\ \xi_{max}(n) & < & \xi_{max}(n+1) & < & \infty \\ 0 & \leq & f_{p,min}(n+1) & < & f_{p,min}(n) \\ f_{p,max}(n) & < & f_{p,max}(n+1) & < & F_s \end{array} \tag{5.40}$$

5.5 BASIC DESIGN ALTERNATIVES

This section presents our case studies of a comprehensive set of design alternatives based on the design space. It is understood that the rational elliptic function can be readily constructed for a given set of design parameters [52, 53]. The advantages of the various designs are discussed.

Usually, the designer selects the minimal order $n = n_{min}$. The design alternatives that follow are general and valid for any n from the design space. We assume that a specification, S, has been mapped into the form S_K.

5.5.1 Design D1

The design D1 sets the three design parameters, $\xi = F_s/F_p$, $\epsilon = K_p$, $f_p = F_p$, directly from the specification (Fig. 5.10).

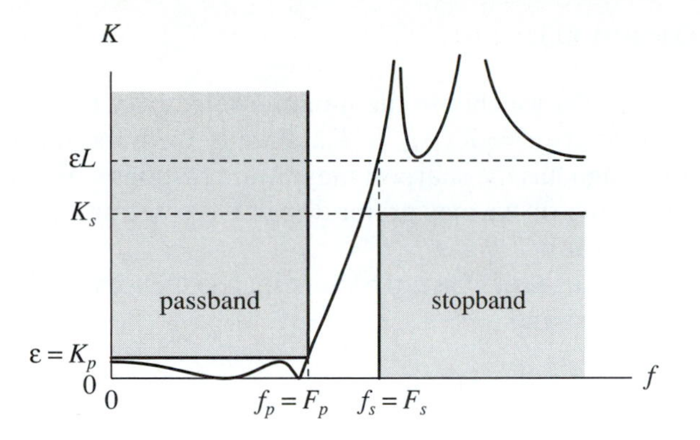

Figure 5.10 Design D1.

This design has higher attenuation in the stopband than is required by the specification. We choose this design when we prefer to achieve as large an attenuation as possible in the stopband—that is, $\epsilon L > K_s$.

5.5.2 Design D2

The design D2 sets the two design parameters, $\xi = F_s/F_p$, $f_p = F_p$, directly from the specification (Fig. 5.11). The ripple factor is computed from $\epsilon = K_s/L(n, \xi)$.

This design has lower attenuation in the passband than is required by the specification. We choose this design when we prefer to achieve as low an attenuation as possible in the passband—that is, $\epsilon < K_p$. Also, this design is suitable when filter element imperfections significantly affect the magnitude response in the passband. In that case, we achieve the highest attenuation margin in the passband (the margin is $K_p - \epsilon$; see Fig. 5.11), and we expect that the imperfections of the implemented filter will not violate the specification.

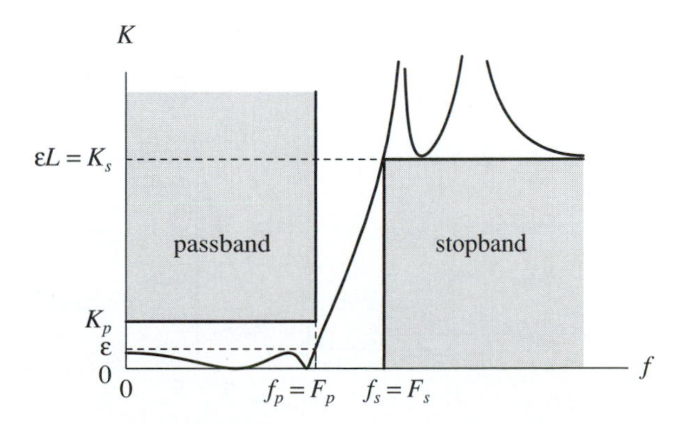

Figure 5.11 Design D2.

5.5.3 Design D3a

In the design D3a we choose the minimal selectivity factor, $\xi = \xi_{min}$, and set the two design parameters, $\epsilon = K_p$, $f_p = F_p$, directly from the specification (Fig. 5.12).

This design has the sharpest magnitude response. When undesired signals exist in the transition region we may prefer design D3a, because it rejects the undesired signals as much as possible.

Disadvantages of D3a can be very high Q-factors and large variation of the group-delay in the passband.

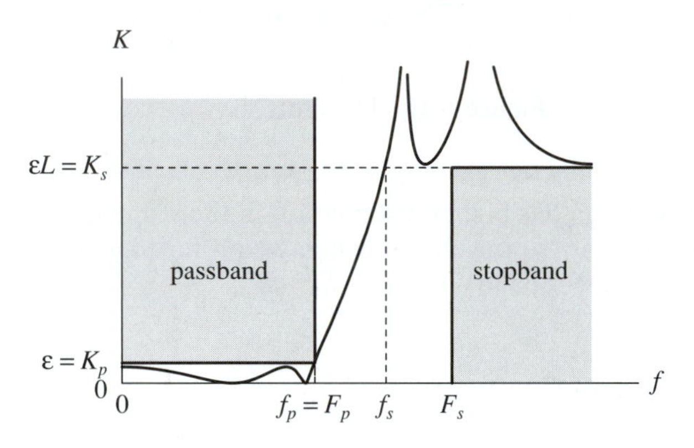

Figure 5.12 Design D3a.

5.5.4 Design D3b

For the design D3b we choose the minimal selectivity factor, $\xi = \xi_{min}$, (the same as in Design 3a) and set the ripple factor, $\epsilon = K_p$, directly from the specification (Fig. 5.13). The actual passband edge is computed from $f_p = f_{p,max} = F_s/\xi$.

This design has the sharpest magnitude response (the same as D3a). When the desired signals exist in the transition region we may prefer the design D3b, because it attenuates the desired signals as low as possible.

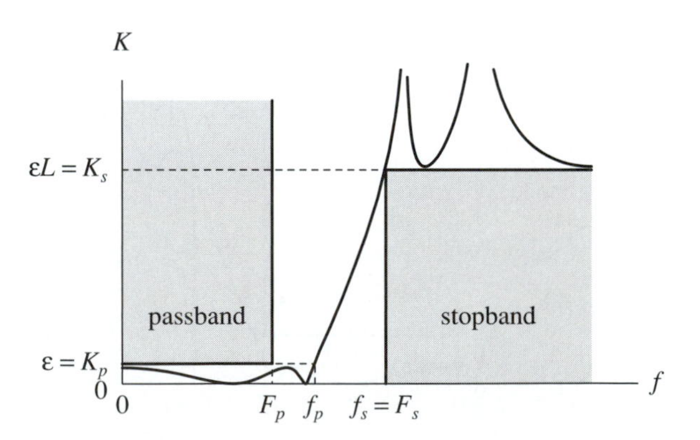

Figure 5.13 Design D3b.

A disadvantage of the design D3b can be very high Q-factors. Although the variation of the group delay can be high, its maximal value can be moved into the transition region, so the group-delay variation can be acceptable in the passband.

5.5.5 Design D4a

In the design D4a we choose the maximal selectivity factor, $\xi = \xi_{max}$, and set the two design parameters, $\epsilon = K_p$ and $f_p = F_p$, directly from the specification (Fig. 5.14).

This design (like the design D1) has higher attenuation in the stopband than is required by the specification, except at the stopband edge frequency. We choose this design when we prefer to achieve as large attenuation as possible in the stopband, i.e. $\epsilon L > K_s$, except at $f = F_s$.

The design D4a has a smoother magnitude response, and that is the main reason for lower Q-factors and smaller variation of the group delay in the passband.

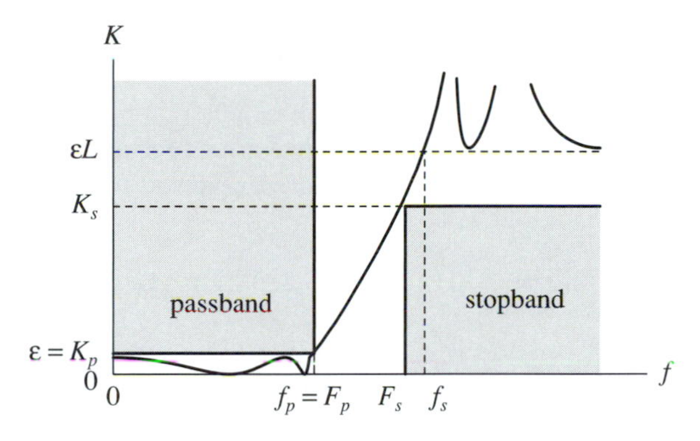

Figure 5.14 Design D4a.

5.5.6 Design D4b

For the design D4b we choose the maximal selectivity factor, $\xi = \xi_{max}$ (the same as in the design D4a) (Fig. 5.15), and calculate the ripple factor from $\epsilon = K_s/L(n, \xi)$. The actual passband edge is computed from $f_p = f_{p,min} = F_s/\xi$.

This design (like the design D2) has lower attenuation in the passband than is required by the specification, except at the passband edge frequency.

We choose this design when we prefer to achieve as low an attenuation as possible in the passband—that is, $\epsilon < K_p$, except at $f = F_p$.

The design D4b has a smoother magnitude response. This design usually yields very low Q-factors and small variation of the group delay in the passband. The design D4b has a very low ripple factor ϵ.

It should be noticed that there exists a straightforward procedure for computing the ripple factor, ϵ, for a given selectivity factor ξ, that yields the minimal Q-factors [47]. We designate this design by D5.

A disadvantage of D3a, D3b, D4a, and D4b is lack of any attenuation margin. Any imperfection, usually in implementation step (such as element tolerances), can violate the specification.

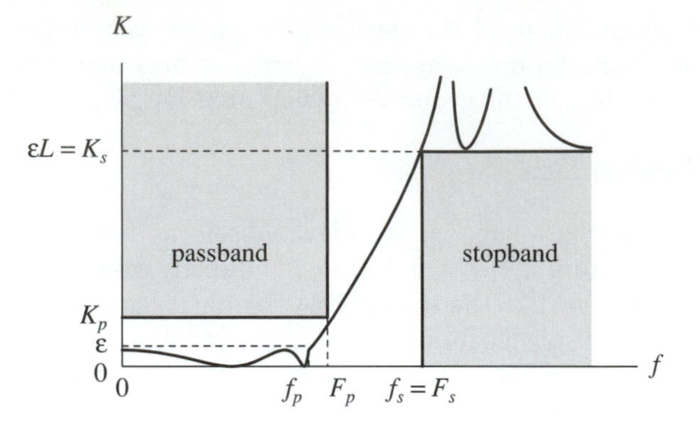

Figure 5.15 Design D4b.

5.5.7 Remarks on Design Alternatives

The approach that we propose in the previous sections we have programmed in *Mathematica* [2]. Several design examples are exercised for an illustrative specification:

$$S_A = \{F_p = 3\,\text{kHz}, F_s = 3.225\,\text{kHz}, A_p = 0.2\,\text{dB}, A_s = 40\,\text{dB}\}.$$

First, the attenuation-limit specification $S_A = \{3,\ 3.225,\ 0.2,\ 40\}$ is transformed into the characteristic-function-limit specification

$$S_K = \{3, 3.225, 0.2171, 100\}.$$

Next, all designs are calculated, and the gain is plotted on Figs. 5.16–5.28. The design parameters, the actual stopband edge, the maximal attenuation in the passband, and the minimal attenuation in the stopband, are summarized in Table 5.1 and Table 5.2. The maximal Q-factor is presented in the last column.

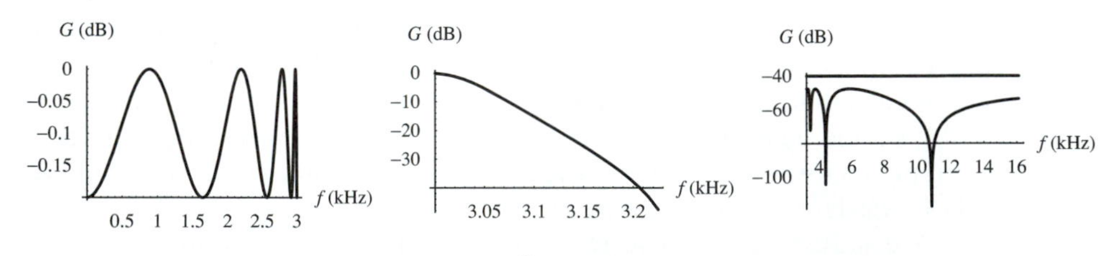

Figure 5.16 Design D1: $D = \{n_{\min}, \frac{F_s}{F_p}, \epsilon_{\max}, F_p\}$, $S_A = \{3, 3.225, 0.2, 40\}$.

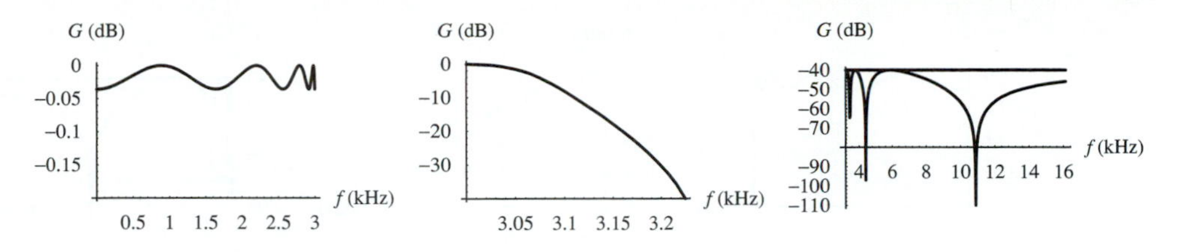

Figure 5.17 Design D2: $D = \{n_{\min}, \frac{F_s}{F_p}, \epsilon < \epsilon_{\max}, F_p\}$, $S_A = \{3, 3.225, 0.2, 40\}$.

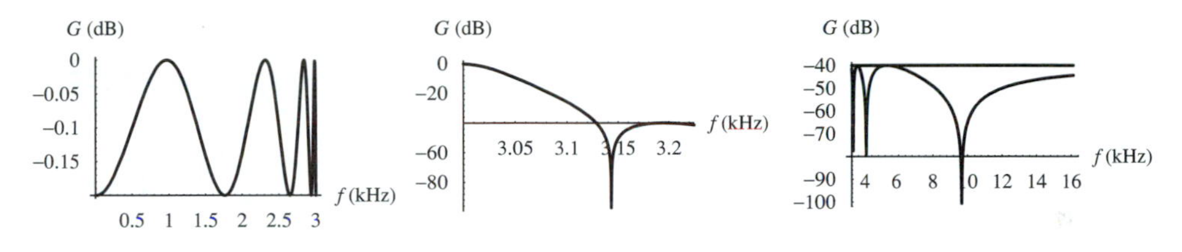

Figure 5.18 Design D3a: $D = \{n_{\min}, \xi_{\min}, \epsilon_{\max}, F_p\}$, $S_A = \{3, 3.225, 0.2, 40\}$.

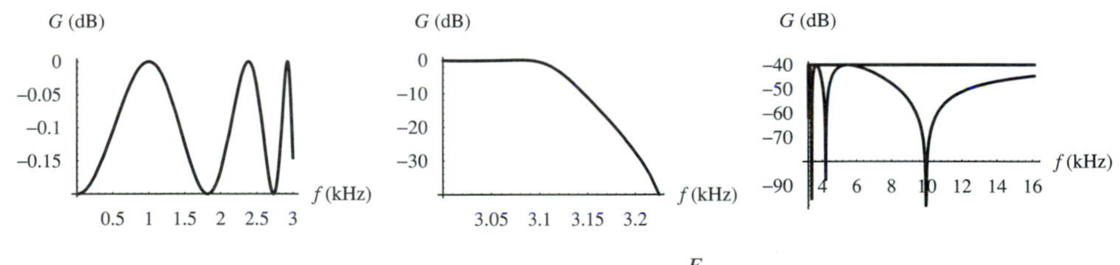

Figure 5.19 Design D3b: $D = \{n_{\min}, \xi_{\min}, \epsilon_{\max}, \frac{F_s}{\xi_{\min}}\}$, $S_A = \{3, 3.225, 0.2, 40\}$.

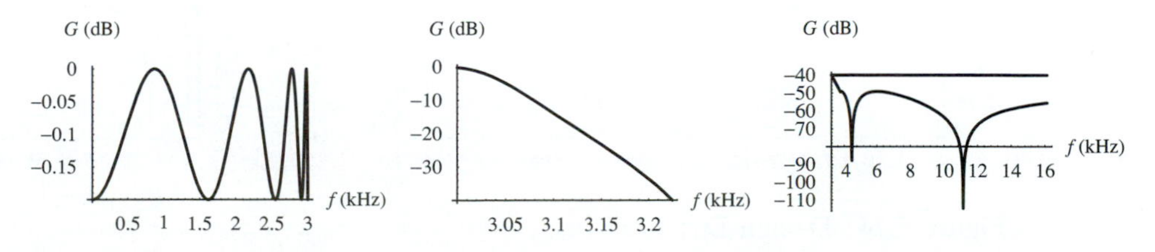

Figure 5.20 Design D4a: $D = \{n_{\min}, \xi_{\max}, \epsilon_{\max}, F_p\}$, $S_A = \{3, 3.225, 0.2, 40\}$.

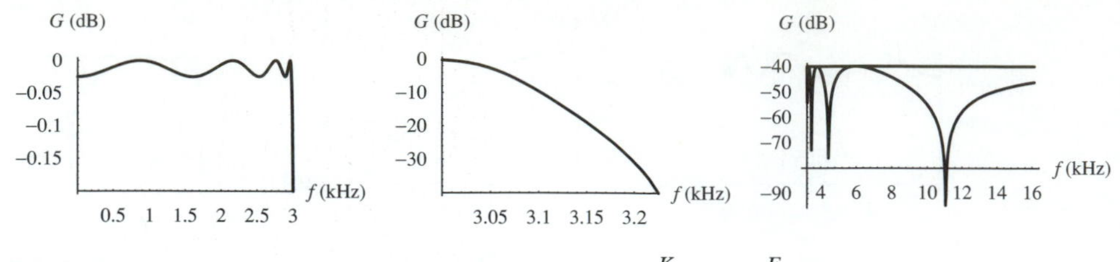

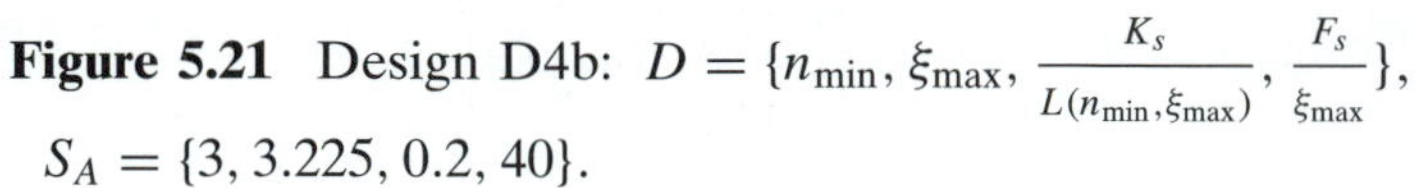

Figure 5.21 Design D4b: $D = \{n_{\min}, \xi_{\max}, \frac{K_s}{L(n_{\min}, \xi_{\max})}, \frac{F_s}{\xi_{\max}}\}$, $S_A = \{3, 3.225, 0.2, 40\}$.

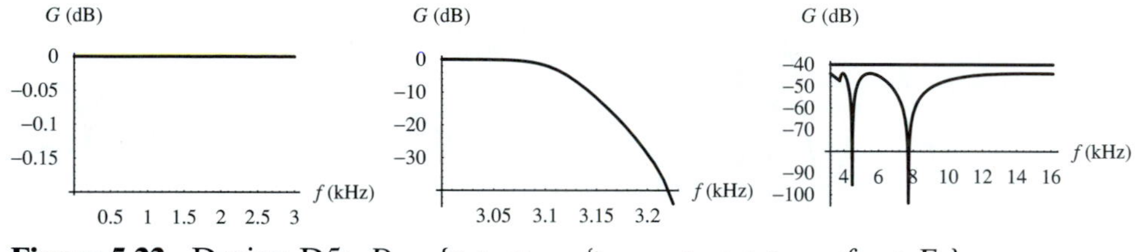

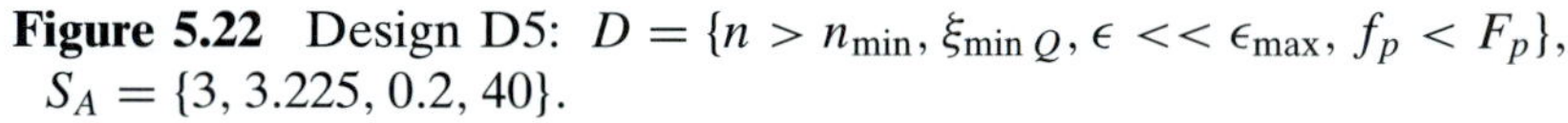

Figure 5.22 Design D5: $D = \{n > n_{\min}, \xi_{\min\,Q}, \epsilon << \epsilon_{\max}, f_p < F_p\}$, $S_A = \{3, 3.225, 0.2, 40\}$.

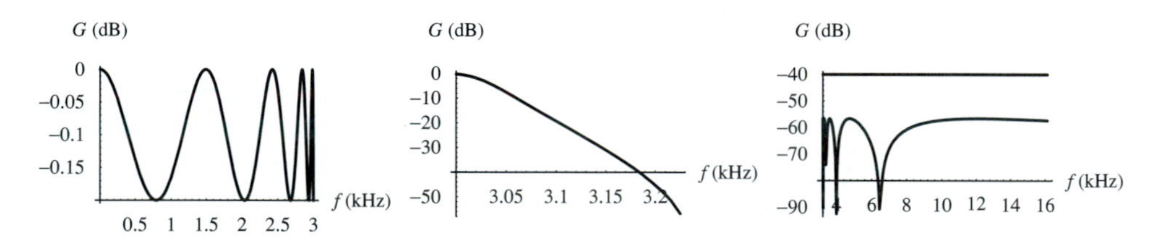

Figure 5.23 Design D1: $D = \{n_{\min} + 1, \frac{F_s}{F_p}, \epsilon_{\max}, F_p\}$, $S_A = \{3, 3.225, 0.2, 40\}$.

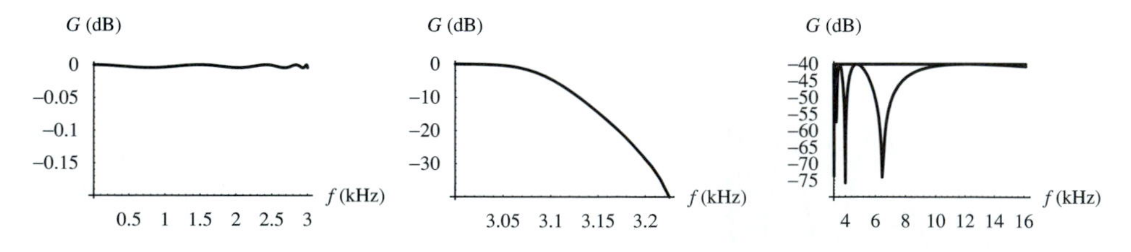

Figure 5.24 Design D2: $D = \{n_{\min} + 1, \frac{F_s}{F_p}, \epsilon < \epsilon_{\max}, F_p\}$, $S_A = \{3, 3.225, 0.2, 40\}$.

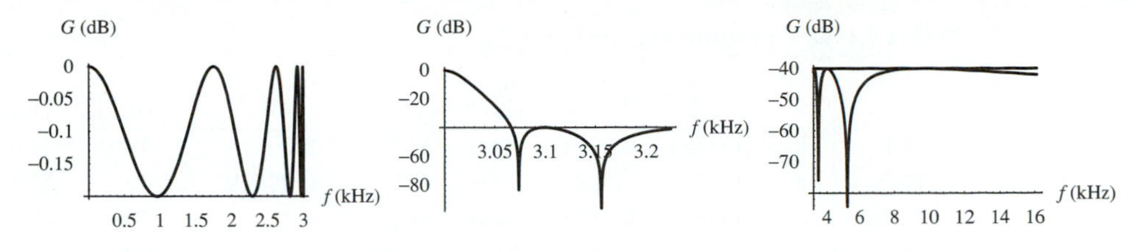

Figure 5.25 Design D3a: $D = \{n_{\min} + 1, \xi_{\min}, \epsilon_{\max}, F_p\}$, $S_A = \{3, 3.225, 0.2, 40\}$.

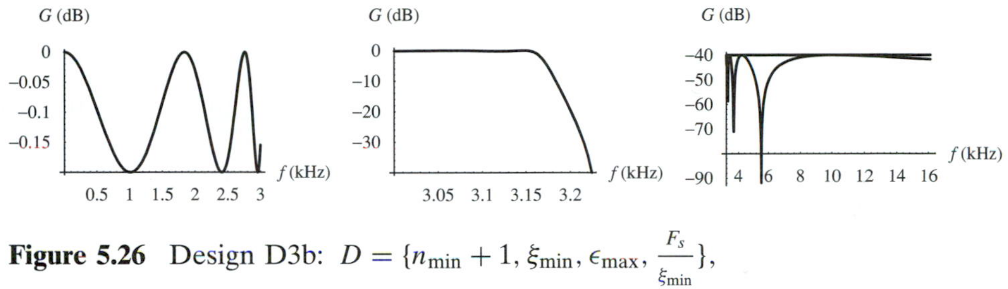

Figure 5.26 Design D3b: $D = \{n_{\min} + 1, \xi_{\min}, \epsilon_{\max}, \frac{F_s}{\xi_{\min}}\}$, $S_A = \{3, 3.225, 0.2, 40\}$.

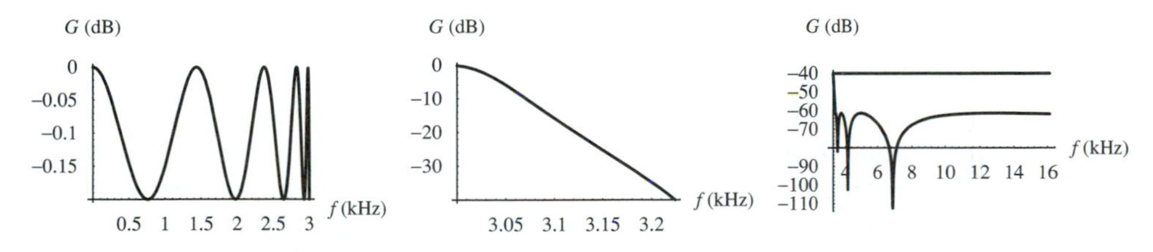

Figure 5.27 Design D4a: $D = \{n_{\min} + 1, \xi_{\max}, \epsilon_{\max}, F_p\}$, $S_A = \{3, 3.225, 0.2, 40\}$.

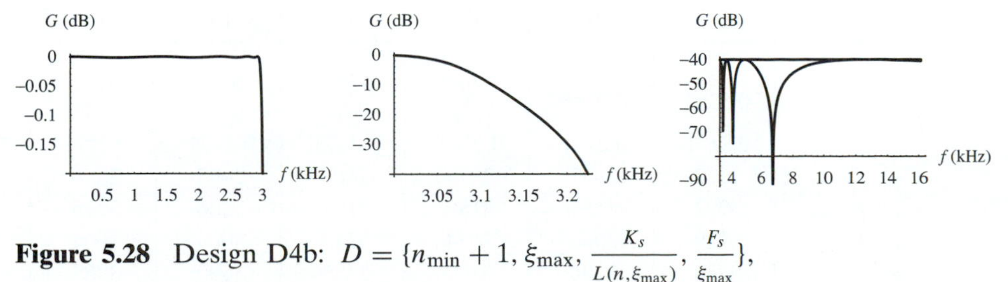

Figure 5.28 Design D4b: $D = \{n_{\min} + 1, \xi_{\max}, \frac{K_s}{L(n, \xi_{\max})}, \frac{F_s}{\xi_{\max}}\}$, $S_A = \{3, 3.225, 0.2, 40\}$.

Table 5.1 Design summary for $n = n_{min}$

	n	ξ	ϵ	f_p (Hz)	f_s (Hz)	a_p (dB)	a_s (dB)	Q_{max}
D 1	8	1.075	0.2171	3000	3225	0.2	47.55	29.9
D 2	8	1.075	0.09097	3000	3225	0.03579	40.	24.2
D 3a	8	1.043	0.2171	3000	3129	0.2	40.	42.1
D 3b	8	1.043	0.2171	3092	3225	0.2	40.	42.1
D 4a	8	1.083	0.2171	3000	3250	0.2	49.14	28.2
D 4b	8	1.083	0.07579	2977	3225	0.02487	40.	22.1
D 5	10	1.079	0.01	2989	3225	$4.343 \cdot 10^{-4}$	40.	21.3

Table 5.2 Design summary for $n = n_{min} + 1$

	n	ξ	ϵ	f_p (kHz)	f_s (kHz)	a_p (dB)	a_s (dB)	Q_{max}
D 1	9	1.075	0.2171	3.	3.225	0.2	56.66	37.4
D 2	9	1.075	0.03188	3.	3.225	0.004412	40.	25.8
D 3a	9	1.022	0.2171	3.	3.066	0.2	40.	81.2
D 3b	9	1.022	0.2171	3.156	3.225	0.2	40.	81.2
D 4a	9	1.098	0.2171	3.	3.294	0.2	61.4	32.1
D 4b	9	1.098	0.01847	2.937	3.225	$1.48 \cdot 10^{-4}$	40.	20.9
D 5	10	1.079	0.01	2.989	3.225	$4.343 \cdot 10^{-4}$	40.	21.3

If technological requirements impose a maximal value of Q-factors (e.g., $Q_{max} = 20$ for active RC filters), Table 5.3 reveals that all six design alternatives fail. The design D4b is the best suboptimal solution.

An advanced design technique, the design D5 [47] with doubled poles [54], achieves the maximum Q-factor lower than 20. This is paid by increasing the filter order; the actual filter order is 12, which is much higher than the minimal order $n = 8$. Although the filter order has been increased, the implementation can be more cost effective [46, 55–58]; the lower tolerance components can be used, and the magnitude response of the implemented filter satisfies the specifications [46].

Table 5.3 Design space for $S_A = \{3000\text{ Hz},\ 3225\text{ Hz},\ 0.2\text{ dB},\ 40\text{ dB}\}$

	Filter Order, n						
	8	9	10	11	12	13	16
ϵ_{min}	0.0758	0.0185	$\frac{3.687}{1000}$	$\frac{5.787}{10^4}$	$\frac{6.76}{10^5}$	$\frac{5.39}{10^6}$	$\frac{2.52}{10^{11}}$
ϵ_{max}	0.2171	0.2171	0.2171	0.2171	0.2171	0.2171	0.2171
$f_{p,min}$	2977	2937	2879	2800	2695	2556	1770
$f_{p,max}$	3092	3156	3189	3101	3206	3215	3225
$f_{s,min}$	3129	3066	3034	3117	3009	3004	3001
$f_{s,max}$	3250	3294	3360	3455	3590	3785	5467

We enlarge the design space by increasing the filter order from $n = 8$ to $n = 9$. The corresponding gain is plotted in Figs. 5.23–5.28, and the design results are summarized in Table 5.4. The Q-factor of the design D4b is 20.9 and is very close to the required maximal value ($Q = 20$).

Table 5.4 Q-factors for $S_A = \{3000\text{ Hz}, 3225\text{ Hz}, 0.2\text{ dB}, 40\text{ dB}\}$

	Filter Order, n								
	8	9	10	11	12	13	14	15	16
Q_{min}	22.1	20.9	20.3	20.0	19.8	19.6	19.6	19.5	19.5
Q_{max}	42.1	81.2	156	301	582	1121	2161	4167	8032

In practice, we choose the most suitable design, D, from the determined design space D_S. Thus, we can try to meet various technological requirements (maximal Q-factors, maximal element tolerances, and so forth) and advanced specifications (maximal group delay variation, maximal rise time, maximal overshoot in step response, maximal settling time).

It should be noticed that the design parameters of the design D5 belong to the design space D_S; D5 and its modification D5a [47] yield, also, elliptic function filters.

Suppose that n_{cheb} designates the order of the Chebyshev-type approximation meeting the specification. For $n = n_{cheb}$ the rational approximation function of D4a is practically equal to the polynomial Chebyshev type I approximation. On the other hand, for the same order $n = n_{cheb}$, the design D4b yields an inverse Chebyshev-type filter.

When the filter order is equal to the order of the Butterworth-type filter, the design D5 practically yields an allpole Butterworth-type filter. This means that the classic filter types—Chebyshev, inverse Chebyshev, and Butterworth—are just special cases of the elliptic function filters and are contained within the design space D_S.

5.6 VISUALIZATION OF DESIGN SPACE

Consider a lowpass filter from Section 5.5 specified by

$$S_A = \{F_p, F_s, A_p, A_s\} = \{3\text{ kHz}, 3225\text{ Hz}, 0.2\text{ dB}, 40\text{ dB}\}$$

with the characteristic-function-limit specification

$$S_K = \{F_p, F_s, K_p, K_s\} = \{3\text{ kHz}, 3225\text{ Hz}, 0.2171, 100\}$$

The minimal filter order has been calculated to be $n = 8$. Next, the range of ξ, ϵ, f_p, and f_s has been determined for $n \geq 8$.

The design subspaces for $n = 8$, $n = 9$, and $n = 13$ are shown in Figs. 5.29–5.31. Notice that the exact design subspaces are nonlinear continuous domains, but we draw rectangular blocks for the sake of simplicity.

Table 5.4 shows that the Q-factor varies from 15 to 156. It should be noticed that for the same specification S_A the MATLAB signal processing toolbox [1] offers an elliptic filter design with $Q = 42$, which might be inappropriate for practical analog implementations. Other classic approximations require very high orders compared to $n_{ellip} = 8$. The order of the Butterworth filter is extremely high ($n_{butt} = 85$), while

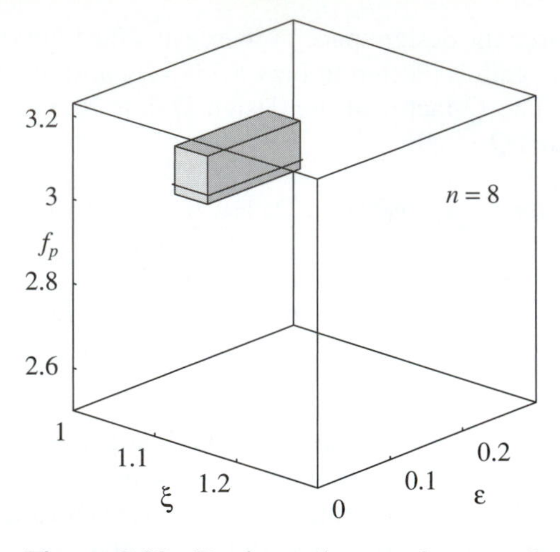

Figure 5.29 Design subspace for $n = 8$.

the order of the Chebyshev and inverse Chebyshev type is also high ($n_{cheb} = 18$) with high Q-factors ($Q_{cheb} = 46$ and $Q_{invcheb} = 20$). Generally, in analog filter design, filter orders that are significantly greater than n_{ellip} are unacceptable because they imply very high implementation complexity and cost.

The range of the design parameters ξ, ϵ, f_p, and f_s are shown in Figs. 5.32 and 5.33. The minimal filter order, $n = n_{min}$, implies a small range for design parameters, and the optimization of the filter behavior can be ineffective.

It is also worth noticing that increasing the filter order, $n > n_{min}$, does not necessarily lead to a better solution. However, there exists an advanced design strategy in which we choose $n > n_{min}$ and obtain robust and selective analog filters with better performance and reduced complexity. For example, the classic filter design failed to meet the specification in the case of an SC filter. However, a design based on our

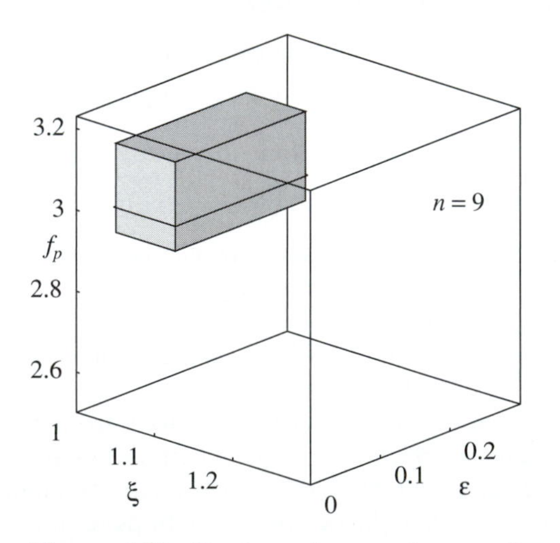

Figure 5.30 Design subspace for $n = 9$.

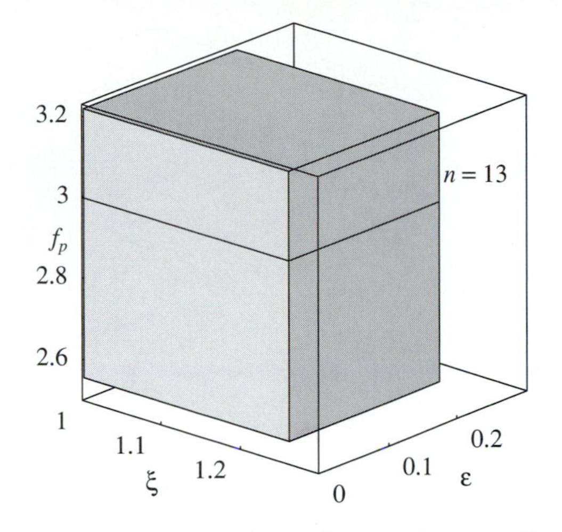

Figure 5.31 Design subspace for $n = 13$.

approach was practically implemented, and the measured filter characteristics showed that our advanced filter design was a successful one [46].

The group delay of the basic designs is plotted in Figs. 5.34 and 5.35.

The maximum group delay is obtained for the minimal transition designs, D3a and D3b, while the maximal transition designs, D4a and D4b, have lower group-delay variation. The design D3b has the minimal variation of the group delay in the passband, while the similar design D3a has the highest overall group-delay variation. The design D5, which is based on the minimal Q-factor design [47], has also a small variation of group delay in the passband.

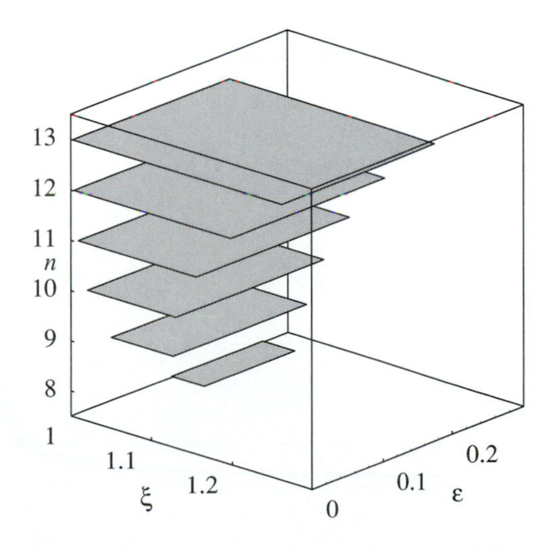

Figure 5.32 Design space of ξ, ϵ and n.

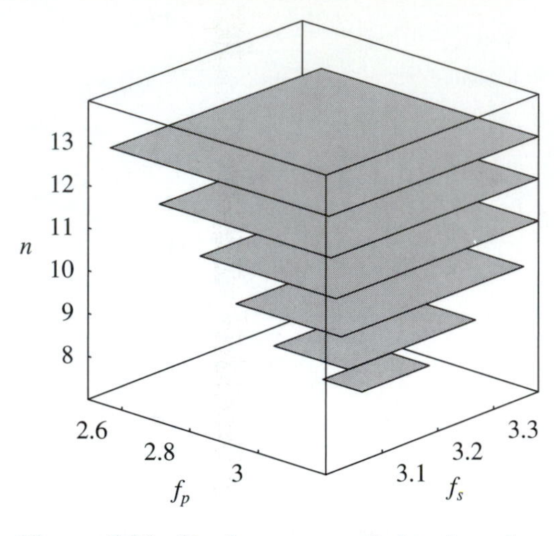

Figure 5.33 Design space of f_p, f_s and n.

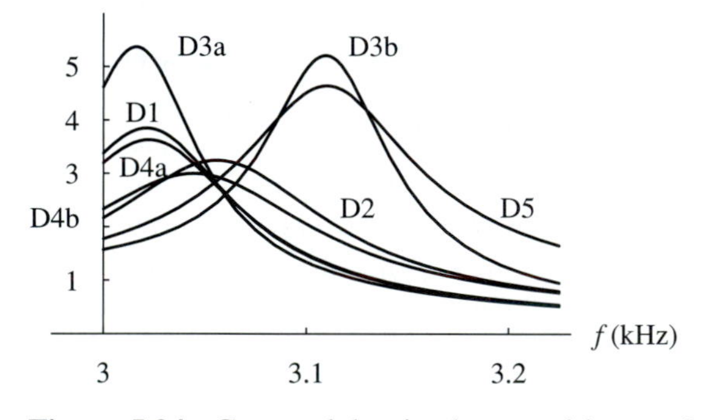

Figure 5.34 Group delay in the transition region.

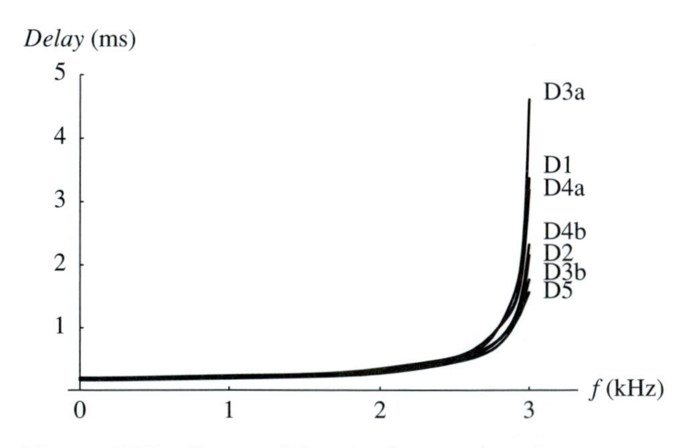

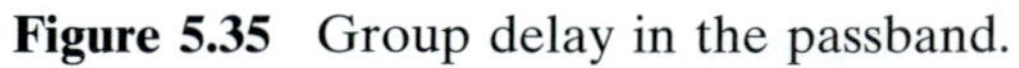

Figure 5.35 Group delay in the passband.

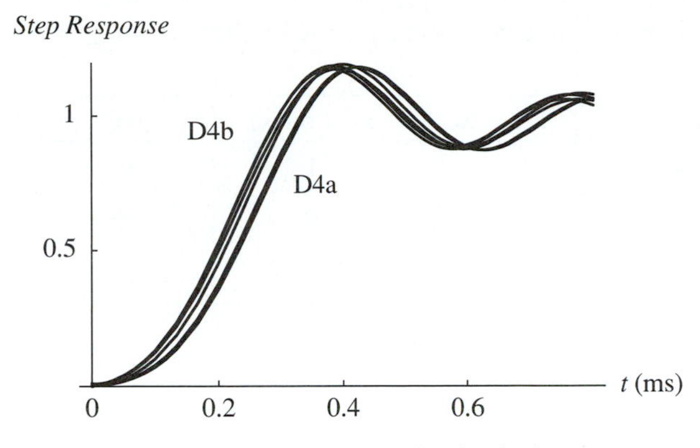

Figure 5.36 Step response of basic designs.

The step response of the basic designs is shown in Fig. 5.36. The shape of all responses is the same with approximately the same amount of overshoots (D3a and D3b have the largest overshoot). As we expect, D4b has the smallest time delay. However, the design D4a, which has also a good group-delay characteristic, has the worst time delay.

5.7 SC FILTER ADVANCED DESIGN EXAMPLE

Advanced filter design techniques can be efficiently exploited in designing robust and selective filters based on commercially available integrated switched capacitor (SC) universal building blocks. Application notes released by manufacturers rely on the classic filter design; often, these notes suggest standard modes of operation of the SC integrated chip that are suitable for different filter types and realizations.

In this section we modify a standard mode of operation of a universal SC filter to design very selective low sensitivity filters.

5.7.1 Introduction

The magnitude response of a high Q-factor second-order filter section is $2Q$ times more sensitive to the pole-magnitude variation than to the Q-factor variation. Traditionally, second-order filter sections (also called *biquads*) has been designed in such a way that the pole-magnitude sensitivities are at their theoretical minima. This means that the sensitivities to passive components (resistors and capacitors) have to be smaller than or equal to $\frac{1}{2}$ [59], [60]. Therefore, special care should be taken on pole-magnitude sensitivities.

The magnitude response of an SC filter is much more sensitive to the pole-magnitude variation than to the Q-factor variation as shown in Table 5.5.

Table 5.5 Approximate value of maximal passive sensitivities for $Q >> 1$

		$\lvert S_R^x \rvert$			$\max(\lvert S_R^{\lvert H \rvert} \rvert)$		
x	$\max(\lvert S_x^{\lvert H \rvert} \rvert)$	RC	SC 3a	SC 1 minQ	RC	SC 3a	SC 1 minQ
Q	1	Q	1	1	Q	1	1
ω_0	Q	$\frac{1}{2}$	$\frac{1}{2}$	0	$\frac{Q}{2}$	$\frac{Q}{2}$	0
ω_z	$\frac{1}{\omega_z^2 - 1}$	$\frac{1}{2}$	$\frac{1}{2}$	$\frac{\omega_z^2 - 1}{2\omega_z^2}$	$\frac{1}{2(\omega_z^2 - 1)}$	$\frac{1}{2(\omega_z^2 - 1)}$	$\frac{1}{2}$
Q=20, $\omega_z = 1.078$ rad/s				$\max(\sum \lvert S_R^{\lvert H \rvert} \rvert)$	33.1	14.1	1.5

Passband edge frequency $\omega_p = 1$ rad/s.

Transfer function zero $\omega_z > \omega_p$.

Transfer function pole $\omega_0 > \omega_p$.

R—external resistor.

RC—second order active RC filter.

SC—switched-capacitor filter in mode 3a or mode 1 minQ.

Universal active SC filters can operate in seven modes. In almost all modes, except mode 1 and mode 4 [43, 44] the pole-magnitude sensitivities to passive components are $\frac{1}{2}$. In mode 1, the pole-magnitude sensitivity to passive components is 0. This means that the pole magnitude depends only on the clock frequency.

The pole Q-factor sensitivity to resistors, of universal SC biquads in mode 1, is equal to 1; it is Q times lower than the corresponding sensitivity of the best active RC filter with a single operational amplifier [59]. Nevertheless, mode 1 has been rarely recommended by manufacturers for practical realizations. The principal reason is that mode 1, usually, requires different clock frequencies for each biquad; thus its complexity increases.

Our target is to design a cost-effective filter with reduced complexity, and we want to use one clock frequency. If we prefer to use mode 1, we have to design a filter whose biquads have identical pole magnitudes.

One solution is the Butterworth filter because it has all poles on a circle; however, its order might be extremely high for very selective filter applications.

The minimal Q-factor filter design, D5, inherently has all poles on a circle; also, very selective specifications could be fulfilled with the order that is slightly higher than the minimal order. In addition, the design D5 features low Q-factors and reduced overall sensitivity. The maximal magnitude-response deviation, due to element tolerances, is in the transition region. Practically, the magnitude response in the passband is insensitive to variations of external resistor values.

5.7.2 Mode 1 Operation

The mode 1 biquad realization [44] is shown in Fig. 5.37.

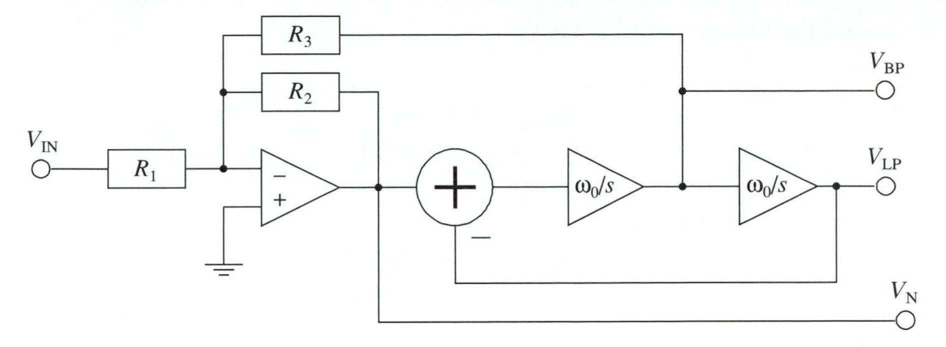

Figure 5.37 The mode 1 of an SC biquad.

The corresponding lowpass (LP), bandpass (BP), and notch (N) transfer functions are

$$H_{LP}(s) = \frac{V_{LP}}{V_{IN}} = -\frac{R_2}{R_1} \frac{\omega_0^2}{s^2 + \dfrac{R_2}{R_3}\omega_0 s + \omega_0^2} \tag{5.41}$$

$$H_{BP}(s) = \frac{V_{BP}}{V_{IN}} = -\frac{R_3}{R_1} \frac{\dfrac{R_2}{R_3}\omega_0 s}{s^2 + \dfrac{R_2}{R_3}\omega_0 s + \omega_0^2} \tag{5.42}$$

$$H_N(s) = \frac{V_N}{V_{IN}} = -\frac{R_2}{R_1} \frac{s^2 + \omega_0^2}{s^2 + \dfrac{R_2}{R_3}\omega_0 s + \omega_0^2} \tag{5.43}$$

where

$$\omega_0 = 2\pi \frac{f_{CLK}}{N} \tag{5.44}$$

$$N = 50 \qquad \text{or} \qquad N = 100 \tag{5.45}$$

The mode 1 operation is recommended for implementation of allpole lowpass and bandpass filters, such as Butterworth-, Chebyshev-, or Bessel-type filters [43, 44]. The mode 1 operation supports the highest clock frequencies because the input summing amplifier is outside the filter's resonant loop [44].

5.7.3 Modification to Mode 1 Operation

The lowpass minimal Q-factor design is essentially an elliptic design, and it yields imaginary transfer function zeros

$$\begin{aligned} s_z &= \pm j\omega_z \\ \omega_z &> \omega_0 \end{aligned} \tag{5.46}$$

and the biquad transfer function

$$H_{\min Q}(s) = -g \frac{s^2 + \omega_z^2}{s^2 + \dfrac{R_2}{R_3}\omega_0 s + \omega_0^2} \tag{5.47}$$

where g is a constant.

Can we modify the realization of Fig. 5.37 to satisfy the condition $\omega_z > \omega_0$? A possible modification is shown in Fig. 5.38, and we call it "mode 1 minQ". The biquad transfer function becomes

$$H_{\min Q}(s) = \frac{V_{N1}}{V_{IN}} = \frac{R_g R_2}{R_1 R_h} \frac{s^2 + \left(\omega_0\sqrt{1 + \dfrac{R_h}{R_l}}\right)^2}{s^2 + \dfrac{R_2\omega_0}{R_3}s + \omega_0^2} \tag{5.48}$$

and its zeros are determined by the clock frequency and the resistors R_h and R_l:

$$\omega_z = 2\pi \frac{f_{CLK}}{N}\sqrt{1 + \frac{R_h}{R_l}} \tag{5.49}$$

The pole Q-factor is

$$Q = \frac{R_3}{R_2} \tag{5.50}$$

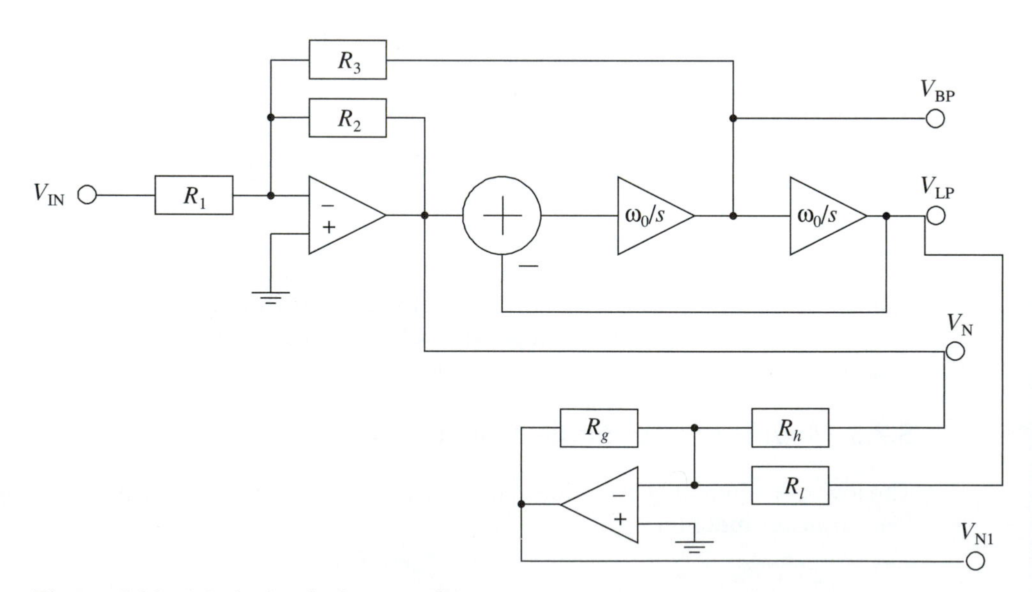

Figure 5.38 Mode 1 minQ operation.

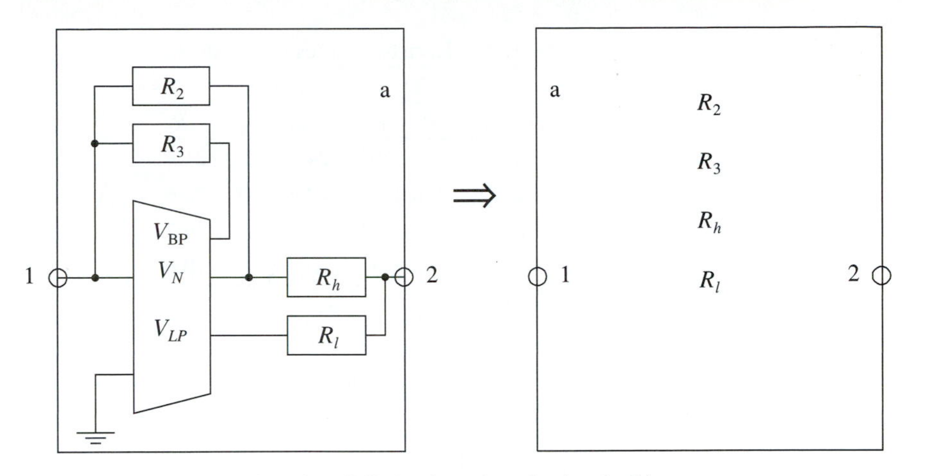

Figure 5.39 Second-order SC section (mode 1 minQ).

The performance of the implemented filter mostly depends on the accuracy of the clock frequency f_{CLK}, which is higher than the accuracy of passive components.

5.7.4 Cascaded Filter Realization

We implement a higher-order filter by cascading second-order sections. Each section is made out of an integrated SC building block and external resistors R_2, R_3, R_h, and R_l as shown in Figs. 5.39 and 5.40.

5.7.5 Practical Implementation

Consider a filter attenuation-limit specification

$$S_A = \{F_p = 2500 \text{ Hz}, \quad F_s = 2690 \text{ Hz}, \quad A_p = 0.5 \text{ dB}, \quad A_s = 40 \text{ dB}\}$$

The manufacturer's application note [43, 44] suggests the mode 3a operation for the design of elliptic filters; the minimal filter order is $n_{\min} = 8$ (Table 5.6).

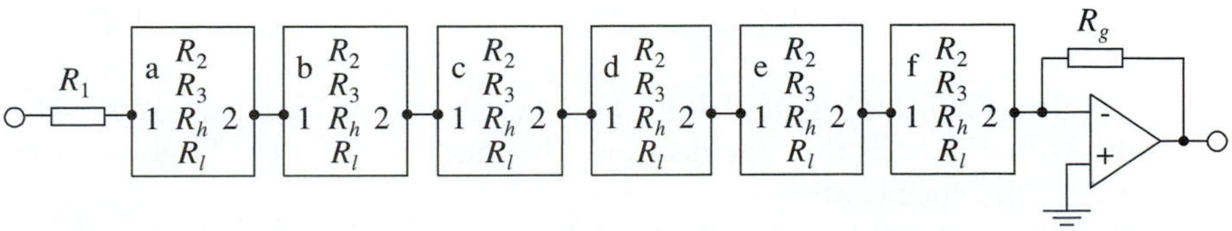

Figure 5.40 Filter implementation by cascading second-order SC sections.

Table 5.6 Resistor values for mode 3a

$R[\mathrm{k}\Omega]$, Mode 3a, n=8					
Section	R_2	R_3	R_4	R_h	R_l
a	10.9	10.0	24.7	262.	21.4
b	10.0	19.5	13.6	41.3	20.6
c	10.0	57.7	11.0	48.7	39.2
d	10.0	254.	10.4	10.9	10.0
$R_1 = 15.1\ \mathrm{k}\Omega$, $R_g = 67.8\ \mathrm{k}\Omega$					

We choose the mode 1 minQ operation and the order $n_{\min Q} = 12$. The element values are summarized in Table 5.7.

Table 5.7 Resistor values for mode 1 minQ

$R[\mathrm{k}\Omega]$, Mode 1 minQ, n=12				
Section	R_2	R_3	R_h	R_l
a	11.2	10.0	25.8	10.0
b	3.96	7.21	7.67	10.0
c	4.34	17.3	15.0	50.0
d	5.77	54.4	20.0	146.
e	4.40	151.	4.96	61.2
f	55.1	29.7	25.7	1.00
$R_1 = 10.0\ \mathrm{k}\Omega$, $R_g = 1.00\ \mathrm{k}\Omega$				

The sensitivity of the transfer function zeros, for mode 3a and mode 1 minQ operations, are shown in Table 5.8. The mode 1 minQ has significantly lower sensitivity.

Table 5.8 Sensitivity of transfer function zeros

	$\left\|S_{R_h}^{\|H\|}\right\| + \left\|S_{R_l}^{\|H\|}\right\|$		
Section/Mode	3a, $n = 8$	3a, $n = 12$	1 minQ, $n = 12$
a	0.08	0.04	0.04
b	0.87	0.35	0.26
c	2.97	1.11	0.53
d	5.86	2.52	0.72
e		4.50	0.82
f		6.17	0.86

The measured gain of both filters is shown in Figs. 5.41 and 5.42. The mode 1 minQ filter satisfies the specification. The mode 3a filter has larger passband ripple and violates the specification.

The measured passband attenuation is larger than the calculated attenuation. This disagreement is caused by the imperfections of the actual device (chip) that is used in

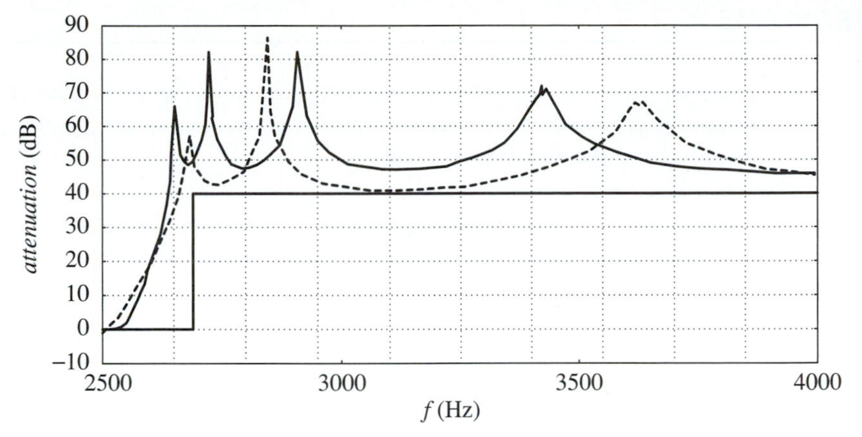

Figure 5.41 Measured attenuation in the stopband: mode 1 minQ (solid line), mode 3a (dashed line).

filter implementation. For a given f_{CLK} the frequency ω_0 varies from chip to chip, typically $\pm 0.2\%$.

5.7.6 Concluding Remarks

The transfer function poles of an integrated SC filter depend only on the clock frequency and are insensitive to variations of external resistors. The sensitivity of Q-factors to external resistors is equal to 1. The sensitivity of transfer function zeros to external resistors is lower than $\frac{1}{2}$.

Practically, our design is insensitive to external resistors. We can manufacture robust filters from standard resistors with looser tolerances. At the same time, we can lower the net price of the product and increase the product yield in mass production.

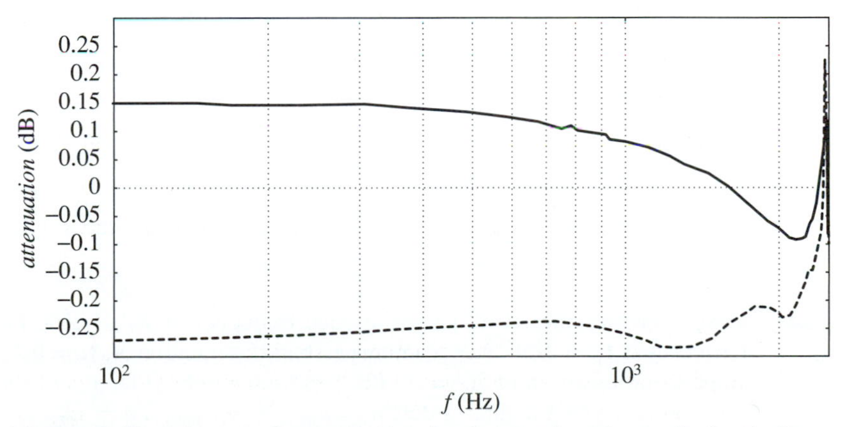

Figure 5.42 Measured attenuation in the passband: mode 1 minQ (solid line), mode 3a (dashed line).

PROBLEMS

5.1 Consider a lowpass filter with passband edge frequency at $F_p = 2000$ Hz, stopband edge frequency at $F_p = 2150$ Hz, maximum passband attenuation of $A_p = 0.2$ dB, and minimum stopband attenuation of $A_s = 40$ dB.

(a) Find the magnitude limits specification, the magnitude tolerances specification, the magnitude ripple tolerances specification, and the gain limits specification.

(b) Determine the minimum order of the filter. Determine the order of the Butterworth, Chebyshev type I, Chebyshev type II, and elliptic transfer function.

(c) Determine the range of the selectivity factor for the orders $n = n_{\min}$ and $n = n_{\min+1}$.

(d) Determine the range of the ripple factor for the orders $n = n_{\min}$ and $n = n_{\min+1}$.

(e) Determine the range of the actual passband edge and actual stopband edge frequencies for the orders $n = n_{\min}$ and $n = n_{\min+1}$.

5.2 Design a minimal Q-factor lowpass filter with passband edge frequency at $F_p = 4000$ Hz, stopband edge frequency at $F_p = 4300$ Hz, maximum passband attenuation of $A_p = 0.2$ dB, and minimum stopband attenuation of $A_s = 40$ dB. Determine the minimum order of the filter, n_{mq}, and the filter order of the Butterworth transfer function, n_{but}. Plot the magnitude response of all minimal Q-factor lowpass filters for $n_{mq} \leq n \leq n_{but}$.

5.3 A filter shall be designed having a maximum passband attenuation $A_p = 0.1$ dB within the passband $0 \leq f \leq F_p = 10$ kHz and a minimum stopband attenuation $A_s = 40$ dB within the stopband $30 \leq f \leq F_s = 1000$ kHz. In addition, the filter shall meet the following:

(a) The attenuation shall be equiripple in the passband and monotonic in the stopband, reaching the attenuation $a = 40$ dB at $f = 30$ kHz.

(b) The attenuation shall be equiripple in the stopband and monotonic in the passband, reaching the attenuation $a = 0.1$ dB at $f = 10$ kHz.

(c) The attenuation shall be monotonic in the passband and stopband, reaching the attenuation $a = 0.1$ dB at $f = 10$ kHz.

(d) The attenuation shall be equiripple in the passband and stopband, reaching the attenuation $a = 0.1$ dB at $f = 10$ kHz.

(e) The attenuation shall be equiripple in the passband and stopband, reaching the attenuation $a = 40$ dB at $f = 30$ kHz.

5.4 Design lowpass filters with passband edge frequency at $F_p = 2000$ Hz, stopband edge frequency at $F_p = 2150$ Hz, maximum passband attenuation of $A_p = 0.2$ dB, and minimum stopband attenuation of $A_s = 40$ dB. Determine pole Q-factors of the elliptic transfer function. Assume the filter orders $n = n_{\min}$, $n = n_{\min+1}$, $n = n_{\min+2}$, and $n = n_{\min+3}$. Determine pole Q-factors of the Butterworth, Chebyshev type I, and Chebyshev type II transfer functions.

5.5 Design lowpass filters with passband edge frequency at $F_p = 2000$ Hz, stopband edge frequency at $F_p = 2150$ Hz, maximum passband attenuation of $A_p = 0.2$ dB, and minimum stopband attenuation of $A_s = 40$ dB.

(a) Determine the minimum order of the filter, $n_{\min}$. Determine the minimum order of the Butterworth, n_{but}, Chebyshev type I, n_{c1}, Chebyshev type II, n_{c2}, and elliptic transfer function, n_e.

(b) Determine the range of the design parameters: selectivity factor, ripple factor, actual passband edge, and actual stopband edge frequencies, for $n_{\min} \leq n \leq n_{c1}$.

(c) Sketch the characteristic function of the designs D1, D2, D3a, D3b, D4a, and D4b, for $n_{\min} \leq n \leq n_{c1}$.

5.6 Design lowpass filters with passband edge frequency at $F_p = 3000$ Hz, stopband edge frequency at $F_p = 3225$ Hz, maximum passband attenuation of $A_p = 0.2$ dB, and minimum stopband attenuation of $A_s = 40$ dB.

(a) Determine the minimum order of the filter, $n_{\min}$. Determine the minimum order of the Butterworth transfer function, n_{but}.

(b) Sketch the characteristic function of the minimal Q-factor elliptic design (design D5) for $n_{\min} \leq n \leq n_{but}$.

(c) Compare the characteristic function of the minimal Q-factor elliptic design with the Butterworth design.

5.7 Design lowpass filters with passband edge frequency at $F_p = 3000$ Hz, stopband edge frequency at $F_p = 3225$ Hz, maximum passband attenuation of $A_p = 0.2$ dB, and minimum stopband attenuation of $A_s = 40$ dB.

(a) Determine the minimum order of the filter, $n_{\min}$. Determine the minimum order of the Chebyshev type I transfer function, n_{c1}.

(b) Sketch the characteristic function of the design D4a for $n_{\min} \leq n \leq n_{c1}$.

(c) Compare the characteristic function of the design D4a with the Chebyshev type I design.

5.8 Consider a lowpass filter specified by

$$S_K = \{F_p = 3 \text{ kHz}, \quad F_s = 3.225 \text{ kHz}, \quad K_p = 0.2171, \quad K_s = 100\}$$

(a) Determine the minimum order of the filter, $n_{\min}$.

(b) Sketch the transfer function, step response, and group delay for designs D1, D2, D3, and D4, for $n = n_{\min}$.

(c) Compare the group delay and the rise time of the designs.

5.9 Design highpass filters with passband edge frequency at $F_p = 2150$ Hz, stopband edge frequency at $F_s = 2$ kHz, maximum passband attenuation of $A_p = 0.2$ dB, and minimum stopband attenuation of $A_s = 40$ dB.

(a) Determine the minimum order of the filter, $n_{\min}$. Determine the minimum order of the Butterworth, n_{but}, Chebyshev type I, n_{c1}, Chebyshev type II, n_{c2}, and elliptic transfer function, n_e.

(b) Determine the range of the design parameters: selectivity factor, ripple factor, actual passband edge, and actual stopband edge frequencies for $n_{\min} \leq n \leq n_{\min+2}$.

(c) Sketch the attenuation of the designs D1, D2, D3a, D3b, D4a, and D4b for $n_{\min} \leq n \leq n_{\min+2}$.

5.10 Design bandpass filters with passband edge frequencies at $F_{p1} = 3007$ Hz, $F_{p2} = 3217$ Hz, stopband edge frequencies at $F_{s1} = 3000$ Hz, $F_{s2} = 3225$ Hz, maximum passband attenuation of $A_p = 0.2$ dB, and minimum stopband attenuation of $A_s = 40$ dB.

- **(a)** Determine the minimum order of the filter, $n_{\min}$. Determine the minimum order of the Butterworth, n_{but}, Chebyshev type I, n_{c1}, Chebyshev type II, n_{c2}, and elliptic transfer function, n_e.
- **(b)** Determine the range of the design parameters: selectivity factor, ripple factor, actual passband edge, and actual stopband edge frequencies for $n_{\min} \leq n \leq n_{\min+2}$.
- **(c)** Sketch the gain of the designs D1, D2, D3a, D3b, D4a, and D4b for $n_{\min} \leq n \leq n_{\min+2}$.

5.11 Design bandreject filters with stopband edge frequencies at $F_{s1} = 3007$ Hz, $F_{s2} = 3217$ Hz, passband edge frequencies at $F_{p1} = 3000$ Hz, $F_{p2} = 3225$ Hz, maximum passband attenuation of $A_p = 0.2$ dB, and minimum stopband attenuation of $A_s = 40$ dB.

- **(a)** Determine the minimum order of the filter, $n_{\min}$. Determine the minimum order of the Butterworth, n_{but}, Chebyshev type I, n_{c1}, Chebyshev type II, n_{c2}, and elliptic transfer function, n_e.
- **(b)** Determine the range of the design parameters: selectivity factor, ripple factor, actual passband edge, and actual stopband edge frequencies, for $n_{\min} \leq n \leq n_{\min+2}$.
- **(c)** Sketch the attenuation of the designs D1, D2, D3a, D3b, D4a, and D4b for filter order $n_{\min} \leq n \leq n_{\min+2}$.

MATLAB EXERCISES

5.1 Write a MATLAB program that computes the magnitude limits specification, the magnitude tolerances specification, the magnitude ripple tolerances specification, and the gain limits specification. Assume the specification given in Problem 5.1. Determine the minimum order of the Butterworth, Chebyshev type I, Chebyshev type II, and elliptic transfer function. Plot the magnitude responses for filter order $n = n_{\min}$ and $n = n_{\min+1}$. Compute the range of the selectivity factor, the range of the ripple factor, and the range of the actual passband edge and actual stopband edge frequencies for filter order $n = n_{\min}$, $n = n_{\min+1}$, and $n = n_{\min+2}$.

5.2 Write a MATLAB program that plots the magnitude response of a minimal-Q-factor lowpass filter. Assume the specification given in Problem 5.2. Compute the magnitude of all poles and show that all poles of a minimal Q-factor transfer function lie on a circle. Plot poles and zeros of the transfer function for $n_{mq} \leq n \leq n_{but}$. Compute the minimum order of this filter, n_{mq}, and the minimal filter order of the Butterworth transfer function, n_{but}.

5.3 Write a MATLAB program that computes a lowpass transfer function with the specification given in Problem 5.4, the passband edge frequency at $F_p = 2000$ Hz, stopband edge frequency at $F_p = 2150$ Hz, maximum passband attenuation of $A_p = 0.2$ dB, and minimum stopband attenuation of $A_s = 40$ dB. Compute Q-factors of the elliptic transfer function for the filter orders $n = n_{\min}$, $n = n_{\min+1}$, $n = n_{\min+2}$, and $n = n_{\min+3}$. Compute the Q-factors of the Butterworth, Chebyshev type I, and Chebyshev type II transfer functions.

5.4 Write a MATLAB program that computes the characteristic function of the designs D1, D2, D3a, D3b, D4a, and D4b for filter order $n_{\min} \leq n \leq n_{c1}$. Assume the specification given in Problem 5.5. Compute the range of the selectivity factor, the ripple factor, the actual passband edge and actual stopband edge frequencies for filter order $n_{\min} \leq n \leq n_{c1}$.

5.5 Write a MATLAB program that computes the characteristic function of the minimal Q-factor elliptic and Butterworth transfer function for filter order $n_{\min} \leq n \leq n_{but}$. Assume the specification given in Problem 5.6.

5.6 Write a MATLAB program that computes the characteristic function of the design D4a and the Chebyshev type I transfer function for filter order $n_{\min} \leq n \leq n_{c1}$. Assume the specification given in Problem 5.7.

5.7 Write a MATLAB program that computes the transfer function, step response, and group delay for filter order $n = n_{\min}$. Assume the specification given in Problem 5.8.

5.8 Write a MATLAB program that computes the attenuation of the designs D1, D2, D3a, D3b, D4a, and D4b for highpass filter. Assume the specification given in Problem 5.9 and the filter order $n_{\min} \leq n \leq n_{\min+2}$.

5.9 Write a MATLAB program that plots the gain of the designs D1, D2, D3a, D3b, D4a, and D4b for bandpass filter. Assume the specification given in Problem 5.10 and the filter order $n_{\min} \leq n \leq n_{\min+2}$.

5.10 Write a MATLAB program that plots the gain of the designs D1, D2, D3a, D3b, D4a, and D4b for bandreject filter. Assume the specification given in Problem 5.11 and the filter order $n_{\min} \leq n \leq n_{\min+2}$.

MATHEMATICA EXERCISES

5.1 Write a *Mathematica* program that computes the magnitude limits specification, the magnitude tolerances specification, the magnitude ripple tolerances specification, and the gain limits specification. Assume the specification given in Problem 5.1. Determine the minimum order of the Butterworth, Chebyshev type I, Chebyshev type II, and elliptic transfer function. Plot the magnitude response for filter order $n = n_{\min}$ and $n = n_{\min+1}$. Compute the range of the selectivity factor, the range of the ripple factor, and the range of the actual passband edge and actual stopband edge frequencies for filter order $n = n_{\min}$, $n = n_{\min+1}$, and $n = n_{\min+2}$.

5.2 Write a *Mathematica* program that plots the magnitude response of a minimal-Q-factor lowpass filter. Assume the specification given in Problem 5.2. Compute the magnitude of all poles and show that all poles of a minimal Q-factor transfer function lie on a circle. Plot poles and zeros of the transfer function for $n_{mq} \leq n \leq n_{but}$. Compute the minimum order of the filter, n_{mq}, and the minimal filter order of the Butterworth transfer function, n_{but}.

5.3 Write a *Mathematica* program that computes lowpass transfer function with the specification given in Problem 5.4, the passband edge frequency at $F_p = 2000$ Hz, stopband edge frequency at $F_p = 2150$ Hz, maximum passband attenuation of $A_p = 0.2$ dB, and minimum stopband attenuation of $A_s = 40$ dB. Compute the Q-factors of elliptic transfer function for the filter orders $n = n_{\min}$, $n = n_{\min+1}$, $n = n_{\min+2}$, and $n = n_{\min+3}$. Compute the Q-factors of the Butterworth, Chebyshev type I, and Chebyshev type II transfer functions.

5.4 Write a *Mathematica* program that computes the characteristic function of the designs D1, D2, D3a, D3b, D4a, and D4b for filter order $n_{\min} \leq n \leq n_{c1}$. Assume the specification given in Problem 5.5. Compute the range of the selectivity factor, the ripple factor, the actual passband edge, and actual stopband edge frequencies for filter order $n_{\min} \leq n \leq n_{c1}$.

5.5 Write a *Mathematica* program that computes the characteristic function of the minimal Q-factor elliptic and Butterworth transfer function for filter order $n_{\min} \leq n \leq n_{but}$. Assume the specification given in Problem 5.6.

5.6 Write a *Mathematica* program that computes the characteristic function of the design D4a and the Chebyshev type I transfer function for filter order $n_{\min} \leq n \leq n_{c1}$. Assume the specification given in Problem 5.7.

5.7 Write a *Mathematica* program that computes the transfer function, step response, and group delay for filter order $n = n_{\min}$. Assume the specification given in Problem 5.8.

5.8 Write a *Mathematica* program that computes the attenuation of the designs D1, D2, D3a, D3b, D4a, and D4b for highpass filter. Assume the specification given in Problem 5.9 and the filter order $n_{\min} \leq n \leq n_{\min+2}$.

5.9 Write a *Mathematica* program that plots the gain of the designs D1, D2, D3a, D3b, D4a, and D4b for bandpass filter. Assume the specification given in Problem 5.10 and the filter order $n_{\min} \leq n \leq n_{\min+2}$.

5.10 Write a *Mathematica* program that plots the gain of the designs D1, D2, D3a, D3b, D4a, and D4b for bandreject filter. Assume the specification given in Problem 5.11 and the filter order $n_{\min} \leq n \leq n_{\min+2}$.

CHAPTER 6

ADVANCED ANALOG FILTER DESIGN ALGORITHMS

Classical analog filter design techniques return only one design from an infinite collection of alternative designs, or fail to design filters when solutions exist. These classic techniques hide a wealth of alternative filter designs that are more robust when implemented in analog circuits. In this chapter, we present (1) case studies of optimal analog filters that cannot be designed with classic techniques and (2) the formal, mathematical framework that underlies their solutions. We have automated the advanced filter design techniques in software, so we present detailed step-by-step design algorithms.

6.1 INTRODUCTION

In designing analog filters, one generally relies on canned software routines or mechanical table-oriented procedures. The primary reason for these "black box" approaches is that the approximation theory that underlies filter design includes complex mathematics. Unfortunately, conventional approaches return only one design, thereby hiding a wealth of alternative filter designs that are more robust when implemented in analog circuits. In addition, conventional approaches may fail to find a filter when in fact one exists.

We develop advanced design techniques to find a comprehensive set of optimal designs to represent the infinite solution space. The optimal designs include filters that have minimal order, minimal quality factors, minimal complexity, minimal sensitivity to pole-zero locations, minimal deviation from a specified group delay, approximate linear phase response, and minimized peak overshoot of the step response. We base our approach on formal, mathematical properties of Jacobi elliptic functions [47, 53]. We automate these advanced filter design techniques in software [2, 51].

The key observations underlying advanced analog filter design are that

1. many designs satisfy the same user specification;
2. Butterworth and Chebyshev filters are special cases of elliptic filters; and
3. minimum-order filters may not be as efficient to implement as some higher-order filters.

The filter optimization problem is a mixed-integer linear programming problem so the classical techniques break down. Instead of using iterative numerical techniques, we solve these problems using closed-form algebraic expressions. Then, we present several new case studies of optimal analog filters that cannot be designed with classical techniques, and the formal, mathematical framework that underlies their solutions.

6.2 NOTATION

We review the list of symbols that we use in formulas and procedures when designing analog filters. Often, we append a suffix to designate a quantity related to a specific filter type. For example, we add h to designate highpass filter; thus, A_{ph} is A_p of a highpass filter.

$A(\omega)$—attenuation (dB)

A_p—maximum passband attenuation in specification (dB)

A_s—minimum stopband attenuation in specification (dB)

$\mathrm{cd}(u, k)$—Jacobi elliptic cd function

$\mathrm{cd}^{-1}(v, k)$—inverse Jacobi elliptic cd function

f—frequency (Hz)

f_{nQ}—normalized frequency in minimal Q-factor design

f_p—passband edge frequency of designed filter (Hz)

F_p—passband edge frequency in specification (Hz)

f_s—stopband edge frequency of designed filter (Hz)

F_s—stopband edge frequency in specification (Hz)

$G(\omega)$—gain (dB)

$h(H(s), t)$—impulse response

$h_s(H(s), t)$—step response

$\mathcal{H}(n, \xi, \epsilon, p)$—normalized lowpass transfer function

$\mathcal{H}_{\min Q}(n, \xi, p)$—minimal Q-factor normalized lowpass transfer function

$H(s)$—transfer function

i—index ($i = 1, 2, \ldots, n$)

j—the imaginary unit ($j = \sqrt{-1}$)

k—modulus of elliptic functions

$K_e(n, \xi, \epsilon, x)$—elliptic characteristic function

$K_J(k)$—complete elliptic integral of first kind
K_p—characteristic function passband specification
K_s—characteristic function stopband specification
$L(n, \xi)$—discrimination factor
$\mathcal{L}^{-1}(H(s))$—the inverse Laplace transform of $H(s)$
$M(\omega)$—magnitude response, $M(\omega) = |H(j\omega)|$
n—transfer function order (order for short)
$n_{but}(F_p, F_s, K_p, K_s)$—minimum Butterworth order
$n_{cheb}(F_p, F_s, K_p, K_s)$—minimum Chebyshev order
$n_{ellip}(F_p, F_s, K_p, K_s)$—minimum elliptic order
$n_{\max}$—maximum order
$n_{\min}$—minimum order
$n_{minQ}(F_p, F_s, K_p, K_s)$—minimum order of minimal Q-factor design
p—normalized complex frequency
$q(k)$—modular constant
$Q(s_i)$—quality factor of pole/zero s_i
$Q_{\min Q}(n, \xi, i)$—quality factor of ith pole of minimal Q-factor design
$R(n, \xi, x)$—elliptic rational function
s—complex frequency, Laplace operator (rad/s)
$\mathrm{sn}(u, k)$—Jacobi elliptic sn function
$\mathrm{sn}^{-1}(v, k)$—inverse Jacobi elliptic sn function
$S(n, \xi, \epsilon, i)$—ith pole of normalized lowpass transfer function
S_A—attenuation-limit specification
S_G—gain-limit specification
S_K—characteristic-function-limit specification
S_M—magnitude-limit specification
S_r—magnitude-ripple specification
S_δ—magnitude-tolerance specification
$S_{minQ}(n, \xi, i)$—ith pole of normalized lowpass transfer function for minimal Q-factor design
t—time (s)
x—dimensionless variable
$X(n, \xi, i)$—ith zero of elliptic rational function
δ_1—passband magnitude ripple
δ_2—stopband magnitude ripple
δ_p—passband magnitude tolerance
δ_s—stopband magnitude tolerance
$\zeta(n, \xi, \epsilon)$—auxiliary function
ϵ—ripple factor
ξ—selectivity factor

$\tau_{GD}(H(s), \omega)$—group delay (s)

$\Phi(\omega)$—phase response, $\Phi(\omega) = \arg(H(j\omega))$

ω—angular frequency (rad/s), $\omega = 2\pi f$

$\lfloor x \rfloor$—integer, $x \leq \lfloor x \rfloor < x + 1$

$\underset{x_1 < x < x_2}{\text{FindRoot}} \; \{ F(x) = G(x) \}$ find real x over interval $x_1 < x < x_2$ by solving $F(x) = G(x)$

$\underset{\substack{x_1 < x < x_2 \\ y_1 < y < y_2}}{\text{FindRoot}} \begin{Bmatrix} F_1(x) = G_1(x) \\ F_2(x) = G_2(x) \end{Bmatrix}$ find real x over interval $x_1 < x < x_2$ and real y over interval $y_1 < y < y_2$ by solving set of equations $\{F_1(x) = G_1(x),\; F_2(x) = G_2(x)\}$

6.3 DESIGN EQUATIONS AND PROCEDURES

In this section we summarize all design equations, formulas, and procedures that are based on Jacobi elliptic functions. We use this relations in the purely numerical design.

6.3.1 Specification

A design specification can be given in different ways:

$$\boxed{S_A = \{F_p, F_s, A_p, A_s\}} \tag{6.1}$$

$$\boxed{S_G = \{F_p, F_s, G_p, G_s\}} \tag{6.2}$$

$$\boxed{S_\delta = \{F_p, F_s, \delta_p, \delta_s\}} \tag{6.3}$$

$$\boxed{S_M = \{F_p, F_s, M_p, M_s\}} \tag{6.4}$$

$$\boxed{S_r = \{F_p, F_s, \delta_1, \delta_2\}} \tag{6.5}$$

$$\boxed{S_K = \{F_p, F_s, K_p, K_s\}} \tag{6.6}$$

We use a set of functions to convert one form of specification into another.

$$\boxed{\begin{aligned} K_p(A_p) &= \frac{\sqrt{1 - 10^{-A_p/10}}}{10^{-A_p/20}} \\ K_s(A_s) &= \frac{\sqrt{1 - 10^{-A_s/10}}}{10^{-A_s/20}} \end{aligned}} \tag{6.7}$$

$$\begin{aligned} K_p(G_p) &= \frac{\sqrt{1 - 10^{G_p/10}}}{10^{G_p/20}} \\ K_s(G_s) &= \frac{\sqrt{1 - 10^{G_s/10}}}{10^{G_s/20}} \end{aligned} \tag{6.8}$$

$$\begin{aligned} K_p(\delta_p) &= \frac{\sqrt{\delta_p(2 - \delta_p)}}{1 - \delta_p} \\ K_s(\delta_s) &= \frac{\sqrt{1 - \delta_s^2}}{\delta_s} \end{aligned} \tag{6.9}$$

$$\begin{aligned} K_p(M_p) &= \frac{\sqrt{1 - M_p^2}}{M_p} \\ K_s(M_s) &= \frac{\sqrt{1 - M_s^2}}{M_s} \end{aligned} \tag{6.10}$$

$$\begin{aligned} K_p(\delta_1) &= \frac{2\sqrt{\delta_1}}{1 - \delta_1} \\ K_s(\delta_1, \delta_2) &= \frac{\sqrt{(1 + \delta_1)^2 - \delta_2^2}}{\delta_2} \end{aligned} \tag{6.11}$$

6.3.2 Special and Auxiliary Functions

The following special mathematical functions are used in the design of elliptic filters:

$$K_J(k) = \int_0^{\pi/2} \frac{d\theta}{\sqrt{1 - k^2 \sin^2 \theta}} \tag{6.12}$$

$$\begin{aligned} v &= \mathrm{sn}^{-1}\left(\frac{1}{\sqrt{1 + \epsilon^2}}, \sqrt{1 - \frac{1}{L^2(n, \xi)}}\right) \\ \zeta(n, \xi, \epsilon) &= \mathrm{sn}\left(\frac{K_J\left(\sqrt{1 - \frac{1}{\xi^2}}\right)}{K_J\left(\sqrt{1 - \frac{1}{L^2(n, \xi)}}\right)} v, \sqrt{1 - \frac{1}{\xi^2}}\right) \end{aligned} \tag{6.13}$$

The procedure given by Eq. (6.13) comes from Eq. (12.302).

$$
\begin{aligned}
& t = \frac{1}{2}\frac{1-\sqrt[4]{1-k^2}}{1+\sqrt[4]{1-k^2}} \\
& q' = t + 2t^5 + 15t^9 + 150t^{13} + 1\,707t^{17} \\
& \quad + 20{,}910t^{21} + 268{,}616t^{25} + 3{,}567{,}400t^{29}, \qquad k \le \frac{1}{\sqrt{2}} \\
& \quad + 48{,}555{,}069t^{33} + 673{,}458{,}874t^{37} \\
& q(k) = q' \\
& t = \frac{1}{2}\frac{1-\sqrt{k}}{1+\sqrt{k}} \\
& q' = t + 2t^5 + 15t^9 + 150t^{13} + 1{,}707t^{17} \\
& \quad + 20{,}910t^{21} + 268{,}616t^{25} + 3{,}567{,}400t^{29}, \qquad \frac{1}{\sqrt{2}} < k < 1 \\
& \quad + 48{,}555{,}069t^{33} + 673{,}458{,}874t^{37} \\
& q(k) = e^{\pi^2/\ln q'}
\end{aligned}
\tag{6.14}
$$

The procedure given by Eq. (6.14) comes form Eqs. (12.60), (12.64)–(12.67).

6.3.3 Transfer Function Order

For a given specification we compute the minimal order of the approximation functions.

$$
\begin{aligned}
k &= \frac{F_p}{F_s} \\
L &= \frac{K_s}{K_p} \\
N &= \frac{K_J\left(\sqrt{1-\frac{1}{L^2}}\right)}{K_J\left(\frac{1}{L}\right)} \\
D &= \frac{K_J\left(\sqrt{1-k^2}\right)}{K_J\,(k)} \\
n_{ellip}\left(F_p, F_s, K_p, K_s\right) &= \left\lfloor \frac{N}{D} \right\rfloor
\end{aligned}
\tag{6.15}
$$

The procedure given by Eq (6.15) comes from Eq. (12.338).

$$\boxed{\begin{aligned} \xi &= \frac{F_p}{F_s} \\ L &= \frac{K_s}{K_p} \\ N &= \cosh^{-1} L \\ D &= \cosh^{-1} \frac{1}{\xi} \\ n_{cheb}\left(F_p, F_s, K_p, K_s\right) &= \left\lfloor \frac{N}{D} \right\rfloor \end{aligned}} \tag{6.16}$$

$$\boxed{\begin{aligned} \xi &= \frac{F_p}{F_s} \\ L &= \frac{K_s}{K_p} \\ N &= \log_{10} L \\ D &= \log_{10} \frac{1}{\xi} \\ n_{but}\left(F_p, F_s, K_p, K_s\right) &= \left\lfloor \frac{N}{D} \right\rfloor \end{aligned}} \tag{6.17}$$

The procedures given by Eqs. (6.16) and (6.17) are based on formulas given in reference 15.

$$\boxed{\begin{aligned} n_{\min} &= n_{ellip}\left(F_p, F_s, K_p, K_s\right) \\ n_{\max} &= 2\, n_{ellip}\left(F_p, F_s, K_p, K_s\right) \end{aligned}} \tag{6.18}$$

$$\boxed{\begin{aligned} i &= n_{ellip}\left(F_p, F_s, K_p, K_s\right) \\ \xi_i &= \xi_{min}(i, 1, K_s^2) \\ & \boxed{\begin{aligned} &\text{While } K_e\left(i, \xi_i, \frac{1}{K_s}, \frac{F_s}{F_p}\xi_i\right) > K_p \\ &i = i + 1 \\ &\xi_i = \xi_{min}(i, 1, K_s^2) \end{aligned}} \\ n_{\min Q}\left(F_p, F_s, K_p, K_s\right) &= i \end{aligned}} \tag{6.19}$$

The procedure given by Eq. (6.19) comes from Eqs. (12.338) and (12.367)–(12.369).

6.3.4 Zeros, Poles, and Q-Factors

The basic functions for the elliptic design are $X(n,\xi,i)$, $S(n,\xi,\epsilon,i)$, and $Q(s)$, as follows:

$$\boxed{\begin{aligned} &X(n,\xi,i) &&= -\mathrm{cd}\left(\frac{2i-1}{n}K_J(\frac{1}{\xi}),\frac{1}{\xi}\right) \\ &X(n,\xi,(n+1)/2) &&= 0, \quad n \text{ odd} \\ &X(n,\xi,1) < X(n,\xi,2) < &&\ldots < X(n,\xi,n) \end{aligned}} \tag{6.20}$$

The procedure given by Eq. (6.20) comes from Eqs. (12.134) and (12.323).

$$\boxed{\begin{aligned} \zeta &= \zeta(n,\xi,\epsilon) \\ x &= X(n,\xi,i) \\ N_{re} &= -\zeta\sqrt{1-\zeta^2}\sqrt{1-x^2}\sqrt{1-\frac{x^2}{\xi^2}} \\ N_{im} &= x\sqrt{1-\left(1-\frac{1}{\xi^2}\right)\zeta^2} \\ N &= N_{re} + jN_{im} \\ D &= 1-\left(1-\frac{x^2}{\xi^2}\right)\zeta^2 \\ S(n,\xi,\epsilon,i) &= \frac{N}{D} \end{aligned}} \tag{6.21}$$

The procedure given by Eq. (6.21) comes from Eq. (12.266).

$$\boxed{\begin{aligned} N_{re} &= -\sqrt{1-X(n,\xi,i)^2}\sqrt{\xi^2-X(n,\xi,i)^2} \\ N_{im} &= X(n,\xi,i)(\xi+1) \\ N &= N_{re} + jN_{im} \\ D &= \xi + X(n,\xi,i)^2 \\ S_{\min Q}(n,\xi,i) &= \sqrt{\xi}\frac{N}{D} \end{aligned}} \tag{6.22}$$

The procedure given by Eq. (6.22) comes from Eq. (12.370).

$$\boxed{Q(s) = -\frac{|s|}{2\mathrm{Re}(s)}} \tag{6.23}$$

$$\boxed{Q_{\min Q}(n,\xi,i) = \frac{\xi + X(n,\xi,i)^2}{2\sqrt{1-X(n,\xi,i)^2}\sqrt{\xi^2-X(n,\xi,i)^2}}} \tag{6.24}$$

Equation (6.24) comes from Eq. (12.372).

6.3.5 Discrimination Factor, Elliptic Rational Function, and Characteristic Function

The essential functions for the elliptic approximation are $L(n,\xi)$, $R(n,\xi,x)$, and $K_e(n,\xi,\epsilon,x)$:

$$L(n,\xi) = \frac{1}{\xi^n} \frac{\prod_{i=1}^{n/2} \left(\xi^2 - X^2(n,\xi,i)\right)^2}{\prod_{i=1}^{n/2} \left(1 - X^2(n,\xi,i)\right)^2} \qquad n \text{ even } n=2,4,\ldots$$

$$L(n,\xi) = \frac{1}{\xi^{n-2}} \frac{\prod_{i=1}^{(n-1)/2} \left(\xi^2 - X^2(n,\xi,i)\right)^2}{\prod_{i=1}^{(n-1)/2} \left(1 - X^2(n,\xi,i)\right)^2} \qquad n \text{ odd } n=1,3,\ldots \tag{6.25}$$

The procedure given by Eq. (6.25) comes from Eqs. (12.359) and (12.360).

$$r_0 = \frac{\prod_{i=1}^{n/2} \left(1 - X^2(n,\xi,i)\right)}{\prod_{i=1}^{n/2} \left(1 - \dfrac{\xi^2}{X^2(n,\xi,i)}\right)} \qquad n \text{ even } n=2,4,\ldots$$

$$R(n,\xi,x) = \frac{1}{r_0} \frac{\prod_{i=1}^{n/2} \left(x^2 - X^2(n,\xi,i)\right)}{\prod_{i=1}^{n/2} \left(x^2 - \dfrac{\xi^2}{X^2(n,\xi,i)}\right)}$$

$$r_0 = \frac{\prod_{i=1}^{(n-1)/2} \left(1 - X^2(n,\xi,i)\right)}{\prod_{i=1}^{(n-1)/2} \left(1 - \dfrac{\xi^2}{X^2(n,\xi,i)}\right)} \qquad n \text{ odd } n=1,3,\ldots$$

$$R(n,\xi,x) = \frac{1}{r_0} \frac{x \prod_{i=1}^{(n-1)/2} \left(x^2 - X^2(n,\xi,i)\right)}{\prod_{i=1}^{(n-1)/2} \left(x^2 - \dfrac{\xi^2}{X^2(n,\xi,i)}\right)} \tag{6.26}$$

The procedure given by Eq. (6.26) comes from Eqs. (12.141)–(12.144).

$$\boxed{K_e(n,\xi,\epsilon,x) = \epsilon\,|R(n,\xi,x)|} \tag{6.27}$$

6.3.6 Normalized Lowpass Elliptic Transfer Function

Our goal is to find the transfer function that meets a lowpass specification.

$$g = \frac{1}{\sqrt{1+\epsilon^2}} \frac{\prod_{i=1}^{n/2} |S(n,\xi,\epsilon,i)|^2}{\prod_{i=1}^{n/2} \frac{\xi^2}{X^2(n,\xi,i)}} \qquad n \text{ even},\ n=2,4,\ldots$$

$$\mathcal{H}(n,\xi,\epsilon,p) = \frac{g \prod_{i=1}^{n/2} p^2 + \frac{\xi^2}{X^2(n,\xi,i)}}{\prod_{i=1}^{n/2} p^2 - 2p\mathrm{Re}\,(S(n,\xi,\epsilon,i)) + |S(n,\xi,\epsilon,i)|^2}$$

$$g = -S(n,\xi,\epsilon,\frac{n+1}{2}) \frac{\prod_{i=1}^{\frac{n-1}{2}} |S(n,\xi,\epsilon,i)|^2}{\prod_{i=1}^{\frac{n-1}{2}} \frac{\xi^2}{X^2(n,\xi,i)}} \qquad n \text{ odd},\ n=1,3,\ldots$$

$$\mathcal{H}(n,\xi,\epsilon,p) = \frac{\frac{g}{p - S(n,\xi,\epsilon,\frac{n+1}{2})} \prod_{i=1}^{\frac{n-1}{2}} p^2 + \frac{\xi^2}{X^2(n,\xi,i)}}{\prod_{i=1}^{\frac{n-1}{2}} p^2 - 2p\mathrm{Re}\,S(n,\xi,\epsilon,i) + |S(n,\xi,\epsilon,i)|^2} \tag{6.28}$$

The procedure given by Eq. (6.28) comes from Eq. (13.196).

$$g = \frac{\prod_{i=1}^{n/2} X(n, \xi, i)^2}{\sqrt{\xi^n}\sqrt{1 + \frac{1}{L(n, a)}}} \qquad n \text{ even}, \; n=2,4,\ldots$$

$$\mathcal{H}_{\min Q}(n, \xi, \epsilon, p) = g \frac{\prod_{i=1}^{n/2} p^2 + \frac{\xi^2}{X^2(n, \xi, i)}}{\prod_{i=1}^{n/2} p^2 + p \frac{\sqrt{\xi}}{Q_{\min Q}(n, \xi, i)} + \xi}$$

$$g = \frac{\prod_{i=1}^{\frac{n-1}{2}} X(n, \xi, i)^2}{\sqrt{\xi^{n-2}}\sqrt{1 + \frac{1}{L(n, a)}}} \qquad n \text{ odd}, \; n=1,3,\ldots$$

$$\mathcal{H}_{\min Q}(n, \xi, \epsilon, p) = g \frac{1}{p + \sqrt{\xi}} \frac{\prod_{i=1}^{\frac{n-1}{2}} p^2 + \frac{\xi^2}{X^2(n, \xi, i)}}{\prod_{i=1}^{\frac{n-1}{2}} p^2 + p \frac{\sqrt{\xi}}{Q_{\min Q}(n, \xi, i)} + \xi} \tag{6.29}$$

The procedure given by Eq. (6.29) comes from Eqs. (12.373) and (12.374).

6.3.7 Selectivity Factor, Ripple Factor, and Edge Frequencies

We compute the boundary values of the design space from

$$\boxed{\begin{aligned} L &= \frac{K_s}{K_p} \\ g &= \left(q\left(\frac{1}{L}\right)\right)^{1/n} \\ N &= 1 + 2\sum_{m=1}^{9}(-1)^m g^{m^2} \\ D &= 1 + 2\sum_{m=1}^{9} g^{m^2} \\ \xi_{min}\left(n, K_p, K_s\right) &= \frac{1}{\sqrt{1 - \left(\frac{N}{D}\right)^4}} \end{aligned}} \tag{6.30}$$

The procedure given by Eq. (6.30) comes from Eq. (12.354).

$$\boxed{\xi_{\max}\left(n, F_p, F_s, K_p, K_s\right) = x \Big| \underset{F_s/F_p<x<10F_s/F_p}{\text{FindRoot}}\ R(n, x, \frac{F_s}{F_p}) = \frac{K_s}{K_p}} \tag{6.31}$$

$$\boxed{1 < \xi_{\min} < \frac{F_s}{F_p} < \xi_{\max} < \infty} \tag{6.32}$$

$$\boxed{\epsilon_{\min}\left(n, F_p, F_s, K_p, K_s\right) = \frac{K_s}{L\left(n, \xi_{\max}\left(n, F_p, F_s, K_p, K_s\right)\right)}} \tag{6.33}$$

$$\boxed{\epsilon_{\max}\left(K_p\right) = K_p} \tag{6.34}$$

$$\boxed{f_{p,min}\left(n, F_p, F_s, K_p, K_s\right) = \frac{F_s}{\xi_{\max}\left(n, F_p, F_s, K_p, K_s\right)}} \tag{6.35}$$

$$\boxed{f_{p,\max}\left(n, F_s, K_p, K_s\right) = \frac{F_s}{\xi_{min}\left(n, K_p, K_s\right)}} \tag{6.36}$$

$$\xi_h = \xi_{\max}\left(n, F_p, F_s, K_p, K_s\right)$$

$$\xi_l = x \left| \underset{\sqrt{\xi_h \frac{F_s}{F_p}} < x < \xi_h}{\text{FindRoot}}\ K_e\left(n, x, \frac{1}{\sqrt{L(n,x)}}, x\frac{F_s}{F_p}\right) = K_p \right.$$

$$f_l = x \left| \underset{\sqrt[4]{\xi_l} < x < \sqrt{\xi_l}}{\text{FindRoot}}\ K_e\left(n, \xi_l, \frac{1}{\sqrt{L(n,\xi_l)}}, x\right) = K_p \right.$$

$$f_h = x \left| \underset{\sqrt[4]{\xi_h} < x < \sqrt{\xi_h}}{\text{FindRoot}}\ K_e\left(n, \xi_h, \frac{1}{\sqrt{L(n,\xi_h)}}, x\right) = K_p \right.$$

$$\begin{aligned} \xi_{\min Q}\left(n, F_p, F_s, K_p, K_s\right) &= x \\ f_{nQ}\left(n, F_p, F_s, K_p, K_s\right) &= y \end{aligned} \left| \underset{\substack{\xi_l, < x < \xi_h \\ f_l < y < f_h}}{\text{FindRoot}} \left\{ \begin{aligned} K_e\left(n, x, \frac{1}{\sqrt{L(n,x)}}, y\right) &= K_p \\ K_e\left(n, x, \frac{1}{\sqrt{L(n,x)}}, y\frac{F_s}{F_p}\right) &= K_s \end{aligned} \right\} \right. \tag{6.37}$$

6.4 DESIGN D1

1. Start from a specification and convert it into the characteristic-function-limit specification

$$\left.\begin{aligned} S_A &= \{F_p, F_s, A_p, A_s\} \\ S_\delta &= \{F_p, F_s, \delta_p, \delta_s\} \\ S_M &= \{F_p, F_s, M_p, M_s\} \\ S_r &= \{F_p, F_s, \delta_1, \delta_2\} \\ S_G &= \{F_p, F_s, G_p, G_s\} \end{aligned}\right\} \longmapsto S_K = \{F_p, F_s, K_p, K_s\} \tag{6.38}$$

2. Compute the minimal order $n_{\min} = n_{ellip}(F_p, F_s, K_p, K_s)$.
3. Choose the order

$$n \geq n_{\min} \tag{6.39}$$

4. Compute the selectivity factor

$$\xi = \frac{F_s}{F_p} \tag{6.40}$$

5. Choose the ripple factor

$$\epsilon = K_p \tag{6.41}$$

6. Choose the actual passband edge

$$f_p = F_p \tag{6.42}$$

7. Construct the normalized lowpass transfer function

$$\mathcal{H}(n, \xi, \epsilon, p) \tag{6.43}$$

8. Construct the lowpass transfer function

$$H(s) = \mathcal{H}\left(n, \xi, \epsilon, \frac{s}{2\pi f_p}\right) \tag{6.44}$$

6.5 DESIGN D2

1. Start from a specification and convert it into the characteristic-function-limit specification

$$\left.\begin{aligned} S_A &= \{F_p, F_s, A_p, A_s\} \\ S_\delta &= \{F_p, F_s, \delta_p, \delta_s\} \\ S_M &= \{F_p, F_s, M_p, M_s\} \\ S_r &= \{F_p, F_s, \delta_1, \delta_2\} \\ S_G &= \{F_p, F_s, G_p, G_s\} \end{aligned}\right\} \longmapsto S_K = \{F_p, F_s, K_p, K_s\} \tag{6.45}$$

2. Compute the minimal order $n_{\min} = n_{ellip}(F_p, F_s, K_p, K_s)$.
3. Choose the order

$$n \geq n_{\min} \tag{6.46}$$

4. Compute the selectivity factor

$$\xi = \frac{F_s}{F_p} \tag{6.47}$$

5. Compute the ripple factor

$$\epsilon = \frac{K_s}{L(n, \xi)} \tag{6.48}$$

6. Choose the actual passband edge

$$f_p = F_p \tag{6.49}$$

7. Construct the normalized lowpass transfer function

$$\mathcal{H}(n, \xi, \epsilon, p) \tag{6.50}$$

8. Construct the lowpass transfer function

$$H(s) = \mathcal{H}\left(n, \xi, \epsilon, \frac{s}{2\pi f_p}\right) \tag{6.51}$$

6.6 DESIGN D3A

1. Start from a specification and convert it into the characteristic-function-limit specification

$$\left.\begin{aligned} S_A &= \{F_p, F_s, A_p, A_s\} \\ S_\delta &= \{F_p, F_s, \delta_p, \delta_s\} \\ S_M &= \{F_p, F_s, M_p, M_s\} \\ S_r &= \{F_p, F_s, \delta_1, \delta_2\} \\ S_G &= \{F_p, F_s, G_p, G_s\} \end{aligned}\right\} \longmapsto S_K = \{F_p, F_s, K_p, K_s\} \tag{6.52}$$

2. Compute the minimal order $n_{\min} = n_{ellip}(F_p, F_s, K_p, K_s)$.
3. Choose the order

$$n \geq n_{\min} \tag{6.53}$$

4. Compute the selectivity factor

$$\xi = \xi_{min}\left(n, K_p, K_s\right) \tag{6.54}$$

5. Choose the ripple factor

$$\epsilon = K_p \tag{6.55}$$

6. Choose the actual passband edge

$$f_p = F_p \tag{6.56}$$

7. Construct the normalized lowpass transfer function

$$\mathcal{H}(n, \xi, \epsilon, p) \tag{6.57}$$

8. Construct the lowpass transfer function

$$H(s) = \mathcal{H}\left(n, \xi, \epsilon, \frac{s}{2\pi f_p}\right) \tag{6.58}$$

6.7 DESIGN D3B

1. Start from a specification and convert it into the characteristic-function-limit specification

$$\left.\begin{aligned} S_A &= \{F_p, F_s, A_p, A_s\} \\ S_\delta &= \{F_p, F_s, \delta_p, \delta_s\} \\ S_M &= \{F_p, F_s, M_p, M_s\} \\ S_r &= \{F_p, F_s, \delta_1, \delta_2\} \\ S_G &= \{F_p, F_s, G_p, G_s\} \end{aligned}\right\} \longmapsto S_K = \{F_p, F_s, K_p, K_s\} \tag{6.59}$$

2. Compute the minimal order $n_{\min} = n_{ellip}(F_p, F_s, K_p, K_s)$.
3. Choose the order

$$n \geq n_{\min} \tag{6.60}$$

4. Compute the selectivity factor

$$\xi = \xi_{min}\left(n, K_p, K_s\right) \tag{6.61}$$

5. Choose the ripple factor

$$\epsilon = K_p \tag{6.62}$$

6. Compute the actual passband edge

$$f_p = \frac{F_s}{\xi} \tag{6.63}$$

7. Construct the normalized lowpass transfer function

$$\mathcal{H}(n, \xi, \epsilon, p) \tag{6.64}$$

8. Construct the lowpass transfer function

$$H(s) = \mathcal{H}\left(n, \xi, \epsilon, \frac{s}{2\pi f_p}\right) \tag{6.65}$$

6.8 DESIGN D4A

1. Start from a specification and convert it into the characteristic-function-limit specification

$$\left.\begin{aligned} S_A &= \{F_p, F_s, A_p, A_s\} \\ S_\delta &= \{F_p, F_s, \delta_p, \delta_s\} \\ S_M &= \{F_p, F_s, M_p, M_s\} \\ S_r &= \{F_p, F_s, \delta_1, \delta_2\} \\ S_G &= \{F_p, F_s, G_p, G_s\} \end{aligned}\right\} \longmapsto S_K = \{F_p, F_s, K_p, K_s\} \tag{6.66}$$

2. Compute the minimal order $n_{\min} = n_{ellip}(F_p, F_s, K_p, K_s)$.
3. Choose the order

$$n \geq n_{\min} \tag{6.67}$$

4. Compute the selectivity factor,

$$\xi = \xi_{\max}\left(n, F_p, F_s, K_p, K_s\right) \tag{6.68}$$

5. Choose the ripple factor

$$\epsilon = K_p \tag{6.69}$$

6. Choose the actual passband edge

$$f_p = F_p \tag{6.70}$$

7. Construct the normalized lowpass transfer function

$$\mathcal{H}(n, \xi, \epsilon, p) \tag{6.71}$$

8. Construct the lowpass transfer function

$$H(s) = \mathcal{H}\left(n, \xi, \epsilon, \frac{s}{2\pi f_p}\right) \tag{6.72}$$

6.9 DESIGN D4B

1. Start from a specification and convert it into the characteristic-function-limit specification

$$\left.\begin{aligned} S_A &= \{F_p, F_s, A_p, A_s\} \\ S_\delta &= \{F_p, F_s, \delta_p, \delta_s\} \\ S_M &= \{F_p, F_s, M_p, M_s\} \\ S_r &= \{F_p, F_s, \delta_1, \delta_2\} \\ S_G &= \{F_p, F_s, G_p, G_s\} \end{aligned}\right\} \longmapsto S_K = \{F_p, F_s, K_p, K_s\} \tag{6.73}$$

2. Compute the minimal order $n_{\min} = n_{ellip}(F_p, F_s, K_p, K_s)$.
3. Choose the order

$$n \geq n_{\min} \tag{6.74}$$

4. Compute the maximal selectivity factor

$$\xi = \xi_{\max}\left(n, F_p, F_s, K_p, K_s\right) \tag{6.75}$$

5. Compute the ripple factor

$$\epsilon = \frac{K_s}{L(n, \xi)} \tag{6.76}$$

6. Compute the actual passband edge

$$f_p = \frac{F_s}{\xi} \tag{6.77}$$

7. Construct the normalized lowpass transfer function

$$\mathcal{H}(n, \xi, \epsilon, p) \tag{6.78}$$

8. Construct the lowpass transfer function

$$H(s) = \mathcal{H}\left(n, \xi, \epsilon, \frac{s}{2\pi f_p}\right) \tag{6.79}$$

6.10 DESIGN D5

1. Start from a specification and convert it into the characteristic-function-limit specification

$$\left.\begin{aligned} S_A &= \{F_p, F_s, A_p, A_s\} \\ S_\delta &= \{F_p, F_s, \delta_p, \delta_s\} \\ S_M &= \{F_p, F_s, M_p, M_s\} \\ S_r &= \{F_p, F_s, \delta_1, \delta_2\} \\ S_G &= \{F_p, F_s, G_p, G_s\} \end{aligned}\right\} \longmapsto S_K = \{F_p, F_s, K_p, K_s\} \tag{6.80}$$

2. Compute the minimal order $n_{\min Q}\left(F_p, F_s, K_p, K_s\right)$

3. Choose the order

$$n \geq n_{\min Q}\left(F_p, F_s, K_p, K_s\right) \tag{6.81}$$

4. Compute the maximal selectivity factor

$$\xi = \xi_{\min Q}\left(n, F_p, F_s, K_p, K_s\right) \tag{6.82}$$

and normalized frequency

$$f_{nQ}\left(n, F_p, F_s, K_p, K_s\right) \tag{6.83}$$

5. Compute the ripple factor

$$\epsilon = \frac{1}{\sqrt{L(n, \xi)}} \tag{6.84}$$

6. Compute the actual passband edge

$$f_p = \frac{F_p}{f_{nQ}\left(n, F_p, F_s, K_p, K_s\right)} \tag{6.85}$$

7. Construct the normalized lowpass transfer function

$$\mathcal{H}(n, \xi, \epsilon, p) \tag{6.86}$$

8. Construct the lowpass transfer function

$$H(s) = \mathcal{H}\left(n, \xi, \epsilon, \frac{s}{2\pi f_p}\right) \tag{6.87}$$

Summary of design parameters:

	$n_{\min}$	ξ	ϵ	f_p
D1	n_{ellip}	$\frac{F_s}{F_p}$	K_p	F_p
D2	n_{ellip}	$\frac{F_s}{F_p}$	$\frac{K_s}{L(n, \xi)}$	F_p
D3a	n_{ellip}	$\xi_{min}\left(n, K_p, K_s\right)$	K_p	F_p
D3b	n_{ellip}	$\xi_{min}\left(n, K_p, K_s\right)$	K_p	$\frac{F_s}{\xi}$
D4a	n_{ellip}	$\xi_{\max}\left(n, F_p, F_s, K_p, K_s\right)$	K_p	F_p
D4b	n_{ellip}	$\xi_{\max}\left(n, F_p, F_s, K_p, K_s\right)$	$\frac{K_s}{L(n, \xi)}$	$\frac{F_s}{\xi}$
D5	$n_{\min Q}$	$\xi_{\min Q}\left(n, F_p, F_s, K_p, K_s\right)$	$\frac{1}{\sqrt{L(n, \xi)}}$	$\frac{F_p}{f_{nQ}\left(n, F_p, F_s, K_p, K_s\right)}$

(6.88)

6.11 TIME RESPONSE AND FREQUENCY RESPONSE

From the known transfer function, $H(s)$, we compute the step response

$$h_s(t) = \mathcal{L}^{-1}\left(\frac{1}{s}H(s)\right) \tag{6.89}$$

and the impulse response

$$h(t) = \mathcal{L}^{-1}(H(s)) \tag{6.90}$$

The frequency response can be found as follows: the magnitude response

$$M(\omega) = |H(j\omega)| \tag{6.91}$$

the phase response

$$\Phi(\omega) = \arg(H(j\omega)) \tag{6.92}$$

the group delay

$$\tau_{GD}(\omega) = -\frac{\mathrm{d}\Phi(\omega)}{\mathrm{d}\omega} \tag{6.93}$$

the gain in dB

$$G(\omega) = 20\log_{10}|H(j\omega)| \tag{6.94}$$

the attenuation in dB

$$A(\omega) = -20\log_{10}|H(j\omega)| \tag{6.95}$$

6.12 HIGHPASS FILTER

A highpass filter can be specified by its edge frequencies and attenuation limits S_{Ah}; F_{ph} designates the passband edge frequency (Hz), F_{sh} is the stopband edge frequency (Hz), $F_{sh} < F_{ph}$, the passband attenuation (dB) is designated by A_{ph}, and A_{sh} stands for the stopband attenuation (dB).

First, we map the highpass filter specification, S_{Ah}, into the lowpass filter specification, S_A,

$$\boxed{S_{Ah} = \{F_{sh}, F_{ph}, A_{ph}, A_{sh}\} \longmapsto S_A = \left\{\begin{array}{rcl} F_p &=& F_{sh} \\ F_s &=& F_{ph} \\ A_p &=& A_{ph} \\ A_s &=& A_{sh} \end{array}\right\}} \tag{6.96}$$

Next, we construct the normalized lowpass transfer function, $\mathcal{H}(n, \xi, \epsilon, p)$, that meets the specification S_A according to the design procedures D1, D2, D3a, D3b, D4a, D4b, and D5.

Finally, the transfer function, $H_{HP}(s)$, of the highpass filter is constructed from the normalized lowpass transfer function according to the transformation:

Design	Step 6	Step 7
D1	$f_p = F_{ph}$	$p = \dfrac{2\pi f_p}{s}$
D2	$f_p = F_{ph}$	$p = \dfrac{2\pi f_p}{s}$
D3a	$f_p = F_{ph}$	$p = \dfrac{2\pi f_p}{s}$
D3b	$f_p = F_{sh}\,\xi_{min}\left(n, K_p, K_s\right)$	$p = \dfrac{2\pi f_p}{s}$
D4a	$f_p = F_{ph}$	$p = \dfrac{2\pi f_p}{s}$
D4b	$f_p = F_{sh}\,\xi_{\max}\left(n, F_p, F_s, K_p, K_s\right)$	$p = \dfrac{2\pi f_p}{s}$
D5	$f_p = F_{ph}\,f_{nQ}\left(n, F_p, F_s, K_p, K_s\right)$	$p = \dfrac{2\pi f_p}{s}$

(6.97)

which yields

$$H_{HP}(s) = \mathcal{H}\left(n, \xi, \epsilon, \frac{2\pi f_p}{s}\right) \tag{6.98}$$

6.13 BANDPASS FILTER

A bandpass filter can be specified by its edge frequencies and attenuation limits S_{Ab}; F_{p1} and F_{p2} designate the passband edge frequencies (Hz), F_{s1} and F_{s2} are the stopband edge frequencies (Hz), $F_{s1} < F_{p1} < F_{p2} < F_{s2}$, the passband attenuation (dB) is designated by A_{pb}, and A_{s1} and A_{s2} stand for the stopband attenuations (dB). We assume that the specification satisfies

$$F_{p1}F_{p2} = F_{s1}F_{s2} \tag{6.99}$$

First, we map the bandpass filter specification, S_{Ab}, into the lowpass filter specification, S_A,

$$S_{Ab} = \{F_{s1}, F_{p1}, F_{p2}, F_{s2}, A_{s1}, A_{pb}, A_{s2}\} \longmapsto S_A = \left\{\begin{array}{l} F_p = F_{s1} \\ F_s = F_{s1}\dfrac{F_{s2} - F_{s1}}{F_{p2} - F_{p1}} \\ A_p = A_{pb} \\ A_s = \max(A_{s1}, A_{s2}) \end{array}\right\} \tag{6.100}$$

Next, we construct the normalized lowpass transfer function, $\mathcal{H}(n, \xi, \epsilon, p)$, that meets the specification S_A according to the design procedures D1, D2, D3a, D3b, D4a, D4b, and D5.

Finally, the transfer function, $H_{BP}(s)$, of the bandpass filter is constructed from the normalized lowpass transfer function according to the transformation:

Design	Step 6	Step 7
D1	$f_p = F_{p2} - F_{p1}$	$p = \dfrac{s^2 + 4\pi^2 F_{p1} F_{p2}}{2\pi f_p s}$
D2	$f_p = F_{p2} - F_{p1}$	$p = \dfrac{s^2 + 4\pi^2 F_{p1} F_{p2}}{2\pi f_p s}$
D3a	$f_p = F_{p2} - F_{p1}$	$p = \dfrac{s^2 + 4\pi^2 F_{p1} F_{p2}}{2\pi f_p s}$
D3b	$f_p = \dfrac{\frac{F_s}{F_p}(F_{p2} - F_{p1})}{\xi_{min}\left(n, K_p, K_s\right)}$	$p = \dfrac{s^2 + 4\pi^2 F_{p1} F_{p2}}{2\pi f_p s}$
D4a	$f_p = F_{p2} - F_{p1}$	$p = \dfrac{s^2 + 4\pi^2 F_{p1} F_{p2}}{2\pi f_p s}$
D4b	$f_p = \dfrac{\frac{F_s}{F_p}(F_{p2} - F_{p1})}{\xi_{\max}\left(n, F_p, F_s, K_p, K_s\right)}$	$p = \dfrac{s^2 + 4\pi^2 F_{p1} F_{p2}}{2\pi f_p s}$
D5	$f_p = \dfrac{F_{p2} - F_{p1}}{f_{nQ}\left(n, F_p, F_s, K_p, K_s\right)}$	$p = \dfrac{s^2 + 4\pi^2 F_{p1} F_{p2}}{2\pi f_p s}$

(6.101)

which yields

$$H_{BP}(s) = \mathcal{H}\left(n, \xi, \epsilon, \frac{s^2 + 4\pi^2 F_{p1} F_{p2}}{2\pi f_p s}\right) \tag{6.102}$$

This type of bandpass filter is said to be the *symmetrical bandpass filter*.

6.14 BANDREJECT FILTER

A bandreject filter can be specified by its edge frequencies and attenuation limits S_{Ar}; F_{p1} and F_{p2} designate the passband edge frequencies (Hz), F_{s1} and F_{s2} are the stopband edge frequencies (Hz), $F_{p1} < F_{s1} < F_{s2} < F_{p2}$, the stopband attenuation (dB) is designated by A_{sr}, and A_{p1} and A_{p2} stand for the passband attenuations (dB). We assume that the specification satisfies

$$F_{p1} F_{p2} = F_{s1} F_{s2} \tag{6.103}$$

First, we map the bandreject filter specification, S_{Ar}, into the lowpass filter specification, S_A,

$$S_{Ar} = \{F_{p1}, F_{s1}, F_{s2}, F_{p2}, A_{p1}, A_{sr}, A_{p2}\} \longmapsto S_A = \left\{ \begin{aligned} F_p &= F_{p1} \\ F_s &= F_{p1}\frac{F_{p2} - F_{p1}}{F_{s2} - F_{s1}} \\ A_p &= \min(A_{p1}, A_{p2}) \\ A_s &= A_{sr} \end{aligned} \right\} \tag{6.104}$$

Next, we construct the normalized lowpass transfer function, $\mathcal{H}(n, \xi, \epsilon, p)$, that meets the specification S_A according to the design procedures D1, D2, D3a, D3b, D4a, D4b, and D5.

Finally, the transfer function, $H_{BR}(s)$, of the bandreject filter is constructed from the normalized lowpass transfer function according to the transformation:

Design	Step 6	Step 7
D1	$f_p = F_{s2} - F_{s1}$	$p = \dfrac{2\pi f_p s}{s^2 + 4\pi^2 F_{s1} F_{s2}}$
D2	$f_p = F_{s2} - F_{s1}$	$p = \dfrac{2\pi f_p s}{s^2 + 4\pi^2 F_{s1} F_{s2}}$
D3a	$f_p = F_{s2} - F_{s1}$	$p = \dfrac{2\pi f_p s}{s^2 + 4\pi^2 F_{s1} F_{s2}}$
D3b	$f_p = \dfrac{F_p}{F_s}(F_{s2} - F_{s1})\,\xi_{min}\left(n, K_p, K_s\right)$	$p = \dfrac{2\pi f_p s}{s^2 + 4\pi^2 F_{s1} F_{s2}}$
D4a	$f_p = F_{s2} - F_{s1}$	$p = \dfrac{2\pi f_p s}{s^2 + 4\pi^2 F_{s1} F_{s2}}$
D4b	$f_p = \dfrac{F_p}{F_s}(F_{s2} - F_{s1})\,\xi_{\max}\left(n, F_p, F_s, K_p, K_s\right)$	$p = \dfrac{2\pi f_p s}{s^2 + 4\pi^2 F_{s1} F_{s2}}$
D5	$f_p = (F_{s2} - F_{s1})\, f_{nQ}\left(n, F_p, F_s, K_p, K_s\right)$	$p = \dfrac{2\pi f_p s}{s^2 + 4\pi^2 F_{s1} F_{s2}}$

(6.105)

which yields

$$H_{BR}(s) = \mathcal{H}\left(n, \xi, \epsilon, \frac{2\pi f_p s}{s^2 + 4\pi^2 F_{s1} F_{s2}}\right) \tag{6.106}$$

This type of bandreject filter is said to be the *symmetrical bandreject filter*.

6.15 CONCLUDING REMARKS

We can improve the computational efficiency by using the closed-form expressions. We prefer to exploit analytical formulas involving simple algebraic manipulation, instead of using expressions in which Jacobi elliptic functions appear.

CHAPTER 7

MULTICRITERIA OPTIMIZATION OF ANALOG FILTER DESIGNS

This chapter presents an extensible framework for designing analog filters that exhibit several desired behavioral properties after being realized in circuits. In the framework, we model the constrained nonlinear optimization problem as a sequential quadratic programming (SQP) problem. SQP requires real-valued constraints and objective function that are differentiable with respect to the free parameters (pole-zero locations). We derive the differentiable constraints and a weighted differentiable objective function for simultaneously optimizing the behavioral properties of magnitude response, phase response, and peak overshoot and the implementation property of quality factors. We use *Mathematica* to define the algebraic equations for the constraints and objective function, compute their gradients symbolically, and generate standalone MATLAB programs to perform the multicriteria optimization. Providing closed-form gradients prevents divergence in the SQP procedure. The automated approach avoids errors in algebraic calculations and errors in transcribing equations into software. The key contributions are

- an extensible, automated, multicriteria filter optimization framework and
- an analytic approximation for peak overshoot.

7.1 INTRODUCTION

Classic elliptic, Chebyshev, Butterworth, and Bessel analog filter designs yield desirable behavioral properties subject to constraints on the magnitude response. For example, the step response of Bessel filters exhibits low overshoot, and its phase response is nearly linear over the passband. Bessel filters have been used as antialiasing filters since antialiasing filters require a minimum deviation in the phase response from linear phase, subject to a set of magnitude specifications [61]. Classic elliptic filter designs have minimal order, but for a given implementation technology, minimal order filters either may not be realizable or may not have minimal complexity [51]. In designing analog filters for implementation, multiple behavioral properties (e.g., magnitude response, phase response, peak overshoot, rise time, and settling time) and implementation properties (e.g., quality factors and capacitance value spread ratio) may be important.

In this chapter, we present a formal extensible framework for simultaneously optimizing analog filter designs for multiple behavioral and implementation properties. We demonstrate the framework using the behavioral properties of magnitude response, phase response, and peak overshoot, as well as the implementation property of quality factors. The framework takes an initial filter design, e.g., one designed using a classic numeric approach or a modern symbolic approach [51], and finds the pole-zero locations that optimize a weighted combination of properties subject to constraints on the properties. Some of the previous multicriteria filter optimization techniques, such as those discussed in reference [62], only optimize for one property subject to constraints on multiple properties. Another technique applies Sequential Quadratic Programming (SQP) methods to optimize loss and delay in digital filter designs [62].

Our framework models the constrained nonlinear optimization problem as a SQP problem. SQP requires that the objective function [63] and the constraints [64] be real-valued and twice continuously differentiable with respect to the free parameters. The free parameters are the pole and zero locations. When closed-form formulas for the gradients of the objective function and constraints are not provided for SQP routines, the SQP routines must approximate the gradient which often leads to divergence. We develop *Mathematica* [65] software to compute the gradients and translate the entire SQP formulation into working MATLAB [66] programs that optimize analog filter designs. The generated MATLAB code is combined with the SQP procedure in the MATLAB Optimization Toolbox [67] to produce standalone programs [68] that solve the constrained nonlinear optimization problem.

The framework is flexible because it is formulated at an algebraic level. At the algebraic level, a designer can use a symbolic mathematics environment such as *Mathematica* to change the objective measure for a given property or add, delete, and change constraints. Our symbolic software will then recompute the gradient and regenerate the numerical optimization code. We have bridged the gap between the symbolic work that designers often do on paper and the working computer implementation, thereby eliminating algebraic errors in hand calculations and bugs in coding the equations in software.

Section 7.2 reviews notation. Section 7.3 derives a family of weighted, differentiable objective functions to measure the deviation in magnitude response, deviation in linear phase response, quality factors, and peak overshoot of the step response, of an analog filter. In the derivation, we find a new analytic approximation for the peak overshoot. Section 7.4 converts filter specifications on the magnitude response, quality factors, and peak overshoot into differentiable constraints. Section 7.5 reports design examples found by our filter optimization framework. Section 7.6 describes the process by which we verified the formulas in Section 7.3, generated MATLAB code for the objective function and constraints as well as their gradients, and validated the generated MATLAB code. Section 7.7 concludes the chapter.

7.2 NOTATION

Without lack of generality, we focus on an analog filter represented by its n complex conjugate pole pairs and r complex conjugate zero pairs such that $r \leq n$. We denote the kth pole pair as $p_k = a_k \pm jb_k$, where $a_k < 0$ for stability, and we denote the lth zero pair as $z_l = c_l \pm jd_l$. The even-order transfer function of the filter is

$$H(s) = K \frac{\prod_{l=1}^{r} (s - (c_l + jd_l))(s - (c_l - jd_l))}{\prod_{k=1}^{n} (s - (a_k + jb_k))(s - (a_k - jb_k))}$$

and for allpole filters it simplifies to

$$H_{\text{allpole}}(s) = K_p \frac{1}{\prod_{k=1}^{n} (s - (a_k + jb_k))(s - (a_k - jb_k))}$$

We assume that K and K_p are real positive constants.

The magnitude response $M_{\text{allpole}}(\omega) = |H_{\text{allpole}}(j\,\omega)|$ and unwrapped phase response $\Phi_{\text{allpole}}(\omega) = \arg H_{\text{allpole}}(j\,\omega)$ of an allpole filter, expressed as real-valued differentiable functions, are

$$\begin{aligned} M_{\text{allpole}}(\omega) &= \prod_{k=1}^{n} \frac{a_k^2 + b_k^2}{\sqrt{a_k^2 + (\omega + b_k)^2}\sqrt{a_k^2 + (\omega - b_k)^2}} \\ &= \prod_{k=1}^{n} \frac{a_k^2 + b_k^2}{\sqrt{(\omega^2 + 2(a_k^2 - b_k^2))\omega^2 + (a_k^2 + b_k^2)^2}} \end{aligned} \tag{7.1}$$

$$\Phi_{\text{allpole}}(\omega) = -\sum_{k=1}^{n} (\arg(j\omega - (a_k + jb_k)) + \arg(j\omega - (a_k - jb_k)))$$

$$\Phi_{\text{allpole}}(\omega) = \lambda_p \pi + \sum_{k=1}^{n} \left(\arctan\left(\frac{\omega - b_k}{a_k}\right) + \arctan(\frac{\omega + b_k}{a_k}) \right) \tag{7.2}$$

where an integer λ_p takes into account the proper quadrant of the complex plane. Note that for real x and y, $\arg(x + jy) = \arctan(y/x)$ only for $x > 0$. We factor the polynomial under the square root in (7.1) into Horner's form because it has better numerical properties. Together with the zero pairs, the magnitude and unwrapped phase responses, respectively, are

$$M(\omega) = M_{\text{allpole}}(\omega) \prod_{l=1}^{r} \frac{\sqrt{(\omega^2 + 2(c_l^2 - d_l^2))\omega^2 + (c_l^2 + d_l^2)^2}}{c_l^2 + d_l^2} \tag{7.3}$$

$$\Phi(\omega) = \Phi_{\text{allpole}}(\omega) + \sum_{l=1}^{r} (\arg(s - (c_l + jd_l)) + \arg(s - (c_l - jd_l)))$$

or

$$\Phi(\omega) = \Phi_{\text{allpole}}(\omega) + \lambda_z \pi - \sum_{l=1}^{r} \left(\arctan\left(\frac{\omega - d_l}{c_l}\right) + \arctan\left(\frac{\omega + d_l}{c_l}\right)\right) \quad (7.4)$$

where an integer λ_z takes into account the proper quadrant of the complex plane.

We use the following notation: Q represents quality factors, ϵ represents a small positive number, σ denotes deviation, m represents slope of a line, t is time, and W is a weighting factor. Unless otherwise stated, we consider lowpass filters with $M(0) = 1$.

7.3 OBJECTIVE FUNCTION

In this section, we derive an objective function that is real and twice continuously differentiable to match an SQP formulation. The objective function is a weighted combination of the deviations from the desired filter properties. Properties are quantified by objective measures. Based on the objective measures for magnitude and phase responses given in the previous section, Sections 7.3.1 and 7.3.2 define deviation from the desired magnitude response and deviation from a phase response that is linear in the passband, respectively. Section 7.3.3 defines quality factors and deviation from desired quality factors. Section 7.3.4 derives an analytic approximation to measure peak overshoot in the step response and defines deviation from the desired filter overshoot. Section 7.3.5 defines the objective function. The objective function is non-negative so that a value of zero represents the desired filter. Using the automated SQP-based framework, a designer may change the objective measures and distance measures to form new objective functions, as well as regenerate the new MATLAB programs to perform the optimization.

7.3.1 Deviation from a Desired Magnitude Response

We use (7.3) as the real, differentiable measure of magnitude response. We measure the deviation from the desired magnitude response separately in the passband, transition band, and stopband. Based on the regions of the desired (sometimes called *ideal*) magnitude response shown in Fig. 7.1, the components of the objective function relate to the deviation from an ideal magnitude response in the least squares sense in each region. Assuming that the ideal filter is lowpass with the passband located on $\omega \in (0, \omega_p)$ and the stopband located on $\omega \in (\omega_s, \infty)$, we obtain

$$\sigma_{pb} = \int_0^{\omega_p} F_p(\omega)\,(M(\omega) - 1)^2\,\mathrm{d}\omega \quad (7.5)$$

$$\sigma_{tb} = \int_{\omega_p}^{\omega_s} F_t(\omega)\,(M(\omega) - m(\omega - \omega_s))^2\,\mathrm{d}\omega \quad (7.6)$$

$$\sigma_{sb} = \int_{\omega_s}^{\infty} F_s(\omega) M(\omega)^2\,\mathrm{d}\omega \quad (7.7)$$

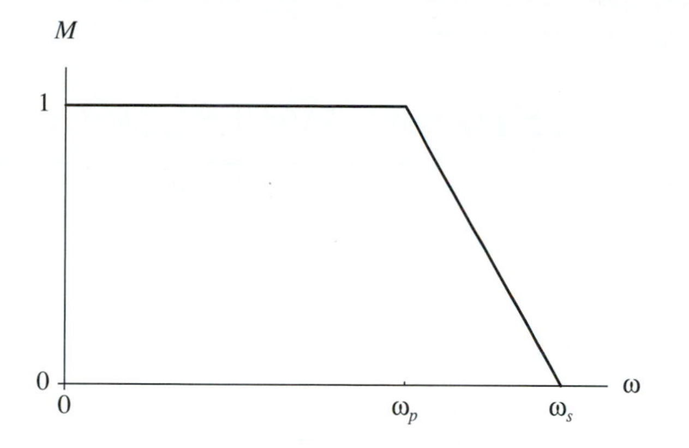

Figure 7.1 The desired magnitude response.

where $F_p(\omega)$, $F_t(\omega)$, and $F_s(\omega)$ are integrable weighting functions, and m is the slope of the ideal response in the transition region defined as $m = 1/(\omega_p - \omega_s)$. These integral quantities represent deviation from the ideal magnitude response in the least-squares sense over the regions shown in Fig. 7.1.

7.3.2 Deviation from an Ideal Phase Response

We use (7.4) as a real, differentiable measure of the phase response. We measure the deviation of a filter from linear phase over some range of frequencies (usually over the passband):

$$\sigma_{lp} = \int_{\omega_1}^{\omega_2} \left(\Phi(\omega) - m_{lp}\,\omega\right)^2 d\omega \tag{7.8}$$

Here, m_{lp} is the slope of the linear phase response. Unfortunately, one does not know the value of m_{lp} *a priori*. We can compute it as the slope of the line in ω that minimizes (7.8):

$$\min_{m_{lp}} \int_{\omega_1}^{\omega_2} \left(\Phi(\omega) - m_{lp}\,\omega\right)^2 d\omega \tag{7.9}$$

In (7.9), $\Phi(\omega)$ does not depend on m_{lp}, so the integrand is quadratic in m_{lp}. To find the minimum, we take the derivative with respect to m_{lp}, set it to zero, and solve for m_{lp}:

$$m_{lp} = \frac{\int_{\omega_1}^{\omega_2} \Phi(\omega)\;\omega\;d\omega}{\int_{\omega_1}^{\omega_2} \omega^2\;d\omega} \tag{7.10}$$

After evaluating the integrals,

$$m_{lp} = \frac{3}{2} \frac{\sum_{k=1}^{n} \left[f_{lp1}(\omega_2) - f_{lp1}(\omega_1) \right] - \sum_{l=1}^{r} \left[f_{lp2}(\omega_2) - f_{lp2}(\omega_1) \right]}{\omega_2^3 - \omega_1^3} \tag{7.11}$$

where $f_{lp1}(\omega)$ is

$$\begin{aligned} f_{lp1}(\omega) &= 2\omega a_k + (b_k^2 - a_k^2 - \omega^2) \left(\arctan\left(\frac{\omega - b_k}{a_k}\right) + \arctan\left(\frac{\omega + b_k}{a_k}\right) \right) \\ &+ a_k b_k \left(\log\left(1 + \frac{(\omega - b_k)^2}{a_k^2}\right) - \log\left(1 + \frac{(\omega + b_k)^2}{a_k^2}\right) \right) \end{aligned}$$

and $f_{lp2}(\omega)$ is

$$\begin{aligned} f_{lp2}(\omega) &= 2\omega c_l + (d_l^2 - c_l^2 - \omega^2) \left(\arctan\left(\frac{\omega - d_l}{c_l}\right) + \arctan\left(\frac{\omega + d_l}{c_l}\right) \right) \\ &+ c_l d_l \left(\log\left(1 + \frac{(\omega - d_l)^2}{c_l^2}\right) - \log\left(1 + \frac{(\omega + d_l)^2}{c_l^2}\right) \right) \end{aligned}$$

Using *Mathematica*, we compute the definite integrals in (7.10) to verify the answers. Now that we have a closed-form solution for m_{lp}, we can substitute (7.11) into (7.8) to obtain a rather complicated but differentiable expression for the deviation from linear phase.

7.3.3 Deviation from Desired Quality Factor

The quality factor Q_k for the kth pole pair $a_k \pm jb_k$ (with $a_k < 0$) is defined by

$$Q_k = \frac{\sqrt{a_k^2 + b_k^2}}{-2a_k}$$

where $Q_k \geq \frac{1}{2}$. $Q_k = \frac{1}{2}$ corresponds to a double real-valued pole ($b_k = 0$), and $Q_k \to \infty$ corresponds to a purely imaginary pole pair ($a_k \to 0$).

Quality factor measures the relative distance of a filter pole from the imaginary frequency axis. For the effective overall measure of the quality factor, Q_{eff}, we use a geometric mean of the individual pole-pair quality factors

$$Q_{\text{eff}} = \left(\prod_{k=1}^{n} Q_k \right)^{\frac{1}{n}} \tag{7.12}$$

where $Q_{\text{eff}} \geq \frac{1}{2}$. Other objectives measures could be used. To measure the distance from the minimal quality factor of $\frac{1}{2}$, we simply use

$$\sigma_Q = Q_{\text{eff}} - \frac{1}{2}$$

This distance measure rewards low quality factors because they are essential in damping oscillatory behavior in the time response of the filter.

7.3.4 Deviation from Desired Peak Overshoot in the Step Response

From the step response, we can numerically compute the peak overshoot and the time t_{peak} at which it occurs. In order to make the peak overshoot calculation differentiable for the SQP-based framework, we derive an analytic expression that approximates t_{peak} in terms of the pole-zero locations. The derivation assumes that the poles are not multiple. The assumption of simple poles is enforced by means of constraints, as explained in Section 7.4.

The Laplace transform of the step response is

$$\frac{H(s)}{s} = \frac{1}{s}\left[\prod_{k=1}^{n} \frac{a_k^2 + b_k^2}{s^2 - 2a_k s + a_k^2 + b_k^2}\right]\left[\prod_{l=1}^{r} \frac{s^2 - 2c_l s + c_l^2 + d_l^2}{c_l^2 + d_l^2}\right] \tag{7.13}$$

Assuming simple poles, partial fractions expansion yields

$$\frac{H(s)}{s} = \left[\frac{A}{s} + \sum_{k=1}^{n} \frac{C_k s + D_k}{s^2 - 2a_k s + a_k^2 + b_k^2}\right] \tag{7.14}$$

$$A = \lim_{s \to 0} H(s) = 1$$

$$B_k = \lim_{s \to (a_k + jb_k)} \left[(s - (a_k + jb_k))\frac{H(s)}{s}\right] = |B_k| e^{j\beta_k}$$

$$C_k = 2|B_k|\cos(\beta_k)$$

$$D_k = -2|B_k|\,(a_k \cos(\beta_k) + b_k \sin(\beta_k))$$

where $|B_k|$ and β_k can be expressed as real-valued differentiable functions of the pole and zero locations.

After inverse transforming (7.14), the step response is

$$h_{\text{step}}(t) = A + \sum_{k=1}^{n} e^{a_k t}\left[C_k \cos(b_k t) + \left(\frac{D_k + C_k a_k}{b_k}\right)\sin(b_k t)\right] \tag{7.15}$$

By substituting the definitions for A, D_k, and C_k from (7.14) into (7.15), the step response simplifies to

$$h_{\text{step}}(t) = 1 + 2\sum_{k=1}^{n} |B_k|\, e^{a_k t} \cos(b_k t + \beta_k) \tag{7.16}$$

The overall step response, given by either (7.15) or (7.16), is one plus a sum of n terms corresponding to pole pairs. Each term is a function of all of the filter poles and zeros. By analyzing the kth term in the summation in (7.15), the kth peak overshoot occurs at time

$$t_{\text{peak}}^{k} = -\frac{1}{b_k}\left[\arctan\left(\frac{(D_k + 2C_k a_k)b_k}{C_k(a_k^2 - b_k^2) + D_k a_k}\right) + \lambda_k \pi\right] \tag{7.17}$$

provided that $b_k \neq 0$; λ_k is an integer such that t_{peak}^{k} is positive and represents the first local maximum. The sign of the second derivative of the kth term must be negative for $t = t_{\text{peak}}^{k}$, and the following must hold:

$$-\text{sign}(b_k)\frac{\sin(b_k t_{\text{peak}}^{k})}{2a_k C_k + D_k} < 0$$

The least and greatest peak times provide a bound on the possible times at which the peak overshoot occurs, that is,

$$t_{\text{peak}} \in \left[\min_k t_{\text{peak}}^{k},\ \max_k t_{\text{peak}}^{k}\right] \tag{7.18}$$

In the SQP-based framework, (7.18) helps in two important areas. First, at each iteration, (7.18) gives a range of time over which to perform a one-dimensional search for the actual t_{peak} value. In the implementation, we search a broader interval. Second, from the observation that (7.18) states that t_{peak} is dependent on the values of t_{peak}^{k}, we construct the following differentiable function to approximate t_{peak} for the sole purpose of computing gradients for the peak overshoot of the filter:

$$t_{\text{peak}} = \alpha \frac{1}{n} \sum_{k=1}^{n} t_{\text{peak}}^{k} \tag{7.19}$$

This analytic approximation is inferred from (7.15), in which the step response is written as a sum of a constant plus an equal additive contribution from each pole pair. At each iteration of the optimization procedure, α is set to the computed value of t_{peak} divided by $\frac{1}{n}\sum_{k=1}^{n} t_{\text{peak}}^{k}$. Section 7.6.3 validates the accuracy of the analytic approximation.

The deviation in peak overshoot is measured by a differentiable function that measures overshoot. For lowpass filters, one measure of deviation in peak overshoot is the percent overshoot defined as $100\% \times \left(h_{\text{step}}(t_{\text{peak}}) - 1\right)$. This formula, however, assumes that the step response will rise above 1. We measure the deviation in step response amplitude when the peak overshoot occurs from the ideal amplitude of one which corresponds to a peak overshoot of zero:

$$\sigma_t = \left(h_{\text{step}}(t_{\text{peak}}) - 1\right)^2$$

7.3.5 The Complete Objective Function

The complete objective function is a weighted sum of the distance measures developed earlier in this section. Since the objective function will ultimately be handed off to a numerical optimizer, each infinity that appears in the limit of the definite integrations, such as in (7.7), must be approximated. We approximate ∞ as 10^d multiplied by the highest frequency specified (e.g., ω_s), where d represents the number of decades of beyond the highest specified frequency. In assembling the composite objective function, we normalize the integrals by dividing by the length of the integration interval and scale the distance measure for the peak overshoot, so that when the weights are equal, each distance measure will contribute more equally. Note that the weighted objective function is non-negative and twice differentiable:

$$\begin{aligned}\sigma = {} & W_{pb}\frac{1}{\omega_p}\sigma_{pb} + W_{tb}\frac{1}{\omega_s - \omega_p}\sigma_{tb} + W_{sb}\frac{1}{10^d\omega_s - \omega_s}\sigma_{sb} \\ & + W_{lp}\frac{1}{\omega_p}\sigma_{lp} + W_Q\sigma_Q + W_t\,1000\sigma_t\end{aligned} \tag{7.20}$$

7.4 CONSTRAINTS

The first set of constraints are on the magnitude response, peak overshoot, and quality, and the second set prevents numerical instabilities in the calculations and enforces assumptions about the poles and zeros. For the magnitude response constraints, we sample the magnitude response given by (7.3) at a set of passband frequencies $\{\omega_k : \omega_k \leq \omega_p\}$ and stopband frequencies $\{\omega_l : \omega_l \geq \omega_s\}$:

$$\begin{aligned}1 - \delta_p \leq |H(j\,\omega_k)| \leq 1 + \delta_p, \qquad \forall k \\ |H(j\,\omega_l)| \leq \delta_s, \qquad \forall l\end{aligned} \tag{7.21}$$

where δ_p, ω_p, δ_s, and ω_s are the magnitude specifications. For the peak overshoot constraint, we compute the peak overshoot by searching a larger interval than $t \in [\min_k t^k_{\text{peak}}, \max_k t^k_{\text{peak}}]$ to find the maximum value of the step response in (7.15). Before finding the gradient of this constraint, we substitute $t = t_{\text{peak}}$ in (7.15) by using the analytic approximation for t_{peak} given by (7.19).

Lower quality factors often mean better noise immunity and lower sensitivity. The implementation technology imposes an upper limit on the quality factors, $Q_{\max}$, for each second-order section:

$$\frac{\sqrt{a_k^2 + b_k^2}}{-2a_k} < Q_{\max} \qquad \text{for} \quad k = 1\ldots, n \tag{7.22}$$

The designer is free to specify the value of $Q_{\max}$ to the optimizer. Section 7.6.2 explains how to set $Q_{\max}$ in the synthesized MATLAB code.

Since a_k appears in the denominators in (7.1) and (7.11), b_k appears in the denominators in (7.15) and (7.17), and c_l appears in the denominators in (7.4) and (7.11), we constrain these parameters to be negative-valued and a neighborhood away from zero:

$$a_k < -\epsilon_{div} < 0 \quad \text{for} \quad k = 1, \ldots, n$$

$$b_k < -\epsilon_{div} < 0 \quad \text{for} \quad k = 1, \ldots, n$$

$$c_l < -\epsilon_{div} < 0 \quad \text{for} \quad l = 1, \ldots, r$$

Here, ϵ_{div} is the distance from 1.0 to the next largest floating point number. In MATLAB, it is defined by the eps constant, which is 2.2204×10^{-16} on an Sun Ultra–2 workstation. To ensure the numerical stability of the denominators of $|B_k|$ and β_k in (7.14),

$$|a_k - a_m| + |b_k - b_m| > 2\epsilon_{div} \quad \text{for} \quad k = 1, \ldots, n \quad \text{and} \quad m = k + 1, \ldots, n$$

This set of constraints also prevents multiple poles and poles from becoming too close to one another relative to the available numerical precision.

It is possible that an initial filter design does not meet all of the constraints. For example, classic filter design algorithms may yield filters with extraordinarily large quality factors which may exceed $Q_{\max}$. When an initial guess is infeasible, the SQP procedure in MATLAB will update the free parameters until the constraint or constraints that were initially violated are satisfied. This relaxation occurs in the design example discussed in Section 7.5.2.

7.5 EXAMPLE FILTER DESIGNS

This section presents three lowpass filter designs found by the automated filter optimization framework. In Sections 7.5.1 and 7.5.2, the filters have minimized deviation from linear phase over the passband and peak overshoot in the step response. In Section 7.5.3, the filter has minimized magnitude response in the stopband, phase response in the passband, peak overshoot, and quality factors. All execution times are given for MATLAB 5 on a 167-MHz Sun Ultra workstation.

7.5.1 Allpole Filter with Near Linear Phase Response and Minimal Peak Overshoot

We apply the framework to optimize an allpole filter to obtain near-linear phase response and minimal peak overshoot. The magnitude specifications are $\omega_p = 20$ rad/s, $\delta_p = 0.21$, $\omega_s = 30$ rad/s, and $\delta_s = 0.31$. The initial design is a fourth-order Butterworth filter. We jointly optimize the pole locations to achieve minimal peak overshoot and minimal deviation from linear phase in the passband phase response. In the objective function, we weight the deviation from linear phase by 0.1 and weight the deviation in peak overshoot by 1. All other weights are zero. We set $Q_{\max}$ to 10. The non-negative objective function is reduced from 1.17 to 4.7×10^{-5}. The peak overshoot of the resulting filter is reduced from 16% to 8%, and the phase becomes approximately linear in the passband. Since the magnitude response was not optimized, it is traded off for better phase response and lower overshoot, but kept within specifications. Table 7.1 lists the initial and optimized pole locations. Figures 7.2–7.4 plot the magnitude, phase,

Table 7.1 Quality factors and pole locations for the two second-order sections of the initial and optimized filters.

Q	Poles
	Initial Butterworth Filter
0.54	$-20.3153 \pm j\,8.4149$
1.31	$-8.4149 \pm j\,20.3153$
	Optimized Filter
0.50	$-19.5623 \pm j\,0.6255$
1.55	$-7.7918 \pm j\,22.8984$

The optimized filter has minimal deviation from linear phase in the passband and peak overshoot. Fig. 7.2 compares the behavior of the two filters.

and step responses for the initial and optimized filters. The plots illustrate that the optimization procedure effectively traded off magnitude response in the passband for a more linear phase response in the passband and a lower overshoot. The optimization takes 13 seconds to run.

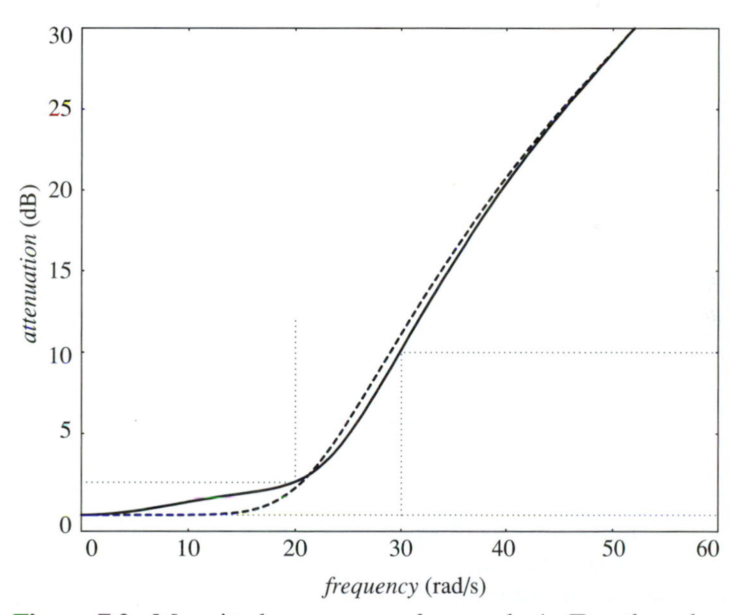

Figure 7.2 Magnitude response of example 1. Fourth-order lowpass filter with optimized phase and step responses. The specification of the magnitude response is $\omega_p = 20$ rad/s and $\delta_p = 0.21$ for the passband and $\omega_s = 30$ rad/s and $\delta_s = 0.31$ for the stopband. The dashed lines represent the initial Butterworth filter, and the solid lines represent the filter optimized for linear phase response in the passband and for overshoot of the step response.

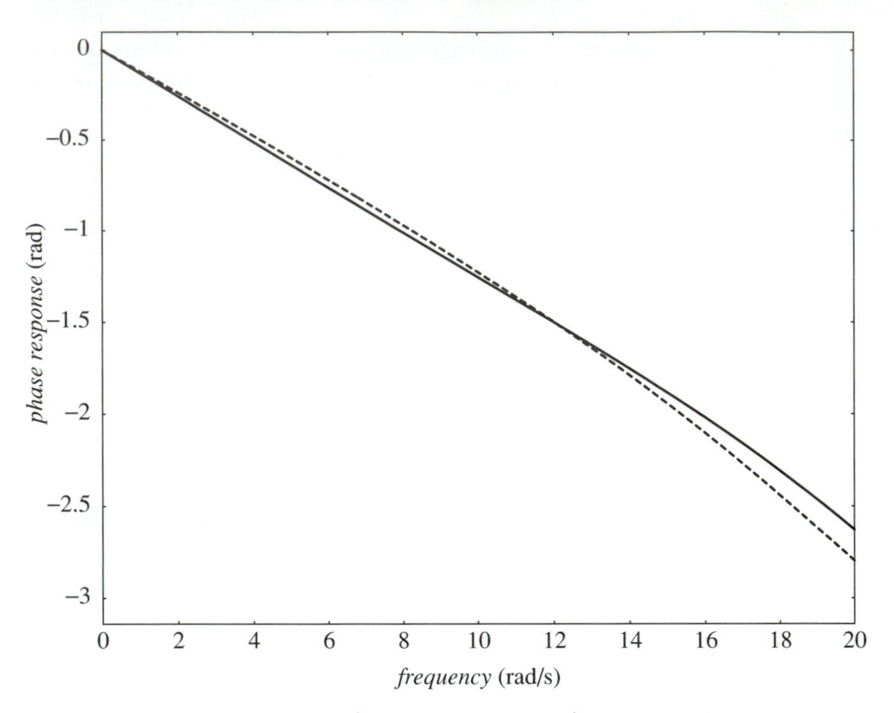

Figure 7.3 Phase response (over the passband) of example 1.

7.5.2 Filter with Near-Linear Phase Response and Minimal Peak Overshoot

As in the previous example, we minimize the peak overshoot and deviation from linear phase of a lowpass filter given the same magnitude specifications $\omega_p = 20$ rad/s, $\delta_p = 0.21$, $\omega_s = 30$ rad/s, and $\delta_s = 0.31$, except that we allow poles and zeros. We use a fourth-order elliptic filter as the initial guess. In the objective function, we weight the deviation from linear phase by 0.1 and deviation in peak overshoot by 1. All other weights are zero. We set $Q_{\max}$ to 10. The non-negative objective function is reduced from 2.87 to 4.33×10^{-5}. Table 7.2 lists the initial and final poles and zeros. Figures 7.5–7.7 plot the magnitude, phase, and step responses for the initial and final filters. Figure 7.5 illustrates that the optimization procedure effectively trades off transition bandwidth in the magnitude response for more linear phase in the passband and a lower overshoot. The peak overshoot is reduced from 25% to 10%. The gradient of the objective function with respect to the poles $\{a_1, b_1, a_2, b_2\}$ is $\{-1.99, -0.81, -17.74, 2.11\} \times 10^{-5}$ and with respect to the zeros $\{c_1, d_1, c_2, d_2\}$ is $\{2.07, 2.38, 0.55, 1.97\} \times 10^{-5}$. Since the second filter section is more sensitive to perturbations in the pole locations, better components should be used for the second section. The optimization takes 20 seconds to run.

7.5.3 Filter Simultaneously Optimized for Four Criteria

We optimize three behavioral properties and one implementation property simultaneously. The specifications on the magnitude response are $\omega_p = 30$ rad/s, $\delta_p = 0.2$,

Table 7.2 Quality factors and pole-zero locations for the two second-order sections of the initial and optimized filters.

Q	Poles	Zeros
	Initial Elliptic Filter	
1.7	$-5.3553 \pm j16.9547$	$0 \pm j20.2479$
61.1	$-0.1636 \pm j19.9899$	$0 \pm j28.0184$
	Optimized Filter	
0.68	$-11.4343 \pm j10.5092$	$-3.4232 \pm j28.6856$
10	$-1.0926 \pm j21.8241$	$-1.2725 \pm j35.5476$

The filter was minimized for deviation from linear phase in the passband and peak overshoot.

$\omega_s = 50$ rad/s, and $\delta_s = 0.3$. In the objective function, we weight the deviation from linear phase by 1, ideal peak overshoot by 1, ideal filter quality by 0.5, and ideal stopband magnitude response by 1. All other weights are zero. We set $Q_{\max}$ to 10. The initial guess is a sixth-order elliptic filter. It was designed using a stricter stopband criterion than specified to illustrate the tradeoffs obtained by searching the infinite design space. The non-negative objective function is reduced from 4.25 to 0.34. The peak overshoot is reduced from 20% to 15%. Table 7.3 lists the initial and final poles and zeros.

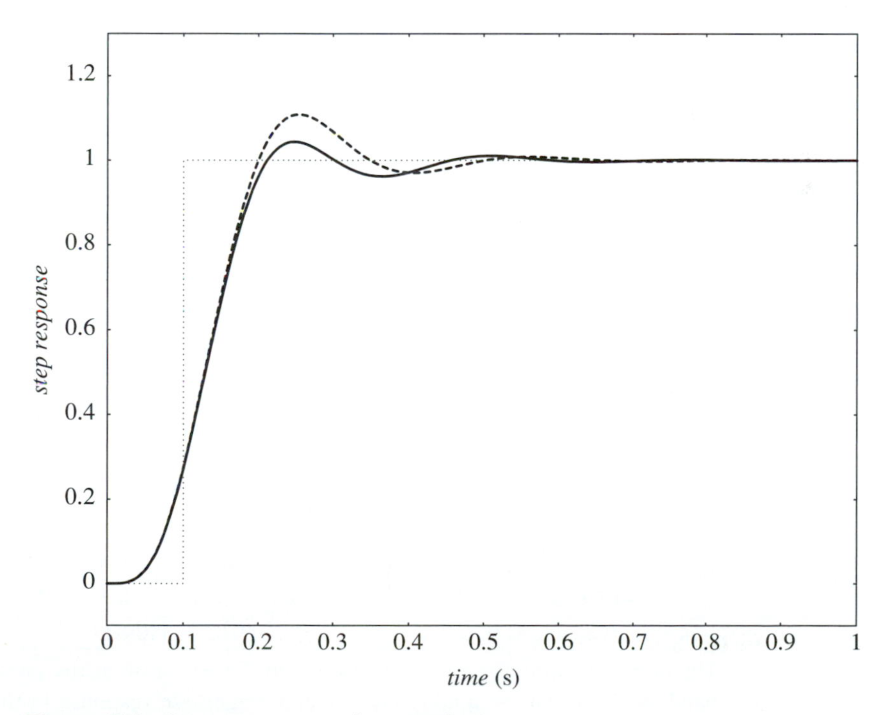

Figure 7.4 Step response of example 1.

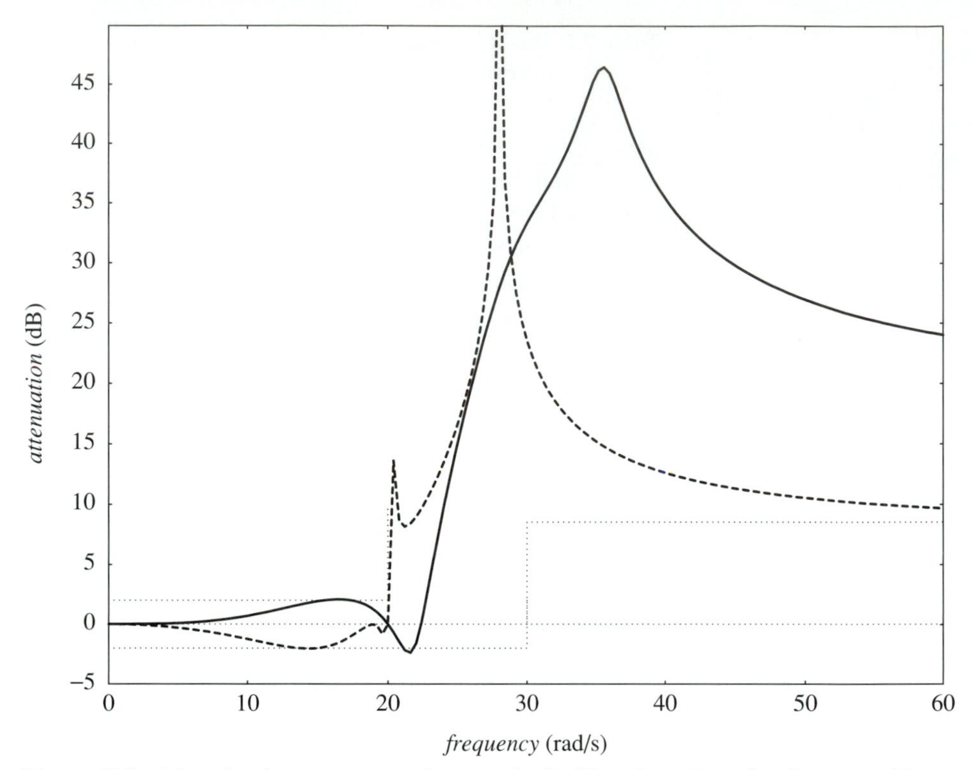

Figure 7.5 Magnitude response of example 2. Two fourth-order lowpass filters to meet the magnitude specifications $\omega_p = 20$ rad/s, $\delta_p = 0.21$, $\omega_s = 30$ rad/s, and $\delta_s = 0.31$. The initial filter is an elliptic filter (shown in dashed lines), and the final filter is optimized for phase and step response (shown in solid lines). We are trading linear phase response over the passband and peak overshoot in the step response for magnitude response while keeping the magnitude response within specification.

Table 7.3 Quality factors and pole-zero locations for the three second-order sections of the initial and optimized filters.

Q	Poles	Zeros
	Initial Elliptic Filter	
0.63	$-16.821 \pm j12.881$	$0 \pm j37.082$
1.93	$-7.3348 \pm j27.291$	$0 \pm j110.07$
9.05	$-1.7224 \pm j31.139$	$0 \pm j45.114$
	Optimized Filter	
0.51	$-38.21 \pm j5.8056$	$-16.407 \pm j45.045$
1.57	$-13.30 \pm j39.507$	$-4.8726 \pm j52.405$
1.74	$-9.84 \pm j32.906$	$-0.09162 \pm j109.78$

The filter was minimized for deviation from linear phase in the passband, peak overshoot, quality factors, and magnitude response in the stopband.

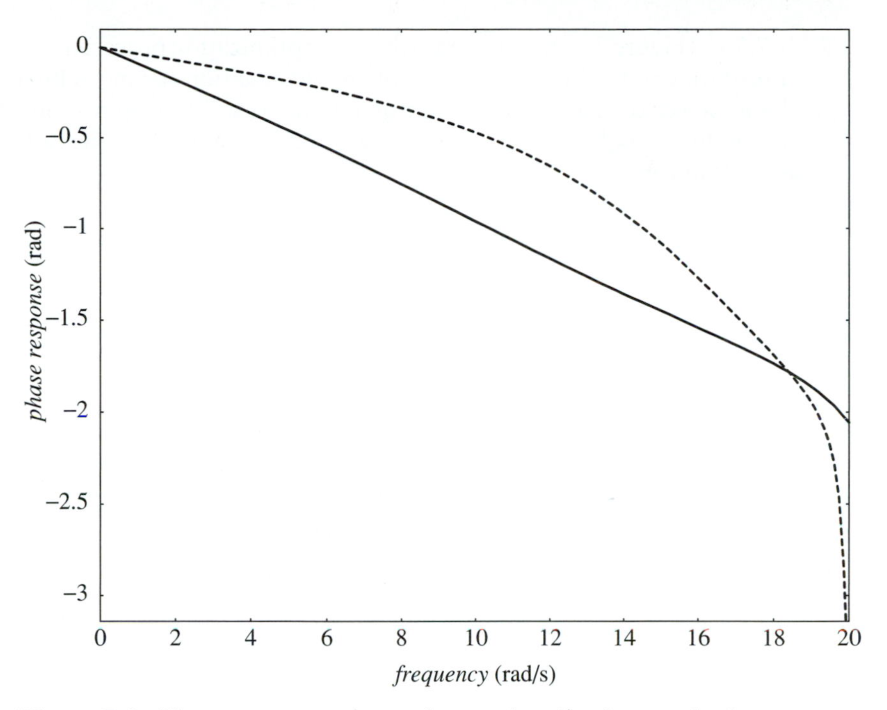

Figure 7.6 Phase response (over the passband) of example 2.

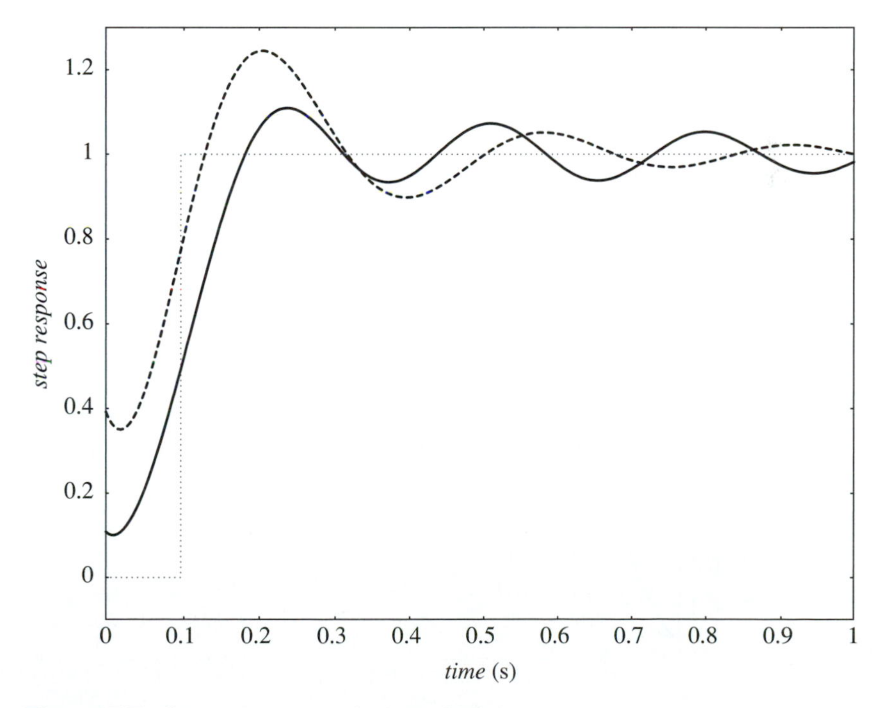

Figure 7.7 Step response of example 2.

Table 7.3 and Figure 7.8–7.10 illustrate that the optimization procedure effectively trades off transition bandwidth and extra stopband attenuation for more linear phase in the passband, lower overshoot, and lower quality factors. The step responses of the terms corresponding to pole pairs are shown in Figures 7.11 and 7.12. The optimization takes 65 seconds to run.

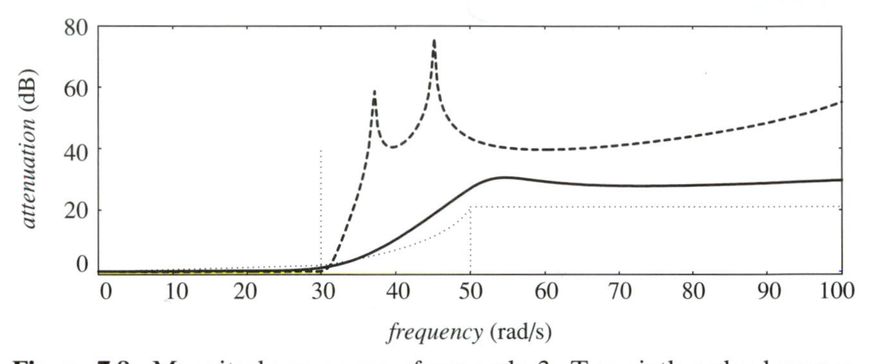

Figure 7.8 Magnitude response of example 3. Two sixth-order lowpass filters to meet the magnitude specifications $\omega_p = 30$ rad/s, $\delta_p = 0.2$, $\omega_s = 50$ rad/s, and $\delta_s = 0.095$. The initial filter is an elliptic filter (shown in dashed lines), and the final filter is optimized for phase, step response magnitude response in the stopband, and quality factor (shown in solid lines).

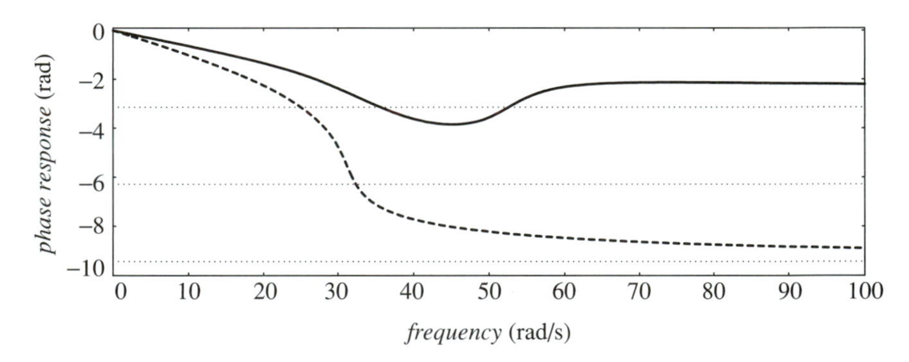

Figure 7.9 Phase response of example 3.

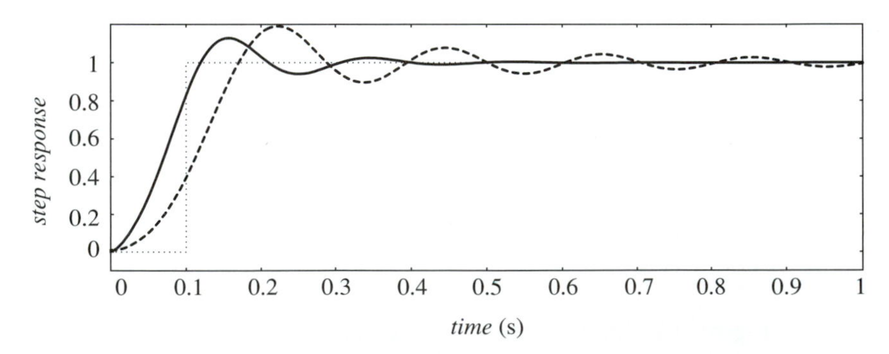

Figure 7.10 Step response of example 3.

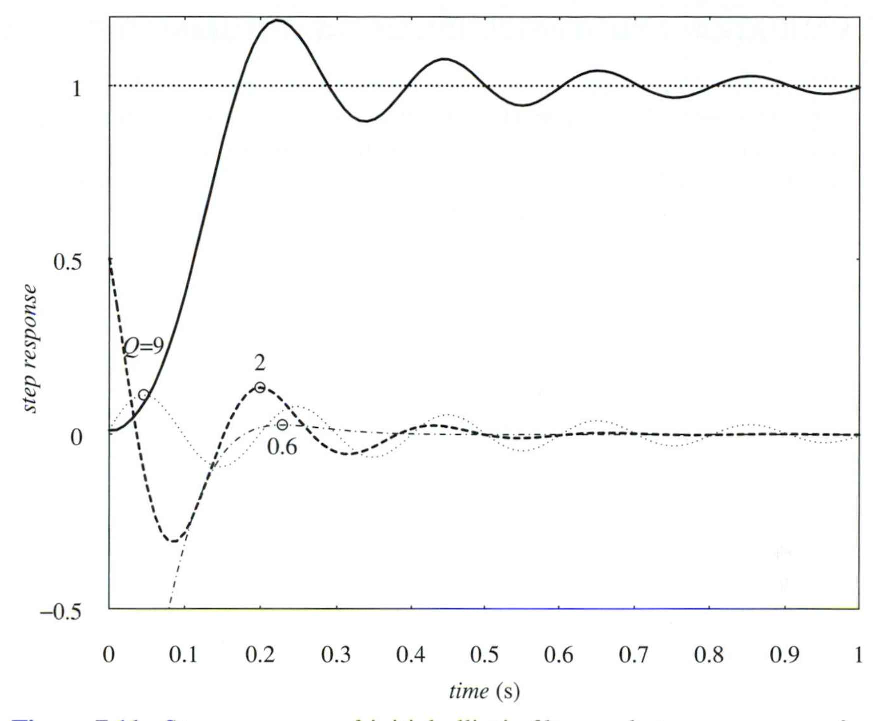

Figure 7.11 Step response of initial elliptic filter and step responses of three terms corresponding to pole pairs.

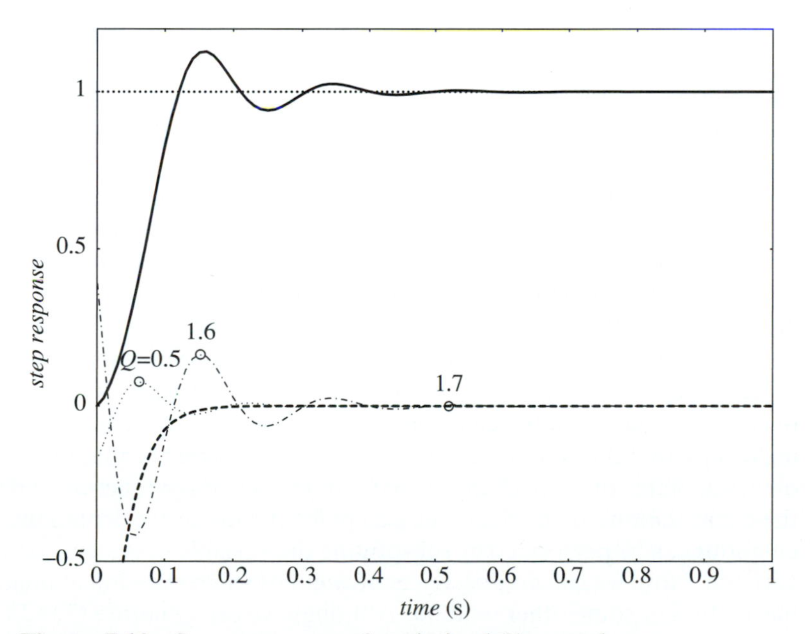

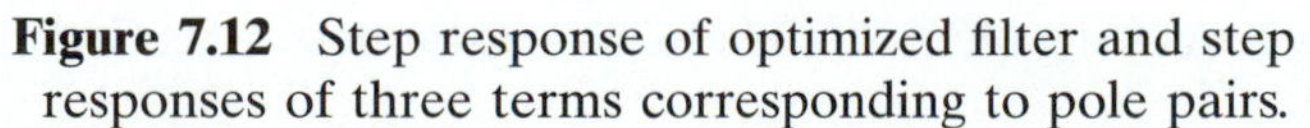

Figure 7.12 Step response of optimized filter and step responses of three terms corresponding to pole pairs.

7.6 VALIDATION AND VERIFICATION OF THE AUTOMATED FRAMEWORK

We use the *Mathematica* symbolic mathematics environment to collect the formulas for objective measures of properties, distance measures, objective function, and constraints, and to automate the conversion of the formulas to working filter optimization programs in MATLAB. Section 7.6.1 describes the steps in using *Mathematica* to verify the formulas we entered into it for the equations in Section 7.3. It also introduces our approach to converting the objective function and constraints, as well as their gradients, into source code. Section 7.6.2 discusses the automatic synthesis of complete MATLAB programs to optimize analog filter designs. Finally, Section 7.6.3 summarizes the ways in which we validate the generated code.

7.6.1 Verification of Formulas and Code Generation

We have written functions in *Mathematica* to return the formulas in Section 7.3 given the names of the free parameters of the pole-zero locations, the number of conjugate pole pairs, and the number of conjugate zero pairs (i.e., `a`, `b`, `c`, `d`, `n`, and `r`). We then use the symbolic mathematics environment to verify that we encoded the key equations properly as follows. For (7.1) and (7.1), we validated them by comparing their formulas to the absolute value and argument, respectively, applied directly to several example frequency responses $H(j\,\omega)$. We verified the formula for the partial fractions coefficients in (7.14) by comparing the formula to the partial fractions decomposition obtained by the symbolic mathematics environment. In validating the encoding of the step response in (7.15), we compared the formula with the inverse Laplace transform of s multiplied by several transfer functions.

Now that the formulas as entered into the symbolic mathematics environment have been verified, we encode the objective function given in (7.20). The next step is to compute the gradient of the objective function and the constraints so that the problem formulation can be passed onto optimization software. In this step, we discovered the need for an "inert" integral operator (one that does not try to evaluate the integrand) to represent equations such as (7.5). The reason is that we want the optimization software to compute the integral rather than the symbolic mathematics environment. Therefore, we introduce an inert integration operator called `integrate`. By using `integrate`, we were able to automate the calculation of gradients for the objective function and constraints by augmenting *Mathematica*'s built-in derivative operator. The gradients are computed with respect to the free parameters (pole-zero locations).

Next, we convert the objective function and constraints and their gradients into source code. In order to generate efficient source code, we reuse the intermediate results of code that has already been generated. Before we can reuse computations, we must make sure that the computations are context-free—that is, that they are not dependent on a parameter that can change value. For the objective function and constraints, all of the computations are context-free except for the integral calculations. The integral calculations can be performed by substituting the variable of integration with some unique variable. Now, we can employ the equivalent of subexpression elimination by means of hash tables or some other method. Although we can generate C and FORTRAN code, we focus on generating MATLAB code to take advantage of its Optimization Toolbox.

7.6.2 Generating Working MATLAB Programs

The first step in generating MATLAB programs is to find the analogous operations in MATLAB. The MATLAB functions corresponding to the *Mathematica* functions Cos, Log, Max, Min, Sign, Sin, and Tan are the same except that the MATLAB names are in lowercase. Similarly, the constant Pi becomes *pi*, but the constant E becomes *exp(1)*. Although Arg maps to *angle*, MATLAB does not have an arctangent function. Therefore, we map ArcTan[x, y], which takes the sign of x and y into account, to *angle(x + j*y)*. Likewise, ArcTan[z] becomes *angle(1 + z)*. We map symbolic definite integrals into numerical integrations via MATLAB's *trapz* function.

MATLAB's syntax is different from *Mathematica*'s, but the conversion is possible by a direct mapping of characters. In delimiting functions, *Mathematica* uses square brackets whereas MATLAB uses parentheses. Unlike *Mathematica*, which performs its computations in terms of formulas, MATLAB is based on matrix–vector operations. In *Mathematica*, the multiplication, division, and power operators work in a pointwise fashion and are therefore mapped into MATLAB's ".* ‘', "./" and ".^" operators. The complete mapping of an expression in *Mathematica* to MATLAB is carried out by first generating a string representing the internal form of the *Mathematica* expression via InputForm and then performing the substitutions listed above via StringReplace.

Now that expressions in *Mathematica* can be mapped directly into MATLAB code, we can add the syntax for function definitions and create a MATLAB script that directs the design procedure. In the synthesis of MATLAB programs, our programs generate four files as shown in Table 7.4. The "go*n*.m" file directs the optimization procedure. First, it initializes several constant optimization parameters as global variables. Then, it calls the constrained nonlinear optimization routine *constr* in the MATLAB Optimization Toolbox [67] to minimize the objective function given by (7.20). The *constr* function calls the filcost*n* function to compute the objective function and constraints, and the

Table 7.4 MATLAB files generated by our programs.

Filename	File Type	Purpose
stepresp.m	Function	Returns the value of the step response at time t (works for all filter orders)
filcost*n*.m	Function	Returns the value of objective function (a scalar) and values of the constraints (a vector)
costder*n*.m	Function	Returns the gradient of the objective function (a vector) and gradient of constraints (a matrix)
go*n*.m	Script	Runs numerical optimization to design "best" analog filter

Here, n is the number of conjugate pole pairs in the analog filter. All functions take the pole-zero locations as arguments.

costder$n$ function to compute the gradients. As stated previously, the *constr* routine is sequential quadratic programming method [64], and not a conjugate-gradient technique.

The designer may change the constant optimization parameters in the MATLAB code. The maximum quality factor $Q_{\max}$ is the `qmax1` variable in the "filcost$n$.m" file. The other constant parameters—weights of the objective function, magnitude response specifications, and maximum overshoot—are defined at the beginning of the "go$n$.m" file.

The "stepresp.m" file computes the step response. It is independent of the number of conjugate pole pairs n, whereas the other three files are not. In order for *Mathematica* to compute the gradient of the objective function, we unravel the product and summation terms in the objective function for fixed values of n and r. The effect on the generated MATLAB code is that the loops that would have depended on n and r have been unrolled (i.e., flattened). In the freely distributable release of the framework, we have generated the MATLAB programs for 4 poles and 0 zeroes, 4 poles and 2 zeroes, 4 poles and 4 zeros, 6 poles and 6 zeroes, and 8 poles and 8 zeros.

7.6.3 Validating the MATLAB Programs

This section discusses the verification of synthesized MATLAB programs. Because we generated the step response as a separate MATLAB file, we compare it directly to the step response generated in *Mathematica* for sampled values. MATLAB implements the step response according to (7.15), whereas *Mathematica* computes the step response by using the inverse Laplace transform. We compare the objective function generated in *Mathematica* directly with the objective function generated in MATLAB as it is also a separate file. The *Mathematica* and MATLAB versions agree.

MATLAB's *constr* SQP routine enables the checking of the symbolic form of the gradients. We expect strong agreement because symbolic differentiation by a symbolic mathematics environment is highly reliable. In numerically computing the gradients, the SQP routine uses finite difference techniques. The finite difference techniques only use the generated objective function, which we have already validated. If the analytic and finite difference gradients agree, then the symbolic form of the gradient is correct because we already know that the code generation is working properly. For the design examples in Section 7.5, the maximum deviation in the components of the gradient is less than 0.06%. The symbolic form of the gradients are important for the stability of the SQP procedure. We found design examples in MATLAB's *constr* routine diverged without using the symbolic gradients but converged when using the symbolic gradients.

By validating the symbolic form of the gradients, we validate the analytic approximation of t_{peak} by (7.19) and (7.17). The analytic approximation is solely used to compute the gradient of the objective measure of the peak overshoot of a filter. We validate the accuracy of the analytic approximation by running 10 design examples requiring at least 40 iterations of the SQP procedure each. At each iteration, we have the SQP routine compare gradients. In all cases, the gradient of the objective measure of the peak overshoot given by (7.19) and (7.17) never varied more than 0.06% of the numerically approximated value.

7.7 CONCLUSION

We have developed a formal, extensible framework for optimizing multiple behavioral and implementation properties of analog filter designs. We have implemented the framework as a set of *Mathematica* programs that generate MATLAB programs to perform the simultaneous optimization of magnitude response, phase response, peak overshoot, and quality factors. In developing the framework, we derive an analytic approximation for peak overshoot, which we validate using the MATLAB implementation. We demonstrate the framework by finding lowpass filter designs optimized for multiple criteria.

In the framework, both the algebraic derivations and programming tasks would be nearly impossible for a human to carry out correctly. By performing both processes together, we can validate that the assumptions in the algebraic derivations are legitimate and that the source code is generated properly. Furthermore, the algebraic abstraction empowers the researcher to create new filter design programs by simply redefining the objective function: Our software will take care of recomputing the derivatives and regenerating the source code.

CHAPTER 8

CLASSIC DIGITAL FILTER DESIGN

This chapter is intended to review the basics of classic digital IIR filter design. Classification, salient properties, and sensitivity of transfer functions in the z-domain are given. The most important digital filter realizations are presented including direct form, transpose direct form, parallel-of-two-allpass, and wave digital filter (WDF) realizations. For each realization we provide complete design equations and procedures that make the design easily applicable to a broad variety of digital filter design problems.

8.1 INTRODUCTION TO DIGITAL FILTERS

Digital filter is a discrete-time system that alters the spectral information contained in some discrete-time signal x producing a new discrete-time signal y. Sampled signals are represented digitally as sequences of numbers. In this book we consider linear time-invariant systems used for smoothing, predicting, differentiating, integrating, and separating signals, as well as for removing noise from the signal x. In fact, we use linear operations on some data sequences. For the purposes of this book, we define a *DSP system* as a system used for digital signal processing.

A key quantity of a digital filter (or, generally, of a DSP system) is its *sampling frequency*, F_{sampling}, also called the *sampling rate*. We assume that a sequence of numbers x_k is obtained from equally spaced samples of some quantity $x(t)$, where k is an integer and t is a continuous variable. Typically, t represents time or space distance, but not necessarily so. If t denotes time, then the samples can be obtained from $x(t)$ at equidistant time instants

$$x(t),\ \ x(t+T),\ \ x(t+2T),\ \ \ldots,\ \ x(t+kT),\ \ \ldots$$

where we observe the signal from the reference time $t = t_0$, and $k = 0, 1, 2, 3, \ldots$. The sampling frequency is the reciprocal value of the time interval T; that is, $F_{\text{sampling}} = 1/T$. In fact, the sampling frequency shows the number of samples per unit time. If no ambiguity can arise, we denote a sequence of numbers x_k by $x(k)$ interchangeably.

Usually, DSP systems use only one sampling rate. *Multirate DSP systems* use more than one sampling rate.

A digital filter can be analyzed in the time or frequency domain. Let us consider a digital filter, DF, with an input sequence $x(n)$ $(n = 0, 1, 2, \ldots)$. The output sequence $y(n)$ results from a set of elementary algebraic operations. The intermediate results are stored in some auxiliary sequences $y_i(n)$ $(i = 0, 1, 2, \ldots)$.

The table to follow illustrates a typical processing that is performed by a digital filter. The intermediate results are stored in the auxiliary sequences $y_1(n)$, $y_2(n)$, $y_3(n)$, $y_4(n)$, $y_5(n)$, $y_6(n)$, $y_7(n)$, $y_8(n)$, and $y_9(n)$.

Input	Processing	Output
	$y_1(n) = x(n)$	
$n = 0, 1, 2, \ldots$	$y_7(n) = y_1(n-1)$	
	$y_5(n) = y_6(n-2)$	
$x(n)$	$y_2(n) = y_1(n) - y_5(n)$	
$y_1(-1) = 0$	$y_3(n) = \frac{1}{2} y_2(n)$	$y(n) = y_9(n)$
$y_6(-1) = 0$	$y_4(n) = y_3(n) + y_5(n)$	
$y_6(-2) = 0$	$y_6(n) = y_1(n) + y_3(n)$	
	$y_8(n) = y_4(n) + y_7(n)$	
	$y_9(n) = \frac{1}{2} y_8(n)$	

(8.1)

This digital filter can be represented pictorially by a block diagram as shown in Fig. 8.1, where $a = \frac{1}{2}$ and $b = \frac{1}{2}$.

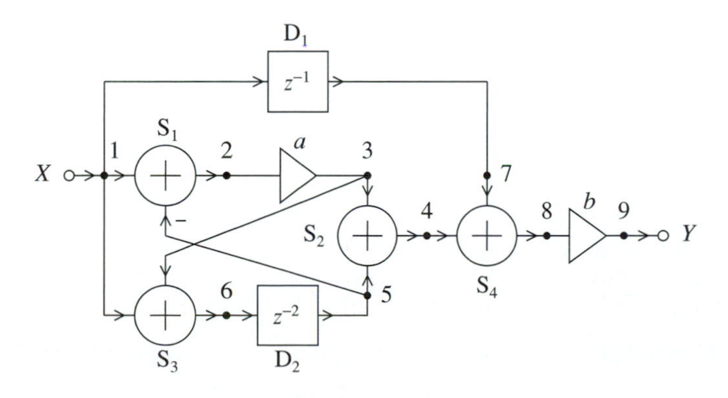

Figure 8.1 Third-order digital filter: $a = \frac{1}{2}$ and $b = \frac{1}{2}$.

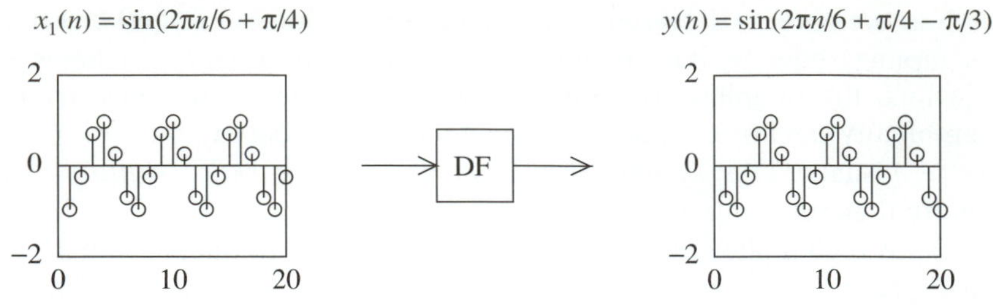

Figure 8.2 Input and output sequences of a digital filter described by (8.1): $x_1(n) = \sin(2\pi n f_1 + \frac{\pi}{4})$ and $y(n) = \sin(2\pi n f_1 + \frac{\pi}{4} - \frac{\pi}{3})$.

Multiplication by a constant, $a = \frac{1}{2}$ or $b = \frac{1}{2}$, is visualized by a triangle. Operation of shifting a sequence, by one or by two, is represented by a box with z^{-1} or z^{-2} inscribed. Addition is depicted by a circle with the plus sign inscribed. The nodes 1 to 9 are related to the auxiliary sequences. The input and the output are designated by smaller circles, labeled by X and Y, respectively.

Assume that the input sequence is a sinusoidal sequence of the digital frequency $f_1 = \frac{1}{6}$ and the unit amplitude. We use "frequency" as a short form of dimensionless "digital frequency" throughout this chapter. Then the steady-state output sequence is again a sinusoidal sequence of the same frequency $f_1 = \frac{1}{6}$, the unit amplitude, but with a different phase shift $-\frac{\pi}{3}$, as shown in Fig. 8.2.

If the input sequence is a sinusoidal sequence of the frequency $f_2 = \frac{1}{3}$ and the unit amplitude, then the steady-state output sequence is a sequence of zeros, as shown in Fig. 8.3.

If the input sequence is a sum of two sinusoidal sequences of the frequencies $f_1 = \frac{1}{6}$ and $f_2 = \frac{1}{3}$ and the unit amplitudes, then the steady-state output sequence is

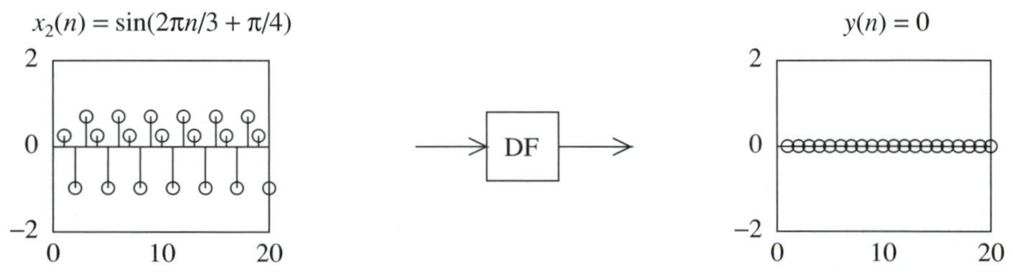

Figure 8.3 Input and output sequences of a digital filter described by (8.1): $x_2(n) = \sin(2\pi n f_2 + \frac{\pi}{4})$ and $y(n) = 0$.

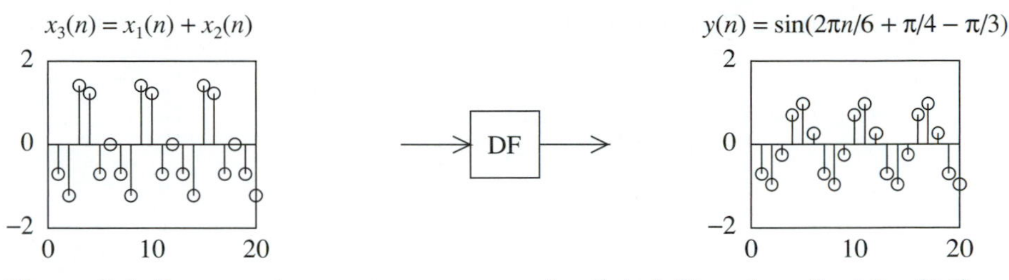

Figure 8.4 Input and output sequences of a digital filter described by (8.1): $x_3(n) = \sin(2\pi n f_1 + \frac{\pi}{4}) + \sin(2\pi n f_2 + \frac{\pi}{4})$ and $y(n) = \sin(2\pi n f_1 + \frac{\pi}{4} - \frac{\pi}{3})$.

only a sinusoidal sequence of the frequency $f_1 = \frac{1}{6}$, the unit amplitude, and the phase shift $-\frac{\pi}{3}$, as shown in Fig. 8.4.

If the input sequence is a sinusoidal sequence of the frequency $f_4 = \frac{1}{4}$ and the unit amplitude, then the steady-state output sequence is again a sinusoidal sequence of the same frequency $f_1 = \frac{1}{4}$, the amplitude $\sqrt{2}/2$, and the phase shift $-\frac{3\pi}{4}$.

$$
\begin{array}{lll}
f & \text{Input Sequence} & \text{Output Sequence} \\
f_1 = \frac{1}{6} & x_1(n) = \sin(2\pi n f_1 + \frac{\pi}{4}) & y(n) = \sin(2\pi n f_1 + \frac{\pi}{4} - \frac{\pi}{3}) \\
f_2 = \frac{1}{3} & x_2(n) = \sin(2\pi n f_2 + \frac{\pi}{4}) & y(n) = 0 \\
f_1 = \frac{1}{6} & x_3(n) = \sin(2\pi n f_1 + \frac{\pi}{4}) & \\
f_2 = \frac{1}{3} & \qquad + \sin(2\pi n f_2 + \frac{\pi}{4}) & y(n) = \sin(2\pi n f_1 + \frac{\pi}{4} - \frac{\pi}{3}) \\
f_4 = \frac{1}{4} & x_4(n) = \sin(2\pi n f_4) & y(n) = \frac{\sqrt{2}}{2}\sin(2\pi n f_4 - \frac{3\pi}{4})
\end{array}
\tag{8.2}
$$

We conclude that this digital filter, described by the set of equations 8.1, passes the sinusoidal sequence of frequency $f_1 = \frac{1}{6}$ without affecting its amplitude and rejects (stops) the sinusoidal sequence of frequency $f_2 = \frac{1}{3}$.

Once when we know the block diagram representing a digital filter (Fig. 8.1), we can formulate the required equations in the complex domain of the z-transform, and we can find the filter transfer function

$$H(z) = \frac{Y(z)}{X(z)} = \frac{1}{2}\frac{\frac{1}{2} + z^{-1} + z^{-2} + \frac{1}{2}z^{-3}}{1 + \frac{1}{2}z^{-2}}$$

We have automated in *Mathematica* the derivation of a transfer function from a given block diagram[69].

The frequency response of the filter is obtained for $z = e^{j2\pi f}$ and can be written as follows:

$$H(e^{j2\pi f}) = \frac{(2\cos\pi f + \cos 3\pi f)(\cos\pi f - j\sin\pi f)}{3\cos 2\pi f + j\sin 2\pi f}$$

The magnitude response of the filter

$$M(f) = \left|H(e^{j2\pi f})\right| = \frac{|2\cos\pi f + \cos 3\pi f|}{\sqrt{1 + 8\cos^2 2\pi f}}$$

is shown in Fig. 8.5.

From the known frequency response and the given input sinusoidal sequence $x(n) = \sin(2\pi n f_x + \phi_x)$, we can find the steady-state output sequence as

$$y(n) = M(f_x)\sin(2\pi n f_x + \phi_x + \arg H(e^{j2\pi f_x}))$$

For example, if the frequency is $f_1 = \frac{1}{6}$ the corresponding magnitude is $M(f_1) = 1$; next, for the frequency $f_2 = \frac{1}{3}$ we compute the magnitude $M(f_2) = 0$; also, for $f_3 = \frac{1}{4}$ we find $M(f_3) = \frac{\sqrt{2}}{2} \approx 0.707$.

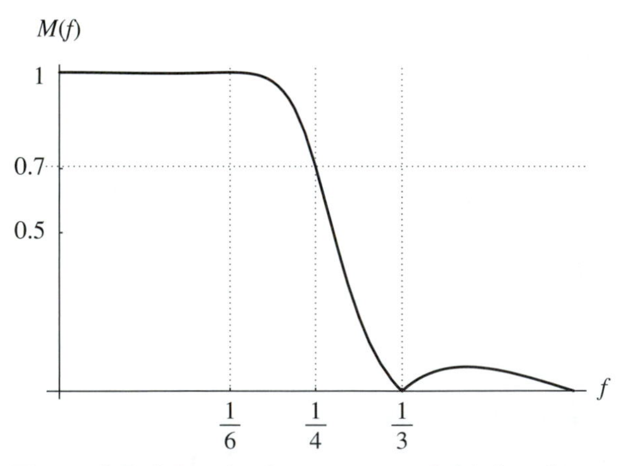

Figure 8.5 Magnitude response of third-order digital filter shown in Fig. 8.1.

It is interesting to examine what happens in the processing (8.1) if we simply change the summing operation $y_8(n) = y_4(n) + y_7(n)$ into the subtracting operation $y_8(n) = y_4(n) - y_7(n)$. We find a new transfer function

$$H(z) = \frac{1}{2}\frac{\frac{1}{2} - z^{-1} + z^{-2} - \frac{1}{2}z^{-3}}{1 + \frac{1}{2}z^{-2}}$$

Notice the change of the signs in the numerator.

Let us observe the output sequence $y(n)$ when this filter is excited by the sequences $x(n)$ according to the table

$$\begin{array}{lll}
f & \text{Input Sequence} & \text{Output Sequence} \\
f_1 = \frac{1}{6} & x_1(n) = \sin(2\pi n f_1) & y(n) = 0 \\
f_2 = \frac{1}{3} & x_2(n) = \sin(2\pi n f_2) & y(n) = \sin(2\pi n f_2 + \frac{\pi}{3}) \\
f_1 = \frac{1}{6} & x_3(n) = \sin(2\pi n f_1) & \\
f_2 = \frac{1}{3} & \qquad\quad + \sin(2\pi n f_2) & y(n) = \sin(2\pi n f_2 + \frac{\pi}{3}) \\
f_4 = \frac{1}{4} & x_4(n) = \sin(2\pi n f_4) & y(n) = \frac{\sqrt{2}}{2}\sin(2\pi n f_4 + \frac{3\pi}{4})
\end{array} \tag{8.3}$$

The table in (8.3) is similar to the table in (8.2), but the following differences in the magnitude response should be observed:

$$\begin{array}{lll}
f & (8.2) & (8.3) \\
f_1 = \frac{1}{6} & M(f_1) = 1 & M(f_1) = 0 \\
f_2 = \frac{1}{3} & M(f_2) = 0 & M(f_2) = 1 \\
f_4 = \frac{1}{4} & M(f_1) = \frac{\sqrt{2}}{2} & M(f_1) = \frac{\sqrt{2}}{2}
\end{array} \tag{8.4}$$

showing that only one change in the sign of a summing operation can significantly affect the filter transfer function, thereby substantially changing the behavior of the filter.

If $x(n)$ is a sum of sinusoidal sequences, then a digital filter can be used to amplify or attenuate a single sinusoidal sequence or a portion of the signal frequency spectrum. The range of frequencies in which the sinusoidal sequences are amplified or passed without considerable attenuation is called the *passband*. The frequency range in which the sinusoidal sequences are significantly attenuated is called the *stopband*. The required minimum and maximum of the attenuation or amplification, along with the corresponding edge frequencies of the passbands and stopbands, are called the *specification*.

A *digital filter design* is a process in which we construct a digital hardware or a program (software) that meets the given specification. The design starts with the specification, and it consists of four basic steps: *approximation*, *realization*, *study of imperfections*, and *implementation*. There is an infinite number of digital hardware and programs that meet the specification; therefore, the filter design is by no means unique.

The example to follow details on the four design steps. A typical digital filter design is shown in Fig. 8.6.

The *angular digital frequency* is designated by θ, and we consider that $0 \le \theta \le 2\pi$.

In this chapter we define the *digital frequency* as a dimensionless quantity

$$f = \frac{\theta}{2\pi}, \qquad 0 \le f \le 1$$

In this book we are concerned with linear constant-coefficient digital filters. Also, we assume that the coefficients are real, thus yielding a real impulse response and the symmetric frequency spectrum. Therefore, we are mainly interested in the frequency interval $0 \le f \le \frac{1}{2}$. Note that the symbol f in this context does not designate the frequency in Hz.

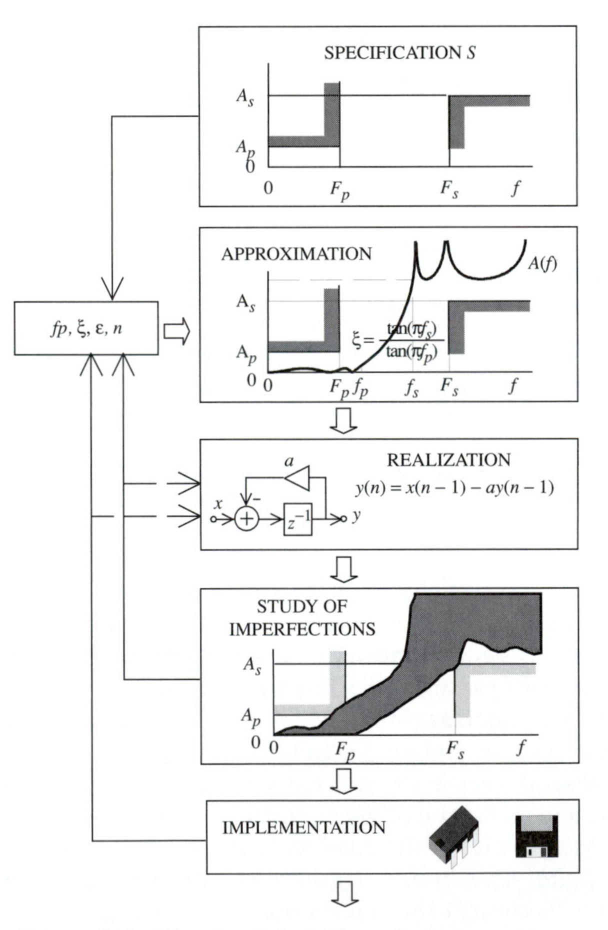

Figure 8.6 Classic digital filter design.

Assume that we want to design a filter that passes sinusoidal sequences over the frequency range $0 \leq f \leq F_p$, but attenuates the sequences for $f \geq F_s > F_p$, where $F_p < F_s < \frac{1}{2}$. This type of filters is called the *digital lowpass filter*. The frequency range $0 \leq f \leq F_p$ is the passband, and the range $F_s \leq f \leq \frac{1}{2}$ is the stopband; F_p is the *passband edge frequency* and F_s is the *stopband edge frequency*. We require that the attenuation in passband must not exceed A_p dB, and that the attenuation in stopband should be no less than A_s dB. Obviously, A_p is the *maximum passband attenuation*, and A_s is the *minimum stopband attenuation*. The four quantities that specify our lowpass filter requirements can be written in the form of the list $S = \{F_p, F_s, A_p, A_s\}$, which we simply call the *digital lowpass specification*.

In the first design step, the *approximation step*, we construct the filter transfer function $H(z)$ which is a rational function in z. The attenuation, $A(f) = -20 \log_{10} |H(e^{j2\pi f})|$, must satisfy the specification S, that is,

$$0 \leq A(f) \leq A_p, \qquad 0 \leq f \leq F_p < \frac{1}{2} \tag{8.5}$$

$$A_s \leq A(f), \qquad F_s \leq f \leq \frac{1}{2} \tag{8.6}$$

The function $A(f)$ is called the *attenuation approximation function*, or simply *attenuation*.

Suppose that we choose the elliptic transfer function which depends on four parameters: the order n, the actual passband edge frequency f_p, the actual stopband edge frequency f_s, and the ripple factor ϵ. It is convenient to define the ratio $\xi = \tan(\pi f_s)/\tan(\pi f_p)$, which is called the *selectivity factor*. We adjust the transfer-function parameters f_p, ξ, ϵ, and n to meet the specification; usually, we set $f_p = F_p$ and $\xi = \tan(\pi F_s)/\tan(\pi F_p)$. The maximal passband attenuation is controlled by the passband ripple factor ϵ, and we traditionally choose $\epsilon = \sqrt{10^{A_p/10} - 1}$. Generally, we prefer the minimum order n of the transfer function.

The *realization step* of a digital filter is the process of converting the transfer function into a block diagram or program (software); this block diagram or software is called the *realization*. The designer is interested in realizations which are economical, simple, and cheap, with short wordlength and high dynamic range. In Fig. 8.6 a realization of a first-order digital IIR filter is presented. Numerical values of the coefficients are calculated from known $H(z)$.

In practice, the filter is implemented with nonideal elements and the designer must accomplish the *study of imperfections* which includes filter coefficient quantization effects, product quantization effects (uncorrelated roundoff or truncation noise) and dynamic range constrains. If the specification can be satisfied only with high-precision coefficients and very long wordlength, then the designer has to choose another transfer function $H(z)$ and reevaluate the realization or approximation step.

In the *implementation step* a device called the product prototype, also called the *implementation*, is constructed and tested. The cost of the mass production depends on the type of components, packaging, methods of manufacturing, and testing. If the requirements are not met, then the realization step (new digital filter network or new software program) or the approximation step (new f_p, ξ, ϵ or n) must be redone.

The classic digital filters can be classified according to the frequency range they pass or reject:

- *Lowpass filter* passes sinusoidal sequences over the range $0 \leq f \leq F_p < \frac{1}{2}$, but attenuates the sequences for $F_p < F_s \leq f < \frac{1}{2}$.
- *Highpass filter* passes sinusoidal sequences for $F_p \leq f < \frac{1}{2}$, but attenuates the sequences over the range $0 \leq f \leq F_s < F_p < \frac{1}{2}$.
- *Bandpass filter* passes sinusoidal sequences over the range $F_{p1} \leq f \leq F_{p2}$, but attenuates the sequences for $0 \leq f \leq F_{s1} < F_{p1} < \frac{1}{2}$ and $F_{p2} < F_{s2} \leq f < \frac{1}{2}$.
- *Bandreject* or *bandstop filter* passes sinusoidal sequences for $0 \leq f \leq F_{p1} < F_{s1} < \frac{1}{2}$ and $F_{s2} < F_{p2} \leq f < \frac{1}{2}$, but attenuates the sequences over the range $F_{s1} \leq f \leq F_{s2} < \frac{1}{2}$.
- *Allpass filter* or *phase equalizer* passes sinusoidal sequences without attenuation, and it shapes the phase response.
- *Lowpass-notch filter* rejects sinusoidal sequences at frequencies $f \approx f_z$, but it passes sequences at high frequencies ($f_z < f$) with some attenuation.
- *Highpass-notch filter* rejects sinusoidal sequences at frequencies are $f \approx f_z$, but it passes sequences at low frequencies ($f < f_z$) with some attenuation.

We mention here several, among many, excellent books on classic digital filter theory, analysis, and design: references [16] and [23].

8.2 BASIC FILTER TRANSFER FUNCTIONS

Filter transfer function is a rational function in z and can be written in the form

$$H(z) = K\frac{(z - z_{z1})(z - z_{z2})\cdots(z - z_{zm})}{(z - z_{p1})(z - z_{p2})\cdots(z - z_{pn})} \tag{8.7}$$

where K is a real constant, $z_{z1}, \ldots, z_{zm}$ are zeros, and $z_{p1}, \cdots, z_{pn}$ are poles of the transfer function. Poles and zeros can be real or complex.

Complex poles or zeros, z_i, occur in complex-conjugate pairs:

$$\begin{aligned} z_i &= \mathrm{Re}\,(z_i) + j\mathrm{Im}\,(z_i) \\ z_{i+1} &= \mathrm{Re}\,(z_i) - j\mathrm{Im}\,(z_i) \end{aligned} \tag{8.8}$$

The corresponding factors of the transfer function can be expressed as

$$(z - z_i)(z - z_{i+1}) = z^2 - (z_i + z_{i+1})\,z + z_i z_{i+1} \tag{8.9}$$

or as a second-order polynomial with real coefficients a_i and b_i:

$$(z - z_i)(z - z_{i+1}) = z^2 + b_i z + a_i, \quad b_i = -2\ \mathrm{Re}\,(z_i), \quad a_i = |z_i|^2 \tag{8.10}$$

Since complex-conjugate poles must occur inside the unit circle, the following must hold:

$$\begin{aligned} 0 < a_i < 1 \\ -2\sqrt{a_i} \le b_i \le 2\sqrt{a_i} \end{aligned} \tag{8.11}$$

A complex pole-zero pair can be represented by the *second-order transfer function*

$$H_i(z) = \frac{e_i z^2 + d_i z + c_i}{z^2 + b_i z + a_i} \tag{8.12}$$

also called the *biquad*, for short.

Usually, the coefficients of a biquad do not give much insight into the corresponding frequency response. Therefore, for practical reasons, we would like to define a single quantity that can be simply calculated from the transfer function coefficients and that can give us a better insight into the magnitude and phase response, such as the maximal value of the magnitude response, the sharpness of the magnitude response, the frequency width of bandpass filter, or the sharpness of the phase response.

In this book we find convenient to introduce the quantity Q_i that we call the *Q-factor* of the complex-conjugate pair z_i and z_{i+1}. The Q-factor is used to characterize a complex-conjugate pair of poles or zeros, and we define it by

$$Q_i = \frac{\sqrt{(1 + a_i)^2 - b_i^2}}{2\,(1 - a_i)} \tag{8.13}$$

A pole Q-factor is always positive because $0 < a_i < 1$, and it reaches the minimal value for $b_i = \pm 2\sqrt{a_i}$:

$$Q_{\min} = Q_i|_{b_i = \pm 2\sqrt{a_i}} = \frac{1}{2} \tag{8.14}$$

If complex-conjugate poles are placed on the imaginary axes (i.e., $b_i = 0$), the Q-factor is

$$Q_i|_{b_i=0} = \frac{1}{2}\,\frac{1 + a_i}{1 - a_i} \tag{8.15}$$

and for $a_i \approx 1$ it is

$$Q_i|_{b_i=0,\ a_i\approx 1} \approx \frac{1}{1 - a_i} \tag{8.16}$$

Formally, for $a_i = 1$ the poles occur on the unity circle and the Q-factor becomes infinity. Therefore, we can consider the Q-factor as a measure of closeness of complex-conjugate poles (or zeros) to the unit circle; the higher the Q-factor, the closer the poles are to the unit circle.

The Q-factor better reflects the transfer function features than the coefficients a_i. A high Q-factor shows that the coefficient a_i is very close to the unity, but if a_i approaches 1, it still does not mean that the Q-factor is extremely large as shown in Fig. 8.7.

The coefficient a_i can be expressed in terms of b_i and Q_i:

$$a_i = \frac{1 + 4Q_i^2 - \sqrt{b_i^2 + 16Q_i^2 - 4b_i^2 Q_i^2}}{-1 + 4Q_i^2}, \quad Q_i > \frac{1}{2} \tag{8.17}$$

To reduce the sensitivity of a transfer function with respect to deviations of the coefficient values and to simplify realizations, it is preferable to realize a digital filter by a cascade or parallel connection of the first-order and second-order filter sections. The cascade approach consists of implementing each of the biquads by an appropriate digital hardware or software. The overall transfer function, $H(z)$, can be expressed as a product of the first-order and second-order transfer functions $H_i(z)$:

$$H(z) = \prod_{i=1}^{N} H_i(z) \tag{8.18}$$

The advantage of the cascade approach is that the realization of a higher-order transfer function is reduced to the much simpler design of first-order and second-order filters. The individual low-order filters are isolated so that any change in one filter does not affect any other filter in the cascade. This property is useful for simpler filter implementations.

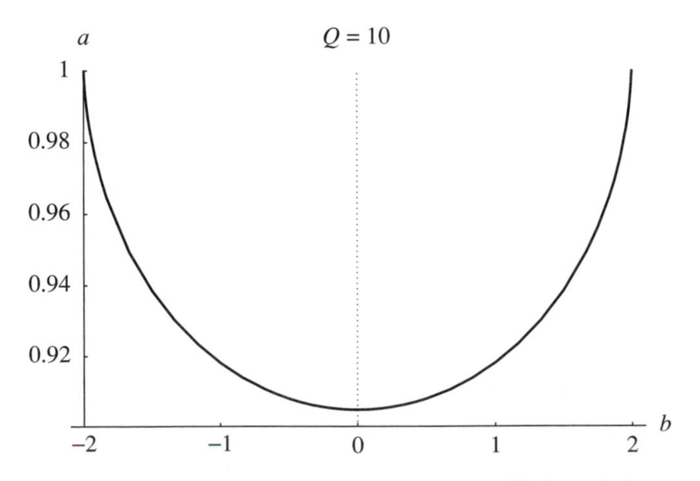

Figure 8.7 Coefficient a in terms of coefficient b for $Q = 10$.

We characterize each second-order filter section by the frequencies of the magnitude response extremes and by the 3-dB frequencies. The *3-dB frequencies*, denoted by $f_{3dB} = \theta_{3dB}/(2\pi)$, are frequencies at which the magnitude response $M(f) = |H(e^{j2\pi f})|$ is $\sqrt{2}$ times smaller than the magnitude response at some reference frequency f_r (3 dB only approximately corresponds to $\sqrt{2}$, strictly speaking it is $10^{3/20}$):

$$\frac{M(f_{3dB})}{M(f_r)} = \frac{|H(e^{j2\pi f_{3dB}})|}{|H(e^{j2\pi f_r})|} = \frac{1}{\sqrt{2}} \tag{8.19}$$

For example, for lowpass filters, the reference frequency is $f_r = 0$.

In filter design, we prefer to use the *normalized transfer function*, $H_n(z)$, defined by

$$H_n(z) = \frac{H(z)}{\max\limits_{f} M(f)} \tag{8.20}$$

The transfer function can be obtained by scaling the normalized transfer function by a constant:

$$H(z) = kH_n(z) \tag{8.21}$$

Quite generally, the normalization constant k can take any real value.

8.2.1 Second-Order Transfer Functions

In this section we analyze the properties of the basic second-order transfer functions. We examine the magnitude response $M(f) = |H(e^{j2\pi f})|$ for real positive digital angular frequencies $\theta = 2\pi f$. The angular frequencies of the magnitude response local extrema are designated by $\theta_e = 2\pi f_e$. We find it convenient to define the frequency f_j at which the frequency response becomes purely imaginary, $\mathrm{Re}(H(e^{j2\pi f_j})) = 0$.

Lowpass Transfer Function. The second-order *lowpass transfer function* is defined as

$$H_{LP}(z) = \frac{1+a+b}{4}\,\frac{(z+1)^2}{z^2+bz+a} = \frac{1+a+b}{4}\,\frac{\left(1+z^{-1}\right)^2}{1+bz^{-1}+az^{-2}} \tag{8.22}$$

At higher frequencies ($f > f_j > f_e$), where

$$f_j = \frac{1}{2\pi}\cos^{-1}\frac{-\,b}{1+a} \tag{8.23}$$

the magnitude response $M(f) = |H_{LP}(e^{j2\pi f})|$ decreases and, thus, high-frequency sinusoidal sequences are rejected.

The key properties of the lowpass transfer function are summarized below:

$$
\begin{aligned}
M(0) &= H_{LP}(1) = 1, && z = 1, && f = 0 \\
M(0.5) &= H_{LP}(-1) = 0, && z = -1, && f = 0.5 \\
M(f_j) &= |H_{LP}(e^{j2\pi f_j})| = \left|jQ_p\right| = Q_p, && z = e^{j2\pi f_j}, && f = f_j \\
M_e &= \max_{0 \le f \le 0.5}(M(f)) = |H_{LP}(e^{j2\pi f_e})| = \frac{Q_p}{\sqrt{1 - \dfrac{1}{4Q_p^2}}}, && z = e^{j2\pi f_e}, && f = f_e
\end{aligned}
\tag{8.24}
$$

where f_e is the frequency at which $M(f)$ has its maximal value M_e:

$$f_e = \frac{1}{2\pi} \cos^{-1} \frac{(1-a)^2 - b(1+a-b)}{4a - b - ab} \tag{8.25}$$

and the pole Q-factor is

$$Q_p = \frac{\sqrt{(1+a)^2 - b^2}}{2(1-a)} \tag{8.26}$$

The maximal value of the magnitude response is approximately equal to Q_p for $Q_p \gg 1$ (i.e., for $a \approx 1$), as shown in Fig. 8.8:

$$\max_{0 \le f \le 0.5} \left| H_{LP}\left(e^{j2\pi f}\right) \right| = \left| H_{LP}\left(e^{j2\pi f_e}\right) \right| \approx Q_p, \quad f_e \approx f_j, \quad Q_p \gg 1 \tag{8.27}$$

This fact is very important for digital filters implemented in fixed point arithmetic. Suppose that the filter output sequence, y_k, must be bounded to $-1 \le y_k \le 1$. Then, the amplitude of the sinusoidal sequence of frequency f_e, at the input of the lowpass second-order filter, must be smaller than $1/Q_p$, so that, after filtering, the amplitude of the output sequence remains within the prescribed range, $-1 \le y_k \le 1$.

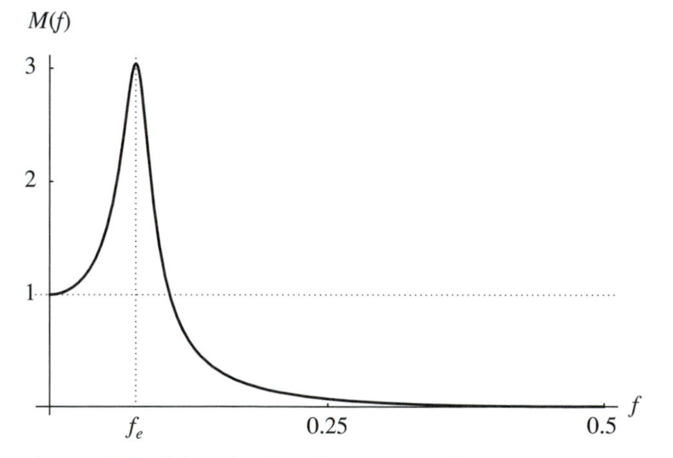

Figure 8.8 Magnitude of second-order lowpass transfer function: $Q_p = 3$, $a = 0.85117$, and $b = -1.621545$.

The maximal value of the magnitude response, M_e, can be expressed in terms of the coefficients a and b:

$$M_e = \max_{0 \le f \le 0.5} (M(f)) = \frac{(1+a)^2 - b^2}{2(1-a)\sqrt{4a - b^2}} \tag{8.28}$$

The magnitude of the normalized transfer function, $H_{LPn}(z)$, is shown in Fig. 8.9. This normalized transfer function is defined as

$$H_{LPn}(z) = \frac{H_{LP}(z)}{M_e} = \frac{(1-a)\sqrt{4a-b^2}}{2(1+a-b)} \frac{\left(z^{-1}+1\right)^2}{1+bz^{-1}+az^{-2}} \tag{8.29}$$

and it has the maximal magnitude, equal to 1, at the frequency f_e.

For $Q_p \le 1/\sqrt{2}$ we find $f_e = 0$ as shown in Fig. 8.10. Therefore, for $1/2 < Q_p \le 1/\sqrt{2}$ the maximal magnitude function is at frequency $f_e = 0$.

The maximal value of the magnitude of second-order lowpass transfer functions, M_e, in terms of the coefficients is shown in Fig. 8.11. M_e dramatically increases when a approaches to 1, while the influence of b is negligible.

Highpass Transfer Function. The second-order *highpass transfer function* is defined as

$$H_{HP}(z) = \frac{1+a-b}{4} \frac{(z-1)^2}{z^2+bz+a} = \frac{1+a-b}{4} \frac{\left(z^{-1}-1\right)^2}{1+bz^{-1}+az^{-2}} \tag{8.30}$$

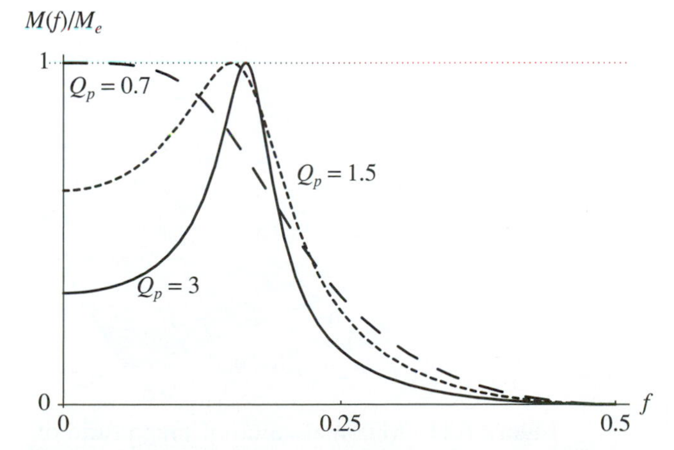

Figure 8.9 Magnitude of second-order normalized lowpass transfer functions $M(f)/M_e$: $Q_p = 3$, 1.5, 0.7.

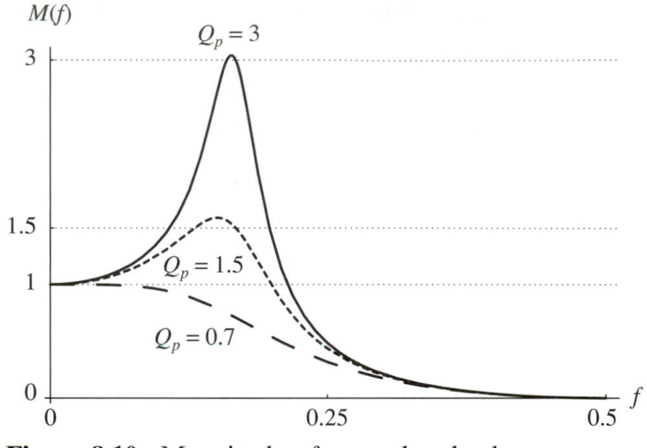

Figure 8.10 Magnitude of second-order lowpass transfer functions: $Q_p = 3,\ 1.5,\ 0.7$.

At lower frequencies, $f < f_j < f_e$, where

$$f_j = \frac{1}{2\pi}\cos^{-1}\frac{-\,b}{1+a} \tag{8.31}$$

the magnitude response $M(f) = |H_{HP}(e^{j2\pi f})|$ decreases when f approaches zero and, thus, low-frequency sinusoidal sequences are rejected, while the high-frequency sinusoidal sequences, $f \geq f_p$, pass without attenuation.

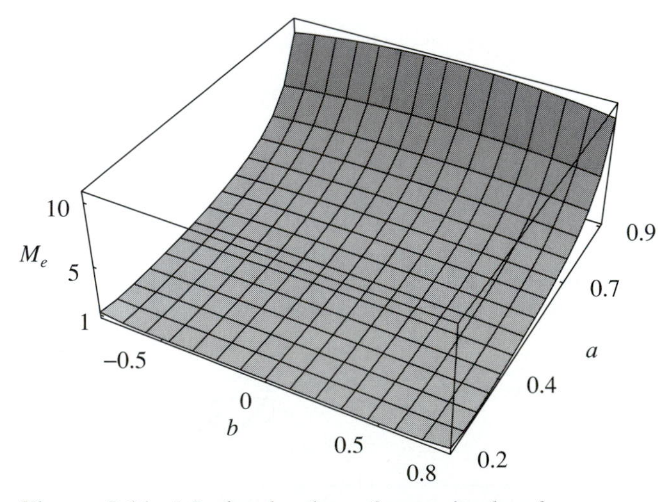

Figure 8.11 Maximal value of magnitude of second-order lowpass transfer functions in terms of filter coefficients.

The key properties of the highpass transfer function are summarized below:

$$
\begin{aligned}
M(0) &= H_{HP}(1) = 0, && z = 1, && f = 0 \\
M(0.5) &= H_{HP}(-1) = 1, && z = -1, && f = 0.5 \\
M(f_j) &= |H_{HP}(e^{j2\pi f_j})| = \left|jQ_p\right| = Q_p, && z = e^{j2\pi f_j}, && f = f_j \\
M_e &= \max_{0 \le f \le 0.5}(M(f)) = |H_{HP}(e^{j2\pi f_e})| = \frac{Q_p}{\sqrt{1 - \dfrac{1}{4Q_p^2}}}, && z = e^{j2\pi f_e}, && f = f_e
\end{aligned}
\tag{8.32}
$$

where f_e is the frequency at which $M(f)$ has its maximal value M_e

$$
f_e = -\frac{1}{2\pi}\cos^{-1} - \frac{(1-a)^2 + b(1+a+b)}{4a + b + ab} \tag{8.33}
$$

and the pole Q-factor is

$$
Q_p = \frac{\sqrt{(1+a)^2 - b^2}}{2(1-a)} \tag{8.34}
$$

The maximal value of the magnitude response is approximately equal to Q_p for $Q_p \gg 1$ (i.e., for $a \approx 1$), as shown in Fig. 8.12:

$$
\max_{0 \le f \le 0.5}\left|H_{HP}\left(e^{j2\pi f}\right)\right| = \left|H_{HP}\left(e^{j2\pi f_e}\right)\right| \approx Q_p, \quad f_e \approx f_j, \quad Q_p \gg 1 \tag{8.35}
$$

As in the case of the lowpass transfer function, the maximal value of the magnitude response is approximately equal to Q_p, as shown in Fig. 8.12. For $Q_p \le 1/\sqrt{2}$ we find that $f_e = 0.5$, as shown in Fig. 8.13.

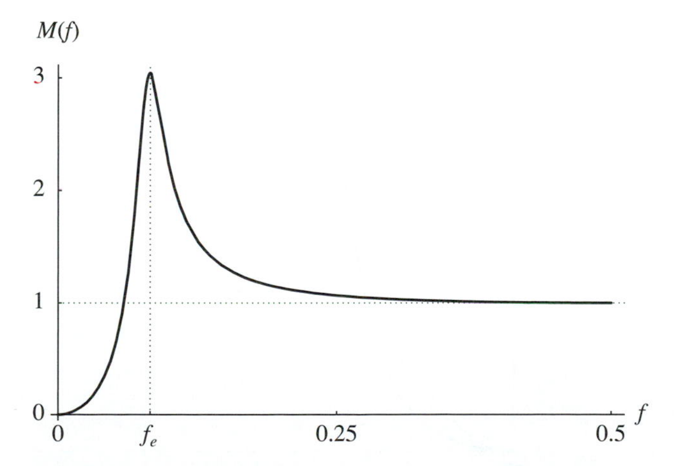

Figure 8.12 Magnitude of second-order highpass transfer function: $Q_p = 3$, $a = 0.85117$, and $b = -1.621545$.

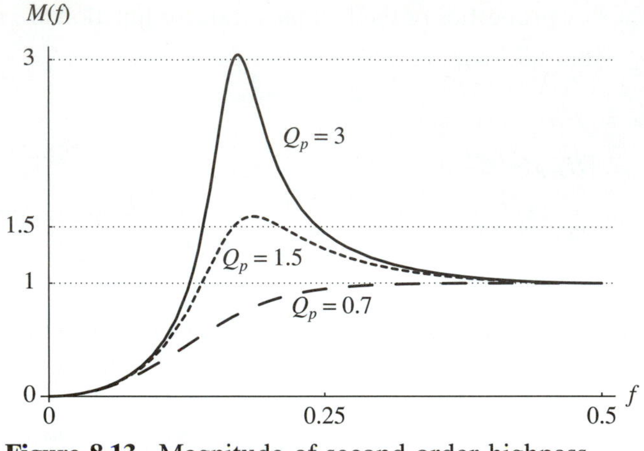

Figure 8.13 Magnitude of second-order highpass transfer functions: $Q_p = 3$, 1.5, 0.7.

The maximal value of the magnitude response, M_e, can be expressed in terms of the coefficients a and b:

$$M_e = \frac{(1+a)^2 - b^2}{2\,(1-a)\,\sqrt{4a - b^2}} \tag{8.36}$$

The magnitude of the normalized transfer function, $H_{HPn}(z)$, is shown in Fig. 8.14. This normalized transfer function

$$H_{HPn}(z) = \frac{H_{HP}(z)}{M_e} = \frac{(1-a)\,\sqrt{4a-b^2}}{2\,(1+a+b)}\,\frac{\left(z^{-1}-1\right)^2}{1+bz^{-1}+az^{-2}} \tag{8.37}$$

has the maximal magnitude, equal to 1, at the frequency f_e.

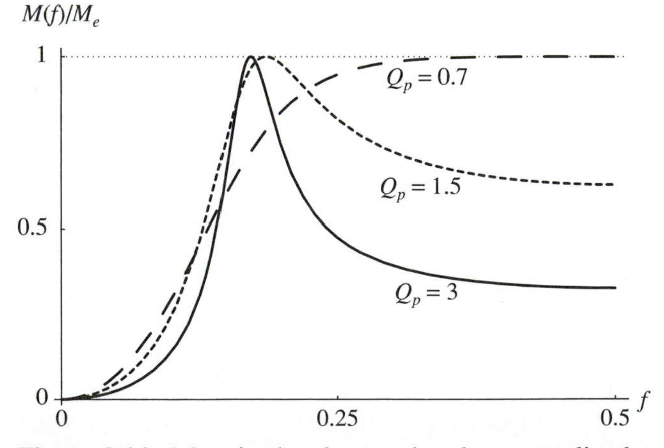

Figure 8.14 Magnitude of second-order normalized highpass transfer functions $M(f)/M_e$: $Q_p = 3$, 1.5, 0.7.

The maximal value of the magnitude of second-order highpass transfer functions, M_e, in terms of the coefficients is the same as for the second-order lowpass transfer functions shown in Fig. 8.11.

Bandpass Transfer Function. The second-order *bandpass transfer function* is defined as

$$H_{BP}(z) = \frac{1-a}{2} \frac{z^2-1}{z^2+bz+a} = \frac{1-a}{2} \frac{1-z^{-2}}{1+bz^{-1}+az^{-2}} \tag{8.38}$$

The key properties of the bandpass transfer function are summarized below:

$$\begin{aligned}
M(0) &= H_{BP}(1) = 0, & z &= 1, & f &= 0 \\
M(0.5) &= H_{BP}(-1) = 0, & z &= -1, & f &= 0.5 \\
M_e &= \max_{0 \le f \le 0.5}(M(f)) = |H_{BP}(e^{j2\pi f_e})| = 1, & z &= e^{j2\pi f_e}, & f &= f_e
\end{aligned} \tag{8.39}$$

where

$$f_e = \frac{1}{2\pi}\cos^{-1}\frac{-b}{1+a} \tag{8.40}$$

The frequency at which the magnitude response reaches its maximum, f_e, is sometimes called the *resonant frequency* or the *central frequency*.

Second-order bandpass filters pass sinusoidal sequences from the band of frequencies $f_{low,3\text{dB}} < f < f_{high,3\text{dB}}$ with insignificant attenuation (less than $20\log_{10}\sqrt{2} \approx$ 3dB), but reject sinusoidal sequences whose frequencies are on either side of this band:

$$\begin{aligned}
&\frac{1}{\sqrt{2}} \le M(f) = |H_{BP}(e^{j2\pi f})| \le 1, \quad f_{low,3\text{dB}} \le f \le f_{high,3\text{dB}} \\
&|H_{BP}(e^{j2\pi f_{low,3\text{dB}}})| = |H_{BP}(e^{j2\pi f_{high,3\text{dB}}})| = \frac{1}{\sqrt{2}} \\
&f_{low,3\text{dB}} = \frac{1}{2\pi}\cos^{-1}\frac{-b(1+a)-(1-a)\sqrt{2+2a^2-b^2}}{2(1+a^2)} \\
&f_{high,3\text{dB}} = \frac{1}{2\pi}\cos^{-1}\frac{-b(1+a)+(1-a)\sqrt{2+2a^2-b^2}}{2(1+a^2)}
\end{aligned} \tag{8.41}$$

The pole Q-factor of the bandpass filter is

$$Q_p = \frac{\sqrt{(1+a)^2-b^2}}{2(1-a)} \tag{8.42}$$

The maximum of the magnitude response is 1. The 3-dB bandwidth, $f_{high,3\text{dB}} - f_{low,3\text{dB}}$, is affected by Q_p (Fig. 8.15). Higher Q-factors produce narrower bandwidths (Fig. 8.16).

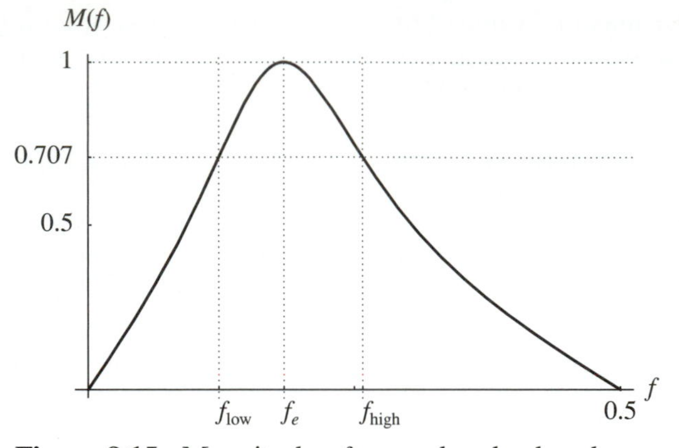

Figure 8.15 Magnitude of second-order bandpass transfer function: $Q_p = 3$, $a = 0.3727$, and $b = -0.5573$.

The reference frequency for f_{3dB}, for bandpass filters, is $f_r = f_e$.

The 3-dB bandwidth mostly depends on the Q-factor, but it also depends on the coefficient b as shown in Fig. 8.17. It should be noticed that for known Q and b the coefficient a is uniquely determined.

Bandreject Transfer Function. The second-order *bandreject transfer function*, also called the *bandstop transfer function*, is defined as

$$H_{BR}(z) = \frac{1}{2}\,\frac{(1+a)\,z^2 + 2bz + (1+a)}{z^2 + bz + a} = \frac{1}{2}\,\frac{(1+a) + 2bz^{-1} + (1+a)\,z^{-2}}{1 + bz^{-1} + az^{-2}} \tag{8.43}$$

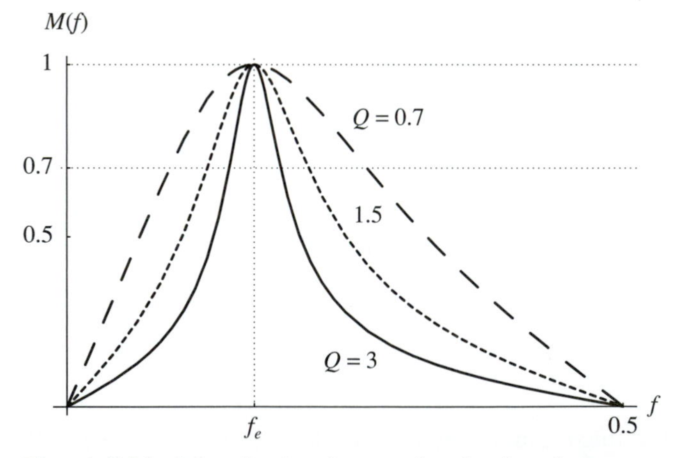

Figure 8.16 Magnitude of second-order bandpass transfer functions: $Q_p = 3$, 1.5, 0.7.

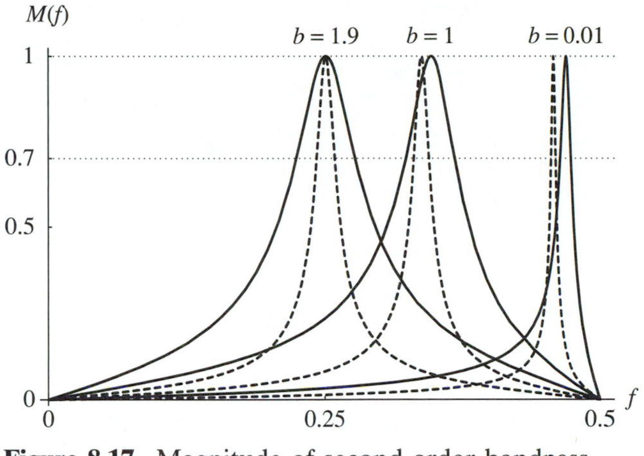

Figure 8.17 Magnitude of second-order bandpass transfer functions: $Q_p = 3$ (solid line), $Q_p = 10$ (dashed line).

The key properties of the bandreject transfer function are summarized below:

$$\begin{aligned} M(0) &= H_{BR}(1) = 1, & z &= 1, & f &= 0 \\ M(0.5) &= H_{BR}(-1) = 1, & z &= -1, & f &= 0.5 \\ M_e &= |H_{BR}(e^{j2\pi f_e})| = 0, & z &= e^{j2\pi f_e}, & f &= f_e \end{aligned} \tag{8.44}$$

where

$$f_e = \frac{1}{2\pi} \cos^{-1} \frac{-b}{1+a} \tag{8.45}$$

The frequency at which the magnitude response reaches its minimum, f_e, is sometimes called the *antiresonant frequency*.

Second-order bandreject filters pass sinusoidal sequences from the two bands of frequencies $0 \leq f < f_{low,3\text{dB}}$ and $f_{high,3\text{dB}} < f \leq 0.5$ with insignificant attenuation (less than $20 \log_{10} \sqrt{2} \approx 3\text{dB}$), but reject sinusoidal sequences whose frequencies are from the band of frequencies

$$\begin{aligned} \frac{1}{\sqrt{2}} &< |H_{BR}(e^{j2\pi f})| \leq 1, & 0 &\leq f < f_{low,3\text{dB}} \\ \frac{1}{\sqrt{2}} &< |H_{BR}(e^{j2\pi f})| \leq 1, & f_{high,3\text{dB}} &< f \leq 0.5 \end{aligned} \tag{8.46}$$

$$|H_{BR}(e^{j2\pi f_{low,3\text{dB}}})| = |H_{BR}(e^{j2\pi f_{high,3\text{dB}}})| = \frac{1}{\sqrt{2}}$$

$$\begin{aligned} f_{low,3\text{dB}} &= \frac{1}{2\pi} \cos^{-1} \frac{-b(1+a)-(1-a)\sqrt{2+2a^2-b^2}}{2(1+a^2)} \\ f_{high,3\text{dB}} &= \frac{1}{2\pi} \cos^{-1} \frac{-b(1+a)+(1-a)\sqrt{2+2a^2-b^2}}{2(1+a^2)} \end{aligned} \tag{8.47}$$

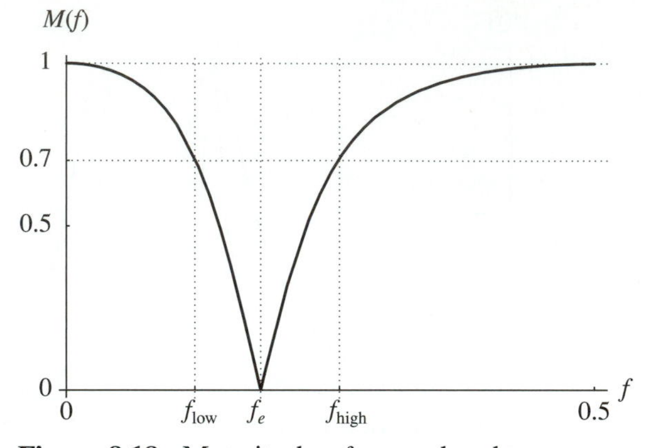

Figure 8.18 Magnitude of second-order bandreject transfer function: $Q_p = 3$, $a = 0.3727$, and $b = -0.5573$.

The pole Q-factor of the bandreject filter is

$$Q_p = \frac{\sqrt{(1+a)^2 - b^2}}{2(1-a)} \tag{8.48}$$

The maximum of the magnitude response is 1. The 3-dB bandwidth of the stopband, $f_{high,3\text{dB}} - f_{low,3\text{dB}}$, is affected by Q_p (Fig. 8.18). Higher Q-factors produce narrower bandwidths (Fig. 8.19).

The reference frequency for $f_{3\text{dB}}$, for bandreject filters, is $f_r = 0$.

The 3-dB bandwidth depends mostly on the Q-factor but also it depends on the coefficient b. It should be noticed that for known Q and b the coefficient a is uniquely determined.

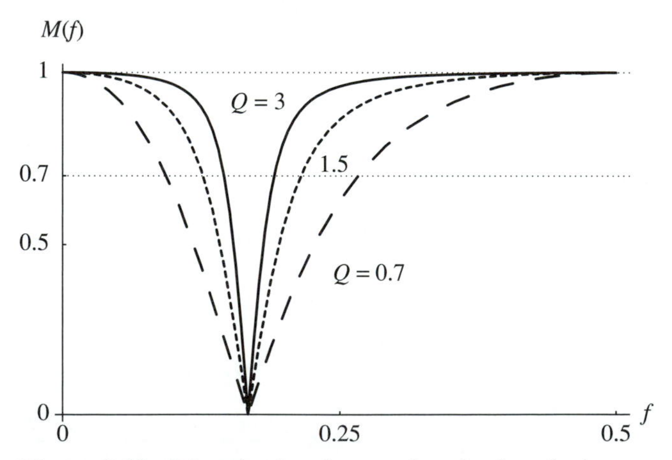

Figure 8.19 Magnitude of second-order bandreject transfer functions: $Q_p = 3$, 1.5, 0.7.

Lowpass-Notch and Highpass-Notch Transfer Function. The second-order *lowpass-notch* and *highpass-notch transfer functions* are defined as

$$
\begin{aligned}
H_{LPN}(z) &= \frac{1+a-b}{2\,(1+a-bc)} \; \frac{(1+a)\,z^2 + 2bcz + (1+a)}{z^2+bz+a} \\
&= \frac{1+a-b}{2\,(1+a-bc)} \; \frac{(1+a) + 2bcz^{-1} + (1+a)\,z^{-2}}{1+bz^{-1}+az^{-2}}, \quad 0 < c \le c_1 < 1 \\
H_{HPN}(z) &= \frac{1+a-b}{2\,(1+a-bc)} \; \frac{(1+a)\,z^2 + 2bcz + (1+a)}{z^2+bz+a} \\
&= \frac{1+a-b}{2\,(1+a-bc)} \; \frac{(1+a) + 2bcz^{-1} + (1+a)\,z^{-2}}{1+bz^{-1}+az^{-2}}, \quad 1 < c_2 \le c < 2
\end{aligned}
\tag{8.49}
$$

where

$$
c_1 = \frac{(1+a)\left(1-2a+a^2+b+ab+b^2\right)}{b\,(4a+b+ab)} \tag{8.50}
$$

$$
c_2 = \frac{(1+a)\left(1-2a+a^2-b-ab+b^2\right)}{b\,(-4a+b+ab)} \tag{8.51}
$$

The key properties of the lowpass-notch and highpass-notch transfer functions are summarized below:

$$
\begin{aligned}
&M(0) = H_{LPN}(1) = H_{HPN}(1) = \frac{(-1-a+b)\;(1+a+bc)}{(1+a+b)\;(-1-a+bc)}, && z = 1, && f = 0 \\
&M(0.5) = H_{LPN}(-1) = H_{HPN}(-1) = 1, && z = -1, && f = 0.5 \\
&M_e = M(f_e) = |H_{LPN}(e^{j2\pi f_e})| = |H_{HPN}(e^{j2\pi f_e})|, && z = e^{j2\pi f_e}, && f = f_e \\
&M_z = M(f_z) = |H_{LPN}(e^{j2\pi f_z})| = |H_{HPN}(e^{j2\pi f_z})| = 0, && z = e^{j2\pi f_z} && f = f_z
\end{aligned}
\tag{8.52}
$$

where

$$
f_e = \frac{1}{2\pi}\cos^{-1}\frac{(1+a)\;\left(-1+2a-a^2-b^2+b^2c\right)}{b\;\left(1+2a+a^2-4ac\right)} \tag{8.53}
$$

$$
f_z = \frac{1}{2\pi}\cos^{-1}\left(\frac{-\,bc}{1+a}\right) \tag{8.54}
$$

The maximum of the magnitude response occurs at f_e and is given below:

$$
M_e = \frac{(1+a-b)}{(1-a)\;(1+a-bc)}\sqrt{\left|\frac{(1+a)^2\left(1-2a+a^2+b^2-2b^2c\right)+4ab^2c^2}{4a-b^2}\right|} \tag{8.55}
$$

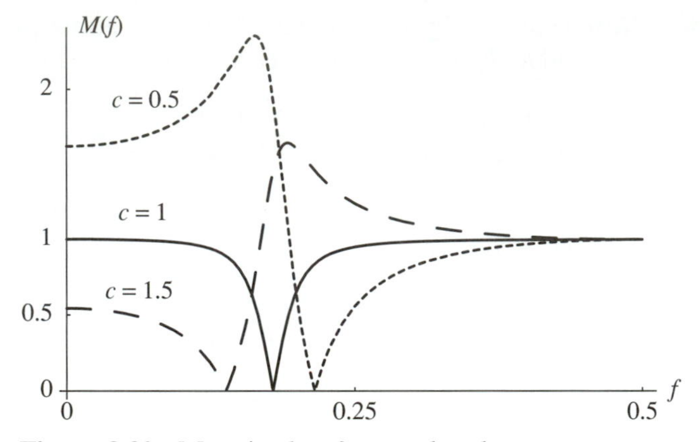

Figure 8.20 Magnitude of second-order lowpass-notch transfer function ($c = 0.5$), highpass-notch transfer function ($c = 1.5$), and bandreject transfer function ($c = 1$).

The transfer functions $H_{LPN}(z)$ and $H_{HPN}(z)$ become purely imaginary at frequency f_j, for $z = e^{j2\pi f_j}$:

$$H_{LPN}(e^{j2\pi f_j}) = H_{HPN}(e^{j2\pi f_j}) = j\frac{b\,(1+a)\,(1-c)\,\sqrt{1+2a+a^2-b^2}}{(1-a)\,(1+a+b)\,(1+a-bc)}$$
$$f_j = \frac{1}{2\pi}\cos^{-1}\frac{-b}{1+a} \tag{8.56}$$

Lowpass-notch filters reject sinusoidal sequences whose frequencies are $f \approx f_z$, but they pass sequences at high frequencies ($f > f_z$) with some attenuation. Highpass-notch filters reject sinusoidal sequences whose frequencies are $f \approx f_z$, but they pass sequences at low frequencies ($f < f_z$) with some attenuation.

The reference frequency for $f_{3\text{dB}}$, for notch filters, is $f_r = 0.5$, as shown in Fig. 8.20.

The maximum of the magnitude response in terms of the coefficient c is plotted in Fig. 8.21; notice that it reaches the minimum for $c = 1$. The magnitude of the second-order notch transfer function is shown in Fig. 8.22 and the magnitude of the corresponding second-order normalized notch transfer function is plotted in Fig. 8.23.

Allpass Transfer Function. Thus far, we have discussed the magnitude response, only. However, for allpass transfer functions the phase response is of key interest. Allpass filter sections can be used to modify the phase response of a filter without changing its magnitude response. On the other hand, allpass second-order sections can be used as basic building blocks for realizations of elliptic transfer functions. The second-order *allpass transfer function* is defined as

$$H_{AP}(z) = \frac{az^2+bz+1}{z^2+bz+a} = \frac{a+bz^{-1}+z^{-2}}{1+bz^{-1}+az^{-2}} \tag{8.57}$$

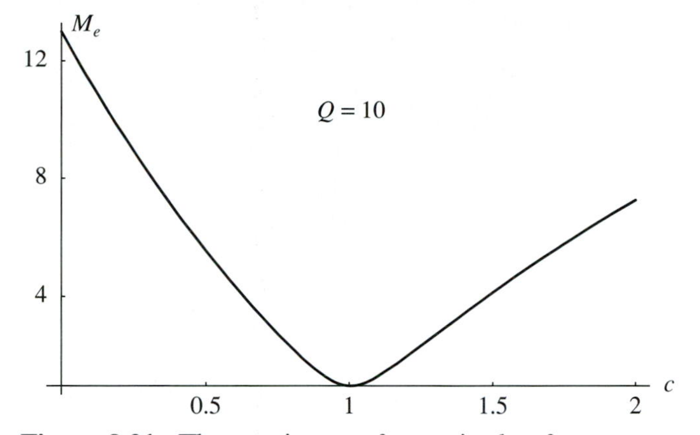

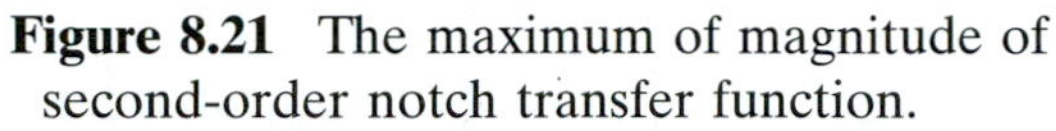
Figure 8.21 The maximum of magnitude of second-order notch transfer function.

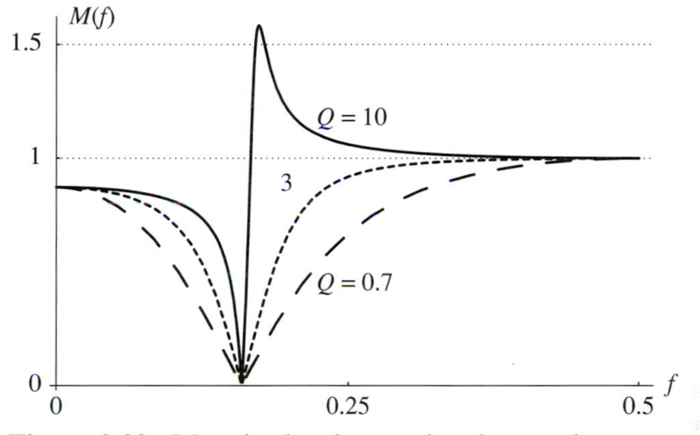

Figure 8.22 Magnitude of second-order notch transfer function: $Q \in \{0.7, 3, 10\}$.

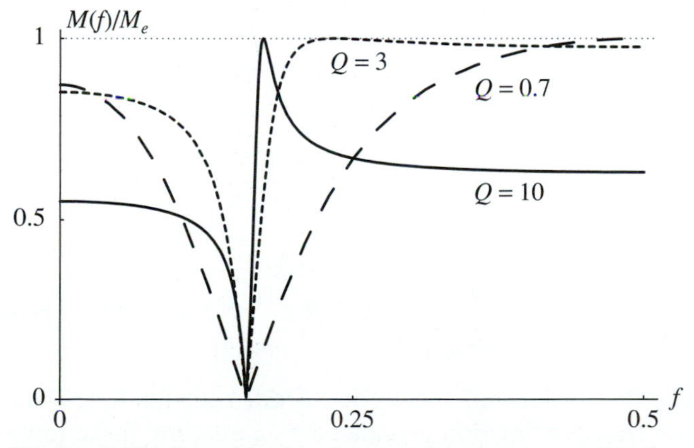

Figure 8.23 Magnitude of second-order normalized notch transfer function: $Q \in \{0.7, 3, 10\}$.

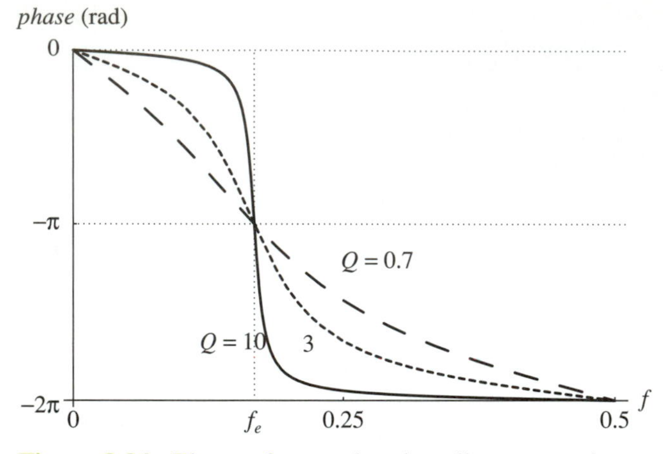

Figure 8.24 Phase of second-order allpass transfer functions, $Q_p = 10$, 1.

The magnitude of the allpass transfer function is constant for all f:

$$M(f) = |H_{AP}(e^{j2\pi f})| = 1 \tag{8.58}$$

The phase response of second-order allpass filters is shown in Fig. 8.24. If the Q-factor is smaller, then the phase response of an allpass filter is more linear.

The pole Q-factor is given by

$$Q = \frac{\sqrt{(1+a)^2 - b^2}}{2(1-a)} \tag{8.59}$$

The zeros of an allpass transfer function are mirror images of the transfer function poles, in the complex z-plane, with respect to the unit circle.

8.3 DECOMPOSITION OF TRANSFER FUNCTIONS

Any higher-order transfer function can be expressed as a product of the first-order and the second-order transfer functions. In practice, when designing a filter, we prefer transfer functions with complex-conjugate poles; however, if the transfer-function order is odd, we prefer transfer functions with only one real pole:

$$H_{even}(z) = \prod_i \frac{e_i + d_i z^{-1} + c_i z^{-2}}{1 + b_i z^{-1} + a_i z^{-2}} \tag{8.60}$$

$$H_{odd}(z) = \frac{e_0 + d_0 z^{-1}}{1 + b_0 z^{-1}} \prod_i \frac{e_i + d_i z^{-1} + c_i z^{-2}}{1 + b_i z^{-1} + a_i z^{-2}} \tag{8.61}$$

The first-order and the second-order transfer functions are realized separately. Next, the transfer function is realized by cascading these low-order filter section. The cascade design is attractive for various reasons:

- The filter design is straightforward.
- The case study for selecting the appropriate low-order filter realizations is facilitated.
- The effort for quantization analyzes is simplified.

8.4 POLE-ZERO PAIRING

The *pole-zero pairing* is (a) a decomposition of the numerator and the denominator of a transfer function into products of constant terms, the first-order and the second-order polynomials in z, and, then, (b) constructing the first-order and the second-order rational functions in z by pairing the numerator and denominator terms.

The pole-zero pairing is not unique and depends on implementation, as well as on selected low-order transfer functions.

The most frequently used pole-zero pairing procedure is based on pairing the poles with higher a (i.e., higher Q-factors) with zeros that are as close as possible to those poles (c closest to 1). This pairing can achieve the maximal dynamic range of the second-order sections.

The most important criteria that are used in practice, as criteria for pole-zero pairing, are as follows:

1. *Maximal dynamic range*. The dynamic range of a filter can be determined as a ratio between the largest and smallest signals (amplitudes of sequences) that can be represented in a filter. A ratio between the maximum magnitude response computed over the whole frequency range and the minimum magnitude response in the passband can be used as a figure-of-merit of dynamic range.
2. *Minimal sensitivity*. The overall sensitivity of the filter can be reduced by appropriate pole-zero pairing.

The choice of the scaling factors of the first-order and the second-order sections is called the *gain distribution*. There exist procedures for optimizing the gain distribution for maximal dynamic range or for minimal sensitivity.

8.5 OPTIMUM CASCADING SEQUENCE

After a higher-order transfer function is decomposed into, say, m first-order and second-order functions, with appropriate scaling factors, the designer has to choose the *optimal cascading sequence*. The number of possible cascade realizations is $m!$, where m is, also, the number of cascaded filter sections.

For example, a sixth-order transfer function can be decomposed into the product $H(z) = H_1(z)H_2(z)H_3(z)$, and realized as a cascade of three second-order sections. The number of possible realizations is $m! = 3! = 6$ as shown in Fig. 8.25.

Frequently, for the maximal dynamic range, the optimal sequence of the second-order filter sections is the sequence in which the preceding section has lower Q-factor than the following section.

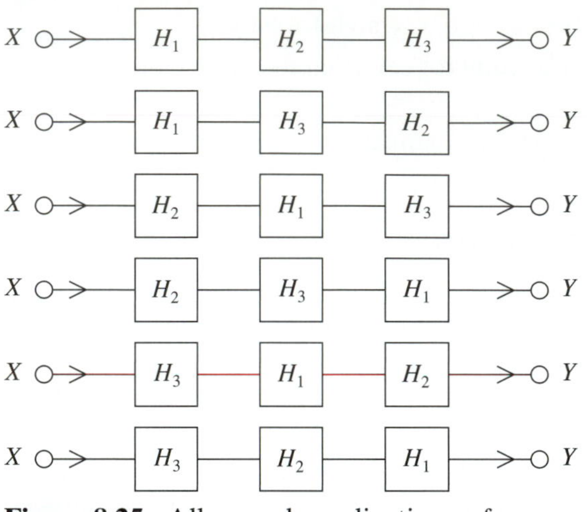

Figure 8.25 All cascade realizations of sixth-order transfer function.

8.6 FINITE WORDLENGTH EFFECTS

8.6.1 Coefficient Quantization

In custom fixed-point implementations of digital filters, such as VLSI or FPGA implementations, the reduction of accuracy saves in chip area and increases processing speed. This reduction requires a thorough analysis and optimization of the finite wordlength properties: coefficient quantization, rounding or truncation of intermediate results, and arithmetic overflows.

Quantized filter coefficients make the transfer function different from the ideal one (with infinite-precision coefficients). Those errors can be predicted by the coefficient sensitivity analysis of a filter. We use "coefficient sensitivity" to denote the sensitivity of the transfer function (or frequency response, or magnitude response, or phase response) to its coefficients or parameters.

8.6.2 Basic Definitions of Sensitivity

The key property of digital filters is that the performances of manufactured filters can be guaranteed to correspond exactly to the designed filter performances. Manufactured filters always operate with finite-precision coefficients, either in floating-point or in fixed-point arithmetic. Thus, in the design step "study of imperfections" we have to check that the designed filter with quantized coefficients meets the specification. The coefficient sensitivity analysis is the simplest method for predicting the quantization effects.

Many filter realizations can be found for a given transfer function. If we assume that the coefficients are exact (of infinite accuracy), all existing realizations must exhibit the same performances. In practice, coefficients are quantized and their values can differ from the exact values; therefore, the performances of the realized filter can differ from the performances of the filter with exact coefficients. Various filter realizations can have rather different amount of degradation after coefficient quantization.

The simplest way to predict the quantization effects is to use the concept of sensitivity, assuming that the coefficient changes are small.

The *single-parameter sensitivity* of a function $F = F(x_1, \ldots, x_i, \ldots, x_n)$, due to a change in quantity x_i, is

$$S_{x_i}^F = \frac{\partial F}{\partial x_i} \tag{8.62}$$

and, for a small change in x_i, the variation of the function F is expected to be

$$\Delta F \approx S_{x_i}^F \Delta x_i \tag{8.63}$$

The expected *variation* in function $F = F(x_1, \ldots, x_i, \ldots, x_n)$, due to the changes in all quantities $x_1, \ldots, x_i, \ldots, x_n$, is given by

$$\Delta F \approx \sum_{i=1}^{n} S_{x_i}^F \Delta x_i \tag{8.64}$$

The function $F(x_1, \ldots, x_i, \ldots, x_n)$ can be a transfer function, magnitude response, phase response, attenuation, group delay, or any other function derived from the transfer function.

Consider the third-order lowpass transfer function

$$H(z) = \frac{1}{2}\frac{a + z^{-1} + z^{-2} + az^{-3}}{1 + az^{-2}}, \qquad a = \frac{1}{2}$$

that can be realized by the block diagram shown in Fig. 8.1, which we call the *allpass-type realization* for short.

This transfer function can be decomposed into a product of two lower-order functions:

$$H(z) = H_1(z)H_2(z)$$

$$H_1(z) = \frac{1 + z^{-1}}{4}, \qquad H_2(z) = \frac{1 + z^{-1} + z^{-2}}{1 + az^{-2}} \qquad a = \frac{1}{2}$$

and it can be realized as a cascade connection of two filter sections; we call this realization the *cascade realization* for short. The sensitivity of the magnitude response, $M(f) = |H(e^{j2\pi f})|$, to the coefficient a, for both realizations, is shown in Figs. 8.26 and 8.27.

The extreme values of the sensitivity for the cascade realization are greater than the extreme values corresponding to the allpass-type realization. However, the sensitivity in the passband ($0 < f < 0.2$) of the allpass-type realization is 10 times smaller. Since the sensitivity in the transition region ($0.2 < f < 0.3$) is of minor interest, we conclude that the allpass-type realization is significantly better design solution, in this example.

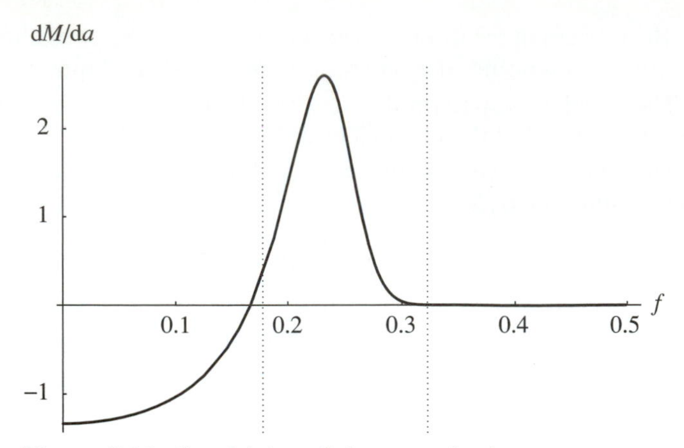

Figure 8.26 Sensitivity of the magnitude response to a for cascade realization.

The magnitude response of the two realizations is shown in Figs. 8.28 and 8.29. From a practical standpoint, the cascade design is a far inferior solution for such a range of coefficient values.

The upper limit of the variation of a function, ΔF, can be estimated by the *worst-case* method that gives the absolute variation:

$$\Delta F|_{\text{worst case}} = \sum_{i=1}^{n} |S_{x_i}^F \Delta x_i| \tag{8.65}$$

Some other methods are based on the formula

$$\Delta F = \sqrt{\sum_{i=1}^{n} |S_{x_i}^F \Delta x_i|^2} \tag{8.66}$$

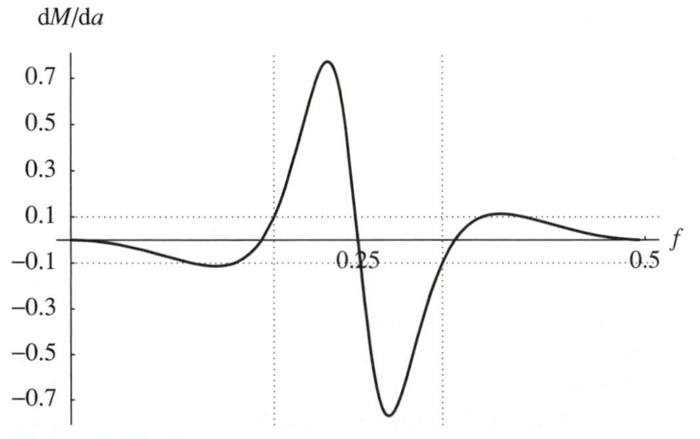

Figure 8.27 Sensitivity of magnitude response to a for allpass-type realization.

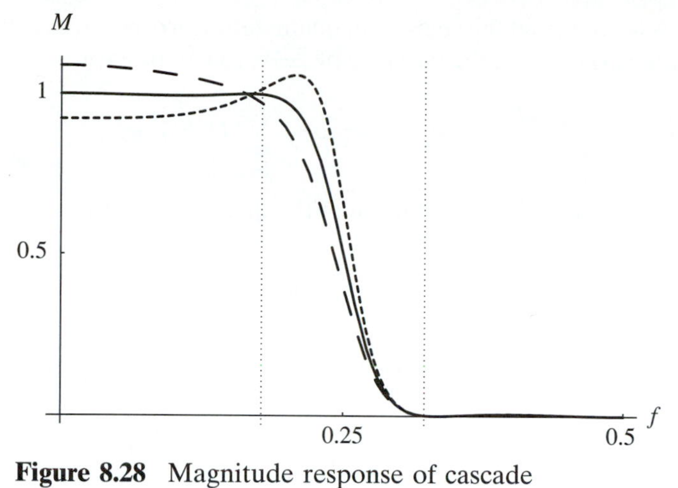

Figure 8.28 Magnitude response of cascade

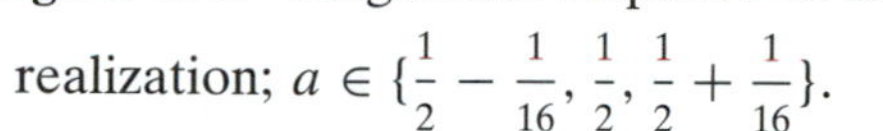

realization; $a \in \{\frac{1}{2} - \frac{1}{16}, \frac{1}{2}, \frac{1}{2} + \frac{1}{16}\}$.

Let us consider the magnitude response sensitivity of a filter with m multipliers with coefficients $x_1, \ldots, x_m$. The magnitude response variation, $\Delta M(f)$, due to small changes Δx_i, is

$$\Delta M(f) = \Delta |H(e^{j2\pi f})| \approx \sum_{i=1}^{m} S_{x_i}^{|H|} \Delta x_i \tag{8.67}$$

where

$$S_{x_i}^{|H|} = \frac{\partial |H(e^{j2\pi f})|}{\partial x_i} \tag{8.68}$$

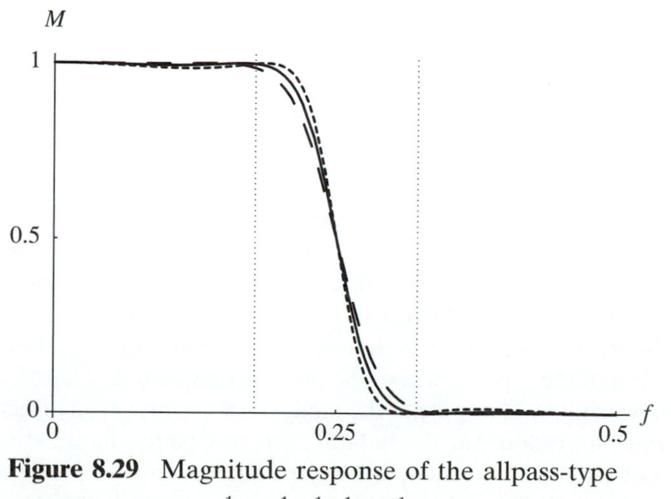

Figure 8.29 Magnitude response of the allpass-type realization; $a \in \{\frac{1}{2} - \frac{1}{16}, \frac{1}{2}, \frac{1}{2} + \frac{1}{16}\}$.

Assuming that the coefficient quantization errors Δx_i are statistically independent with the variance σ_i^2, the variance of $\Delta M(f)$ can be expressed as

$$\sigma_{\Delta M}^2 \approx \sum_{i=1}^{m} \left(S_{x_i}^{|H|}\right)^2 \sigma_i^2 \tag{8.69}$$

For the uniform distribution of the quantization error, with the quantization step size q, the coefficient variance is

$$\sigma_x^2 = \frac{q^2}{12} \tag{8.70}$$

For equal coefficient variances, $\sigma_1^2 = \cdots = \sigma_m^2 = \sigma_x^2$, we obtain

$$\sigma_{\Delta M}^2 \approx \sigma_x^2 \sum_{i=1}^{m} \left(S_{x_i}^{|H|}\right)^2 \tag{8.71}$$

In general, we prefer to realize a transfer function with first-order and second-order sections; the sections are cascaded or connected in parallel. Direct higher-order realizations are inferior from the viewpoint of the coefficient quantization. This is particularly pronounced in the case of very selective digital filters. For this reason, in the next section we consider the sensitivity of the second-order transfer functions.

8.6.3 Sensitivity of Second-Order Transfer Function

For a given quantization step the transfer-function poles can occur only at certain locations in the z-plane inside the unit circle. In Figs. 8.30 and 8.31, the pole locations of a second-order transfer function are shown for the four-bit quantization. The transfer function is expressed in two ways, in terms of parameters a and b or parameters α and β:

$$H_1(z) = \frac{1}{z^2 + bz + a} \tag{8.72}$$

$$H_2(z) = \frac{1}{z^2 + \beta(1+\alpha)z + \alpha}, \qquad a = \alpha,\ b = \beta(1+\alpha) \tag{8.73}$$

Without lack of generality we assume that the numerator is constant.

The poles of $H_1(z)$ are restricted to 765 different values, while the number of pole values of $H_2(z)$ is 961. We can find plenty of different realizations of the second-order transfer functions, each having different pole locations.

In practice, we prefer to choose a realization which has the densest pole locations in the region of the z-plane where the poles are expected. Therefore, in an advanced filter design we should incorporate the analysis of quantization effects in the approximation step; thus, we reduce the degradation effects due to quantization.

Let us consider the sensitivity of the second-order transfer function $H_1(z) = 1/(z^2 + bz + a)$. The sensitivity of the magnitude response to a and to b are shown in Figs. 8.32 and 8.33.

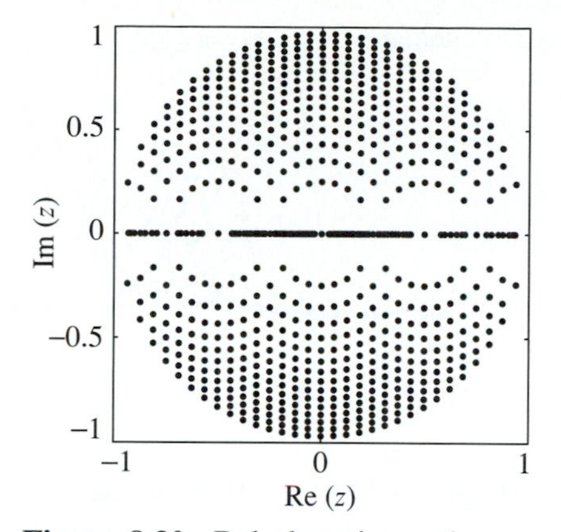

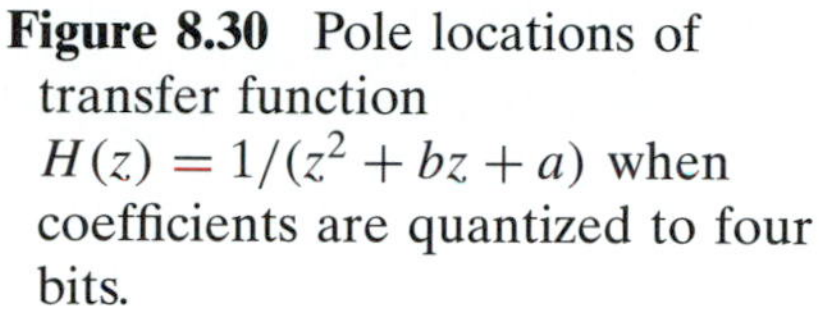

Figure 8.30 Pole locations of transfer function $H(z) = 1/(z^2 + bz + a)$ when coefficients are quantized to four bits.

The maximal value of $S_a^{|H|} = \dfrac{\partial |H_1(e^{j2\pi f})|}{\partial a}$ has the minimum for $b = 0$:

$$\begin{aligned} S_{a_i}^{|H|}\Big|_{b=0} &= \frac{2}{(1-a)^3} \\ S_{a_i}^{|H|} &\geq \frac{2}{(1-a)^3} \end{aligned} \tag{8.74}$$

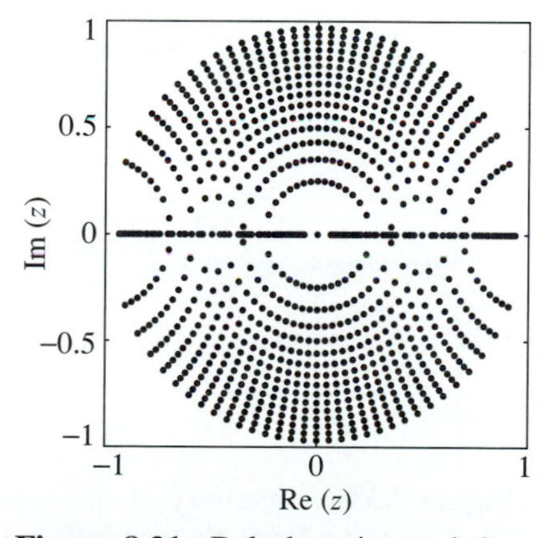

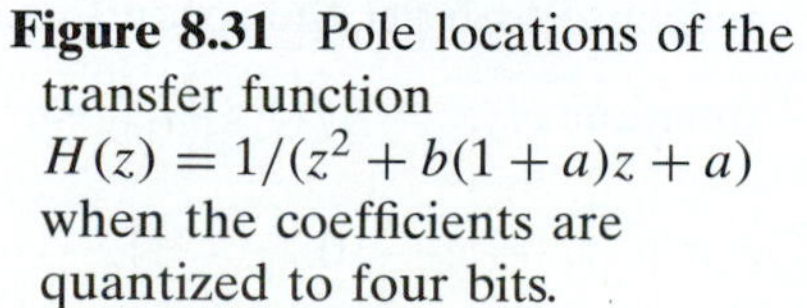

Figure 8.31 Pole locations of the transfer function $H(z) = 1/(z^2 + b(1 + a)z + a)$ when the coefficients are quantized to four bits.

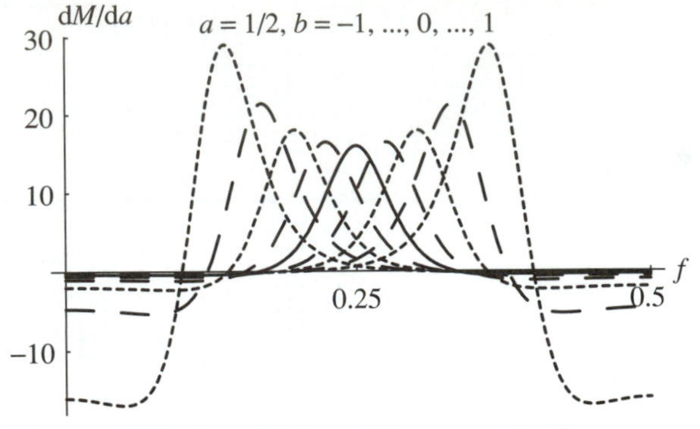

Figure 8.32 Sensitivity of the magnitude response to a for the second-order transfer function $H(z) = 1/(z^2 + bz + a)$; $a = \frac{1}{2}$, $b \in \{-1, -\frac{3}{4}, -\frac{1}{2}, 0, \frac{1}{2}, \frac{3}{4}, 1\}$.

We conclude that it is very important to keep the coefficient a as small as possible $(0 < a < 1)$ in order to obtain lower sensitivity.

Phase Sensitivity of Second-Order Transfer Functions. The first-order and the second-order allpass transfer functions can be the basic functions for realization of the most economical elliptic filters because the minimal number of multipliers is required. For that reason, we examine the phase response and the sensitivity of the phase response for several transfer functions.

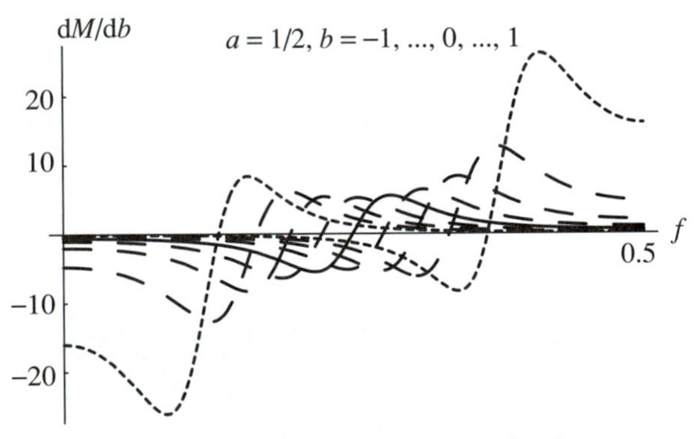

Figure 8.33 Sensitivity of the magnitude response to b for the second-order transfer function $H(z) = 1/(z^2 + bz + a)$; $a = \frac{1}{2}$, $b \in \{-1, -\frac{3}{4}, -\frac{1}{2}, 0, \frac{1}{2}, \frac{3}{4}, 1\}$.

A first-order transfer function, the corresponding phase response, and the phase-response sensitivity are summarized below:

$$H_1(z) = \frac{a_1 + z^{-1}}{1 + a_1 z^{-1}}$$

$$\phi = \tan^{-1} \frac{-\left(1 - a_1^2\right)\sin\omega}{2a_1 + (1 + a_1^2)\cos\omega}$$

$$\frac{\partial\phi}{\partial a_1} = \frac{2\sin\omega}{1 + a_1^2 + 2a_1\cos\omega}$$

Second-order transfer functions, the corresponding phase response, and the phase-response sensitivity are as follows:

$$H_{2a}(z) = \frac{a_i + b_i(1+a_i)z^{-1} + z^{-2}}{1 + b_i(1+a_i)z^{-1} + a_i z^{-2}}, \qquad b_i = b, \quad a_i = a$$

$$\phi = \tan^{-1} \frac{-2\left(b_i - b_i a_i^2 + \left(1 - a_i^2\right)\cos\omega\right)\sin\omega}{b_i^2(1+a_i)^2 + 2b_i(1+a_i)^2\cos\omega + (1+a_i)^2\cos^2\omega - (1-a_i)^2\sin^2\omega}$$

$$\frac{\partial\phi}{\partial b_i} = \frac{2\left(1 - a_i^2\right)\sin\omega}{(1-a_i)^2 + b_i^2(1+a_i)^2 + 2b_i(1+a_i)^2\cos\omega + 4a_i\cos^2\omega}$$

$$\frac{\partial\phi}{\partial a_i} = \frac{4(b_i + \cos\omega)\sin\omega}{(1-a_i)^2 + 4a_i\cos^2\omega + b_i(1+a_i)^2(b_i + 2\cos\omega)}$$

$$H_{2b}(z) = \frac{a_i + b_i z^{-1} + z^{-2}}{1 + b_i z^{-1} + a_i z^{-2}}, \qquad b_i = \frac{b}{1+a}, \quad a_i = a$$

$$\phi = \tan^{-1} \frac{2(1-a_i)(b_i + (1+a_i)\cos\omega)\sin\omega}{b_i^2 - (1-a_i)^2 + 2b_i(1+a_i)\cos\omega + 2\left(1+a_i^2\right)\cos\omega^2}$$

$$\frac{\partial\phi}{\partial b_i} = \frac{2(1-a_i)\sin\omega}{b_i^2 + (1-a_i)^2 + 2b_i(1+a_i)\cos\omega + 4a_i\cos\omega^2}$$

$$\frac{\partial\phi}{\partial a_i} = \frac{2(b_i + 2\cos\omega)\sin\omega}{b_i^2 + (1-a_i)^2 + 2b_i(1+a_i)\cos\omega + 4a_i\cos\omega^2}$$

$$H_{2c}(z) = \frac{b_i a_i + b_i z^{-1} + z^{-2}}{1 + b_i z^{-1} + b_i a_i z^{-2}}, \qquad b_i = \frac{b}{1+a}, \quad a_i = \frac{a(1+a)}{b}$$

$$\phi = \tan^{-1} \frac{2(1 - b_i a_i)(b_i + (1 + b_i a_i)\cos\omega)\sin\omega}{b_i^2 - (1 - b_i a_i)^2 + 2b_i(1 + b_i a_i)\cos\omega + 2\left(1 + b_i^2 a_i^2\right)\cos\omega^2}$$

$$\frac{\partial\phi}{\partial b_i} = \frac{2(1 + 2a_i\cos\omega)\sin\omega}{b_i^2 + (1 - b_i a_i)^2 + 2b_i(1 + b_i a_i)\cos\omega + 4b_i a_i\cos\omega^2}$$

$$\frac{\partial\phi}{\partial a_i} = \frac{2b_i(b_i + 2\cos\omega)\sin\omega}{b_i^2 + (1 - b_i a_i)^2 + 2b_i(1 + b_i a_i)\cos\omega + 4b_i a_i\cos\omega^2}$$

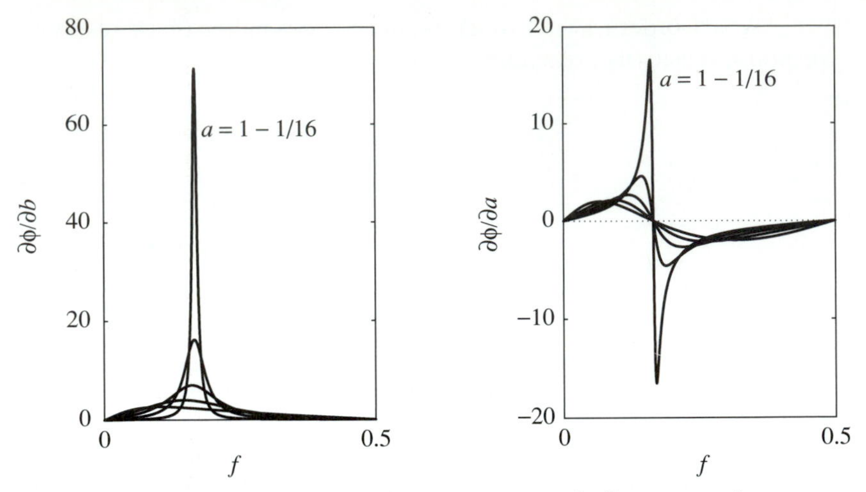

Figure 8.34 Sensitivity of the phase response of allpass transfer function $\frac{a_i+b_i(1+a_i)z^{-1}+z^{-2}}{1+b_i(1+a_i)z^{-1}+a_iz^{-2}}$; $b = -\frac{1}{2}$, $a \in \{0, \frac{1}{4}, \frac{1}{2}, \frac{3}{4}, 1 - \frac{1}{16}\}$.

The plots of the sensitivity functions are shown in Figs. 8.34–8.36.

8.6.4 Overflow Arithmetic Operations

In fixed-point arithmetic the result of an arithmetic operation (for example, adding) can exceed the allowed range of numeric values. For example, the result can exceed the magnitude of unity and thus it is not representable. Since the new result cannot be adequately represented, a new value from allowed range of numbers must be used for

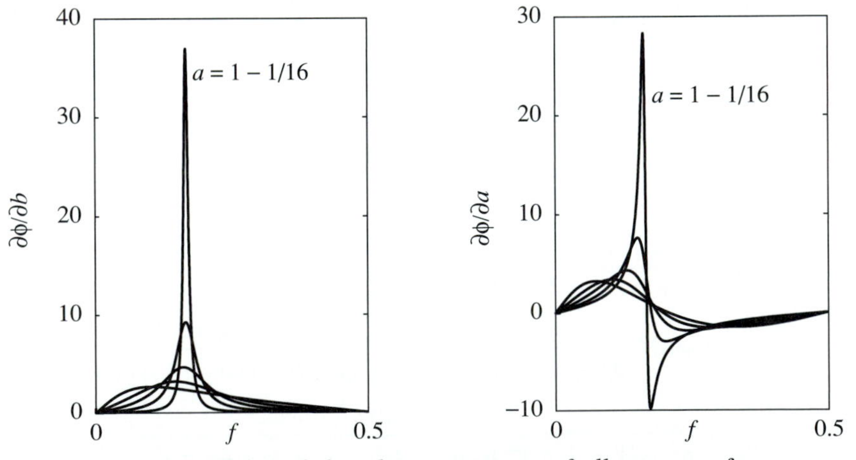

Figure 8.35 Sensitivity of the phase response of allpass transfer function $\frac{a_i+b_iz^{-1}+z^{-2}}{1+b_iz^{-1}+a_iz^{-2}}$; $b = -\frac{1}{2}$, $a \in \{0, \frac{1}{4}, \frac{1}{2}, \frac{3}{4}, 1 - \frac{1}{16}\}$.

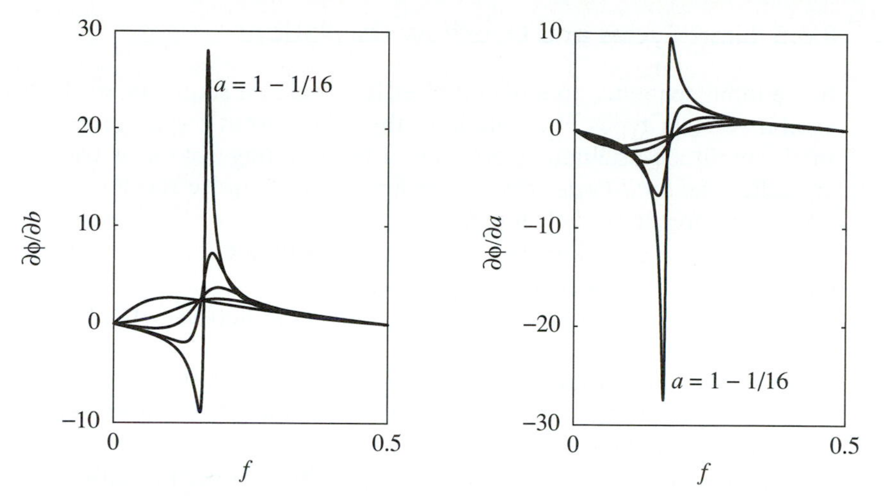

Figure 8.36 Sensitivity of the phase response of allpass transfer function $\frac{b_i a_i + b_i z^{-1} + z^{-2}}{1 + b_i z^{-1} + b_i a_i z^{-2}}$; $b = -\frac{1}{2}$, $a \in \{0, \frac{1}{4}, \frac{1}{2}, \frac{3}{4}, 1 - \frac{1}{16}\}$.

representing the result. For instance, if the set of numbers for representing the results is $\Sigma_i k/2^i$ $(i = 1, 2, \ldots)$, $k \in \{0, 1\}$, and $\left|\Sigma_i k/2^i\right| \leq 1$, then the result of $\frac{3}{4} + \frac{1}{2} = \frac{5}{4}$ exceeds the upper limit 1; the results can be represented as upper limit 1, or in two's complement type as $-\frac{5}{4} + 1 = -\frac{1}{4}$.

If the result of an arithmetic operation, x, has exceeded the permitted range, which we call the *overflow*, then this result should be replaced with another number, y, that belongs to the permitted range. In practice, two rules of replacements are used:

1. *Saturation arithmetic*

$$\text{If } 1 < x, \text{ then } y = 1$$
$$\text{If } x < -1 \text{ then } y = -1$$

2. *Two's complement overflow*

$$\text{If } 1 \leq x \text{ then } y = x - 2$$
$$\text{If } x < -1 \text{ then } y = x + 2$$

It should be noticed that if the summation of more than two numbers is performed and the final result is from the permitted range, but the intermediate results are not from the permitted range, then the two's complement overflow gives the exact result while saturation arithmetic does not. Therefore, in practice, only the final result is forced to saturate at $+1$ or -1, since intermediate overflows in two's complement arithmetic do not affect the accuracy of the final result.

8.6.5 Limit Cycles and Overflow Oscillations

If the input sequence to a digital filter is a constant sequence or if it varies slowly or exhibits certain types of oscillations, then the output sequence may oscillate. Those small-amplitude oscillations are caused by rounding errors, or truncating errors, and are called the *limit-cycle effects*. The limit cycles require recursion to exist and do not occur in nonrecursive FIR filters.

Overflows generated by additions can result in undesirable large-amplitude oscillations that are called the *overflow oscillations*.

Both limit-cycle oscillations and overflow oscillations settle the filter into the nonlinear mode of operation.

As a result of roundoff operations, sometimes the output sequence can become constant; thus, the output sequence does not follow the input sequence. This phenomenon is called the *dead band effect*.

In practice, we tend to reduce the probability of overflow rather than to completely avoid overflow.

8.6.6 Signal Quantization Error

In practical digital systems the input sequence can be of higher precision compared to the precision of the DSP system. For example, we can generate a high-precision sequence in MATLAB and want to process it in a DSP system which operates with B bits. First, we have to preprocess the high-precision sequence—that is, to quantize it to B bits.

The *quantization error* is the difference between the higher precision sequence value and the value obtained after quantization. If it is bounded as

$$-\frac{\Delta}{2} \leq \text{quantization error} \leq \frac{\Delta}{2}$$

then the variance of the error sequence can be obtained as

$$\sigma^2_{\text{error}} \approx \frac{\Delta^2}{12}$$

Notice that $\Delta = 2^{-B}$, where B is the number of fractional bits. If we have $B + 1$ bits (one bit for sign) representing a sinusoidal sequence of unit amplitude, then the signal-to-noise ratio (SNR) becomes

$$\text{SNR} = 10 \log_{10} \frac{\frac{1}{2}}{\frac{\Delta^2}{12}} \approx 6.02B + 7.8 \text{ dB}$$

If we find that the overflow may occur, we can scale the input sequence and use more bits for representation. Adding one bit to the representation increases the SNR by approximately 6 dB. In general, the input sequence should be scaled so that the probability of reaching overflow range becomes small enough.

In fixed-point representations the SNR of sinusoidal sequence decreases as the amplitude of the sinusoidal sequence decreases. This happens with scaled sinusoidal sequence (to avoid overflow it is necessary to restrict the amplitude of input sequence). Therefore, the scaling can reduce the overflow effects, but unfortunately it decreases the SNR.

Suppose that we have to scale a zero-mean random sequence with Gaussian distribution of values and the variance σ^2. Assume that the probability of overflow is 10^{-3}. It could be shown that $\sigma/3$ level of the sequence has to be scaled to correspond to the maximal level of one and then SNR becomes

$$\text{SNR} \approx 6.02B + 0.4 \text{ dB}$$

8.6.7 Roundoff Noise

When fixed-point arithmetic is used, and quantization to B bits is performed after each multiplication, the variance of output noise is

$$\sigma^2 = \frac{2^{-2B}}{12} \tag{8.75}$$

This type of quantization is called the *product quantization*.

In order to determine the roundoff noise at the output of a digital filter we assume that the noise due to a quantization is stationary and uncorrelated with the filter input, output, and internal variables and that it is zero-mean white noise. Also, we assume that the product of two noise samples is negligible.

The error due to product quantization produced by a multiplier a can be considered as a noise sequence superimposed on the signal at the output of the multiplier. The steady-state value of the variance of the noise sequence at the output of the filter is given by

$$\sigma_a^2 = \frac{2^{-2B}}{12} \frac{1}{2\pi j} \oint_{|z|=1} H_a(z) H_a(z^{-1}) z^{-1} \, dz$$

where $H_a(z)$ is the transfer function from the output of the multiplier a to the filter output that we call the *noise transfer function*.

If $H_{a1}(z)$ is the transfer function from the output of the multiplier a_1 to the filter output, $H_{a2}(z)$ is the transfer function from the output of the multiplier a_2 to the filter output, and $H_{ai}(z)$ is the transfer function from the output of the multiplier a_i to the filter output, then the steady-state output noise is given as

$$\sigma_a^2 = \sigma_{a1}^2 + \sigma_{a2}^2 + \cdots + \sigma_{ai}^2$$

where

$$\sigma_{ai}^2 = \frac{2^{-2B}}{12} \frac{1}{2\pi j} \oint_{|z|=1} H_{ai}(z) H_{ai}(z^{-1}) z^{-1} \, dz$$

Product Quantization in Cascade and Parallel Realizations. Let us consider the cascade connection of a filter containing the multiplier a whose noise transfer function is $H_a(z)$, followed by another filter with transfer function $G(z)$. The noise at the output of the second filter due to product quantization of multiplier a is given by

$$\sigma_{ai}^2 = \frac{2^{-2B}}{12} \frac{1}{2\pi j} \oint_{|z|=1} H_a(z)H_a(z^{-1})G(z)G(z^{-1})z^{-1}\,dz \tag{8.76}$$

If the second transfer function is allpass, then $G(z)G(z^{-1}) = 1$, which reduces (8.76) to

$$\sigma_a^2 = \frac{2^{-2B}}{12} \frac{1}{2\pi j} \oint_{|z|=1} H_a(z)H_a(z^{-1})z^{-1}\,dz \tag{8.77}$$

Consider now the parallel connection of (a) a filter containing the multiplier a whose noise transfer function is $H_a(z)$ and (b) a filter with transfer function $G(z)$. The noise at the output of summer due to product quantization of multiplier a is again σ_a^2.

This implies that the noise variance at the output of the filters that are followed by allpass filters is the same as that obtained at the output of the each lower-order filter section. Therefore, the total noise at the output of the cascade connection of an allpass filter is given by the sum of the variances of each section in the cascade.

Let us consider the variance of uncorrelated noise due to rounding the output of the multiplier for the coefficient a, whose noise transfer function is of the form

$$H_a(z) = \frac{c_2z^2 + c_1z + c_0}{z^2 + d_1z + d_0} = \frac{c_2z^2 + c_1z + c_0}{(z - z_1)(z - z_2)}, \qquad |z_1| = |z_2| = \sqrt{d_0} < 1$$

Then $H_a(z^{-1})$ is of the form

$$H_a(z^{-1}) = \frac{c_2 + c_1z + c_0z^2}{1 + d_1z + d_0z^2}$$

The noise variance can be calculated from

$$\sigma_a^2 = \frac{2^{-2B}}{12} \frac{1}{2\pi j} \oint_{|z|=1} H_a(z)H_a(z^{-1})z^{-1}\,dz \tag{8.78}$$

where

$$\frac{1}{2\pi j} \oint_{|z|=1} H_a(z)H_a(z^{-1})z^{-1}\,dz = R_0 + R_1 + R_2 \tag{8.79}$$

and

$$R_0 = c_2 \frac{c_0}{d_0}$$

$$R_1 = \frac{1}{z_1} \frac{c_2 z_1^2 + c_1 z_1 + c_0}{z_1 - z_2} \frac{c_2 + c_1 z_1 + c_0 z_1^2}{1 + d_1 z_1 + d_0 z_1^2}$$

$$R_2 = \frac{1}{z_2} \frac{c_2 z_2^2 + c_1 z_2 + c_0}{z_2 - z_1} \frac{c_2 + c_1 z_2 + c_0 z_2^2}{1 + d_1 z_2 + d_0 z_2^2}$$

$$\sigma_a^2 = \frac{2^{-2B}}{12} (R_0 + R_1 + R_2) \tag{8.80}$$

It is convenient to express the noise variance in terms of the pole magnitude, r, and the pole argument, Θ:

$$d_0 = r^2$$

$$d_1 = -2r \cos \Theta$$

If $c_0 = 0$ or $c_2 = 0$, the integral becomes

$$\oint_{|z|=1} H_a(z) H_a(z^{-1}) z^{-1} \mathrm{d}z = \oint_{|z|=1} \frac{c_2 z + c_1}{z^2 + d_1 z + d_0} \frac{c_2 + c_1 z}{1 + d_1 z + d_0 z^2}$$

$$\oint_{|z|=1} H_a(z) H_a(z^{-1}) z^{-1} \mathrm{d}z = \oint_{|z|=1} \frac{c_1 z + c_0}{z^2 + d_1 z + d_0} \frac{c_1 + c_0 z}{1 + d_1 z + d_0 z^2}$$

Some second-order sections that are implemented with a nonminimal number of delays have the noise transfer function with one or two poles at the origin

$$H_a(z) = \frac{N(z)}{z\left(z^2 + d_1 z + d_0\right)}$$

$$H_a(z) = \frac{N(z)}{z^2\left(z^2 + d_1 z + d_0\right)}$$

where $N(z)$ is the numerator of the noise transfer function. For that case, we derive a new procedure for calculating the variance.

The *Mathematica* code for computing the variance of uncorrelated noise due to rounding the output of the multiplier is shown below:

```
VQNR[H_,z_Symbol,a_Symbol,b_Symbol
    ,r_Symbol,theta_Symbol] := Module[
   {ax, bx, d0, d1, d2, denH2, H0, H0inv, H2, numH2
   ,res0, res1, res2, rtheta2ab, sumres, var=Infinity
   ,z1, z2},
   H0 = Together[H] /. {a -> ax, b -> bx};
   H0inv = Together[H0 /. z->1/z];
   H2 = Together[Cancel[z^2*H0]];
   numH2 = Collect[Numerator[H2],z];
   denH2 = Collect[Denominator[H2],z];
   If[Exponent[denH2,z] == 2
   ,{d0,d1,d2} = CoefficientList[denH2,z];
    {z1,z2} = Flatten[Solve[denH2==0,z]];
    res0 = D[H2*H0inv,{z,2}]/2 /. z->0;
    res1 = numH2*H0inv/(d2*(z-(z/.z2))*z^3) /. z1;
    res2 = numH2*H0inv/(d2*(z-(z/.z1))*z^3) /. z2;
    sumres = Simplify[res0 + res1 + res2];
    rtheta2ab = Solve[{d0 == d2*(r^2),
                       d1 == d2*(-2*r*Cos[theta])}
                  ,{ax,bx}] //Flatten;
    var = Together[sumres /. rtheta2ab];
   ,Print[''Error in denominator!   '', denH2];
   ];
  var]

VQNR[H_,z_Symbol,a_Symbol] := Module[
   {d0, d1, denH2, H0, H0inv, H2, numH2
   ,res0, res1, res2, rtheta2ab, sumres, var=Infinity, z1},
   H0 = Together[H];
   H0inv = Together[H0 /. z->1/z];
   H2 = Together[Cancel[z^2*H0]];
   numH2 = Collect[Numerator[H2],z];
   denH2 = Collect[Denominator[H2],z];
   If[Exponent[denH2,z] == 1
   ,{d0,d1} = CoefficientList[denH2,z];
    {z1} = Flatten[Solve[denH2==0,z]];
    res0 = D[H2*H0inv,{z,2}]/2 /. z->0;
    res1 = numH2*H0inv/(d1*z^3) /. z1;
    var = Together[Simplify[res0 + res1 ]];
   ,Print[''Error in denominator!   '', denH2];
   ];
  var]
```

In the following table we compute noise variances, σ_a^2, for different noise transfer functions, $H_a(z)$:

	$H_a(z)$	σ_a^2
1.	$\dfrac{(-1+c)\,(-1+z^2)}{c+(1+c)\,dz+z^2}$	$2\,(1-r^2)$
2.	$\dfrac{1+2dz+z^2}{c+(1+c)\,dz+z^2}$	$\dfrac{2}{1+r^2}$
3.	$\dfrac{1-z^2}{c+(1+c)\,dz+z^2}$	$\dfrac{2}{1-r^2}$
4.	$\dfrac{(1-c)\,(-1+z)}{c+(1+c)\,dz+z^2}$	$\dfrac{2\,(1-r^2)}{1+r^2+2r\cos\Theta}$
5.	$\dfrac{(1-c)\,(1+z)}{c+(1+c)\,dz+z^2}$	$\dfrac{2\,(1-r^2)}{1+r^2-2r\cos\Theta}$

where c and d are the multiplier coefficients.

In the first case, when the pole approaches the unit circle, the noise variance tends to zero and it does not depend on the angle Θ. In the third case, the noise variance can take extremely large values when the pole approaches the unit circle. In the fourth case, the noise variance is, again, small for $\Theta = 0$, but it can take very large values for $\Theta = \pi$. The fifth case is quite different: The noise variance is small for $\Theta = \pi$, but it can take very large values for $\Theta = 0$.

8.7 DIGITAL FILTER REALIZATIONS

After having accomplished the approximation step, the filter transfer function is known, and the designer must choose a realization of the digital filter: A block diagram or a set of equations that indicates the sequence of computations to be performed on the input sequence of numbers. Digital filters can be classified on the basis of their structure as follows:

- Direct-form I filters
- Direct-form II (or Direct canonic) filters
- Cascade filters
- Parallel filters
- Wave digital filters (WDF)
- Parallel-of-two-allpass filters
- Ladder filters
- Filters based on structural passivity

Direct-form I digital filters are realized directly from the difference equation. The filter coefficients can be identified directly from the difference equation describing the digital filter. Those realizations can be implemented in a straightforward and efficient way on programmable processors.

Direct-form II digital filters, also referred to as *direct canonic* filters, have the minimal number of delays. By definition, a *canonic realization* is a realization with the minimal number of delays.

Cascade filters are realized as a cascade of first-order and second-order sections. Each section can be realized as direct-form I, direct-form II, or any other type.

Parallel filters are realized as a parallel connection of first-order and second-order sections, that is, the outputs of the lower-order sections are connected to an adder. Each section can be realized as direct-form I, direct-form II, or any other type.

Wave digital filters are derived from microwave filters. Analysis and design of these filters rely on microwave theory and techniques.

Parallel-of-two-allpass filters are realized as a parallel connection of two allpass filters, each filter being realized as a cascade of allpass lower-order sections (first-order and second-order sections). Although they are not derived from the microwave filters, they have identical properties as wave digital filters.

Ladder filters may be preferable to direct, cascade, or parallel realizations in filter applications where coefficient sensitivity is of primary concern. However, its implementation nearly doubles the computational cost.

Filters based on structural passivity are based on the zero sensitivity of magnitude response to multiplier coefficients at the frequencies of maximal magnitude response. The key property of these filters is that the magnitude response cannot be larger than 1, and small changes of the filter coefficient values only change the frequency at which the maximum occurs. Those filters can be designed in the z-domain with no need for any analog prototype. In fact, the filters based on the structural passivity, wave digital filters, and parallel-of-two-allpass filters have the zero sensitivity at the frequencies at which magnitude response maxima occur. This class of filters also comprises FIR filters with low passband sensitivity [70].

In this book we focus on realizations of direct-form II filters, cascade filters, parallel filters, wave digital filters, and parallel-of-two-allpass filters.

8.7.1 Direct-Form I Realization

Consider a second-order transfer function of the form

$$H(z) = \frac{b_0 + b_1 z^{-1} + b_2 z^{-2}}{1 - a_1 z^{-1} - a_2 z^{-2}} \tag{8.81}$$

or, equivalently, the corresponding difference equation

$$y(n) = a_1 y(n-1) + a_2 y(n-2) + b_0 x(n) + b_1 x(n-1) + b_2 x(n-2) \tag{8.82}$$

For the given transfer function of the form (8.81) we can directly derive the difference equation of the form (8.82), and vice versa. Next, we can construct an algorithm (sequence of arithmetic operations) which directly follows the evaluations in (8.82).

The algorithm represents a digital filter realization (for software implementation) that is called the *direct-form I* realization.

The direct-form I realization is a noncanonic realization because it requires a number of delays that are greater than the transfer-function order; for example, the second-order filter requires four delays: $y(n-1)$, $y(n-2)$, $x(n-1)$, $x(n-2)$.

8.7.2 Direct-Form II Realization

A second-order transfer function $H(z)$ can be expressed as a product of two transfer functions, $H_1(z)$ and $H_2(z)$, as follows:

$$\begin{aligned} H(z) &= H_1(z)\ H_2(z) \\ &= \frac{1}{1 - a_1 z^{-1} - a_2 z^{-2}} \ \frac{b_0 + b_1 z^{-1} + b_2 z^{-2}}{1} \\ H_1(z) &= \frac{1}{1 - a_1 z^{-1} - a_2 z^{-2}} \\ H_2(z) &= b_0 + b_1 z^{-1} + b_2 z^{-2} \end{aligned} \tag{8.83}$$

where $H_1(z)$ is the reciprocal of the denominator of $H(z)$, while $H_2(z)$ is the numerator of $H(z)$.

Consider a realization which is a cascade of two sections with transfer functions $H_1(z)$ and $H_2(z)$. We denote the output of the first section by $d(n)$. The output of the first section is an input to the second section. We can directly write the corresponding difference equations by inspection of the transfer functions $H_1(z)$ and $H_2(z)$

$$\begin{aligned} H_1(z) \rightarrow d(n) &= x(n) + a_1 d(n-1) + a_2 d(n-2) \\ H_2(z) \rightarrow y(n) &= b_0 d(n) + b_1 d(n-1) + b_2 d(n-2) \end{aligned} \tag{8.84}$$

We can construct an algorithm (sequence of arithmetic operations) which directly follows the evaluations in (8.84). The algorithm represents a digital filter realization (for software implementation) that is called the *direct-form II* realization.

Alternatively, a digital filter realization can be represented by a block diagram, as shown in Figs. 8.37 and 8.38.

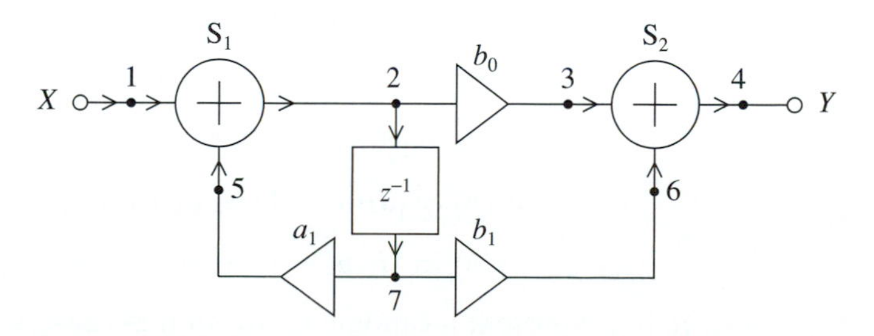

Figure 8.37 Direct-form II first-order realization.

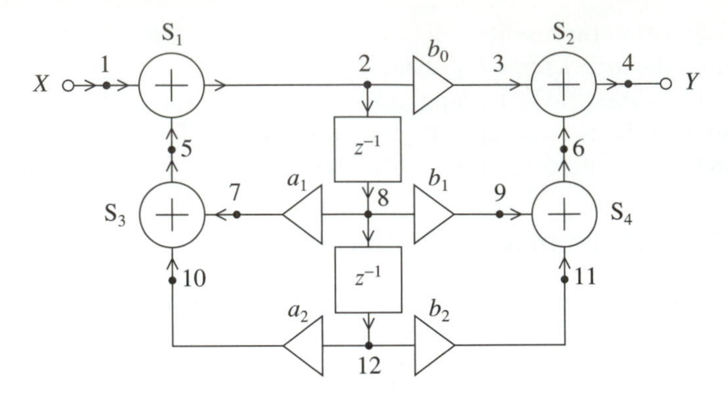

Figure 8.38 Direct-form II biquad.

In both difference equations the same intermediate samples are used, $d(n-1)$ and $d(n-2)$. Therefore, the direct-form II realization is a canonic realization because the second-order filter requires two delays: $d(n-1)$ and $d(n-2)$. Notice that $d(n-2)$ does not require two delays but only one: $d(n-2) = d_1(n-1)$, where $d_1(n-1) = d(n-1)$.

8.7.3 Transposed Direct-Form II Realization

The difference equation (8.82) (realized as direct-form I) was rewritten as two difference equations (8.84) (realized as direct-form II). There are a variety of rearrangements of a difference equation. We can group the samples that have the same delay

$$y(n) = a_2 y(n-2) + b_2 x(n-2) + a_1 y(n-1) + b_1 x(n-1) + b_0 x(n) \quad (8.85)$$

and introduce the intermediate samples $d_1(n)$ and $d_2(n)$ as

$$d_1(n) = a_1 y(n) + b_1 x(n)$$

$$d_2(n) = a_2 y(n) + b_2 x(n)$$

Equation (8.85) becomes

$$y(n) = d_2(n-2) + d_1(n-1) + b_0 x(n) \quad (8.86)$$

Another intermediate sample can be defined as

$$d_3(n) = d_2(n-1) + d_1(n)$$

so the difference equation (8.85) transforms into

$$\begin{aligned} d_2(n) &= a_2 y(n) + b_2 x(n) \\ d_1(n) &= a_1 y(n) + b_1 x(n) \\ d_3(n) &= d_2(n-1) + d_1(n) \\ y(n) &= d_3(n-1) + b_0 x(n) \end{aligned} \quad (8.87)$$

Equations (8.87) represent a canonic realization because the second-order filter requires two delays: $d_2(n-1)$ and $d_3(n-1)$.

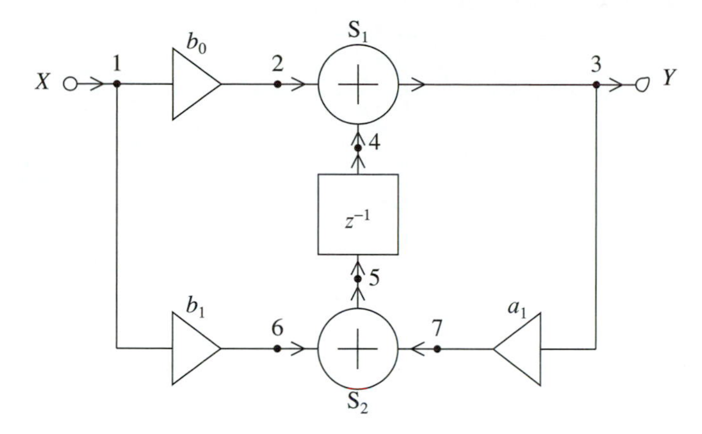

Figure 8.39 Transposed direct-form II first-order realization.

The same realization can be obtained from the direct-form II realization by using the *flow-graph reversal rules*:

- Reverse the input and output of delay.
- Reverse the input and output of multiplier.
- The adder becomes a node.
- The node, from which two or more branches go out, is replaced by adder.
- Swap input and output.

This realization is called a *transposed realization* because it is based on the *flow-graph reversal* or *transposition*.

The block diagram of the transposed direct-form II first-order realization is shown in Fig. 8.39. The block diagram of the transposed direct-form II biquad is shown in Fig. 8.40.

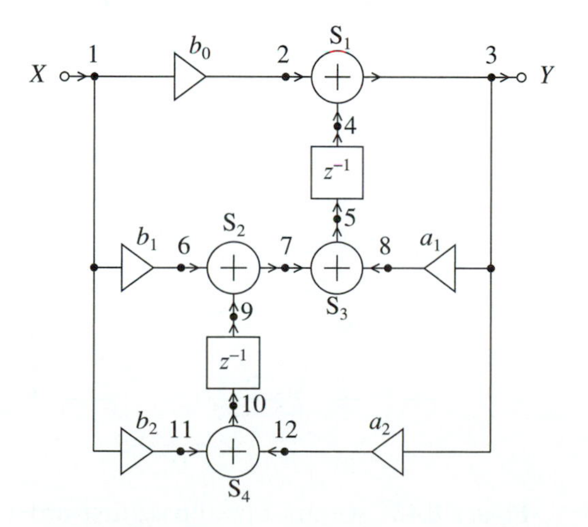

Figure 8.40 Transposed direct-form II biquad.

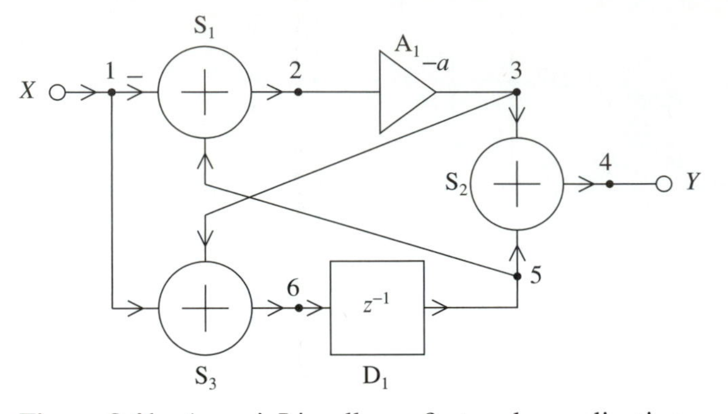

Figure 8.41 Ansari–Liu allpass first-order realization type A.

8.7.4 Ansari–Liu Allpass Realizations

Ansari-Liu allpass realizations have very small noise variance due to the product quantization. Also, the realization parameters can be directly controlled by design parameters of elliptic-type transfer function, as will be shown in subsequent sections.

The Ansari–Liu allpass realizations realize the transfer functions of the form

$$H_1(z) = \frac{a + z^{-1}}{1 + az^{-1}} \tag{8.88}$$

$$H_2(z) = \frac{a + b(1 + a)z^{-1} + z^{-2}}{1 + b(1 + a)z^{-1} + az^{-2}} \tag{8.89}$$

The block diagram of the Ansari–Liu allpass first-order realizations is shown in Figs. 8.41–8.43. The block diagram of the Ansari–Liu allpass biquads is shown in Figs. 8.44–8.46.

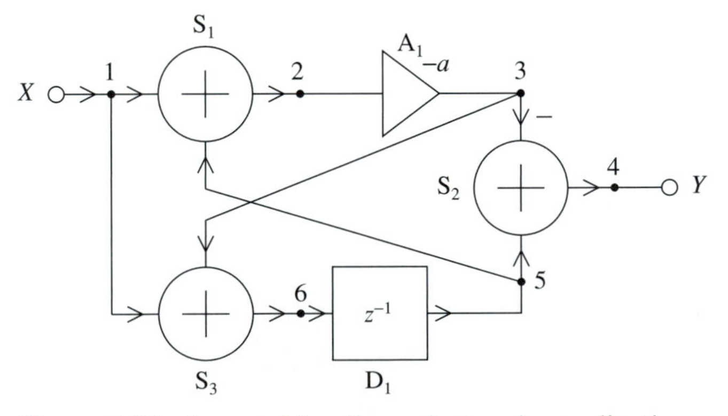

Figure 8.42 Ansari–Liu allpass first-order realization type B.

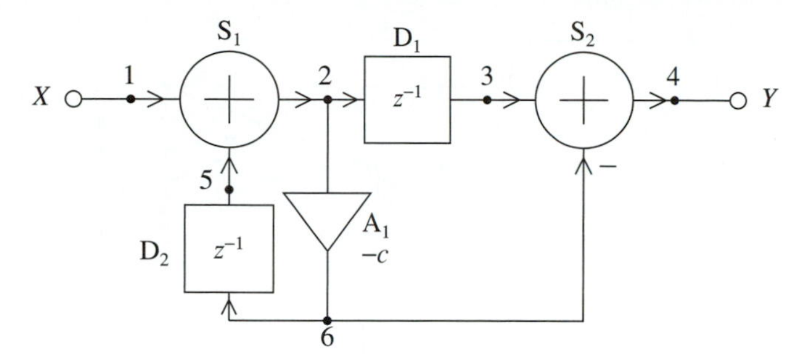

Figure 8.43 Ansari–Liu allpass first-order realization type C.

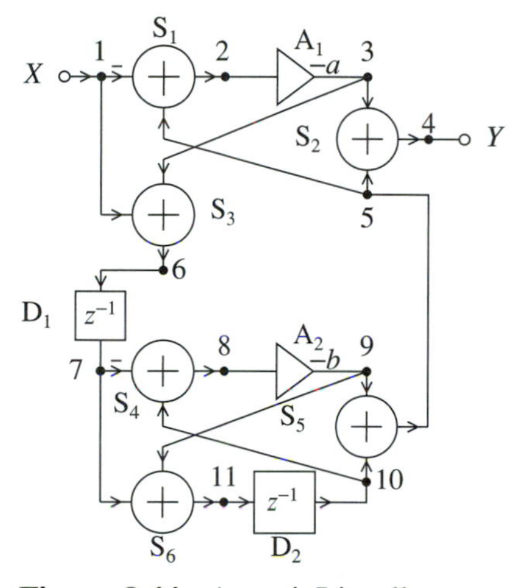

Figure 8.44 Ansari–Liu allpass biquad type A.

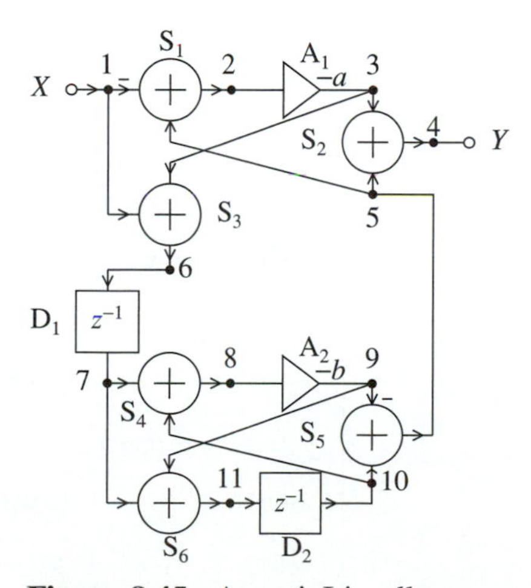

Figure 8.45 Ansari–Liu allpass biquad type B.

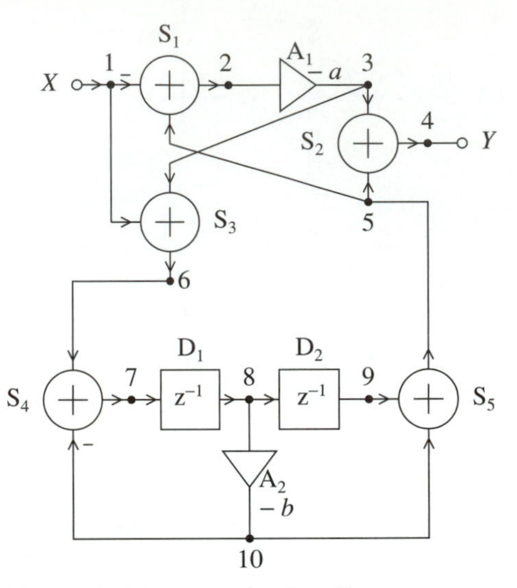

Figure 8.46 Ansari–Liu allpass biquad type C.

The set of difference equations describing each realization is shown below:

Figure 8.41

$$
\begin{aligned}
y_1(n) &= x(n) \\
y_5(n) &= y_6(n-1) \\
y_2(n) &= -y_1(n) + y_5(n) \\
y_3(n) &= -a y_2(n) \\
y_4(n) &= y_3(n) + y_5(n) \\
y_6(n) &= y_1(n) + y_3(n) \\
y(n) &= y_4(n)
\end{aligned}
$$

Figure 8.42

$$
\begin{aligned}
y_1(n) &= x(n) \\
y_5(n) &= y_6(n-1) \\
y_2(n) &= y_1(n) + y_5(n) \\
y_3(n) &= -a y_2(n) \\
y_4(n) &= -y_3(n) + y_5(n) \\
y_6(n) &= y_1(n) + y_3(n) \\
y(n) &= y_4(n)
\end{aligned}
$$

Figure 8.43

$$
\begin{aligned}
y_1(n) &= x(n) \\
y_3(n) &= y_2(n-1) \\
y_5(n) &= y_6(n-1) \\
y_2(n) &= y_1(n) + y_5(n) \\
y_6(n) &= -ay_2(n) \\
y_4(n) &= y_3(n) - y_6(n) \\
y(n) &= y_4(n)
\end{aligned}
$$

Figure 8.44

$$
\begin{aligned}
y_1(n) &= x(n) \\
y_7(n) &= y_6(n-1) \\
y_{10}(n) &= y_{11}(n-1) \\
y_8(n) &= -y_7(n) + y_{10}(n) \\
y_9(n) &= -by_8(n) \\
y_5(n) &= y_9(n) + y_{10}(n) \\
y_2(n) &= -y_1(n) + y_5(n) \\
y_3(n) &= -ay_2(n) \\
y_4(n) &= y_3(n) + y_5(n) \\
y_6(n) &= y_1(n) + y_3(n) \\
y_{11}(n) &= y_7(n) + y_9(n) \\
y(n) &= y_4(n)
\end{aligned}
$$

Figure 8.45

$$
\begin{aligned}
y_1(n) &= x(n) \\
y_7(n) &= y_6(n-1) \\
y_{10}(n) &= y_{11}(n-1) \\
y_8(n) &= y_7(n) + y_{10}(n) \\
y_9(n) &= -by_8(n) \\
y_5(n) &= -y_9(n) + y_{10}(n) \\
y_2(n) &= -y_1(n) + y_5(n) \\
y_3(n) &= -ay_2(n) \\
y_4(n) &= y_3(n) + y_5(n) \\
y_6(n) &= y_1(n) + y_3(n) \\
y_{11}(n) &= y_7(n) + y_9(n) \\
y(n) &= y_4(n)
\end{aligned}
$$

Figure 8.46

$$
\begin{aligned}
y_1(n) &= x(n)\\
y_8(n) &= y_7(n-1)\\
y_9(n) &= y_8(n-1)\\
y_{10}(n) &= by_8(n)\\
y_5(n) &= y_9(n) + y_{10}(n)\\
y_2(n) &= -y_1(n) + y_5(n)\\
y_3(n) &= -ay_2(n)\\
y_4(n) &= -y_3(n) + y_5(n)\\
y_6(n) &= y_1(n) + y_3(n)\\
y_7(n) &= y_6(n) - y_{10}(n)\\
y(n) &= y_4(n)
\end{aligned}
$$

Notice that the first-order realization type C is not a canonic realization since it requires two delays.

8.7.5 Wave Digital Filter Realizations

Wave digital filters (WDFs) are derived from passive lossless microwave filters and, if properly designed, retain completely inherent advantages of the passive microwave filters, such as low sensitivities. The key advantages of WDFs are excellent stability properties even under nonlinear operating conditions resulting from overflow and roundoff effects, low coefficients wordlength requirements, and inherently good dynamic range. For a proper design of WDFs a solid mathematical basis developed for classic synthesis techniques, including those for microwave filters, is at one's disposal.

Many different WDF realizations exist according to the microwave filters from which the WDFs are derived. Design of lattice WDFs where both lattice branches are realized by cascaded first-order and second-order allpass sections has been presented in reference [71]. The number of multipliers in a lattice WDF is equal to the filter order, which is the minimum in comparison with the number of multipliers required for other possible realizations. The principles of WDF design have been described in references [71]–[73].

WDFs are derived from lossless microwave filters using the voltage wave quantities and the scattering-parameter characterization of two-port networks [74]. The lattice WDF is derived from a real symmetric doubly terminated two-port network [72], [75].

Figure 8.47 shows the block diagram of a lattice WDF. The blocks labeled by S_1 and S_2 are reflectances of reactances [71] and are characterized by allpass transfer functions.

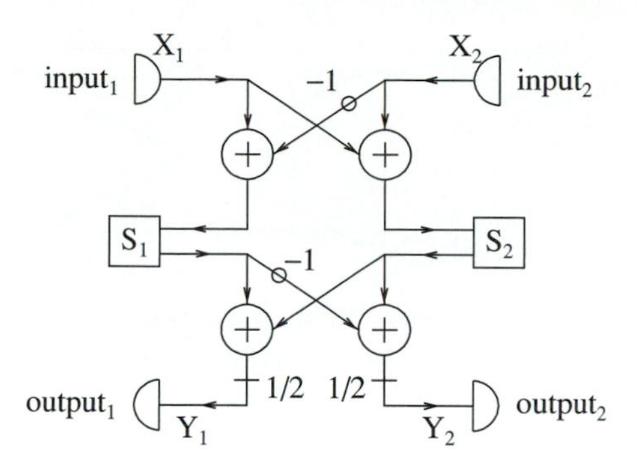

Figure 8.47 Block diagram of a lattice wave digital filter.

From the viewpoint of a digital filter designer, WDF realizations are based on two-input two-output blocks called *two-port adaptors*. A two-port adaptor is realized with adders and one multiplier; no delays appear in the adaptor. In this book we consider five types of adaptors, shown in Figs. 8.48–8.52, according to Gazsi [71]; the symbol α is the multiplier coefficient, $|\alpha| \leq 1/2$.

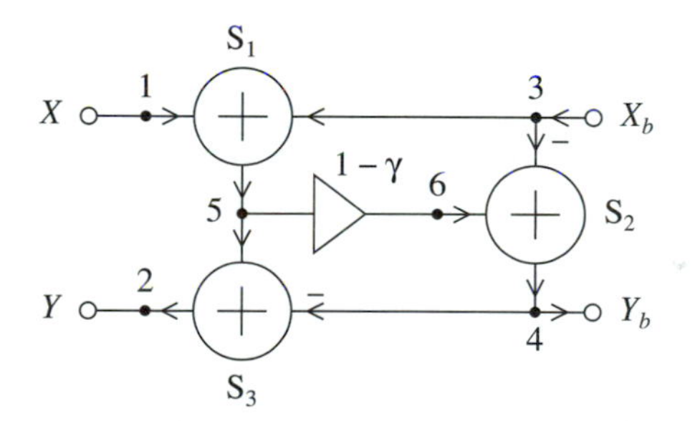

Figure 8.48 Two-port adaptor with multiplier coefficient $\alpha = 1 - \gamma$.

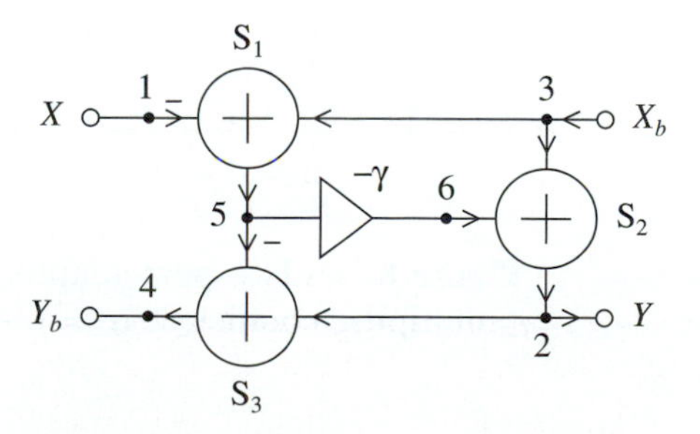

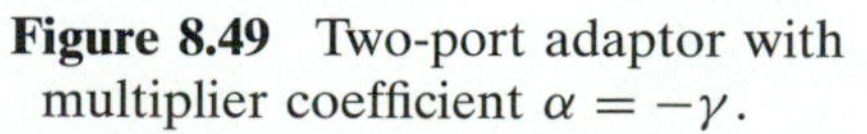
Figure 8.49 Two-port adaptor with multiplier coefficient $\alpha = -\gamma$.

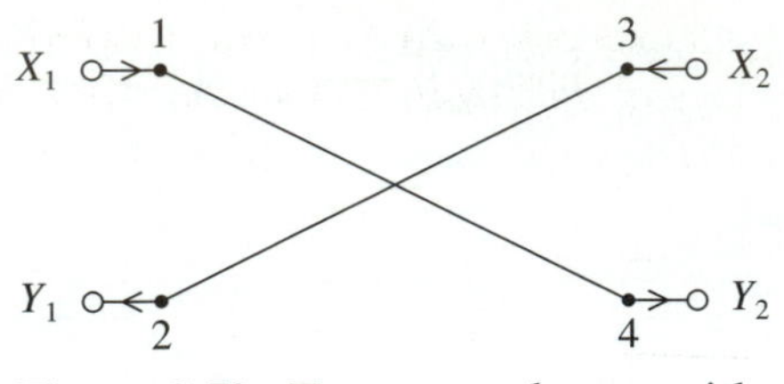

Figure 8.50 Two-port adaptor with multiplier coefficient $\alpha = 0$.

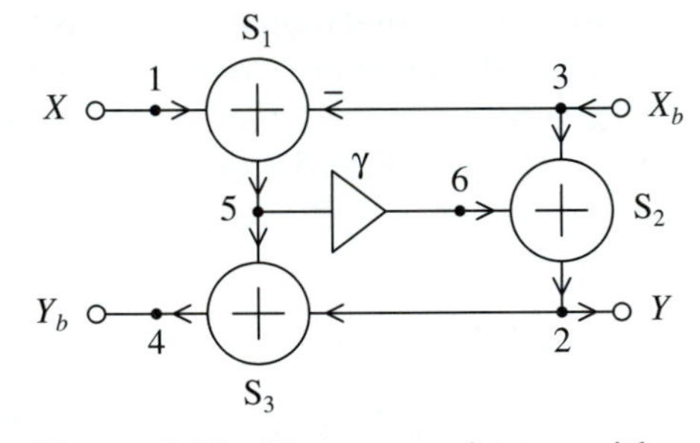

Figure 8.51 Two-port adaptor with multiplier coefficient $\alpha = \gamma$.

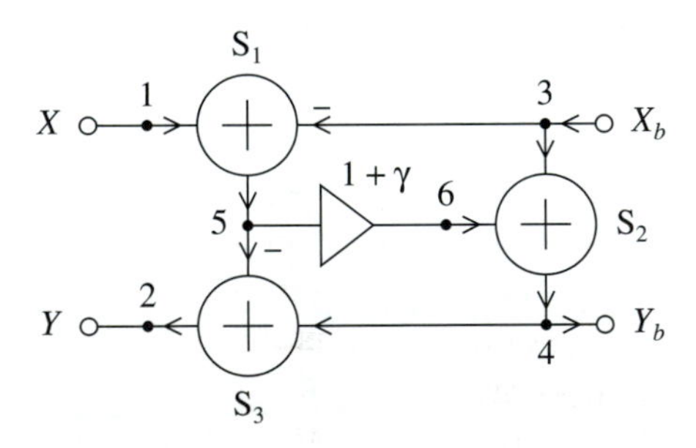

Figure 8.52 Two-port adaptor with multiplier coefficient $\alpha = 1 + \gamma$.

The five two-port adaptors can be described by difference equations as follows:

Figure 8.48

$$
\begin{aligned}
y_1(n) &= x(n)\\
y_2(n) &= -y_4(n) + y_5(n)\\
y_3(n) &= x_b(n)\\
y_4(n) &= -y_3(n) + y_6(n)\\
y_5(n) &= y_1(n) + y_3(n)\\
y_6(n) &= \alpha y_5(n)\\
y_b(n) &= y_4(n)\\
y(n) &= y_2(n)\\
\alpha &= 1 - \gamma
\end{aligned}
$$

Figure 8.49

$$
\begin{aligned}
y_1(n) &= x(n)\\
y_2(n) &= y_3(n) + y_6(n)\\
y_3(n) &= x_b(n)\\
y_4(n) &= y_2(n) - y_5(n)\\
y_5(n) &= -y_1(n) + y_3(n)\\
y_6(n) &= \alpha y_5(n)\\
y_b(n) &= y_4(n)\\
y(n) &= y_2(n)\\
\alpha &= -\gamma
\end{aligned}
$$

Figure 8.50

$$
\begin{aligned}
y_1(n) &= x(n)\\
y_2(n) &= y_3(n)\\
y_3(n) &= x_b(n)\\
y_4(n) &= y_1(n)\\
y_b(n) &= y_4(n)\\
y(n) &= y_2(n)\\
\alpha &= 0
\end{aligned}
$$

Figure 8.51

$$y_1(n) = x(n)$$
$$y_2(n) = y_3(n) + y_6(n)$$
$$y_3(n) = x_b(n)$$
$$y_4(n) = y_2(n) + y_5(n)$$
$$y_5(n) = y_1(n) - y_3(n)$$
$$y_6(n) = \alpha y_5(n)$$
$$y_b(n) = y_4(n)$$
$$y(n) = y_2(n)$$
$$\alpha = \gamma$$

Figure 8.52

$$y_1(n) = x(n)$$
$$y_2(n) = y_4(n) - y_5(n)$$
$$y_3(n) = x_b(n)$$
$$y_4(n) = y_3(n) + y_6(n)$$
$$y_5(n) = y_1(n) - y_3(n)$$
$$y_6(n) = \alpha y_5(n)$$
$$y_b(n) = y_4(n)$$
$$y(n) = y_2(n)$$
$$\alpha = 1 + \gamma$$

Notice that the adaptor in Fig. 8.50 is the simplest one and that it contains no adder or multiplier.

Any of the five two-port adaptors can be represented by a graphical symbol, as shown in Fig. 8.53.

We use two-port adaptors and delays to construct realizations of first-order and second-order allpass transfer functions. In this way, we generate first-order and second-order allpass WDF sections.

A first-order allpass WDF section that consists of one two-port adaptor and one delay is shown in Fig. 8.54. It is described by (1) the difference equations that characterize the two-port adaptor and (2) the equation for the delay and interconnections:

$$x_b(n) = y_b(n - 1)$$

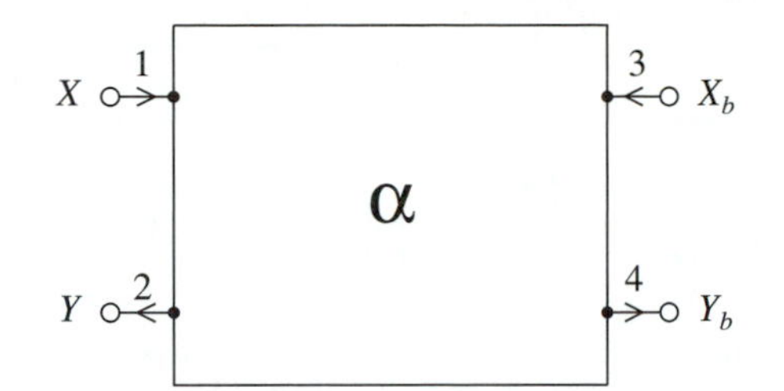

Figure 8.53 Two-port adaptor graphical symbol with multiplier coefficient α.

The corresponding transfer function of the first-order allpass WDF section is

$$H_{WDF1}(z) = \frac{z^{-1} + \gamma}{1 + \gamma z^{-1}}$$

A second-order allpass WDF section consists of two two-port adaptors and two delays (see Fig. 8.55). It is described by (1) the difference equations that characterize the two two-port adaptors A and B and (2) the equations for the two delays and interconnections:

$$x_A(n) = x(n)$$

$$y(n) = y_A(n)$$

$$x_{Ab}(n) = y_B(n-1)$$

$$x_B(n) = y_{Ab}(n)$$

$$x_{Bb}(n) = y_{Bb}(n-1)$$

The multiplier coefficient of the adaptor A is α, and the multiplier coefficient of the adaptor B is β.

The multiplier coefficient α is of the form γ or $1-\gamma$, where $0 < \gamma < 1$. The multiplier coefficient β is of the form $\vartheta, 1-\vartheta, 0, -\vartheta, 1+\vartheta$, where $-1 < \vartheta < 1$. The corresponding transfer function of the second-order allpass WDF section is

$$H_{WDF2}(z) = \frac{z^{-2} + \vartheta(1+\gamma)z^{-1} + \gamma}{1 + \vartheta(1+\gamma)z^{-1} + \gamma z^{-2}}$$

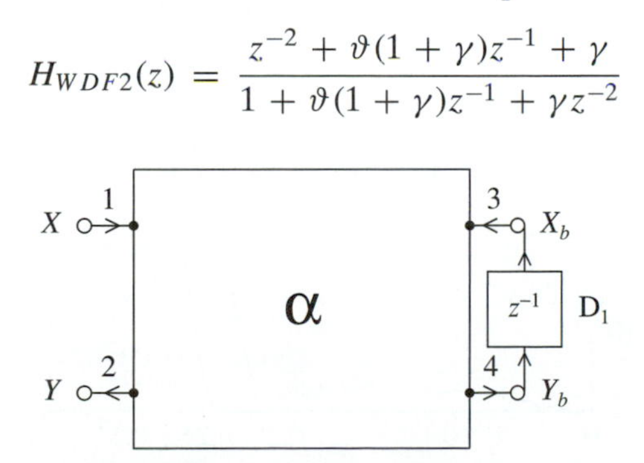

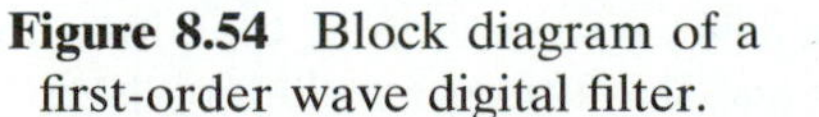

Figure 8.54 Block diagram of a first-order wave digital filter.

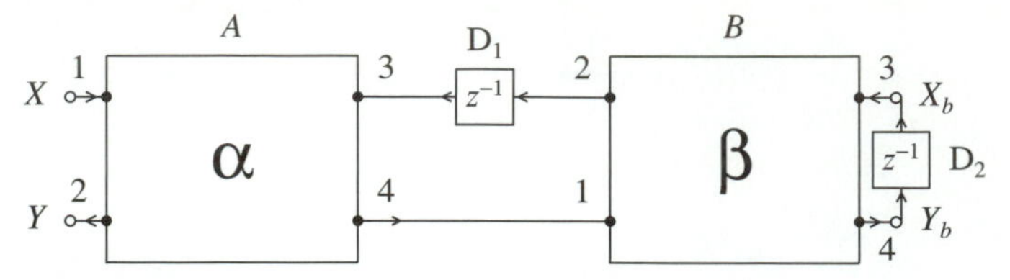

Figure 8.55 Block diagram of a second-order wave digital filter.

It should be noted that the Ansari–Liu allpass realizations have the transfer functions of the form identical to $H_{WDF1}(z)$ and $H_{WDF2}(z)$. Therefore, we can use the same procedure for calculating multiplier coefficients for both the WDF realizations and the Ansari–Liu realizations.

We can always choose an appropriate two-port adaptor in a WDF realization, such that $0 \leq \alpha < \frac{1}{2}$ and $0 < \beta < \frac{1}{2}$, which is very important in implementation.

8.8 COMPARISON OF REALIZATIONS

Consider a digital filter specification: $F_p = 0.22$, $F_a = 0.28$, $A_p = 2.5$ dB, $A_a = 57$ dB. The specification can be fulfilled by a fifth-order IIR elliptic filter and by a ninth-order half-band IIR elliptic filter, as shown in Fig. 8.56.

The fifth-order filter achieves 57 dB minimum attenuation in the stopband, provided that the maximal attenuation in the passband is $A_p = 2.5$ dB. The ninth-order half-band filter achieves the same attenuation in the stopband, but it has extremely

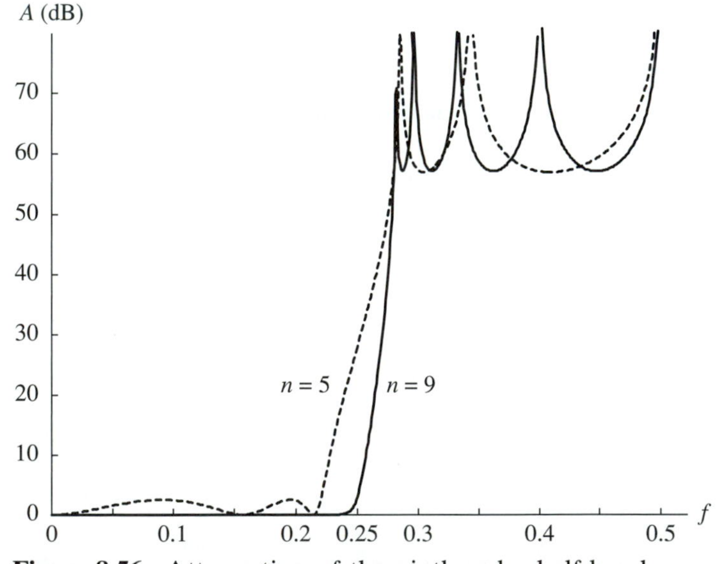

Figure 8.56 Attenuation of the ninth-order half-band filter and the fifth-order elliptic filter.

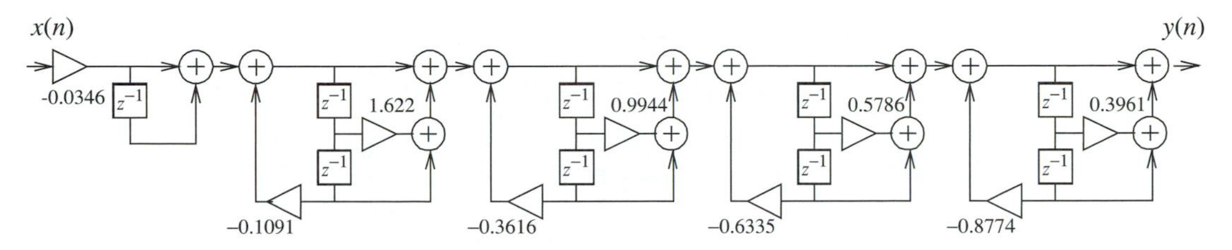

Figure 8.57 Cascade realization of the ninth-order elliptic half-band IIR filter.

small attenuation in the passband, $A_p = 0.00001$ dB. Also, the ninth-order half-band filter has smaller pole magnitudes (0.9367) compared to the pole magnitudes of the fifth-order filter (0.9563).

Half-band IIR filters are special types of lowpass or highpass filters with the 3-dB frequency at $\frac{1}{4}$. The transfer-function poles of these filters occur on the imaginary axis in the z-plane. A pair of purely imaginary complex-conjugate poles can be uniquely represented by only one parameter, which leads to simpler realizations.

We examine, in more detail, various realizations of the ninth-order elliptic half-band IIR filter.

Cascade realization. Since the transfer function poles lie on the imaginary axis, and the zeros are located on the unit circle in the z-plane, the second-order filter sections can be realized with only two multipliers as shown in Fig. 8.57. Namely, since the poles are purely imaginary, some multipliers in the feedback branch do not exist. The transfer function zeros are on the unit circle, which enables the realization of a feedforward branch with only one multiplier with a nonunity coefficient. The cascade realization with the reduced number of multipliers can realize a transfer functions of any order—odd or even.

Parallel realization. Consider an odd-order elliptic half-band IIR filter and assume that we know its transfer function. It can be shown that the residues of the transfer-function are either purely real or purely imaginary. In that case, the feedforward branches of the second-order sections of parallel realization (formed by decomposing the transfer function into partial fractions) have only one multiplier with a nonzero coefficient. Also, some multipliers in the feedback branches do not exist. So, in a parallel realization, biquads with only two multipliers appear, as shown in Fig. 8.58.

Parallel connection of two allpass filters. This realization exists only for the odd-order transfer functions. We form a parallel connection of two allpass filters. Each of the allpass filters is realized as a cascade connection of first-order and second-order allpass sections, as shown in Fig. 8.59. The parallel-of-two-allpass realization is the most economical realization for half-band filters, because the second-order filter sections are realized with only one multiplier, and the first-order section has no multipliers at all. It must be kept in mind that this realization has larger magnitude response sensitivity in the stopband when compared to the cascade or parallel realization.

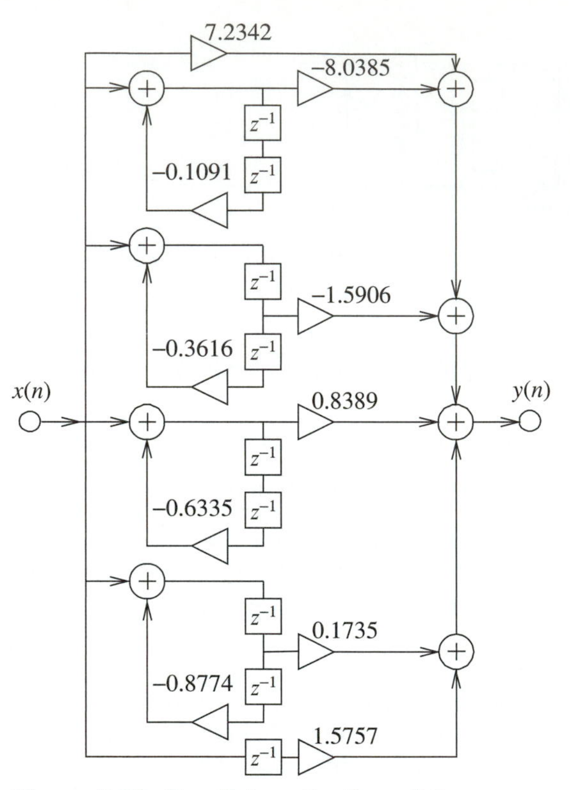

Figure 8.58 Parallel realization of the ninth-order elliptic half-band IIR filter.

Wave digital filters. This realization exists for the odd-order transfer functions. It is based on a set of four-port building blocks derived from the corresponding microwave theory. We form, again, a parallel connection of two allpass filters. Each of the allpass filters is realized as a cascade connection of first-order and second-order allpass sections, as shown in Fig. 8.60. Wave digital realizations exhibit the same sensitivity features as parallel-of-two-allpass realizations.

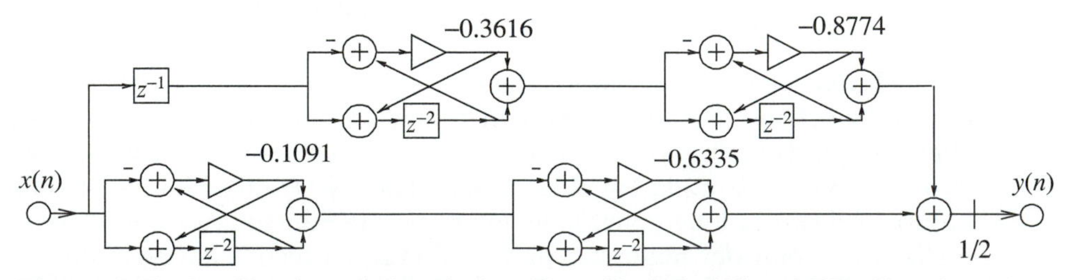

Figure 8.59 Realization of the ninth-order elliptic half-band IIR filter based on a parallel connection of two allpass filters.

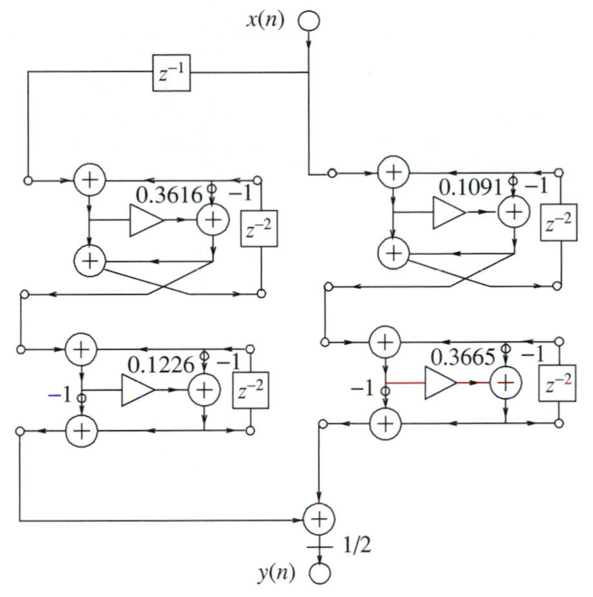

Figure 8.60 Wave digital lattice filter realization of the ninth-order elliptic half-band IIR filter.

The minimal number of multipliers and adders for the above realizations of the fifth-order elliptic IIR filter, and the ninth-order elliptic half-band IIR filter, is summarized below:

Filter type →	Half-band	Ordinary	Half-band	Ordinary
Filter order →	$n = 9$	$n = 5$	$n = 9$	$n = 5$
Realization ↓	Number of multipliers	Number of multipliers	Number of adders	Number of adders
Cascade	9	8	13	10
Parallel	9	10	9	10
Parallel-of-two Allpass	4	5	13	14
Wave digital	4	5	13	16

We conclude that the elliptic half-band IIR filter of higher order can have fewer multipliers and adders than the lower-order elliptic IIR filter, for the same specification. In other words, the half-band filter can be designed to have better performance for the given complexity. A very sharp magnitude response and a flat attenuation characteristic in the passband are achieved. The pole magnitude is smaller in comparison with lower-order elliptic filters, which is important for the effects occurring due to the finite wordlength.

PROBLEMS

8.1 Consider a processing that is performed by a digital filter. The intermediate results are stored in the auxiliary sequences $y_1(n)$, $y_2(n)$, $y_3(n)$, $y_4(n)$, $y_5(n)$, $y_6(n)$, $y_7(n)$, $y_8(n)$, and $y_9(n)$.

input	processing	output
$n = 0, 1, 2, \ldots$	$y_1(n) = x(n)$	
	$y_7(n) = y_1(n-2)$	
$x(n)$	$y_5(n) = y_6(n-4)$	
$y_1(-1) = 0$	$y_2(n) = y_1(n) - y_5(n)$	
$y_1(-2) = 0$	$y_3(n) = ay_2(n)$	$y(n) = y_9(n)$
$y_6(-1) = 0$	$y_4(n) = y_3(n) + y_5(n)$	
$y_6(-2) = 0$	$y_6(n) = y_1(n) + y_3(n)$	
$y_6(-3) = 0$	$y_8(n) = y_4(n) + y_7(n)$	
$y_6(-4) = 0$	$y_9(n) = by_8(n)$	

Find the transfer function and sketch a block diagram of this digital filter.

8.2 A digital filter is described by a set of algebraic equations in the z-domain

$$Y_1(z) = X(z)$$

$$Y_2(z) = Y_1(z) - Y_5(z)$$

$$Y_3(z) = aY_2(z)$$

$$Y_4(z) = Y_3(z) + Y_5(z)$$

$$Y_5(z) = v^3 Y_6(z)$$

$$Y_6(z) = Y_1(z) + Y_3(z)$$

$$Y_7(z) = vY_1(z)$$

$$Y_8(z) = Y_4(z) + Y_7(z)$$

$$Y(z) = bY_8(z)$$

where $v = z^{-1}$. Find the transfer function and sketch a block diagram of the filter.

8.3 A digital filter is specified by a set of algebraic equations in the z-domain

$$\begin{aligned}
Y_1(z) &= X(z) \\
Y_2(z) &= Y_1(z) - Y_5(z) \\
Y_3(z) &= aY_2(z) \\
Y_4(z) &= Y_3(z) + Y_5(z) \\
Y_5(z) &= v^2 Y_6(z) \\
Y_6(z) &= Y_1(z) + Y_3(z) \\
Y_7(z) &= vY_1(z) \\
Y_8(z) &= Y_4(z) + Y_7(z) \\
Y(z) &= bY_8(z)
\end{aligned}$$

Find the transfer function and sketch a block diagram of this digital filter if (a) $v = z^{-1}$, (b) $v = (z^{-1} - \frac{1}{2})/(1 - \frac{1}{2}z^{-1})$.

8.4 For a given c, determine the coefficients d and e to make $H_2(z)$ a second-order lowpass transfer function

$$H_2(z) = \frac{e + dz^{-1} + cz^{-2}}{1 + bz^{-1} + az^{-2}}$$

Find the normalized transfer function, $H_{2n}(z)$, that is, $H_{2n}(1) = 1$. Find the maximum of the magnitude response of $H_{2n}(z)$ for $a = 0.95$, $b = 1.5$, $c = e = 1$, and $d = 2$. Compare this maximum to the value of the pole Q-factor ($Q = \dfrac{\sqrt{(1+a)^2 - b^2}}{2\,(1-a)}$).

8.5 Consider transfer functions of some digital filters

$$H_1(z) = \frac{1 + 2z^{-1} + z^{-2}}{1 + bz^{-1} + az^{-2}}$$

$$H_2(z) = \frac{\left(1 - z^{-1}\right)^2}{1 + bz^{-1} + az^{-2}}$$

$$H_3(z) = \frac{1 - z^{-2}}{1 + bz^{-1} + az^{-2}}$$

$$H_4(z) = \frac{\left(1 - z^{-1}\right)\left(1 + z^{-1}\right)}{1 + bz^{-1} + az^{-2}}$$

Find the maximum of the magnitude response of the normalized transfer functions,

$$\frac{H_1(z)}{H_1(1)}$$

$$\frac{H_2(z)}{H_2(-1)}$$

$$\frac{H_1(z)}{QH_1(1)}$$

$$\frac{H_2(z)}{QH_2(-1)}$$

where $a = 0.93333$, $b = 1.4$, $Q = \dfrac{\sqrt{(1+a)^2 - b^2}}{2(1-a)}$. Construct the fourth-order transfer functions

$$H_{12}(z) = \frac{H_1(z)H_2(z)}{Q^2 H_1(1) H_2(-1)}$$

$$H_{34}(z) = \frac{H_3(z)H_4(z)}{Q^2 H_1(1) H_2(-1)}$$

Compare $H_{12}(z)$ with $H_{34}(z)$. Are the two functions identical?

Find scaling constants k_3 and k_4 to obtain $\max_f(|k_3 H_3(e^{j2\pi f})|) = 1$ and $\max_f(|k_4 H_4(e^{j2\pi f})|) = 1$.

8.6 Given $a = 0.93333$, $b = 1.4$, sketch the magnitude response of

$$H_2(z) = \frac{1 + cz^{-1} + z^{-2}}{1 + bz^{-1} + az^{-2}}$$

for (a) $c = 0.8$, (b) $c = 1.2$, (c) $c = 1.44828$, (d) $c = 1.8$.

8.7 Transfer function of a digital filter can be written as

$$H_2(z) = \frac{1 + a + b}{4} \frac{(1 + z^{-1})^2}{1 + bz^{-1} + az^{-2}}$$

Find and plot the sensitivity of the magnitude response maximum to the coefficients a and b. (a) Assume $b = 1.98$ and plot the sensitivity as a function of a over the range $0.981 < a < 0.991$. (b) Assume $a = 0.99$ and plot the sensitivity as a function of b. What do you conclude about relationship between the sensitivity functions, coefficient a, and Q-factor?

8.8 Consider a class of transfer functions

$$H_2(z) = \frac{(1 + z^{-1})^2}{D(z)}$$

for (a) $D(z) = 1 + bz^{-1} + az^{-2}$,

(a) $D(z) = 1 + b(1 + a)z^{-1} + az^{-2}$,

(b) $D(z) = 1 + bz^{-1} + (1 - a)z^{-2}$,

(c) $D(z) = 1 - 2bz^{-1} + az^{-2}$,

(d) $D(z) = 1 - bz^{-1} + az^{-2}$,

(e) $D(z) = 1 + abz^{-1} + az^{-2}$,

(f) $D(z) = 1 + 2b(1 - a)z^{-1} + az^{-2}$,

(g) $D(z) = 1 + 4b(1-a)z^{-1} + az^{-2}$,

(h) $D(z) = 1 + bz^{-1} + abz^{-2}$,

(i) $D(z) = 1 + 2bz^{-1} + abz^{-2}$,

(j) $D(z) = 1 + (2-b)z^{-1} + (1-a)z^{-2}$.

Find the poles of the transfer function, and plot only the complex poles occurring inside the unit z-plane circle, $|z| < 1$. Assume

$$a \in \{-15/16, -14/16, \cdots, 0, \cdots, 14/16, 15/16\},$$

$$b \in \{-15/16, -14/16, \cdots, 0, \cdots, 14/16, 15/16\}.$$

8.9 Sketch the transposed realization of (1) allpass Ansari–Liu first-order realization type A, (2) allpass Ansari–Liu first-order realization type B, (3) allpass Ansari–Liu first-order realization type C, (4) allpass Ansari–Liu second-order realization type A.

8.10 Find the maximal value of the magnitude response of the transfer function of (1) transformed allpass Ansari–Liu first-order realization type A, (2) transformed allpass Ansari–Liu first-order realization type B,and (3) transformed allpass Ansari–Liu first-order realization type C. The transformation replaces z^{-1} with z^{-2}. Plot the magnitude response of all partial transfer functions. Suggest the best realization that is suitable for fixed-point implementation, and give reasons for your choice.

MATLAB EXERCISES

8.1 Write a MATLAB program to plot the magnitude response of the transfer function given in Problem 8.1. Assume $a = \frac{1}{2}$ and $b = \frac{1}{2}$. Verify your result with the function `freqz`. Write a MATLAB program that implements this filter.

8.2 Write a MATLAB program to plot the magnitude response of the transfer function given in Problem 8.2. Assume $a = \frac{1}{2}$ and $b = \frac{1}{2}$. Verify your result with the function `freqz`. Write MATLAB program that implements this filter.

8.3 Write a MATLAB program to plot the magnitude response of the transfer function given in Problem 8.3. Assume $a = \frac{1}{2}$ and $b = \frac{1}{2}$. Write MATLAB program that implements this filter.

8.4 Write a MATLAB program to plot the magnitude and phase response of the transfer function calculated in Problem 8.4. Assume $a = 0.95$, $b = 1.5$, $c = e = 1$, $d = 2$. Verify your result with the function `freqz`.

8.5 Write a MATLAB program to plot the magnitude response of the transfer functions given in Problem 8.5. Assume $a = 0.93333$, $b = 1.4$. Find the maximal value of the magnitude response of the normalized transfer functions

$$\frac{1+a+b}{4}H_1(z), \quad \frac{1+a-b}{4}H_2(z), \qquad \frac{1+a+b}{4Q}H_1(z), \quad \frac{1+a-b}{4Q}H_2(z),$$

$$\frac{\sqrt{(1+a+b)(1+a-b)}}{4Q}H_3(z), \qquad \frac{\sqrt{(1+a+b)(1+a-b)}}{4Q}H_4(z)$$

Compare the fourth-order transfer function

$$\frac{(1+a+b)\,(1+a-b)}{16Q^2}H_1(z)H_2(z) \text{ with } \frac{(1+a+b)\,(1+a-b)}{16Q^2}H_3(z)H_4(z)$$

8.6 Write a MATLAB program to plot the magnitude response of the transfer function given in Problem 8.6 scaled by constants $1/Q$, $1/H(1)$ and $1/H(-1)$. Which ones of the transfer functions are lowpass-notch transfer functions. Find the maximal value of the magnitude response.

8.7 Write a MATLAB program to plot the sensitivity of the magnitude response of the transfer function given in Problem 8.7.

8.8 Write a MATLAB program to plot the complex poles in the z-plane of the transfer function given in Problem 8.8. How many different complex poles can be realized?

8.9 Write a MATLAB program to draw the schematic of the transposed realization of (1) allpass Ansari–Liu first-order realization type A, (2) allpass Ansari–Liu first-order realization type B, (3) allpass Ansari–Liu first-order realization type C, and (4) allpass Ansari–Liu second-order realization type A.

8.10 Write MATLAB programs to draw the schematic of the transposed realization of (1) allpass Ansari–Liu first-order realization type A, (2) allpass Ansari–Liu first-order realization type B, (3) allpass Ansari–Liu first-order realization type C, and (4) allpass Ansari–Liu second-order realization type A. Assume z^{-1} is replaced with z^{-2}. Plot the magnitude response of the partial transfer functions.

8.11 Write MATLAB programs to (1) draw the schematic, (2) filter with and without quantization, (3) verify the filter realization, and plot the frequency response of the following filters:

(a) Direct form first-order realization,

(b) Direct form second-order realization,

(c) Transpose direct form first-order realization,

(d) Transpose direct form II second-order realization,

(e) Allpass Ansari–Liu first-order realization type A,

(f) Allpass Ansari–Liu first-order realization type B,

(g) Allpass Ansari–Liu first-order realization type C,

(h) Allpass Ansari–Liu second-order realization type A,

(i) Allpass Ansari–Liu second-order realization type B,

(j) Allpass Ansari–Liu second-order realization type C,

(k) Lowpass halfband IIR 3rd-order realization,

(l) Highpass halfband IIR 3rd-order realization.

MATHEMATICA EXERCISES

8.1 Write a *Mathematica* program to plot the magnitude response of the transfer function calculated in Problem 8.1. Assume $a = \frac{1}{2}$ and $b = \frac{1}{2}$. Write a *Mathematica* program that implements this filter. Verify your result using the functions `InverseFourier` and `Fourier`.

8.2 Write a *Mathematica* program to plot the magnitude response of the transfer function calculated in Problem 8.2. Assume $a = \frac{1}{2}$ and $b = \frac{1}{2}$. Verify your result using the functions `InverseFourier` and `Fourier`.

8.3 Write a *Mathematica* program to plot the magnitude response of the transfer function calculated in Problem 8.3. Assume $a = \frac{1}{2}$ and $b = \frac{1}{2}$.

8.4 Write a *Mathematica* program to plot the magnitude and phase response of the transfer function calculated in Problem 8.4. Assume $a = 0.95$, $b = 1.5$, $c = e = 1$, $d = 2$.

8.5 Write *Mathematica* programs to plot the magnitude response of the transfer functions calculated in Problem 8.5. Assume $a = 0.93333$, $b = 1.4$. Find the maximal value of the magnitude response of the normalized transfer functions

$$\frac{1+a+b}{4}H_1(z), \quad \frac{1+a-b}{4}H_2(z), \quad \frac{1+a+b}{4Q}H_1(z),$$

$$\frac{1+a-b}{4Q}H_2(z),$$

$$\frac{\sqrt{(1+a+b)(1+a-b)}}{4Q}H_3(z),$$

and

$$\frac{\sqrt{(1+a+b)(1+a-b)}}{4Q}H_4(z).$$

Compare the fourth-order transfer functions

$$\frac{(1+a+b)(1+a-b)}{16Q^2}H_1(z)H_2(z)$$

$$\frac{(1+a+b)(1+a-b)}{16Q^2}H_3(z)H_4(z).$$

8.6 Write a *Mathematica* program to plot the magnitude response of the transfer function given in Problem 8.6 scaled by constants $1/Q$, $1/H(1)$ and $1/H(-1)$. Which ones of the transfer functions are lowpass-notch transfer functions. Find the maximal value of the magnitude response.

8.7 Write a *Mathematica* program to plot the sensitivity of the magnitude response of the transfer function given in Problem 8.7.

8.8 Write a *Mathematica* program to plot the complex poles in the z-plane of the transfer function given in 8.8. How many different complex poles can be realized? Derive the coefficient sensitivity.

8.9 Write *Mathematica* programs to draw the schematic of the transposed realization of (1) allpass Ansari–Liu first-order realization type A, (2) allpass Ansari–Liu first-order realization type B, (3) allpass Ansari–Liu first-order realization type C, and (4) allpass Ansari–Liu second-order realization type A. Derive the transfer function and noise transfer functions.

8.10 Write *Mathematica* programs to draw the schematic of the transposed realization of (1) allpass Ansari–Liu first-order realization type A, (2) allpass Ansari–Liu first-order realization type B, (3) allpass Ansari–Liu first-order realization type C, and (4) allpass Ansari–Liu second-order realization type A. Assume z^{-1} is replaced with z^{-2}. Derive and plot the magnitude response of the partial transfer functions.

8.11 Write *Mathematica* programs to (1) draw the schematic, (2) derive the transfer function and noise transfer functions, poles, zeros, and Q-factors in terms of coefficients, (3) find coefficients in terms of design parameters, (4) filter with and without quantization, (5) verify the filter realization, and plot the frequency response of the following filters:

(a) Direct form first-order realization,
(b) Direct form second-order realization,
(c) Transpose direct form first-order realization,
(d) Transpose direct form II second-order realization,
(e) Allpass Ansari–Liu first-order realization type A,
(f) Allpass Ansari–Liu first-order realization type B,
(g) Allpass Ansari–Liu first-order realization type C,
(h) Allpass Ansari–Liu second-order realization type A,
(i) Allpass Ansari–Liu second-order realization type B,
(j) Allpass Ansari–Liu second-order realization type C,
(k) Lowpass halfband IIR third-order realization,
(l) Highpass halfband IIR third-order realization.

CHAPTER 9

ADVANCED DIGITAL FILTER DESIGN CASE STUDIES

This chapter reviews basic definitions of digital IIR filter design. It introduces straightforward procedures to map the filter specification into a design space—that is, a set of ranges for parameters that we use in the filter design. We search this design space for the optimum solution according to given criteria, such as minimal quantization error.

The principal drawback of the classical digital filter design is in returning only one solution, which can be unacceptable for many practical implementations. We propose an advanced approach to the filter design by using a mixture of symbolic and numeric computation and discrete nonlinear optimization. This approach should provide reduced filter complexity for a desired performance, or better performances for the required complexity.

We conclude this chapter by several important application examples in which we design low-sensitivity selective multiplierless IIR filters, power-of-two IIR filters, half-band IIR filters, 1/3-band filters, narrow-band IIR filters, Hilbert transformers, and zero-phase IIR filters. Each example design is followed by a comprehensive step-by-step procedure for computing the filter coefficients.

9.1 BASIC DEFINITIONS

A *filter* is a system that can be used to modify or reshape the frequency spectrum of a signal according to some prescribed requirements.

A *digital filter* takes an input sequence of numbers and produces an output sequence of numbers. Usually, the input sequence of numbers are samples of a continuous function of time, but it can be any kind of numbers such as (a) prices from the daily stock market or (b) pixels of an image.

Generally, the purpose of most filters is to separate the desired signals from undesired signals or noise. Often, the descriptions of the signals and noise are given in terms of their frequency content or the energy of the signals in the frequency bands. For this reason, the filter specifications are usually given in the frequency domain as magnitude response, or by gain, or by attenuation.

The range of frequencies in which the sinusoidal signals are rejected is called a *stopband*. The range of frequencies in which the sinusoidal signals pass with tolerated distortion is called a *passband*. A region between the passband and stopband, where neither desired nor undesired signals exist or the spectra of the desired and undesired signals are overlapped, can be defined as a *transition region*.

In this chapter we will consider a filter with single passband referred to as a *lowpass* filter. All other types of filters (*highpass*, *bandpass*, and *bandstop*) can be easily obtained by simple transformation from the lowpass filter.

Once the filter requirements are known, the filter specification can be established; for example, we specify the passband and stopband edge frequencies and tolerances. Next, we proceed with the filter design.

The *design* is a set of processes that starts with the specification, and ends with the implementation, of a (product) filter prototype. It comprises four general steps, as follows:

- Approximation
- Realization
- Study of imperfections
- Implementation

The *approximation* step is the process of generating a transfer function that satisfies the desired specification.

The *realization* step is the process of converting the transfer function of the filter into a block diagram (also called the digital filter network) or a set of equations that indicates the sequence of computations to be performed on the input sequence of numbers.

The *study of imperfections* investigates the effects of imperfections, such as (a) finite wordlength for storing the samples and coefficients or (b) the largest quantization step that can be tolerated without violating the specification of the filter.

The *implementation* step is constructing the product prototype of the filter in hardware (DSP processors, dedicated hardware, custom VLSI chips) or software that executes on general purpose computers, specialized computers, or array processors. Decisions to be made involve (a) the type of components and (b) the methods to be used for the manufacture, the data wordlength, coefficient wordlengths, and so on.

Usually, those four design steps are considered separately, although they are not independent of each other. The main goal is to find the most economical solution in short time. Which filter is better depends on the hardware used for the implementation. Many different constrains have to be fulfilled. The finite wordlength effects have significant influence on fulfilling the specification. In this case, classic approaches are not adequate for optimizing both the behavior (performance) and implementation (complexity and cost).

We propose an advanced approach to the digital filter design by using a mixture of symbolic and numeric computation and discrete nonlinear optimization. Opposite to the conventional approaches, which return only one design and hide a wealth of alternative filter designs, the advanced design techniques (that we introduce) find a comprehensive set of optimal designs to represent the infinite solution space.

9.2 SPECIFICATION OF A DIGITAL FILTER

Usually, a digital-filter specification is given, but in many cases the designer has to establish the specification by himself. This is the most important prerequisite for the filter design; that is, if the specification is too restrictive (e.g., very low passband and stopband tolerances, narrow transition region), the filter may not be feasible.

The proper selection of the specification must be done according to the nature of signals (i.e. frequency bands and the corresponding levels of the desired and undesired signals or noise) and the available hardware or software (floating-point or fixed-point arithmetic, coefficient wordlength, etc.).

In this chapter we assume that the specification is given. Next, we examine the feasibility of the practical filter design. Finally, if the filter is not feasible, we propose the minimum changes in the given specification to make the filter design possible.

In practice, there are several ways to present the digital filter specification. To provide a unified and consistent design, we adopt one form of presenting the specification.

Let us consider a *discrete-time linear time-invariant system* (DTLTI), with the input sine sequence $x_{in}(n)$

$$x_{in}(n) = X_m \sin(2\pi f n + \zeta) \tag{9.1}$$

and the steady-state output sequence $y_{out}(n)$,

$$y_{out}(n) = Y_m \sin(2\pi f n + \eta) \tag{9.2}$$

as shown in Fig. 9.1. The ratio $M(f) = Y_m/X_m$ describes the change in the amplitude of the input sine sequence; we call $M(f)$ the *magnitude response*. With $\Phi(f) = \eta - \zeta$ we designate the change in the phase of the input sine sequence; we call $\Phi(f)$ the *phase response*. Both quantities, $M(f)$ and $\Phi(f)$, are defined at the *digital frequency*, f, which is a dimensionless quantity. Some authors prefer to express $M(f)$ and $\Phi(f)$ in terms of the *digital angular frequency*, $\theta = 2\pi f$. The *frequency response* of the system is defined as $M(f)e^{j\Phi(f)}$.

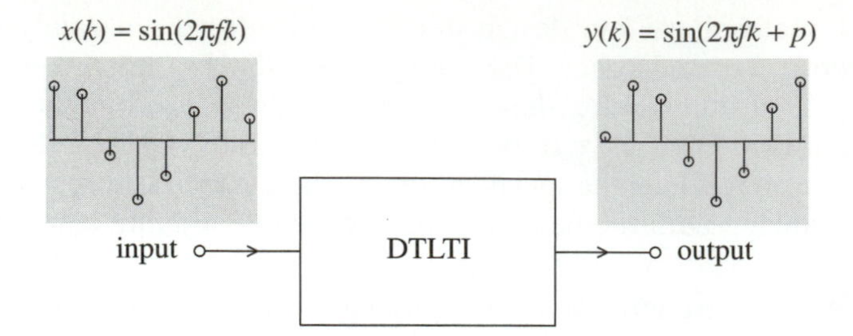

Figure 9.1 Discrete-time linear time-invariant system.

In other words, $M(f)e^{j\Phi(f)}$ shows how the input signal is transferred through the system at the specific frequency f. From the frequency response we can derive the *transfer function*. The transfer function of a DTLTI, $H(z)$, is a rational function in complex variable z; for $z = e^{j2\pi f}$ the transfer function becomes the frequency response, $H(e^{j2\pi f}) = M(f)e^{j\Phi(f)}$.

Several functions are derived from the magnitude response, $M(f) = |H(e^{j2\pi f})|$. The reciprocal of the squared magnitude response is called the *loss function*:

$$L_F(f) = \frac{1}{M^2(f)} \tag{9.3}$$

We call the function $\sqrt{L_F(f) - 1}$ the *characteristic function*:

$$K(f) = \sqrt{\frac{1}{M^2(f)} - 1} \tag{9.4}$$

Attenuation (in dB) or *loss characteristic* is defined by

$$A(f) = 20 \log_{10} \frac{1}{M(f)} \tag{9.5}$$

Gain (in dB) is the negative of the attenuation:

$$G(f) = -A(f) = 20 \log_{10} M(f) \tag{9.6}$$

We assume that the maximal value of the magnitude response is 1:

$$\max_f M(f) = \max_f |H(e^{j2\pi f})| = 1$$

If this is not the case, the required attenuation or gain can be easily compensated for by multiplying $H(z)$ by a constant.

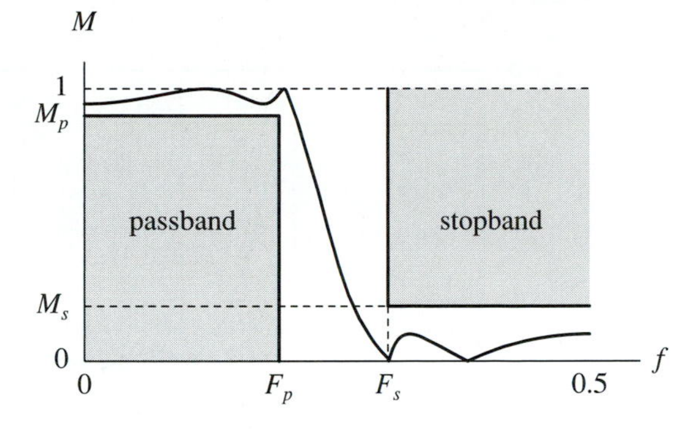

Figure 9.2 Magnitude-limit specification.

The magnitude response $M(f)$, of a lowpass digital filter, against digital frequency f, is plotted in Fig. 9.2. The maximum digital frequency of interest equals 1. However, if we focus on transfer functions with real coefficients, then the upper limit for the digital frequency of interest is $\frac{1}{2}$.

The filter specification can be expressed in several ways:

1. The *magnitude limits* (Fig. 9.2), define the minimum magnitude in passband, M_p, and the maximum magnitude in stopband, M_s.
2. The *magnitude tolerances* (Fig. 9.3) specify the maximum magnitude decrease in passband, $\delta_p = 1 - M_p$, and the maximum magnitude in stopband, $\delta_s = M_s$.
3. The *magnitude ripple tolerances* (Fig. 9.4) describe the maximum magnitude variation in passband, δ_1, and in stopband, δ_2.
4. The *attenuation limits* in dB (Fig. 9.5) specify the maximum attenuation in passband, A_p, and the minimum attenuation in stopband, A_s.
5. The *gain limits* in dB (Fig. 9.6) specify the minimum gain in passband, $G_p = -A_p$, and the maximum gain in stopband, $G_s = -A_s$.

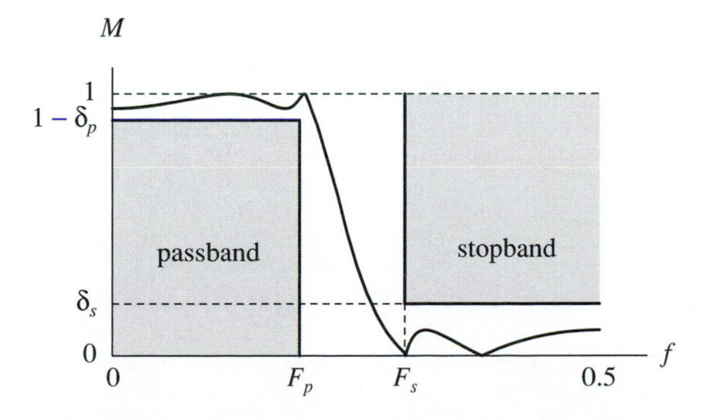

Figure 9.3 Magnitude-tolerance specification.

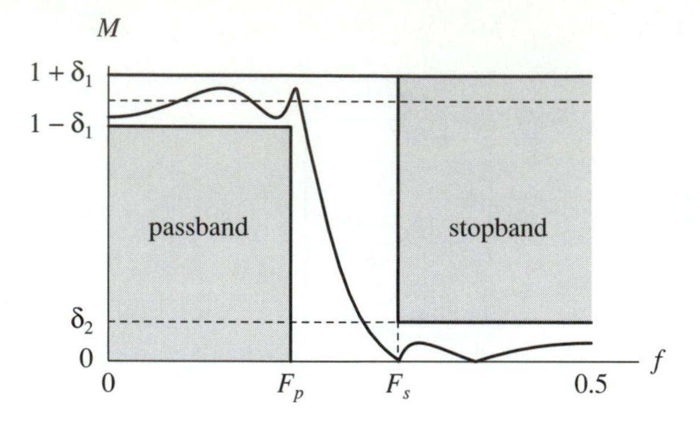

Figure 9.4 Magnitude-ripple specification.

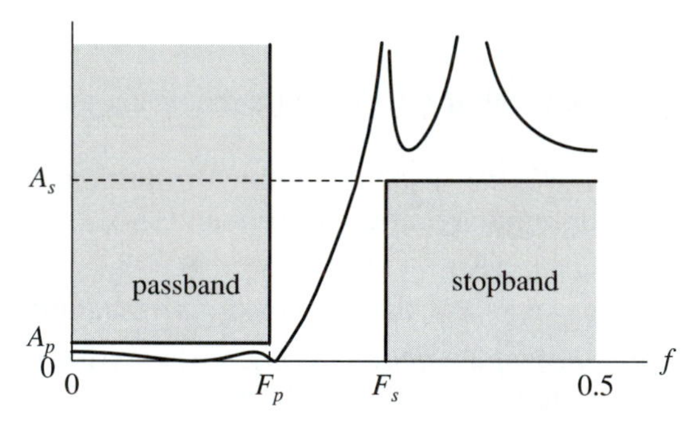

Figure 9.5 Attenuation-limit specification.

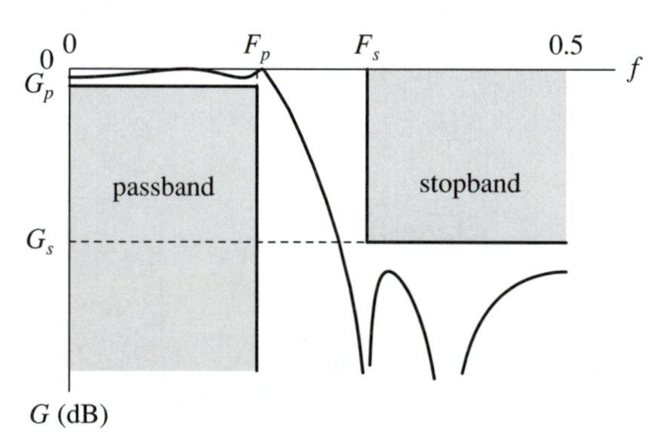

Figure 9.6 Gain-limit specification.

Relations between the specification quantities are summarized in Eqs. (9.7)–(9.12)

$$
\begin{array}{|ccccccc|}
\hline
\delta_p & = & 1 - M_p & = & \dfrac{2\delta_1}{1+\delta_1} & = & \dfrac{-1+\sqrt{1+K_p^2}}{\sqrt{1+K_p^2}} \\
1-\delta_p & = & M_p & = & \dfrac{1-\delta_1}{1+\delta_1} & = & \dfrac{1}{\sqrt{1+K_p^2}} \\
\dfrac{\delta_p}{2-\delta_p} & = & \dfrac{1-M_p}{1+M_p} & = & \delta_1 & = & \dfrac{-1+\sqrt{1+K_p^2}}{1+\sqrt{1+K_p^2}} \\
\dfrac{\sqrt{\delta_p(2-\delta_p)}}{1-\delta_p} & = & \dfrac{\sqrt{1-M_p^2}}{M_p} & = & \dfrac{2\sqrt{\delta_1}}{1-\delta_1} & = & K_p \\
-20\log_{10}(1-\delta_p) & = & -20\log_{10} M_p & = & 20\log_{10}\dfrac{1+\delta_1}{1-\delta_1} & = & 10\log_{10}(1+K_p^2) \\
20\log_{10}(1-\delta_p) & = & 20\log_{10} M_p & = & 20\log_{10}\dfrac{1-\delta_1}{1+\delta_1} & = & -10\log_{10}(1+K_p^2) \\
\hline
\end{array}
\tag{9.7}
$$

$$
\begin{array}{|ccccc|}
\hline
\dfrac{-1+\sqrt{1+K_p^2}}{\sqrt{1+K_p^2}} & = & 1 - 10^{-A_p/20} & = & 1 - 10^{G_p/20} \\
\dfrac{1}{\sqrt{1+K_p^2}} & = & 10^{-A_p/20} & = & 10^{G_p/20} \\
\dfrac{-1+\sqrt{1+K_p^2}}{1+\sqrt{1+K_p^2}} & = & \dfrac{1-10^{-A_p/20}}{1+10^{-A_p/20}} & = & \dfrac{1-10^{G_p/20}}{1+10^{G_p/20}} \\
K_p & = & \dfrac{\sqrt{1-10^{-A_p/10}}}{10^{-A_p/20}} & = & \dfrac{\sqrt{1-10^{G_p/10}}}{10^{G_p/20}} \\
10\log_{10}(1+K_p^2) & = & A_p & = & -G_p \\
-10\log_{10}(1+K_p^2) & = & -A_p & = & G_p \\
\hline
\end{array}
\tag{9.8}
$$

$$
\begin{array}{ccccccc}
\delta_s & = & M_s & = & \dfrac{\delta_2}{1+\delta_1} & = & \dfrac{1}{\sqrt{1+K_s^2}} \\
\dfrac{2\delta_s}{2-\delta_p} & = & \dfrac{2M_s}{1+M_p} & = & \delta_2 & = & 2\dfrac{\sqrt{\dfrac{1+K_p^2}{1+K_s^2}}}{1+\sqrt{1+K_p^2}} \\
\dfrac{\sqrt{1-\delta_s^2}}{\delta_s} & = & \dfrac{\sqrt{1-M_s^2}}{M_s} & = & \dfrac{\sqrt{(1+\delta_1)^2-\delta_2^2}}{\delta_2} & = & K_s \\
-20\log_{10}(\delta_s) & = & -20\log_{10} M_s & = & -20\log_{10}\dfrac{\delta_2}{1+\delta_1} & = & 10\log_{10}(1+K_s^2) \\
20\log_{10}(\delta_s) & = & 20\log_{10} M_s & = & 20\log_{10}\dfrac{\delta_2}{1+\delta_1} & = & -10\log_{10}(1+K_s^2)
\end{array}
\tag{9.9}
$$

$$
\begin{array}{ccccc}
\dfrac{1}{\sqrt{1+K_s^2}} & = & 10^{-A_s/20} & = & 10^{G_s/20} \\
2\dfrac{\sqrt{\dfrac{1+K_p^2}{1+K_s^2}}}{1+\sqrt{1+K_p^2}} & = & 2\dfrac{10^{-A_s/20}}{1+10^{-A_p/20}} & = & 2\dfrac{10^{G_p/20}}{1+10^{G_p/20}} \\
K_s & = & \dfrac{\sqrt{1-10^{-A_s/10}}}{10^{-A_s/20}} & = & \dfrac{\sqrt{1-10^{G_s/10}}}{10^{G_s/20}} \\
10\log_{10}(1+K_s^2) & = & A_s & = & -G_s \\
-10\log_{10}(1+K_s^2) & = & -A_s & = & G_s
\end{array}
\tag{9.10}
$$

$$
\begin{array}{lll}
\delta_p = 1 - \dfrac{\sqrt{2}}{2} & M_p = \dfrac{\sqrt{2}}{2} & K_p = 1 \\
\delta_s \approx 0.1 & M_s \approx 0.1 & K_s = 10 \\
\delta_s \approx 0.01 & M_s \approx 0.01 & K_s = 10^2 \\
\delta_s \approx 0.0001 & M_s \approx 0.0001 & K_s = 10^4 \\
\delta_s \approx 0.00001 & M_s \approx 0.00001 & K_s = 10^5 \\
\delta_p \approx 0.005 & M_p \approx 0.995 & \delta_1 \approx 0.0099 \;\; K_p = 1/10 \\
\delta_p \approx 0.00005 & M_p \approx 0.99995 & \delta_1 \approx 0.00001 \;\; K_p = 1/100
\end{array}
\tag{9.11}
$$

$$\begin{array}{|lll|} K_p = 1 & A_p \approx 3 & G_p \approx -3 \\ K_s = 10 & A_s \approx 20 & G_s \approx -20 \\ K_s = 10^2 & A_s \approx 40 & G_s \approx -40 \\ K_s = 10^4 & A_s \approx 80 & G_s \approx -80 \\ K_s = 10^5 & A_s \approx 100 & G_s \approx -100 \\ K_p = 1/10 & A_p \approx 0.043 & G_p \approx -0.043 \\ K_p = 1/100 & A_p \approx 0.0004 & G_p \approx -0.0004 \end{array} \tag{9.12}$$

The underlining idea of the design that we propose is to map any specification into a new one, expressed in terms of the *characteristic-function limits* (Fig. 9.7), specifying the maximum value in passband, K_p, and the minimum value in stopband, K_s. This provides a unified start for the subsequent design steps.

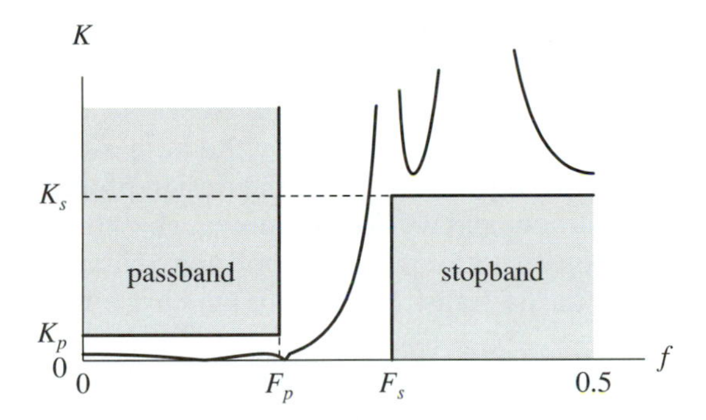

Figure 9.7 Characteristic-function-limit specification.

9.3 APPROXIMATION STEP

In this section we consider the approximation step of a digital filter design.

Let us consider the specification shown in Fig. 9.7. The first step is to generate the characteristic function $K(f)$. A function that is a candidate for $K(f)$ is known as the *approximation function*, also called the *approximation*. From the approximation the transfer function that meets the specification can be derived. Besides several classic approximations there are numerous closed-form expressions and numerical procedures for generating approximation functions.

The *Butterworth* approximation is smooth and monotonically increases with respect to frequency. It is maximally flat at $f = 0$.

The *Chebyshev* approximation, sometimes called the *Chebyshev type I* approximation, gives the smallest ripple over the entire passband. In the stopband, this approximation monotonically increases with respect to frequency.

The *Chebyshev type II* approximation, also called the *inverse Chebyshev* approximation, is smooth and monotonically increases with respect to frequency in the passband. It is maximally flat at $f = 0$ like the Butterworth approximation. This approximation gives the smallest ripple over the entire stopband.

The *elliptic function approximation*, also called the *elliptic approximation*, *Cauer approximation*, or *Darlington approximation*, gives the smallest ripple over the entire passband and stopband.

Other types of approximations exist, and they exhibit good properties of group delay or impulse response. There is a class of approximations that combines the properties of the classic ones; we call this class of approximation functions the *transitional approximations*.

In digital filter theory there are additional approaches for solving the approximation problem (least-squares error method, windows approach) [50].

Which approximation should the designer choose?

One approach is to analyze all known approximations and to use all known procedures for calculating the magnitude response. The examination of all known approximations has a very high computational cost. Such an approach can be time-consuming, too.

One straightforward approach is to specify a desired magnitude response and to try to find, numerically, an approximation closest to the response [50]. We also must add some constraints that are required by the numerical optimizer; the purpose of the additional constraints is just to simplify the design procedure. The numerical approach gives only one solution, and we do not known whether a better solution exists.

It is interesting to notice [50] that good approximations are obtained by specifying some constraints in the transition region, which we normally treat as a "don't care" region.

In this chapter we focus on only one approximation, the elliptic approximation. Next, we find the design space—that is, the range of design parameters that satisfy the specification—and keep the design parameters as symbols.

9.4 DESIGN SPACE

In this section we define the design space. First, we map the specification into a standard form. Next, we identify the design parameters. Finally, we calculate the limits of the design parameters.

In the previous section we have shown several ways of presenting required specifications. Any digital lowpass filter can be specified by a set of four quantities

$$S_\delta = \{F_p, F_s, \delta_p, \delta_s\} \tag{9.13}$$

$$S_M = \{F_p, F_s, M_p, M_s\} \tag{9.14}$$

$$S_r = \{F_p, F_s, \delta_1, \delta_2\} \tag{9.15}$$

$$S_K = \{F_p, F_s, K_p, K_s\} \tag{9.16}$$

$$S_A = \{F_p, F_s, A_p, A_s\} \tag{9.17}$$

$$S_G = \{F_p, F_s, G_p, G_s\} \tag{9.18}$$

and the relations between them are summarized in Eqs. (9.7)–(9.12). The symbol F_p designates the passband edge frequency, and F_s stands for the stopband edge frequency. We use the term "frequency" as a short form for "digital frequency."

It is more convenient to transform a given specification S into the specification S_K because it provides a clearer relationship between the design parameters and the specification. Since we have to find a characteristic function $K(f)$ in the approximation step, S_K is the most suitable way of presenting the specification.

An infinite number of characteristic functions that fit S exists. We consider the *elliptic approximation*, because it fulfills the requirements with the minimal transfer function order. The minimal order can often lead to the most economical solution (the minimal number of components, the minimal number of multiplications, and so forth). Also, it will be shown that some other classic approximations are special cases of the elliptic approximation.

The *prototype elliptic approximation*, K_e, is an nth-order rational function in the real variable x:

$$K_e(x) = \epsilon |R(n, \xi, x)| \tag{9.19}$$

where R, referred to as the *elliptic rational function*, satisfies the conditions

$$0 \leq |R(n, \xi, x)| \leq 1, \qquad |x| \leq 1 \tag{9.20}$$

$$L(n, \xi) \leq |R(n, \xi, x)|, \qquad \xi \leq |x| \leq +\infty \tag{9.21}$$

and L is the *discrimination factor*,—that is, the minimal value of the magnitude of R for $|x| \geq \xi$—and can be calculated as

$$L(n, \xi) = |R(n, \xi, \xi)| \tag{9.22}$$

The normalized transition band $1 < x < \xi$ is defined by

$$1 < |R(n, \xi, x)| < L(n, \xi), \qquad 1 < |x| < \xi \tag{9.23}$$

The parameter ξ is called the *selectivity factor*.

The parameter ϵ determines the maximal variation of K_e in the normalized passband $0 \leq x \leq 1$

$$0 \leq K_e(x) \leq \epsilon, \qquad |x| \leq 1 \tag{9.24}$$

and is called the *ripple factor*.

The *elliptic approximation*, $K(f)$, is a function in frequency f (Figs. 9.8 and 9.9):

$$K(f) = K_e(x), \qquad x = \frac{\tan(\pi f)}{\tan(\pi f_p)} \tag{9.25}$$

where f_p represents a design parameter that we call the *actual passband edge*. Traditionally, it has been set to $f_p = F_p$.

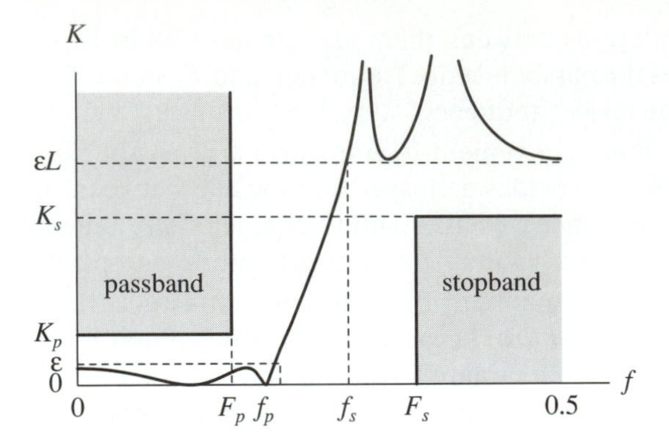

Figure 9.8 Elliptic approximation.

The four dimensionless quantities, n, ξ, ϵ, and f_p, are collectively referred to as *design parameters* and can be expressed as a list of the form

$$D = \{n, \xi, \epsilon, f_p\} \tag{9.26}$$

Quantities ξ, ϵ, and f_p, take a value over a continuous range of numbers. The order n can take a value from a discrete range of numbers and is referred to as the *filter order* or *transfer function order*.

It is known [50] that the elliptic approximation provides the minimal order, $n_{min} = n_{ellip}$, for a given specification. The maximal order, from the practical viewpoint, can be assumed to be $n_{max} = 2n_{min}$.

The selectivity factor, ξ, falls within the limits which are found by solving the equations [51]

$$R(n, \xi, \xi) = \frac{K_s}{K_p} \quad \Rightarrow \quad \xi_{min} = \xi_{min}(n) \tag{9.27}$$

$$R\left(n, \xi, \frac{\tan(\pi F_s)}{\tan(\pi F_p)}\right) = \frac{K_s}{K_p}, \quad \xi > \frac{\tan(\pi F_s)}{\tan(\pi F_p)} \quad \Rightarrow \quad \xi_{max} = \xi_{max}(n) \tag{9.28}$$

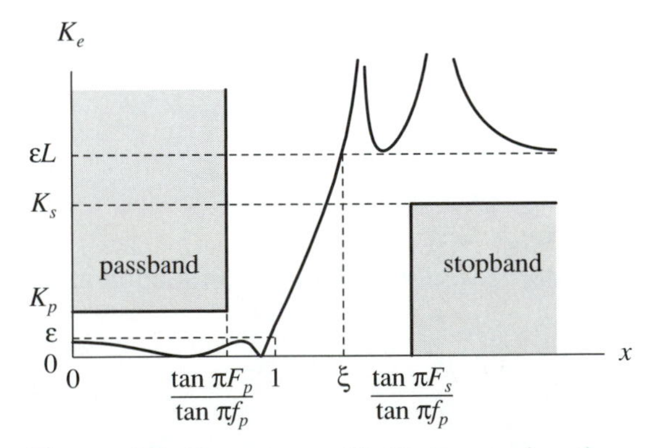

Figure 9.9 Prototype elliptic approximation.

It follows from Eqs. (9.19) and (9.19) that the maximal value of ϵ must be equal to or less than K_p:

$$\epsilon \leq K_p \tag{9.29}$$

The ripple factor quantifies the output signal amplitude, Y_m, with respect to the input signal amplitude, X_m. When $K(f) = 0$, both amplitudes have the same value, that is,

$$Y_m = X_m, \qquad K(f) = 0 \tag{9.30}$$

when $K(f) = \epsilon$ the amplitudes are different:

$$Y_m = \frac{X_m}{\sqrt{1+\epsilon^2}}, \qquad K(f) = \epsilon \tag{9.31}$$

From the previous equations it follows that the smaller the value of ϵ, the smaller the difference between the input and output amplitude.

What is the lower limit of ϵ?

The ripple factor, for $x > \xi$, must meet another condition $K_e(x) \geq K_s$ (Fig. 9.9), thus

$$\epsilon L(n, \xi) \geq K_s \tag{9.32}$$

Therefore, the maximal and minimal values of ϵ has to be determined:

$$\epsilon_{min} \leq \epsilon \leq \epsilon_{max} \tag{9.33}$$

From Eq. (9.29) we find the upper bound:

$$\epsilon_{max} = K_p \tag{9.34}$$

From Eq. (9.32) we determine the lower bound:

$$\epsilon_{min} = \frac{K_s}{L(n, \xi)} = \epsilon_{min}(n, \xi) \tag{9.35}$$

The maximal value of ϵ directly follows from specification, while the minimal value of ϵ depends on the order n and the selectivity factor ξ.

The actual passband edge, f_p, can take a value from the interval

$$\begin{aligned} f_{p,min} &= \frac{1}{\pi}\tan^{-1}\left(\frac{\tan(\pi F_s)}{\xi_{max}}\right) \\ f_{p,max} &= \frac{1}{\pi}\tan^{-1}\left(\frac{\tan(\pi F_s)}{\xi_{min}}\right) \\ f_{p,min} &\leq f_p \leq f_{p,max} \end{aligned} \tag{9.36}$$

Obviously, the actual passband edges are functions of filter order $f_{p,min} = f_{p,min}(n)$ and $f_{p,max} = f_{p,max}(n)$.

The set of all quadruples $D = \{n, \xi, \epsilon, f_p\}$ satisfying the constraints $\{n_{min} \le n \le n_{max}, \xi_{min} \le \xi \le \xi_{max}, \epsilon_{min} \le \epsilon \le \epsilon_{max}, f_{p,min} \le f_p \le f_{p,max}\}$ is called the *design space*

$$D_S = \{D_{S,n}\}_{n=n_{min},n_{min}+1,\ldots,n_{max}} \tag{9.37}$$

$$D_{S,n} = \left\{ \begin{array}{ccc} & n & \\ \xi_{min}(n) & \le \xi \le & \xi_{max}(n) \\ \epsilon_{min}(n,\xi) & \le \epsilon \le & K_p \\ f_{p,min}(n) & \le f_p \le & f_{p,max}(n) \end{array} \right\} \tag{9.38}$$

The order n is an integer, and it takes only the discrete numeric values, so it is more convenient to express the design space, D_S, as a list of subspaces, $D_{S,n}$

$$\left\{ \begin{array}{l} \left\{ \begin{array}{ccc} & n = n_{min} & \\ \xi_{min}(n) & \le \xi \le & \xi_{max}(n) \\ \epsilon_{min}(n,\xi) & \le \epsilon \le & K_p \\ f_{p,min}(n) & \le f_p \le & f_{p,max}(n) \end{array} \right\} \\ \left\{ \begin{array}{ccc} & n = n_{min}+1 & \\ \xi_{min}(n) & \le \xi \le & \xi_{max}(n) \\ \epsilon_{min}(n,\xi) & \le \epsilon \le & K_p \\ f_{p,min}(n) & \le f_p \le & f_{p,max}(n) \end{array} \right\} \\ \ldots \\ \left\{ \begin{array}{ccc} & n = n_{max} & \\ \xi_{min}(n) & \le \xi \le & \xi_{max}(n) \\ \epsilon_{min}(n,\xi) & \le \epsilon \le & K_p \\ f_{p,min}(n) & \le f_p \le & f_{p,max}(n) \end{array} \right\} \end{array} \right\} \tag{9.39}$$

where

$$\begin{array}{ccccc} 0 & < & \epsilon_{min}(n+1) & < & \epsilon_{min}(n) \\ 1 & < & \xi_{min}(n+1) & < & \xi_{min}(n) \\ \xi_{max}(n) & < & \xi_{max}(n+1) & < & \infty \\ 0 & \le & f_{p,min}(n+1) & < & f_{p,min}(n) \\ f_{p,max}(n) & < & f_{p,max}(n+1) & < & F_s \end{array} \tag{9.40}$$

9.5 BASIC DESIGN ALTERNATIVES

This section presents our case studies of a comprehensive set of design alternatives based on the design space. It is understood that the rational elliptic function can be readily constructed for a given set of design parameters [52, 53]. The advantages of the various designs are discussed.

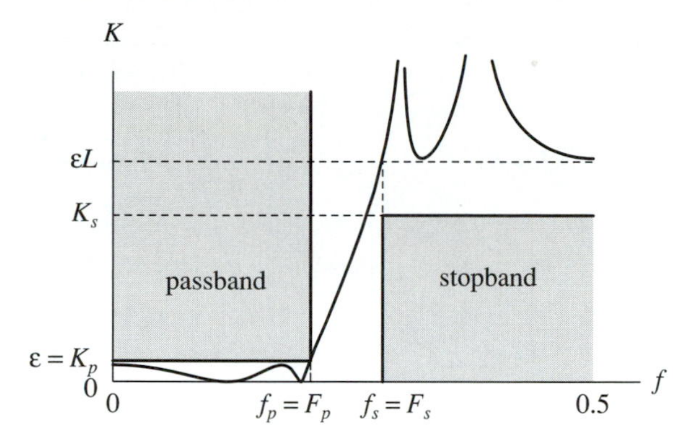

Figure 9.10 Design D1.

Usually, the designer selects the minimal order $n = n_{min}$. The design alternatives that follow are general and valid for any n from the design space. We assume that a specification, S, has been mapped into the form S_K.

9.5.1 Design D1

The design D1 sets the three design parameters, $\xi = \tan(\pi F_s)/\tan(\pi F_p)$, $\epsilon = K_p$, and $f_p = F_p$, directly from the specification (Fig. 9.10).

This design has higher attenuation in the stopband than is required by S_A. We choose this design when we prefer to achieve as large an attenuation as possible in the stopband; that is, $\epsilon L > K_s$.

9.5.2 Design D2

The design D2 sets the two design parameters, $\xi = \tan(\pi F_s)/\tan(\pi F_p)$, $f_p = F_p$, directly from the specification (Fig. 9.11). The ripple factor is computed from $\epsilon = K_s/L(n, \xi)$.

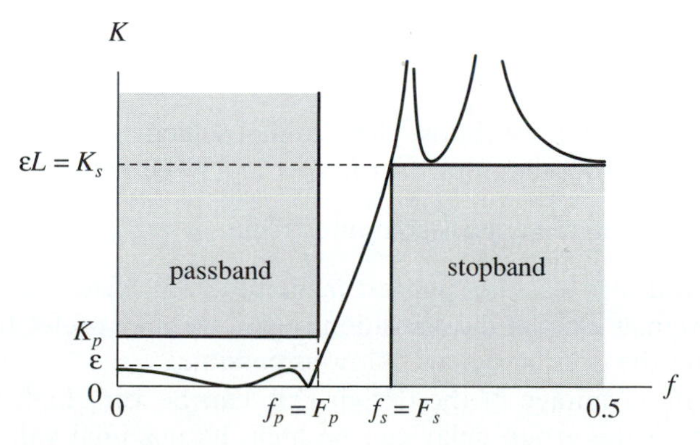

Figure 9.11 Design D2.

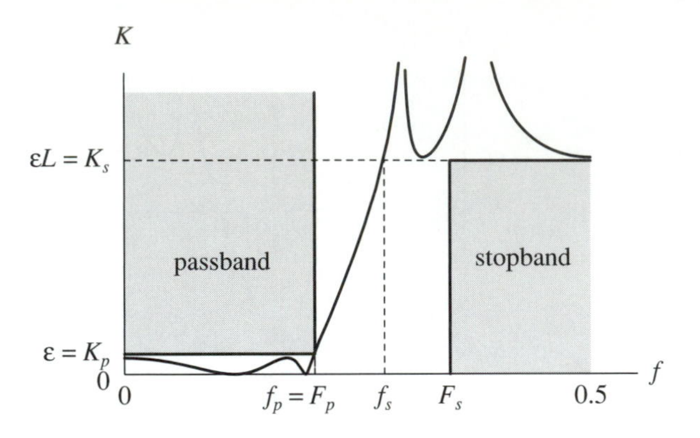

Figure 9.12 Design D3a.

This design has lower attenuation in the passband than is required by S_A. We choose this design when we prefer to achieve as low an attenuation as possible in the passband—that is, $\epsilon < K_p$. Also, this design is suitable when filter finite wordlength effects significantly affect the magnitude response in the passband. In that case, we achieve the highest attenuation margin in the passband (the margin is $K_p - \epsilon$, Fig. 9.11), and we expect that the imperfections of the implemented filter will not violate the specification.

9.5.3 Design D3a

In the design D3a we choose the minimal selectivity factor, $\xi = \xi_{min}$, and set the two design parameters, $\epsilon = K_p$, $f_p = F_p$, directly from the specification (Fig. 9.12).

This design has the sharpest magnitude response. When undesired signals exist in the transition region we may prefer design D3a, because it rejects the undesired signals as much as possible.

Disadvantages of D3a can be very high Q-factors and large variation of the group delay in the passband. It should be noticed that higher Q-factors imply higher sensitivity of the magnitude response to the filter parameters.

9.5.4 Design D3b

For the design D3b we choose the minimal selectivity factor, $\xi = \xi_{min}$ (the same as in Design 3a) and set the ripple factor, $\epsilon = K_p$, directly from the specification (Fig. 9.13). The actual passband edge is computed from $f_p = f_{p,max} = \frac{1}{\pi} \tan^{-1}\left(\frac{\tan(\pi F_s)}{\xi_{min}}\right)$.

This design has the sharpest magnitude response (the same as D3a). When the desired signals exist in the transition region we may prefer the design D3b, because it attenuates the desired signals as low as possible.

A disadvantage of the design D3b can be very high Q-factors. Although the variation of the group delay can be high, its maximal value can be moved into the transition region, so the group-delay variation can be acceptable in the passband.

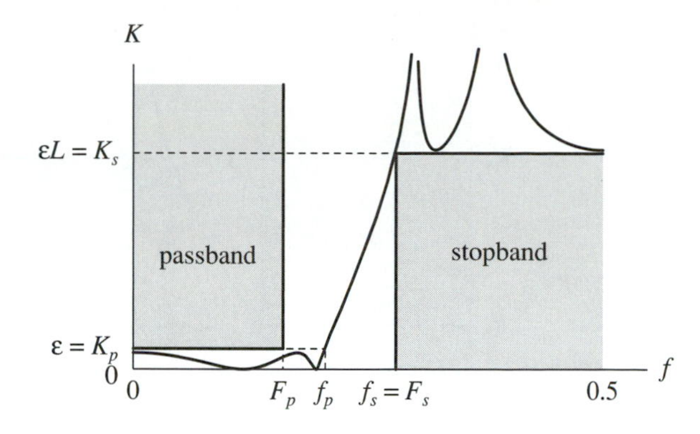

Figure 9.13 Design D3b.

9.5.5 Design D4a

In the design D4a we choose the maximal selectivity factor, $\xi = \xi_{max}$, and set the two design parameters, $\epsilon = K_p$, $f_p = F_p$, directly from the specification (Fig. 9.14).

This design (like the design D1) has higher attenuation in the stopband than is required by the specification, except at the stopband edge frequency. We choose this design when we prefer to achieve as large an attenuation as possible in the stopband; that is $\epsilon L > K_s$, except at $f = F_s$.

The design D4a has a smoother magnitude response.

9.5.6 Design D4b

For the design D4b we choose the maximal selectivity factor, $\xi = \xi_{max}$ (the same as in the design D4a) (Fig. 9.15), and calculate the ripple factor from $\epsilon = K_s/L(n, \xi_{max})$. The actual passband edge is computed from $f_p = f_{p,min} = \dfrac{1}{\pi} \tan^{-1}\left(\dfrac{\tan(\pi F_s)}{\xi_{max}}\right)$

This design (like the design D2) has lower attenuation in the passband than is required by the specification, except at the passband edge frequency.

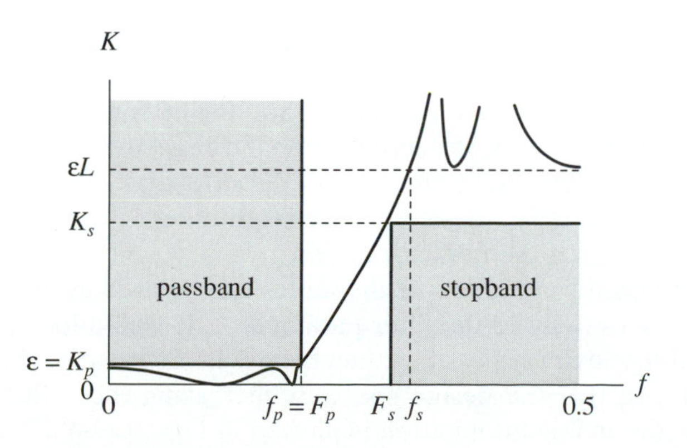

Figure 9.14 Design D4a.

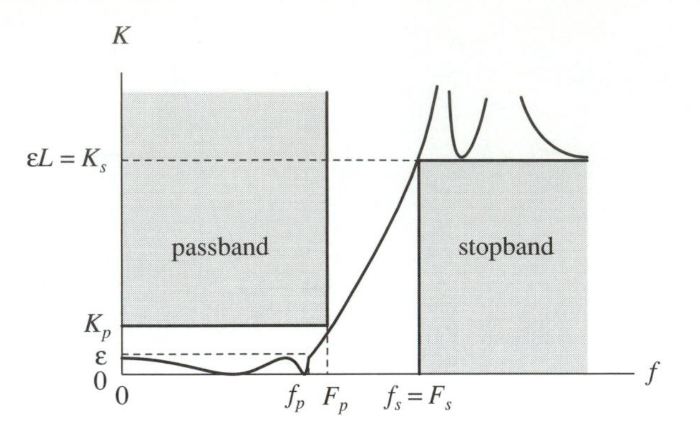

Figure 9.15 Design D4b.

We choose this design when we prefer to achieve as low an attenuation as possible in the passband; that is, $\epsilon < K_p$, except at $f = F_p$.

The design D4b has a smoother magnitude response. This design usually yields very low Q-factors. The design D4b has a very low ripple factor ϵ.

It should be noticed that there exists a straightforward procedure for computing the ripple factor, ϵ, for a given selectivity factor ξ, which yields the minimal Q-factors [47]. We designate this design by D5.

A disadvantage of D3a, D3b, D4a, and D4b is lack of any attenuation margin. Any imperfection, usually in implementation step (like coefficient quantization), can violate the specification.

9.5.7 Remarks on Design Alternatives

The approach that we propose in the previous sections we have programmed in *Mathematica* [2, 76]. Several design examples are exercised for an illustrative specification:

$$S_A = \{F_p = 0.2,\ F_s = 0.212,\ A_p = 0.2\ \text{dB},\ A_s = 40\ \text{dB}\}$$

First, the attenuation-limit specification $S_A = \{0.2, 0.212, 0.2, 40\}$ is transformed into the characteristic-function-limit specification

$$S_K = \{0.2, 0.212, 0.2171, 100\}$$

Next, all designs are calculated, and the attenuation is plotted in Figs. 9.16–9.27. The design parameters, the actual stopband edge, the maximal attenuation in the passband, and the minimal attenuation in the stopband, are summarized in Tables 9.1 and 9.2. The symbol $|z_i|_{\max}$ designates the magnitude of the pole with the largest magnitude, and $Q_{\max}$ denotes the Q-factor of $|z_i|_{\max}$.

The last two columns of the tables can be used as an estimate of the magnitude response sensitivity to the filter parameters. If technological requirements impose as minimal sensitivity as possible, then the design D4b can be the best solution (Table 9.3).

We enlarge the design space by increasing the order from $n = 8$ to $n = 9$. The corresponding attenuation is plotted in Figs. 9.22–9.27, and the design results are summarized in Table 9.4.

In practice, we choose the most suitable design, D, from the determined design space D_S. Thus, we can try to meet various technological requirements (prescribed values of coefficients of digital IIR filter [77–81]), and advanced specifications, such as minimal group-delay variation, and minimal overshoot in step response.

By further increasing the filter order the design D4a arrives at the Chebyshev-type approximation, for $f_s = \frac{1}{2}$. Alternatively, for the same order and $f_p = 0$, the design D4b yields an inverse Chebyshev-type filter. When the filter order is equal to the order of the Butterworth-type filter, with $f_p = 0$ and $f_s = \frac{1}{2}$, the elliptic approximation transforms into the Butterworth approximation. This means that the classic filter types, Chebyshev, inverse Chebyshev, and Butterworth, are just special cases of the elliptic filters and are contained within the design space D_S.

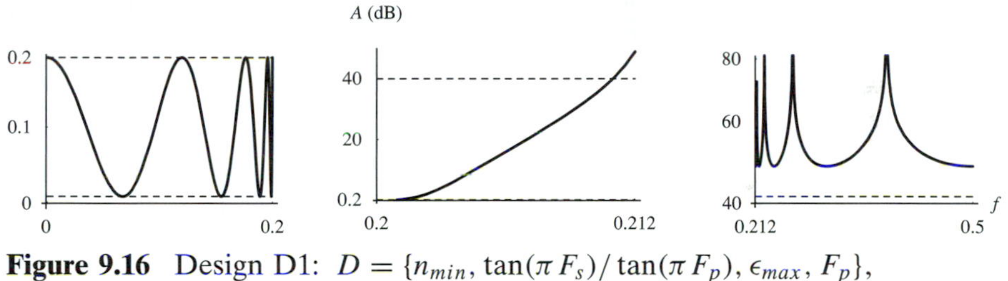

Figure 9.16 Design D1: $D = \{n_{min}, \tan(\pi F_s)/\tan(\pi F_p), \epsilon_{max}, F_p\}$, $S_A = \{0.2, 0.212, 0.2, 40\}$.

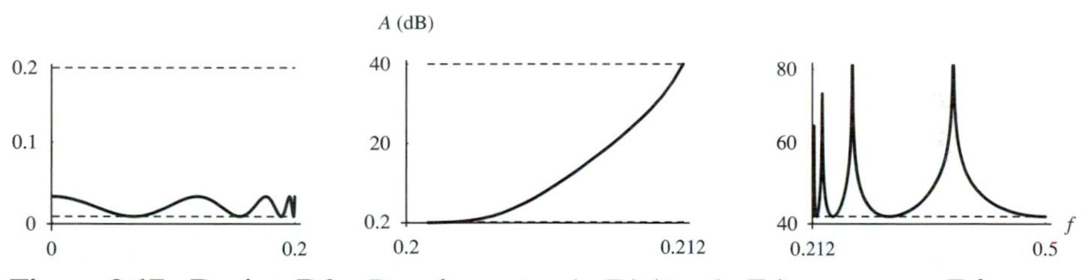

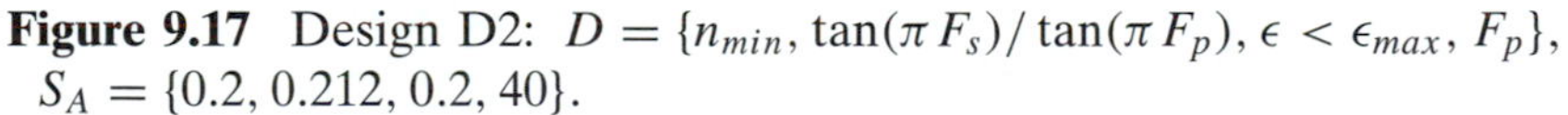

Figure 9.17 Design D2: $D = \{n_{min}, \tan(\pi F_s)/\tan(\pi F_p), \epsilon < \epsilon_{max}, F_p\}$, $S_A = \{0.2, 0.212, 0.2, 40\}$.

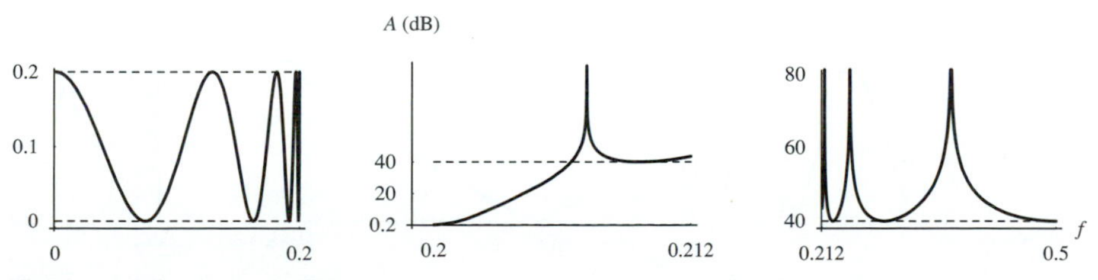

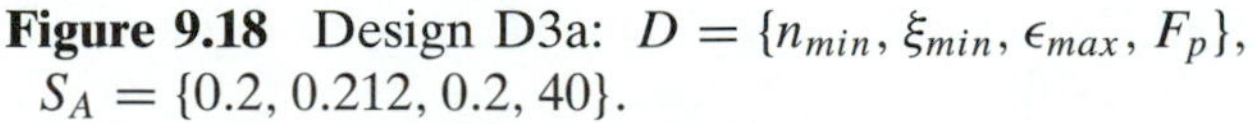

Figure 9.18 Design D3a: $D = \{n_{min}, \xi_{min}, \epsilon_{max}, F_p\}$, $S_A = \{0.2, 0.212, 0.2, 40\}$.

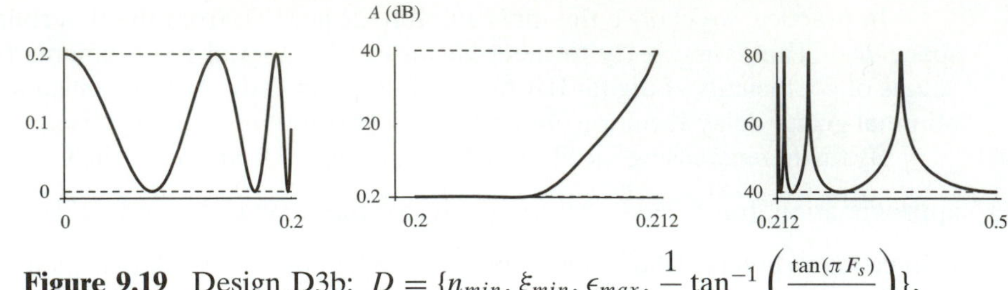

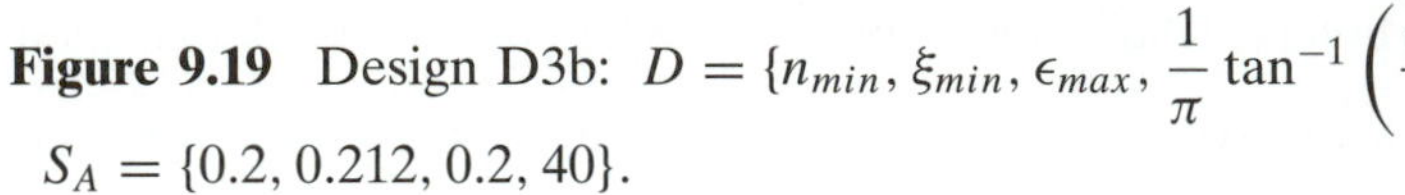

Figure 9.19 Design D3b: $D = \{n_{min}, \xi_{min}, \epsilon_{max}, \frac{1}{\pi}\tan^{-1}\left(\frac{\tan(\pi F_s)}{\xi_{\min}}\right)\}$, $S_A = \{0.2, 0.212, 0.2, 40\}$.

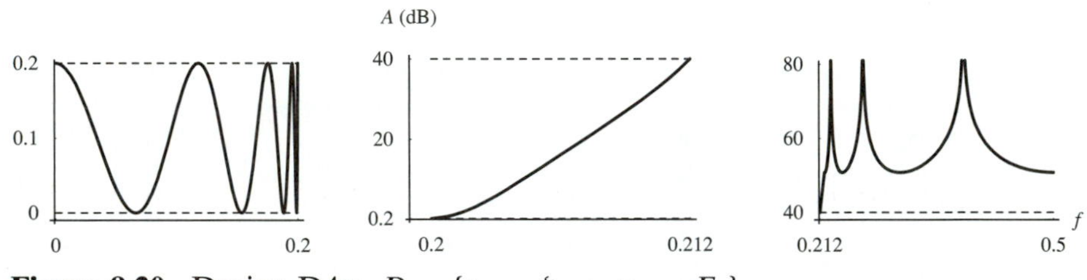

Figure 9.20 Design D4a: $D = \{n_{min}, \xi_{max}, \epsilon_{max}, F_p\}$, $S_A = \{0.2, 0.212, 0.2, 40\}$.

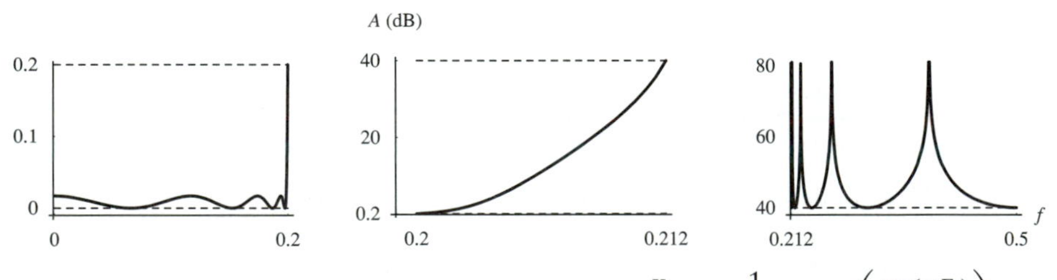

Figure 9.21 Design D4b: $D = \{n_{min}, \xi_{max}, \frac{K_s}{L(n_{\min}, \xi_{\max})}, \frac{1}{\pi}\tan^{-1}\left(\frac{\tan(\pi F_s)}{\xi_{max}}\right)\}$, $S_A = \{0.2, 0.212, 0.2, 40\}$.

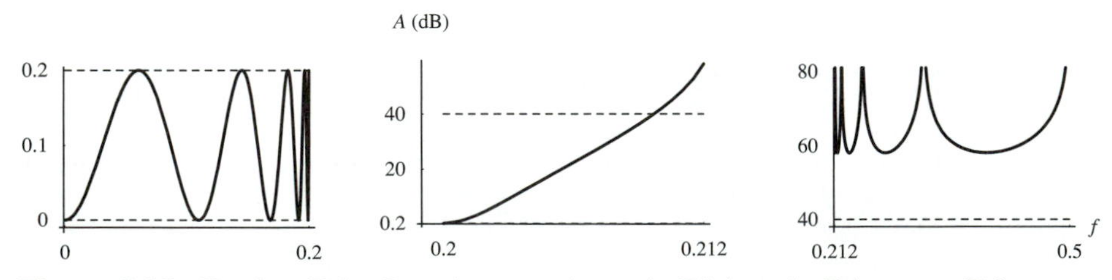

Figure 9.22 Design D1: $D = \{n_{min} + 1, \tan(\pi F_s)/\tan(\pi F_p), \epsilon_{max}, F_p\}$, $S_A = \{0.2, 0.212, 0.2, 40\}$.

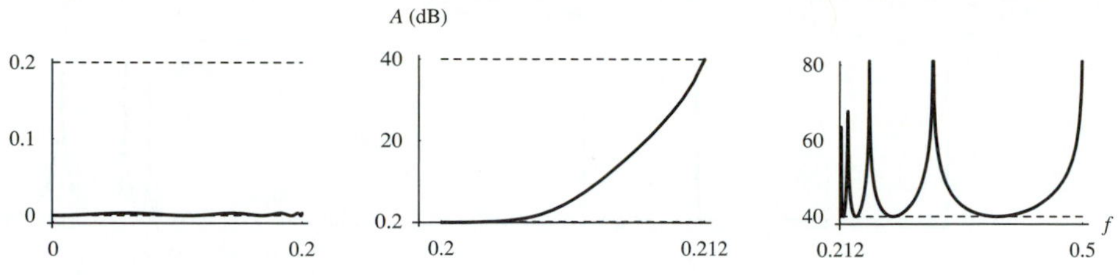

Figure 9.23 Design D2: $D = \{n_{min} + 1, \tan(\pi F_s)/\tan(\pi F_p), \epsilon < \epsilon_{max}, F_p\}$, $S_A = \{0.2, 0.212, 0.2, 40\}$.

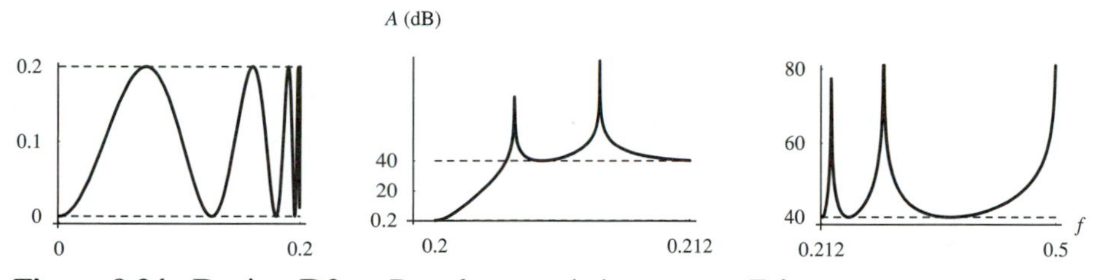

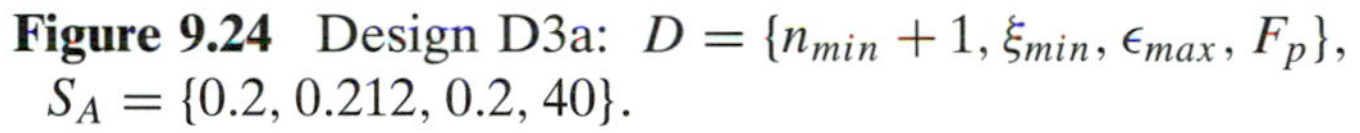

Figure 9.24 Design D3a: $D = \{n_{min} + 1, \xi_{min}, \epsilon_{max}, F_p\}$, $S_A = \{0.2, 0.212, 0.2, 40\}$.

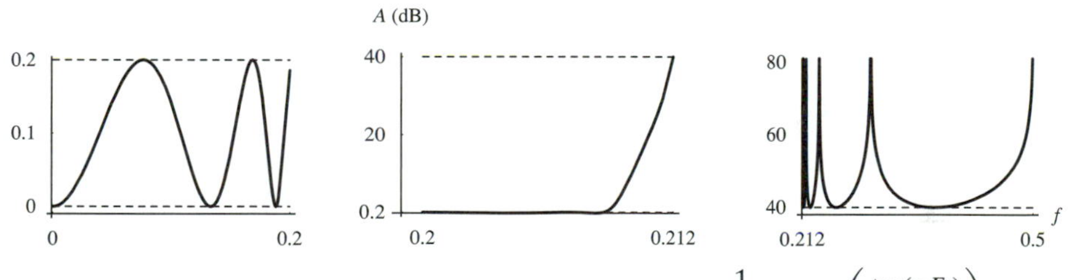

Figure 9.25 Design D3b: $D = \{n_{min} + 1, \xi_{min}, \epsilon_{max}, \frac{1}{\pi} \tan^{-1}\left(\frac{\tan(\pi F_s)}{\xi_{\min}}\right)\}$, $S_A = \{0.2, 0.212, 0.2, 40\}$.

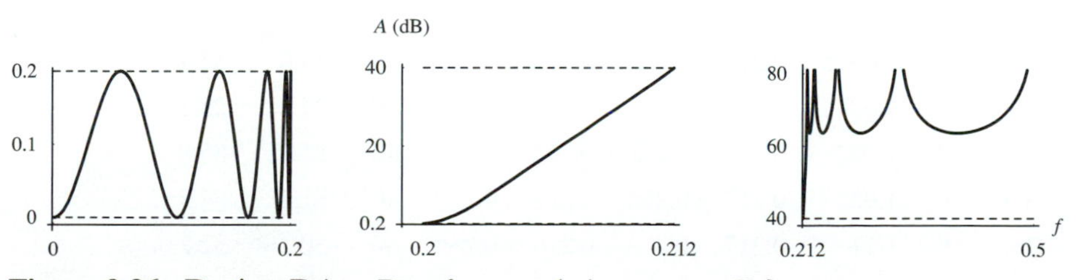

Figure 9.26 Design D4a: $D = \{n_{min} + 1, \xi_{max}, \epsilon_{max}, F_p\}$, $S_A = \{0.2, 0.212, 0.2, 40\}$.

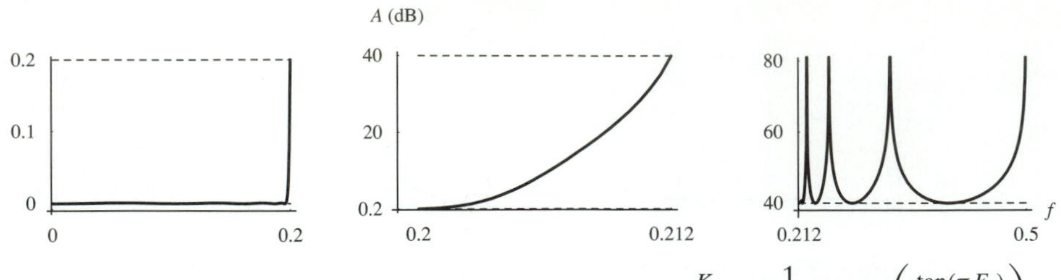

Figure 9.27 Design D4b: $D = \{n_{min}+1, \xi_{max}, \frac{K_s}{L(n,\xi_{\max})}, \frac{1}{\pi}\tan^{-1}\left(\frac{\tan(\pi F_s)}{\xi_{max}}\right)\}$, $S_A = \{0.2, 0.212, 0.2, 40\}$.

Table 9.1 Digital filter design summary for $n = n_{min}$

	n	ξ	ϵ	f_p	f_s	a_p dB	a_s dB	Q_{max}	$\frac{1}{1-\lvert z_i\rvert^2_{max}}$	$\frac{1}{(1-\lvert z_i\rvert^2_{max})^3}$
D 1	8	1.08155	0.217	0.2	0.212	0.2	49	28	30.4	28113
D 2	8	1.08155	0.079	0.2	0.212	0.03	40	22	24.0	13783
D 3a	8	1.04285	0.217	0.2	0.206	0.2	40	42	44.7	89094
D 3b	8	1.04285	0.217	0.205	0.212	0.2	40	42	44.3	86661
D 4a	8	1.09245	0.217	0.2	0.214	0.2	51	27	28.4	22797
D 4b	8	1.09245	0.063	0.198	0.212	0.02	40	20	21.6	10028

Table 9.2 Digital filter design summary for $n = n_{min}+1$

	n	ξ	ϵ	f_p	f_s	a_p dB	a_s dB	Q_{max}	$\frac{1}{1-\lvert z_i\rvert^2_{max}}$	$\frac{1}{(1-\lvert z_i\rvert^2_{max})^3}$
D 1	9	1.0815	0.217	0.2	0.212	0.2	58	35.6	37.9	54328
D 2	9	1.0815	0.027	0.2	0.212	0.003	40	24.1	25.6	16747
D 3a	9	1.022	0.217	0.2	0.203	0.2	40	81.2	85.8	631677
D 3b	9	1.022	0.217	0.208	0.212	0.2	40	81.2	84.5	602459
D 4a	9	1.11	0.217	0.2	0.216	0.2	60	30.2	32.2	33309
D 4b	9	1.11	0.014	0.196	0.212	$\frac{1}{10^3}$	40	19.1	20.5	8599

Table 9.3 Design space for $S_A = \{0.2,\ 0.212,\ 0.2\ \text{dB},\ 40\ \text{dB}\}$

	Filter Order, n					
	8	9	10	11	12	13
ϵ_{min}	0.063	0.0143	$\frac{6}{10^5}$	$\frac{1.2}{10^6}$	$\frac{1.1}{10^8}$	$\frac{4.3}{10^{11}}$
ϵ_{max}	0.2171	0.2171	0.2171	0.2171	0.2171	0.2171
$f_{p,min}$	0.198	0.196	0.193	0.188	0.181	0.171
$f_{p,max}$	0.206	0.209	0.210	0.211	0.212	0.212
$f_{s,min}$	0.206	0.203	0.202	0.201	0.200	0.200
$f_{s,max}$	0.214	0.216	0.220	0.225	0.232	0.243

9.6 VISUALIZATION OF DESIGN SPACE

Consider a lowpass filter from Section 9.5 specified by

$$S_A = \{F_p, F_s, A_p, A_s\} = \{0.2, 0.212, 0.2\,\text{dB}, 40\,\text{dB}\}$$

with the characteristic-function-limit specification

$$S_K = \{F_p, F_s, K_p, K_s\} = \{0.2, 0.212, 0.2171, 100\}$$

The minimal filter order has been calculated to be $n_{\text{min}} = 8$. Next, the range of ξ, ϵ, f_p, and f_s has been determined for $n \geq 8$.

The design subspaces for $n = 8$, $n = 9$, and $n = 13$ are shown in Figs. 9.28–9.30. Notice that the exact design subspaces are nonlinear continuous domains, but we draw rectangular blocks for the sake of simplicity.

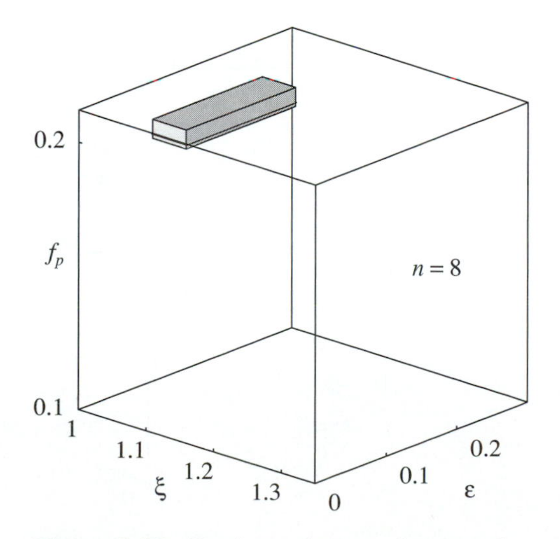

Figure 9.28 Design subspace for $n = 8$.

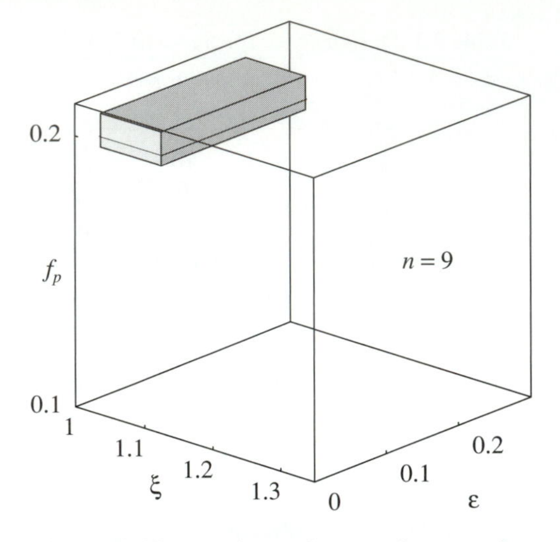

Figure 9.29 Design subspace for $n = 9$.

Tables 9.4 and 9.5 show that the Q-factor varies from 18 to 1121. It should be noticed that for the same specification S_A the MATLAB signal processing toolbox [1] offers an elliptic filter design with $Q = 42$, which might be inappropriate for practical implementations. Other classic approximations require very high orders compared to $n_{ellip} = 8$. The order of the Butterworth filter is extremely high ($n_{butt} = 79$), while the order of the Chebyshev and inverse Chebyshev type is also high ($n_{cheb} = 18$).

By increasing the order of the design D4b, we decrease the largest pole magnitude $|z_i|$ and decrease the magnitude response sensitivity. In contrast, by increasing the filter order of the design D3a and D3b, the magnitude response sensitivity increases.

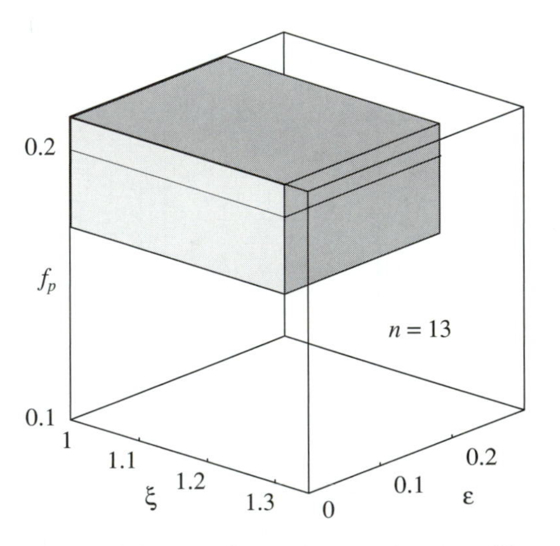

Figure 9.30 Design subspace for $n = 13$.

Table 9.4 Q-factors for D4b, $S_A = \{0.2,\ 0.212,\ 0.2\ \text{dB},\ 40\ \text{dB}\}$

	Filter Order, n					
	8	9	10	11	12	13
Q	20.14	19.14	18.65	18.37	18.20	18.19
$\lvert z_i\rvert^2_{max}$	0.9486	0.9536	0.9512	0.9499	0.9491	0.9486

Table 9.5 Q-factors for D3a, $S_A = \{0.2,\ 0.212,\ 0.2\ \text{dB},\ 40\ \text{dB}\}$

	Filter Order, n					
	8	9	10	11	12	13
Q	42.12	81.20	156.5	301.7	581.7	1121.
$\lvert z_i\rvert^2_{max}$	0.9772	0.9881	0.9938	0.9968	0.9983	0.9991

The range of the design parameters ξ, ϵ, f_p, and f_s are shown in Figs. 9.31 and 9.32. The minimal filter order, $n = n_{min}$ implies a small range for design parameters, and the optimization of the filter behavior can be ineffective.

It is also worth noticing that increasing the filter order, $n > n_{min}$, does not necessarily lead to a better solution. However, there exists an advanced design strategy in which we choose $n > n_{min}$ and obtain robust and selective digital filters with better performance and reduced complexity.

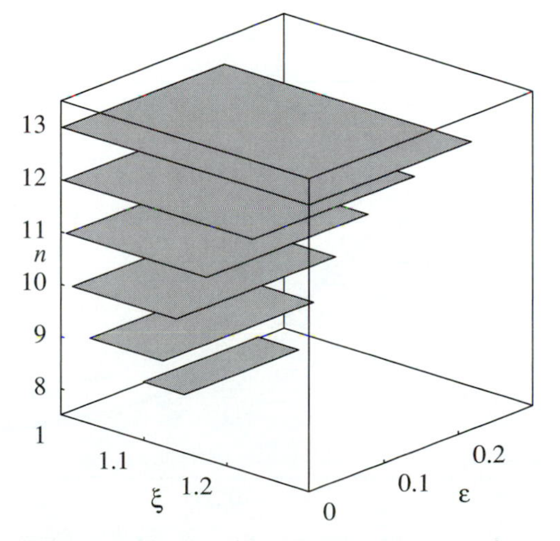

Figure 9.31 Design space of ξ, ϵ and n.

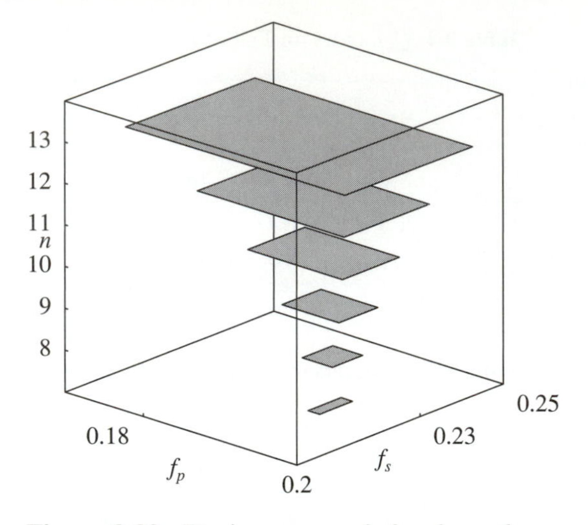

Figure 9.32 Design space of f_p, f_s, and n.

The group delay of the designs D1–D4 is plotted in Figs. 9.33 and 9.34. The maximum delay is obtained for minimal transition designs, D3a and D3b, while the maximal transition designs, D4a and D4b, have lower group-delay variation. The design D3b has the minimal variation of the group delay in the passband, while the similar design D3a has the highest overall group-delay variation.

The impulse response of the designs D1–D4 is shown in Fig. 9.35. The shape of all responses is the same with approximately the same amount of overshoots. As we expect, D4b has the smallest time delay.

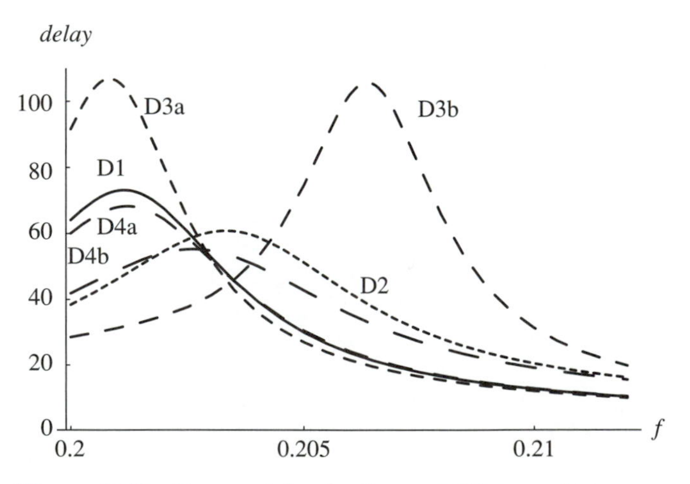

Figure 9.33 Group delay in the transition region.

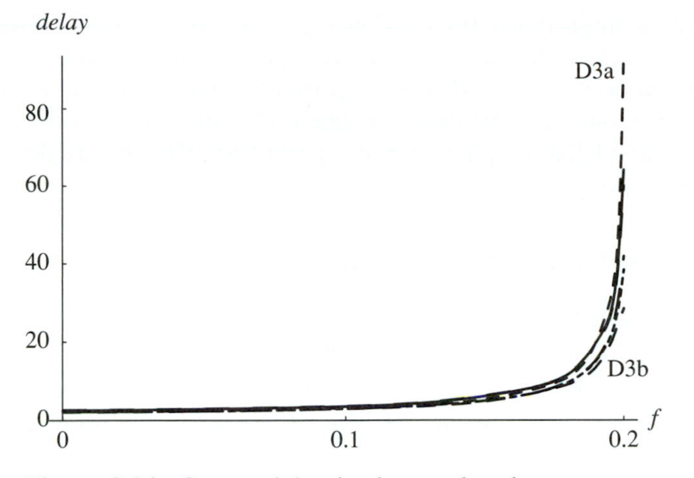

Figure 9.34 Group delay in the passband.

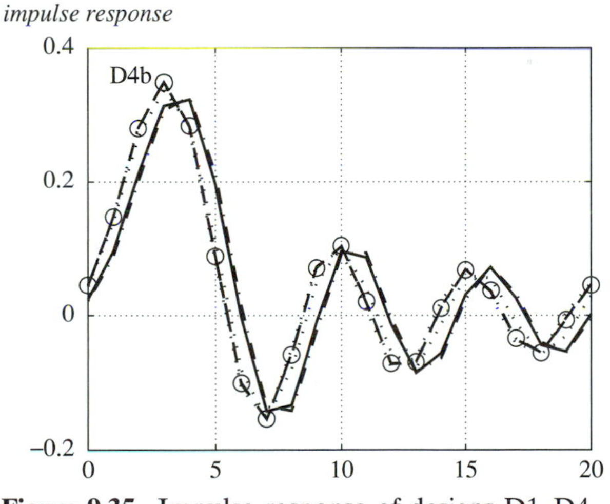

Figure 9.35 Impulse response of designs D1–D4.

9.7 ELLIPTIC HALF-BAND IIR FILTERS

Multirate digital signal processing is a very important part of modern digital telecommunications because digital transmission systems are required to handle data at several rates, such as teletype, facsimile, low bit-rate speech, and video. A special class of digital filters for multirate systems, known as the half-band filters, provides improved efficiency in some fairly general applications [82]. Half-band filters are also used for decimation in A/D conversion.

Filters that divide the operating frequency range of a discrete-time system into two equal parts are known as *half-band filters*. The FIR filters are most often used as half-band filters due to the suitability in saving the number of arithmetic operations and, also, due to their linear phase response [82]. The main disadvantage of FIR filters is that the number of filter coefficients required for sharp-cutoff filters is generally quite large [82]. The main disadvantage of classic IIR filters is their non-linear phase response. Powell and Chau [83] have reported real-time IIR filter realizations with exact linear phase response [83].

9.7.1 Definition of Half-Band Filters

Initially, half-band filters have been realized as FIR filters possessing the frequency symmetry in the passband and stopband [82].

Let us denote the edge frequency of the passband by F_p and denote the stopband edge frequency by F_s. Half-band filters, by definition, must fulfill the frequency symmetry condition

$$F_s = \frac{1}{2} - F_p \tag{9.41}$$

For the half-band IIR filter the magnitude ripple condition must be met:

$$\Delta_p = \Delta_s \tag{9.42}$$

where

$$\begin{aligned} M_p &= |H(e^{j2\pi F_p})| \\ M_s &= |H(e^{j2\pi F_s})| \\ \Delta_p &= 1 - M_p^2 \\ \Delta_s &= M_s^2 \end{aligned} \tag{9.43}$$

(see Fig. 9.36).

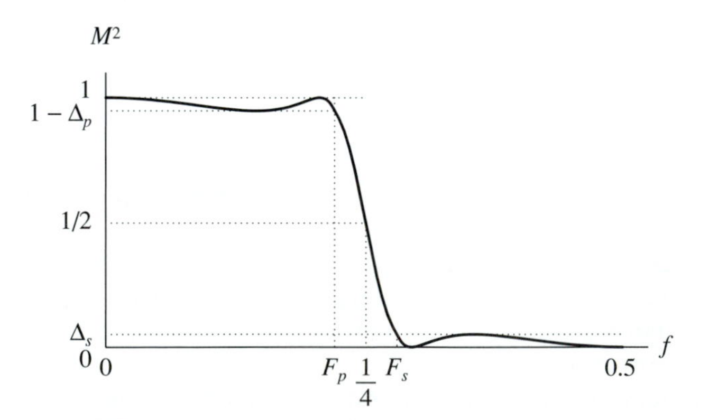

Figure 9.36 Squared magnitude response of half-band filter.

In the middle of the operating frequency range $0 \leq f \leq \frac{1}{2}$, at $f_{3\text{dB}} = \frac{1}{4}$, the squared magnitude response is $\frac{1}{2}$:

$$M^2(f_{3\text{dB}}) = \left|H(e^{j2\pi f_{3\text{dB}}})\right|^2 = \frac{1}{2} \tag{9.44}$$

We call $f_{3\text{dB}}$ the 3-dB frequency because the attenuation at the frequency $f_{3\text{dB}}$ is $a(f_{3\text{dB}}) = -10\log\left(M^2(f_{3\text{dB}})\right) \approx 3$ dB.

We can express the condition (9.42) in terms of δ_p and δ_s (see Fig. 9.37):

$$\begin{aligned} \delta_p &= 1 - \sqrt{1 - \delta_s^2} \\ \delta_s &= \sqrt{\delta_p\,(2 - \delta_p)} \end{aligned} \tag{9.45}$$

where

$$\begin{aligned} \delta_p &= 1 - M_p \\ \delta_s &= M_s \end{aligned} \tag{9.46}$$

Let us denote by A_p the maximal attenuation in the passband in dB, and by A_s the minimal attenuation in the stopband in dB. If δ_p and δ_s satisfy relation (9.45), then the attenuations in the passband and the stopband must be such that

$$\begin{aligned} A_p &= -10\log\left(1 + \frac{1}{10^{A_s/10} - 1}\right) \\ A_s &= -10\log\left(1 - 10^{A_p/10}\right) \end{aligned} \tag{9.47}$$

It should be noticed that the specification for a half-band IIR filter is not the same as the specification for a half-band FIR filter [82]: the ripple of the magnitude response, $M(f) = |H(e^{j2\pi f})|$, in the passband and in the stopband is

$$\text{FIR [82]}: \quad \delta_s = 2\delta_p \tag{9.48}$$

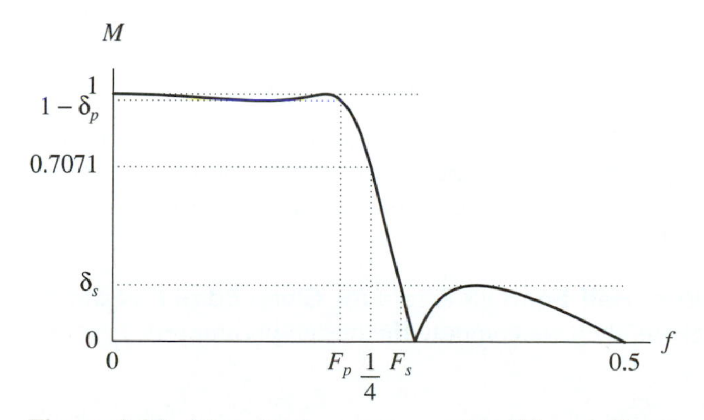

Figure 9.37 Magnitude response of half-band filter.

It is known that a special class of the elliptic transfer functions, the minimal Q-factor elliptic transfer functions [47], satisfies the relation (9.47). If we choose the edge frequencies to satisfy the frequency symmetry condition, Eq. (9.41), the minimal Q-factor elliptic transfer functions can be used for the design of the elliptic half-band IIR filters.

9.7.2 Elliptic Minimal Q-Factor Transfer Functions

Elliptic minimal Q-factor transfer functions have a property [47]

$$\epsilon^2 = \frac{1}{L} \tag{9.49}$$

where ϵ is the ripple factor and L can be computed by using equations derived in reference 53:

$$L = \sqrt{\frac{|H(e^{j2\pi F_s})|^2 - 1}{|H(e^{j2\pi F_p})|^2 - 1}} \tag{9.50}$$

It can be shown that the passband and stopband ripple are equal:

$$\Delta_p = \Delta_s = \frac{1}{1 + L} \tag{9.51}$$

From (9.49) and (9.51) it follows that for this class of filters the relation (9.47) is satisfied.

The selectivity factor ξ of elliptic half-band IIR filters can be calculated from one of the following expressions:

$$\begin{aligned} \xi &= \frac{\tan(\pi F_s)}{\tan(\pi F_p)} \\ \xi &= \frac{1}{\tan^2(\pi F_p)} \\ \xi &= \tan^2(\pi F_s) \end{aligned} \tag{9.52}$$

All three equations give the same result that can be proved by using the known trigonometry identity $\tan(\frac{\pi}{2} - \pi F_p) = 1/\tan(\pi F_p)$, $\tan(\frac{\pi}{2} - \pi F_s) = 1/\tan(\pi F_s)$, and the condition Eq. (9.41), that is, $\pi F_s = \frac{\pi}{2} - \pi F_p$.

For a given filter order n, and F_p or F_s, the selectivity factor ξ can be calculated from Eq. (9.52). L is a function of the filter order, n, and the selectivity factor, ξ, only; therefore, Δ_p and Δ_s are uniquely determined by n and ξ and cannot be arbitrarily specified.

Design of Half-Band IIR Filters for Given Edge Frequency. If the edge frequency F_s is given, then we compute the design parameters f_p and f_s from

$$\begin{aligned} f_p &= \frac{1}{2} - F_s \\ f_s &= F_s \end{aligned} \tag{9.53}$$

If F_p is given instead, then

$$\begin{aligned} f_p &= F_p \\ f_s &= \frac{1}{2} - F_p \end{aligned} \tag{9.54}$$

The selectivity factor is

$$\xi = \tan^2(\pi f_s) \tag{9.55}$$

Next, we find the attenuation design parameters a_s and a_p as follows:

$$\xi \geq \sqrt{2} \rightarrow \begin{cases} t = \dfrac{1}{2}\dfrac{1 - \sqrt[4]{1 - \dfrac{1}{\xi^2}}}{1 + \sqrt[4]{1 - \dfrac{1}{\xi^2}}} \\ q = t + 2t^5 + 15t^9 + 150t^{13} \end{cases}$$

$$\xi < \sqrt{2} \rightarrow \begin{cases} t = \dfrac{1}{2}\dfrac{1 - \dfrac{1}{\sqrt{\xi}}}{1 + \dfrac{1}{\sqrt{\xi}}} \\ q_p = t + 2t^5 + 15t^9 + 150t^{13} \\ q = e^{\pi^2/\log(q_p)} \end{cases} \tag{9.56}$$

$$L = \frac{1}{4}\sqrt{\frac{1}{q^n} - 1} \tag{9.57}$$

$$\begin{aligned} a_p &= 10\log_{10}\left(1 + \frac{1}{L}\right) \\ a_s &= 10\log_{10}(1 + L) \end{aligned} \tag{9.58}$$

Finally, we can use MATLAB for calculating the transfer function zeros, poles, and the gain constant:

```
[z,p,k]=ellip(n,ap,as,2*fp)
```

Design of Half-Band IIR Filters for Given Passband or Stopband Attenuation. If the stopband attenuation, A_s, is known, we calculate L from

$$L = 10^{A_s/10} - 1 \tag{9.59}$$

and the attenuation design parameters as

$$\begin{aligned} a_p &= -10\log\left(1 + \frac{1}{10^{A_s/10} - 1}\right) \\ a_s &= A_s \end{aligned} \tag{9.60}$$

If the passband attenuation, A_p, is known, we calculate L from

$$L = \frac{1}{10^{A_p/10} - 1} \tag{9.61}$$

and the attenuation design parameters as

$$\begin{aligned} a_p &= A_p \\ a_s &= -10\log\left(1 - 10^{A_p/10}\right) \end{aligned} \tag{9.62}$$

The selectivity factor, ξ, in terms of n and L, can be calculated by using approximate relations from reference 84 that we have adapted as follows:

$$t = \frac{1}{2}\frac{1 - \sqrt[4]{1 - \frac{1}{L^2}}}{1 + \sqrt[4]{1 - \frac{1}{L^2}}} \tag{9.63}$$

$$q = t + 2t^5 + 15t^9 + 150t^{13} \tag{9.64}$$

$$g = e^{\log(q)/n} \tag{9.65}$$

$$q_0 = \frac{g + g^9 + g^{25} + g^{49} + g^{81} + g^{121} + g^{169}}{1 + 2(g^4 + g^{16} + g^{36} + g^{64} + g^{100} + g^{144})} \tag{9.66}$$

$$\xi = \frac{1}{\sqrt{1 - \left(\frac{1 - 2q_0}{1 + 2q_0}\right)^4}} \tag{9.67}$$

The frequency design parameters are

$$\begin{aligned} f_p &= \frac{1}{\pi}\tan^{-1}\frac{1}{\sqrt{\xi}} \\ f_s &= \frac{1}{2} - \frac{1}{\pi}\tan^{-1}\frac{1}{\sqrt{\xi}} \end{aligned} \tag{9.68}$$

Finally, we can use MATLAB for calculating the transfer function zeros, poles, and the gain constant:

```
[z,p,k]=ellip(n,ap,as,2*fp)
```

Figure 9.38 displays the family of curves calculated for filters from the second to the twelfth order. The minimal stopband attenuation, A_s, is expressed as a function of the passband edge frequency, F_p.

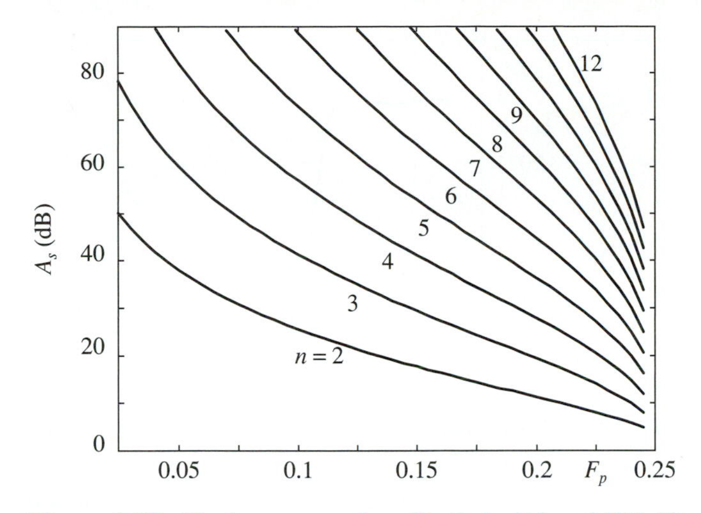

Figure 9.38 Design curves for elliptic half-band IIR filter.

9.7.3 1/3-Band IIR Filters

This section shows that very economical digital filters separating 1/3 of the frequency band can be designed. The magnitude ripple condition is the same as in the case of the half-band filter, $\Delta_p = \Delta_s$ (see Figs. 9.39 and 9.40).

The filter attenuation at $f_{3\text{dB}} = 0.5/3$ is 3 dB,

$$M^2(\frac{1}{6}) = \left|H(e^{j\pi/3})\right|^2 = \frac{1}{2} \tag{9.69}$$

and the edge frequencies of the passband and stopband are related by

$$\tan(\pi F_p)\tan(\pi F_s) = \tan^2(\pi f_{3\text{dB}}) = \tan^2\frac{\pi}{3} = \frac{1}{3} \tag{9.70}$$

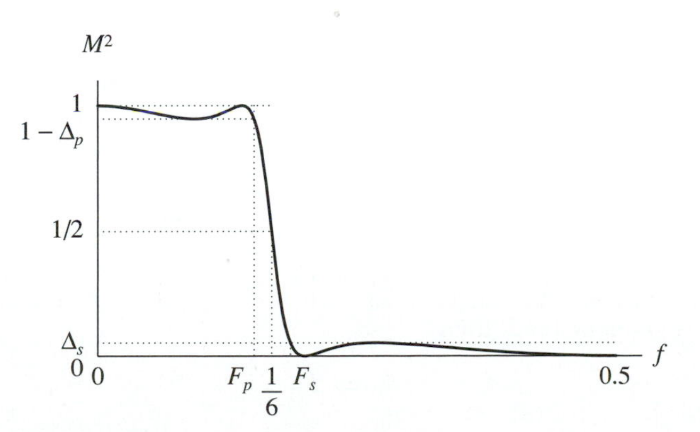

Figure 9.39 Squared magnitude response of 1/3-band IIR filter.

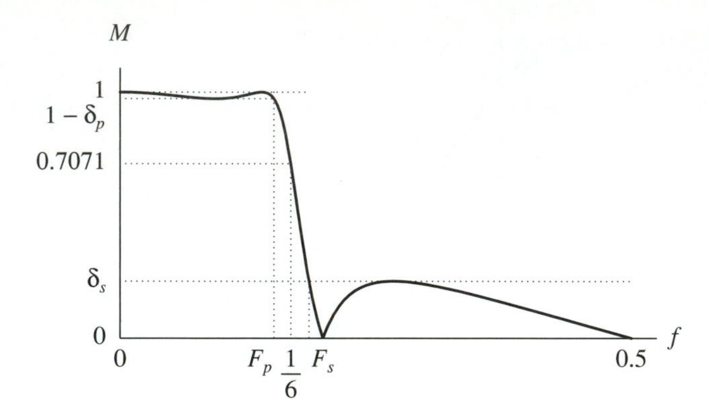

Figure 9.40 Magnitude response of 1/3-band IIR filter.

When this filter is realized as a parallel connection of two allpass networks, or as a lattice wave digital filter, considerable savings in the number of multipliers could be achieved; that is, it has been shown that the half of multipliers have the value $\alpha_i = \frac{1}{2}$ [78, 85, 86]

$$\alpha_i = -\cos(2\pi f_{3\text{dB}}) = -\cos\frac{\pi}{3} = -\frac{1}{2} \tag{9.71}$$

The remaining coefficients, β_i, for the second-order sections are calculated as squared pole magnitudes,

$$\beta_i = |z_i|^2 \tag{9.72}$$

while the coefficient of the first-order section is

$$\alpha_1 = -2 + \sqrt{3} \approx -\left(\frac{1}{2^2} + \frac{1}{2^6}\right) \tag{9.73}$$

For $\alpha_i = -1/2$, we realize $(n-1)/2$ multiplications by a binary shifting operation because the filter has $(n-1)/2$ second-order sections.

Design of 1/3-Band IIR Filters for Given Edge Frequency. If the edge frequency F_s is given, then we compute the design parameters f_p and f_s from

$$\begin{aligned} f_p &= \frac{1}{\pi}\tan^{-1}\frac{1}{3\tan(\pi F_s)} \\ f_s &= F_s \end{aligned} \tag{9.74}$$

If F_p is given instead, then

$$\begin{aligned} f_p &= F_p \\ f_s &= \frac{1}{\pi}\tan^{-1}\frac{1}{3\tan(\pi F_p)} \end{aligned} \tag{9.75}$$

The selectivity factor is

$$\xi = 3\tan^2(\pi f_s)$$
$$\xi = \frac{1}{3\tan^2(\pi f_p)} \tag{9.76}$$

Next, we proceed as for half-band filters, Eqs. (9.56)–(9.58).

Design of 1/3-Band IIR Filters for Given Passband or Stopband Attenuation. If the passband or stopband attenuation is known, we proceed as for half-band filters up to the step in which we calculate the frequency design parameters, Eqs. (9.59)–(9.67):

$$f_p = \frac{1}{\pi}\tan^{-1}\frac{1}{\sqrt{3\xi}}$$
$$f_s = \frac{1}{\pi}\tan^{-1}\sqrt{\frac{\xi}{3}} \tag{9.77}$$

Finally, we can use MATLAB for calculating the transfer function zeros, poles, and the gain constant:

```
[z,p,k]=ellip(n,ap,as,2*fp)
```

9.7.4 Comparison of FIR and IIR Half-Band Filters

A linear phase FIR filter designed to meet the half-band specification has the transfer function with every second coefficient reduced to zero. In multirate systems, if a half-band FIR filter is realized in a direct form, all computations can be evaluated at the lower sampling rate. On the other hand, the problem of computation savings in half-band IIR filters is more complex.

Usually, when we design half-band FIR filters, F_p and δ_p have to be specified. If we design elliptic half-band IIR filters, F_p and the filter order n have to be specified; δ_p is calculated from n and F_p [78, 85, 86]. For the half-band filter design, it is usually convenient to represent the stopband attenuation as a function of the transition bandwidth ΔF and the filter order n, as is given for FIR filters in reference 82. From the passband and stopband symmetry we obtain

$$\Delta F = F_s - F_p = \frac{1}{2} - 2F_p \tag{9.78}$$

Performance of half-band FIR filters reported in reference [82] is compared with the performance of half-band IIR filters (see Fig. 9.41). The design curves show relations between the stopband attenuation A_s and ΔF for several half-band FIR and elliptic IIR filters.

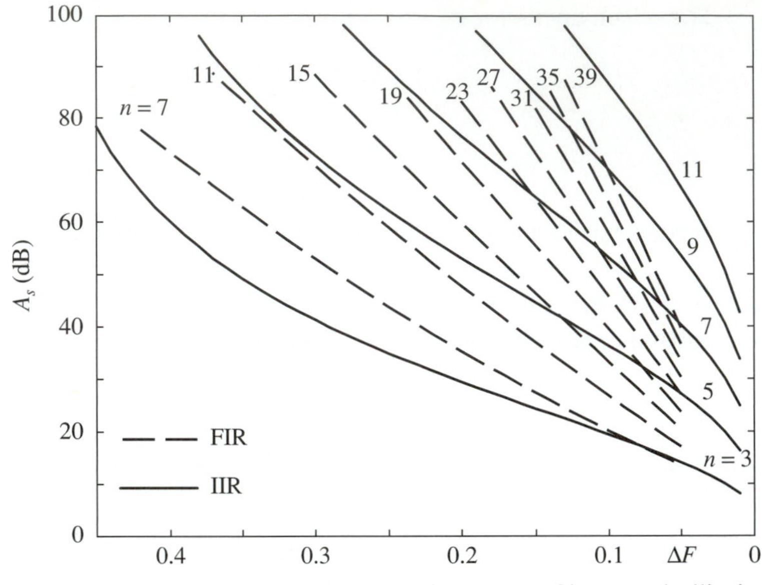

Figure 9.41 Design curves for half-band FIR filters and elliptic half-band IIR filters.

Notice that for the linear phase elliptic IIR filter [83], n and A_s in Fig. 9.41 have to be multiplied by 2. Figure 9.41 displays the efficiency of elliptic half-band IIR filters, particularly if we take into account the possibility of reducing the number of multipliers for odd-order filters.

Although the presented design for a reasonable stopband attenuation gives an elliptic IIR filter with a very low passband ripple, the realization of such a filter can be simpler than the realization of a lower-order IIR filter with a higher δ_p as shown in the following example.

The linear phase IIR filter [83], with the transfer function $H_{LPF}(z)$, has the magnitude response equal to the squared magnitude response of an elliptic IIR filter, with the transfer function $H(z)$:

$$|H_{LPF}(e^{j2\pi f})| = |H(e^{j2\pi f})|^2 \tag{9.79}$$

The nth-order elliptic half-band IIR filter transfer function is a minimal Q-factor elliptic transfer function [78, 86, 87] with the poles on the imaginary axis and can be written in the form

$$H(z) = c\,(z+1) \prod_{i=1}^{(n-1)/2} \frac{1 + \alpha_i z^{-1} + z^{-2}}{1 + \beta_i^2 z^{-2}}, \qquad n \text{ is odd} \tag{9.80}$$

9.7.5 Parallel Realization of Elliptic Half-Band IIR Filters

A parallel realization of elliptic half-band IIR filters is based on the partial fraction expansion of the odd-order transfer function

$$H(z) = c_0 + c_1 z^{-1} + \sum_{i=1}^{(n-1)/2} 2\frac{\mathrm{Re}(R_i) + z^{-1}\beta_i \,\mathrm{Im}(R_i)}{1 + \beta_i^2 z^{-2}} \tag{9.81}$$

where n is an odd integer, and R_i are the residues of the transfer function $H(z)$ at $z = j\beta_i$:

$$R_i = \lim_{z \to j\beta_i} H(z)\,(1 - j\beta_i z^{-1}) \tag{9.82}$$

It has been reported [78] that R_i can be either real, Re $R_i = 0$, or imaginary, Im $R_i = 0$, for n odd. The proof follows from the representation of $H(z)$ as a sum of two allpass transfer functions. It has been shown [78, 85, 86] that an elliptic transfer function can be expressed as a sum or difference of two allpass transfer functions, $H_a(z)$ and $H_b(z)$:

$$H(z) = \frac{H_a(z) + H_b(z)}{2} \tag{9.83}$$

The poles are on the imaginary axis, so $H_a(z)$ and $H_b(z)$ can be written in the form

$$\begin{aligned} H_a(z) &= z \prod_{1}^{[(n+3)/4]} \frac{\beta_i^2 + z^{-2}}{(1 + j\beta_i z^{-1})(1 - j\beta_i z^{-1})} \\ H_b(z) &= \prod_{[(n+7)/4]}^{(n+1)/2} \frac{\beta_i^2 + z^{-2}}{(1 + j\beta_i z^{-1})(1 - j\beta_i z^{-1})} \end{aligned} \tag{9.84}$$

where β_i are the transfer function poles (including $\beta_i = 0$), and $[x]$ gives the smallest integer greater than or equal to x.

When the residues R_i are computed according to (9.82), one of the addends $(H_a(z))\,(1 - j\beta_i z^{-1})$ or $(H_b(z))\,(1 - j\beta_i z^{-1})$ is zero, and the other, which contains the pole $z_i = j\beta_i$, has all real multiplication factors except for the factor $z = j\beta_i$, which is imaginary. Therefore, if the pole β_i belongs to $H_a(z)$, the residue R_i is imaginary, and if β_i belongs to $H_b(z)$, R_i is real.

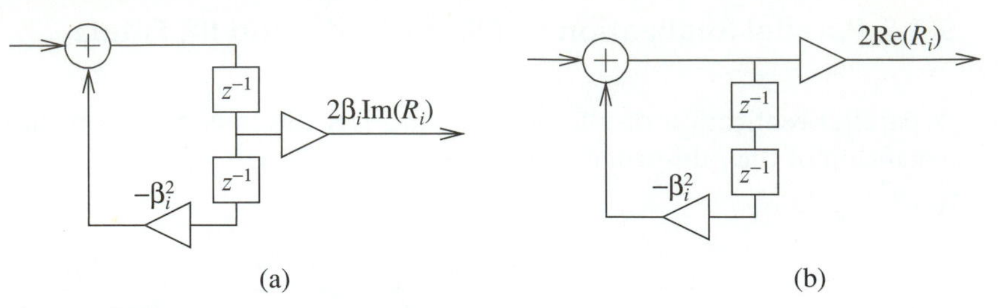

Figure 9.42 Basic second-order sections for parallel realization of elliptic half-band IIR filters: (a) Re $R_i = 0$, (b) Im $R_i = 0$.

Each second-order section from Fig. 9.42 has only one nonzero multiplier in the direct branches, instead of two. Consequently, a second-order section of a parallel realization is implemented with only two multipliers: one in the feedback branch for the complex-conjugate pole pair, and one in the direct branch for the corresponding residues. In summary, if the filter order n is an odd integer, an elliptic half-band IIR filter can be realized with only $n + 1$ multipliers.

Example 9.1

In Fig. 9.43 the attenuation curves of two IIR filters that have a practically equal number of multiplication constants are compared. The specification is $F_p = 0.22$, $F_s = 0.28$, and $A_s = 57$ dB. The parallel realization of the ninth-order elliptic half-band IIR filter requires 10 multipliers, as shown in Fig. 9.44. A fifth-order elliptic IIR filter, which is not a half-band filter, requires 11

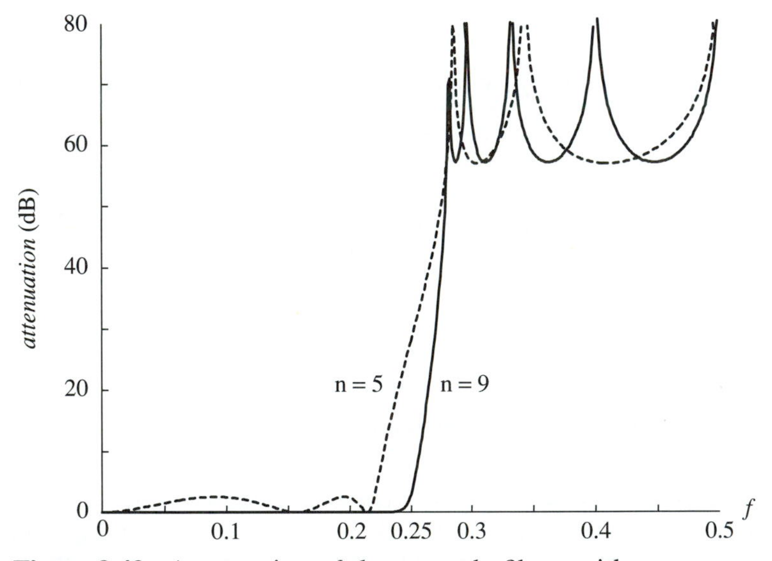

Figure 9.43 Attenuation of the example filters with $F_p = 0.22$, $F_s = 0.28$, $A_s = 57$ dB: ninth-order elliptic half-band IIR filter with $A_p = 0.000008316$ dB, solid line; fifth-order elliptic half-band IIR filter with $A_p = 2.5$ dB, dashed line.

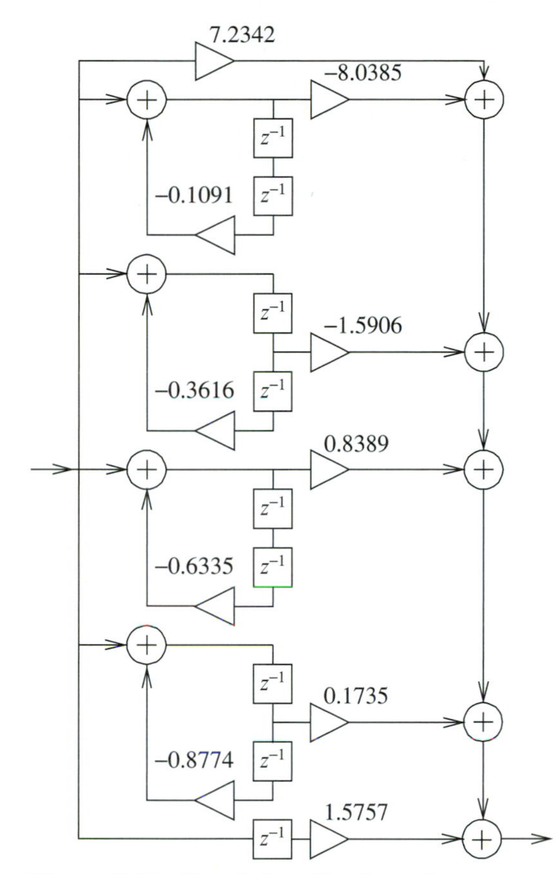

Figure 9.44 Parallel realization of ninth-order elliptic half-band IIR example filter.

multipliers for the parallel realization (Fig. 9.45). The magnitude of the pole closest to the unit circle for the ninth-order filter is $r_{9,1}$=0.9367, and for the fifth-order filter $r_{5,1}$=0.9563. Evidently, the ninth-order filter provides a sharper magnitude response, smaller pole magnitudes, and a very low passband attenuation. ◆

Conclusion. It is possible to design an elliptic half-band IIR filter with every second multiplication constant reduced to zero. The reduced number of multipliers can be achieved with odd-order elliptic minimal Q-factor transfer functions. The obtained IIR filter transfer function has two useful properties: (1) The poles are on the imaginary axis of the z-plane, and (2) the residues of the poles are either purely real or purely imaginary. Therefore, in the parallel realization based on the second-order sections, the number of multipliers is reduced by half. The parallel realization also permits the computations in the direct branches to be evaluated at the lower sampling rate. In addition, the obtained filter has a very low passband attenuation, and the pole magnitudes are smaller compared to other elliptic filters. A nearly flat magnitude response in the passband, the

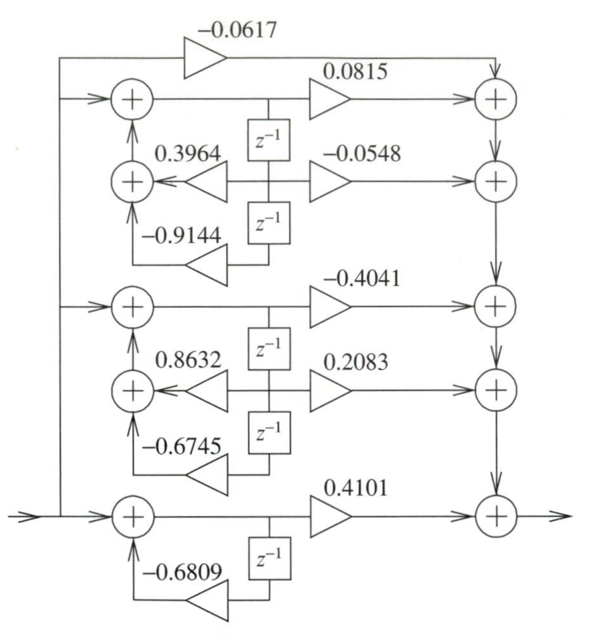

Figure 9.45 Parallel realization of fifth-order elliptic IIR example filter.

relatively small pole magnitudes, and considerable computation savings recommend this class of half-band filters for multirate applications.

9.8 HILBERT TRANSFORMER

The Hilbert transform is a set of mathematical equations relating the real and imaginary parts, or the magnitude and phase, of the Fourier transform of certain signals. Based on the properties of the discrete Hilbert transform, an *ideal Hilbert transformer* is defined as a DTLTI system with the frequency response

$$H_{HT}(e^{j2\pi f}) = \begin{cases} -j, & 0 \le f < \frac{1}{2} \\ +j, & -\frac{1}{2} \le f < 0 \end{cases}$$

An ideal Hilbert transformer is also called a 90-*degree phase-shifter*. It is non-causal and cannot be realized. Any realization that approximates the ideal Hilbert transformer is referred to as a *Hilbert transformer*. In this section we present a realization of Hilbert transformers using half-band filters.

9.8.1 Half-Band Filter and Hilbert Transformer

A half-band filter is defined by the passband-stopband symmetry related to $f = \frac{1}{4}$. If $H(z)$ is the transfer function of the half-band filter, then $H_H(z) = H(z/e^{j\pi/2}) =$

$H(-jz)$ is the transfer function of a complex filter with passband-stopband symmetry related to $f = \frac{1}{2}$. The squared magnitude responses of the Hilbert transformer and the half-band filter are plotted in Fig. 9.46. If the input sequence to a filter with $H_H(z)$ is real, say $x(n)$, then the output sequence is complex, say $y(n) = y_r(n) + jy_i(n)$, where $y_r(n)$ and $y_i(n)$ are real. The signal $y(n)$ is called a *complex analytic signal.* A filter with one input and two outputs, one for $y_r(n)$ and one $y_i(n)$, is the Hilbert transformer. The real and imaginary outputs of the complex filter form a Hilbert transform pair over the specified frequency range [88].

The key properties of the elliptic half-band IIR filter are

$$\begin{aligned} F_p &= \frac{1}{2} - F_s \\ \left|H(e^{j\pi/2})\right|^2 &= \frac{1}{2} \\ \Delta_p = \Delta_s \quad &\Leftrightarrow \quad a_p = 10\log_{10}\left(1 + \frac{1}{10^{a_s/10} - 1}\right) \end{aligned} \tag{9.85}$$

where

$$\begin{aligned} \Delta_p &= 1 - |H(e^{j2\pi F_p})|^2 \\ \Delta_s &= |H(e^{j2\pi F_s})|^2 \end{aligned} \tag{9.86}$$

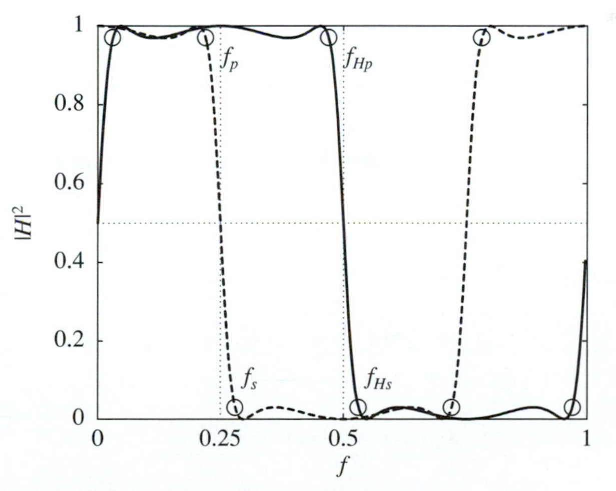

Figure 9.46 Squared magnitude response of Hilbert transformer (solid line) and half-band filter (dashed line).

The key properties of the Hilbert transformer are

$$\begin{aligned} F_{Hp} &= 1 - F_{Hs} \\ \left|H_H(e^{j\pi})\right|^2 &= \frac{1}{2} \\ \Delta_{Hp} = \Delta_{Hs} \quad &\Leftrightarrow \quad a_{Hp} = 10\log_{10}\left(1 + \frac{1}{10^{a_{Hs}/10} - 1}\right) \end{aligned} \tag{9.87}$$

where

$$\begin{aligned} \Delta_{Hp} &= 1 - |H_H(e^{j2\pi F_{Hp}})|^2 \\ \Delta_{Hs} &= |H_H(e^{j2\pi F_{Hs}})|^2 \end{aligned} \tag{9.88}$$

Hilbert transformers can be designed from half-band filters in a straightforward manner, by replacing z with $-jz$.

The transfer function poles of the half-band IIR filter are on the imaginary axis of the z-plane. The realization of the filter is based on the sum of two allpass transfer functions, $H_a(z)$ and $H_b(z)$ (Fig. 9.47):

$$H(z) = \frac{1}{2}(H_a(z) + H_b(z)) \tag{9.89}$$

with

$$H_a(z) = \prod_{i=[(n+7)/4]}^{(n+1)/2} \frac{\beta_i + z^{-2}}{1 + \beta_i z^{-2}} \tag{9.90}$$

$$H_b(z) = z^{-1} \prod_{i=3}^{[(n+1)/4]} \frac{\beta_i + z^{-2}}{1 + \beta_i z^{-2}} \tag{9.91}$$

where β_i is the square magnitude of a pole z_i, such that

$$0 \le \beta_i = |z_i|^2 < 1 \tag{9.92}$$

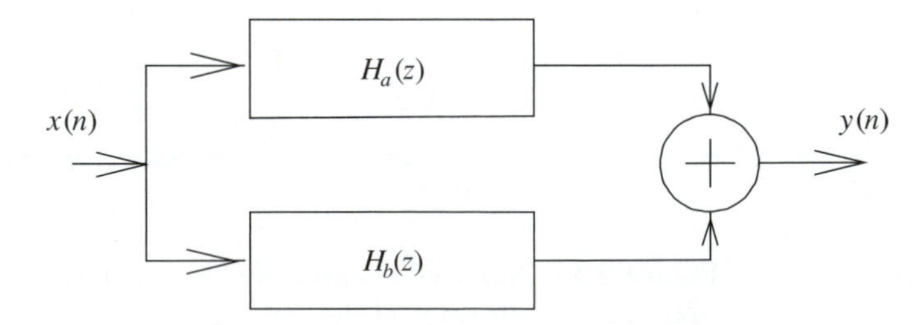

Figure 9.47 Realization of a half-band filter.

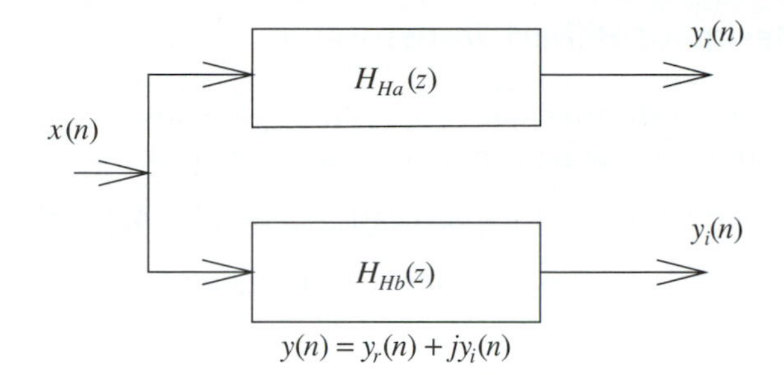

Figure 9.48 Realization of Hilbert transformer (complex filter).

The poles of the Hilbert transformer are on the real axis of the z plane. Replacing z with $-jz$ in $H_a(z)$ and $H_b(z)$, we obtain the transfer function of the complex filter (Fig. 9.48):

$$H_H(z) = H_{Ha}(z) + jH_{Hb}(z) \tag{9.93}$$

$$H_{Ha}(z) = \prod_{i=[(n+7)/4]}^{(n+1)/2} \frac{\beta_i - z^{-2}}{1 - \beta_i z^{-2}} \tag{9.94}$$

$$H_{Hb}(z) = z^{-1} \prod_{i=3}^{[(n+1)/4]} \frac{\beta_i - z^{-2}}{1 - \beta_i z^{-2}} \tag{9.95}$$

The allpass transfer functions $H_{Ha}(z)$ and $H_{Hb}(z)$ correspond to branches of a 90-degree phase-shifter (also called the $\frac{\pi}{2}$ *phase splitter*) whose outputs approximate the Hilbert transform pair [89].

We can rewrite Eqs. (9.94) and (9.95) as follows:

$$H_{Ha}(z) = \prod_{i=[(n+7)/4]}^{(n+1)/2} \frac{(-\beta_i) + z^{-2}}{1 + (-\beta_i)z^{-2}} \tag{9.96}$$

$$H_{Hb}(z) = (-1)^{(n+1)/2} z^{-1} \prod_{i=3}^{[(n+1)/4]} \frac{(-\beta_i) + z^{-2}}{1 + (-\beta_i)z^{-2}} \tag{9.97}$$

Eqations (9.96) and (9.97) show that the coefficients of the Hilbert transformer are the negative coefficients of the half-band filter. Therefore, the procedure for designing multiplierless elliptic IIR filters can be used for the design of multiplierless Hilbert transformers.

9.8.2 Design of Hilbert Transformer

The Hilbert transformer can be specified by the lower stopband edge frequency, F_{Hs}, and the minimum stopband attenuation, A_s, that is, $S_H = \{F_{Hs}, A_s\}$. The corresponding half-band filter has the specification $S = \{F_s, A_s\}$ with $F_s = F_{Hs} - \frac{1}{4}$. Next, the coefficients of the half-band filter are calculated to meet $S = \{F_s, A_s\}$. If the coefficients of the half-band filter are β_i, then the coefficients of the Hilbert transformer are equal to $-\beta_i$.

Example 9.2

Assume a half-band filter specification $F_s = 0.28$, along with $A_s = 46$ dB and $S = \{F_s, A_s\} = \{0.28, 46\}$. For $n = 9$ the two cases exist:

1. $a_s = A_s$, $f_s = 0.275 < F_s$;
2. $A_s < a_s = 57.2$ and $f_s = F_s$.

The specification is satisfied if $46 \leq a_s \leq 57.2$ and $0.275 \leq f_s \leq 0.28$.

For the two cases we find:

1. $n = 9$, $a_s = A_s = 46$ dB, $a_p = 0.00011$, $\xi = 1.217$, $f_s = 0.2656$, $f_p = 0.2344$ and $\beta_2 = 0.1532$, $\beta_3 = 0.7336$, $\beta_4 = 0.4646$, $\beta_5 = 0.9202$;
2. $n = 9$, $a_s = 57.18$ dB, $a_s = 0.0000083$, $\xi = 1.461$, $f_s = 0.28$, $f_p = 0.2216$, and $\beta_2 = 0.1091$, $\beta_3 = 0.3616$, $\beta_4 = 0.6335$, $\beta_5 = 0.8774$. ◆

Figs. 9.49 and 9.50 show attenuation: solid lines for $\beta_2 = 0.1206$, $\beta_3 = 0.6628$, $\beta_4 = 0.3900$, $\beta_5 = \beta_{\max} = 1 - 1/2^3 + 1/2^6$, dashed lines for quantized coefficients $\beta_{2q} = 1/2^3 - 1/2^8$, $\beta_{3q} = 1/2 + 1/2^3 + 1/2^5 + 1/2^7 = (1 + 1/2^2)(1/2 + 1/2^5)$, $\beta_{4q} = 1/2^2 + 1/2^3 + 1/2^6$, $\beta_5 = \beta_{\max} = 1 - 1/2^3 + 1/2^6$.

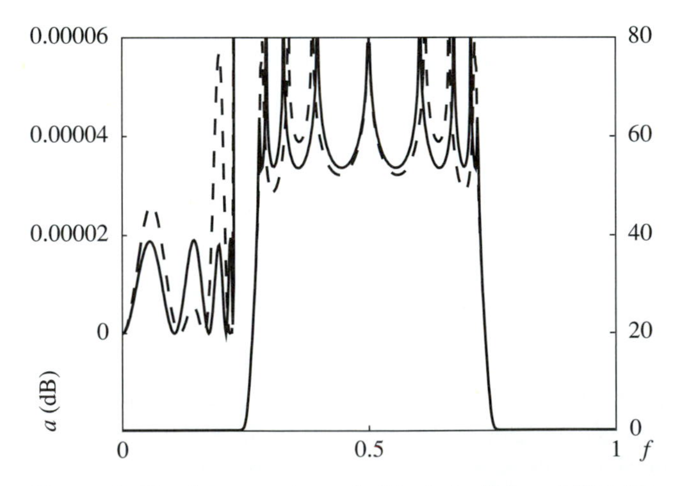

Figure 9.49 Attenuation of elliptic half-band IIR filter.

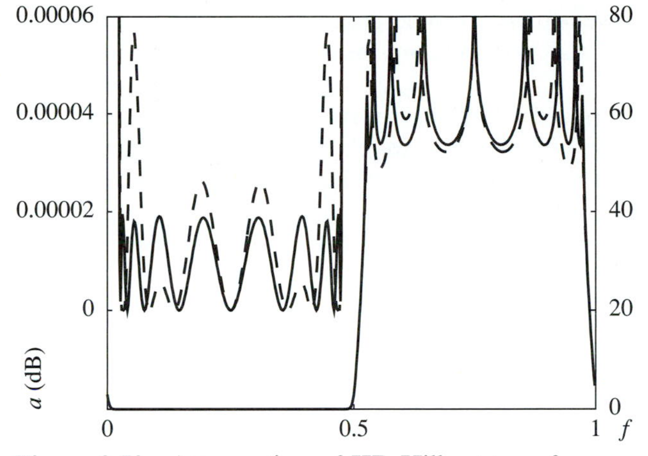

Figure 9.50 Attenuation of IIR Hilbert transformer.

A realization of the ninth-order half-band filter is shown in Fig. 9.51, and the corresponding complex filter is presented in Fig. 9.52, for $f_s = 0.2751$ and $a_s = 53.64$ dB. The corresponding magnitude responses are shown in Figs. 9.49 and 9.50.

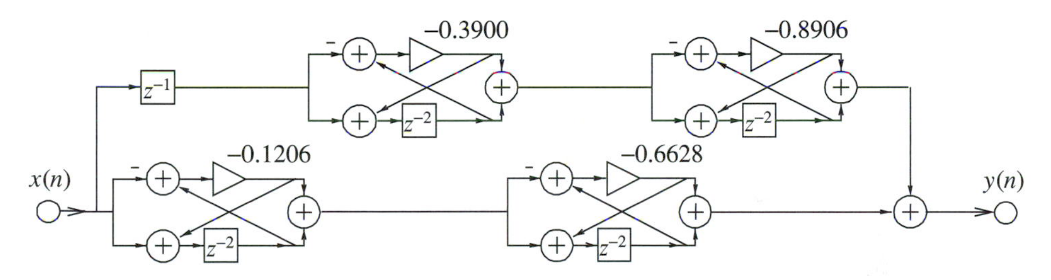

Figure 9.51 Ninth-order elliptic half-band IIR filter.

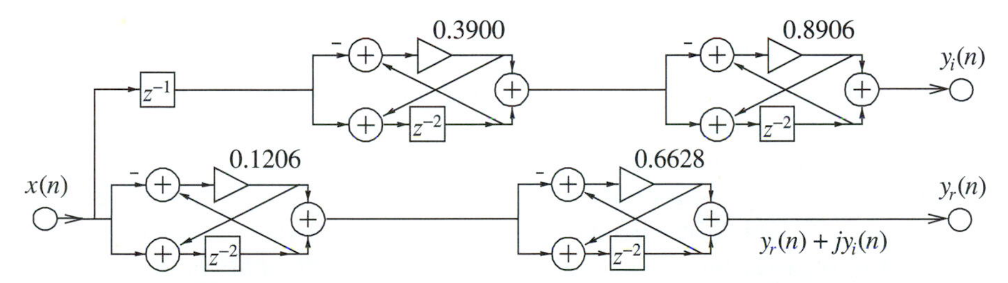

Figure 9.52 Realization of Hilbert transformer by using ninth-order elliptic complex IIR filter.

9.9 MULTIPLIERLESS ELLIPTIC IIR FILTERS

An analytical expression for amplitude response sensitivity is derived for the filter realizations consisting of two allpass subfilters in parallel. It is shown that the amplitude response sensitivity to some parameter x can be expressed as a product of the filter reflectance function and the phase sensitivity of the allpass section that implements the parameter.

The $(n+1)/2$ most sensitive parameters can be directly controlled by the transfer function parameters if we choose the elliptic minimal Q-factor (EMQF) transfer function. This way, $(n+1)/2$ multiplier coefficients can be implemented without quantization leaving the filter characteristic strictly elliptic. This is achieved for a class of low-noise allpass sections and for the wave lattice digital filter, as well.

The quantization of the remaining $(n-1)/2$ less sensitive parameters is performed using the phase tolerance scheme and phase sensitivity functions. Our design technique is straightforward, and consequently very fast. The application is demonstrated on the examples of narrowband, wideband, and half-band filters.

9.9.1 Introduction

A digital filter whose multiplier coefficients are implemented with a small number of shift-and-add operations is called the *multiplierless filter*. The multiplierless filter complexity is considerably reduced in implementation.

Design of multiplierless IIR filters is particularly difficult because of the high sensitivity of IIR filter frequency responses to multiplier coefficients. An approach is given in references [90] and [91] where the application of a multiplier block is proposed for the implementation of a group of multipliers sharing the common input. This approach can be applied in the direct, cascade, and parallel realizations of digital filters. The multiplier block concept is inapplicable for the wave digital filters and other realizations based on allpass sections.

Multiplierless elliptic IIR filters are considered in references 92–94. It has been shown that the elliptic minimal Q-factor (EMQF) transfer function [47] provides half of multipliers to be implemented with a small number of shift-and-add operations. The multiplierless filters in references 92 and 93 rely on parallel-of-two-allpass realizations [75, 95] and wave digital filters [71].

It is well known that the variable coefficient wordlength implementations are superior to the uniform wordlength case [96–100]. Thus, for the multiplierless design, we investigate the individual coefficient sensitivities.

It follows that the amplitude response sensitivity to an arbitrary multiplier coefficient is determined by the phase sensitivity of the allpass section that implements the coefficient. The characteristics of the phase sensitivity functions of allpass sections provide an easy way for the computation of the range of permitted values for each multiplier coefficient. It has been discovered that by the appropriate design of the nth-order EMQF transfer function, $(n+1)/2$ multiplier coefficients of the highest sensitivity can be implemented as exact values with a minimal number of shift-and-add operations. This feature offers the whole design margin to the quantization process of less sensitive parameters.

Using the fact that the amplitude response sensitivity is simply expressed by the phase sensitivity, the design method considers the amplitude response approximation problem as a phase approximation problem. Using this assumption, we quantize remaining $(n-1)/2$ multiplier coefficients on the basis of the phase tolerance scheme.

The tolerance scheme is restricted to the stopband, since the passband sensitivity is low and the selected class of elliptic transfer functions gives a very small passband ripple. Our design technique can be used for arbitrary filter specifications.

9.9.2 Amplitude Response Sensitivity for Allpass Realization of IIR Filters

The transfer function of an odd-order IIR filter, $H(z)$, can be represented as a difference or sum of two allpass transfer functions $H_a(z)$ and $H_b(z)$ [75]:

$$H(z) = \frac{1}{2}\left(H_a(z) \pm H_b(z)\right) \tag{9.98}$$

where $+$ is for a lowpass and $-$ is for a complementary highpass filter.

The transfer function $H(z)$ has one real pole and $(n-1)/2$ complex-conjugate pole pairs, so $H_a(z)$ and $H_b(z)$ can be written in the form of products of the first- and second-order factors:

$$\begin{aligned} H_a(z) &= z \prod_{1}^{[(n+3)/4]} \frac{\beta_i + \alpha_i(1+\beta_i)z^{-1} + z^{-2}}{1 + \alpha_i(1+\beta_i)z^{-1} + \beta_i z^{-2}} \\ H_b(z) &= \prod_{[(n+7)/4]}^{(n+1)/2} \frac{\beta_i + \alpha_i(1+\beta_i)z^{-1} + z^{-2}}{1 + \alpha_i(1+\beta_i)z^{-1} + \beta_i z^{-2}} \end{aligned} \tag{9.99}$$

where $[x]$ returns the integer value such that $x \leq [x] < x + 1$. With $\beta_1 = 0$, the first-order section corresponding to the real pole of the transfer function becomes

$$z\,\frac{\alpha_1 z^{-1} + z^{-2}}{1 + \alpha_1 z^{-1}} = \frac{\alpha_1 + z^{-1}}{1 + \alpha_1 z^{-1}} \tag{9.100}$$

Equation (9.99) represents the cascade realization of allpass transfer functions $H_a(z)$ and $H_b(z)$. From the variety of existing first-order and second-order sections we can use the sections reported in reference 95.

A transfer function pole pair, z_i and z_{i+1}, can be represented by

$$\begin{aligned} z_i &= r_i\, e^{j\theta_i} \\ z_{i+1} &= r_i\, e^{-j\theta_i} \end{aligned} \tag{9.101}$$

and the parameters α_i and β_i are determined from [95]

$$\begin{aligned} \alpha_1 &= -r_1 \\ \beta_1 &= 0 \\ \alpha_i &= -2\frac{r_i \cos\theta_i}{1 + r_i^2}, \qquad i > 1 \\ \beta_i &= r_i^2, \qquad i > 1 \end{aligned} \tag{9.102}$$

This is the most economical implementation because it requires only n multiplications for an odd nth-order filter.

The first-order sensitivity of the magnitude response to a parameter x is defined as a partial derivative of the magnitude response with respect to x. Since the magnitude response is not differentiable, it is better to consider the real-valued *amplitude response* defined by

$$\mathcal{A}(f) = \pm\left|H(e^{j2\pi f})\right|$$

which may be positive or negative [50]. From (9.98) we can write the frequency response of an arbitrary lowpass filter as

$$H\left(e^{j2\pi f}\right) = \frac{1}{2}\left(e^{j\varphi_a(f)} + e^{j\varphi_b(f)}\right) \tag{9.103}$$

where $e^{j\varphi_a(f)}$ and $e^{j\varphi_b(f)}$ are the frequency responses of allpass branches.

The expression for $\mathcal{A}(f)$ is directly obtained from (9.103) as

$$\mathcal{A}(f) = \cos(\psi(f)) \tag{9.104}$$

where $\psi(f)$ is a *phase difference function* defined by

$$\psi(f) = \frac{\varphi_a(f) - \varphi_b(f)}{2} \tag{9.105}$$

Evidently, passband is obtained for $\psi(f) \approx 0$, while $\psi(f) \approx \pi/2$ gives a stopband. If $H_a(z)$ and $H_b(z)$ are realized by the cascade connection of a single first-order and $(n-1)/2$ second-order allpass sections, we have

$$\psi(f) = \frac{1}{2}\sum_{i=1}^{(n+1)/2} \pm\varphi_i(f) \tag{9.106}$$

where $\varphi_i(f)$ is the phase of the ith section, and the plus sign "+" refers to the branch a, while the minus sign "−" refers to the branch b.

The *amplitude sensitivity function* to an arbitrary multiplier coefficient x, $S_x^{\mathcal{A}}$, is defined as a partial derivative:

$$S_x^{\mathcal{A}}(f) = \frac{\partial\mathcal{A}(f)}{\partial x} \tag{9.107}$$

Applying (9.107) to Eq. (9.104) and using the substitutions from (9.105) and (9.106), we obtain

$$S_x^{\mathcal{A}}(f) = -\sin(\psi(f))\frac{\partial\psi(f)}{\partial x} = \pm\frac{1}{2}\sin(\psi(f))\frac{\partial\varphi_i}{\partial x} \tag{9.108}$$

where $\varphi_i(f)$ is the phase of the section i which contains the multiplier coefficient x. The plus sign "+" refers to the branch b, while the minus sign "−" refers to the branch a.

9.9.3 Sensitivity Functions

From the sensitivity analysis several useful properties can be brought out:

- From Eq. (9.108), it follows that the amplitude response sensitivity can be computed as the product of the filter *reflectance amplitude function*, defined as $\sin(\psi(f))$, and the phase sensitivity of the corresponding first-order or second-order section, $\dfrac{\partial \varphi_i(f)}{\partial x}$.
- From Eqs. (9.104) and (9.108) it follows that the amplitude response sensitivity in the passband is very low because $\sin(\psi(f)) \approx \psi(f) \approx 0$. The amplitude response sensitivity in the stopband is nearly equal to $\dfrac{1}{2}\dfrac{\partial \varphi_i(f)}{\partial x}$ because $\psi(f) \approx \pm\frac{\pi}{2}$ and $\sin(\psi(f)) \approx 1$. The above analysis proves that the parallel allpass realization has a low passband sensitivity [101].
- The sensitivity to α is higher in comparison with the sensitivities to β.
- The sensitivity to α increases with a decrease in the filter cutoff frequency, while the sensitivity to β does not have this type of change.
- The increase in β produces the increase in the extrema of the sensitivity functions.

9.9.4 Phase Tolerance Scheme

Digital filter specifications are usually given as the attenuation-limit specification

$$S_A = \{F_p, F_s, A_p, A_s\}$$

where F_p is the passband edge frequency, F_s is the stopband edge frequency, A_p stands for the maximum passband attenuation in dB, and A_s designates the minimal stopband attenuation in dB.

An infinite number of elliptic transfer functions that fit the specification exists. We can always find the transfer function order that provides a sufficient safety margin in attenuation characteristics, as shown in Fig. 9.53. The *attenuation tolerance scheme* S_a is defined for elliptic filters as

$$S_a = \left\{a_s - A_s,\ \ A_p - a_p,\ \ F_s - f_s,\ \ f_p - F_p\right\} \tag{9.109}$$

where a_s stands for the actual minimal stopband attenuation, a_p is the actual maximal passband attenuation, f_s is the actual stopband edge frequency, and f_p is the actual passband edge.

Since the amplitude response sensitivity is expressed in terms of the phase sensitivity (9.108), it is useful to transform the attenuation tolerance scheme (9.109) into a corresponding phase tolerance scheme. This way, the amplitude response approximation problem may be considered as a phase approximation problem. The desired transformation directly follows from (9.104)

$$\psi(f) = \cos^{-1}(\mathcal{A}(f)) \tag{9.110}$$

The phase difference function of a fifth-order filter, $\psi(f)$, with the phase tolerance scheme, is displayed in Fig. 9.54. The function $\psi(f)$ oscillates in an equal-ripple manner

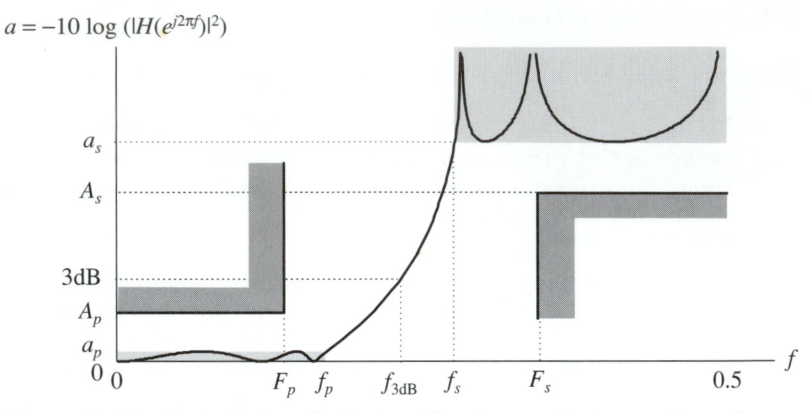

Figure 9.53 Attenuation-limit specification and attenuation of elliptic filter.

between ψ_p and $-\psi_p$ in the passband and between $\frac{\pi}{2} - \psi_s$ and $\frac{\pi}{2} + \psi_s$ in the stopband.

For the frequency $f_{3\text{dB}}$, at which the filter attenuation is 3 dB, we find $\psi(f_{3\text{dB}}) = \frac{\pi}{4}$. D_p and D_s denote the permitted variations of $\psi(f)$ in passband and stopband, respectively. For a given A_p and A_s, D_p and D_s can be determined from Eq. (9.110) and Fig. 9.54 as

$$D_p = \cos^{-1}\left(1 - 10^{-A_p/20}\right) \tag{9.111}$$

and

$$D_s = \frac{\pi}{2} - \cos^{-1}\left(10^{-A_s/20}\right) \tag{9.112}$$

This way, instead of an attenuation tolerance scheme, S_a, we use the phase tolerance scheme, S_ψ, defined as

$$S_\psi = \left\{D_s - \psi_s, \quad D_p - \psi_p, \quad F_s - f_s, \quad f_p - F_p\right\} \tag{9.113}$$

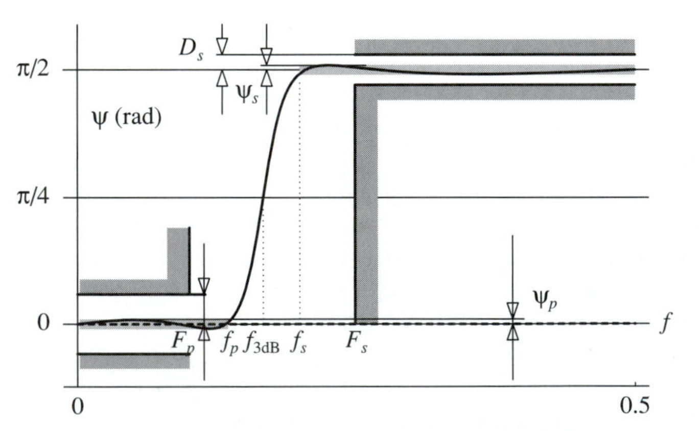

Figure 9.54 Phase tolerance scheme.

9.9.5 EMQF Transfer Function

In this section, we use elliptic minimal Q-factor (EMQF) transfer functions [47, 92] and directly control the values of the most sensitive parameters of the filter.

EMQF filters have two important properties: (1) The square magnitude response has equal ripples in passband and stopband:

$$\begin{aligned} \Delta_p &= \Delta_s \\ \Delta_p &= 1 - |H(e^{j2\pi f_p})|^2 \\ \Delta_s &= |H(e^{j2\pi f_s})|^2 \end{aligned} \tag{9.114}$$

(2) The passband attenuation is very small.

For the known passband edge frequency, f_p, and the known stopband edge frequency, f_s, the selectivity factor ξ is computed from

$$\xi = \frac{\tan \pi f_s}{\tan \pi f_p} \tag{9.115}$$

and f_p and f_s satisfy the relation

$$\tan^2(\pi f_{3\text{dB}}) = \tan(\pi f_s) \tan(\pi f_p) \tag{9.116}$$

where

$$|H(e^{j2\pi f_{3\text{dB}}})|^2 = \frac{1}{2} \tag{9.117}$$

An EMQF transfer function is uniquely defined by the order n, the selectivity factor ξ, and the passband edge frequency f_p [47, 92].

We define L as

$$L = \sqrt{\frac{|H(e^{j2\pi f_s})|^2 - 1}{|H(e^{j2\pi f_p})|^2 - 1}} \tag{9.118}$$

and compute as shown in reference 53.

The ripple factor ϵ of a EMQF transfer functions is [47]

$$\epsilon = \frac{1}{\sqrt{L}} \tag{9.119}$$

It can be shown that the passband and stopband ripple are equal:

$$\Delta_p = \Delta_s = \frac{1}{1 + L} \tag{9.120}$$

The actual passband and stopband attenuations are

$$\begin{aligned} a_p &= 10 \log_{10}\left(1 + \frac{1}{L}\right) \\ a_s &= 10 \log_{10}(1 + L) \end{aligned} \tag{9.121}$$

Note that L is uniquely determined by the selectivity factor ξ and the filter order n:

$$\xi \geq \sqrt{2} \quad \rightarrow \quad \begin{cases} t = \dfrac{1}{2}\dfrac{1-\sqrt[4]{1-\dfrac{1}{\xi^2}}}{1+\sqrt[4]{1-\dfrac{1}{\xi^2}}} \\ q = t + 2t^5 + 15t^9 + 150t^{13} \end{cases}$$

$$\xi < \sqrt{2} \quad \rightarrow \quad \begin{cases} t = \dfrac{1}{2}\dfrac{1-\dfrac{1}{\sqrt{\xi}}}{1+\dfrac{1}{\sqrt{\xi}}} \\ q_p = t + 2t^5 + 15t^9 + 150t^{13} \\ q = e^{\pi^2/\log(q_p)} \end{cases} \tag{9.122}$$

$$L = \frac{1}{4}\sqrt{\frac{1}{q^n} - 1} \tag{9.123}$$

For known L the selectivity factor ξ can be calculated by using approximate relations from reference 84 that have been adapted as follows

$$\begin{aligned} t &= \frac{1}{2}\frac{1-\sqrt[4]{1-\frac{1}{L^2}}}{1+\sqrt[4]{1-\frac{1}{L^2}}} \\ q &= t + 2t^5 + 15t^9 + 150t^{13} \\ g &= e^{\log(q)/n} \\ q_0 &= \frac{g + g^9 + g^{25} + g^{49} + g^{81} + g^{121} + g^{169}}{1 + 2(g^4 + g^{16} + g^{36} + g^{64} + g^{100} + g^{144})} \\ \xi &= \frac{1}{\sqrt{1 - \left(\frac{1-2q_0}{1+2q_0}\right)^4}} \end{aligned} \tag{9.124}$$

where t, q, g, and q_0 are auxiliary variables.

The poles of EMQF transfer functions, in the z-plane, occur on a circle with the center on the real axis, which is orthogonal with the unit circle, as shown in the Fig. 9.55

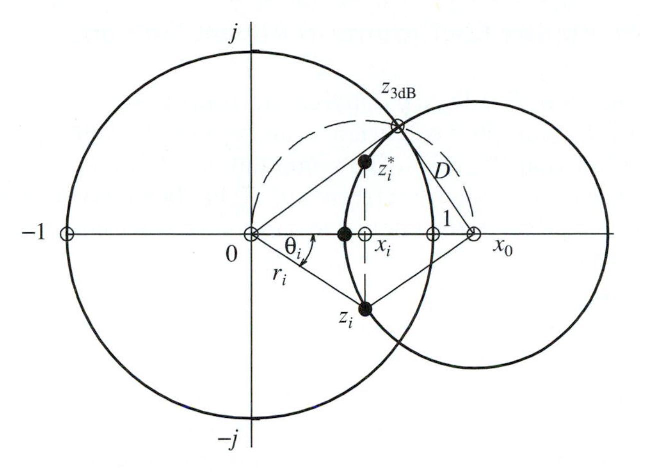

Figure 9.55 Poles loci of EMQF transfer function.

(for the third-order EMQF transfer function). The center of the circle on which the poles reside, x_0, is always placed outside the unit circle.

The pole distribution of EMQF transfer function among $H_a(z)$ and $H_b(z)$ is shown in Fig. 9.56. The following simple procedure can be applied for the distribution of the poles among two groups in order to form $H_a(z)$ and $H_b(z)$: Order the poles according to the increasing moduli, $H_a(z)$ encloses the real pole and then every second complex-conjugate pair, and $H_b(z)$ encloses the remaining poles.

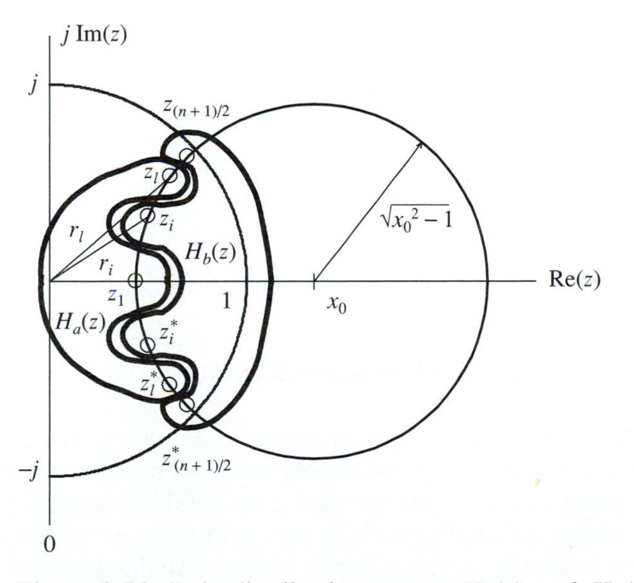

Figure 9.56 Pole distribution among $H_a(z)$ and $H_b(z)$.

9.9.6 Multiplier Coefficients in Allpass Sections

For IIR filters realized by a parallel connection of two allpass networks (9.98), the values of $(n-1)/2$ multiplier coefficients can be directly controlled if we choose an EMQF transfer function [92, 93]. The computation of coefficients of allpass sections from reference 95 is described in reference 92. It has been shown that all allpass second-order sections of an EMQF IIR filter have one common parameter whose value depends only on the frequency f_{3dB} at which the filter attenuation is 3 dB. This common parameter enables the direct control over the values of $(n-1)/2$ multiplier coefficients. It has been proved [93] that the same property holds for adaptor coefficients γ_i of a lattice wave digital filter (WDF) [71, 72].

Table 9.6 presents the expressions for computing the coefficients in allpass sections of an EMQF transfer function. The expressions from Table 9.6 are derived in reference 92 and [93]. Evidently, $(n-1)/2$ multiplier coefficients are directly controlled by f_{3dB}. The phase difference function has the value $\pi/4$ at f_{3dB}. Usually, f_{3dB} is placed in the transition band.

The common coefficients of the second-order sections can be adjusted for the multiplierless implementation by adjusting f_{3dB} [92, 93]. It is important to notice that the filter is still elliptic; therefore, $(n-1)/2$ multiplier coefficients are to be implemented as exact values eliminating the influence of high sensitivities to coefficients α.

The coefficients for two types of allpass sections differ only in sign, as is evident from Table 9.6. Hence, the sensitivity analysis can also be applied for wave lattice digital filters.

Table 9.6 Coefficients in allpass sections of EMQF filter

i	Coefficients [92]	Adaptor coefficients [93]	Comment
$i=1$ real pole	$\alpha_1 = -\frac{1-\tan(\pi f_{3dB})}{1+\tan(\pi f_{3dB})}$	$\gamma_1 = \frac{1-\tan(\pi f_{3dB})}{1+\tan(\pi f_{3dB})}$	First-order section
$2 \le i \le \frac{n+1}{2}$	$\alpha_i = \alpha = -\cos(2\pi f_{3dB})$	$\gamma_{2i-1} = \gamma = \cos(2\pi f_{3dB})$	Second-order
Conjugate complex pole pair	$\beta_i = r_i^2$	$\gamma_{2i-2} = -r_i^2$	section

9.9.7 β_{max} Coefficient in Allpass Sections

The deviations in the amplitude response are mostly produced by the section which implements the transfer function pole pair closest to the unit circle. Thus, the coefficient quantization error may be reduced if we establish a relation between the amplitude response and the coefficient β_{max}. β_{max} is determined by the magnitude of the transfer function pole, $r_{max} = |z_i|_{max}$, which is closest to the unit circle:

$$\beta_{max} = r_{max}^2 \tag{9.125}$$

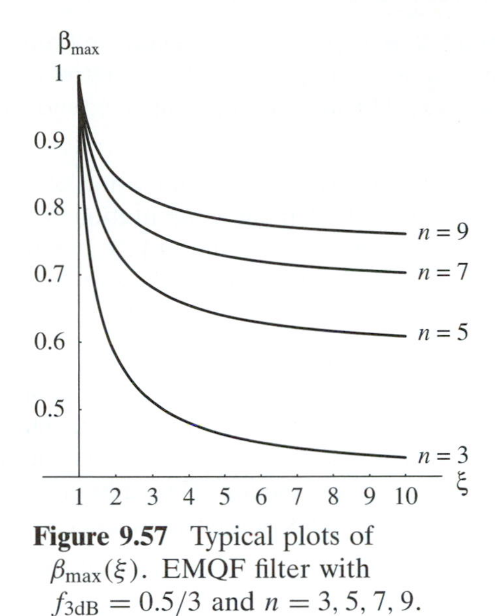

Figure 9.57 Typical plots of $\beta_{\max}(\xi)$. EMQF filter with $f_{3\text{dB}} = 0.5/3$ and $n = 3, 5, 7, 9$.

It is useful to examine whether by the adjustment of the filter transition band $(f_s - f_p)$ a suitable value for $\beta_{\max}$ can be found. The value of $\beta_{\max}$ can be controlled only by ξ for a known $f_{3\text{dB}}$.

A plot of $\beta_{\max}$ versus ξ is displayed in Fig. 9.57 for the third-, fifth-, seventh-, and ninth-order EMQF filters. $\beta_{\max}$ is a monotonically decreasing function of ξ and is suitable for performance optimization.

The multiplierless design begins with an initial EMQF transfer function which fits the specification and provides suitable values for α (or γ), by optimizing $f_{3\text{dB}}$. After this step $f_{3\text{dB}}$ remains constant. Next, we optimize ξ to obtain desired $\beta_{\max}$. We conclude with an EMQF transfer function, $H(z)$, whose multiplier coefficients $\alpha_2, \cdots, \alpha_{(n+1)/2}$ and $\beta_{\max}$ can be implemented as exact values with a minimal number of shift-and-add operations.

9.9.8 Multiplierless Elliptic IIR Filter Design

Our design procedure uses allpass sections from reference 95. The same procedure can also be developed for WDFs [71].

Multiplierless filter design means that each coefficient x from the set

$$\left\{\alpha_1,\ \alpha_2, \cdots,\ \alpha_{(n+1)/2},\ \beta_2, \cdots,\ \beta_{(n+1)/2}\right\} \tag{9.126}$$

should have a value suitable for a minimum shift-and-add implementation, that is,

$$x \in \left\{\pm 1/2^m,\ \pm\left(1 - 1/2^m\right)\right\} \tag{9.127}$$

or

$$x \in \left\{\pm 1/2^m \pm 1/2^p,\ \pm\left(1 - 2^m \pm 2^p\right)\right\} \tag{9.128}$$

or

$$x \in \left\{\pm 1/2^m \pm 1/2^p \pm 1/2^q,\ \pm\left(1 - 1/2^m \pm 1/2^p \pm 1/2^q\right)\right\} \tag{9.129}$$

with m, p, q integers, $m < p < q$. This signed digit representation is used in this section for the sake of simplicity. The use of other multiplier realizations [102] increases the number of shift-and-add combinations, which extends the choice of possible shift-and-add values.

Our algorithm is based on the computation of the range of permitted values for each multiplier coefficient. For the computation of the range of permitted values a design margin is to be used [(9.109) and (9.113)]. Afterwards, we have to verify which value from sets (9.127)–(9.129) belongs to the computed range. The value with minimal number of shift-and-add operations is chosen.

The multiplier coefficients can be divided into two groups that can be considered separately. The first group includes the coefficients that are controlled by f_{3dB} and ξ; that is, $\alpha_i = \alpha$ $(i > 1)$ and β_{max} (see Table 9.6). The second group includes α_1 and the remaining coefficients β_i. For the second group, new values α_{1q} and β_{iq} have to be selected using the sensitivity functions. The design proceeds through three main steps $(A, B$, and $C)$ where the values of the coefficients are successively determined.

Below, the process will be explained in detail and each design step will be demonstrated on an example specification: $F_p = 0.135$, $F_s = 0.2$, $A_p = 0.2$ dB, $A_s = 30$ dB.

9.9.9 The Choice of α

The value of α has to be coordinated with the parameters of the EMQF transfer function—that is, with f_{3dB} (see Table 9.6). For the given specification the range of permitted values of α is determined from F_p and F_s according to reference 92:

$$-\frac{1 - \tan^2 \pi F_p}{1 + \tan^2 \pi F_p} < \alpha < -\frac{1 - \tan^2 \pi F_s}{1 + \tan^2 \pi F_s} \tag{9.130}$$

Table 9.7 gives the pairs α, f_{3dB} arising from (9.127) and representing the simplest implementation. If there are several values from Table 9.7 that satisfy (9.130) and given that $F_p < f_{3dB} < F_s$, we then select α as

$$\alpha \approx -\frac{1 - \tan \pi F_p \tan \pi F_s}{1 + \tan \pi F_p \tan \pi F_s} \tag{9.131}$$

If not a single value from Table 9.7 satisfies (9.130), α has to be represented with more operations as in (9.128) or (9.129). Figures 9.58 and 9.59 illustrate densities of f_{3dB} for all possible values of α from (9.127) and (9.128). Evidently, f_{3dB} is uniformly distributed over the whole frequency range $(0 < f < 0.5)$, which confirms that a satisfactory value for α can be found without difficulty.

Table 9.7 List of the corresponding values α, f_{3dB}

α	f_{3dB}
$-1+1/2^8$	$0.014072 \approx 0.5/35$
$-1+1/2^7$	$0.019907 \approx 0.5/25$
$-1+1/2^6$	0.028172
$-1+1/2^5$	0.039893
$-1+1/2^4$	0.056567
$-1+1/2^3$	0.080431
$-1+1/2^2$	0.115027
$-1/2$	$0.166667 = 0.5/3$
$-1/2^2$	0.209785
$-1/2^3$	0.230053
$-1/2^4$	0.240046
$-1/2^5$	0.245026
$-1/2^6$	0.247513
$-1/2^7$	0.248757
$-1/2^8$	0.249378
0	$0.25 = 0.5/2$
$1/2^8$	0.250622
$1/2^7$	0.251243
$1/2^6$	0.252487
$1/2^5$	0.254974
$1/2^4$	0.259954
$1/2^3$	0.269947
$1/2^2$	0.290215
$1/2$	$1/3 = 0.5(1-1/3)$
$1-1/2^2$	0.384973
$1-1/2^3$	0.419569
$1-1/2^4$	0.443433
$1-1/2^5$	0.460107
$1-1/2^6$	0.471828
$1-1/2^7$	$0.48 \approx 0.5(1-1/25)$
$1-1/2^8$	$0.486 \approx 0.5(1-1/35)$

The procedure of this design step is shown below:

Step	Equation	
$A1$	$-0.6613 < \alpha < -0.309$	(9.130)
$A2$	$\alpha \approx -0.506$	(9.131)
$A3$	$\alpha \in [\ldots, -0.75, -0.5, -0.25, \ldots]$	(9.127)
$A4$	$\alpha = -0.5$	We choose
$A5$	$f_{3dB} = \dfrac{0.5}{3}$	$f_{3dB} = \dfrac{1}{\pi}\tan^{-1}\sqrt{\frac{1+\alpha}{1-\alpha}}$

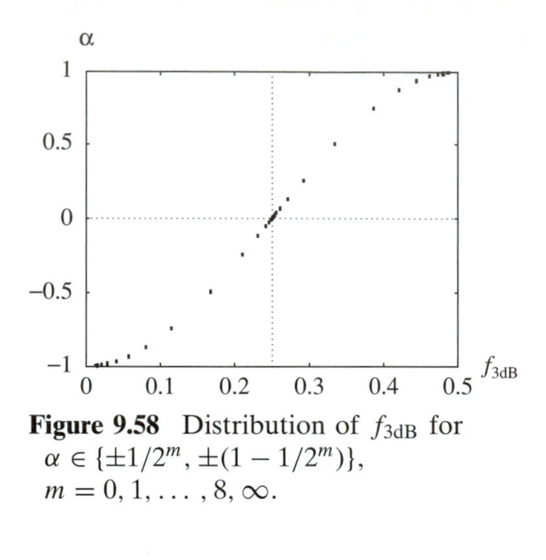

Figure 9.58 Distribution of f_{3dB} for $\alpha \in \{\pm 1/2^m, \pm(1 - 1/2^m)\}$, $m = 0, 1, \ldots, 8, \infty$.

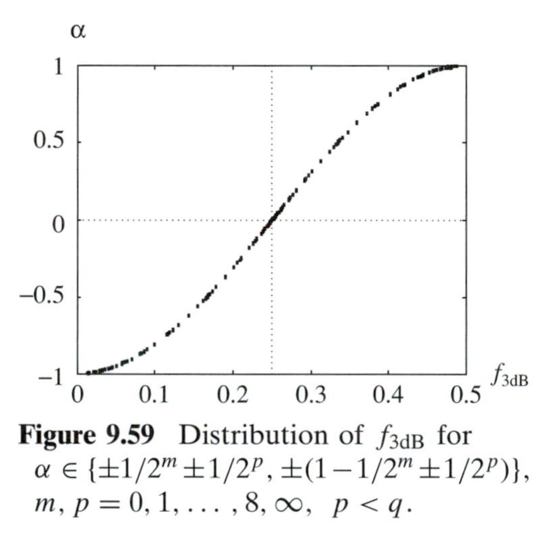

Figure 9.59 Distribution of f_{3dB} for $\alpha \in \{\pm 1/2^m \pm 1/2^p, \pm(1 - 1/2^m \pm 1/2^p)\}$, $m, p = 0, 1, \ldots, 8, \infty,\ p < q$.

9.9.10 The Choice of $\beta_{\max}$

Keeping the value of f_{3dB} from step A unchanged, the transition bandwidth is to be adjusted to provide the suitable value for $\beta_{\max}$.

In the first place, two transfer functions as boundary cases have to be determined: $H_1(z)$ with the design parameters f_{p1}, f_{3dB}, f_s, a_{p1}, and a_{s1}; and $H_2(z)$ with f_{p2}, f_{3dB}, f_{s2}, a_{p2}, and a_{s2}.

$H_1(z)$ is an EMQF transfer function with $f_{s1} = F_s$, and f_{p1} is calculated from (9.116). The filter order n is selected to ensure $a_{s1} > A_s$. From a practical standpoint, we always have $a_{p1} << A_p$.

$H_2(z)$ is also an nth-order EMQF transfer function, but for $a_{s2} = A_s$. The stopband edge frequency f_{s2} ($f_{s2} < f_{s1}$) is then computed, and the corresponding f_{p2} follows from (9.116).

The poles of $H_1(z)$ and $H_2(z)$ define the two coefficients $\beta_{\max 1}$ and $\beta_{\max 2}$:

$$\begin{aligned}\beta_{\max 1} &= r_{\max 1}^2 \\ \beta_{\max 2} &= r_{\max 2}^2\end{aligned} \tag{9.132}$$

where $r_{\max 1}$ and $r_{\max 2}$ are the magnitudes of the poles closest to the unit circle. A value for $\beta_{\max}$, implemented with minimal number of shift-and-add operations, has to be selected from the range

$$\beta_{\max 1} < \beta_{\max} < \beta_{\max 2} \tag{9.133}$$

For the chosen $\beta_{\max}$, the corresponding ξ of the EMQF transfer function has to be determined according to Section 9.9.7. The resulting $H(z)$ is the nth-order elliptic transfer function adjusted to provide the exact implementation of totally $(n+1)/2$ multiplier coefficients with a minimal number of shifters and adders.

Below, we demonstrate the design step B:

Step	Find n	Equation [ref.]
$B1$	$n \geq \text{ellipord}(2 * F_p, 2 * F_a, A_p, A_a) = 4$	MATLAB [1]
$B2$	$n = 5$	n is odd

Step	Find $\beta_{\max 1}$	Equation [ref.]
$B3$	$f_{p1} = 0.1369$	(9.116), $f_{a1} = F_a$
$B4$	$\xi = 1.5836$	(9.115)
$B5$	$L = 1391.$	(9.122)–(9.123)
$B6$	$a_{p1} = 0.00312$ dB	(9.121)
$B7$	$a_{a1} = 31.44$ dB	(9.121)
$B8$	$[z, p, c] = \text{ellip}(n, a_{p1}, a_{a1}, 2 * f_{p1});$	MATLAB [1]
$B9$	$\beta_{\max 1} = \max(\lvert p\rvert^2) = 0.7871$	(9.125)

Step	Find $\beta_{\max 2}$	Equation [ref.]
$B10$	$L = 999.$	(9.121)
$B11$	$\xi = 1.506$	(9.122), (9.123)
$B12$	$f_{p2} = 0.140$	(9.115), (9.116)
$B13$	$a_{p2} = 0.004345$ dB	(9.121)
$B14$	$a_{s2} = A_s = 30$ dB	(9.121)
$B15$	$[z, p, c] = \text{ellip}(n, a_{p2}, a_{s2}, 2 * f_{p2});$	MATLAB [1]
$B16$	$\beta_{\max 2} = \max(\lvert p\rvert^2) = 0.8001$	(9.125)

Step	Find β_{max}	Equation [ref.]
$B17$	$0.7871 < \beta_{max} < 0.8001$	$\beta_{max\,1} < \beta_{max} < \beta_{max\,2}$
$B18$	$\beta_{max} \in [\ldots, 0.75, 0.875, \ldots], \ldots$	(9.127)–(9.129)
$B19$	$\beta_{max} = 1 - 1/2^2 + 1/2^5 + 1/2^7 = 0.7891$	We choose

Step	Find β_i and α_1	Equation [ref.]
$B20$	$f_s = 0.1994$	$\beta_{max}(f_s) - 0.7891 = 0$, fzero.m, MATLAB
$B21$	$f_p = 0.1374$	(9.116)
$B22$	$\xi = 1.571$	(9.115)
$B23$	$L = 1322.$	(9.122)–(9.123)
$B24$	$a_p = 0.003283$ dB	(9.121)
$B25$	$a_s = 31.22$ dB	(9.121)
$B26$	$[z, p, c] = \text{ellip}(n, a_p, a_s, 2 * f_p);$	MATLAB [1]
$B27$	$\beta_{max} = \max(\lvert p\rvert^2) = 0.7891$	(9.125)
	$\beta_3 = 0.3467$	Table 9.6
	$\alpha_1 = -0.2679$	Table 9.6

9.9.11 The Choice of α_1 and β_i

The procedure for adjusting the second group of coefficients is based on the sensitivity analysis. The range of permitted values for α_1 and β_i (excluding β_{max}) can be computed using the phase sensitivity functions and the specified design margin. Within the range of permitted values, a new α_{1q} and β_{iq} have to be selected while keeping the filter specifications satisfied. The simplest way is to use the phase tolerance scheme (9.113) (Fig. 9.54). Since the passband sensitivity is very low, and EMQF filters yield a very small a_p, it is sufficient to consider the stopband margin only.

The choice of α_{1q} and β_{iq} is bounded by the condition

$$\left|\psi(f) - \frac{\pi}{2}\right| < D_s, \quad F_s \le f < \frac{1}{2} \tag{9.134}$$

where $\psi(f)$ is the phase difference function defined in (9.105), D_s is the permitted stopband tolerance of $\psi(f)$ computed from (9.112), and F_s is the required stopband edge frequency.

The errors in $\psi(f)$, produced by the replacements of α_1 or β_i with the quantized values α_{1q} and β_{iq} may be computed from the phase sensitivities of the first-order and second-order sections:

$$\Delta\psi(f) = \pm\frac{1}{2}\frac{\partial\varphi_1(f)}{\partial\alpha_1}\Delta\alpha_1,$$
$$\Delta\psi(f) = \pm\frac{1}{2}\frac{\partial\varphi_i(f)}{\partial\beta_i}, \qquad i > 1 \tag{9.135}$$

where + is used for branch a, and − is used for branch b.

The range of permitted values for α_1 and β_i can be computed from Eqs. (9.134) and (9.135) applied to the set of critical frequencies in the stopband. The *critical frequencies* are the specified stopband edge F_s, and the extremal frequencies of $\psi(f)$. The process should start from the coefficient having the highest sensitivity (excluding $\beta_{\max}$) and proceed to the lower sensitivity coefficients [96, 97].

Figure 9.60 displays the sensitivity functions for the example specification. The shaded area indicates the stopband. Dotted lines represent the sensitivity functions for the coefficients determined in steps A and B ($\alpha_2 = \alpha$, $\alpha_3 = \alpha$, $\beta_2 = \beta_{\max}$). Evidently, exact implementation of these coefficients avoids the influence of their high sensitivities on the filter response. The sensitivity of the remaining coefficients, α_1 and β_3, is considerably lower, as depicted in Fig. 9.60. Moreover, the sensitivity functions of α_1 and β_3 are relatively uniform, which guarantees with a good probability the success of the quantization process [96].

The quantization procedure starts by defining the phase tolerance scheme. Since only the stopband is to be considered, the phase tolerance scheme (9.113) reduces to

$$S_\psi = \{D_s - \psi_s, \quad F_s - f_s\} \tag{9.136}$$

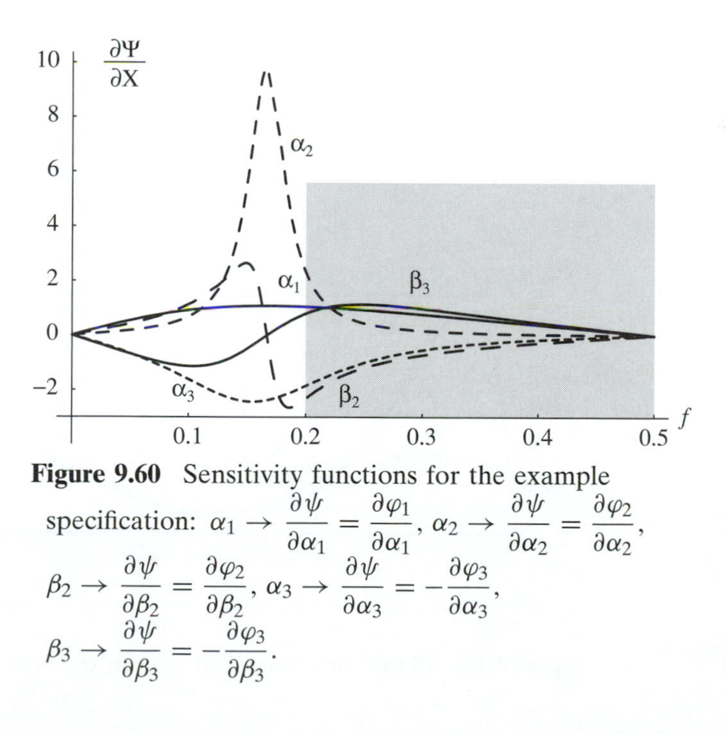

Figure 9.60 Sensitivity functions for the example specification: $\alpha_1 \to \frac{\partial\psi}{\partial\alpha_1} = \frac{\partial\varphi_1}{\partial\alpha_1}$, $\alpha_2 \to \frac{\partial\psi}{\partial\alpha_2} = \frac{\partial\varphi_2}{\partial\alpha_2}$, $\beta_2 \to \frac{\partial\psi}{\partial\beta_2} = \frac{\partial\varphi_2}{\partial\beta_2}$, $\alpha_3 \to \frac{\partial\psi}{\partial\alpha_3} = -\frac{\partial\varphi_3}{\partial\alpha_3}$, $\beta_3 \to \frac{\partial\psi}{\partial\beta_3} = -\frac{\partial\varphi_3}{\partial\beta_3}$.

Figure 9.61 illustrates the design margin and behavior of a fifth-order filter in the stopband. The solid line represents the phase difference function $\psi_0(f) = \psi(f, \alpha_1, \alpha, \beta_3, \beta_{max})$ computed for the EMQF filter designed in step B. The quantization procedure develops sequentially starting from $\psi_0(f)$. The process ends when all multiplier coefficients are quantized. As shown below, we first quantize β_3:

Step	Find β_{3q}			Equation
$C1$	$F_a = 0.2$	$f_x = 0.2256$	$f_y = 0.3584$	Extremum (Fig. 9.61)
$C2$	$\psi_0(F_a) = 1.547$	$\psi_0(f_x) = 1.598$	$\psi_0(f_y) = 1.543$	
$C3$	$\Delta_1 = -0.024$	$\Delta_2 = 0.02748$	$\Delta_3 = -0.02751$	$\Delta = \psi_0(f) - \frac{\pi}{2}$
$C4$	$D_a = 0.03163$			(9.112)
$C5$	$\Delta_1 + D_a = 0.0076$	$\Delta_2 - D_a = -0.0041$	$\Delta_3 + D_a = 0.0041$	$\Delta_i \pm D_a$
$C6$	$\frac{\partial \psi}{\partial \beta_3} = 0.80$	$\frac{\partial \psi}{\partial \beta_3} = 1.08$	$\frac{\partial \psi}{\partial \beta_3} = 0.68$	$\frac{\partial \psi(2\pi f)}{\partial \beta_3}$
$C7$	$\Delta_{\beta 3} = 0.0094$	$\Delta_{\beta 3} = -0.0038$	$\Delta_{\beta 3} = 0.0060$	$\Delta_{\beta 3} = \frac{\Delta_i \pm D_a}{\frac{\partial \psi}{\partial \beta_3}}$
$C8$	0.3429	$< \beta_{3q} <$	0.3527	$\beta_3 + \Delta_{\beta 3}$
$C9$	$\beta_{3q} =$	$1/2 - 1/2^3 - 1/2^5$	$= 0.3438$	We choose

When β_{3q} is selected, we compute a new phase difference function $\psi_1(f) = \psi(f, \alpha_1, \alpha, \beta_{3q}, \beta_{max})$, depicted in Fig. 9.61. Evidently, the margin is decreased, but condition (9.134) is satisfied, and some margin is still left for the next quantization. Finally, in the second quantization step, we quantize α_1.

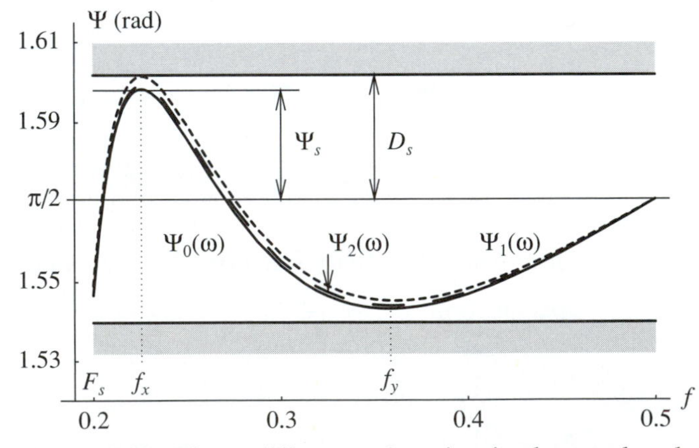

Figure 9.61 Phase difference function in the stopband.

Step	Find β_{1q}			Equation
$C10$	$F_a = 0.2$	$f_x = 0.2256$	$f_y = 0.3613$	Extremum (Fig. 9.61)
$C11$	$\psi_1(F_a) = 1.549$	$\psi_1(f_x) = 1.601$	$\psi_1(f_y) = 1.545$	
$C12$	$\Delta_1 = -0.0216$	$\Delta_2 = 0.03069$	$\Delta_3 = -0.02549$	$\Delta = \psi_1(f) - \frac{\pi}{2}$
$C13$	$D_a = 0.03163$			(9.112)
$C14$	$\Delta_1 + D_a = 0.01$	$\Delta_2 - D_a = -0.0009$	$\Delta_3 + D_a = 0.00614$	$\Delta_i \pm D_a$
$C15$	$\frac{\partial\psi}{\partial\alpha_1} = 1.05$	$\frac{\partial\psi}{\partial\alpha_1} = 1.00$	$\frac{\partial\psi}{\partial\alpha_1} = 0.54$	$\frac{\partial\psi(2\pi f)}{\partial\alpha_1}$
$C16$	$\Delta_{\alpha 1} = 0.00954$	$\Delta_{\alpha 1} = -0.001$	$\Delta_{\alpha 1} = 0.0113$	$\Delta_{\alpha 1} = \frac{\Delta_i \pm D_a}{\frac{\partial\psi}{\partial\alpha_1}}$
$C17$	-0.2689	$< \alpha_{1q} <$	-0.2584	$\alpha_1 + \Delta_{\alpha 1}$
$C18$	$\alpha_{1q} =$	$-1/2^2 - 1/2^6$	-0.2656	We choose

The new phase difference function $\psi_2(f) = \psi(f, \alpha_{1q}, \alpha, \beta_{3q}, \beta_{\max})$ is computed and depicted in Fig. 9.61. Since $\psi_2(f)$ satisfies the condition defined in (9.134), we conclude that the quantization process is terminated.

The filter attenuation characteristic, displayed in Fig. 9.62, shows that the required specification has been met.

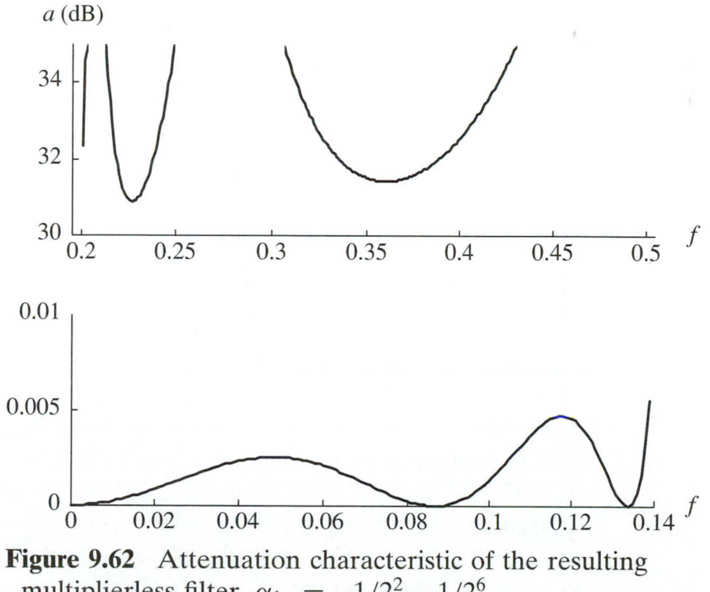

Figure 9.62 Attenuation characteristic of the resulting multiplierless filter, $\alpha_{1q} = -1/2^2 - 1/2^6$, $\alpha_2 = \alpha_3 = \alpha = -1/2$, $\beta_2 = \beta_{\max} = 1 - 1/2^2 + 1/2^5 + 1/2^7$, $\beta_{3q} = 1/2 - 1/2^3 - 1/2^5$.

From Fig. 9.61, we notice some increase of the margin in the second quantization step, that is, $|D_s - |\psi_2(f)||_{\min} > |D_s - |\psi_1(f)||_{\min}$. This is because the sensitivities of α_1 and β_3 behave uniformly in the stopband (Fig. 9.60), and we quantize α_1 and β_3 in the opposite directions.

9.9.12 Applications

Our method can be equally applied in the design of wideband and narrowband filters. It is illustrated in Figs. 9.58 and 9.59 that for values of α from (9.127)–(9.129), $f_{3\text{dB}}$ is practically uniformly distributed over the range [0, 0.5]. In Section 9.9.8, the example illustrates the case $f_{3\text{dB}} = 1/3$. This section presents an example of a half-band filter, $f_{3\text{dB}} = 1/2$, and an example of a narrowband filter with $f_{3\text{dB}} = 1/35$.

Example 9.3 Half-band filter example

Specification: $F_s = 0.28$, $A_s = 46$ dB.

The specifications are fulfilled with a ninth-order EMQF filter. The selectivity factor is adjusted to give $\beta_{\max} = 1 - 1/2^3 + 1/2^5$, which is achieved with $\xi = 1.3709$ resulting in $F_p = 0.2250$ and $F_s = 0.2750$. Figure 9.63 displays two attenuation characteristics: (1) The elliptic half-band IIR filter with exact coefficients and (2) The multiplierless

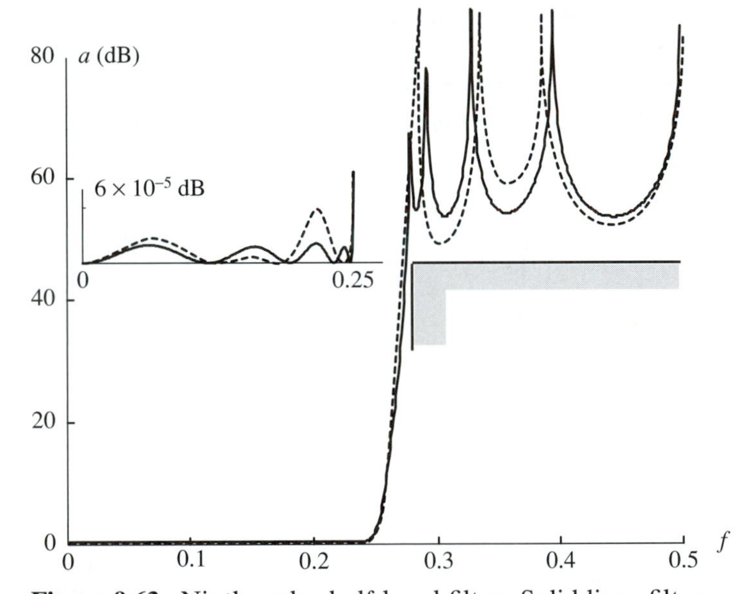

Figure 9.63 Ninth-order half-band filter. Solid line: filter with exact multiplier coefficients $\beta_2 = 0.1206$, $\beta_3 = 0.6629$, $\beta_4 = 0.3900$, $\beta_5 = \beta_{\max} = 1 - 1/2^3 + 1/2^6$. Dotted line: multiplierless filter with $\beta_{2q} = 1/2^3 - 1/2^8$, $\beta_{3q} = 1/2 + 1/2^3 + 1/2^5 + 1/2^7 = (1 + 1/2^2)(1/2 + 1/2^5)$, $\beta_{4q} = 1/2^2 + 1/2^3 + 1/2^6$, $\beta_5 = \beta_{\max} = 1 - 1/2^3 + 1/2^6$.

filter obtained by the quantization of β_2, β_3, and β_4. Using the design margin in the stopband, we reach values of β_{2q}, β_{3q}, and β_{4q} such that each individual coefficient can be implemented with only one or two additions. The realization of the half-band filter is based on the first-order allpass sections where a delay element z^{-1} is replaced by z^{-2}. ◆

EXAMPLE 9.4 Narrowband lowpass filter example

Design a lowpass filter with $f_{3dB} = 1/35$.

The third-order filter with $f_{3dB} = 1/35$ is designed giving $\alpha_{1q} = -(1 - 1/2^3 + 1/2^5)$, $\alpha_2 = -(1-1/2^8)$, and $\beta_2 = \beta_{max} = 1-1/2^4$. The attenuation characteristic is shown in Fig. 9.64. In this example only α_1 is quantized, since α_2 is determined by f_{3dB}, and $\beta_2 = \beta_{max}$ by the choice of the filter selectivity factor. All coefficients are of the type $1 - c, c << 1$. This type of coefficients always requires one extra addition if compared with coefficients from the range $[-1/2, 1/2]$. The problem is eliminated by the application of the first-order and second-order sections of the wave lattice digital filter [71]. In the WDF, the multipliers c and $1-c$ are realized in the same way and the only difference is in the sign of the adder [71]. Although the adaptor coefficients γ lie in the range $[-1,\ 1]$, the multiplier coefficients of a WDF are limited to the range $[-1/2,\ 1/2]$. The resulting block diagram of the multiplierless third-order WDF is shown in Fig. 9.65.

The multiplier coefficients in the examples provide implementations with at most two additions justifying the name "multiplierless." A very small passband ripple, which is a general property of EMQF filters, offers the possibility to design high-order filters by a cascade connection of lower-order filters. This is particularly useful for the design of linear phase IIR filters [83]. ◆

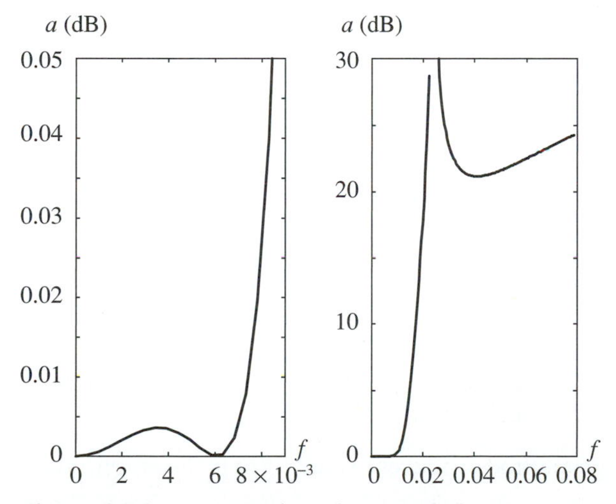

Figure 9.64 Attenuation characteristic of multiplierless narrowband lowpass third-order filter.

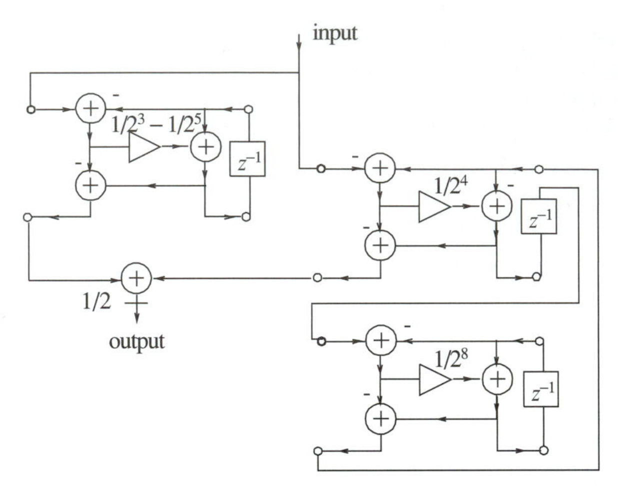

Figure 9.65 Block diagram of multiplierless narrowband lowpass third-order filter.

9.9.13 Conclusion

We present a straightforward procedure for designing multiplierless elliptic IIR filters in the form of a parallel connection of two allpass sections. The procedure is based on the first-order sensitivity analysis.

The closed-form expressions for the amplitude response sensitivities to multiplier coefficients are derived. It is shown that the amplitude sensitivity to a particular coefficient can be expressed as a product of the amplitude reflectance function and the phase sensitivity of the allpass section implementing the coefficient.

By using elliptic minimal Q-factor transfer functions, $(n + 1)/2$ highly sensitive multiplier coefficients are implemented as exact values with a minimal number of shifters and adders.

For the quantization of the remaining coefficients, the application of sensitivity analysis to phase tolerance scheme is proposed. This approach is computationally efficient for several reasons:

1. A deviation in the phase difference function produced by coefficient quantization is a linear combination of the sensitivity functions of the allpass sections.
2. The explicit phase sensitivity functions for allpass sections can be used.
3. A total of $(n - 1)/2$ coefficients, all of them of lower sensitivity, have to be quantized.
4. Since the passband sensitivity and passband ripple are small, it is sufficient to consider only the stopband.

This design method produces a small error in the amplitude response for several reasons:

- $(n + 1)/2$ coefficients are implemented as exact values.
- The remaining $(n - 1)/2$ coefficients are of lower sensitivity and their influence is directly controlled by the explicit sensitivity functions.
- The effect of the quantization on the passband is negligible.

Our design technique works well with most specifications including half-band filters and filters with a very low cutoff frequency. As demonstrated through examples, a sharp multiplierless filter can be obtained with a very small passband ripple. Using this property, high-order multiplierless filters can be formed by a cascade connection of lower-order filters while still exhibiting a small passband ripple.

9.10 LINEAR PHASE IIR FILTERS

In previous sections we have considered the design of digital elliptic IIR filters with sharp magnitude response. The phase response of these filters is highly nonlinear. The advanced filter design techniques were efficiently exploited in designing robust and selective low complexity filters. In this section we use the same design techniques for designing low-complexity digital elliptic IIR filters with sharp magnitude response and linear phase response. The approximation step is the same, and we focus on magnitude specifications only, but in the realization step we achieve the linear phase response by using the double filtering technique.

9.10.1 Filtering Finite Length Sequences

Consider a finite input sequence $x_0(l)$ $(l = 1, \ldots, L)$, and assume that we may process that sequence off-line. Our goal is to use a low-complexity filter with sharp magnitude response and zero phase response. One solution to the problem is the *double filtering* technique, also known as the *two-pass filtering* [83].

First, we extend the input sequence with K zeros on both sides and create a new sequence $x(k)$ (see Figs. 9.66 and 9.67):

$$\begin{aligned} x(k) &= 0, & k &= 1, \ldots, K \\ x(k) &= x_0(l), & k &= K + l \\ x(k) &= 0, & k &= K + L, \ldots, 2K + L \end{aligned} \tag{9.137}$$

Second, we process the sequence $x(k)$, by an IIR filter with the transfer function $H_{\text{forward}}(z)$, processing all samples in the forward direction from $x(1)$ to $x(2K + L)$. Obviously, for a relaxed filter the output sequence is $y(k) = 0$ $(k = 1, \ldots, K)$. We assume that the IIR filter is stable; theoretically, the filtered sequence $y(k)$ tends to become zero after infinite number of operations, $k \to \infty$. Practically, after a finite number of operations the quantization effects always cause the output sequence to become constant or to exhibit certain type of oscillations. Therefore, after some finite number of operations, say K, we can stop filtering.

Third, we process the sequence $y(k)$ in the reversed direction from $y(2K + L)$ to $y(1)$, by the same IIR filter with the transfer function $H_{\text{forward}}(z)$. Since $y(k) = 0$, $(k = 1, \ldots, K)$, the last K samples of the reversed sequence are zero. After a finite number of operations, K, the quantization effects mostly affect the output sequence and we stop filtering.

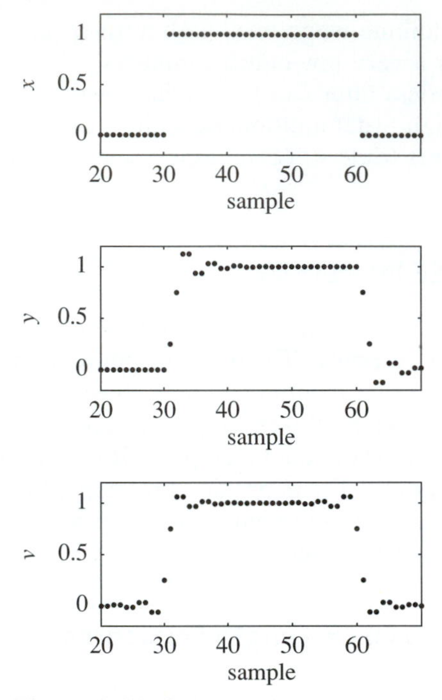

Figure 9.66 Input sequence x, output sequence after filtering in the forward direction, y, and output sequence after filtering in both the forward and reverse direction, v.

As an illustrative example of the double filtering technique, consider a filter whose transfer function is

$$H_3(z) = \frac{1}{2}\frac{\frac{1}{2} + z^{-1} + z^{-2} + \frac{1}{2}z^{-3}}{1 + \frac{1}{2}z^{-2}}$$

Excite the filter by the sequence $x(k)$, defined by Eq. (9.137) with $L = K = 30$ and $x_0(l) = 1$ $(l = 1, \ldots, 30)$.

Without sample quantization we find that $|y(k)| < 1/2^{16}$ for $k > 90$. If we quantize samples $y(k)$ to 15 bits all samples for $k > 90$ are represented as zeros. Therefore, we stop filtering at $k = 90$.

Next, we reverse the sequence $y(k)$ and input the reversed sequence to the same filter with $H_3(z)$. The output sequence is called the *double filtered sequence.* We find that the quantized double filtered sequence is approximately 0 for $k > 90$. Again, we stop filtering at $k = 90$. We denote the reversed double filtered sequence by $v(k)$, and this sequence is the output of the double filtering system. The three sequences are plotted in Figs. 9.66 and 9.67.

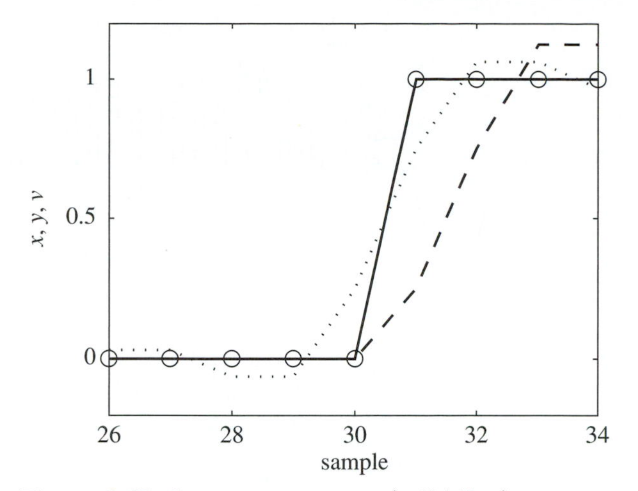

Figure 9.67 Input sequence x (solid line), output sequence after causal filtering, y (dashed line), and output sequence after double filtering, v (dotted line).

From Figs. 9.66 and 9.67 we conclude that the sequence y is delayed approximately one sample with respect to the sequence x. The sequence v is delayed approximately one sample with respect to the sequence y, but in the reversed direction. Therefore, the sequence v is not delayed with respect to the input sequence x.

The amplitude spectra of all three sequences are shown in Fig. 9.68. The higher-frequency components in the spectrum of the sequence y are more than 20 dB below the corresponding components in the spectrum of x. The higher-frequency spectral components of v are more than 40 dB below the corresponding components of x.

If the frequency response of the filter with $H_{\text{forward}}(z)$ is

$$H_{\text{forward}}(e^{j2\pi f}) = M(f)e^{j\Phi(f)}$$

then the frequency response of an equivalent system for reversed filtering is $M(f)e^{-j\Phi(f)}$. If we denote the equivalent transfer function of the system for reversed filtering with $H_{\text{reversed}}(z)$, then $H_{\text{reversed}}(e^{j2\pi f}) = M(f)e^{-j\Phi(f)}$ and

$$H_{\text{reversed}}(e^{j2\pi f}) = H_{\text{forward}}(e^{-j2\pi f})$$

$$H_{\text{reversed}}(z) = H_{\text{forward}}(\frac{1}{z})$$

The transfer function of the system for double filtering is

$$H(z) = H_{\text{forward}}(z)H_{\text{reversed}}(z) = H_{\text{forward}}(z)H_{\text{forward}}(\frac{1}{z})$$

with the frequency response

$$H(e^{j2\pi f}) = M(f)e^{j\Phi(f)}M(f)e^{-j\Phi(f)} = M^2(f)$$

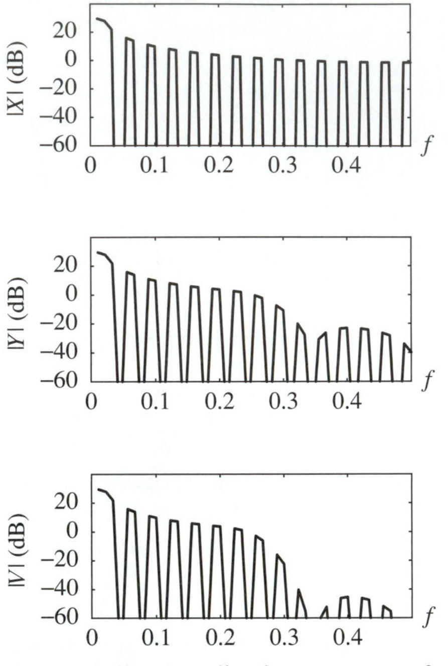

Figure 9.68 Amplitude spectrum of input sequence x, single filtered sequence y, and double filtered sequence v.

The magnitude response of the filter that performs double filtering is equal to the squared magnitude response of the filter that performs forward filtering. Notice that $H(e^{j2\pi f})$ is a real positive function in f that implies the zero phase response.

If the transfer function $H_{\text{forward}}(z)$ refers to a stable IIR filter with real coefficients (with the poles inside the unit circle), then the poles of $H_{\text{reversed}}(z)$ occur outside the unit circle. An IIR filter with poles outside the unit circle is unstable; fortunately, we realize $H_{\text{reversed}}(z)$ by a stable filter with the transfer function $H_{\text{forward}}(z)$, and we "reverse the time"; in other words, the input sequence is filtered from the last sample to the first one.

The error of truncating the infinite filtered sequence to the finite length is negligible in the previous example. Both the truncating error and the quantization of the samples cause the nonlinear distortion. The distortion of the sequence v due to sample quantization is shown in Fig. 9.69. The effect of the sample quantization to $B = 8$ bits is pronounced at higher frequencies. The quantization to $B = 5$ bits considerably distorts the spectrum of the sequence.

The nonlinear distortion due to truncating the filtered sequence can be reduced by increasing the number of zero samples, K. Similarly, we can reduce the nonlinear quantization distortion by increasing the number of bits, B, for storing the samples. In practice, we optimize K and B in such a way that the nonlinear distortion in the stopband is smaller than the required passband or stopband attenuation.

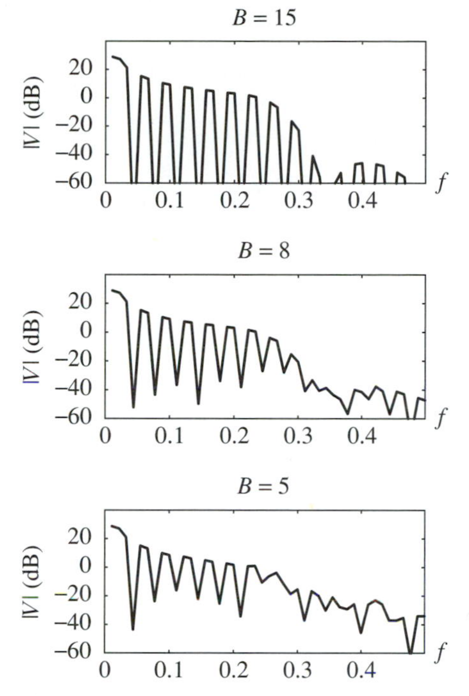

Figure 9.69 Amplitude spectrum of double filtered sequence, v, after quantization to B bits.

We search for minimal K such that the frequency response of the double filtering system is not noticeably different from the ideal frequency response $M^2(f)$ [83].

Real-time linear phase IIR filters, that process infinite length input sequences, can be realized by using block processing techniques. An input sequence is divided into finite length segments and each segment is filtered separately as it has been shown in this section. The resulting output sequences are combined using overlap-add methods. Realization of a linear phase IIR filter with minimal truncation distortion has been reported in [83].

9.10.2 Filtering Infinite Sequences

Powell and Chau [83] have devised an efficient method for the design of real-time linear phase IIR filters using suitable modification of the well-known time-reversing technique [103]. In their realization, shown in Fig. 9.70, the input signal is divided into L-sample segments, time-reversed, and twice filtered using two IIR filters whose transfer functions are the same, for example, $H_1(z) = H_2(z) = H(z)$. The transfer function $H(z)$ is usually of elliptic type, giving the best selectivity. It has been shown that the proposed procedure is much faster than previous methods, having at the same time low distortions of the magnitude response, $M(f) = |H(e^{j2\pi f})|$, and the phase response $\Phi(f) = \arg(H(e^{j2\pi f}))$.

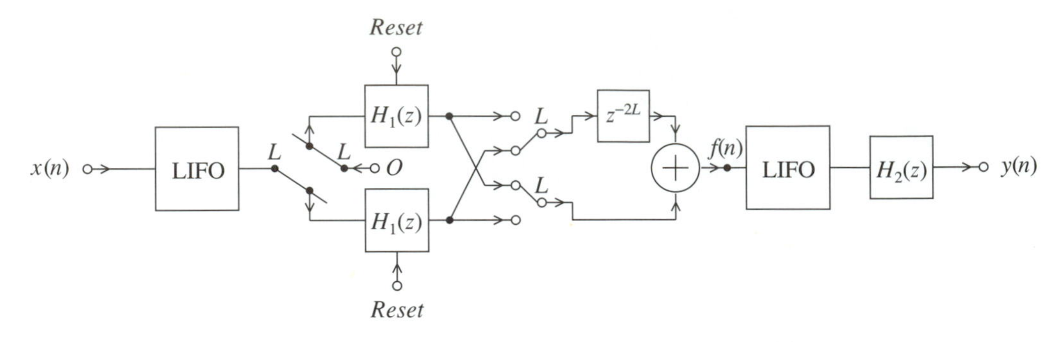

Figure 9.70 Powell–Chau implementation of linear phase IIR filter.

The performance of the realization shown in Fig. 9.70 has been further improved by Willson and Orchard [104], by separating the double zeros on the unit circle, which gives a better magnitude response in the stopband and slightly smaller magnitude and phase response distortions. The transfer functions $H_1(z)$ and $H_2(z)$ are different, but have the same denominator.

We further improve the Powell–Chau and Willson–Orchard methods by reordering the polynomials in the numerators of $H_1(z)$ and $H_2(z)$.

Assume that the transfer functions of the two IIR filters shown in Fig. 9.70 are written as a ratio of two polynomials as follows:

$$H_1(z) = \frac{A(z)}{D(z)} \tag{9.138}$$

$$H_2(z) = \frac{B(z)}{D(z)} \tag{9.139}$$

The polynomials $A(z)$ and $B(z)$ have zeros on the unit circle and may be equal [83], or different [104]. The polynomial $A(z)$ has no influence on the phase response $\Phi_1(f) = \arg(H_1(e^{j2\pi f}))$; similarly, $B(z)$ does not affect the phase response $\Phi_2(f) = \arg(H_2(e^{j2\pi f}))$. But, the phase response $\Phi(f)$ depends on $A(z)$ and $B(z)$ because of the segmentation of input sequence into length-L segments. Also, it may be noticed that the effects of the segmentation of the input signal are concentrated into the time-reversed part of the realization where infinite impulse responses of $H_1(z)$ filters are interrupted after L samples.

Following the results obtained by Willson and Orchard [104] we examine whether some other rearrangements of the polynomials $A(z)$ and $B(z)$ yield smaller truncation noise, smaller magnitude, and smaller phase response distortion.

We define the *truncation noise* as the noise caused by the truncation of the input sequence to L-sample segments. The truncation noise depends only on the realization of $H_1(z)$, therefore, the spectrum of the signal $f(n)$ at the output of the time-reversing part of Fig. 9.70 should be analyzed. As in reference 83, the *total harmonic distortion* (THD) of the signal $f(n)$ is measured in response to a sinusoidal input $x(n) = \sin(2\pi f_0 n)$, $0 \leq n \leq N-1$. The input sequence frequency f_0 is chosen so that an integer number

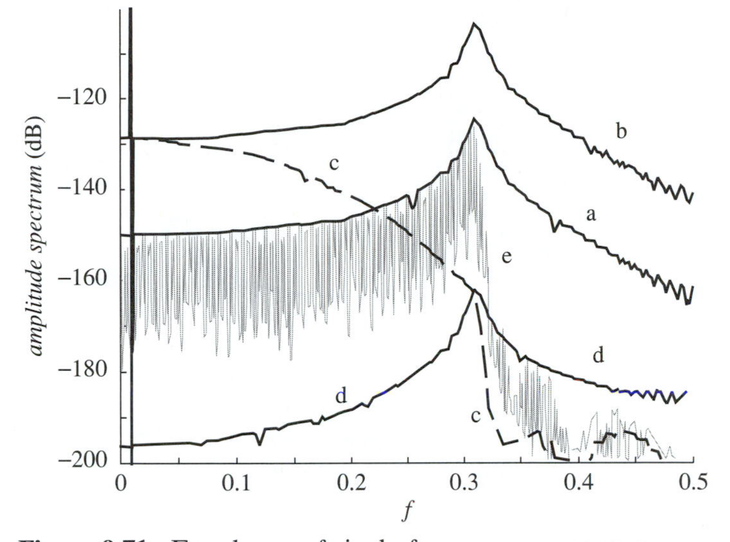

Figure 9.71 Envelopes of single frequency responses: $l = 37$, $N = 4096$, $L = 200$. Time-reversed sections are designated by z^{-1}, and direct sections are designated by z. (a) $A(z^{-1})/D(z^{-1})$, (b) $B(z)\ [A(z^{-1})/D(z^{-1})]$, (c) $[A(z^{-1})/D(z^{-1})]\ B(z)$, (d) $[A(z^{-1})B(z^{-1})]/D(z^{-1})$, (e) entire single frequency responses $[A(z^{-1})/D(z^{-1}]B(z)/D(z)$.

of periods is contained in the N input samples, $f_0 = l/N$. In our example, $l = 37$, $N = 4096$, and $L = 200$. The THD of the input sequence alone is approximately −150 dB. Some representative results are shown in Fig. 9.71, where effects of placing the polynomial $B(z)$ before or after the $H_1(z)$ filter, or included in $H_1(z)$, are examined. In all examples the polynomials $A(z)$, $B(z)$, and $D(z)$ are synthesized to satisfy the attenuation-limit specification from Example 6 in reference 83:

$$F_p = 0.3,\ F_s = 0.325,\ A_p = 0.01 \text{ dB},\ A_s = 70 \text{ dB} \tag{9.140}$$

Figure 9.71 shows the envelope of the output sequence spectra for all examined realizations, assuming the sinusoidal input sequence of frequency $f_0 = l/N$. Curve **a** represents the spectrum of the sequence at output of the $H_1(z) = A(z)/D(z)$ filter as reported in reference 83. If $B(z)$ is realized as a FIR filter before the $H_1(z)$ filter, the input signal in $H_1(z)$ is increased and consequently the truncation noise at the output of $H_1(z)$ is increased at all frequencies (curve **b**). On the contrary, if $B(z)$ is realized after the $H_1(z)$ filter, the truncation noise is filtered by $B(z)$ and the noise at higher frequencies is more attenuated (curve **c**).

When $B(z)$ is included into the $H_1(z)$ filter, the truncation noise is considerably decreased at lower frequencies (curve **d**). As can be seen in Fig. 9.71, this realization produces the lowest truncation noise in the passband and slightly larger noise in the stopband than the solution represented by curve **c**.

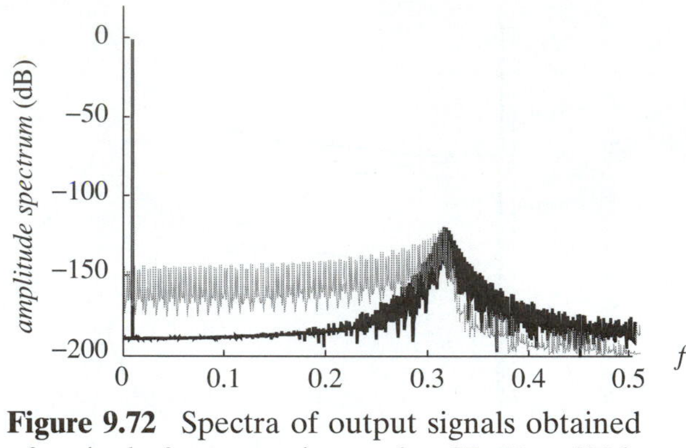

Figure 9.72 Spectra of output signals obtained for single frequency input: $l = 37$, $N = 4096$, $L = 200$; new realization—thick line, Powell–Chau realization—thin line.

If our goal is to minimize the truncation noise, we choose

$$H_1(z) = \frac{A(z)B(z)}{D(z)} \tag{9.141}$$

$$H_2(z) = \frac{1}{D(z)} \tag{9.142}$$

where $A(z)$, $B(z)$, and $D(z)$ are the same as in references 83 or 104.

Figure 9.72 shows the amplitude spectrum of the output sequence $y(n)$ for the original realization (thin line) using Eqs. (9.138) and (9.139), and for our realization (thick line) using Eqs. (9.141) and (9.142). The input sequence $x(n)$ and the polynomials $A(z)$, $B(z)$, and $D(z)$ are the same as in the previous example. The harmonic distortion produced by our realization is smaller by 20–40 dB in most parts of the passband. This means that the finite section length produces smaller harmonic distortions when both numerator polynomials are concentrated in the time-reversed section as in (9.141).

Consider Example 3 in reference 83:

$$F_p = 0.15, \qquad F_s = 0.2, \qquad A_p = 0.005 \text{ dB}, \qquad A_s = 82 \text{ dB} \tag{9.143}$$

We found $A(z)$, $B(z)$, and $D(z)$ meeting the above specification according to reference 104. Our realization has better performance than the Willson–Orchard realization [104, 105].

This choice of the transfer functions $H_1(z)$ and $H_2(z)$ considerably affects the decrease of the impulse response, as shown in Fig. 9.73. As can be seen, after few samples of the impulse response, which are larger than the corresponding samples in the Powell-Chau realization [83], the remaining samples become significantly smaller. This fact is a consequence of an increased number of transfer function zeros in $H_1(z)$. In

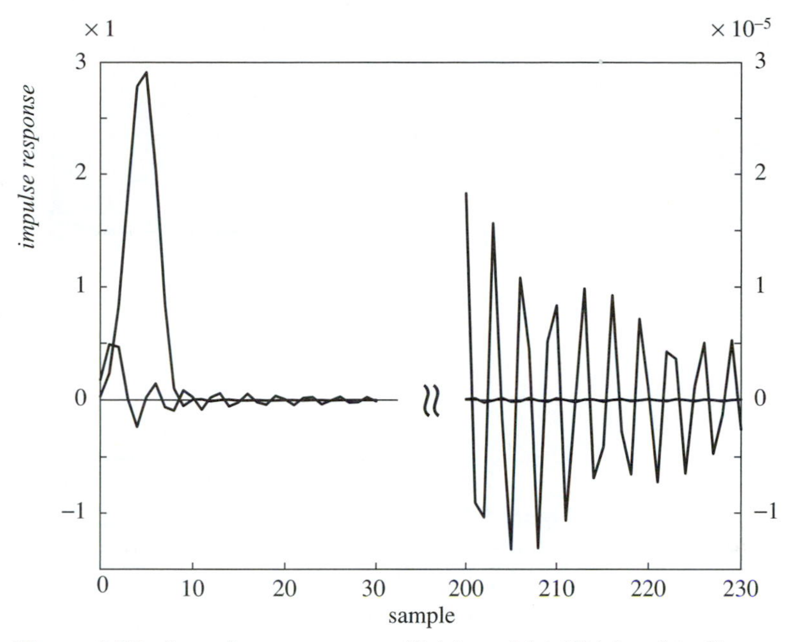

Figure 9.73 Impulse responses: $H_1(z) = A(z)/D(z)$—thin line, $H_1(z) = [A(z)B(z)]/D(z)$—thick line.

order to better explain this behavior of the impulse response we can decompose $H_1(z)$ into FIR and IIR parts as follows:

$$\begin{aligned} H_1(z) &= \frac{A(z)}{D(z)} = A_F(z) + \frac{A_I(z)}{D(z)} \\ A_I(z) &= \text{remainder}\left(\frac{A(z)}{D(z)}\right) \end{aligned} \tag{9.144}$$

Consequently, the impulse response is the sum of the finite impulse response $h_{FIR}(n)$ and infinite impulse response $h_{IIR}(n)$, where $h_{FIR}(n)$ and $h_{IIR}(n)$ are impulse responses of $A_F(z)$ and $A_I(z)/D(z)$, respectively. The response $h_{FIR}(n)$ is very short. The rates of decrease of the infinite impulse responses $h_{IIR}(n)$ of realizations according to (9.138) and (9.141) are the same due to the same denominators $D(z)$ in both realizations. Our realization has smaller amplitude of impulse response $h_{IIR}(n)$ that is caused by the numerator $A_I(z)$ [the numerators of realizations according to (9.138) and (9.141) are different]. The faster decrease of the impulse response of our realization makes less significant negative effects of the truncation to L-sample segments.

Having reduced the truncation noise in the passband, we can also expect that the phase response will be improved in the passband. In the second numerical experiment the phase response distortion of our realization and the realizations reported in references 83 and 104 are compared. We apply a unit impulse sequence of 4096 samples and plot the phase response in Fig. 9.74; thick line refers to our realization, and thin line refers to the Powell–Chau realization [83]. The phase response of our realization, in the

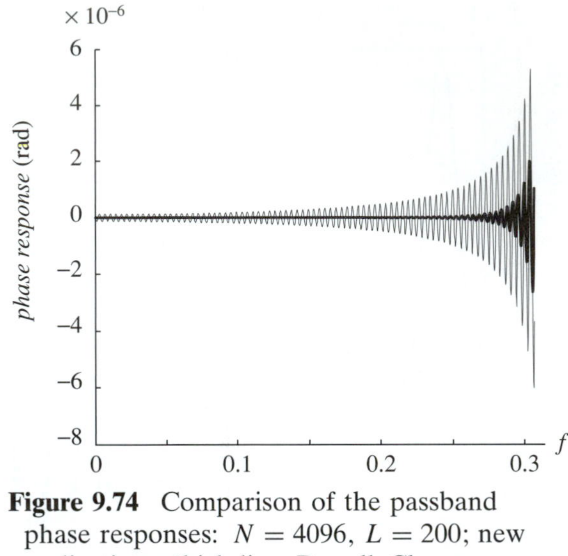

Figure 9.74 Comparison of the passband phase responses: $N = 4096$, $L = 200$; new realization—thick line, Powell–Chau realization—thin line.

passband, is very small and has the maximum of about 2×10^{-6} rad, which is 2–3 times smaller than the corresponding maximum in [83]. This experiment was performed again using the impulse response of a high-order allpass filter as an input signal; similar results were obtained. Also, the comparison with the Willson–Orchard realization [104], based on the specification (9.140), shows similar results in favor of our realization.

PROBLEMS

9.1 Consider a lowpass filter with passband edge frequency at $F_p = 0.2$, stopband edge frequency at $F_p = 0.212$, maximum passband attenuation of $A_p = 0.2$ dB, and minimum stopband attenuation of $A_s = 40$ dB.

(a) Find the magnitude limits specification, the magnitude tolerances specification, the magnitude ripple tolerances specification, and the gain limits specification.

(b) Determine the minimum order of the Butterworth, Chebyshev type I, Chebyshev type II, and elliptic transfer function.

(c) Determine the range of the selectivity factor for the orders $n = n_{\min}$, and $n = n_{\min+1}$.

(d) Determine the range of the ripple factor for the orders $n = n_{\min}$, and $n = n_{\min+1}$.

(e) Determine the range of the actual passband edge and actual stopband edge frequencies for the orders $n = n_{\min}$, and $n = n_{\min+1}$.

9.2 Design a minimal Q-factor lowpass filter with passband edge frequency at $F_p = 0.1$, stopband edge frequency at $F_p = 0.106$, maximum passband attenuation of $A_p = 0.2$ dB, and minimum stopband attenuation of $A_s = 40$ dB. Determine the minimum order of this filter, n_{mq}, and the minimum order of the Butterworth transfer function, n_{but}. Plot the magnitude response for all minimal Q-factor lowpass filters and for $n_{mq} \leq n \leq n_{but}$.

9.3 A filter shall be designed having a maximum passband attenuation $A_p = 0.1$ dB within the passband $0 \le f \le F_p = 0.1$ and a minimum stopband attenuation $A_s = 40$ dB within the stopband $0.2 \le f \le 0.5$. In addition, the filter shall meet the following:

- **(a)** The attenuation shall be equiripple in the passband and monotonic in the stopband, reaching the attenuation $a = 40$ dB at $f = 0.2$.
- **(b)** The attenuation shall be equiripple in the stopband and monotonic in the passband, reaching the attenuation $a = 0.1$ dB at $f = 0.1$.
- **(c)** The attenuation shall be monotonic in the passband and stopband, reaching the attenuation $a = 0.1$ dB at $f = 0.1$.
- **(d)** The attenuation shall be equiripple in the passband and stopband, reaching the attenuation $a = 0.1$ dB at $f = 0.1$.
- **(e)** The attenuation shall be equiripple in the passband and stopband, reaching the attenuation $a = 40$ dB at $f = 0.2$.

9.4 Design lowpass filters with passband edge frequency at $F_p = 0.2$, stopband edge frequency at $F_p = 0.212$, maximum passband attenuation of $A_p = 0.2$ dB, and minimum stopband attenuation of $A_s = 40$ dB. Determine pole Q-factors of the elliptic transfer function. Assume the filter orders $n = n_{\min}$, $n = n_{\min+1}$, $n = n_{\min+2}$, and $n = n_{\min+3}$. Determine pole Q-factors of the Butterworth, Chebyshev type I, and Chebyshev type II transfer functions.

9.5 Design lowpass filters operating at sampling rate of 10 kHz with the passband edge frequency at $F_p = 2$ kHz, stopband edge frequency at $F_p = 2.120$ kHz, maximum passband attenuation of $A_p = 0.2$ dB, and minimum stopband attenuation of $A_s = 40$ dB.

- **(a)** Determine the minimum order of the filter, $n_{\min}$. Determine the minimum order of the Butterworth, n_{but}, Chebyshev type I, n_{c1}, Chebyshev type II, n_{c2}, and elliptic transfer function, n_e.
- **(b)** Determine the range of the selectivity factor, the ripple factor, the actual passband edge and actual stopband edge frequencies for filter order $n_{\min} \le n \le n_{c1}$.
- **(c)** Sketch the characteristic function of the designs D1, D2, D3a, D3b, D4a, D4b, for $n_{\min} \le n \le n_{c1}$.

9.6 Design lowpass filters with passband edge frequency at $F_p = 0.2$, stopband edge frequency at $F_p = 0.212$, maximum passband attenuation of $A_p = 0.2$ dB, and minimum stopband attenuation of $A_s = 40$ dB.

- **(a)** Determine the minimum order of the filter, $n_{\min}$. Determine the minimum order of the Butterworth transfer function, n_{but}.
- **(b)** Sketch the characteristic function of the minimal Q-factor elliptic design (design D5) for $n_{\min} \le n \le n_{but}$.
- **(c)** Compare the characteristic function of the minimal Q-factor elliptic design with the Butterworth design.

9.7 Design lowpass filters with passband edge frequency at $F_p = 0.2$, stopband edge frequency at $F_p = 0.212$, maximum passband attenuation of $A_p = 0.2$ dB, and minimum stopband attenuation of $A_s = 40$ dB.

- **(a)** Determine the minimum order of the filter, $n_{\min}$. Determine the minimum order of the Chebyshev type I transfer function, n_{c1}.

(b) Sketch the characteristic function of the design D4a for $n_{\min} \leq n \leq n_{c1}$.

(c) Compare the characteristic function of the design D4a with the Chebyshev type I design.

9.8 Consider a lowpass filter operating at sampling rate of 10 kHz and specified by

$$S_K = \{F_p = 3 \text{ kHz}, \ F_s = 3.225 \text{ kHz}, \ K_p = 0.2171, \ K_s = 100\}$$

(a) Determine the minimum order of the filter, $n_{\min}$.

(b) Sketch the transfer function, step response and group delay for designs D1, D2, D3, D4, for $n = n_{\min}$.

(c) Compare the group delay and rise time of the designs.

9.9 Design highpass filters operating at sampling rate of 20 kHz and with passband edge frequency at $F_p = 2120$ Hz, stopband edge frequency at $F_s = 2$ kHz, maximum passband attenuation of $A_p = 0.2$ dB, and minimum stopband attenuation of $A_s = 40$ dB.

(a) Determine the minimum order of the filter, $n_{\min}$. Determine the minimum order of the Butterworth, n_{but}, Chebyshev type I, n_{c1}, Chebyshev type II, n_{c2}, and elliptic transfer function, n_e.

(b) Determine the range of the design parameters: selectivity factor, ripple factor, actual passband edge and actual stopband edge frequencies, for $n_{\min} \leq n \leq n_{\min+2}$.

(c) Sketch the attenuation of the designs D1, D2, D3a, D3b, D4a, D4b, for $n_{\min} \leq n \leq n_{\min+2}$.

9.10 Design bandpass filters with passband edge frequencies at $F_{p1} = 0.151$, $F_{p2} = 0.159$, stopband edge frequencies at $F_{s1} = 0.15$, $F_{s2} = 0.16$, maximum passband attenuation of $A_p = 0.2$ dB, and minimum stopband attenuation of $A_s = 45$ dB.

(a) Determine the minimum order of the filter, $n_{\min}$. Determine the minimum order of the Butterworth, n_{but}, Chebyshev type I, n_{c1}, Chebyshev type II, n_{c2}, and elliptic transfer function, n_e.

(b) Determine the range of the design parameters: selectivity factor, ripple factor, actual passband edge and actual stopband edge frequencies, for $n_{\min} \leq n \leq n_{\min+3}$.

(c) Sketch the gain of the designs D1, D2, D3a, D3b, D4a, D4b, for $n_{\min} \leq n \leq n_{\min+3}$.

9.11 Design bandreject filters with stopband edge frequencies at $F_{s1} = 0.151$, $F_{s2} = 0.159$, passband edge frequencies at $F_{p1} = 0.15$, $F_{p2} = 0.16$, maximum passband attenuation of $A_p = 0.2$ dB, and minimum stopband attenuation of $A_s = 45$ dB.

(a) Determine the minimum order of the filter, $n_{\min}$. Determine the minimum order of the Butterworth, n_{but}, Chebyshev type I, n_{c1}, Chebyshev type II, n_{c2}, and elliptic transfer function, n_e.

(b) Determine the range of the design parameters: selectivity factor, ripple factor, actual passband edge and actual stopband edge frequencies, for $n_{\min} \leq n \leq n_{\min+3}$.

(c) Sketch the attenuation of the designs D1, D2, D3a, D3b, D4a, D4b, for $n_{\min} \leq n \leq n_{\min+3}$.

9.12 Consider a IIR half-band lowpass filter operating at sampling rate of 10 kHz and specified by

$$S_\delta = \{F_p = 2 \text{ kHz}, \ F_p = 3 \text{ kHz}, \ \delta_p = 0.0001201322, \ \delta_s = 0.0155\}$$

(a) Determine the minimum order of the elliptic filter, $n_{\min}$.

(b) Sketch the magnitude response, step response and group delay.

(c) Verify that the transfer function zeros lie on the unit circle and the real part of the transfer function poles are zero.

9.13 Design a lowpass multiplierless filter with passband edge frequency at $F_p = 0.137$, stopband edge frequency at $F_p = 0.2$, maximum passband attenuation of $A_p = 0.03$ dB, and minimum stopband attenuation of $A_s = 60$ dB.

(a) Determine the minimum order of the elliptic filter. Find the minimal coefficient wordlength.

(b) Determine the minimum order of the filter implemented as a cascade connection of two lower odd-order elliptic filters, $H_1(z)$ and $H_2(z)$. Find the minimal coefficient wordlength. Implement the multiplier coefficients with minimal number of shift-and-add operations (the coefficients are represented by a sign bit and b binary bits). Implement each lower-order filter as a parallel connection of two allpass filters.

(c) Find the group-delay variation and phase variation of the filter implemented with two-pass filtering technique; $H_1(z)$ is used for filtering in the forward direction and $H_2(z)$ is used for filtering in the reverse direction.

9.14 Design a Hilbert transformer with stopband edge frequencies at $F_{s1} = 0.53$, and $F_{s2} = 0.97$, the phase tolerance of $D = 0.005$ rad. Determine the minimum order of the corresponding elliptic filter. Find the minimal coefficient wordlength.

9.15 Consider three transfer functions

$$H_1(z) = \frac{\frac{1}{2} + z^{-4}}{1 + \frac{1}{2}z^{-4}} + z^{-2}$$

$$H_2(z) = \frac{\frac{1}{2} + z^{-4} + z^{-2} + \frac{1}{2}z^{-6}}{1 + \frac{1}{2}z^{-4}}$$

$$H_3(z) = \left(1 + z^{-2}\right) \frac{0.51076788\left(1 - z^{-1} + z^{-2}\right)}{1 - 1.1892z^{-1} + 0.7071z^{-2}} \frac{3.91567\left(1 + z^{-1} + z^{-2}\right)}{1 + 1.1892z^{-1} + 0.7071z^{-2}}$$

Determine the order of the transfer functions. Suggest the best transfer function for implementation and give reasons for your choice.

MATLAB EXERCISES

9.1 Write a MATLAB program that computes the magnitude limits specification, the magnitude tolerances specification, the magnitude ripple tolerances specification, and the gain limits specification. Assume the specification given in Problem 9.1. Determine the minimum order of the Butterworth, Chebyshev type I, Chebyshev type II, and elliptic transfer function. Plot the magnitude responses for filter order $n = n_{\min}$ and $n = n_{\min+1}$. Compute the range of the selectivity factor, the range of the ripple factor, and the range of the actual passband edge and actual stopband edge frequencies for filter order $n = n_{\min}$, $n = n_{\min+1}$, and $n = n_{\min+2}$.

9.2 Write a MATLAB program that plots the magnitude response of a minimal-Q-factor lowpass filter. Assume the specification given in Problem 9.2. Compute the magnitudes of all poles and show that all poles of a minimal Q-factor transfer function lie on a circle. Plot poles and zeros of the transfer function for $n_{mq} \le n \le n_{but}$. Compute the minimum order of the filter, n_{mq}, and the minimal filter order of the Butterworth transfer function, n_{but}.

9.3 Write a MATLAB program that computes lowpass transfer function with the specification given in Problem 9.4. Compute pole Q-factors of the elliptic transfer function for the filter orders $n = n_{\min}$, $n = n_{\min+1}$, $n = n_{\min+2}$, and $n = n_{\min+3}$. Compute the Q-factors of the Butterworth, Chebyshev type I, and Chebyshev type II transfer functions.

9.4 Write a MATLAB program that computes the characteristic function of the designs D1, D2, D3a, D3b, D4a, D4b, for filter order $n_{\min} \le n \le n_{c1}$. Assume the specification given in Problem 9.5. Compute the range of the selectivity factor, the ripple factor, the actual passband edge and actual stopband edge frequencies for filter order $n_{\min} \le n \le n_{c1}$.

9.5 Write a MATLAB program that computes the characteristic function of the minimal Q-factor elliptic and Butterworth transfer function for filter order $n_{\min} \le n \le n_{but}$. Assume the specification given in Problem 9.6.

9.6 Write a MATLAB program that computes the characteristic function of the design D4a and the Chebyshev type I transfer function for filter order $n_{\min} \le n \le n_{c1}$. Assume the specification given in Problem 9.7.

9.7 Write a MATLAB program that computes the transfer function, step response and group delay for $n = n_{\min}$ with the specification given in Problem 9.8.

9.8 Write a MATLAB program that computes the attenuation of the designs D1, D2, D3a, D3b, D4a, D4b, for highpass filter. Assume the specification given in Problem 9.9 and the filter order $n_{\min} \le n \le n_{\min+2}$.

9.9 Write a MATLAB program that plots the gain of the designs D1, D2, D3a, D3b, D4a, D4b, for bandpass filter. Assume the specification given in Problem 9.10 and the filter order $n_{\min} \le n \le n_{\min+2}$.

9.10 Write a MATLAB program that plots the gain of the designs D1, D2, D3a, D3b, D4a, D4b, for bandreject filter. Assume the specification given in Problem 9.11 and the filter order $n_{\min} \le n \le n_{\min+2}$.

9.11 Write a MATLAB program that plots the attenuation of the half-band lowpass filter. Assume the specification given in Problem 9.12. Verify the results using the MATLAB programs `ellipord` and `ellip`.

9.12 Write a MATLAB program that plots the attenuation of the lowpass multiplierless filters. Assume the specification given in Problem 9.13.

9.13 Write a MATLAB program that plots the attenuation of a Hilbert transformer. Assume the specification given in Problem 9.14.

9.14 Write a MATLAB program that plots the attenuation of the filters given in Problem 9.15. Write a MATLAB program that implements the filters.

MATHEMATICA EXERCISES

9.1 Write a *Mathematica* program that computes the magnitude limits specification, the magnitude tolerances specification, the magnitude ripple tolerances specification, and the gain limits specification. Assume the specification given in Problem 9.1. Determine the minimum order of the Butterworth, Chebyshev type I, Chebyshev type II, and elliptic transfer function. Plot the magnitude responses for filter order $n = n_{\min}$ and $n = n_{\min+1}$. Compute the range of the selectivity factor, the range of the ripple factor, and the range of the actual passband edge and actual stopband edge frequencies for filter order $n = n_{\min}$, $n = n_{\min+1}$, and $n = n_{\min+2}$.

9.2 Write a *Mathematica* program that plots the magnitude response of a minimal-Q-factor lowpass filter. Assume the specification given in Problem 9.2. Compute the magnitude of all poles and show that all poles of a minimal Q-factor transfer function lie on a circle. Plot poles and zeros of the transfer function for $n_{mq} \leq n \leq n_{but}$. Compute the minimum order of the filter, n_{mq}, and the minimal filter order of the Butterworth transfer function, n_{but}.

9.3 Write a *Mathematica* program that computes lowpass transfer function with the specification given in Problem 9.4. Compute pole Q-factors of the elliptic transfer function for the filter orders $n = n_{\min}$, $n = n_{\min+1}$, $n = n_{\min+2}$, and $n = n_{\min+3}$. Compute pole Q-factors of the Butterworth, Chebyshev type I, and Chebyshev type II transfer functions.

9.4 Write a *Mathematica* program that computes the characteristic function of the designs D1, D2, D3a, D3b, D4a, and D4b, for filter order $n_{\min} \leq n \leq n_{c1}$. Assume the specification given in Problem 9.5. Compute the range of the selectivity factor, the ripple factor, the actual passband edge and actual stopband edge frequencies for filter order $n_{\min} \leq n \leq n_{c1}$.

9.5 Write a *Mathematica* program that computes the characteristic function of the minimal Q-factor elliptic and Butterworth transfer function for filter order $n_{\min} \leq n \leq n_{but}$. Assume the specification given in Problem 9.6.

9.6 Write a *Mathematica* program that computes the characteristic function of the design D4a and the Chebyshev type I transfer function for filter order $n_{\min} \leq n \leq n_{c1}$. Assume the specification given in Problem 9.7.

9.7 Write a *Mathematica* program that computes the transfer function, step response and group delay for $n = n_{\min}$ with the specification given in Problem 9.8.

9.8 Write a *Mathematica* program that computes the attenuation of the designs D1, D2, D3a, D3b, D4a, and D4b, for highpass filter. Assume the specification given in Problem 9.9 and the filter order $n_{\min} \leq n \leq n_{\min+2}$.

9.9 Write a *Mathematica* program that plots the gain of the designs D1, D2, D3a, D3b, D4a, and D4b, for bandpass filter. Assume the specification given in Problem 9.10 and the filter order $n_{\min} \leq n \leq n_{\min+2}$.

9.10 Write a *Mathematica* program that plots the gain of the designs D1, D2, D3a, D3b, D4a, and D4b, for bandreject filter. Assume the specification given in Problem 9.11 and the filter order $n_{\min} \leq n \leq n_{\min+2}$.

9.11 Write a *Mathematica* program that plots the attenuation of the IIR half-band lowpass filter. Assume the specification given in Problem 9.12.

9.12 Write a *Mathematica* program that plots the attenuation of the lowpass multiplierless filters. Assume the specification given in Problem 9.13.

9.13 Write a *Mathematica* program that plots the attenuation of a Hilbert transformer. Assume the specification given in Problem 9.14.

9.14 Write a *Mathematica* program that plots the attenuation of the filters given in Problem 9.15. Write a *Mathematica* program that implements the filters.

Chapter 10

ADVANCED DIGITAL FILTER DESIGN ALGORITHMS

Classic digital filter design techniques return only one design from an infinite collection of alternative designs, or they fail to design filters when solutions exist. These classic techniques hide a wealth of alternative filter designs that are more robust when implemented in digital hardware and embedded software. In this chapter, we present (1) case studies of optimal digital filters that cannot be designed with classic techniques and (2) the formal, mathematical framework that underlies their solutions. We have automated the advanced filter design techniques in software, so we present detailed step-by-step design algorithms.

10.1 INTRODUCTION

In designing digital IIR filters, one generally relies on canned software routines or mechanical table-oriented procedures. The primary reason for these "black box" approaches is that the approximation theory that underlies filter design includes complex mathematics. Unfortunately, conventional approaches return only one design, thereby hiding a wealth of alternative filter designs that are more robust when implemented in digital hardware and embedded software. In addition, conventional approaches may fail to find a filter when in fact one exists.

We develop advanced design techniques to find a comprehensive set of optimal designs to represent the infinite solution space. The optimal designs include filters that have minimal order, minimal complexity, minimal sensitivity to pole-zero locations, minimal deviation from a specified group delay, approximate linear phase response, and minimized peak overshoot of the step response. The design space also includes digital filters with power-of-two coefficients. We base our approach on formal, mathematical properties of Jacobi elliptic functions [47, 53]. We automate these advanced filter design techniques in software [2, 51].

The key observations underlying advanced digital filter design are as follows:

1. Many designs satisfy the same user specification.
2. Butterworth and Chebyshev IIR filters are special cases of elliptic IIR filters.
3. Minimum-order filters may not be as efficient to implement as some higher-order filters.

The filter optimization problem is a mixed-integer linear programming problem so the classic techniques break down. Instead of using iterative numerical techniques, we solve these problems using closed-form algebraic expressions. Then, we present (a) several new case studies of optimal digital IIR filters that cannot be designed with classic techniques and (b) the formal, mathematical framework that underlies their solutions.

10.2 NOTATION

We review the list of symbols that we use in formulas and procedures when designing digital IIR filters. Often, we append a suffix to designate a quantity related to a specific filter type. For example, we add h to designate highpass filter; thus, A_{ph} is A_p of a highpass filter.

a—parameter of second-order section
$A(f)$—attenuation (dB)
A_p—maximum passband attenuation in specification (dB)
A_s—minimum stopband attenuation in specification (dB)
b—parameter of second-order section
$\mathrm{cd}(u, k)$—Jacobi elliptic cd function
$\mathrm{cd}^{-1}(v, k)$—inverse Jacobi elliptic cd function
f—digital frequency (frequency for short), $0 \le f \le 0.5$
f_{nQ}—normalized frequency of minimal Q-factor design
f_p—passband edge frequency of designed filter
F_p—passband edge frequency in specification
f_s—stopband edge frequency of designed filter
F_s—stopband edge frequency in specification
$G(f)$—gain
$\mathcal{H}(n, \xi, \epsilon, f_p, z)$—lowpass elliptic transfer function
$\mathcal{H}_{\min Q}(n, \xi, f_p, z)$—minimal Q-factor lowpass elliptic transfer function
$H(z)$—transfer function
i—index ($i = 1, 2, \ldots, n$)

j—the imaginary unit ($j = \sqrt{-1}$)

k—modulus of elliptic functions

$K_e(n, \xi, \epsilon, x)$—elliptic characteristic function

$K_J(k)$—complete elliptic integral of first kind

K_p—characteristic function passband specification

K_s—characteristic function stopband specification

$L(n, \xi)$—discrimination factor

$M(f)$—magnitude response, $M(f) = |H(e^{j2\pi f})|$

n—transfer function order (order for short)

$n_{but}(F_p, F_s, K_p, K_s)$—minimum Butterworth order

$n_{cheb}(F_p, F_s, K_p, K_s)$—minimum Chebyshev order

$n_{ellip}(F_p, F_s, K_p, K_s)$—minimum elliptic order

$n_{\max}$—maximum order

$n_{\min}$—minimum order

$n_{minQ}(F_p, F_s, K_p, K_s)$—minimum order of minimal Q-factor design

$q(k)$—modular constant

$Q(a, b)$—quality factor of $H(z) = z^2/(z^2 + bz + a)$

$Q_{\min Q}(n, \xi, i)$—quality factor of ith pole of minimal Q-factor design

$R(n, \xi, x)$—elliptic rational function

$\mathrm{sn}(u, k)$—Jacobi elliptic sn function

$\mathrm{sn}^{-1}(v, k)$—inverse Jacobi elliptic sn function

$S(n, \xi, \epsilon, i)$—ith pole of elliptic rational function

S_A—attenuation-limit specification

S_G—gain-limit specification

S_K—characteristic-function-limit specification

S_M—magnitude-limit specification

S_r—magnitude-ripple specification

S_δ—magnitude-tolerance specification

$S_{minQ}(n, \xi, i)$—ith pole of minimal Q-factor design

x—dimensionless variable

$X(n, \xi, i)$—ith zero of elliptic rational function

z—complex variable in the z-plane; $z = e^{j2\pi f}$ refers to the unit circle

Z—auxiliary complex variable in the z-plane

$Z_{BL}(s, f_p)$—bilinear transformation

δ_1—passband magnitude ripple
δ_2—stopband magnitude ripple
δ_p—passband magnitude tolerance
δ_s—stopband magnitude tolerance
$\zeta(n, \xi, \epsilon)$—auxiliary function
ϵ—ripple factor
ξ—selectivity factor
$\tau_{GD}(H(z), f)$—group delay (in samples)
$\Phi(f)$—phase response, $\Phi(f) = \arg(H(e^{j2\pi f}))$
θ—angular digital frequency, $\theta = 2\pi f$
$\lfloor x \rfloor$—integer, $x \leq \lfloor x \rfloor < x + 1$

$\underset{x_1<x<x_2}{\text{FindRoot}} \; \{ \; F(x) = G(x) \; \}$ — find real x over interval $x_1 < x < x_2$ by solving $F(x) = G(x)$

$\underset{\substack{x_1<x<x_2 \\ y_1<y<y_2}}{\text{FindRoot}} \begin{Bmatrix} F_1(x) = G_1(x) \\ F_2(x) = G_2(x) \end{Bmatrix}$ — find real x over interval $x_1 < x < x_2$ and real y over interval $y_1 < y < y_2$ by solving set of equations $\{F_1(x) = G_1(x), \;\; F_2(x) = G_2(x)\}$

10.3 DESIGN EQUATIONS AND PROCEDURES

In this section we summarize all design equations, formulas, and procedures that are based on Jacobi elliptic functions. We use these relations in the purely numerical design.

10.3.1 Specification

A design specification can be given in different ways:

$$\boxed{S_A = \{F_p, F_s, A_p, A_s\}} \tag{10.1}$$

$$\boxed{S_G = \{F_p, F_s, G_p, G_s\}} \tag{10.2}$$

$$\boxed{S_\delta = \{F_p, F_s, \delta_p, \delta_s\}} \tag{10.3}$$

$$\boxed{S_M = \{F_p, F_s, M_p, M_s\}} \tag{10.4}$$

$$\boxed{S_r = \{F_p, F_s, \delta_1, \delta_2\}} \tag{10.5}$$

$$\boxed{S_K = \{F_p, F_s, K_p, K_s\}} \tag{10.6}$$

We use a set of functions to convert one form of specification into another.

$$\boxed{\begin{aligned} K_p(A_p) &= \frac{\sqrt{1 - 10^{-A_p/10}}}{10^{-A_p/20}} \\ K_s(A_s) &= \frac{\sqrt{1 - 10^{-A_s/10}}}{10^{-A_s/20}} \end{aligned}} \tag{10.7}$$

$$\boxed{\begin{aligned} K_p(G_p) &= \frac{\sqrt{1 - 10^{G_p/10}}}{10^{G_p/20}} \\ K_s(G_s) &= \frac{\sqrt{1 - 10^{G_s/10}}}{10^{G_s/20}} \end{aligned}} \tag{10.8}$$

$$\boxed{\begin{aligned} K_p(\delta_p) &= \frac{\sqrt{\delta_p(2 - \delta_p)}}{1 - \delta_p} \\ K_s(\delta_s) &= \frac{\sqrt{1 - \delta_s^2}}{\delta_s} \end{aligned}} \tag{10.9}$$

$$\boxed{\begin{aligned} K_p(M_p) &= \frac{\sqrt{1 - M_p^2}}{M_p} \\ K_s(M_s) &= \frac{\sqrt{1 - M_s^2}}{M_s} \end{aligned}} \tag{10.10}$$

$$\boxed{\begin{aligned} K_p(\delta_1) &= \frac{2\sqrt{\delta_1}}{1 - \delta_1} \\ K_s(\delta_1, \delta_2) &= \frac{\sqrt{(1 + \delta_1)^2 - \delta_2^2}}{\delta_2} \end{aligned}} \tag{10.11}$$

10.3.2 Special and Auxiliary Functions

The following special mathematical functions are used in the design of elliptic filters:

$$K_J(k) = \int_0^{\pi/2} \frac{d\theta}{\sqrt{1 - k^2 \sin^2 \theta}} \tag{10.12}$$

$$\boxed{\begin{aligned} v &= \text{sn}^{-1}\left(\frac{1}{\sqrt{1+\epsilon^2}}, \sqrt{1 - \frac{1}{L^2(n,\xi)}}\right) \\ \zeta(n,\xi,\epsilon) &= \text{sn}\left(\frac{K_J\left(\sqrt{1-\frac{1}{\xi^2}}\right)}{K_J\left(\sqrt{1-\frac{1}{L^2(n,\xi)}}\right)} v, \sqrt{1-\frac{1}{\xi^2}}\right) \end{aligned}} \tag{10.13}$$

The procedure given by Eq. (10.13) comes from Eq. (12.302).

$$\boxed{\begin{aligned} t &= \frac{1}{2}\frac{1-\sqrt[4]{1-k^2}}{1+\sqrt[4]{1-k^2}} \\ q' &= t + 2t^5 + 15t^9 + 150t^{13} + 1{,}707t^{17} \\ &\quad + 20{,}910t^{21} + 268{,}616t^{25} + 3{,}567{,}400t^{29}, \qquad k \le \frac{1}{\sqrt{2}} \\ &\quad + 48{,}555{,}069t^{33} + 673{,}458{,}874t^{37} \\ q(k) &= q' \\ t &= \frac{1}{2}\frac{1-\sqrt{k}}{1+\sqrt{k}} \\ q' &= t + 2t^5 + 15t^9 + 150t^{13} + 1{,}707t^{17} \\ &\quad + 20{,}910t^{21} + 268{,}616t^{25} + 3{,}567{,}400t^{29}, \qquad \frac{1}{\sqrt{2}} < k < 1 \\ &\quad + 48{,}555{,}069t^{33} + 673{,}458{,}874t^{37} \\ q(k) &= e^{\pi^2/\ln q'} \end{aligned}} \tag{10.14}$$

The procedure given by Eq. (10.14) comes from Eqs. (12.60) and (12.64)–(12.67).

10.3.3 Transfer Function Order

For a given specification we compute the minimal order of the approximation functions for $(0 \le F_p < F_s \le \frac{1}{2})$.

$$
\begin{aligned}
k &= \frac{\tan\left(\pi F_p\right)}{\tan\left(\pi F_s\right)} \\
L &= \frac{K_s}{K_p} \\
N &= \frac{K_J\left(\sqrt{1-\dfrac{1}{L^2}}\right)}{K_J\left(\dfrac{1}{L}\right)} \\
D &= \frac{K_J\left(\sqrt{1-k^2}\right)}{K_J\,(k)} \\
n_{ellip}\left(F_p, F_s, K_p, K_s\right) &= \left\lfloor \frac{N}{D} \right\rfloor
\end{aligned}
\tag{10.15}
$$

The procedure given by Eq. (10.15) comes from Eq. (12.338).

$$
\begin{aligned}
\xi &= \frac{\tan\left(\pi F_p\right)}{\tan\left(\pi F_s\right)} \\
L &= \frac{K_s}{K_p} \\
N &= \cosh^{-1} L \\
D &= \cosh^{-1} \xi \\
n_{cheb}\left(F_p, F_s, K_p, K_s\right) &= \left\lfloor \frac{N}{D} \right\rfloor
\end{aligned}
\tag{10.16}
$$

$$
\begin{aligned}
\xi &= \frac{\tan\left(\pi F_p\right)}{\tan\left(\pi F_s\right)} \\
L &= \frac{K_s}{K_p} \\
N &= \log_{10} L \\
D &= \log_{10} \xi \\
n_{but}\left(F_p, F_s, K_p, K_s\right) &= \left\lfloor \frac{N}{D} \right\rfloor
\end{aligned}
\tag{10.17}
$$

The procedure given by Eqs. (10.16) and (10.17) are based on formulas given in reference [15].

$$\boxed{\begin{aligned} n_{\min} &= n_{ellip}\left(F_p, F_s, K_p, K_s\right) \\ n_{\max} &= 2\, n_{ellip}\left(F_p, F_s, K_p, K_s\right) \end{aligned}} \tag{10.18}$$

$$\boxed{\begin{aligned} i &= n_{ellip}\left(F_p, F_s, K_p, K_s\right) \\ \xi_1 &= \frac{\tan\left(\pi F_p\right)}{\tan\left(\pi F_s\right)} \\ \xi_2 &= \xi_1^2 \\ \xi_i &= x \,\Big|\, \underset{\xi_1 < x < \xi_2}{\text{FindRoot}} \sqrt{L(i, x)} = K_s \\ & \boxed{\begin{aligned} &\text{While } K_e\left(i, \xi_i, \frac{1}{K_s}, \frac{\tan\left(\pi F_p\right)}{\tan\left(\pi F_s\right)} \xi_i\right) > K_p \\ &i = i + 1 \\ &\xi_2 = \xi_i \\ &\xi_i = x \,\Big|\, \underset{\xi_1 < x < \xi_2}{\text{FindRoot}} \sqrt{L(i, x)} = K_s \end{aligned}} \\ n_{\min Q}\left(F_p, F_s, K_p, K_s\right) &= i \end{aligned}} \tag{10.19}$$

The procedure given by Eq. (10.19) comes from Eqs. (12.338) and (12.367)–(12.369).

10.3.4 Bilinear Transformation

We define the bilinear transformation by

$$\boxed{z_{BL}\left(s, f_p\right) = \frac{1 + s \tan\left(\pi f_p\right)}{1 - s \tan\left(\pi f_p\right)}} \tag{10.20}$$

10.3.5 Zeros, Poles, and Q-Factors

The basic functions for the elliptic design are $X(n,\xi,i)$, $S(n,\xi,\epsilon,i)$ and $Q(s)$, as follows:

$$\begin{aligned} X(n,\xi,i) &= -\mathrm{cd}\left(\frac{2i-1}{n}K_J(\frac{1}{\xi}),\frac{1}{\xi}\right) \\ X(n,\xi,(n+1)/2) &= 0, \quad n \text{ odd} \\ X(n,\xi,1) < X(n,\xi,2) < &\ldots < X(n,\xi,n) \end{aligned} \tag{10.21}$$

The procedure given by Eq. (10.21) comes from Eqs. (12.134) and (12.323).

$$\begin{aligned} \zeta &= \zeta(n,\xi,\epsilon) \\ x &= X(n,\xi,i) \\ N_{re} &= -\zeta\sqrt{1-\zeta^2}\sqrt{1-x^2}\sqrt{1-\frac{x^2}{\xi^2}} \\ N_{im} &= x\sqrt{1-\left(1-\frac{1}{\xi^2}\right)\zeta^2} \\ N &= N_{re}+jN_{im} \\ D &= 1-\left(1-\frac{x^2}{\xi^2}\right)\zeta^2 \\ S(n,\xi,\epsilon,i) &= \frac{N}{D} \end{aligned} \tag{10.22}$$

The procedure given by Eq. (10.22) comes from Eq. (12.266).

$$\begin{aligned} N_{re} &= -\sqrt{1-X(n,\xi,i)^2}\sqrt{\xi^2-X(n,\xi,i)^2} \\ N_{im} &= X(n,\xi,i)(\xi+1) \\ N &= N_{re}+jN_{im} \\ D &= \xi+X(n,\xi,i)^2 \\ S_{\min Q}(n,\xi,i) &= \sqrt{\xi}\frac{N}{D} \end{aligned} \tag{10.23}$$

The procedure given by Eq. (10.23) comes from Eq. (12.370).

$$Q(a,b) = \frac{\sqrt{(1+a)^2 - b^2}}{2(1-a)}, \qquad H(z) = \frac{z^2}{z^2 + bz + a} \tag{10.24}$$

10.3.6 Discrimination Factor, Elliptic Rational Function, and Characteristic Function

The essential functions for the elliptic approximation are $L(n, \xi)$, $R(n, \xi, x)$, and $K_e(n, \xi, \epsilon, x)$.

$$L(n,\xi) = \frac{1}{\xi^n} \frac{\prod_{i=1}^{n/2} \left(\xi^2 - X^2(n,\xi,i)\right)^2}{\prod_{i=1}^{n/2} \left(1 - X^2(n,\xi,i)\right)^2} \qquad n \text{ even} \quad n=2,4,\ldots$$

$$L(n,\xi) = \frac{1}{\xi^{n-2}} \frac{\prod_{i=1}^{(n-1)/2} \left(\xi^2 - X^2(n,\xi,i)\right)^2}{\prod_{i=1}^{(n-1)/2} \left(1 - X^2(n,\xi,i)\right)^2} \qquad n \text{ odd} \quad n=1,3,\ldots \tag{10.25}$$

The procedure given by Eq. (10.25) comes from Eqs. (12.359) and (12.360).

$$r_0 = \frac{\prod_{i=1}^{n/2}\left(1 - X^2(n,\xi,i)\right)}{\prod_{i=1}^{n/2}\left(1 - \frac{\xi^2}{X^2(n,\xi,i)}\right)}$$

n even
n=2,4,...

$$R(n,\xi,x) = \frac{1}{r_0}\frac{\prod_{i=1}^{n/2}\left(x^2 - X^2(n,\xi,i)\right)}{\prod_{i=1}^{n/2}\left(x^2 - \frac{\xi^2}{X^2(n,\xi,i)}\right)}$$

$$r_0 = \frac{\prod_{i=1}^{(n-1)/2}\left(1 - X^2(n,\xi,i)\right)}{\prod_{i=1}^{(n-1)/2}\left(1 - \frac{\xi^2}{X^2(n,\xi,i)}\right)}$$

n odd
n=1,3,...

$$R(n,\xi,x) = \frac{1}{r_0}\frac{x\prod_{i=1}^{(n-1)/2}\left(x^2 - X^2(n,\xi,i)\right)}{\prod_{i=1}^{(n-1)/2}\left(x^2 - \frac{\xi^2}{X^2(n,\xi,i)}\right)} \tag{10.26}$$

The procedure given by Eq. (10.26) comes from Eqs. (12.141) and (12.144).

$$\boxed{K_e\,(n,\xi,\epsilon,x) = \epsilon\,|R\,(n,\xi,x)|} \tag{10.27}$$

10.3.7 Lowpass Elliptic Transfer Function

Our goal is to find the transfer function that meets a lowpass specification.

$$z_{zi} = Z_{BL}\left(j\frac{\xi}{X(n,\xi,i)}, f_p\right)$$

$$z_{pi} = Z_{BL}\left(S(n,\xi,\epsilon,i), f_p\right)$$

$$g = \frac{\sqrt{1+\epsilon^2}\prod_{i=1}^{n/2}(2 - 2\ \mathrm{Re}(z_{zi}))}{\prod_{i=1}^{n/2}\left(1 - 2\ \mathrm{Re}(z_{pi}) + |z_{pi}|^2\right)} \qquad \begin{array}{l} n \text{ even} \\ n = 2, 4, \ldots \end{array}$$

$$\mathcal{H}(n,\xi,\epsilon,f_p,z) = \frac{\frac{1}{g}\prod_{i=1}^{n/2}\left(z^2 - 2z\ \mathrm{Re}(z_{zi}) + 1\right)}{\prod_{i=1}^{n/2}\left(z^2 - 2z\ \mathrm{Re}(z_{pi}) + |z_{pi}|^2\right)}$$

$$z_{zi} = Z_{BL}\left(j\frac{\xi}{X(n,\xi,i)}, f_p\right)$$

$$z_{pi} = Z_{BL}\left(S(n,\xi,\epsilon,i), f_p\right)$$

$$g = \frac{2\prod_{i=1}^{(n-1)/2}(2 - 2\ \mathrm{Re}(z_{zi}))}{(1 - z_{(n+1)/2})\prod_{i=1}^{(n-1)/2}\left(1 - 2\ \mathrm{Re}(z_{pi}) + |z_{pi}|^2\right)} \qquad \begin{array}{l} n \text{ odd} \\ n = 3, 5, \ldots \end{array}$$

$$\mathcal{H}(n,\xi,\epsilon,f_p,z) = \frac{\frac{1}{g}(z+1)\prod_{i=1}^{(n-1)/2}\left(z^2 - 2z\ \mathrm{Re}(z_{zi}) + 1\right)}{(z - z_{(n+1)/2})\prod_{i=1}^{(n-1)/2}\left(z^2 - 2z\ \mathrm{Re}(z_{pi}) + |z_{pi}|^2\right)} \qquad (10.28)$$

The procedure given by Eq. (10.28) comes from Eq. (13.196).

$$
\begin{aligned}
z_{zi} &= Z_{BL}\left(j\frac{\xi}{X(n,\xi,i)}, f_p\right) \\
z_{pi} &= Z_{BL}\left(S_{\min Q}(n,\xi,i), f_p\right) \\
g &= \frac{\sqrt{1 + \dfrac{1}{L(n,\xi)}\displaystyle\prod_{i=1}^{n/2}(2 - 2\,\mathrm{Re}(z_{zi}))}}{\displaystyle\prod_{i=1}^{n/2}\left(1 - 2\,\mathrm{Re}(z_{pi}) + |z_{pi}|^2\right)} && \begin{array}{l} n \text{ even} \\ n = 2, 4, \ldots \end{array} \\
\mathcal{H}_{\min Q}(n,\xi,f_p,z) &= \frac{\dfrac{1}{g}\displaystyle\prod_{i=1}^{n/2}\left(z^2 - 2z\,\mathrm{Re}(z_{zi}) + 1\right)}{\displaystyle\prod_{i=1}^{n/2}\left(z^2 - 2z\,\mathrm{Re}(z_{pi}) + |z_{pi}|^2\right)} \\
z_{zi} &= Z_{BL}\left(j\frac{\xi}{X(n,\xi,i)}, F_p\right) \\
z_{pi} &= Z_{BL}\left(S_{\min Q}(n,\xi,i), F_p\right) \\
g &= \frac{2\displaystyle\prod_{i=1}^{(n-1)/2}(2 - 2\,\mathrm{Re}(z_{zi}))}{(1 - z_{(n+1)/2})\displaystyle\prod_{i=1}^{(n-1)/2}\left(1 - 2\,\mathrm{Re}(z_{pi}) + |z_{pi}|^2\right)} && \begin{array}{l} n \text{ odd} \\ n = 3, 5, \ldots \end{array} \\
\mathcal{H}_{\min Q}(n,\xi,f_p,z) &= \frac{\dfrac{1}{g}(z+1)\displaystyle\prod_{i=1}^{(n-1)/2}\left(z^2 - 2z\,\mathrm{Re}(z_{zi}) + 1\right)}{(z - z_{(n+1)/2})\displaystyle\prod_{i=1}^{(n-1)/2}\left(z^2 - 2z\,\mathrm{Re}(z_{pi}) + |z_{pi}|^2\right)}
\end{aligned}
\tag{10.29}
$$

The procedure given by Eq. (10.29) comes from Eqs. (12.373) and (12.374).

10.3.8 Selectivity Factor, Ripple Factor, and Edge Frequencies

We compute the boundary values of the design space from

$$\boxed{\begin{aligned} L &= \frac{K_s}{K_p} \\ g &= \left(q\left(\frac{1}{L}\right)\right)^{1/n} \\ N &= 1 + 2\sum_{m=1}^{9}(-1)^m g^{m^2} \\ D &= 1 + 2\sum_{m=1}^{9} g^{m^2} \\ \xi_{min}\left(n, K_p, K_s\right) &= \frac{1}{\sqrt{1 - \left(\frac{N}{D}\right)^4}} \end{aligned}} \tag{10.30}$$

The procedure given by Eq. (10.30) comes from Eq. (12.354).

$$\boxed{\xi_{\max}\left(n, F_p, F_s, K_p, K_s\right) = x \left| \underset{\frac{\tan(\pi F_s)}{\tan(\pi F_p)} < x < 10\frac{\tan(\pi F_s)}{\tan(\pi F_p)}}{\text{FindRoot}} \quad R(n, x, \frac{\tan(\pi F_s)}{\tan(\pi F_p)}) = \frac{K_s}{K_p} \right.} \tag{10.31}$$

$$\boxed{1 < \xi_{\min} < \frac{\tan(\pi F_s)}{\tan(\pi F_p)} < \xi_{\max} < \infty} \tag{10.32}$$

$$\boxed{\epsilon_{\min}\left(n, F_p, F_s, K_p, K_s\right) = \frac{K_s}{L\left(n, \xi_{\max}\left(n, F_p, F_s, K_p, K_s\right)\right)}} \tag{10.33}$$

$$\boxed{\epsilon_{\max}\left(K_p\right) = K_p} \tag{10.34}$$

$$\boxed{f_{p,min}\left(n, F_p, F_s, K_p, K_s\right) = \frac{1}{\pi}\tan^{-1}\left(\frac{\tan(\pi F_s)}{\xi_{\max}\left(n, F_p, F_s, K_p, K_s\right)}\right)} \tag{10.35}$$

$$\boxed{f_{p,\max}\left(n, F_s, K_p, K_s\right) = \frac{1}{\pi}\tan^{-1}\left(\frac{\tan\left(\pi F_s\right)}{\xi_{min}\left(n, K_p, K_s\right)}\right)} \tag{10.36}$$

$$\begin{aligned}
\xi_h &= \xi_{\max}\left(n, F_p, F_s, K_p, K_s\right) \\
\xi_l &= x \left| \underset{\sqrt{\xi_h \frac{\tan(\pi F_s)}{\tan(\pi F_p)}} < x < \xi_h}{\text{FindRoot}} \quad K_e\left(n, x, \frac{1}{\sqrt{L(n,x)}}, x\frac{\tan\left(\pi F_s\right)}{\tan\left(\pi F_p\right)}\right) = K_p \right. \\
f_l &= x \left| \underset{\sqrt[4]{\xi_l} < x < \sqrt{\xi_l}}{\text{FindRoot}}\, K_e\left(n, \xi_l, \frac{1}{\sqrt{L(n,\xi_l)}}, x\right) = K_p \right. \\
f_h &= x \left| \underset{\sqrt[4]{\xi_h} < x < \sqrt{\xi_h}}{\text{FindRoot}}\, K_e\left(n, \xi_h, \frac{1}{\sqrt{L(n,\xi_h)}}, x\right) = K_p \right. \\
\begin{matrix}\xi_{\min Q}\left(n, F_p, F_s, K_p, K_s\right) = x \\ f_{nQ}\left(n, F_p, F_s, K_p, K_s\right) = y\end{matrix} &\left| \underset{\substack{\xi_l < x < \xi_h \\ f_l < y < f_h}}{\text{FindRoot}} \left\{ \begin{matrix} K_e\left(n, x, \frac{1}{\sqrt{L(n,x)}}, y\right) = K_p \\ K_e(n, x, \frac{1}{\sqrt{L(n,x)}}, y\frac{\tan(\pi F_s)}{\tan(\pi F_p)}) = K_s \end{matrix} \right\} \right.
\end{aligned} \tag{10.37}$$

10.4 DESIGN D1

1. Start from a specification and convert it into the characteristic-function-limit specification:

$$\left.\begin{aligned}
S_A &= \{F_p, F_s, A_p, A_s\} \\
S_\delta &= \{F_p, F_s, \delta_p, \delta_s\} \\
S_M &= \{F_p, F_s, M_p, M_s\} \\
S_r &= \{F_p, F_s, \delta_1, \delta_2\} \\
S_G &= \{F_p, F_s, G_p, G_s\}
\end{aligned}\right\} \longmapsto S_K = \{F_p, F_s, K_p, K_s\} \tag{10.38}$$

2. Compute the minimal order: $n_{\min} = n_{ellip}(F_p, F_s, K_p, K_s)$.
3. Choose the order:

$$n \geq n_{\min} \tag{10.39}$$

4. Compute the selectivity factor:

$$\xi = \frac{\tan(\pi F_s)}{\tan(\pi F_p)} \tag{10.40}$$

5. Choose the ripple factor:

$$\epsilon = K_p \tag{10.41}$$

6. Choose the actual passband edge:

$$f_p = F_p \tag{10.42}$$

7. Construct the lowpass transfer function:

$$H(z) = \mathcal{H}(n, \xi, \epsilon, f_p, z) \tag{10.43}$$

10.5 DESIGN D2

1. Start from a specification and convert it into the characteristic-function-limit specification:

$$\left.\begin{aligned} S_A &= \{F_p, F_s, A_p, A_s\} \\ S_\delta &= \{F_p, F_s, \delta_p, \delta_s\} \\ S_M &= \{F_p, F_s, M_p, M_s\} \\ S_r &= \{F_p, F_s, \delta_1, \delta_2\} \\ S_G &= \{F_p, F_s, G_p, G_s\} \end{aligned}\right\} \longmapsto S_K = \{F_p, F_s, K_p, K_s\} \tag{10.44}$$

2. Compute the minimal order: $n_{\min} = n_{ellip}(F_p, F_s, K_p, K_s)$.
3. Choose the order:

$$n \geq n_{\min} \tag{10.45}$$

4. Compute the selectivity factor:

$$\xi = \frac{\tan(\pi F_s)}{\tan(\pi F_p)} \tag{10.46}$$

5. Compute the ripple factor:

$$\epsilon = \frac{K_s}{L(n, \xi)} \tag{10.47}$$

6. Choose the actual passband edge:

$$f_p = F_p \tag{10.48}$$

7. Construct the lowpass transfer function:

$$H(z) = \mathcal{H}(n, \xi, \epsilon, f_p, z) \tag{10.49}$$

10.6 DESIGN D3A

1. Start from a specification and convert it into the characteristic-function-limit specification:

$$\left.\begin{aligned} S_A &= \{F_p, F_s, A_p, A_s\} \\ S_\delta &= \{F_p, F_s, \delta_p, \delta_s\} \\ S_M &= \{F_p, F_s, M_p, M_s\} \\ S_r &= \{F_p, F_s, \delta_1, \delta_2\} \\ S_G &= \{F_p, F_s, G_p, G_s\} \end{aligned}\right\} \longmapsto S_K = \{F_p, F_s, K_p, K_s\} \tag{10.50}$$

2. Compute the minimal order: $n_{\min} = n_{ellip}(F_p, F_s, K_p, K_s)$.
3. Choose the order:

$$n \geq n_{\min} \tag{10.51}$$

4. Compute the selectivity factor:

$$\xi = \xi_{min}\left(n, K_p, K_s\right) \tag{10.52}$$

5. Choose the ripple factor:

$$\epsilon = K_p \tag{10.53}$$

6. Choose the actual passband edge:

$$f_p = F_p \tag{10.54}$$

7. Construct the lowpass transfer function:

$$H(z) = \mathcal{H}(n, \xi, \epsilon, f_p, z) \tag{10.55}$$

10.7 DESIGN D3B

1. Start from a specification and convert it into the characteristic-function-limit specification:

$$\left.\begin{aligned} S_A &= \{F_p, F_s, A_p, A_s\} \\ S_\delta &= \{F_p, F_s, \delta_p, \delta_s\} \\ S_M &= \{F_p, F_s, M_p, M_s\} \\ S_r &= \{F_p, F_s, \delta_1, \delta_2\} \\ S_G &= \{F_p, F_s, G_p, G_s\} \end{aligned}\right\} \longmapsto S_K = \{F_p, F_s, K_p, K_s\} \tag{10.56}$$

2. Compute the minimal order: $n_{\min} = n_{ellip}(F_p, F_s, K_p, K_s)$.
3. Choose the order:

$$n \geq n_{\min} \tag{10.57}$$

4. Compute the selectivity factor:

$$\xi = \xi_{min}\left(n, K_p, K_s\right) \tag{10.58}$$

5. Choose the ripple factor:

$$\epsilon = K_p \tag{10.59}$$

6. Compute the actual passband edge:

$$f_p = \frac{1}{\pi}\tan^{-1}\left(\frac{\tan(\pi F_s)}{\xi}\right) \tag{10.60}$$

7. Construct the lowpass transfer function:

$$H(z) = \mathcal{H}(n, \xi, \epsilon, f_p, z) \tag{10.61}$$

10.8 DESIGN D4A

1. Start from a specification and convert it into the characteristic-function-limit specification:

$$\left.\begin{aligned} S_A &= \{F_p, F_s, A_p, A_s\} \\ S_\delta &= \{F_p, F_s, \delta_p, \delta_s\} \\ S_M &= \{F_p, F_s, M_p, M_s\} \\ S_r &= \{F_p, F_s, \delta_1, \delta_2\} \\ S_G &= \{F_p, F_s, G_p, G_s\} \end{aligned}\right\} \longmapsto S_K = \{F_p, F_s, K_p, K_s\} \tag{10.62}$$

2. Compute the minimal order: $n_{\min} = n_{ellip}(F_p, F_s, K_p, K_s)$.
3. Choose the order:

$$n \geq n_{\min} \tag{10.63}$$

4. Compute the selectivity factor:

$$\xi = \xi_{\max}\left(n, F_p, F_s, K_p, K_s\right) \tag{10.64}$$

5. Choose the ripple factor:

$$\epsilon = K_p \tag{10.65}$$

6. Choose the actual passband edge:

$$f_p = F_p \tag{10.66}$$

7. Construct the lowpass transfer function:

$$H(z) = \mathcal{H}(n, \xi, \epsilon, f_p, z) \tag{10.67}$$

10.9 DESIGN D4B

1. Start from a specification and convert it into the characteristic-function-limit specification:

$$\left.\begin{aligned} S_A &= \{F_p, F_s, A_p, A_s\} \\ S_\delta &= \{F_p, F_s, \delta_p, \delta_s\} \\ S_M &= \{F_p, F_s, M_p, M_s\} \\ S_r &= \{F_p, F_s, \delta_1, \delta_2\} \\ S_G &= \{F_p, F_s, G_p, G_s\} \end{aligned}\right\} \longmapsto S_K = \{F_p, F_s, K_p, K_s\} \tag{10.68}$$

2. Compute the minimal order: $n_{\min} = n_{ellip}(F_p, F_s, K_p, K_s)$.
3. Choose the order:

$$n \geq n_{\min} \tag{10.69}$$

4. Compute the selectivity factor:

$$\xi = \xi_{\max}\left(n, F_p, F_s, K_p, K_s\right) \tag{10.70}$$

5. Compute the ripple factor:

$$\epsilon = \frac{K_s}{L(n, \xi)} \tag{10.71}$$

6. Compute the actual passband edge:

$$f_p = \frac{1}{\pi}\tan^{-1}\left(\frac{\tan(\pi F_s)}{\xi}\right) \tag{10.72}$$

7. Construct the lowpass transfer function:

$$H(z) = \mathcal{H}(n, \xi, \epsilon, f_p, z) \tag{10.73}$$

10.10 DESIGN D5

1. Start from a specification and convert it into the characteristic-function-limit specification:

$$\left.\begin{aligned} S_A &= \{F_p, F_s, A_p, A_s\} \\ S_\delta &= \{F_p, F_s, \delta_p, \delta_s\} \\ S_M &= \{F_p, F_s, M_p, M_s\} \\ S_r &= \{F_p, F_s, \delta_1, \delta_2\} \\ S_G &= \{F_p, F_s, G_p, G_s\} \end{aligned}\right\} \longmapsto S_K = \{F_p, F_s, K_p, K_s\} \tag{10.74}$$

2. Compute the minimal order: $n_{\min Q}\left(F_p, F_s, K_p, K_s\right)$
3. Choose the order:

$$n \geq n_{\min Q}\left(F_p, F_s, K_p, K_s\right) \tag{10.75}$$

4. Compute the selectivity factor:

$$\xi = \xi_{\min Q}\left(n, F_p, F_s, K_p, K_s\right) \tag{10.76}$$

and normalized frequency:

$$f_{nQ} = f_{nQ}\left(n, F_p, F_s, K_p, K_s\right) \tag{10.77}$$

5. Compute the ripple factor:

$$\epsilon = \frac{1}{\sqrt{L(n,\xi)}} \tag{10.78}$$

6. Compute the actual passband edge:

$$f_p = \frac{1}{\pi}\tan^{-1}\left(\frac{\tan(\pi F_s)}{f_{nQ}}\right) \tag{10.79}$$

7. Construct the lowpass transfer function:

$$H(z) = \mathcal{H}(n, \xi, \epsilon, f_p, z) \tag{10.80}$$

Summary of the design parameters:

	$n_{\min}$	ξ	ϵ	f_p
D1	n_{ellip}	$\frac{\tan(\pi F_s)}{\tan(\pi F_p)}$	K_p	F_p
D2	n_{ellip}	$\frac{\tan(\pi F_s)}{\tan(\pi F_p)}$	$\frac{K_s}{L}$	F_p
D3a	n_{ellip}	ξ_{min}	K_p	F_p
D3b	n_{ellip}	ξ_{min}	K_p	$\frac{1}{\pi}\tan^{-1}\left(\frac{\tan(\pi F_s)}{\xi_{\min}}\right)$
D4a	n_{ellip}	$\xi_{\max}$	K_p	F_p
D4b	n_{ellip}	$\xi_{\max}$	$\frac{K_s}{L}$	$\frac{1}{\pi}\tan^{-1}\left(\frac{\tan(\pi F_s)}{\xi_{\max}}\right)$
D5	$n_{\min Q}$	$\xi_{\min Q}$	$\frac{1}{\sqrt{L}}$	$\frac{1}{\pi}\tan^{-1}\left(\frac{\tan(\pi F_s)}{f_{nQ}}\right)$

(10.81)

where $\xi_{\min} = \xi_{\min}(n, K_p, K_s)$, $\xi_{\max} = \xi_{\max}(n, K_p, K_s)$, $f_{nQ} = f_{nQ}(n, F_p, F_s, K_p, K_s)$ and $L = L(n, \xi)$.

10.11 FREQUENCY RESPONSE

From the known transfer function, $H(z)$, we compute the magnitude response

$$M(f) = |H(e^{j2\pi f})| \tag{10.82}$$

the phase response

$$\Phi(f) = \arg(H(e^{j2\pi f})) \tag{10.83}$$

the group delay

$$\tau_{DGD}(f) = -\frac{d\Phi(f)}{df} \tag{10.84}$$

the gain (in dB)

$$G(f) = 20\log_{10}|H(e^{j2\pi f})| \tag{10.85}$$

and the attenuation (in dB)

$$A(f) = -20\log_{10}|H(e^{j2\pi f})| \tag{10.86}$$

10.12 HIGHPASS FILTER

A highpass filter can be specified by its edge frequencies and attenuation limits S_{Ah}; F_{ph} designates the passband edge frequency, F_{sh} is the stopband edge frequency, $F_{sh} < F_{ph}$, the passband attenuation (dB) is designated by A_{ph}, and A_{sh} stands for the stopband attenuation (dB).

First, we map the highpass filter specification, S_{Ah}, into the lowpass filter specification, S_A,

$$\boxed{S_{Ah} = \{F_{sh}, F_{ph}, A_{ph}, A_{sh}\} \longmapsto S_A = \left\{\begin{array}{c} F_p = 0.5 - F_{sh} \\ F_s = 0.5 - F_{ph} \\ A_p = A_{ph} \\ A_s = A_{sh} \end{array}\right\}} \tag{10.87}$$

Next, we construct the lowpass transfer function, $\mathcal{H}(n, \xi, \epsilon, f_p, Z)$, which meets the specification S_A according to the design procedures D1, D2, D3a, D3b, D4a, D4b, and D5.

Finally, the transfer function, $H_{HP}(z)$, of the highpass filter is constructed from the lowpass transfer function according to the transformation:

$$Z = -z \tag{10.88}$$

which yields

$$\boxed{H_{HP}(z) = \mathcal{H}(n, \xi, \epsilon, f_p, -z)} \tag{10.89}$$

10.13 BANDPASS FILTER

A bandpass filter can be specified by its edge frequencies and attenuation limits S_{Ab}; F_{p1} and F_{p2} designate the passband edge frequencies, F_{s1} and F_{s2} are the stopband edge frequencies, $0 < F_{s1} < F_{p1} < F_{p2} < F_{s2} < 0.5$, the passband attenuation (dB) is designated by A_{pb}, and A_{s1} and A_{s2} stand for the stopband attenuations (dB). We assume that the specification satisfies

$$\tan(\pi F_{p1})\ \tan(\pi F_{p2}) = \tan(\pi F_{s1})\ \tan(\pi F_{s2}) \tag{10.90}$$

If Eq. (10.90) is not satisfied, then we use new F_{s1} or F_{s2}:

$$F'_{s2} = \frac{1}{\pi} \tan^{-1}\left(\frac{\tan(\pi F_{p1})\ \tan(\pi F_{p2})}{\tan(\pi F_{s1})}\right) < F_{s2} \tag{10.91}$$

or

$$F'_{s1} = \frac{1}{\pi} \tan^{-1}\left(\frac{\tan(\pi F_{p1})\ \tan(\pi F_{p2})}{\tan(\pi F_{s2})}\right) > F_{s1} \tag{10.92}$$

First, we map the bandpass filter specification, S_{Ab}, into the lowpass filter specification, S_A:

$$S_{Ab} = \{F_{s1}, F_{p1}, F_{p2}, F_{s2}, A_{s1}, A_{pb}, A_{s2}\}$$
$$\downarrow$$
$$S_A = \left\{\begin{array}{lcl} F_p & = & F_{s1} \\ F_s & = & \dfrac{1}{\pi}\tan^{-1}\left(\tan(\pi F_{s1})\,\dfrac{\tan(\pi F_{s2}) - \tan(\pi F_{s1})}{\tan(\pi F_{p2}) - \tan(\pi F_{p1})}\right) \\ A_p & = & A_{pb} \\ A_s & = & \max(A_{s1}, A_{s2}) \end{array}\right\} \tag{10.93}$$

Next, we construct the lowpass transfer function, $\mathcal{H}(n, \xi, \epsilon, f_p, Z)$, which meets the specification S_A according to the design procedures D1, D2, D3a, D3b, D4a, D4b, and D5.

Finally, the transfer function, $H_{BP}(z)$, of the bandpass filter is constructed from the lowpass transfer function according to the transformation:

$$Z = -\frac{z^2 + \beta z + \gamma}{1 + \beta z + \gamma z^2}$$

where

$$k = \tan(\pi F_p)\ \cot(\pi(F_{p2} - F_{p1}))$$

$$\gamma = \frac{k - 1}{k + 1}$$

$$\beta = -\frac{2k\cos(\pi(F_{p2} + F_{p1}))}{(1 + k)\cos(\pi(F_{p2} - F_{p1}))}$$

	f_p
D1, D2, D3a, D4a	F_{p1}
D3b	$\frac{1}{\pi}\tan^{-1}\left(\frac{\tan(\pi F_s)\ \tan(\pi F_{p1})}{\xi_{min}\ \tan(\pi F_p)}\right)$
D4b	$\frac{1}{\pi}\tan^{-1}\left(\frac{\tan(\pi F_s)\ \tan(\pi F_{p1})}{\xi_{\max}\ \tan(\pi F_p)}\right)$
D5	$\frac{1}{\pi}\tan^{-1}\left(\frac{\tan(\pi F_{p1})}{f_{nQ}}\right)$

(10.94)

which yields

$$H_{BP}(z) = \mathcal{H}\left(n, \xi, \epsilon, F_{p1}, -\frac{z^2 + \beta z + \gamma}{1 + \beta z + \gamma z^2}\right) \tag{10.95}$$

This type of bandpass filter is said to be the *symmetrical bandpass filter*.

10.14 BANDREJECT FILTER

A bandreject filter can be specified by its edge frequencies and attenuation limits S_{Ar}; F_{p1} and F_{p2} designate the passband edge frequencies, F_{s1} and F_{s2} are the stopband edge frequencies, $0 < F_{p1} < F_{s1} < F_{s2} < F_{p2} < 0.5$, the stopband attenuation (dB) is designated by A_{sr}, and A_{p1} and A_{p2} stand for the passband attenuations (dB). We assume that the specification satisfies

$$\tan(\pi F_{p1})\ \tan(\pi F_{p2}) = \tan(\pi F_{s1})\ \tan(\pi F_{s2}) \tag{10.96}$$

If Eq. (10.96) is not satisfied, than we use new F_{s1} or F_{s2}

$$F'_{p2} = \frac{1}{\pi}\tan^{-1}\left(\frac{\tan(\pi F_{s1})\ \tan(\pi F_{s2})}{\tan(\pi F_{p1})}\right) < F_{p2} \tag{10.97}$$

or

$$F'_{p1} = \frac{1}{\pi}\tan^{-1}\left(\frac{\tan(\pi F_{s1})\ \tan(\pi F_{s2})}{\tan(\pi F_{p2})}\right) > F_{p1} \tag{10.98}$$

First, we map the bandreject filter specification, S_{Ar}, into the corresponding low-pass filter specification, S_A:

$$\begin{array}{|l|}\hline S_{Ar} = \{F_{p1}, F_{s1}, F_{s2}, F_{p2}, A_{p1}, A_{sr}, A_{p2}\} \\ \qquad\qquad\qquad \downarrow \\ S_A = \left\{\begin{array}{l} F_p = F_{p1} \\ F_s = \frac{1}{\pi}\tan^{-1}\left(\tan(\pi F_{p1})\,\frac{\tan(\pi F_{p2}) - \tan(\pi F_{p1})}{\tan(\pi F_{s2}) - \tan(\pi F_{s1})}\right) \\ A_p = \min(A_{p1}, A_{p2}) \\ A_s = A_{sr} \end{array}\right\} \\ \hline \end{array} \tag{10.99}$$

Next, we construct the lowpass transfer function, $\mathcal{H}(n, \xi, \epsilon, f_p, Z)$, which meets the specification S_A according to the design procedures D1, D2, D3a, D3b, D4a, D4b, and D5.

Finally, the transfer function, $H_{BR}(z)$, of the bandreject filter is constructed from the normalized transfer function according to the transformation:

$$Z = -\frac{z^2 + \beta z + \gamma}{1 + \beta z + \gamma z^2}$$

where

$$k = \tan(\pi F_p)\ \tan\left(\pi(F_{p2} - F_{p1})\right)$$

$$\gamma = \frac{1 - k}{k + 1}$$

$$\beta = -\frac{2\cos\left(\pi(F_{p2} + F_{p1})\right)}{(1 + k)\cos\left(\pi(F_{p2} - F_{p1})\right)}$$

Design	f_p
D1, D2, D3a, D4a	F_{p1}
D3b	$\frac{1}{\pi}\tan^{-1}\left(\frac{\tan(\pi F_s)\ \tan(\pi F_{p1})}{\xi_{min}\ \tan(\pi F_p)}\right)$
D4b	$\frac{1}{\pi}\tan^{-1}\left(\frac{\tan(\pi F_s)\ \tan(\pi F_{p1})}{\xi_{\max}\ \tan(\pi F_p)}\right)$
D5	$\frac{1}{\pi}\tan^{-1}\left(\frac{\tan(\pi F_s)\ \tan(\pi F_{p1})}{f_{nQ}\ \tan(\pi F_p)}\right)$

(10.100)

which yields

$$H_{BR}(z) = \mathcal{H}\left(n, \xi, \epsilon, F_{p1}, -\frac{z^2 + \beta z + \gamma}{1 + \beta z + \gamma z^2}\right) \tag{10.101}$$

This type of bandreject filter is said to be the *symmetrical bandreject filter*.

10.15 CONCLUDING REMARKS

We can improve the computational efficiency by using the closed-form expressions. We prefer to exploit analytical formulas involving simple algebraic manipulation, instead of using expressions in which Jacobi elliptic functions appear.

CHAPTER 11

MULTICRITERIA OPTIMIZATION OF DIGITAL FILTER DESIGNS

This chapter presents an extensible framework for the simultaneous constrained optimization of multiple properties of digital IIR filters. The framework optimizes the pole-zero locations for behavioral properties of magnitude and phase response, as well as the implementation property of quality factors, subject to constraints on the same properties. We formulate the constrained nonlinear optimization problem as a sequential quadratic programming (SQP) problem. SQP solvers are robust when provided formulas for the gradients of the cost function and constraints. We program *Mathematica* to compute the gradient formulas and convert the formulas into MATLAB programs to perform the optimization. The automated approach eliminates errors in manipulating the algebraic equations and transcribing equations into software. The key contribution is an automated, extensible, multicriteria digital filter optimization framework.

11.1 INTRODUCTION

Classic digital infinite impulse response (IIR) filters introduce significant phase distortion in output signals. This phase distortion is generally acceptable in single-speaker voice processing applications, such as speech coding and plain old telephone service, but generally unacceptable in digital communications, digital audio, and digital image processing. A conventional method to linearize the phase is to cascade a sequence of allpass filters. This structure is clearly inefficient when compared to a unified design. Digital IIR filters may also suffer from numerical instability when implemented. Quality factors provide a technology-independent measure of the potential numerical instability.

Elliptic, Chebyshev, and Butterworth digital IIR filter designs yield desirable behavioral properties subject to constraints on the magnitude response. Classic elliptic filter designs have minimal order, but for a given implementation technology, minimal order filters either may not be realizable or may not have minimal complexity [51]. Modern methods for design of filters satisfying multiple criteria sometimes are capable of handling only a particular class of filters [62, 106] . For example, in reference 62, the filters are designed as a parallel arrangement of allpass sections. Quality factors are not considered in the optimization. We instead provide a framework to optimize and improve the filters generated by conventional methods.

This chapter develops a framework for the simultaneous optimization of multiple user-specified criteria for lowpass digital IIR filters. We use the framework to design a digital IIR filter with near-linear phase response over the passband and minimized quality factors, subject to constraints on the magnitude response. The framework requires an initial filter design, which it will automatically generate if not supplied by the user. New properties may be added to the framework. The framework is an extension of our multicriteria analog IIR filter optimization framework [68].

We model the constrained nonlinear optimization problem as a sequential quadratic programming (SQP) problem. SQP requires that the objective function and constraints be real-valued and twice differentiable with respect to the free parameters [107]. The free parameters are the pole-zero locations. In order to avoid divergence that may occur when an SQP solver approximates the gradient [68], we supply closed-form gradients of the objective function and constraints with respect to the free parameters. We develop *Mathematica* software [65] to compute the gradients and translate the SQP formulation into MATLAB programs to perform the optimization [66, 67]. Constraints and objective functions may be added or modified in the framework. The *Mathematica* software would regenerate the MATLAB code.

Section 11.2 defines notation. Section 11.3 defines differentiable measures of the magnitude response, phase response, and quality factors. Section 11.4 describes the constraints. Section 11.5 presents a design example. Section 11.6 concludes the chapter.

11.2 NOTATION

Without lack of generality we focus on a lowpass even-order digital IIR filter represented by its n complex conjugate pole pairs and μ complex conjugate zero pairs. We denote the kth pole pair as $p_k = a_k \pm jb_k$, where $(a_k^2 + b_k^2) < 1$ for stability, and we denote the lth zero pair as $q_l = c_l \pm jd_l$. The transfer function of the filter in the z-transform domain is

$$H(z) = K \frac{\prod_{l=1}^{\mu} \left(1 - z^{-1}(c_l + jd_l)\right)\left(1 - z^{-1}(c_l - jd_l)\right)}{\prod_{k=1}^{n} \left(1 - z^{-1}(a_k + jb_k)\right)\left(1 - z^{-1}(a_k - jb_k)\right)} \tag{11.1}$$

or, equivalently,

$$H(z) = K \frac{\prod_{l=1}^{\mu} \left(1 - 2c_l z^{-1} + \left(c_l^2 + d_l^2\right) z^{-2}\right)}{\prod_{k=1}^{n} \left(1 - 2a_k z^{-1} + \left(a_k^2 + b_k^2\right) z^{-2}\right)}$$

where K is a real positive constant.

The magnitude response $M(f) = |H(e^{j2\pi f})|$ as a real-valued differentiable function is

$$M(f) = \sqrt{\frac{\prod_{l=1}^{\mu} \left(4c_l^2 + (c_l^2 + d_l^2 - 1)^2 - 4c_l(1 + c_l^2 + d_l^2)\cos(2\pi f) + 4(c_l^2 + d_l^2)\cos(2\pi f)^2\right)}{\prod_{k=1}^{n} \left(4a_k^2 + (a_k^2 + b_k^2 - 1)^2 - 4a_k(1 + a_k^2 + b_k^2)\cos(2\pi f) + 4(a_k^2 + b_k^2)\cos(2\pi f)^2\right)}}$$

where f denotes the digital frequency. The closed-form expression for the unwrapped phase response is complicated, so we write the phase response as the sum of phase components from the poles and zeros. First, we represent the transfer function as

$$H(z) = K z^{n-\mu} \prod_{l=1}^{\mu} \frac{1 - 2c_l z^{-1} + \left(c_l^2 + d_l^2\right) z^{-2}}{z^{-1}} \prod_{k=1}^{n} \frac{z^{-1}}{1 - 2a_k z^{-1} + \left(a_k^2 + b_k^2\right) z^{-2}}$$

Next, we find the phase component due to a single pole pair, p_k, within the unit circle that can be expressed as a differentiable function:

$$\Phi_{p_k}(f) = -\frac{\pi}{2} + \tan^{-1} \frac{\left(1 + a_k^2 + b_k^2\right)\cos(2\pi f) - 2a_k}{\left(1 - (a_k^2 + b_k^2)\right)\sin(2\pi f)} \tag{11.2}$$

The phase component due to a single zero pair, q_l, is

$$\Phi_{q_l}(f) = \begin{cases} \frac{\pi}{2} - \tan^{-1} \frac{(1 + c_l^2 + d_l^2)\cos(2\pi f) - 2c_l}{\left(1 - (c_l^2 + d_l^2)\right)\sin(2\pi f)} & \text{for } c^2 + d^2 < 1 \\ -\frac{\pi}{2} - \tan^{-1} \frac{(1 + c_l^2 + d_l^2)\cos(2\pi f) - 2c_l}{\left(1 - (c_l^2 + d_l^2)\right)\sin(2\pi f)} & \text{for } c^2 + d^2 > 1 \end{cases} \tag{11.3}$$

The total phase response is

$$\Phi(f) = -2\pi(n - \mu)f + \sum_{k=1}^{n} \Phi_{p_k}(f) + \sum_{l=1}^{\mu} \Phi_{q_l}(f) \tag{11.4}$$

In this chapter, Q represents quality factor, σ represents deviation, and m represents slope of a line.

11.3 OBJECTIVE FUNCTION

In this section, we derive cost functions to be used in the minimization problem to measure the deviation from desired (often called *ideal*) magnitude response, phase response, and quality factors. The final objective function is formed by weighting these various measures.

11.3.1 Deviation in Magnitude Response

Since we are optimizing only lowpass filters, the ideal magnitude response is shown in Fig. 11.1. Given integrable weighting functions $F_p(f)$, $F_t(f)$, and $F_s(f)$, the deviation from the ideal magnitude responses in the passband, transition band, and stopband, respectively, are given by

$$\begin{aligned}
\sigma_{pb} &= \int_0^{f_p} F_p(f)(M(f) - 1)^2 \mathrm{d}f \\
\sigma_{tb} &= \int_{f_p}^{f_s} F_t(f)\left(M(f) - \frac{f_s - f}{f_s - f_p}\right)^2 \mathrm{d}f \\
\sigma_{sb} &= \int_{f_s}^{1/2} F_s(f)M(f)^2 \mathrm{d}f
\end{aligned} \tag{11.5}$$

11.3.2 Deviation in the Phase Response

We could specify a differentiable expression as the desired phase response, and we could measure the deviation from the desired phase response in the objective function. In this section, we measure the deviation from linear phase over the passband. When using (11.4), the phase response is wrapped, and the denominators of the arctan functions could become zero. To prevent these problems, we apply constraints, as explained in

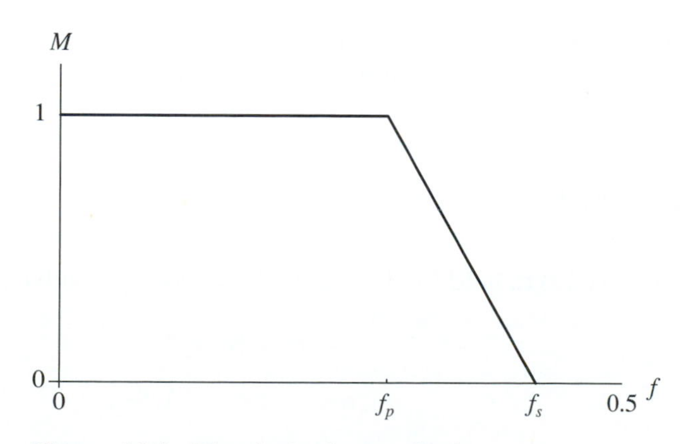

Figure 11.1 The desired magnitude response.

Section 11.4. The cost function is given by

$$\sigma_{lp} = \int_0^{f_p} (\Phi(f) - mf)^2 \,\mathrm{d}f \tag{11.6}$$

where m is the slope of the best line through the phase response over the passband given by

$$m = \frac{\int_0^{f_p} f\Phi(f)\mathrm{d}f}{\int_0^{f_p} f^2\mathrm{d}f} \tag{11.7}$$

We require a differentiable expression for m to use in the SQP framework. Symbolic integration of (11.7) is not possible. We use a first-order approximation of $\Phi(f)$ to determine m, even though this approximation is unlikely to be valid over a wide passband. Since the transfer function of a filter with real coefficients is conjugate symmetric, the phase response is an odd function. We write the phase response as a three-term Taylor series around $f = 0$:

$$\tilde{\Phi}(f) = 2\pi f\left(h_1 + h_3\left(\frac{f}{f_p}\right)^2 + h_5\left(\frac{f}{f_p}\right)^4\right)$$

After substituting $\Phi(f) = \tilde{\Phi}(f)$ in (11.7) and solving for m, we obtain

$$m = \tilde{m} = h_1 + \frac{3}{5}h_3 + \frac{3}{7}h_5 \tag{11.8}$$

We compute $\tilde{m}$ as the weighted mean of two first-order terms as

$$\tilde{m} = \alpha\frac{\tilde{\Phi}(r_1 f_p)}{r_1 f_p} + \beta\frac{\tilde{\Phi}(r_2 f_p)}{r_2 f_p} \tag{11.9}$$

where $0 < r_1, r_2 < 1$. We compute α, β, r_1, and r_2 such that for every fifth-order polynomial, $\tilde{\Phi}(f)$, the right-hand side of (11.9) is identically equal to $\tilde{m}$. Substituting $\tilde{\Phi}(f)$ and $\tilde{m}$ in (11.9), we obtain

$$h_1 + \frac{3}{5}h_3 + \frac{3}{7}h_5 = (\alpha + \beta)h_1 + (\alpha r_1^2 + \beta r_2^2)h_3 + (\alpha r_1^4 + \beta r_2^4)h_5 \tag{11.10}$$

We require (11.10) to hold for all real h_1, h_3, and h_5. We solve the system of equations

$$\alpha + \beta = 1$$
$$\alpha r_1^2 + \beta r_2^2 = 3/5$$
$$\alpha r_1^4 + \beta r_2^4 = 3/7$$

This system has multiple solutions. The solution that we implement is

$$\begin{bmatrix} \alpha \\ \beta \\ r_1 \\ r_2 \end{bmatrix} = \begin{bmatrix} 0.50000 \\ 0.50000 \\ 0.92836 \\ 0.58150 \end{bmatrix} \tag{11.11}$$

Using the above, we compute $\tilde{m}$ from (11.9) and approximate m as $\tilde{m}$ in (11.6) to obtain the cost function.

A zero outside of or on the unit circle—that is, a zero at $q_l = re^{j\rho}$ for $r \geq 1$—becomes a point of discontinuity in the phase expression in (11.4) at $f = \frac{1}{2\pi}\rho$. At these zero locations, phase is neither wrapped nor differentiable. Since we desire to optimize the phase response in the passband, we want to prevent phase discontinuities in the passband. The passband should not contain zeros; otherwise, the magnitude response will be disrupted. We add constraints to force the zeros to remain outside of the passband to ensure that the phase expression is differentiable and well-behaved in the passband.

11.3.3 Filter Quality Factor

The quality factor is a measure of the sensitivity of the pole locations. A perturbation in a pole location leads to unexpected oscillations and/or more attenuation in the filter response than designed. The quality factor also reflects the sharpness in the phase and magnitude response. We define the quality factor Q_k of a pole pair $p_k = a_k \pm jb_k$ as

$$Q_k = \frac{\sqrt{(1 + a_k^2 + b_k^2)^2 - 4a_k^2}}{2(1 - \sqrt{a_k^2 + b_k^2})} \tag{11.12}$$

where $a_k^2 + b_k^2 < 1$ and $\frac{1}{2} \leq Q < \infty$. Lower-quality factors are desirable.

For the effective overall measure of the quality factor, Q_{eff}, we use a geometric mean of the individual pole-pair quality factors:

$$Q_{\text{eff}} = \left(\prod_{k=1}^{n} Q_k \right)^{\frac{1}{n}} \tag{11.13}$$

Since $Q_{\text{eff}} \geq \frac{1}{2}$, we use

$$\sigma_Q = Q_{\text{eff}} - \frac{1}{2}$$

in the objective measure of deviation in filter quality from the ideal.

11.3.4 The Complete Objective Function

The complete objective function is a weighted sum of the measures developed earlier in this section:

$$\sigma = W_{pb}\frac{1}{f_p}\sigma_{pb} + W_{tb}\frac{1}{f_s - f_p}\sigma_{tb} + W_{sb}\frac{1}{\frac{1}{2} - f_s}\sigma_{sb} + W_{lp}\frac{1}{f_p}\sigma_{lp} + W_Q\sigma_Q \quad (11.14)$$

where W represents a weight factor.

Note that the weighted objective function is non-negative and twice differentiable.

11.4 CONSTRAINTS

We place constraints on magnitude response, quality factors, numerical stability, and filter stability. The magnitude response constraints occur at uniformly spaced passband frequencies f_i and stopband frequencies f_l:

$$1 - \delta_p < M(f_i) < 1 + \delta_p$$
$$M(f_l) < \delta_s$$

We also add a constraint on the maximum value of the quality factor for the poles. Users can set the maximum value of $Q_{\max}$ for a particular implementation technology. Zero locations are constrained to be outside of the passband so as to avoid phase discontinuities in the passband (see Section 11.3.2). Pole locations are constrained to be within the unit circle to ensure the stability of the filter.

11.5 EXAMPLE

We use the framework to optimize an elliptic lowpass filter with four zeros and four poles. The magnitude specifications were $f_p = \frac{1}{2\pi}$, passband ripple of $\delta_p = 0.05$, $f_s = \frac{1}{2\pi}1.8$, and stopband ripple of $\delta_s = 0.01$. The initial elliptic filter has poles $p_1 = 0.5176 \pm j0.3264$, ($Q_1 = 0.72$) and $p_2 = 0.4584 \pm j0.7602$, ($Q_2 = 3.62$). In the optimization, we constrained the maximum quality factor to $Q_{\max} = 2$ and we optimized for linear phase over the passband. For the optimized design, the poles were $p_1 = 0.2470 \pm j0.2399$ ($Q_1 = 0.57$) and $p_2 = 0.2815 \pm j0.7355$ ($Q_2 = 2$). Figures. 11.2 and 11.3 show the magnitude and phase responses of the filters. The optimized filter satisfies the required magnitude specification.

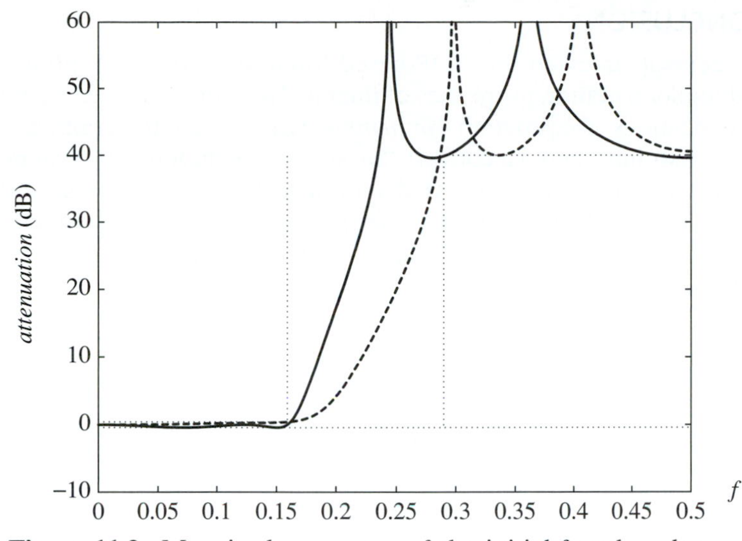

Figure 11.2 Magnitude response of the initial fourth-order lowpass elliptic filter (solid line) and optimized filter (dashed line). Both filters meet the magnitude specifications of $f_p = \frac{1}{2\pi}$, $f_s = \frac{1}{2\pi}1.8$, $\delta_p = 0.05$, and $\delta_s = 0.01$.

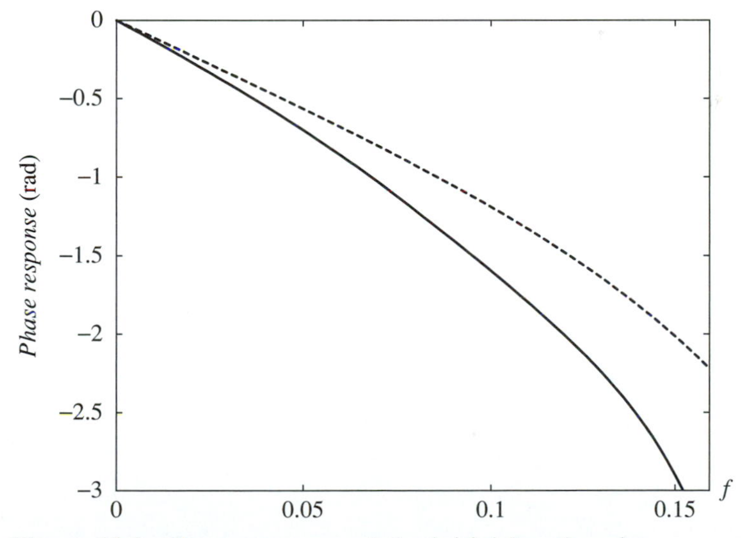

Figure 11.3 Phase response of the initial fourth-order lowpass elliptic filter (solid line) and optimized filter (dashed line). In the optimized filter, the phase response is more linear over the passband.

11.6 CONCLUSION

We develop an extensible SQP-based framework for joint optimization of behavioral and implementation properties of digital IIR filters. The free parameters are the pole-zero locations. We program *Mathematica* to compute the gradients of the cost function and constraints, and we convert the SQP problem into MATLAB programs to perform the optimization. The automated approach eliminates algebraic and programming mistakes. We use the framework to design a digital IIR filter with near-linear phase over the passband and minimized quality factors, subject to constraints on the magnitude response.

CHAPTER 12

ELLIPTIC FUNCTIONS

This chapter introduces the basic Jacobi elliptic functions and reviews the most important relations between them. Several related theorems not found in standard textbooks are presented. Elliptic functions make a background for development of elliptic rational functions which are of central interest for advanced filter design. Various useful approximation formulas are offered to facilitate the derivation of elliptic rational functions. A nesting property of the Jacobi elliptic functions is derived in the strict mathematical sense.

It is not necessary for the filter designer to understand the elliptic theory presented in this chapter. However, the reader may want to know the proofs of the equations that we use throughout this book. Thus, we provide here a brief treatment of the theory of the Jacobi elliptic functions and the Legendre elliptic integral. Those who are not interested in this mathematical treatment may proceed to the next chapter.

Elliptic filters offer the steepest rolloff characteristics and consequently meet an assigned set of filter performance specifications with the lowest filter order. As a consequence of the extreme interest in these filters, many different design procedures were developed relying on iterative numerical algorithms, or approximations, for evaluation of elliptic functions. In this chapter we present a novel approach to the design of elliptic filters in which we use exact closed-form expressions based on the nesting property. The key benefit of the nesting property is a possibility to derive analytically lower-order elliptic rational functions and to use them for building up the higher-order elliptic rational functions.

We conclude this chapter by proving that there exists a special class of elliptic filters, called minimal Q-factor filters, which is of prime importance in many practical applications. The principal feature of the minimal Q-factor filters is that their transfer function poles lay on a circle in the complex s-plane.

12.1 LEGENDRE ELLIPTIC INTEGRAL

The *elliptic integral of the first kind* u is defined by

$$u = u(\phi, k) = \int_0^{\phi} \frac{d\theta}{\sqrt{1 - k^2 \sin^2 \theta}}, \qquad 0 \leq k < 1 \tag{12.1}$$

where k is called the *modulus* and the upper limit of integration ϕ is called the *amplitude*. The function $u(\phi, k)$ is called the *Legendre standard form of the elliptic integral of the first kind.*

The *complete elliptic integral of the first kind* K, also called the *real quarterperiod*, is given by:

$$K = K(k) = u\left(\frac{\pi}{2}, k\right) = \int_0^{\pi/2} \frac{d\theta}{\sqrt{1 - k^2 \sin^2 \theta}} \tag{12.2}$$

A plot of K as a function of k is shown in Fig. 12.1. Figure 12.2 shows $u(\phi, k)/K(k)$ for various values of k.

The *complementary complete elliptic integral of the first kind* K', also called the *imaginary quarterperiod*, is given by

$$K' = K'(k) = u\left(\frac{\pi}{2}, \sqrt{1 - k^2}\right) = \int_0^{\pi/2} \frac{d\theta}{\sqrt{1 - (1 - k^2) \sin^2 \theta}} \tag{12.3}$$

The *complementary modulus* k' is defined in terms of the corresponding modulus k:

$$k' = \sqrt{1 - k^2} \tag{12.4}$$

From Eqs. (12.2), (12.3), and (12.4) we obtain

$$K'(k) = K(k') \tag{12.5}$$

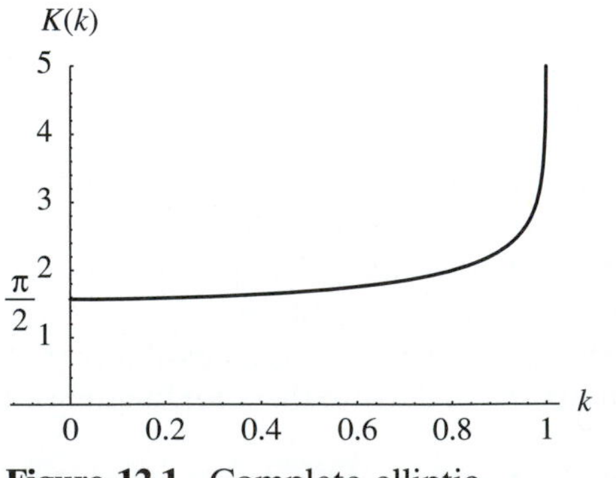

Figure 12.1 Complete elliptic integral of the first kind $K(k)$.

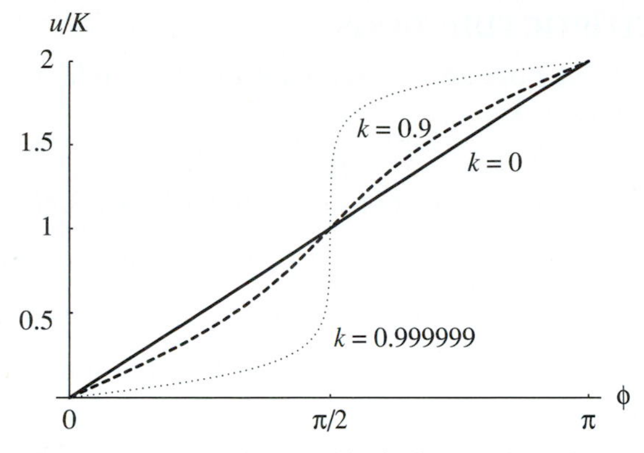

Figure 12.2 Ratio of elliptic integral and complete elliptic integral of the first kind $u(\phi, k)/K(k)$.

The complete elliptic integral K and the corresponding K' are plotted in Fig. 12.3. $K(k)$ and $K'(k)$ are positive and steadily increasing and decreasing functions, respectively, in the interval $0 < k < 1$. Their ratio $K(k)/K'(k)$ is a strictly increasing function in interval $0 < k < 1$. If any one of k, k', $K(k)$, $K'(k)$, or $K(k)/K'(k)$ is given, all the rest are uniquely determined. Thus, there is a one-to-one correspondence between $K(k)$, $K'(k)$, $K(k)/K'(k)$, and k.

The elliptic integrals have the following properties:

$$K(0) = K'(1) = \frac{\pi}{2} \tag{12.6}$$

$$K(1) = K'(0) = \infty \tag{12.7}$$

$$K\left(\frac{\sqrt{2}}{2}\right) = K'\left(\frac{\sqrt{2}}{2}\right) \tag{12.8}$$

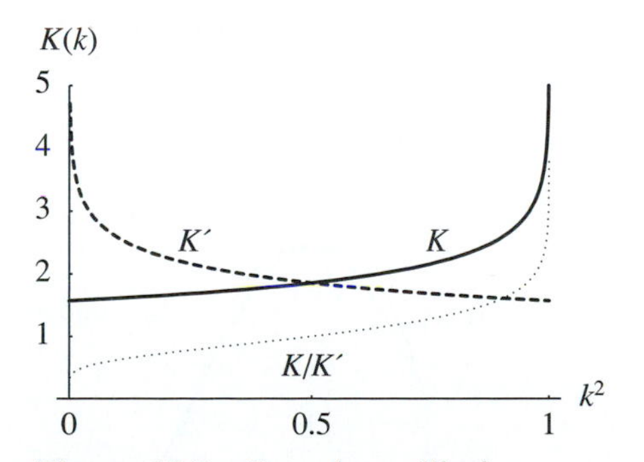

Figure 12.3 Complete elliptic integral $K(k)$, complementary complete elliptic integral $K'(k)$, and their ratio $K(k)/K'(k)$.

12.2 JACOBI ELLIPTIC FUNCTIONS

Jacobi introduced notation for expressing the amplitude ϕ in terms of the elliptic integral u and the modulus k:

$$\phi = \mathrm{am}(u, k) \tag{12.9}$$

The *Jacobi elliptic sine function of modulus k* is defined as

$$\mathrm{sn}(u, k) = \sin\phi = \sin \mathrm{am}(u, k) \tag{12.10}$$

For convenience, the *Jacobi elliptic cosine function* is defined as

$$\mathrm{cn}(u, k) = \cos\phi = \cos \mathrm{am}(u, k) \tag{12.11}$$

The *Jacobi elliptic tangent* is defined in an analogous manner:

$$\mathrm{tn}(u, k) = \frac{\sin\phi}{\cos\phi} = \frac{\mathrm{sn}(u, k)}{\mathrm{cn}(u, k)} \tag{12.12}$$

The *difference function*, which is the derivative of ϕ, is

$$\mathrm{dn}(u, k) = \frac{d\phi}{du} = \sqrt{1 - k^2 \mathrm{sn}^2(u, k)} \tag{12.13}$$

Figures 12.4 and 12.5 show Jacobi elliptic sine and cosine functions for various values of the modulus k. Many of the properties of Jacobi elliptic functions follow from the properties of trigonometric functions:

$$\mathrm{sn}^2(u, k) + \mathrm{cn}^2(u, k) = 1 \tag{12.14}$$

$$k^2\mathrm{sn}^2(u, k) + \mathrm{dn}^2(u, k) = 1 \tag{12.15}$$

$$\mathrm{sn}(u, k) = -\mathrm{sn}(-u, k) \tag{12.16}$$

$$\mathrm{cn}(u, k) = \mathrm{cn}(-u, k) \tag{12.17}$$

$$\mathrm{sn}(0, k) = 0 \tag{12.18}$$

$$\mathrm{sn}(K, k) = 1 \tag{12.19}$$

$$\mathrm{cn}(0, k) = 1 \tag{12.20}$$

$$\mathrm{cn}(K, k) = 0 \tag{12.21}$$

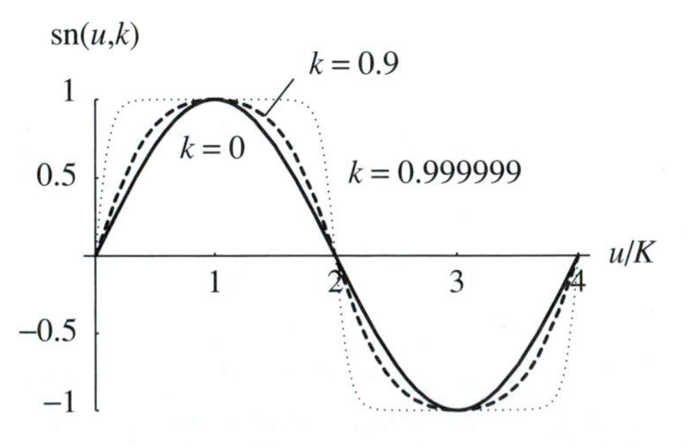

Figure 12.4 Elliptic sine functions $\mathrm{sn}(u, k)$.

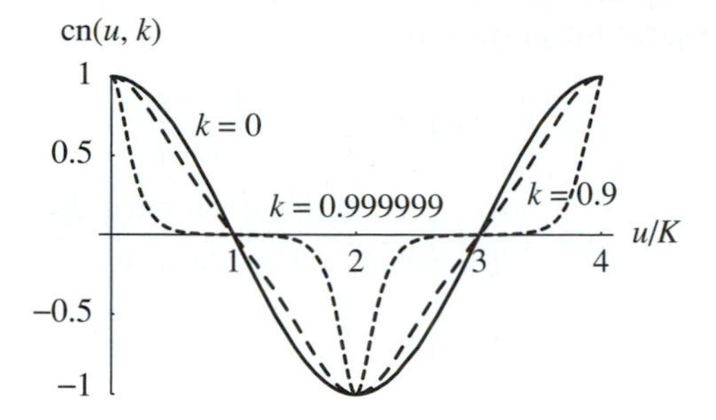

Figure 12.5 Elliptic cosine functions $\mathrm{cn}(u, k)$.

The addition theorems for Jacobi elliptic functions are

$$\mathrm{sn}(u \pm v, k) = \frac{\mathrm{sn}(u, k)\,\mathrm{cn}(v, k)\,\mathrm{dn}(v, k) \pm \mathrm{sn}(v, k)\,\mathrm{cn}(u, k)\,\mathrm{dn}(u, k)}{1 - k^2\mathrm{sn}^2(u, k)\,\mathrm{sn}^2(v, k)} \tag{12.22}$$

$$\mathrm{cn}(u \pm v, k) = \frac{\mathrm{cn}(u, k)\,\mathrm{cn}(v, k) \mp \mathrm{sn}(u, k)\,\mathrm{sn}(v, k)\,\mathrm{dn}(u, k)\,\mathrm{dn}(v, k)}{1 - k^2\mathrm{sn}^2(u, k)\,\mathrm{sn}^2(v, k)} \tag{12.23}$$

$$\mathrm{dn}(u \pm v, k) = \frac{\mathrm{dn}(u, k)\,\mathrm{dn}(v, k) \mp k^2\,\mathrm{sn}(u, k)\,\mathrm{sn}(v, k)\,\mathrm{cn}(u, k)\,\mathrm{cn}(v, k)}{1 - k^2\mathrm{sn}^2(u, k)\,\mathrm{sn}^2(v, k)} \tag{12.24}$$

Applying the Jacobi imaginary transformations

$$\mathrm{sn}(ju, k) = j\frac{\mathrm{sn}(u, k')}{\mathrm{cn}(u, k')} \tag{12.25}$$

$$\mathrm{cn}(ju, k) = \frac{1}{\mathrm{cn}(u, k')} \tag{12.26}$$

$$\mathrm{dn}(ju, k) = \frac{\mathrm{dn}(u, k')}{\mathrm{cn}(u, k')} \tag{12.27}$$

into the addition theorems, we obtain

$$\mathrm{sn}(u \pm jv, k) = \frac{\mathrm{sn}(u, k)\,\mathrm{dn}(v, k') \pm j\mathrm{cn}(u, k)\,\mathrm{dn}(u, k)\,\mathrm{sn}(v, k')\,\mathrm{cn}(v, k')}{\mathrm{cn}^2(v, k') + k^2\mathrm{sn}^2(u, k)\,\mathrm{sn}^2(v, k')} \tag{12.28}$$

$$\mathrm{cn}(u \pm jv, k) = \frac{\mathrm{cn}(u, k)\,\mathrm{cn}(v, k') \mp j\mathrm{sn}(u, k)\,\mathrm{sn}(v, k')\,\mathrm{dn}(u, k)\,\mathrm{dn}(v, k')}{\mathrm{cn}^2(v, k') + k^2\mathrm{sn}^2(u, k)\,\mathrm{sn}^2(v, k')} \tag{12.29}$$

$$\mathrm{dn}(u \pm jv, k) = \frac{\mathrm{dn}(u, k)\,\mathrm{cn}(v, k')\,\mathrm{dn}(v, k') \mp jk^2\,\mathrm{sn}(u, k)\,\mathrm{sn}(v, k')\,\mathrm{cn}(u, k)}{\mathrm{cn}^2(v, k') + k^2\mathrm{sn}^2(u, k)\,\mathrm{sn}^2(v, k')} \tag{12.30}$$

Other useful properties are

$$\mathrm{sn}^2\left(\frac{u}{2}, k\right) = \frac{1 - \mathrm{cn}(u, k)}{1 + \mathrm{dn}(u, k)} \tag{12.31}$$

$$\mathrm{cn}^2\left(\frac{u}{2}, k\right) = \frac{\mathrm{dn}(u, k) + \mathrm{cn}(u, k)}{1 + \mathrm{dn}(u, k)} \tag{12.32}$$

$$\mathrm{dn}^2\left(\frac{u}{2}, k\right) = \frac{1 - k^2 + \mathrm{dn}(u, k) + k^2\mathrm{cn}(u, k)}{1 + \mathrm{dn}(u, k)} \tag{12.33}$$

12.3 PERIODS OF ELLIPTIC FUNCTIONS

Elliptic functions have two periods: a real period, $4K$,

$$\mathrm{sn}(u + 4K, k) = \mathrm{sn}(u, k) \tag{12.34}$$

and an imaginary period, $2K'$,

$$\mathrm{sn}(u + j2K', k) = \mathrm{sn}(u, k) \tag{12.35}$$

Since $u(\phi, 0) = \phi$ it follows that

$$\mathrm{sn}(u, 0) = \sin u \tag{12.36}$$

$$\mathrm{cn}(u, 0) = \cos u \tag{12.37}$$

and

$$K(0) = \frac{\pi}{2} \tag{12.38}$$

Unlike the trigonometric functions, $\mathrm{cn}(u, k)$ is not the same as $\mathrm{sn}(u, k)$ shifted by K [84], except for $k = 0$. The Jacobi function

$$\mathrm{cd}(u, k) = \mathrm{sn}(u + K, k) \tag{12.39}$$

is preferred to $\mathrm{cn}(u, k)$ or $\mathrm{sn}(u, k)$ when we describe elliptic transfer functions with a single formula [84] (Fig. 12.6).

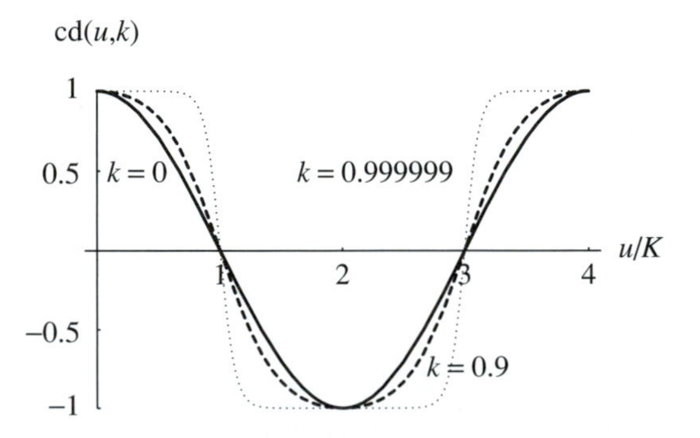

Figure 12.6 Shifted elliptic sine functions $\mathrm{cd}(u, k)$.

The Jacobi elliptic functions $\mathrm{sn}(u, k)$, $\mathrm{cn}(u, k)$, and $\mathrm{cd}(u, k)$ are plotted in Fig. 12.7.

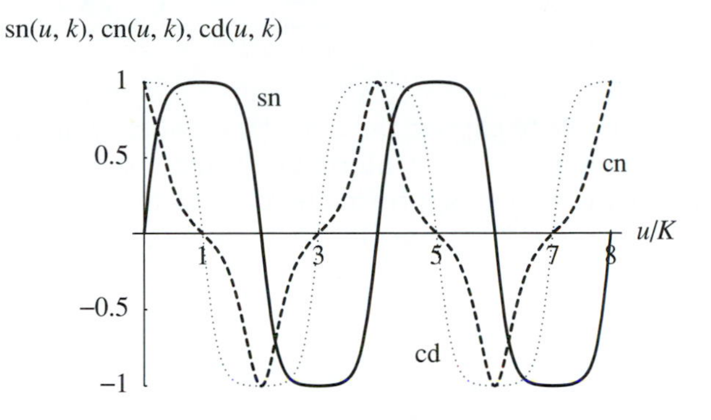

Figure 12.7 Typical elliptic functions.

12.4 SERIES REPRESENTATION AND MODULAR CONSTANT

Elliptic functions can be represented in terms of the elliptic theta functions [108]:

$$\mathrm{sn}(u, k) = \frac{1}{\sqrt{k}} \frac{\theta_1\left(\frac{u}{2K}, q\right)}{\theta_0\left(\frac{u}{2K}, q\right)} \tag{12.40}$$

$$\mathrm{cn}(u, k) = \frac{\sqrt{k'}}{\sqrt{k}} \frac{\theta_2\left(\frac{u}{2K}, q\right)}{\theta_0\left(\frac{u}{2K}, q\right)} \tag{12.41}$$

$$\mathrm{dn}(u, k) = \sqrt{k'} \frac{\theta_3\left(\frac{u}{2K}, q\right)}{\theta_0\left(\frac{u}{2K}, q\right)} \tag{12.42}$$

where parameter q is called the *modular constant* and is given by

$$q = e^{-\pi K'/K} \tag{12.43}$$

The *elliptic theta functions* are defined by rapidly converging series:

$$\theta_0\left(\frac{u}{2K}, q\right) = 1 + 2\sum_{m=1}^{\infty}(-1)^m q^{m^2} \cos\left(2m\frac{\pi u}{2K}\right) \tag{12.44}$$

$$\theta_1\left(\frac{u}{2K}, q\right) = 2q^{1/4}\sum_{m=0}^{\infty}(-1)^m q^{m(m+1)} \sin\left((2m+1)\frac{\pi u}{2K}\right) \tag{12.45}$$

$$\theta_2\left(\frac{u}{2K}, q\right) = 2q^{1/4}\sum_{m=0}^{\infty} q^{m(m+1)} \cos\left((2m+1)\frac{\pi u}{2K}\right) \tag{12.46}$$

$$\theta_3\left(\frac{u}{2K}, q\right) = 1 + 2\sum_{m=1}^{\infty} q^{m^2} \cos\left(2m\frac{\pi u}{2K}\right) \tag{12.47}$$

It is desirable to express the modular constant q in terms of k—that is, to find an explicit function $q = q(k)$. Unfortunately, there is no exact closed-form solution for $q(k)$. In order to find an approximate function for $q(k)$ we substitute $u = 0$ into Eq. (12.42). From Eqs. (12.15) and (12.18), for $u = 0$, we find

$$\mathrm{dn}(0, k) = 1 \tag{12.48}$$

which is the left-hand side of Eq. (12.42). Next, for $u = 0$, from Eqs. (12.44) and (12.47), the right-hand side of Eq. (12.42) becomes

$$\sqrt{k'}\,\frac{\theta_3(0, q)}{\theta_0(0, q)} = \sqrt{k'}\,\frac{1 + 2\sum\limits_{m=1}^{\infty} q^{m^2}}{1 + 2\sum\limits_{m=1}^{\infty} (-1)^m q^{m^2}} \tag{12.49}$$

Equating Eq.(12.48) and Eq. (12.49) yields

$$\sqrt{k'} = \frac{1 + 2\sum\limits_{m=1}^{\infty} (-1)^m q^{m^2}}{1 + 2\sum\limits_{m=1}^{\infty} q^{m^2}} \tag{12.50}$$

which can be expressed in the form

$$\sqrt{k'} = \frac{1 + 2\sum\limits_{m=1}^{\infty} q^{(2m)^2} - 2\sum\limits_{m=1}^{\infty} q^{(2m-1)^2}}{1 + 2\sum\limits_{m=1}^{\infty} q^{(2m)^2} + 2\sum\limits_{m=1}^{\infty} q^{(2m-1)^2}} \tag{12.51}$$

or

$$\sqrt{k'} = \frac{1 - \dfrac{2\sum\limits_{m=1}^{\infty} q^{(2m-1)^2}}{1 + 2\sum\limits_{m=1}^{\infty} q^{(2m)^2}}}{1 + \dfrac{2\sum\limits_{m=1}^{\infty} q^{(2m-1)^2}}{1 + 2\sum\limits_{m=1}^{\infty} q^{(2m)^2}}} \tag{12.52}$$

We introduce a new quantity q_0:

$$q_0 = \frac{\sum_{m=1}^{\infty} q^{(2m-1)^2}}{1 + 2\sum_{m=1}^{\infty} q^{(2m)^2}} \tag{12.53}$$

Now Eq. (12.52) can be expressed in terms of q_0:

$$\sqrt{k'} = \frac{1 - 2q_0}{1 + 2q_0} \tag{12.54}$$

Expanding Eq. (12.53) we obtain

$$q_0 = \frac{q + q^9 + q^{25} + q^{49} + q^{81} + q^{121} + q^{169} + \cdots}{1 + 2q^4 + 2q^{16} + 2q^{36} + 2q^{64} + 2q^{100} + 2q^{144} + \cdots} \tag{12.55}$$

Performing long division, we have [108]

$$\begin{aligned} q_0 = {} & q - 2q^5 + 5q^9 - 10q^{13} + 18q^{17} - 32q^{21} + 55q^{25} - 90q^{29} \\ & + 144q^{33} - 226q^{37} + 346q^{41} - 522q^{45} + 777q^{49} - \cdots \end{aligned} \tag{12.56}$$

By moving q to the left-hand side, and moving q_0 to the right-hand side we find q as a function of q_0 and $q^5, q^9, \ldots$:

$$\begin{aligned} q = {} & q_0 + 2q^5 + 5q^9 + 10q^{13} - 18q^{17} + 32q^{21} - 55q^{25} + 90q^{29} \\ & - 144q^{33} + 226q^{37} - 346q^{41} + 522q^{45} - 777q^{49} + \cdots \end{aligned} \tag{12.57}$$

Now, substituting q from Eq. (12.57) into the right-hand side of the same Eq. (12.57) and repeatedly performing the substitution several times, we find that Eq. (12.57) becomes

$$\begin{aligned} q = {} & q_0 + 2q_0^5 + 15q_0^9 + 150q_0^{13} + 1707q_0^{17} + 20{,}910q_0^{21} + 268{,}616q_0^{25} \\ & + 3{,}567{,}400q_0^{29} + 48{,}555{,}069q_0^{33} + 673{,}458{,}874q_0^{37} + \cdots \end{aligned} \tag{12.58}$$

From Eq. (12.54), which is rewritten as $q_0(k')$, we obtain

$$q_0 = \frac{1}{2}\frac{1 - \sqrt{k'}}{1 + \sqrt{k'}} \tag{12.59}$$

and from Eq. (12.58) we derive an approximate value of q (designated by $\hat{q}$) in terms of k:

$$\begin{aligned} q \approx \hat{q} = {} & q_0 + 2q_0^5 + 15q_0^9 + 150q_0^{13} + 1707q_0^{17} + 20{,}910q_0^{21} + 268{,}616q_0^{25} \\ & + 3{,}567{,}400q_0^{29} + 48{,}555{,}069q_0^{33} + 673{,}458{,}874q_0^{37} \end{aligned}$$

for $$q_0 = \frac{1}{2}\frac{1 - \sqrt[4]{1 - k^2}}{1 + \sqrt[4]{1 - k^2}} \tag{12.60}$$

In practice, k is from range $0 < k \leq \sin 86^o$. For example, for $k = \sin 86^o$ the series is

$$q \approx \hat{q} = 0.291068 + 0.0041783 + 0.000224924 + 0.000016144$$
$$+ 1.31865 \times 10^{-6} + 1.15938 \times 10^{-7} + 1.069 \times 10^{-8}$$
$$+ 1.019 \times 10^{-9} + 9.95476 \times 10^{-11} + 9.91021 \times 10^{-12}$$

$$q_0 = \frac{1}{2} \frac{1 - \sqrt[4]{1 - \sin^2 86^o}}{1 + \sqrt[4]{1 - \sin^2 86^o}}$$

resulting in

$$\hat{q} = 0.2954883855575757$$

The accurate value is

$$q = 0.2954883855586914$$

and therefore the error is

$$\text{error} = q - \hat{q} = 10^{-12}$$

So, we can retain 9 terms and obtain a ten-digit accuracy in the worst case, for the maximum q of interest.

If we consider the edge case for $k = 1$, the values of terms in series are shown in Fig. 12.8 and the error of $q - \hat{q}$ is shown in Fig. 12.9.

The error rapidly increases when k tends to 1. The approximation becomes impractical for $k > \sin 86°$.

Since all terms in the series are positive and with the increasing coefficients, it is obvious that the required accuracy is easier to achieve for smaller values of k. If we consider $k \leq 1/\sqrt{2}$, the first four terms are enough to ensure the 20-digit accuracy. The values of terms in the series are shown in Fig. 12.10 and the error $q - \hat{q}$ is shown in Fig. 12.11.

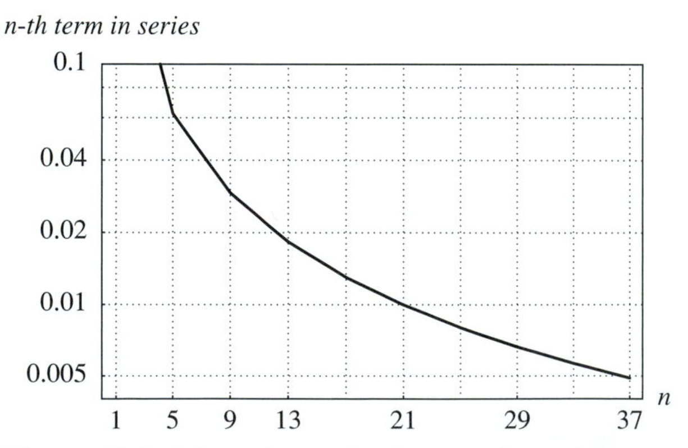

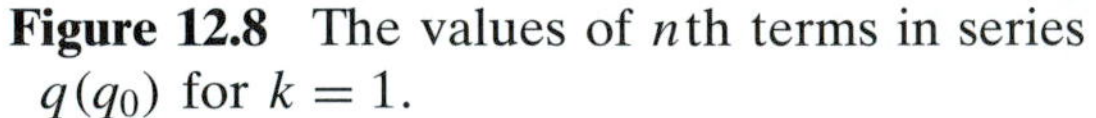

Figure 12.8 The values of nth terms in series $q(q_0)$ for $k = 1$.

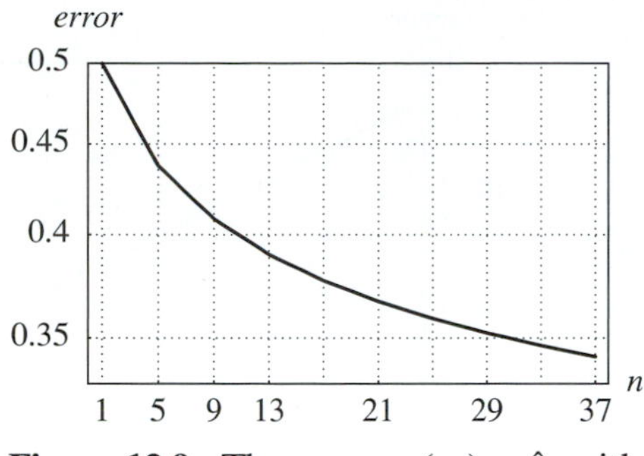

Figure 12.9 The error, $q(q_0) - \hat{q}$, with 10 terms in series for $k = 1$.

Usually, computer programs (for example, in MATLAB) work with 16 digits. In this case it is of no use to keep more than four terms in the series, for $k \leq 1/\sqrt{2}$. If we can control the number of digits—that is, the accuracy of calculations—with the first 10 terms we obtain approximately 50-digit accuracy.

In Table 12.1 a set of particular values is shown for k, k', K, K', q_0, and q.

Theoretically, $0 < k < 1$, thus, $0 < q_0 < 1/2$, $0 < q < 1$. The functions $q_0(k)$ and $q(k)$ are monotonically increasing and bounded. The series in Eq. (12.58) is convergent.

The problem of accuracy can be avoided by modifying the procedure for calculating $q(k)$ [84]. In the case $k \geq 1/\sqrt{2}$ we introduce q':

$$q' = e^{-\pi K/K'} \tag{12.61}$$

According to Eq. (12.5)

$$K'(k) = K(k') \tag{12.62}$$

we find

$$q'(k) = q(k') \tag{12.63}$$

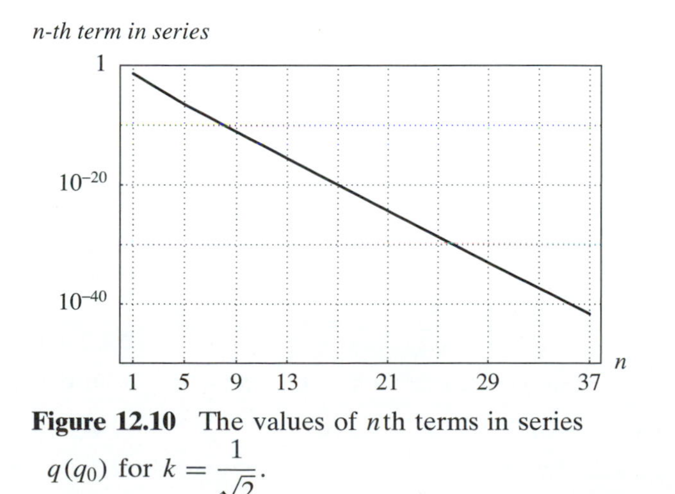

Figure 12.10 The values of nth terms in series $q(q_0)$ for $k = \dfrac{1}{\sqrt{2}}$.

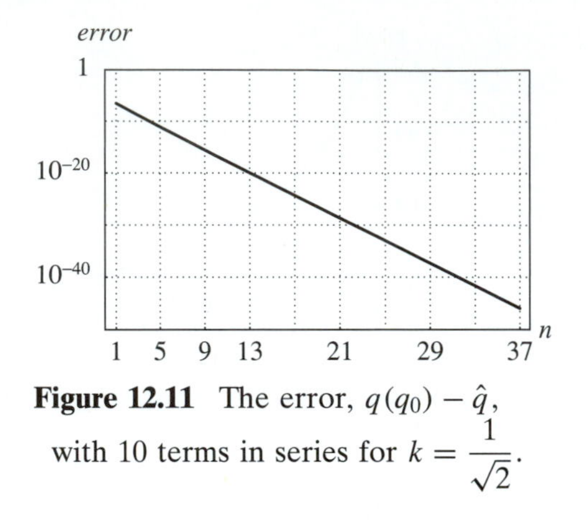

Figure 12.11 The error, $q(q_0) - \hat{q}$, with 10 terms in series for $k = \dfrac{1}{\sqrt{2}}$.

By using the same procedure from Eq. (12.58) we find

$$\begin{aligned} q' &= q_0' + 2(q_0')^5 + 15(q_0')^9 + 150(q_0')^{13} \\ &\quad + 1{,}707(q_0')^{17} + 20{,}910(q_0')^{21} + 268{,}616(q_0')^{25} \\ &\quad + 3{,}567{,}400(q_0')^{29} + 48{,}555{,}069(q_0')^{33} + 673{,}458\,874(q_0')^{37} + \cdots \end{aligned} \tag{12.64}$$

where

$$q_0' = \frac{1}{2}\frac{1 - \sqrt{k}}{1 + \sqrt{k}} \tag{12.65}$$

If we use Eqs. (12.43) and (12.61), we obtain

$$\ln q \; \ln q' = \pi^2 \tag{12.66}$$

and find q as a function of q':

$$q = e^{\pi^2/\ln q'} \tag{12.67}$$

The error is maximal for $k = k' = 1/\sqrt{2}$. The modified procedure gives the exact value for $k = 1$, $q(1) = 1$.

Table 12.1 Special values of k and k'

k	k'	$K(k)$	$K'(k)$	$q_0(k)$	$q(k)$
0	1	$\frac{\pi}{2}$	∞	0	0
$\frac{1}{\sqrt{2}}$	$\frac{1}{\sqrt{2}}$	$K = K'$	$K' = K$	$\frac{1}{2}\frac{\sqrt[4]{2}-1}{\sqrt[4]{2}+1}$	$e^{-\pi}$
$\sin 86^o$	$\cos 86^o$	$K > K'$	$K' < K$	0.2910676	0.2954884
1	0	∞	$\frac{\pi}{2}$	0.5	1

12.5 CHEBYSHEV POLYNOMIALS

The most commonly used definition of the well-known Chebyshev polynomial is

$$C(n, x) = \cos(n \arccos(x)) \tag{12.68}$$

or, by a pair of parametric equations,

$$\begin{aligned} C(n, x) &= \cos nw \\ x &= \cos w \end{aligned} \tag{12.69}$$

where w is the parametric variable and n is the polynomial degree or order. In Figs. 12.12 and 12.13, the plots of the third-order and the fourth-order parametric relations are shown. It should be noted that $C(n, x)$ is, by definition, a real-valued function for any non-negative integer n and any real variable x. The parametric variable w takes only real values for $|x| \leq 1$, but w becomes complex for $|x| > 1$.

If w continuously changes (increases) from $w = \pi$ to $w = 2\pi$, then x monotonically increases from $x = -1$ to $x = 1$ (Figs. 12.12 and 12.13), and $C(n, x)$ oscillates n times between values of ± 1:

$$\pi \leq w \leq 2\pi$$

$$-1 \leq x \leq 1$$

Let us find complex values for w in order to obtain $x > 1$. For $w = u + jv$, u and v real, and $j = \sqrt{-1}$, the equations defining the Chebyshev polynomial take the form

$$\begin{aligned} C(n, x) &= \cos(n(u + jv)) \\ x &= \cos(u + jv) \end{aligned} \tag{12.70}$$

Introducing the transformation

$$\cos(\alpha + \beta) = \cos(\alpha)\cos(\beta) - \sin(\alpha)\sin(\beta)$$

and the hyperbolic cosine function defined by

$$\cos(j\beta) = \frac{e^{\beta} + e^{-\beta}}{2} = \cosh(\beta)$$

we obtain

$$x = \cos(u + jv) = \cos(u)\cosh(v) - \sin(u)\sin(jv)$$

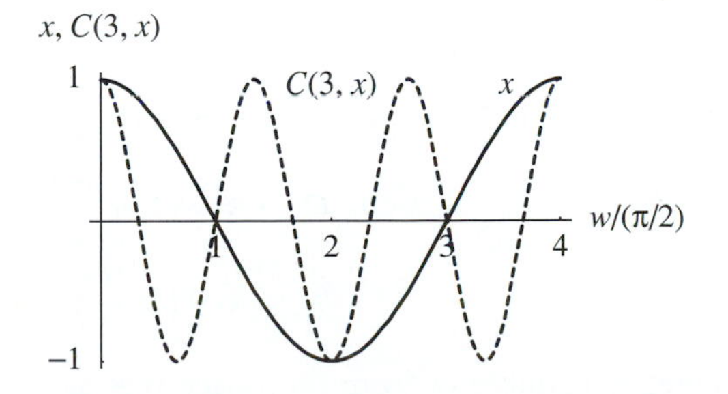

Figure 12.12 $C(3, x) = \cos 3w, \quad x = \cos w.$

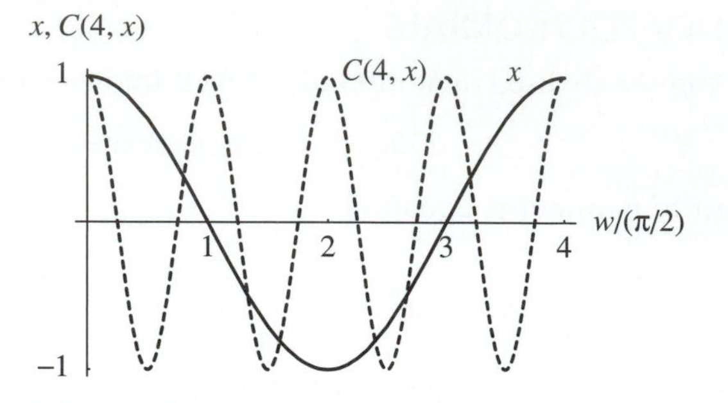

Figure 12.13 $C(4, x) = \cos 4w, \quad x = \cos w.$

If we keep u constant at the edge value $u = 2\pi$, x becomes

$$x = \cosh(v)$$

In a similar way we derive

$$C(n, x) = \cosh(nv)$$

$C(n, x)$ and x are monotonically increasing functions of the real variable v:

$$\begin{aligned} C(n, x) &= \cos\left(n(2\pi + jv)\right) = \cosh(nv) \\ x &= \cos\left(2\pi + jv\right) = \cosh(v), \qquad x \geq 1 \end{aligned} \tag{12.71}$$

We can conclude that the variable x monotonically increases from -1 to 1 for $w = u + j0$ and $\pi < u < 2\pi$; that is, we change only the real part of the parametric variable w; by changing only the imaginary part of w we generate an arbitrary value $1 < x < +\infty$. For negative values of x we use the properties

$$C(2n, -x) = C(2n, x)$$

$$C(2n + 1, -x) = -C(2n + 1, x)$$

to compute the Chebyshev polynomial.

Let us consider only the range $-1 \leq x \leq 1$. Instead of w, we can use the variable μ defined by

$$\mu = \frac{w}{\frac{\pi}{2}} \tag{12.72}$$

After we substitute Eq. (12.70) into Eqs. (12.69) the parametric equations become

$$\begin{aligned} C(n, x) &= \cos\left(n\,\mu\,\frac{\pi}{2}\right) \\ x &= \cos\left(\mu\,\frac{\pi}{2}\right) \end{aligned} \tag{12.73}$$

For the parametric variable μ from the range $0 \leq \mu \leq 4$, $\cos(\mu(\pi/2))$ contains one period while $\cos(n\mu(\pi/2))$ has n periods.

The parametric equations defining the Chebyshev polynomial for any non-negative x are

$$\begin{aligned} C(n,x) &= \cos(nu) \\ x &= \cos u \qquad 0 \le x \le 1, \quad \pi \le u \le 2\pi \\ C(n,x) &= \operatorname{ch} nv \\ x &= \operatorname{ch} v \qquad x \ge 1, \quad 0 \le v < +\infty \end{aligned} \tag{12.74}$$

Expanding $\cos(nu)$, we obtain a polynomial in $\cos u$:

$$\begin{aligned} \cos(1u) &= \cos u \\ \cos(2u) &= 2\cos^2 u - 1 \\ \cos(3u) &= 4\cos^3 u - 3\cos u \\ \cos(4u) &= 8\cos^4 u - 8\cos^2 u + 1 \\ \cos(5u) &= 16\cos^5 u - 20\cos^3 u + 5\cos u \\ &\dots \end{aligned} \tag{12.75}$$

After replacing $\cos u$ with x, we find explicit formulas for $C(n,x)$:

$$\begin{aligned} C(1,x) &= x \\ C(2,x) &= 2x^2 - 1 \\ C(3,x) &= 4x^3 - 3x \\ C(4,x) &= 8x^4 - 8x^2 + 1 \\ C(5,x) &= 16x^5 - 20x^3 + 5x \\ &\dots \end{aligned} \tag{12.76}$$

which verify that $C(n,x)$ is an nth-order polynomial in x.

Higher-order polynomials $C(n,x)$ can be derived from lower-order ones. By using the identity

$$\cos(nu) = \cos(u + (n-1)u) = \cos u \cos((n-1)u) - \sin u \sin((n-1)u) \tag{12.77}$$

and expressing $\sin(\cdot)$ in terms of $\cos(\cdot)$

$$-\sin u \sin((n-1)u) = \frac{1}{2}\cos(nu) - \frac{1}{2}\cos((n-2)u) \tag{12.78}$$

we find

$$\frac{1}{2}\cos(nu) = \cos u \cos((n-1)u) - \frac{1}{2}\cos((n-2)u) \tag{12.79}$$

Substituting $\cos u = x$ and $\cos(nu) = C(n, x)$ into Eq. (12.79), we find the well-known recurrence relation for the Chebyshev polynomials:

$$C(n, x) = 2x\,C(n-1, x) - C(n-2, x) \tag{12.80}$$

In the subsequent paragraphs we demonstrate an interesting property of the Chebyshev polynomials. By definition,

$$\begin{aligned} C(m, X) &= \cos(m(pu)) \\ X &= \cos(pu) \\ C(p, x) &= \cos(pu) \\ x &= \cos u \end{aligned} \tag{12.81}$$

For integer p, assuming $X = C(p, x)$, it follows that

$$\begin{aligned} C(m, C(p, x)) &= \cos(m(pu)) = \cos(m\,p\,u) \\ x &= \cos u \end{aligned} \tag{12.82}$$

According to Eq. (12.74) we obtain

$$\begin{aligned} C(mp, x) &= \cos(m\,p\,u) \\ x &= \cos u \end{aligned} \tag{12.83}$$

and from Eqs. (12.82) and (12.83) we find

$$C(mp, x) = C(m, C(p, x)) \tag{12.84}$$

Equation (12.84) shows a property that we call the *nesting property* of the Chebyshev polynomial; that is, when the independent variable, x, in the mth-order Chebyshev polynomial is replaced by the pth-order Chebyshev polynomial, the resulting polynomial is again the Chebyshev polynomial of the order $n = mp$.

For example, let us consider the second-order Chebyshev polynomial, $n = 2$:

$$C(2, x) = 2x^2 - 1 \tag{12.85}$$

After we replace the independent variable x with $C(2, x)$

$$C(2, C(2, x)) = 2\,(C(2, x))^2 - 1 = 8x^4 - 8x^2 + 1 \tag{12.86}$$

we obtain the fourth-order Chebyshev polynomial:

$$C(4, x) = C(2, C(2, x)) \tag{12.87}$$

The Chebyshev polynomials are plotted in Fig. 12.14 for $|x| \leq 1$. All curves take the unit value for $x = 1$.

By increasing the polynomial order, n, we increase the slope at $x = 1$ as shown in Fig. 12.15.

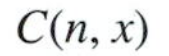

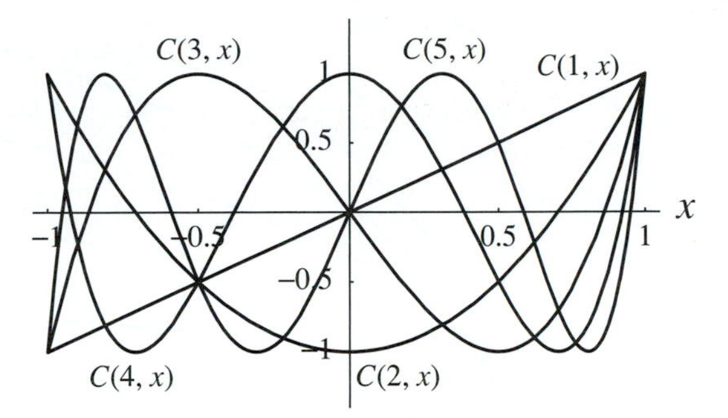

Figure 12.14 Chebyshev polynomials $C(n, x)$, $|x| \leq 1$.

For $x > 1$, the Chebyshev polynomials are monotonically increasing (Fig. 12.16). From Figs. 12.15 and 12.16 we conclude that the slope at $x = 1$ increases when the largest polynomial zero approaches $x = 1$. Chebyshev polynomials have the highest slope of all polynomials oscillating between -1 and $+1$, over the range $0 \leq x \leq 1$. Further increasing the slope at $x = 1$ is possible only by introducing poles for $x > 1$; that is, instead of polynomials we have to use rational functions.

The zeros of the Chebyshev polynomial play an important role in deriving the zeros of the elliptic rational functions.

The zeros of the cosine function are

$$\cos\alpha = 0 \quad \Rightarrow \quad \alpha = (2i - 1)\frac{\pi}{2}, \quad i = 0, \pm 1, \pm 2, \ldots \tag{12.88}$$

The zeros of $C(n, x)$ are obtained from the equation

$$C(n, x) = \cos(n \arccos x) = 0 \tag{12.89}$$

which yields

$$n \arccos x = (2k - 1)\frac{\pi}{2}, \quad i = 0, \pm 1, \pm 2, \ldots \tag{12.90}$$

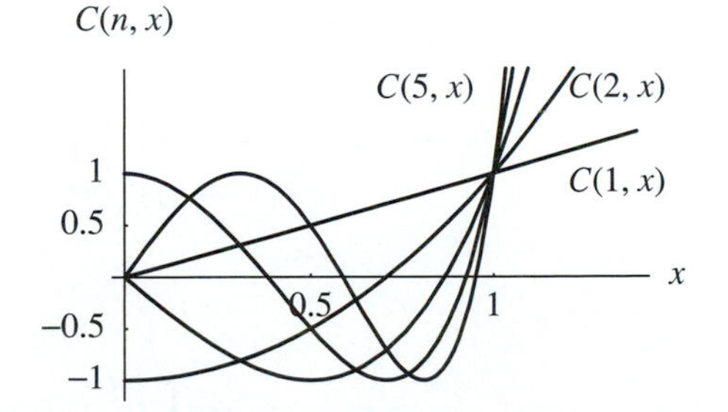

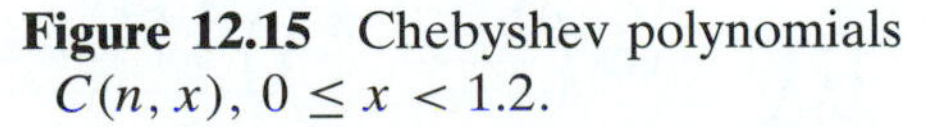
Figure 12.15 Chebyshev polynomials $C(n, x)$, $0 \leq x < 1.2$.

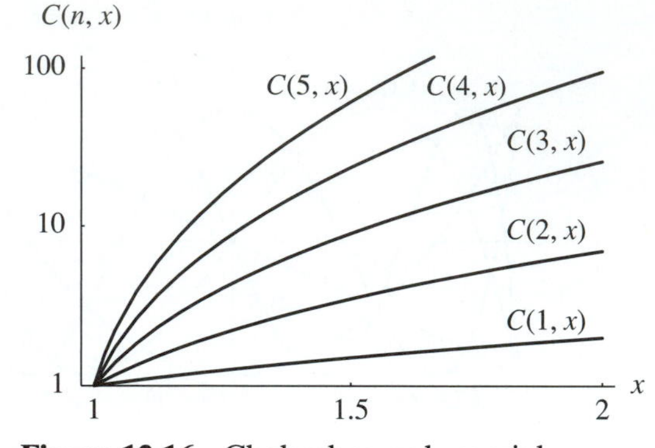

Figure 12.16 Chebyshev polynomials $C(n, x)$, $x \geq 1$.

that is,

$$\arccos x = \frac{2i - 1}{n} \frac{\pi}{2}, \qquad i = 0, \pm 1, \pm 2, \ldots \tag{12.91}$$

We have shown that $C(n, x)$ is a polynomial, so it has exactly n zeros. From Eq. (12.91), $C(n, x)$ has n different zeros within the interval $-1 < x < 1$. Therefore, we can represent the zeros by the formula

$$x_i = \cos\left(\frac{2i - 1}{n} \frac{\pi}{2}\right), \qquad i = 1, 2, \ldots, n \tag{12.92}$$

The following relations between the zeros exist:

$$x_n = -x_1$$

$$x_{n-1} = -x_2$$

$$\ldots$$

$$x_{n+1-i} = -x_i$$

and for odd n we have

$$x_{(n+1)/2} = 0$$

The Chebyshev polynomial can be expressed in terms of its zeros:

$$C(n, x) = \frac{1}{\prod_{i=1}^{n}(1 - x_i)} \prod_{i=1}^{n}(x - x_i) \tag{12.93}$$

that is,

$$C(n, x) = \frac{1}{\prod_{i=1}^{n}\left(1 - \cos\left(\frac{2i - 1}{n} \frac{\pi}{2}\right)\right)} \prod_{i=1}^{n}\left(x - \cos\left(\frac{2i - 1}{n} \frac{\pi}{2}\right)\right) \tag{12.94}$$

The term

$$\frac{1}{\prod_{i=1}^{n}(1 - x_i)}$$

ensures that $C(n, 1) = 1$.

12.6 ELLIPTIC RATIONAL FUNCTION

We examine, now, what happens if we replace the trigonometric cosine function, cos, by the Jacobi cosine function, cd, in Eq. (12.73). We no longer have only one independent variable. The Jacobi cd function is a function of two independent variables: the first variable is μ and the second variable is the modulus k, $0 \leq k < 1$. The periods of the Jacobi functions are functions of modulus, and we assume that each parametric equation has a different modulus. A new function, generated by replacing cos with cd, is a function of the order n, independent variable x (as in the case of the Chebyshev polynomial), and the moduli k, κ:

$$\begin{aligned} \mathcal{R}(n, k, \kappa, x) &= \mathrm{cd}\,(n\,\mu\, K(\kappa), \kappa) \\ x &= \mathrm{cd}\,(\mu\, K(k), k) \end{aligned} \tag{12.95}$$

Since $K(k)$ is the quarter period of the Jacobi elliptic functions of modulus k, and $K(\kappa)$ is the quarter period of modulus κ, x monotonically increases from -1 to $+1$ and $\mathcal{R}(n, k, \kappa, x)$ has n oscillations between ± 1 if μ takes a value from 2 to 4, as in the case of the Chebyshev polynomial. This equiripple property for $-1 \leq x \leq 1$ holds for any k and κ.

In this chapter we consider a special class of $\mathcal{R}(n, k, \kappa, x)$ functions specified by the conditions

$$\begin{aligned} \frac{1}{\kappa} &= \mathrm{cd}\,(n\,\mu\, K(\kappa), \kappa) \\ \frac{1}{k} &= \mathrm{cd}\,(\mu\, K(k), k) \end{aligned} \tag{12.96}$$

that is,

$$\mathcal{R}\left(n, k, \kappa, \frac{1}{k}\right) = \frac{1}{\kappa} \tag{12.97}$$

For given n and k the two equations (12.95) uniquely define $\kappa = \kappa(n, k)$; this particular value of κ we designate by k_n. The corresponding function $\mathcal{R}(n, k, k_n, x)$ becomes a function of three independent variables and is designated by $R(n, \xi, x)$. In filter design we prefer to use the reciprocal value of the modulus k, $\xi = 1/k$, which we call the *selectivity factor*:

$$\begin{aligned} R(n, \xi, x) &= \mathcal{R}\left(n, \frac{1}{\xi}, k_n, x\right) = \mathrm{cd}\,(n\,\mu\, K\,(k_n)\,, k_n) \\ x &= \mathrm{cd}\left(\mu\, K\left(\frac{1}{\xi}\right), \frac{1}{\xi}\right) \end{aligned} \tag{12.98}$$

Substituting $\xi = 1/k$ and Eq. (12.97) into Eq. (12.95), we find $k_n = 1/R(n, \xi, \xi)$ and

$$R(n, \xi, x) = \operatorname{cd}\left(n \, \mu \, K\left(\frac{1}{R(n, \xi, \xi)}\right), \frac{1}{R(n, \xi, \xi)}\right)$$
$$x = \operatorname{cd}\left(\mu \, K\left(\frac{1}{\xi}\right), \frac{1}{\xi}\right) \tag{12.99}$$

We call $R(n, \xi, x)$ the *elliptic rational function.*

In this chapter, for the sake of simplicity, we introduce a new function

$$L(n, \xi) = R(n, \xi, \xi) \tag{12.100}$$

Also, we use the compact notation

$$K = K(k) \tag{12.101}$$

$$K_n = K(k_n) \tag{12.102}$$

$$K' = K(k') = K\left(\sqrt{1 - k^2}\right) \tag{12.103}$$

$$K_n' = K(k_n') = K\left(\sqrt{1 - k_n^2}\right) \tag{12.104}$$

$$k = \frac{1}{\xi} \tag{12.105}$$

$$k_n = \frac{1}{L(n, \xi)} \tag{12.106}$$

With this notation we can write the elliptic rational function as

$$R(n, \xi, x) = \operatorname{cd}(n \, \mu \, K_n, k_n)$$
$$x = \operatorname{cd}(\mu \, K, k), \qquad k = \frac{1}{\xi} \tag{12.107}$$

Substituting $w = \mu K$, we obtain the elliptic rational function in a form that appears in standard textbooks:

$$R(n, \xi, x) = \operatorname{cd}\left(n w \frac{K_n}{K}, \frac{1}{L(n, \xi)}\right)$$
$$x = \operatorname{cd}\left(w, \frac{1}{\xi}\right) \tag{12.108}$$

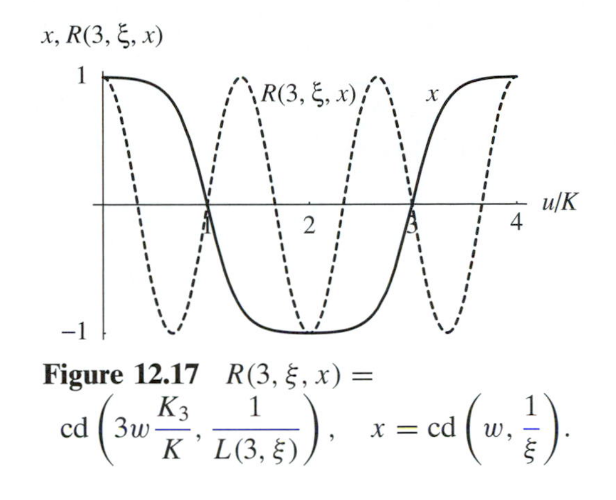

Figure 12.17 $R(3, \xi, x) = \text{cd}\left(3w\frac{K_3}{K}, \frac{1}{L(3, \xi)}\right), \quad x = \text{cd}\left(w, \frac{1}{\xi}\right).$

In Figs. 12.17 and 12.18 the third-order and the fourth-order parametric equations are visualized.

The Jacobi cd function looks like the trigonometric cos function, but it has flat maxima and minima; the smaller the value of ξ, the larger the flatness. Two special cases are derived from Eq. (12.108): (1) For $k = 1/\xi = 0$, cd becomes equal to cos, and (2) letting $k = 1/\xi \to 1$ the cd function looks like a pulse train. The domain in which all the cd functions must occur is bounded by the two special cases (Fig. 12.19).

The well-known properties of the Jacobi sn and cn functions are thoroughly documented in textbooks. For this reason it is convenient to express cd in terms of sn:

$$\text{cd}\left(w, \frac{1}{\xi}\right) = \text{sn}\left(w + K, \frac{1}{\xi}\right) \tag{12.109}$$

The Jacobi cd function is the shifted Jacobi sn function for the quarter period K in the same way as the trigonometric cos function is the shifted trigonometric sin function for the amount of $\pi/2$. Finally, we can write the parametric equations for the elliptic

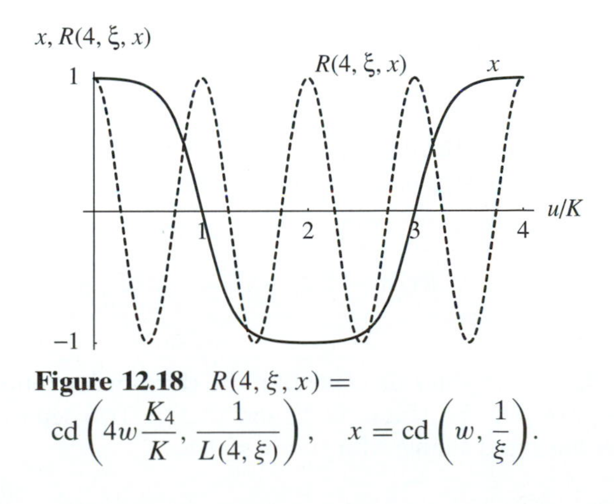

Figure 12.18 $R(4, \xi, x) = \text{cd}\left(4w\frac{K_4}{K}, \frac{1}{L(4, \xi)}\right), \quad x = \text{cd}\left(w, \frac{1}{\xi}\right).$

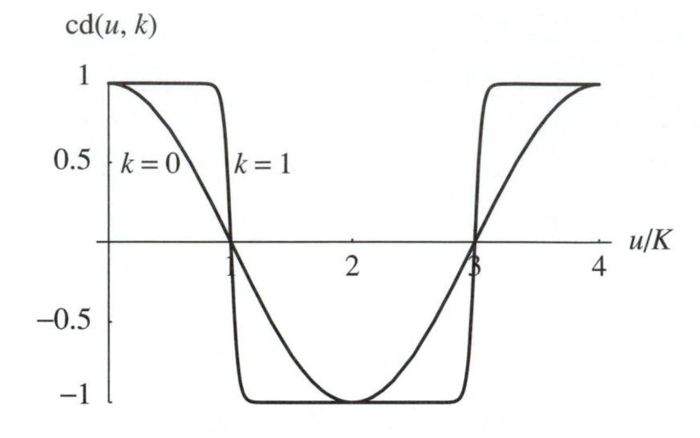

Figure 12.19 $\mathrm{cd}(w, k)$.

rational function as

$$R(n, \xi, x) = \mathrm{sn}\left(nw\frac{K_n}{K} + K_n, \frac{1}{L(n, \xi)}\right)$$
$$x = \mathrm{sn}\left(w + K, \frac{1}{\xi}\right) \tag{12.110}$$

The parametric variable w can be complex, $w = u + jv$, u and v real, $j = \sqrt{-1}$, as in the analysis of the Chebyshev polynomial

$$R(n, \xi, x) = \mathrm{cd}\left(\frac{n(u + jv)K_n}{K}, k_n\right) \tag{12.111}$$

$$x = \mathrm{cd}\,((u + jv), k) \tag{12.112}$$

The boundary values in Eq. (12.113) specify the parametric variable w which yields x in a continuous range from 0 to $+\infty$. Notice the three ranges of x and the corresponding ranges for u and v as shown in Fig. 12.20:

$$3K \leq u \leq 4K \quad \text{and} \quad v = 0 \Rightarrow 0 \leq x \leq 1$$
$$u = 4K \quad \text{and} \quad 0 \leq v \leq K' \Rightarrow 1 \leq x \leq \xi \tag{12.113}$$
$$4K \geq u \geq 3K \quad \text{and} \quad v = K' \Rightarrow \xi \leq x < +\infty$$

It can be shown, as in the case of the Chebyshev polynomial, that R is an even function of x for even n and is an odd function for odd n

$$R(2n, \xi, -x) = R(2n, \xi, x) \tag{12.114}$$

$$R(2n + 1, \xi, -x) = -R(2n + 1, \xi, x) \tag{12.115}$$

For the same values of w we plot the third-order rational function $R(3, \xi, x) = \mathrm{cd}\,(3(u + jv)K_3/K, k_3)$ (Figs. 12.21 and 12.22). Two separate diagrams are required because R has rapid changes for $0 < x < +\infty$.

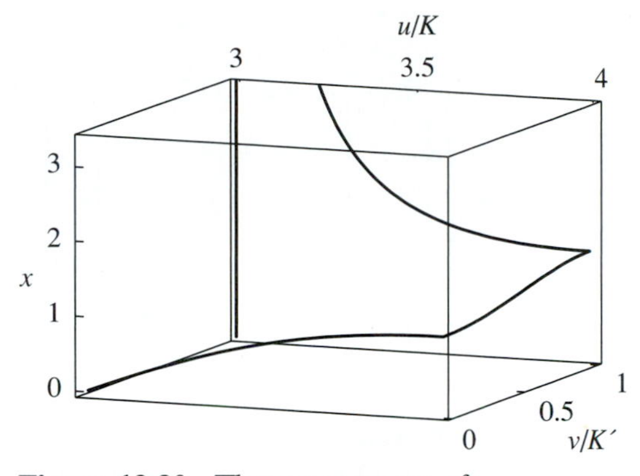

Figure 12.20 Three segments of $x = \mathrm{cd}\,((u + jv), k)$: $u + jv \in \{(3K, 4K), (4K, 4K + jK'), (3K + jK', 4K + jK')\}$, $x \in \{(0, 1), (1, \xi), (\xi, \infty)\}$, $\xi = 10/7$.

In the same way we plot the fourth-order rational function $R(4, \xi, x) = \mathrm{cd}\,(4(u + jv)K_4/K, k_4)$ (Figs. 12.23 and 12.24).

Let us find the value of x for the edge value of $w = 4K(k) + jK'(k)$:

$$w = u + jv = 4K(k) + jK'(k) \tag{12.116}$$

Since $4K$ is a period of the cd function, we find

$$x = \mathrm{cd}\left(4K(k) + jK'(k), k\right) = \mathrm{cd}\left(jK'(k), k\right) \tag{12.117}$$

Next, we use the identity [109, page 572, Eq. (16.7.4)]

$$\mathrm{cd}(jK'(k), k) = \frac{1}{k} \tag{12.118}$$

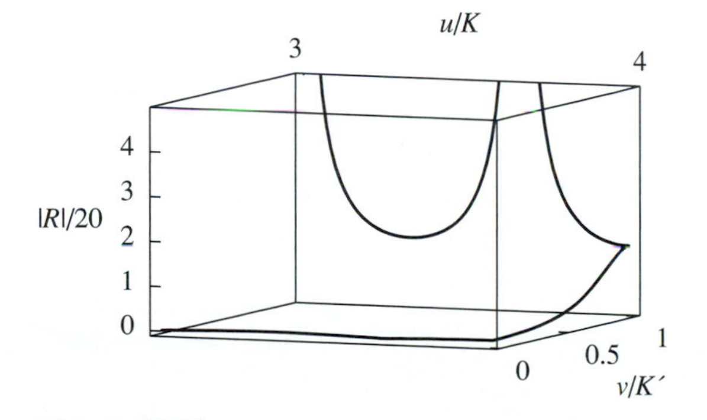

Figure 12.21 $R(3, \xi, x) = \mathrm{cd}\,(3(u + jv)K_3/K, k_3)$ for $u + jv \in \{(3K, 4K), (4K, 4K + jK'), (3K + jK', 4K + jK')\}$, $\xi = 10/7$.

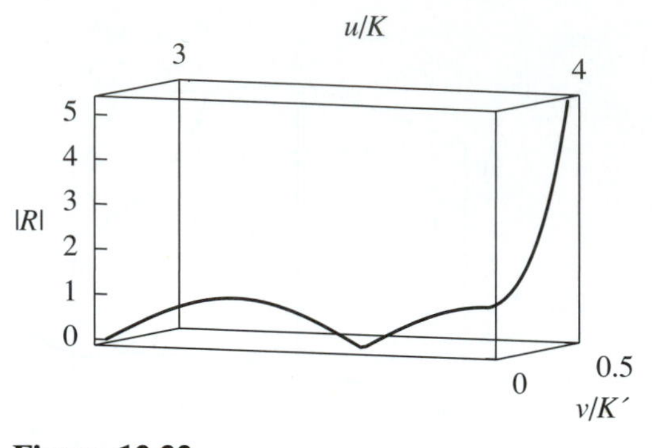

Figure 12.22
$R(3, \xi, x) = \mathrm{cd}\,(3(u + jv)K_3/K, k_3)$ for $u + jv \in \{(3K, 4K), (4K, 4K + jK')\}$, $x < \xi = 10/7$.

and Eq. (12.117) transforms into

$$x = \mathrm{cd}\left(jK'(k), k\right) = \frac{1}{k} = \xi \tag{12.119}$$

A similar procedure can be applied to Eq. (12.111):

$$R(n, \xi, x) = \mathrm{cd}\left(nw\frac{K_n}{K}, k_n\right)$$

Substituting $x = \xi$ from Eq. (12.119) along with the corresponding $w = 4K + jK'$, we obtain

$$R(n, \xi, \xi) = \mathrm{cd}\left(n(4K + jK')\frac{K_n}{K}, k_n\right)$$

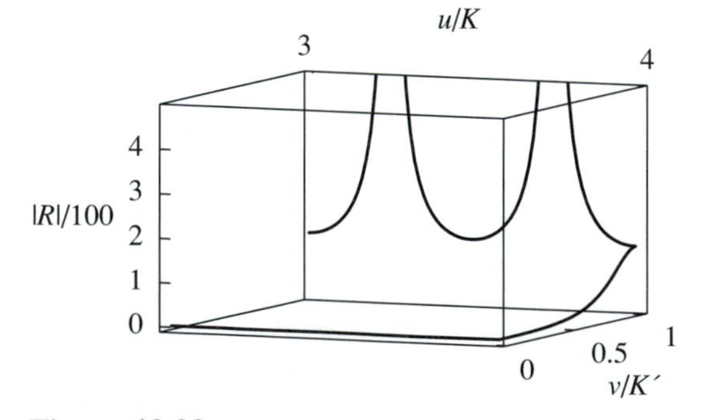

Figure 12.23
$R(4, \xi, x) = \mathrm{cd}\,(4(u + jv)K_4/K, k_4)$ for $u + jv \in \{(3K, 4K), (4K, 4K + jK'), (3K + jK', 4K + jK')\}$, $\xi = 10/7$.

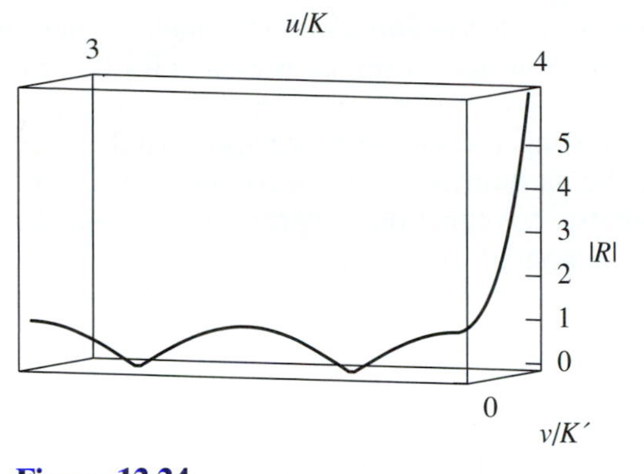

Figure 12.24
$R(4, \xi, x) = \mathrm{cd}\,(4(u + jv)K_4/K, k_4)$ for $u + jv \in \{(3K, 4K), (4K, 4K + jK')\}$, $x < \xi = 10/7$.

which is equivalent to

$$R(n, \xi, \xi) = \mathrm{cd}\left(4nK_n + jnK'\frac{K_n}{K}, k_n\right)$$

Since $4K_n$ is the period of $\mathrm{cd}(w, k_n)$, we obtain

$$R(n, \xi, \xi) = \mathrm{cd}\left(jnK'\frac{K_n}{K}, k_n\right)$$

Multiplying and dividing the first cd argument by K'_n, we obtain

$$R(n, \xi, \xi) = \mathrm{cd}\left(jK'_n(n\frac{K_n}{K}\frac{K'}{K'_n}), k_n\right)$$

By definition of the elliptic rational function, we have

$$R(n, \xi, \xi) = \frac{1}{k_n}$$

implying that

$$\mathrm{cd}\left(jK'_n\left(n\frac{K_n}{K}\frac{K'}{K'_n}\right), k_n\right) = \frac{1}{k_n}$$

must hold. According to Eq. (12.118) we identify that

$$n\frac{K_n}{K}\frac{K'}{K'_n} = 1 \tag{12.120}$$

must be satisfied. This Eq. (12.120) can be rewritten as

$$n = \frac{K}{K'}\frac{K'_n}{K_n} \tag{12.121}$$

and is known as the *degree equation*. The degree equation is a very important relation in the elliptic filter design. It serves as a basis for determining the filter order.

It is interesting to examine what happens if the selectivity factor ξ tends to infinity. In that case the modulus $k = 1/\xi$ becomes zero, and the Jacobi cd function becomes the trigonometric cos function. Therefore, the elliptic rational function becomes the Chebyshev polynomial. In summary,

$$\lim_{\xi \to +\infty} \operatorname{cd}\left(w, \frac{1}{\xi}\right) = \lim_{k \to 0} \operatorname{cd}(w, k) = \operatorname{cd}(w, 0) = \cos(w)$$

$$\lim_{\xi \to +\infty} R(n, \xi, x) = C(n, x)$$

which means that the Chebyshev polynomial is a special case of the elliptic rational function.

Expanding $\operatorname{cd}(nu\frac{K_n}{K}, k_n)$, we obtain a rational function in terms of $\operatorname{cd}(u, k)$:

$$\begin{aligned}
\operatorname{cd}\left(1u\frac{K_1}{K}, k_1\right) &= \operatorname{cd}(u, k) \\
\operatorname{cd}\left(2u\frac{K_2}{K}, k_2\right) &= \frac{\left(\sqrt{1-k^2}+1\right)\operatorname{cd}^2(u, k) - 1}{\left(\sqrt{1-k^2}-1\right)\operatorname{cd}^2(u, k) + 1} \\
\operatorname{cd}\left(3u\frac{K_3}{K}, k_3\right) &= \operatorname{cd}(u, k)\,\frac{(1+\operatorname{dn}(\frac{2K}{3}, k))^2\operatorname{cd}^2(u, k) - (1+2\operatorname{dn}(\frac{2K}{3}, k))}{(-1+\operatorname{dn}^2(\frac{2K}{3}, k))\operatorname{cd}^2(u, k) + 1} \\
\operatorname{cd}\left(4u\frac{K_4}{K}, k_4\right) &= \frac{(1+k')\left(1+\sqrt{k'}\right)^2\operatorname{cd}^4(u, k) - 2(1+k')\left(1+\sqrt{k'}\right)\operatorname{cd}^2(u, k) + 1}{(1+k')\left(1-\sqrt{k'}\right)^2\operatorname{cd}^4(u, k) - 2(1+k')\left(1-\sqrt{k'}\right)\operatorname{cd}^2(u, k) + 1} \\
\operatorname{cd}\left(5u\frac{K_5}{K}, k_5\right) &= \operatorname{cd}(u, k)\,\frac{a_5\operatorname{cd}^4(u, k) - a_3\operatorname{cd}^2(u, k) + a_1}{a_4\operatorname{cd}^4(u, k) - a_2\operatorname{cd}^2(u, k) + 1} \\
\cdots &= \cdots
\end{aligned} \tag{12.122}$$

where

$$a_5 = \left(1 + \mathrm{dn}\left(\frac{2K}{5}, k\right)\right)^2 \left(1 + \mathrm{dn}\left(\frac{4K}{5}, k\right)\right)^2$$

$$a_4 = \left(1 - \mathrm{dn}^2\left(\frac{2K}{5}, k\right)\right)\left(1 - \mathrm{dn}^2\left(\frac{4K}{5}, k\right)\right)$$

$$a_3 = 1 - \mathrm{dn}^2\left(\frac{2K}{5}, k\right)\mathrm{dn}^2\left(\frac{4K}{5}, k\right) + \left(1 + \mathrm{dn}\left(\frac{2K}{5}, k\right)\right)^2\left(1 + \mathrm{dn}\left(\frac{4K}{5}, k\right)\right)^2 + 2\left(\mathrm{dn}\left(\frac{2K}{5}, k\right) + \mathrm{dn}\left(\frac{4K}{5}, k\right)\right)$$

$$a_2 = 2 - \mathrm{dn}^2\left(\frac{2K}{5}, k\right) - \mathrm{dn}^2\left(\frac{4K}{5}, k\right)$$

$$a_1 = \left(1 + 2\left(\mathrm{dn}\left(\frac{2K}{5}, k\right) + \mathrm{dn}\left(\frac{4K}{5}, k\right)\right)\right)$$

Replacing $\mathrm{cd}(u, k)$ with x, we find explicit formulas for $R(n, \xi, x)$:

$$\begin{aligned}
R(1, \xi, x) &= x \\
R(2, \xi, x) &= \frac{\left(\sqrt{1 - k^2} + 1\right)x^2 - 1}{\left(\sqrt{1 - k^2} - 1\right)x^2 + 1} \\
R(3, \xi, x) &= x\,\frac{\left(1 + \mathrm{dn}\left(\frac{2K}{3}, k\right)\right)^2 x^2 - \left(1 + 2\,\mathrm{dn}\left(\frac{2K}{3}, k\right)\right)}{\left(-1 + \mathrm{dn}^2\left(\frac{2K}{3}, k\right)\right)x^2 + 1} \\
R(4, \xi, x) &= \frac{(1 + k')(1 + \sqrt{k'})^2 x^4 - 2(1 + k')(1 + \sqrt{k'})x^2 + 1}{(1 + k')(1 - \sqrt{k'})^2 x^4 - 2(1 + k')(1 - \sqrt{k'})x^2 + 1} \\
R(5, \xi, x) &= x\,\frac{a_5 x^4 - a_3 x^2 + a_1}{a_4 x^4 - a_2 x^2 + 1} \\
\cdots &= \cdots
\end{aligned} \tag{12.123}$$

Note that for $k = 0$ we have

$$\mathrm{dn}(u, 0) = 1$$

and $k' = 1$, and the elliptic rational function becomes the Chebyshev polynomial.

In the subsequent paragraphs we demonstrate an interesting property of the elliptic rational function that is similar to the property of the Chebyshev polynomial. By definition, we have

$$
\begin{aligned}
R(m, \Xi, X) &= \mathrm{cd}\left(m(p\mu)K\left(\frac{1}{L(m, \Xi)}\right), \frac{1}{L(m, \Xi)}\right) \\
X &= \mathrm{cd}\left((p\mu)K\left(\frac{1}{\Xi}\right), \frac{1}{\Xi}\right) \\
R(p, \xi, x) &= \mathrm{cd}\left(p\mu K\left(\frac{1}{L(p, \xi)}\right), \frac{1}{L(p, \xi)}\right) \\
x &= \mathrm{cd}\left(\mu K\left(\frac{1}{\xi}\right), \frac{1}{\xi}\right)
\end{aligned}
\tag{12.124}
$$

For integer p and $\Xi = L(p, \xi) = R(p, \xi, \xi)$, assuming $L(m, \Xi) = L(m, L(p, \xi))$, $L(m, \Xi) = L(mp, \xi)$, and $X = \mathrm{cd}(p\mu K\,(1/\Xi)\,, 1/\Xi)$, it follows that

$$
\begin{aligned}
R(m, R(p, \xi, \xi), R(p, \xi, x)) &= \mathrm{cd}(m\ p\ \mu K(1/L(mp, \xi)), 1/L(mp, \xi)) \\
x &= \mathrm{cd}(\mu K(1/\xi), 1/\xi)
\end{aligned}
\tag{12.125}
$$

According to Eq. (12.74) and notation $k = 1/\xi$, $K = K(1/\xi)$, $k_{mp} = 1/L(mp, \xi)$, and $K_{mp} = K(1/L(mp, \xi))$, we obtain

$$
\begin{aligned}
R(mp, \xi, x) &= \mathrm{cd}\left(m\ p\ u\frac{K_{mp}}{K}, k_{mp}\right) \\
x &= \mathrm{cd}(u, k)
\end{aligned}
\tag{12.126}
$$

and from Eqs. (12.82) and (12.83) we find

$$
R(mp, \xi, x) = R(m, R(p, \xi, \xi), R(p, \xi, x)) \tag{12.127}
$$

Equation (12.127) shows the nesting property of the elliptic rational functions.

For example, let us consider the second-order elliptic rational function, $n = 2$. After we replace the independent variable x with $R(2, \xi, x)$, we obtain the fourth-order elliptic rational function.

The zeros of the elliptic rational function can be derived in a way that resembles the procedure for finding the zeros of the Chebyshev polynomial.

The zeros of the Jacobi cd function are

$$
\mathrm{cd}(w, k) = 0 \quad \Rightarrow \quad w = (2i - 1)K(k), \qquad i = 0, \pm 1, \pm 2, \ldots \tag{12.128}
$$

$$
\mathrm{cd}(w, k_n) = 0 \quad \Rightarrow \quad w = (2i - 1)K(k_n), \quad i = 0, \pm 1, \pm 2, \ldots \tag{12.129}
$$

The zeros of $R(n, \xi, x)$ are obtained from the equation

$$
R(n, \xi, x) = \mathrm{cd}\left(n\ w\,\frac{K(k_n)}{K(k)}, k_n\right) = 0 \tag{12.130}
$$

or

$$R(n,\xi,x) = \text{cd}\left(n\ (\text{cd}^{-1}(x,k))\,\frac{K(k_n)}{K(k)},\, k_n\right) = 0 \tag{12.131}$$

which yields

$$n\,\frac{K(k_n)}{K(k)}\,\text{cd}^{-1}(x,k) = (2i-1)K(k_n), \qquad i = 0, \pm 1, \pm 2, \ldots \tag{12.132}$$

that is,

$$\text{cd}^{-1}(x,k) = \frac{2i-1}{n}\,K(k), \qquad i = 0, \pm 1, \pm 2, \ldots \tag{12.133}$$

We have shown that $R(n,\xi,x)$ is a rational function and that its numerator order is n, so it has exactly n zeros. From Eq. (12.91), $R(n,\xi,x)$ has n different zeros within the interval $-1 < x < 1$. Therefore, we can represent the zeros by the formula

$$x_i = \text{cd}\left(\frac{2i-1}{n}\,K(k)\right), \qquad i = 1, 2, \ldots, n \tag{12.134}$$

or

$$x_i = \text{sn}\left(\frac{2i-1}{n}\,K(k) + K(k)\right), \qquad i = 1, 2, \ldots, n \tag{12.135}$$

The argument u_i which corresponds to the zero x_i is

$$u_i = \frac{2i-1}{n}\,K(k)$$

The following relations between the zeros exist:

$$x_n = -x_1$$

$$x_{n-1} = -x_2$$

$$\ldots$$

$$x_{n+1-i} = -x_i$$

and for odd n we have

$$x_{(n+1)/2} = 0$$

The poles of the elliptic rational function are the zeros of the equation

$$\frac{1}{R(n,\xi,x)} = 0$$

which is equivalent to

$$\frac{1}{\text{cd}\left(n\,w\frac{K(k_n)}{K(k)},\, k_n\right)} = 0 \tag{12.136}$$

This equation has zeros only for complex w:

$$w = u + jK'$$

yielding

$$\frac{1}{\mathrm{cd}\left(n\,u\dfrac{K(k_n)}{K(k)} + jn\,\dfrac{K(k_n)}{K(k)}K', k_n\right)} = 0$$

According to the degree equation we have $n\,\frac{K(k_n)}{K(k)}K' = K'_n$, so we obtain

$$\frac{1}{\mathrm{cd}\left(n\,u\dfrac{K(k_n)}{K(k)} + jK'_n, k_n\right)} = 0$$

Applying the change-of-argument theorem for the cd function [109, page 572, Eq. (16.8.4)], we obtain

$$\frac{1}{\mathrm{cd}\left(n\,u\frac{K(k_n)}{K(k)} + jK'_n, k_n\right)} = k_n\,\mathrm{cd}\left(n\,u\frac{K(k_n)}{K(k)}, k_n\right) = 0$$

that is,

$$\mathrm{cd}\left(n\,u\frac{K(k_n)}{K(k)}, k_n\right) = 0$$

The solution of the above equation is

$$u_i = \frac{2i-1}{n}\,K(k), \qquad i = 1, 2, \ldots, n$$

The corresponding values of w are

$$w_i = u_i + jK' = \frac{2i-1}{n}\,K(k) + jK', \qquad i = 1, 2, \ldots, n$$

Finally, the poles of $R(n, \xi, x)$ are derived from the second parametric equation that defines $R(n, \xi, x)$:

$$x_{p,i} = \mathrm{cd}\,(w_i, k) = \mathrm{cd}\left(u_i + jK', k\right) = \frac{1}{k}\,\frac{1}{\mathrm{cd}\,(u_i, k)}$$

which can be written in terms of ξ, by substituting $k = 1/\xi$:

$$x_{p,i} = \frac{\xi}{\mathrm{cd}(u_i, k)} = \frac{\xi}{x_i}$$

Notice that the poles $x_{p,i}$ are inversely proportional to the zeros x_i of the elliptic rational function.

The elliptic rational function can be expressed in terms of its zeros and poles:

$$R(n, \xi, x) = r_0\,\frac{\prod_{i=1}^{n}(x - x_i)}{\prod_{i=1}^{n}\left(x - \dfrac{\xi}{x_i}\right)} \tag{12.137}$$

where r_0 is a scaling constant which ensures

$$R(n, \xi, 1) = 1 \tag{12.138}$$

that is,

$$r_0 = \frac{\prod_{i=1}^{n}\left(1 - \frac{\xi}{x_i}\right)}{\prod_{i=1}^{n}(1 - x_i)} \tag{12.139}$$

In the case when n is an odd number, one zero is $x_{(n+1)/2} = 0$, and the corresponding pole is $\xi / x_{(n+1)/2} = \infty$; for that reason we use

$$R(n, \xi, x) = r_0\, x \frac{\prod_{i=1}^{(n-1)/2}(x - x_i)}{\prod_{i=1}^{(n-1)/2}\left(x - \frac{\xi}{x_i}\right)} \frac{\prod_{i=(n+3)/2}^{n}(x - x_i)}{\prod_{i=(n+3)/2}^{n}\left(x - \frac{\xi}{x_i}\right)}, \qquad n = 3, 5, \ldots, 2m + 1 \tag{12.140}$$

where r_0 is

$$r_0 = \frac{\prod_{i=1}^{(n-1)/2}\left(1 - \frac{\xi}{x_i}\right)}{\prod_{i=1}^{(n-1)/2}(1 - x_i)} \frac{\prod_{i=(n+3)/2}^{n}\left(1 - \frac{\xi}{x_i}\right)}{\prod_{i=(n+3)/2}^{n}(1 - x_i)}, \qquad n = 3, 5, \ldots, 2m + 1$$

The poles and the zeros appear in pairs with equal magnitudes and opposite signs (except $x_{(n+1)/2}$ for n odd). Therefore, we can write $(x - x_i)(x - (-x_{n-i+1})) = (x^2 - x_i^2)$

$$R(n, \xi, x) = r_0 \frac{\prod_{i=1}^{n/2}(x^2 - x_i^2)}{\prod_{i=1}^{n/2}\left(x^2 - \frac{\xi^2}{x_i^2}\right)}, \qquad n \text{ is even} \tag{12.141}$$

$$r_0 = \frac{\prod_{i=1}^{n/2}\left(1 - \frac{\xi^2}{x_i^2}\right)}{\prod_{i=1}^{n/2}(1 - x_i^2)}, \qquad n \text{ is even} \tag{12.142}$$

and

$$R(n, \xi, x) = r_0\, x \frac{\prod_{i=1}^{(n-1)/2}(x^2 - x_i^2)}{\prod_{i=1}^{(n-1)/2}\left(x^2 - \frac{\xi^2}{x_i^2}\right)}, \qquad n \text{ is odd} \tag{12.143}$$

$$r_0 = \frac{\prod_{i=1}^{(n-1)/2}\left(1 - \frac{\xi^2}{x_i^2}\right)}{\prod_{i=1}^{(n-1)/2}(1 - x_i^2)}, \qquad n \text{ is odd} \tag{12.144}$$

12.7 NESTING PROPERTY OF JACOBI ELLIPTIC FUNCTIONS

In this section we reveal the nesting property of the Jacobi elliptic functions. Firs, let us clearly define the nesting property of a function.

Consider a function $F(p, x)$ of an argument x, and assume that the function depends on at least one parameter p. The domain of the parameter can be any set of numbers $\mathcal{D}$, that is, $p \in \mathcal{D}$. The function is said to have the *nesting property* if the following holds:

$$F(a, F(b, x)) = F(f(a, b), x) \tag{12.145}$$

for an arbitrary argument x, parameters $a, b \in \mathcal{D}$, and $f(a, b) \in \mathcal{D}$, where $f(a, b)$ represents a function of two arguments. Frequently, $f(a, b) = ab$.

For example, consider the second-order Chebyshev polynomial

$$F(a, x) = C(2, x) = 2x^2 - 1 \tag{12.146}$$

After replacing the independent variable x with $F(b, x) = C(2, x)$,

$$F(a, F(b, x)) = C(2, C(2, x)) = 8x^4 - 8x^2 + 1 \tag{12.147}$$

we obtain the fourth-order Chebyshev polynomial

$$F(f(a, b), x) = C(4, x) = 8x^4 - 8x^2 + 1 \tag{12.148}$$

where $f(a, b) = ab$, obviously, a and b are integers, and the domain is $\mathcal{D} = \{2, 4\}$.

The Jacobi elliptic functions have the nesting property, too. First, we review the key relations and the notation we are going to use:

$$K(k) = \int_0^{\pi/2} \frac{d\theta}{\sqrt{1 - k^2 \sin^2\theta}} \tag{12.149}$$

$$K'(k) = \int_0^{\pi/2} \frac{d\theta}{\sqrt{1 - (1 - k^2)\sin^2\theta}} \tag{12.150}$$

$$k' = \sqrt{1 - k^2} \tag{12.151}$$

$$K'(k) = K(k') \tag{12.152}$$

$$n\frac{K(k_n)}{K'(k_n)} = \frac{K(k)}{K'(k)} \tag{12.153}$$

Often, we use the compact notation

$$\begin{aligned} K &= K(k) \\ K' &= K'(k) \\ k'_n &= \sqrt{1-k_n^2} \\ K_n &= K(k_n) \\ K'_n &= K'(k_n) \end{aligned} \tag{12.154}$$

Unless otherwise stated, K and K' are evaluated for the modulus k; similarly, K_n and K'_n are evaluated for the modulus k_n. The symbol k_n will always represent a solution of the degree equation [Eq. (12.153)] for a given n. The degree equation [Eq. (12.153)] must be satisfied in order to generate the elliptic rational function $R(n, \xi, x)$. Throughout this book, by default, k_n will refer to the order n and the modulus k—that is, $k_n = k_n(k)$. When this convention is not obvious or when this notation may be ambiguous, we write $k_n(k)$ to explicitly specify the modulus.

Note that K and K' are functions of k, while K_n and K'_n are functions of k_n. There is a one-to-one correspondence between K/K' and k and, also, between K_n/K'_n and k_n. Therefore, for any n, according to Eq. (12.153), there is a one-to-one correspondence between k and k_n; that is, k_n is a monotonic function of k:

$$k_n = k_n(k) \tag{12.155}$$

The suffix n in Eq. (12.155) means that the ratio $K(k)/K'(k)$ is n times larger than the ratio $K(k_n)/K'(k_n)$.

For $n = 1$ the degree equation simplifies to $K(k)/K'(k) = K(k_1)/K'(k_1)$, yielding

$$k_1 = k_1(k) = k \tag{12.156}$$

The next theorem states the key relation for advanced analysis of the elliptic rational function, and it proves the nesting property of the modulus.

Theorem 12.1 For the moduli of Jacobi elliptic functions the following holds:

$$k_{mp}(k) = k_m(k_p(k)) = k_p(k_m(k)) \tag{12.157}$$

where m and p are positive integers.

Proof: To prove this identity we repeat Eq. (12.153) for the case of the pth-order function

$$p\frac{K(k_p(k))}{K'(k_p(k))} = \frac{K(k)}{K'(k)} \tag{12.158}$$

Considering $k_p(k)$ as another modulus, we repeat Eq. (12.153) for the mth-order function

$$m\frac{K(k_m(k_p(k)))}{K'(k_m(k_p(k)))} = \frac{K(k_p(k))}{K'(k_p(k))} \tag{12.159}$$

Multiplying Eq. (12.159) by p and equating the left-hand parts with Eq. (12.158), one has

$$mp\frac{K(k_m(k_p(k)))}{K'(k_m(k_p(k)))} = \frac{K(k)}{K'(k)} \tag{12.160}$$

On the other hand, assuming $n = mp$, Eq. (12.160) becomes

$$mp\frac{K(k_{mp}(k))}{K'(k_{mp}(k))} = \frac{K(k)}{K'(k)}$$

From the last two equations we find

$$\frac{K(k_m(k_p(k)))}{K'(k_m(k_p(k)))} = \frac{K(k_{mp}(k))}{K'(k_{mp}(k))} \tag{12.161}$$

K and K' in Eq. (12.161) are positive and, according to their derivatives, steadily increasing and decreasing functions, respectively, on the open interval $0 < k < 1$. Thus, their ratio $K(k)/K'(k)$ is a strictly increasing function on $0 < k < 1$. Since there is a one-to-one correspondence between $K(k)/K'(k)$ and k on the interval $0 < k < 1$, Eq. (12.157) is proved. ∎

Theorem 12.2

$$\begin{aligned} \operatorname{sn}\left(u\ K\left(k_{mp}(k)\right), k_{mp}(k)\right) &= \operatorname{sn}\left(u\ K(k_m(k_p)), k_m(k_p)\right) \\ \operatorname{cd}\left(u\ K\left(k_{mp}(k)\right), k_{mp}(k)\right) &= \operatorname{cd}\left(u\ K(k_m(k_p)), k_m(k_p)\right) \end{aligned} \tag{12.162}$$

where

$$k_p = k_p(k) \tag{12.163}$$

Proof: Validity of the theorem is proved by substituting Eqs. (12.157) and (12.163) in Eq. (12.162). ∎

Consider the parametric equations

$$\begin{aligned} \zeta &= \operatorname{sn}(\nu K', k') \\ \frac{1}{\sqrt{1+\epsilon^2}} &= \operatorname{sn}(\nu K'_n, k'_n) \end{aligned} \tag{12.164}$$

in which the integer n, the complementary modulus $k' = \sqrt{1-k^2}$, the modulus k, and the real parameter ϵ are known. Let us determine ζ. Note that ν is also unknown. One possible solution is to calculate ν from the second equation (the inverse Jacobi sine) and, then, to find ζ as the Jacobi sine. In this procedure we have to calculate the complementary complete elliptic integrals, K'_n and K', the Jacobi functions, sn and sn^{-1}, and k'_n:

$$\zeta = \operatorname{sn}\left(\frac{K'}{K'_n}\operatorname{sn}^{-1}\left(\frac{1}{\sqrt{1+\epsilon^2}}, k'_n\right), k'\right) \tag{12.165}$$

Our goal is to find an analytic, closed-form computationally efficient procedure to calculate ζ. We want to avoid evaluation of complete elliptic integrals and Jacobi functions whenever possible. Our target is to find a closed-form expression for ζ in terms of n, k and ϵ; this expression we denote as SN

$$\zeta = \mathrm{SN}(n, k, \epsilon) \Leftrightarrow \begin{cases} \zeta = \mathrm{sn}(\nu K', k') \\ \dfrac{1}{\sqrt{1+\epsilon^2}} = \mathrm{sn}(\nu K'_n, k'_n) \end{cases} \tag{12.166}$$

For example, for $n = 1$, the parametric equations

$$\begin{aligned} \zeta &= \mathrm{sn}(\nu K', k') \\ \frac{1}{\sqrt{1+\epsilon^2}} &= \mathrm{sn}(\nu K'_1, k'_1) \end{aligned} \tag{12.167}$$

becomes

$$\zeta = \frac{1}{\sqrt{1+\epsilon^2}} \tag{12.168}$$

because $k'_1 = k'$ and $K'_1 = K'$. Hence, we derive a particular expression for the $\mathrm{SN}(1, k, \epsilon)$ function

$$\mathrm{SN}(1, k, \epsilon) = \frac{1}{\sqrt{1+\epsilon^2}} \tag{12.169}$$

The derived function $\mathrm{SN}(1, k, \epsilon)$ is a function of ϵ only. Alternatively, we can express ϵ in terms of $\mathrm{sn}(\nu K'_1, k'_1)$:

$$\epsilon = \sqrt{\frac{1}{\mathrm{sn}^2(\nu K'_1, k'_1)} - 1} \tag{12.170}$$

In the case $n = 2$, the parametric equation takes a form

$$\begin{aligned} \zeta &= \mathrm{sn}(\nu K', k') \\ \frac{1}{\sqrt{1+\epsilon^2}} &= \mathrm{sn}(\nu K'_2, k'_2) \end{aligned} \tag{12.171}$$

It can be shown that the following holds [Eq. (12.299)]:

$$\zeta = \frac{2}{(1+k')\sqrt{1+\epsilon^2} + \sqrt{(1-k')^2 + \epsilon^2(1+k')^2}} \tag{12.172}$$

Hence, we derive a particular expression for the function $\mathrm{SN}(2, k, \epsilon)$:

$$\mathrm{SN}(2, k, \epsilon) = \frac{2}{(1+k')\sqrt{1+\epsilon^2} + \sqrt{(1-k')^2 + \epsilon^2(1+k')^2}} \tag{12.173}$$

The function $\mathrm{SN}(2, k, \epsilon)$ is a function of ϵ and k.

From Eq. (12.166) we determine ϵ in terms of $\mathrm{sn}(\nu K'_2, k'_2)$:

$$\epsilon = \sqrt{\frac{1}{\mathrm{sn}^2(\nu K'_n, k'_n)} - 1} \tag{12.174}$$

Also, the parametric equations (12.164) can be expressed in the following form:

$$\begin{aligned}\sqrt{\frac{1}{\zeta^2}-1} &= \sqrt{\frac{1}{\text{sn}^2(\nu K', k')}-1} \\ \epsilon &= \sqrt{\frac{1}{\text{sn}^2(\nu K_n', k_n')}-1}\end{aligned} \tag{12.175}$$

From Eq. (12.166) we find

$$\sqrt{\frac{1}{\zeta^2}-1} = \sqrt{\frac{1}{\text{SN}^2(n, k, \epsilon)}-1} \tag{12.176}$$

We can assume that $\sqrt{1/\zeta^2 - 1}$ is a new function in terms of n, k, and ϵ. Therefore, we define a new function

$$\mathcal{E}(n, k, \epsilon) = \sqrt{\frac{1}{\text{SN}^2(n, k, \epsilon)}-1} \tag{12.177}$$

which is, according to Eqs. (12.169) and (12.173),

$$\mathcal{E}(1, k, \epsilon) = \epsilon \tag{12.178}$$

$$\mathcal{E}(2, k, \epsilon) = \sqrt{\frac{(1+k')\sqrt{1+\epsilon^2} + \sqrt{(1-k')^2 + \epsilon^2\,(1+k')^2}}{4} - 1} \tag{12.179}$$

In the next theorem we prove the nesting property for $\mathcal{E}(n, k, \epsilon)$, that is useful for determining ζ in terms of n, k, and ϵ.

Theorem 12.3 For a given ϵ, integers m, p and $n = mp$, and the modulus k the nesting property holds:

$$\mathcal{E}(n, k, \epsilon) = \mathcal{E}(m, k, \mathcal{E}(p, k_m, \epsilon)) = \mathcal{E}(p, k, \mathcal{E}(m, k_p, \epsilon)) \tag{12.180}$$

Proof: If we substitute Eqs. (12.166) and (12.177) into Eq. (12.175), the parametric equations become

$$\begin{aligned}\mathcal{E}(n, k, \epsilon) &= \sqrt{\frac{1}{\text{sn}^2(\nu K', k')}-1} \\ \epsilon &= \sqrt{\frac{1}{\text{sn}^2(\nu K_n', k_n')}-1}\end{aligned} \tag{12.181}$$

If we substitute ϵ with $\mathcal{E}(m, k_p, \epsilon)$, and n with p, Eq. (12.181) becomes

$$\begin{aligned}\mathcal{E}(p, k, \mathcal{E}(m, k_p, \epsilon)) &= \sqrt{\frac{1}{\text{sn}^2(\nu K', k')} - 1} \\ \mathcal{E}(m, k_p, \epsilon) &= \sqrt{\frac{1}{\text{sn}^2(\nu K'_p, k'_p)} - 1}\end{aligned} \tag{12.182}$$

while if we substitute k with k_p, and n with m, Eq. (12.181) becomes

$$\begin{aligned}\mathcal{E}(m, k_p, \epsilon) &= \sqrt{\frac{1}{\text{sn}^2(\nu K'_p, k'_p)} - 1} \\ \epsilon &= \sqrt{\frac{1}{\text{sn}^2(\nu K'_m(k_p), k'_m(k_p))} - 1}\end{aligned} \tag{12.183}$$

Since $k'_{mp} = k'_{mp}(k) = k'_m(k_p), k'_m = k'_m(k), k'_p = k'_p(k)$, and $K'_{mp}(k) = K'_m(k_p)$, and the second equation from Eq. (12.182) is equivalent to the first equation from Eq. (12.183), we can formulate new parametric equations:

$$\begin{aligned}\mathcal{E}(p, k, \mathcal{E}(m, k_p, \epsilon)) &= \sqrt{\frac{1}{\text{sn}^2(\nu K', k')} - 1} \\ \epsilon &= \sqrt{\frac{1}{\text{sn}^2(\nu K'_m(k_p), k'_m(k_p))} - 1} = \sqrt{\frac{1}{\text{sn}^2(\nu K'_n, k'_n)} - 1}\end{aligned} \tag{12.184}$$

The right-hand sides of the above equations are equivalent to Eq. (12.180), which proves the theorem. ■

Theorem 12.4 For a given ϵ, integers m, p, and $n = mp$, and the modulus k, the parametric equations

$$\begin{aligned}\zeta &= \text{sn}(\nu K', k') \\ \frac{1}{\sqrt{1 + \epsilon^2}} &= \text{sn}(\nu K'_n, k'_n)\end{aligned} \tag{12.185}$$

can be solved:

$$\zeta = \text{SN}(n, k, \epsilon) \tag{12.186}$$

as

$$\zeta = \text{SN}\left(p, k, \sqrt{\frac{1}{\text{SN}^2(m, k_p, \epsilon)} - 1}\right) \tag{12.187}$$

or

$$\zeta = \text{SN}\left(m, k, \sqrt{\frac{1}{\text{SN}^2(p, k_m, \epsilon)} - 1}\right) \tag{12.188}$$

Proof: If we substitute Eqs. (12.166) and (12.177) into Eq. (12.175), the parametric equations become

$$\begin{aligned} \zeta &= \mathrm{SN}(n,k,\epsilon) = \mathrm{sn}(\nu K',k') \\ \frac{1}{\sqrt{1+\epsilon^2}} &= \mathrm{sn}(\nu K'_n,k'_n) \end{aligned} \tag{12.189}$$

or

$$\begin{aligned} \mathrm{SN}(n,k,\epsilon) = \frac{1}{\sqrt{1+\mathcal{E}^2(n,k,\epsilon)}} &= \mathrm{sn}(\nu K',k') \\ \frac{1}{\sqrt{1+\epsilon^2}} &= \mathrm{sn}(\nu K'_n,k'_n) \end{aligned} \tag{12.190}$$

If we substitute ϵ with $\mathcal{E}(m,k_p,\epsilon)$ and substitute n with p, Eq. (12.181) becomes

$$\begin{aligned} \frac{1}{\sqrt{1+\mathcal{E}^2(p,k,\mathcal{E}(m,k_p,\epsilon))}} &= \mathrm{sn}(\nu\, K',k') \\ \frac{1}{\sqrt{1+\mathcal{E}^2(m,k_p,\epsilon)}} &= \mathrm{sn}(\nu\, K'_p,k'_p) \end{aligned} \tag{12.191}$$

while if we substitute k with k_p and substitute n with m, Eq. (12.181) becomes

$$\begin{aligned} \frac{1}{\sqrt{1+\mathcal{E}^2(m,k_p,\epsilon)}} &= \mathrm{sn}(\nu\, K'_p,k'_p) \\ \frac{1}{\sqrt{1+\epsilon^2}} &= \mathrm{sn}(\nu\, K'_m(k_p),k'_m(k_p)) \end{aligned} \tag{12.192}$$

Since $k'_{mp} = k'_{mp}(k) = k'_m(k_p)$, $k'_m = k'_m(k)$, $k'_p = k'_p(k)$, and $K'_{mp}(k) = K'_m(k_p)$, and the second equation from Eq. (12.182) is equivalent to the first equation from Eq. (12.183), we can formulate new parametric equations:

$$\begin{aligned} \mathrm{SN}(n,k,\epsilon) = \frac{1}{\sqrt{1+\mathcal{E}^2(n,k,\epsilon)}} = \frac{1}{\sqrt{1+\mathcal{E}^2(p,k,\mathcal{E}(m,k_p,\epsilon))}} &= \mathrm{sn}(\nu\, K',k') \\ \frac{1}{\sqrt{1+\epsilon^2}} = \mathrm{sn}(\nu\, K'_m(k_p),k'_m(k_p)) &= \mathrm{sn}(\nu\, K'_n,k'_n) \end{aligned} \tag{12.193}$$

The right-hand sides of the above equations are equal to the right-hand sides of Eq. (12.185). From Eq. (12.177) we obtain

$$\frac{1}{\sqrt{1+\mathcal{E}^2(p,k,\mathcal{E}(m,k_p,\epsilon))}} = \mathrm{SN}\left(p,k,\sqrt{\frac{1}{\mathrm{SN}^2(m,k_p,\epsilon)}-1}\right) \tag{12.194}$$

we prove the theorem. ■

The second-order Chebyshev polynomial is obtained from the trigonometric relation

$$\cos(2u) = 2(\cos u)^2 - 1 \tag{12.195}$$

We expect that the substitution of $\cos(2u)$ with $\mathrm{cd}(2u)$ and $\cos(u)$ with $\mathrm{cd}(u)$ yields the elliptic rational function. However, $\mathrm{cd}(2u, k_2)$ and $\mathrm{cd}(u, k)$ are functions of two arguments, and k_2 is a function of k, where k_2 and k are related by the degree equation.

In the next theorem we state the relation between $\mathrm{cd}(2u, k_2)$ and $\mathrm{cd}(u, k)$.

Theorem 12.5

$$\mathrm{cd}\left(\frac{2K(k_2)}{K(k)}\,u, k_2\right) = \frac{\left(\sqrt{1-k^2}+1\right)\ \mathrm{cd}^2(u,k)\ -\ 1}{\left(\sqrt{1-k^2}-1\right)\ \mathrm{cd}^2(u,k)\ +\ 1} \tag{12.196}$$

Proof: From the definition of the cd function, we obtain

$$\mathrm{cd}(u,k) = \frac{\mathrm{cn}(u,k)}{\mathrm{dn}(u,k)} \tag{12.197}$$

By using the Landen transformations of the cn and dn functions [109, Eqs. (16.14.3) and (16.14.4)], which are adapted to our notation, we obtain

$$\mathrm{cn}(v, k_2) = \frac{1+\sqrt{1-k^2}}{k^2}\ \frac{\mathrm{dn}^2(u,k)-\sqrt{1-k^2}}{\mathrm{dn}(u,k)} \tag{12.198}$$

and

$$\mathrm{dn}(v, k_2) = \frac{1-\sqrt{1-k^2}}{k^2}\ \frac{\mathrm{dn}^2(u,k)+\sqrt{1-k^2}}{\mathrm{dn}(u,k)} \tag{12.199}$$

Equation (12.197) becomes

$$\mathrm{cd}(v, k_2) = \frac{1+\sqrt{1-k^2}}{1-\sqrt{1-k^2}}\ \frac{\mathrm{dn}^2(u,k)-\sqrt{1-k^2}}{\mathrm{dn}^2(u,k)+\sqrt{1-k^2}} \tag{12.200}$$

We can eliminate cn from Eq. (12.197) using the known relation between the squares of the elliptic functions:

$$\mathrm{cd}^2(u,k) = \frac{\mathrm{cn}^2(u,k)}{\mathrm{dn}^2(u,k)} = \frac{\dfrac{1}{k^2}\mathrm{dn}^2 - \dfrac{k\prime^2}{k^2}}{\mathrm{dn}^2(u,k)} = \frac{1}{k^2} - \frac{1-k^2}{k^2\,\mathrm{dn}^2(u,k)} \tag{12.201}$$

Finally, substituting $\mathrm{dn}^2(u, k)$ from Eq. (12.201) into Eq. (12.200) gives

$$\mathrm{cd}(v, k_2) = \frac{\left(1+\sqrt{1-k^2}\right)\ \mathrm{cd}^2(u,k) - 1}{\left(1+\sqrt{1-k^2}\right)\ \mathrm{cd}^2(u,k) + 1} \tag{12.202}$$

It remains to determine the argument v in Eq. (12.202). It can be done by means of Landen‘s transformation, which reads as follows:

$$2K(k_2) = (1+k')\,K(k) \tag{12.203}$$

where

$$k_2 = \frac{1 - k'}{1 + k'} \tag{12.204}$$

To prove relation Eq. (12.203) we start with the defining expression for the complete elliptic integral:

$$K(k) = \int_0^{\pi/2} \frac{d\theta}{\sqrt{1 - k^2 \sin^2 \theta}} \tag{12.205}$$

which, after substituting $t = \tan\theta$, gives

$$K(k) = \int_0^{\infty} \frac{dt}{\sqrt{(1 + t^2)(1 + k'^2 t^2)}} \tag{12.206}$$

A new substitution

$$z = \frac{(1 + k')t}{1 - k't^2} \tag{12.207}$$

gives

$$dz = (1 + k')\frac{1 + k't^2}{(1 - k't^2)^2}\, dt \tag{12.208}$$

$$1 + z^2 = \frac{(1 + t^2)(1 + k'^2 t^2)}{(1 - k'^2 t^2)^2} \tag{12.209}$$

$$1 + {k_2'}^2 z^2 = (1 - k_2^2)z^2 = \left(\frac{1 + k't^2}{1 - k't^2}\right)^2 \tag{12.210}$$

Combining these expressions, one obtains

$$\frac{dz}{\sqrt{(1 + z^2)\,(1 + {k_2'}^2 z^2)}} = (1 + k')\, \frac{dt}{\sqrt{(1 + t^2)\,(1 + k'^2 t^2)}} \tag{12.211}$$

Finally,

$$\int_{-\infty}^{\infty} \frac{dz}{\sqrt{(1 + z^2)\,(1 + {k_2'}^2 z^2)}} = (1 + k') \int_0^{\infty} \frac{dt}{\sqrt{(1 + t^2)\,(1 + k'^2 t^2)}} \tag{12.212}$$

Equation (12.203) directly follows from Eq. (12.212).

By means of Eqs. (12.196) and (12.203), we determine the argument v as

$$v = \frac{2K(k_2)}{K(k)}\, u = (1 + k')u \tag{12.213}$$

Repeated application of the above procedure enables the formulation of a more general theorem in terms of k_n, $n = 2^i$, instead of k_2. First, we define a function

$$J(a,b) = \int_0^{\pi/2} \frac{d\theta}{\sqrt{a^2\cos^2\theta + b^2\sin^2\theta}} = \frac{1}{a}K\left(\sqrt{1-\frac{b^2}{a^2}}\right) \qquad (a > b > 0) \tag{12.214}$$

It could be proved that for any positive a_i and b_i satisfying

$$a_i = \frac{a_{i-1} + b_{i-1}}{2}, \qquad b_i = \sqrt{a_{i-1}\,b_{i-1}}, \qquad a_0 = a,\ b_0 = b \tag{12.215}$$

the following holds:

$$J(a,b) = J(a_i, b_i) \tag{12.216}$$

and we identify the relation

$$k_{2^i} = \sqrt{1 - \frac{b_i^2}{a_i^2}} \tag{12.217}$$

■

12.7.1 Nesting Property of Rational Elliptic Functions

By definition, the rational elliptic function of the order $n = 2$ can be expressed as a parametric set of equations:

$$\begin{aligned} R(2,\xi,x) &= \mathrm{cd}\left(\frac{2K_2}{K}\,w, \frac{1}{L(2,\xi)}\right) \\ x &= \mathrm{cd}\left(w, \frac{1}{\xi}\right) \end{aligned} \tag{12.218}$$

From Eq. (12.196) we obtain

$$\mathrm{cd}\left(\frac{2K_2}{K}\,u, k_2\right) = \frac{\left(\sqrt{1-k^2}+1\right)\ \mathrm{cd}^2(u,k)\ -\ 1}{\left(\sqrt{1-k^2}-1\right)\ \mathrm{cd}^2(u,k)\ +\ 1} \tag{12.219}$$

and for

$$k = \frac{1}{\xi} \tag{12.220}$$

$$k_2 = \frac{1}{L(2,\xi)} \tag{12.221}$$

we derive an explicit analytic formula for $R(2, \xi, x)$:

$$R(2, \xi, x) = \frac{\left(\sqrt{1 - \frac{1}{\xi^2}} + 1\right) x^2 - 1}{\left(\sqrt{1 - \frac{1}{\xi^2}} - 1\right) x^2 + 1} \tag{12.222}$$

Notice that the Jacobi elliptic functions do not appear in Eq. (12.222).

Similarly, substituting

$$k = \frac{1}{L(n, \xi)} \tag{12.223}$$

$$k_2 = \frac{1}{L(2n, \xi)} \tag{12.224}$$

into Eq. (12.219), we derive a closed-form expression for $R(2n, \xi, x)$:

$$R(2n, \xi, x) = \frac{\left(\sqrt{1 - \frac{1}{L^2(n, \xi)}} + 1\right) R^2(n, \xi, x) - 1}{\left(\sqrt{1 - \frac{1}{L^2(n, \xi)}} - 1\right) R^2(n, \xi, x) + 1} \tag{12.225}$$

which yields a closed-form expression for $L(2n, \xi)$, too:

$$L(2n, \xi) = R(2n, \xi, \xi) = \frac{\left(\sqrt{1 - \frac{1}{L^2(n, \xi)}} + 1\right) L^2(n, \xi) - 1}{\left(\sqrt{1 - \frac{1}{L^2(n, \xi)}} - 1\right) L^2(n, \xi) + 1} \tag{12.226}$$

Note that the first-order elliptic rational function reduces to

$$R(1, \xi, x) = x \tag{12.227}$$

and, also,

$$L(1, \xi) = R(1, \xi, \xi) = \xi$$

To demonstrate the nesting property of the elliptic rational function, we derive the fourth-order function as

$$\begin{aligned} R(4, \xi, x) &= R\left(2, R(2, \xi, \xi), R\left(2, \xi, x\right)\right) \\ &= \frac{(1 + k')\left(1 + \sqrt{k'}\right)^2 x^4 - 2(1 + k')\left(1 + \sqrt{k'}\right) x^2 + 1}{(1 + k')\left(1 - \sqrt{k'}\right)^2 x^4 - 2(1 + k')\left(1 - \sqrt{k'}\right) x^2 + 1} \end{aligned} \tag{12.228}$$

$$k' = \sqrt{1 - \frac{1}{\xi^2}}$$

The same expression can be obtained by using the Jacobi elliptic functions

$$\operatorname{cd}\left(\frac{4K(k_4)}{K(k)}\,u, k_4\right) \tag{12.229}$$

$$= \frac{(1+k')\left(1+\sqrt{k'}\right)^2 \operatorname{cd}^4(u,k) - 2(1+k')\left(1+\sqrt{k'}\right)\operatorname{cd}^2(u,k) + 1}{(1+k')\left(1-\sqrt{k'}\right)^2 \operatorname{cd}^4(u,k) - 2(1+k')\left(1-\sqrt{k'}\right)\operatorname{cd}^2(u,k) + 1} \tag{12.230}$$

where

$$\frac{4K(k_4)}{K(k)} = \left(1+\sqrt{1-k_2^2}\right)\left(1+\sqrt{1-k^2}\right) \tag{12.231}$$

with

$$k_2 = \left(\frac{1-k'}{k}\right)^2 \tag{12.232}$$

and

$$k_4 = \left(\frac{1-k_2'}{k_2}\right)^2 \tag{12.233}$$

12.8 POLES OF NORMALIZED TRANSFER FUNCTION

The elliptic rational function is an even or odd function with real coefficients. Therefore, $R^2(n,\xi,x)$ is an even function and can be expressed in terms of x^2, as a ratio of two even polynomials with real coefficients:

$$R^2(n,\xi,x) = \frac{N(x^2)}{D(x^2)} \tag{12.234}$$

From now on, let us consider an nth-order filter whose normalized squared magnitude response can be written in the form

$$|\mathcal{H}(s)|^2 = \frac{1}{1+\epsilon^2 R^2(n,\xi,x)} = \frac{D(x^2)}{D(x^2)+\epsilon^2 N(x^2)}, \qquad s^2 = -x^2 \tag{12.235}$$

where ϵ denotes the *ripple factor* ($\epsilon > 0$), and s represents the normalized complex frequency. The transfer function poles are the left half-plane roots of the equation

$$D(-s^2) + \epsilon^2 N(-s^2) = 0, \qquad s^2 = -x^2 \tag{12.236}$$

in which it is assumed that x^2 is replaced by $-s^2$ in the polynomials $N(x^2)$ and $D(x^2)$. Also, the poles of the transfer function are the left half-plane roots of the equation

$$1 + \epsilon^2 R^2(n,\xi,x) = 0, \qquad x = -js \tag{12.237}$$

where $j = \sqrt{-1}$.

We denote by s_i the values of s that satisfy Eq. (12.237), which we can rewrite as follows:

$$R^2(n,\xi,-js_i) = -\frac{1}{\epsilon^2} \tag{12.238}$$

Since ϵ^2 is a positive number, $R^2(n, \xi, -js_i)$ must have a negative value, i.e. $R(n, \xi, -js_i)$ must be a complex (purely imaginary) number

$$R(n, \xi, -js_i) = \pm j\frac{1}{\epsilon} \tag{12.239}$$

Substituting $R(n, \xi, -js_i)$ from Eq. (12.108) into Eq. (12.239), we find

$$\mathrm{sn}\left(nw_i\frac{K_n}{K} + K_n, \frac{1}{L_n}\right) = \pm j\frac{1}{\epsilon} \tag{12.240}$$

and for $w_i = u_i + jv_i$ we obtain

$$\mathrm{sn}\left(\left(nu_i\frac{K_n}{K} + K_n\right) + jnv_i\frac{K_n}{K}, \frac{1}{L_n}\right) = \pm j\frac{1}{\epsilon} \tag{12.241}$$

We split the left-hand side of Eq. (12.241) into a sum of two functions [109, Eq. (16.21.2); 32, Eq. (6.22)]

$$\mathrm{sn}\left(U_i + jV_i, \frac{1}{L_n}\right) = \frac{\mathrm{sn}\left(U_i, \frac{1}{L_n}\right)\mathrm{dn}\left(V_i, \frac{1}{L'_n}\right)}{\mathrm{cn}^2\left(V_i, \frac{1}{L'_n}\right) + \frac{1}{\xi^2}\mathrm{sn}^2\left(U_i, \frac{1}{L_n}\right)\mathrm{sn}^2\left(V_i, \frac{1}{L'_n}\right)}$$
$$+ j\frac{\mathrm{sn}\left(V_i, \frac{1}{L'_n}\right)\mathrm{cn}\left(V_i, \frac{1}{L'_n}\right)\mathrm{cn}\left(U_i, \frac{1}{L_n}\right)\mathrm{dn}\left(U_i, \frac{1}{L_n}\right)}{\mathrm{cn}^2\left(V_i, \frac{1}{L'_n}\right) + \frac{1}{\xi^2}\mathrm{sn}^2\left(U_i, \frac{1}{L_n}\right)\mathrm{sn}^2\left(V_i, \frac{1}{L'_n}\right)} \tag{12.242}$$

where

$$U_i = nu_i\frac{K_n}{K} + K_n \tag{12.243}$$

$$V_i = nv_i\frac{K_n}{K} \tag{12.244}$$

$$\frac{1}{L'_n} = \sqrt{1 - \frac{1}{L_n^2}} \tag{12.245}$$

The real part of the right-hand side of Eq. (12.241) is 0. Consequently, the real part of the left-hand side of Eq. (12.242) is zero, too:

$$\frac{\mathrm{sn}\left(nu_i\frac{K_n}{K} + K_n, \frac{1}{L_n}\right)\mathrm{dn}\left(nv_i\frac{K_n}{K}, \frac{1}{L'_n}\right)}{\mathrm{cn}^2\left(nv_i\frac{K_n}{K}, \frac{1}{L'_n}\right) + \frac{1}{\xi^2}\mathrm{sn}^2\left(nu_i\frac{K_n}{K} + K_n, \frac{1}{L_n}\right)\mathrm{sn}^2\left(nv_i\frac{K_n}{K}, \frac{1}{L'_n}\right)} = 0 \tag{12.246}$$

The Jacobi elliptic function dn is always positive:

$$\operatorname{dn}\left(nv_i\frac{K_n}{K}, \frac{1}{L'_n}\right) > 0 \qquad \text{for} \quad \frac{1}{L'_n} < 1 \tag{12.247}$$

so

$$\operatorname{sn}\left(nu_i\frac{K_n}{K} + K_n, \frac{1}{L_n}\right) = 0 \tag{12.248}$$

Because the Jacobi elliptic sn function equals zero, we obtain

$$\operatorname{cn}\left(nu_i\frac{K_n}{K} + K_n, \frac{1}{L_n}\right) = 1 \tag{12.249}$$

$$\operatorname{dn}\left(nu_i\frac{K_n}{K} + K_n, \frac{1}{L_n}\right) = 1 \tag{12.250}$$

Now, we can simplify the imaginary part of Eq. (12.242). After Eqs. (12.248)–(12.250) are substituted into Eq. (12.242), Eq. (12.241) becomes

$$j\frac{\operatorname{sn}\left(nv_i\frac{K_n}{K}, \frac{1}{L'_n}\right)}{\operatorname{cn}\left(nv_i\frac{K_n}{K}, \frac{1}{L'_n}\right)} = \pm j\frac{1}{\epsilon} \tag{12.251}$$

By using the identity $\operatorname{cn}^2(w, k) = 1 - \operatorname{sn}^2(w, k)$ and after squaring both sides of Eq. (12.251), we find

$$\frac{\operatorname{sn}^2\left(nv_i\frac{K_n}{K}, \frac{1}{L'_n}\right)}{1 - \operatorname{sn}^2\left(nv_i\frac{K_n}{K}, \frac{1}{L'_n}\right)} = \frac{1}{\epsilon^2} \tag{12.252}$$

or

$$\operatorname{sn}\left(nv_i\frac{K_n}{K}, \frac{1}{L'_n}\right) = \frac{1}{\sqrt{1+\epsilon^2}} \tag{12.253}$$

Equation (12.253) shows that v_i depends only on n, ξ, and ϵ (K, K_n, and L'_n are functions of n and ξ). Therefore, v_i has the same value for all i, and we can omit the subscript

$$v_i = v \tag{12.254}$$

We can solve Eq. (12.248) to find u_i. Note that u_i is not a function of ϵ. This means that ϵ has no influence on the real part of poles.

The condition given by Eq. (12.248), for $k_n = 1/L_n$ and $k = 1/\xi$, simplifies to zeros of the elliptic rational function $R(n, \xi, x)$:

$$\operatorname{sn}\left(nu_i\frac{K_n}{K} + K_n, \frac{1}{L_n}\right) = \operatorname{cd}\left(nw\frac{K_n}{K}, k_n\right) = R(n, \xi, x) = 0 \tag{12.255}$$

yielding

$$x_i = \operatorname{sn}\left(\frac{2i-1}{n}K + K, \frac{1}{\xi}\right), \qquad i = 1, 2, \ldots, n \tag{12.256}$$

with

$$u_i = \frac{2i-1}{n}K, \qquad i = 1, 2, \ldots, n \tag{12.257}$$

The poles of the normalized transfer function, $\mathcal{H}(s)$, are

$$s_i = j\mathrm{sn}\left(u_i + K + jv, \frac{1}{\xi}\right) \tag{12.258}$$

Proceeding as described when obtaining Eq. (12.242), Eq. (12.258) can be rewritten as follows:

$$\mathrm{sn}\left((u_i + K) + jv, \frac{1}{\xi}\right) = \frac{\mathrm{sn}\left(u_i + K, \frac{1}{\xi}\right)\mathrm{dn}\left(v, \frac{1}{\xi'}\right)}{\mathrm{cn}^2\left(v, \frac{1}{\xi'}\right) + \frac{1}{\xi^2}\mathrm{sn}^2\left(u_i + K, \frac{1}{\xi}\right)\mathrm{sn}^2\left(v, \frac{1}{\xi'}\right)} + j\frac{\mathrm{sn}\left(v, \frac{1}{\xi'}\right)\mathrm{cn}\left(v, \frac{1}{\xi'}\right)\mathrm{cn}\left(u_i + K, \frac{1}{\xi}\right)\mathrm{dn}\left(u_i + K, \frac{1}{\xi}\right)}{\mathrm{cn}^2\left(v, \frac{1}{\xi'}\right) + \frac{1}{\xi^2}\mathrm{sn}^2\left(u_i + K, \frac{1}{\xi}\right)\mathrm{sn}^2\left(v, \frac{1}{\xi'}\right)} \tag{12.259}$$

where

$$\frac{1}{\xi'} = \sqrt{1 - \frac{1}{\xi^2}} \tag{12.260}$$

Next, we express $\mathrm{sn}(u_i + K, 1/\xi)$, $\mathrm{cn}(u_i + K, 1/\xi)$ and $\mathrm{dn}(u_i + K, 1/\xi)$ in terms of x_i and ξ. Combining Eq. (12.256) with the relations between sn, cn, and dn Jacobi functions, we obtain

$$\mathrm{cn}\left(\frac{2i-1}{n}K + K, \frac{1}{\xi}\right) = \sqrt{1 - \mathrm{sn}^2\left(\frac{2i-1}{n}K + K, \frac{1}{\xi}\right)} = \sqrt{1 - x_i^2} \tag{12.261}$$

$$\mathrm{dn}\left(\frac{2i-1}{n}K + K, \frac{1}{\xi}\right) = \sqrt{1 - \frac{1}{\xi^2}\mathrm{sn}^2\left(\frac{2i-1}{n}K + K, \frac{1}{\xi}\right)} = \sqrt{1 - \frac{x_i^2}{\xi^2}} \tag{12.262}$$

Equation (12.259) becomes

$$\mathrm{sn}\left(\left(\frac{2i-1}{n}K + K\right) + jv, \frac{1}{\xi}\right) = \frac{x_i\mathrm{dn}\left(v, \frac{1}{\xi'}\right) + j\mathrm{sn}\left(v, \frac{1}{\xi'}\right)\mathrm{cn}\left(v, \frac{1}{\xi'}\right)\sqrt{1 - x_i^2}\sqrt{1 - \frac{x_i^2}{\xi^2}}}{\mathrm{cn}^2\left(v, \frac{1}{\xi'}\right) + \frac{x_i^2}{\xi^2}\mathrm{sn}^2\left(v, \frac{1}{\xi'}\right)} \tag{12.263}$$

The Jacobi elliptic functions $\mathrm{cn}(v, 1/\xi')$ and $\mathrm{dn}(v, 1/\xi')$ can be also expressed in terms of $\mathrm{sn}(v, 1/\xi')$. The second argument in those Jacobi functions is $1/\xi'$, from Eq. (12.260), while the second argument in Eq. (12.262) is $1/\xi$. For the sake of simplicity we denote $\mathrm{sn}(v, 1/\xi')$ as ζ:

$$\zeta = \mathrm{sn}\left(v, \frac{1}{\xi'}\right) \tag{12.264}$$

By expressing the Jacobi functions with argument v in terms of ζ, we find that Eq. (12.263) becomes

$$\mathrm{sn}\left(\frac{2i-1}{n}K + K + jv, \frac{1}{\xi}\right) = \frac{x_i\sqrt{1-\left(1-\dfrac{1}{\xi^2}\right)\zeta^2} + j\zeta\sqrt{1-\zeta^2}\sqrt{1-x_i^2}\sqrt{1-\dfrac{x_i^2}{\xi^2}}}{1-\zeta^2+\dfrac{x_i^2}{\xi^2}\zeta^2} \tag{12.265}$$

Substituting Eq. (12.265) into Eq. (12.258), we find the transfer function poles

$$s_i = \frac{-\zeta\sqrt{1-\zeta^2}\sqrt{1-x_i^2}\sqrt{1-\dfrac{x_i^2}{\xi^2}} + j\,x_i\sqrt{1-\left(1-\dfrac{1}{\xi^2}\right)\zeta^2}}{1-\left(1-\dfrac{x_i^2}{\xi^2}\right)\zeta^2}, \qquad i = 1, 2, \ldots, n \tag{12.266}$$

It is interesting to consider two special cases of Eq. (12.266): $\epsilon = 0$ and $\epsilon \to +\infty$. If $\epsilon \to +\infty$, Eq. (12.254) becomes

$$\mathrm{sn}\left(\frac{nvK_n}{K}, \frac{1}{L_n'}\right) = 0 \tag{12.267}$$

that is,

$$v = 0 \tag{12.268}$$

Substituting $v = 0$ into Eq. (12.264), we obtain

$$\zeta = 0 \tag{12.269}$$

With $\zeta = 0$ in Eq. (12.266) the poles become

$$s_i = jx_i, \quad i = 1, 2, \ldots, n \tag{12.270}$$

that is, the poles are equal to the zeros of the elliptic rational function multiplied by j.

In the second case, by formal substitution $\epsilon = 0$, from Eq. (12.254) we find

$$\mathrm{sn}\left(nv\frac{K_n}{K}, \frac{1}{L_n'}\right) = 1 \tag{12.271}$$

and

$$nv\frac{K_n}{K} = K'_n \tag{12.272}$$

According to the degree equation we obtain

$$n\frac{K_n}{K} = \frac{K'_n}{K'} \tag{12.273}$$

so we compute v:

$$v = K' \tag{12.274}$$

After substituting v into Eq. (12.264) we find

$$\zeta = \text{sn}\left(K', \frac{1}{\xi'}\right) = 1 \tag{12.275}$$

Substituting $\zeta = 1$ into Eq. (12.266), the transfer function poles are

$$s_i = j\frac{\xi}{x_i}, \qquad i = 1, 2, \ldots, n \tag{12.276}$$

and are equal to the elliptic rational function poles multiplied by j.

Another important case is when $\xi \to \infty$, that is, when the elliptic rational function becomes the Chebyshev polynomial. Equation (12.266) gives the poles of the Chebyshev-type transfer function.

If $\xi \to \infty$, then $K = \pi/2$, and the Jacobi elliptic sine function becomes the trigonometric sine function; thus,

$$x_i = \sin\left(\frac{2i-1}{2n}\pi + \frac{\pi}{2}\right), \quad i = 1, 2, \ldots, n \tag{12.277}$$

which are the zeros of the Chebyshev polynomial.

If $\xi \to \infty$, then from Eq. (12.260) we have $\xi' = 1$, 1and from Eq. (12.264) and known relations between the Jacobi functions [109, Eq. (16.6.1)] we obtain

$$\zeta = \text{sn}(v, 1) = \tanh v = \frac{\sinh v}{\cosh v} \tag{12.278}$$

Substituting $1/\xi = 0$ in Eq. (12.266), we obtain

$$s_i = \frac{-\zeta\sqrt{1-\zeta^2}\sqrt{1-x_i^2} + jx_i\sqrt{1-\zeta^2}}{1-\zeta^2}, \qquad i = 1, 2, \ldots, n \tag{12.279}$$

or

$$s_i = \frac{-\zeta\sqrt{1-x_i^2} + jx_i}{\sqrt{1-\zeta^2}}, \qquad i = 1, 2, \ldots, n \tag{12.280}$$

and substituting Eq. (12.278) into Eq. (12.280), we find the poles of the Chebyshev-type transfer function

$$s_i = -\sinh v\sqrt{1-x_i^2} + jx_i\cosh v, \qquad i = 1, 2, \ldots, n \tag{12.281}$$

According to Eq. (12.277) we find

$$s_i = -\sinh v \left(-\cos\left(\frac{2i-1}{2n}\pi + \frac{\pi}{2}\right)\right) + j\sin\left(\frac{2i-1}{2n}\pi + \frac{\pi}{2}\right)\cosh v \quad (12.282)$$

Since $\cos(\alpha + \pi/2) = -\sin(\alpha)$ the poles are

$$s_i = \sinh v \sin\left(\frac{2i-1}{2n}\pi\right) + j\cos\left(\frac{2i-1}{2n}\pi\right)\cosh v, \qquad i = 1, 2, \ldots, n \quad (12.283)$$

We can conclude that Eq. (12.266) gives the transfer function poles including the poles of the Chebyshev-type transfer function.

The poles and zeros of the elliptic transfer function are derived in closedform in terms of the Jacobi elliptic functions. In *Mathematica* the Jacobi elliptic functions are standard built-in functions; therefore, the formulas for the transfer function poles and zeros can be directly programmed. However, we can use these formulas only in purely numerical computation.

Sometimes, we prefer to manipulate the elliptic rational function, or the transfer function, in a symbolic way. In that case we try to find closed-form expressions, exact or approximate, free from the Jacobi elliptic functions.

12.8.1 Exact Formulas for Transfer Function Poles

Let us derive explicit exact formulas for the transfer function poles s_i, in terms of the order n, the selectivity factor ξ, and the ripple factor ϵ. Our objective is to find analytic expressions free from the Jacobi elliptic functions.

First, s_i are computed from Eq. (12.266) in terms of the selectivity factor ξ, the zeros of the elliptic rational function x_i, and ζ, that is, $s_i = f_i(\xi, x_i, \zeta)$, for a given n. Second, s_i is obtained as a solution of the equation $1 + \epsilon^2 R^2(n, \xi, -js) = 0$, in terms of ξ, x_i, and the ripple factor ϵ, that is, $s_i = g_i(\xi, x_i, \epsilon)$. Equating the two functions representing the poles, $f_i(\xi, x_i, \zeta) = g_i(\xi, x_i, \epsilon)$, we formulate an equation from which we want to derive ζ in terms of ξ and ϵ.

This derivation of ζ can be simply performed for lower orders $n = 1, 2, 3$. For higher orders we use the nesting property.

In the simplest case of the first-order elliptic rational function, $n = 1$, we obtain

$$R(1, \xi, x) = x \quad (12.284)$$

and it has only one zero:

$$x_1 = 0 \quad (12.285)$$

The equation $1 + \epsilon^2 R^2(1, \xi, -js) = 0$ simplifies to

$$1 - \epsilon^2 s^2 = 0 \quad (12.286)$$

yielding

$$s_1 = -\frac{1}{\epsilon} \quad (12.287)$$

Alternatively, from Eq. (12.266) with $x_1 = 0$ it follows

$$s_1 = -\frac{\zeta\sqrt{1-\zeta^2}}{1-\zeta^2} \tag{12.288}$$

By equating Eqs. (12.287) and (12.288)

$$-\frac{1}{\epsilon} = -\frac{\zeta\sqrt{1-\zeta^2}}{1-\zeta^2} \tag{12.289}$$

we determine ζ in terms of ϵ:

$$\zeta = \frac{1}{\sqrt{1+\epsilon^2}} \tag{12.290}$$

The same result has already been obtained from the parametric equations (12.168).

For the second-order elliptic rational function, $n = 2$, we obtain

$$R(2, \xi, x) = \frac{(1+k')x^2 - 1}{(-1+k')x^2 + 1}, \qquad k' = \sqrt{1 - \frac{1}{\xi^2}} \tag{12.291}$$

the zeros x_i are

$$x_1 = -\xi\sqrt{1-k'}, \qquad x_2 = \xi\sqrt{1-k'} \tag{12.292}$$

The selectivity factor ξ can be expressed in terms of k':

$$\xi = \frac{1}{\sqrt{1-(k')^2}} \tag{12.293}$$

The solution of equation $1 + \epsilon^2 R^2(2, \xi, -js) = 0$ is

$$s_{1,2} = -\sqrt{\frac{-(1-k') - \epsilon^2(1+k') + \sqrt{1+\epsilon^2}\sqrt{(1-k')^2 + \epsilon^2(1+k')^2}}{2\left((1-k')^2 + \epsilon^2(1+k')^2\right)}}$$
$$\pm j\sqrt{\frac{(1-k') + \epsilon^2(1+k') + \sqrt{1+\epsilon^2}\sqrt{(1-k')^2 + \epsilon^2(1+k')^2}}{2\left((1-k')^2 + \epsilon^2(1+k')^2\right)}} \tag{12.294}$$

The squared magnitudes of the poles are

$$|s_i|^2 = \frac{\sqrt{1+\epsilon^2}}{\sqrt{(1-k')^2 + \epsilon^2(1+k')^2}}, \qquad i = 1, 2 \tag{12.295}$$

On the other hand, from Eq. (12.266), with x_i determined by Eq. (12.292), the poles are

$$s_{1,2} = \frac{-k'\zeta\sqrt{1-\zeta^2} \pm j\sqrt{1-(k')^2\zeta^2}}{(1-k'\zeta^2)\sqrt{1+k'}} \tag{12.296}$$

and

$$|s_i|^2 = \frac{1+k'\zeta^2}{(1+k')(1-k'\zeta^2)} \tag{12.297}$$

Equating the right-hand sides of Eq. (12.295) and Eq. (12.297)

$$\frac{\sqrt{1+\epsilon^2}}{\sqrt{(1-k')^2+\epsilon^2(1+k')^2}} = \frac{1+k'\zeta^2}{(1+k')(1-k'\zeta^2)} \tag{12.298}$$

we calculate ζ as a function of ϵ and k'—that is, in terms of ϵ and ξ:

$$\zeta = \frac{2}{(1+k')\sqrt{1+\epsilon^2}+\sqrt{(1-k')^2+\epsilon^2(1+k')^2}}, \qquad k' = \sqrt{1-\frac{1}{\xi^2}} \tag{12.299}$$

We carry out the same procedure for $n = 3$ and find ζ in terms of ξ and ϵ:

$$\begin{aligned}
b &= \sqrt[3]{1-\left(1-\frac{2}{\xi^2}\right)^2} \\
c &= \sqrt{1+b+b^2} \\
r &= \frac{1}{2}\left(\frac{1}{c}\left(1-\frac{2}{\xi^2}\right)+\sqrt{2+b+2c}-1\right) \\
b_1 &= 9\epsilon^2(1+r+r^2)+(1-r)^3 \\
a_1 &= 3\epsilon\sqrt{3(1+\epsilon^2)(\epsilon^2(1+2r)^3+(1+r)(1-r)^3)} \\
E &= 3\frac{\epsilon(1+r)}{\sqrt[3]{a_1+b_1}-\sqrt[3]{a_1-b_1}+1-r} = E(\xi,\epsilon) \\
\zeta &= \frac{1}{\sqrt{1+E(\xi,\epsilon)^2}}
\end{aligned} \tag{12.300}$$

Equations (12.299), (12.290), and (12.300) show that for $n = 1, 2, 3$ we can express ζ in terms of the ripple factor ϵ and the selectivity factor ξ without resorting to the Jacobi elliptic functions. Formally, we can represent this result in the functional form:

$$\begin{aligned}
\zeta &= \mathrm{SN}(1,k,\epsilon) = \frac{1}{\sqrt{1+\epsilon^2}}, \qquad k = \frac{1}{\xi}, \qquad n = 1 \\
\zeta &= \mathrm{SN}(2,k,\epsilon) = \frac{2}{(1+k')\sqrt{1+\epsilon^2}+\sqrt{(1-k')^2+\epsilon^2(1+k')^2}}, \\
&\qquad k = \frac{1}{\xi}, \qquad k' = \sqrt{1-k^2}, \qquad n = 2 \\
\zeta &= \mathrm{SN}(3,k,\epsilon) = \frac{1}{\sqrt{1+E(\xi,\epsilon)^2}}, \qquad k = \frac{1}{\xi}, \qquad n = 3
\end{aligned} \tag{12.301}$$

Consider a more general case $n > 3$. By definition, ζ has been introduced by the parametric equations

$$\zeta = \mathrm{sn}(vK', k')$$
$$\frac{1}{\sqrt{1+\epsilon^2}} = \mathrm{sn}\left(vK'_n, \sqrt{1-\frac{1}{L_n^2}}\right) \tag{12.302}$$

where $k = 1/\xi$, $k_n = 1/L_n$, $L_n = L(n, \xi)$, and v is the parametric variable. The solution of Eq. (12.302) is designated by

$$\zeta = \mathrm{SN}(n, k, \epsilon) \tag{12.303}$$

According to Eq. (12.166), for arbitrary orders m and p, we can use the following relation:

$$\mathrm{SN}(mp, k, \epsilon) = \mathrm{SN}\left(m, k, \sqrt{\frac{1}{\mathrm{SN}^2(p, k_m, \epsilon)} - 1}\right) \tag{12.304}$$

For example, let us consider the order $n = 4$. Choosing $m = 2$ and $p = 2$ (i.e., $mp = 4$) and with $k = 1/\xi$, $k_2 = 1/L(2, \xi)$, ζ becomes

$$\zeta = \mathrm{SN}(4, k, \epsilon) = \mathrm{SN}\left(2, \frac{1}{\xi}, \sqrt{\frac{1}{\mathrm{SN}^2\left(2, \frac{1}{L(2,\xi)}, \epsilon\right)} - 1}\right), \qquad n = 4 \tag{12.305}$$

Consider another example of a high-order elliptic rational function, $n = 18$, and derive $\zeta = \mathrm{SN}(18, k, \epsilon)$. The procedure is as follows: First, for $m = 2$ and $p = 9$ we obtain

$$\mathrm{SN}(18, k, \epsilon) = \mathrm{SN}\left(2, k, \sqrt{\frac{1}{\mathrm{SN}^2(9, k_2, \epsilon)} - 1}\right) \tag{12.306}$$

Next, for $m = 3$, $p = 3$, and $k_6(k) = k_3(k_2)$ we have

$$\mathrm{SN}(9, k_2, \epsilon) = \mathrm{SN}\left(3, k_2, \sqrt{\frac{1}{\mathrm{SN}^2(3, k_6, \epsilon)} - 1}\right) \tag{12.307}$$

Finally,

$$\zeta = \mathrm{SN}(18, k, \epsilon) = \mathrm{SN}\left(2, \frac{1}{\xi}, \sqrt{\frac{1}{\mathrm{SN}^2\left(3, k_2, \sqrt{\frac{1}{\mathrm{SN}^2(3, k_6, \epsilon)} - 1}\right)} - 1}\right) \tag{12.308}$$

Notice that k_2 and k_6 are uniquely determined by $k = 1/\xi$.

12.9 EXACT FORMULAS FOR ZEROS OF ELLIPTIC RATIONAL FUNCTIONS

Our goal in this section is to derive explicit exact formulas for the zeros, x_i, of the elliptic rational function, in terms of the order n and the selectivity factor ξ. We seek a solution that is free from the Jacobi elliptic functions.

We start from Eq. (12.84):

$$R(mp, \xi, x) = R(m, R(p, \xi, \xi), R(p, \xi, x)) \tag{12.309}$$

which holds for arbitrary orders m and p. For $p = 2$ we have

$$R(2m, \xi, x) = R(m, R(2, \xi, \xi), R(2, \xi, x)) \tag{12.310}$$

and with the short notation

$$\begin{aligned} X_{(2)}(x) &= R(2, \xi, x) \\ \Xi_2 &= R(2, \xi, \xi) \end{aligned} \tag{12.311}$$

the nth-order function, $n = 2m$, becomes

$$R(n, \xi, x) = R(m, \Xi_2, X_{(2)}(x)) \tag{12.312}$$

For $x = x_i$ $(i = 1, 2, \ldots, n)$ we obtain

$$R(n, \xi, x_i) = 0$$

and, due to Eq. (12.309), we have

$$R(n, \xi, x_i) = R(m, \Xi_2, X_{(2)}(x_i)) = 0 \tag{12.313}$$

For the sake of simplicity we designate $X_{(2)}(x_i)$ by $X_{(2)I}$:

$$X_{(2)I} = X_{(2)}(x_i) = R(2, \xi, x_i) \tag{12.314}$$

Since

$$R(2, \xi, x) = \frac{(t+1)x^2 - 1}{(t-1)x^2 + 1}, \qquad t = \sqrt{1 - \frac{1}{\xi^2}} \tag{12.315}$$

one has

$$X_{(2)I} = \frac{(t+1)x_i^2 - 1}{(t-1)x_i^2 + 1} \tag{12.316}$$

or

$$x_I = \frac{1}{\sqrt{1 + t\dfrac{1 - X_{(2)I}}{1 + X_{(2)I}}}}, \qquad I = 1, \ldots, m \tag{12.317}$$

and for negative values one has

$$x_{2m+1-I} = -\frac{1}{\sqrt{1 + t\dfrac{1 - X_{(2)I}}{1 + X_{(2)I}}}}, \qquad I = 1, \ldots, m \tag{12.318}$$

To explain the application of the above formulas we solve the following example. Find the zeros of the eighth-order elliptic rational function for $\xi = \sqrt{2}$:

$$\begin{aligned} R(8, \xi, x) &= R(1, \Xi_8, R(8, \xi, x)) = R(1, \Xi_8, R(2, \Xi_4, R(4, \xi, x))) \\ &= R(1, \Xi_8, R(2, \Xi_4, R(2, \Xi_2, R(2, \xi, x)))) \end{aligned} \tag{12.319}$$

where

$$R(8, \xi, x) = R(1, \Xi_8, X_{(8)}(x))$$

Since

$$R(1, \xi, x) = x \tag{12.320}$$

we have

$$\begin{aligned} X_{(8)}(x) &= R(2, \Xi_4, X_{(4)}(x)) \\ X_{(4)}(x) &= R(2, \Xi_2, X_{(2)}(x)) \\ X_{(2)}(x) &= R(2, \xi, x) \end{aligned}$$

while

$$\begin{aligned} \Xi_2 &= R(2, \xi, \xi) \\ \Xi_4 &= R(2, \Xi_2, \Xi_2) \\ \Xi_8 &= R(2, \Xi_4, \Xi_4) \end{aligned} \tag{12.321}$$

From Eq. (12.315) we obtain

$$R(2, \xi, \xi) = \left(\xi + \sqrt{\xi^2 - 1}\right)^2 \tag{12.322}$$

According to Eq. (12.321) we find

$$\Xi_2 = \left(\xi + \sqrt{\xi^2 - 1}\right)^2 = 5.828427$$

$$\Xi_4 = \left(\Xi_2 + \sqrt{\Xi_2^2 - 1}\right)^2 = 133.874781$$

Corresponding values of the parameter t are

$$t = \sqrt{1 - \frac{1}{\xi^2}} = 0.707107$$

$$t_2 = \sqrt{1 - \frac{1}{\Xi_2^2}} = 0.985171$$

$$t_4 = \sqrt{1 - \frac{1}{\Xi_4^2}} = 0.999972$$

The next step is finding the zeros. Starting with the first-order elliptic rational function, $R(1, \xi, x) = X_{(1)}(x)$, with one zero at the origin, $X_{(1)1} = 0$, we can calculate zeros of higher-order functions. From Eqs. (12.317) and (12.318) we can determine the zeros $X_{(2)I}$ of $R(2, \Xi_4, X_{(2)}(x))$:

$$X_{(2)1} = \frac{1}{\sqrt{1 + t_4 \dfrac{1 - X_{(1)1}}{1 + X_{(1)1}}}} = 0.707112$$

$$X_{(2)2} = -X_{(2)1} = -0.707112$$

Similarly, we calculate the zeros of the function $R(2, \Xi_2, X_{(4)}(x))$:

$$X_{(4)I} = \frac{1}{\sqrt{1 + t_2 \dfrac{1 - X_{(2)I}}{1 + X_{(2)I}}}}, \qquad I = 1, 2$$

$$X_{(4)1} = 0.924886$$

$$X_{(4)2} = 0.385125$$

$$X_{(4)3} = -0.385125$$

$$X_{(4)4} = -0.924886$$

The zeros of $R(2, \xi, x)$ can be obtained as follows:

$$x_I = \frac{1}{\sqrt{1 + t \dfrac{1 - X_{(4)I}}{1 + X_{(4)I}}}}, \qquad I = 1, 2, 3, 4$$

$$x_1 = 0.986482$$

$$x_2 = 0.872408$$

$$x_3 = 0.621023$$

$$x_4 = 0.228692$$

$$x_5 = -0.228692$$

$$x_6 = -0.621023$$

$$x_7 = -0.872408$$

$$x_8 = -0.986482$$

It is useful to emphasize that all zeros are determined by a single expression in which no Jacobi elliptic function appears.

Since we know how to derive analytically the zeros of the third-order elliptic rational function, the above procedure is applicable to all elliptic rational functions of the order $n = 2^i 3^j$ $(i, j = 0, 1, 2, \ldots)$. A graphical presentation of our approach is shown in Fig. 12.25.

12.9.1 Approximate Determination of Zeros

The zeros of the rational elliptic function are

$$x_i = \operatorname{sn}\left(u_i, \frac{1}{\xi}\right), \qquad i = 1, 2, \ldots, n \tag{12.323}$$

where

$$u_i = \left(\frac{2i-1}{n} + 1\right) K, \qquad i = 1, 2, \ldots, n \tag{12.324}$$

The series representation of the Jacobi elliptic sine function is given by

$$\operatorname{sn}(u, k) = \frac{1}{\sqrt{k}} \frac{\theta_1\left(\frac{u}{2K}, q\right)}{\theta_0\left(\frac{u}{2K}, q\right)} \tag{12.325}$$

where

$$\theta_0\left(\frac{u}{2K}, q\right) = 1 + 2\sum_{m=1}^{\infty}(-1)^m q^{m^2} \cos\left(2m\frac{\pi u}{2K}\right) \tag{12.326}$$

$$\theta_1\left(\frac{u}{2K}, q\right) = 2q^{1/4}\sum_{m=0}^{\infty}(-1)^m q^{m(m+1)} \sin\left((2m+1)\frac{\pi u}{2K}\right) \tag{12.327}$$

Substituting the selectivity factor ξ

$$k = \frac{1}{\xi} \tag{12.328}$$

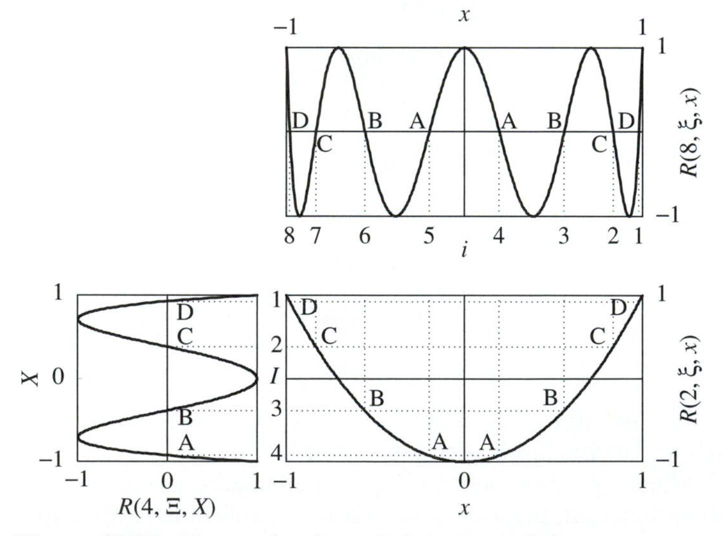

Figure 12.25 Determination of the zeros of the eight-order rational function using the nesting property.

into Eq. (12.325), we obtain

$$\operatorname{sn}\left(u, \frac{1}{\xi}\right) = 2\sqrt{\xi\sqrt{q}}\,\frac{\sum_{m=0}^{\infty}(-1)^m q^{m(m+1)} \sin\left((2m+1)\frac{\pi u}{2K}\right)}{1 + 2\sum_{m=1}^{\infty}(-1)^m q^{m^2} \cos\left(2m\frac{\pi u}{2K}\right)} \tag{12.329}$$

After Eq. (12.324) is substituted into Eq. (12.329), we find

$$\operatorname{sn}\left(u_i, \frac{1}{\xi}\right) = 2\sqrt{\xi\sqrt{q}}\,\frac{\displaystyle\sum_{m=0}^{\infty}(-1)^m q^{m(m+1)} \sin\frac{(2m+1)(n+2i-1)\pi}{2n}}{\displaystyle 1 + 2\sum_{m=1}^{\infty}(-1)^m q^{m^2} \cos\frac{m(n+2i-1)\pi}{n}} \tag{12.330}$$

The expanded form of Eq. (12.330) is

$$x_i = 2\sqrt{\xi\sqrt{q}}\,\frac{\begin{gathered}\sin\frac{(n+2i-1)\pi}{2n} - q^2\sin\frac{3(n+2i-1)\pi}{2n} \\ + q^6\sin\frac{5(n+2i-1)\pi}{2n} - q^{12}\sin\frac{7(n+2i-1)\pi}{2n} \\ + q^{20}\sin\frac{9(n+2i-1)\pi}{2n} - q^{30}\sin\frac{11(n+2i-1)\pi}{2n} \\ + q^{42}\sin\frac{13(n+2i-1)\pi}{2n} + \cdots\end{gathered}}{\begin{gathered}1 - 2q\cos\frac{(n+2i-1)\pi}{n} \\ + 2q^4\cos\frac{2(n+2i-1)\pi}{n} - 2q^9\cos\frac{3(n+2i-1)\pi}{n} \\ + 2q^{16}\cos\frac{4(n+2i-1)\pi}{n} - 2q^{25}\cos\frac{5(n+2i-1)\pi}{n} \\ + 2q^{36}\cos\frac{6(n+2i-1)\pi}{n} - 2q^{49}\cos\frac{7(n+2i-1)\pi}{n} + \cdots\end{gathered}} \tag{12.331}$$

where q is a function of ξ.

12.10 THE DEGREE EQUATION

In the design of elliptic filters the first step is determining the minimal filter order—that is, the minimal degree n of the elliptic rational function $R(n, \xi, x)$. In this section, we consider a specification of a lowpass filter, $S_A = \{F_p, F_s, A_p, A_s\}$, where F_p denotes passband edge frequency, F_s is stopband edge frequency, A_p is maximal passband attenuation in dB, and A_s is minimal stopband attenuation in dB.

The elliptic transfer function depends on three parameters: n, ϵ, and ξ. Traditionally, filter designers choose

$$\xi \le \frac{F_s}{F_p} \tag{12.332}$$

and the following conditions must be met:

$$A_p \ge 10\log_{10}(1+\epsilon^2) \tag{12.333}$$

$$A_s \le 10\log_{10}\left(1+\epsilon^2 L(n,\xi)\right) \tag{12.334}$$

where $L(n,\xi) = R(n,\xi,\xi)$.

From the above set of inequations we have to find the minimal order n.

In this chapter we have already derived the *degree equation*

$$n\frac{K_n}{K} = \frac{K'_n}{K'} \tag{12.335}$$

with $K = K(k)$, $K' = K(k')$, $K_n = K(k_n)$, and $K'_n = K(k'_n)$.

We have already defined the modular constants q_1 and q_n:

$$q_1 = e^{-\pi K'/K} \tag{12.336}$$

$$q_n = e^{-\pi K'_n/K_n} \tag{12.337}$$

Since n can be expressed as

$$n = \frac{\dfrac{K'_n}{K_n}}{\dfrac{K'}{K}} \tag{12.338}$$

we find

$$n = \frac{\log_e q_n}{\log_e q_1} \tag{12.339}$$

The modular constants q_1 and q_n can be determined by using the approximate relation

$$q(k) \approx \hat{q}(k) \tag{12.340}$$

where

$$q_0 = \frac{1}{2}\frac{1-\sqrt[4]{1-k^2}}{1+\sqrt[4]{1-k^2}}$$

$$\begin{aligned}\hat{q} &= q_0 + 2q_0^5 + 15q_0^9 + 150q_0^{13}\\ &+ 1707q_0^{17} + 20{,}910q_0^{21}\\ &+ 268{,}616q_0^{25} + 3{,}567{,}400q_0^{29}\\ &+ 48{,}555{,}069q_0^{33} + 673{,}458{,}874q_0^{37}, \qquad k \le \frac{1}{\sqrt{2}}\end{aligned}$$

$$q_p = \frac{1}{2}\frac{1-\sqrt{k}}{1+\sqrt{k}} \tag{12.341}$$

$$\begin{aligned}q' &= q_p + 2q_p^5 + 15q_p^9 + 150q_p^{13}\\ &+ 1707q_p^{17} + 20{,}910'q_p^{21}\\ &+ 268{,}616q_p^{25} + 3{,}567{,}400q_p^{29}\\ &+ 48{,}555{,}069q_p^{33} + 673{,}458\,874q_p^{37}, \qquad \frac{1}{\sqrt{2}} < k < 1\end{aligned}$$

$$\hat{q} = e^{\pi^2/\ln q'}$$

and

$$q_1 = q\left(\frac{1}{\xi}\right) \tag{12.342}$$

$$q_n = q\left(\frac{1}{L(n,\xi)}\right) \tag{12.343}$$

The inequalities (12.333) and (12.334) determine the upper bound for ϵ:

$$\epsilon_{\max} = \sqrt{10^{A_p/10} - 1} \tag{12.344}$$

and the lower bound for $L(n, \xi)$:

$$L_{\min} = \frac{1}{\epsilon_{\max}}\sqrt{10^{A_s/10} - 1} = \sqrt{\frac{10^{A_s/10} - 1}{10^{A_p/10} - 1}} \tag{12.345}$$

Substituting Eqs. (12.342) and (12.343) into Eq. (12.339), we find the degree n. For an arbitrary specification the degree n from Eq. (12.339) is not an integer, so we have to round up the solution of the degree equation to an integer value:

$$n \geq \frac{\log_e q\left(\dfrac{1}{L_{\min}}\right)}{\log_e q\left(\dfrac{1}{\xi}\right)} \tag{12.346}$$

or, assuming $\xi = F_s/F_p$, we have

$$n \geq \frac{\log_e q\left(\sqrt{\dfrac{10^{A_p/10}-1}{10^{A_s/10}-1}}\right)}{\log_e q\left(\dfrac{F_p}{F_s}\right)} \tag{12.347}$$

Choosing an integer value for n, we obtain a range of edge values

$$\Xi_{min} \leq \xi = \frac{f_s}{f_p} \leq \frac{F_s}{F_p} \tag{12.348}$$

$$a_{p,min} \leq a_p = 10\log_{10}(1+\epsilon^2) \leq A_p \tag{12.349}$$

$$a_{s,max} \geq a_s = 10\log_{10}\left(1+\epsilon^2 L(n,\xi)\right) \geq A_s \tag{12.350}$$

where $\{f_p, f_s, a_p, a_s\}$ are actual parameters of the lowpass filter. The lower bound of ξ, denoted by $\Xi_{\min}$, is computed from the equation $L(n, \Xi_{\min}) = L_{\min}$.

Therefore, we can use any f_p, f_s, a_p, and a_s that satisfy the inequalities (12.349) and (12.350) and the equation

$$n = \frac{\log_e q\left(\sqrt{\dfrac{10^{a_p/10}-1}{10^{a_s/10}-1}}\right)}{\log_e q\left(\dfrac{f_p}{f_s}\right)} \tag{12.351}$$

We can find the minimal value of ξ from

$$q\left(\frac{1}{\Xi_{min}}\right) = q\left(\frac{f_p}{f_s}\right) = \sqrt[n]{q\left(\sqrt{\frac{10^{A_p/10}-1}{10^{A_s/10}-1}}\right)} \tag{12.352}$$

Finally, we calculate Ξ_{min} for $k = f_p/f_s$:

$$\Xi_{min} = \frac{1}{k} = \frac{1}{\sqrt{1 - \sqrt{(k')^4}}} \tag{12.353}$$

$$\Xi_{min} = \frac{1}{\sqrt{1 - \left(\dfrac{1 + 2\displaystyle\sum_{m=1}^{\infty}(-1)^m \left(\sqrt[n]{q\left(\sqrt{\dfrac{10^{A_p/10} - 1}{10^{A_s/10} - 1}}\right)} \right)^{m^2}}{1 + 2\displaystyle\sum_{m=1}^{\infty} \left(\sqrt[n]{q\left(\sqrt{\dfrac{10^{A_p/10} - 1}{10^{A_s/10} - 1}}\right)} \right)^{m^2}} \right)^4}} \tag{12.354}$$

Typically, we take the first 10 terms of the sums.

12.10.1 Discrimination Factor

The reciprocal value of k_n, $L_n = 1/k_n$, is called the *discrimination factor*. One possible procedure to find L_n is to solve the equation

$$n \frac{K\left(\dfrac{1}{L_n}\right)}{K\left(\sqrt{1 - \dfrac{1}{L_n^2}}\right)} = \frac{K\left(\dfrac{1}{\xi}\right)}{K\left(\sqrt{1 - \dfrac{1}{\xi^2}}\right)} \tag{12.355}$$

for given ξ and n, numerically. Such a numeric procedure is computationally intensive and we prefer to have an alternative, much simpler, algorithm to compute the discrimination factor.

Let us now determine L_n in a different way. By definition,

$$L_n = L(n, \xi) = R(n, \xi, \xi) \tag{12.356}$$

where

$$R(n, \xi, \xi) = \frac{\displaystyle\prod_{i=1}^{n/2}\left(1 - \frac{\xi^2}{x_i^2}\right)}{\displaystyle\prod_{i=1}^{n/2}(1 - x_i^2)} \frac{\displaystyle\prod_{i=1}^{n/2}(\xi^2 - x_i^2)}{\displaystyle\prod_{i=1}^{n/2}\left(\xi^2 - \frac{\xi^2}{x_i^2}\right)} \quad \text{for } n \text{ even} \tag{12.357}$$

and

$$R(n,\xi,\xi) = \xi \frac{\prod_{i=1}^{(n-1)/2}\left(1-\frac{\xi^2}{x_i^2}\right)\prod_{i=1}^{(n-1)/2}(\xi^2-x_i^2)}{\prod_{i=1}^{(n-1)/2}(1-x_i^2)\prod_{i=1}^{(n-1)/2}\left(\xi^2-\frac{\xi^2}{x_i^2}\right)} \qquad \text{for } n \text{ odd} \tag{12.358}$$

Simplifying, we obtain

$$L_n = R(n,\xi,\xi) = \frac{1}{\xi^n}\frac{\prod_{i=1}^{n/2}(\xi^2-x_i^2)^2}{\prod_{i=1}^{n/2}(1-x_i^2)^2}, \qquad n \text{ even} \tag{12.359}$$

and

$$L_n = R(n,\xi,\xi) = \frac{1}{\xi^{n-2}}\frac{\prod_{i=1}^{(n-1)/2}(\xi^2-x_i^2)^2}{\prod_{i=1}^{(n-1)/2}(1-x_i^2)^2}, \qquad n \text{ odd} \tag{12.360}$$

Obviously, for given ξ and n it is easier to evaluate Eq. (12.360) than to solve the degree equation, Eq. (12.355).

12.11 ELLIPTIC FILTERS WITH MINIMAL Q-FACTORS

The poles of the nth-order elliptic filter are the poles of the normalized transfer function $\mathcal{H}(s)$:

$$|\mathcal{H}(s)|^2 = \frac{1}{1+\epsilon^2 R^2(n,\xi,x)}, \qquad x = -js \tag{12.361}$$

which can be expressed in terms of the Jacobi elliptic functions $\text{sn}(u,k)$, $\text{cn}(u,k)$, and $\text{dn}(u,k)$:

$$s_i = \frac{-\text{sn}(v,k')\text{cn}(v,k')\text{cn}(u_{oi},k)\text{dn}(u_{oi},k) \pm j\text{sn}(u_{oi},k)\text{dn}(v,k)}{1-\text{sn}^2(v,k')\text{dn}^2(u_{oi},k)} \tag{12.362}$$

The modulus k is inversely proportional to the selectivity factor ξ, $k = 1/\xi$ and $k' = \sqrt{1-k^2}$. The zeros of the elliptic rational function are $x_i = \text{sn}(u_i,k)$. The term $\text{sn}(v,k')$ depends on the order n, the ripple-factor ϵ, and ξ, and can be determined indirectly from the second parametric equation

$$\text{sn}\left(\frac{nK_n}{K}v, \sqrt{1-\frac{1}{L^2}}\right) = \text{sn}\left(\frac{K_n'}{K'}v, \sqrt{1-\frac{1}{L^2}}\right) = \frac{1}{\sqrt{1+\epsilon^2}} \tag{12.363}$$

The quality factor Q_i of a pole s_i is defined as

$$Q_i = -\frac{|s_i|}{2\,\text{Re}\,s_i} \tag{12.364}$$

In fact, the Q-factor is a function of the argument of the pole s_i:

$$Q_i = -\frac{1}{2\cos(\arg(s_i))} \tag{12.365}$$

and it does not depend on the pole magnitude. The minimal value of the Q-factor is 1/2, and it occurs when the pole is real. The Q-factor takes the infinite value for poles on the imaginary axis.

From Eq. (12.362) and (12.363) one can see that the poles s_i become purely imaginary if $\epsilon \to +\infty$ and take a value $s_i = jx_i$; the other extreme is $\epsilon = 0$, and the poles become $s_i = j\xi/x_i$. In both cases the Q-factors of the poles are infinite. Except for these two extremes the poles are complex conjugate.

Analytically, Q_{min} can be determined by equating to zero the imaginary part of the logarithmic derivative of the pole with respect to the independent variable v:

$$\text{Im}\left(\frac{1}{s_i}\frac{\mathrm{d}s_i}{\mathrm{d}v}\right) = \text{Im}\left(\frac{\mathrm{d}}{\mathrm{d}v}\log(s_i)\right) = 0 \tag{12.366}$$

A closer examination shows that Eq. (12.366) is satisfied for

$$v = \frac{K'}{2} \tag{12.367}$$

Substituting this value into Eq. (12.363) gives

$$\text{sn}\left(\frac{K_n'}{2}, \sqrt{1 - \frac{1}{L^2}}\right) = \frac{1}{\sqrt{1 + \frac{1}{L}}} \tag{12.368}$$

Finally, combining Eq. (12.363) and (12.368) we find

$$\epsilon = \epsilon_{minQ} = \frac{1}{\sqrt{L(n, \xi)}} \tag{12.369}$$

Equation (12.369) relates, directly, the ripple-factor required for Q_{min} to the value of $L(n, \xi)$. Since the above relation does not depend on x_i, all Q's are simultaneously minimal.

Substituting Eq. (12.367) into Eq. (12.362), we find a relatively simple expression for poles:

$$s_i = \sqrt{\xi}\frac{\sqrt{1 - x_i^2}\sqrt{\xi^2 - x_i^2} + jx_i(\xi + 1)}{\xi + x_i^2} \tag{12.370}$$

where

$$|s_i| = \sqrt{\xi} \tag{12.371}$$

$$Q_{min,i} = \frac{\xi + x_i^2}{2\sqrt{1 - x_i^2}\sqrt{\xi^2 - x_i^2}} \tag{12.372}$$

By means of Eq. (12.371) and (12.372) it is now possible to express the normalized transfer function with minimal Q-factors, $\mathcal{H}_Q$, as

$$\mathcal{H}_Q(s) = \prod_{i=1}^{n/2} \frac{s^2 + \dfrac{\xi^2}{x_i^2}}{s^2 + \dfrac{\sqrt{\xi}}{Q_{min,i}} s + \xi}, \qquad n \text{ even} \tag{12.373}$$

$$\mathcal{H}_Q(s) = \frac{1}{s + \sqrt{\xi}} \prod_{i=1}^{(n-1)/2} \frac{s^2 + \dfrac{\xi^2}{x_i^2}}{s^2 + \dfrac{\sqrt{\xi}}{Q_{min,i}} s + \xi}, \qquad n \text{ odd} \tag{12.374}$$

As an illustration of the above considerations, the root loci of the sixth-order elliptic transfer function are presented in Fig. 12.26. All curves start at the zeros of the elliptic rational function, at jx_i, for $\epsilon = \infty$, and terminate at the poles of the elliptic rational function, at $j\xi/x_i$, for $\epsilon = 0$. As can be seen in Fig. 12.26, all poles with minimal Q-factors lie on the circle of the radius $\sqrt{\xi}$. The points b, e, and h designate the positions of the $\mathcal{H}_Q$-function poles ($v = K'/2$). For $v_1 = K'/4$ and $v_2 = K' - v_1 = 3K'/4$ the poles with equal Q-factors, but different magnitudes, are $\{a,c\}$ and $\{d,f\}$.

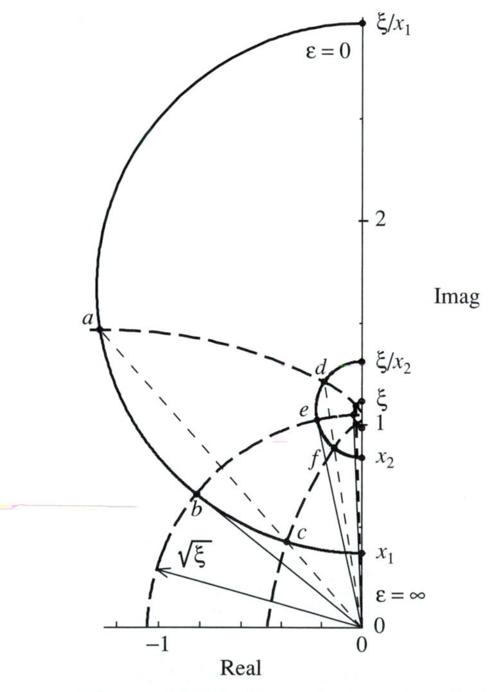

Figure 12.26 Root loci of the sixth-order transfer function poles.

PROBLEMS

12.1 Sketch the elliptic integral of the first kind $u(\phi, k)$ for $0 \le \phi \le \frac{\pi}{2}$ and $k \in \{0, 0.5, 0.9\}$. Find $u(\frac{\pi}{2}, 0)$ $u(\frac{\pi}{2}, 1)$ $u(\frac{\pi}{2}, \frac{\sqrt{2}}{2})/K'(\frac{\sqrt{2}}{2})$.

12.2 Sketch the Jacobi functions $\text{sn}(u, k)$, $\text{cn}(u, k)$, and $\text{cd}(u, k)$. Assume for $0 \le k \le 0.8$, and $0 \le u \le 20$. Find $\text{cd}^{-1}(0, 0.5)$.

12.3 Sketch the Jacobi functions $\text{sn}(yK(k), k)$, $\text{cn}(yK(k), k)$, and $\text{cd}(yK(k), k)$. Assume for $0 \le k \le 0.8$, and $0 \le y \le 8$.

12.4 Find $C(8, x)$, $C(4, C(2, x))$, $C(2, C(4, x))$, and $C(2, C(2, C(2, x)))$, where $C(n, x)$ is the nth-order Chebyshev polynomial.

12.5 Sketch the function $\mathcal{R}(n, k, p, x) = \text{cd}(n\frac{K(p)}{K(k)}\text{cd}^{-1}(x, k), p)$ for $-1 \le x \le 1$, $k = 0.8$, $p = 0.1$, $k = -0.8$, and $n \in \{1, 2\}$.

12.6 Sketch the function $\mathcal{R}(n, k, p, x) = \text{cd}(n\frac{K(p)}{K(k)}\text{cd}^{-1}(x, k), p)$ for $-1 \le x \le 1$, $k = 0.8$, $p = 0.1$, $k = -0.8$, and $n \in \{1, 2, 3, 4, 5\}$.

MATLAB EXERCISES

12.1 Write a MATLAB program that computes $C(8, x)$, $C(4, C(2, x))$, $C(2, C(4, x))$, and $C(2, C(2, C(2, x)))$. Plot the Chebyshev polynomials $C(n, x)$, $\cos(n\cos^{-1}(x))$, and $\cosh(n\cosh^{-1}(x))$ for $0 \le x \le 1.2$ and $1 \le n \le 5$.

12.2 Write a MATLAB program that computes all poles and zeros of the 18th-order elliptic transfer function. Assume the ripple factor $\epsilon = 0.1$, and the selectivity factor $\xi = 1.01$.

MATHEMATICA EXERCISES

12.1 Write a *Mathematica* program that plots the elliptic integral of the first kind $u(\phi, k)$ for $0 \le \phi \le 1.5$ and $0 \le k \le 0.8$. Plot the complete and complementary complete elliptic integral of the first kind $K(k)$, $K'(k)$, and the ratio $K(k)/K'(k)$. Plot the ratio $u(\phi, k)/K'(k)$ for $k = 0$, $k = 0.5$, $k = 0.9$, $k = 0.99999$.

12.2 Write a *Mathematica* program that plots the Jacobi functions $\text{sn}(u, k)$, $\text{cn}(u, k)$, and $\text{cd}(u, k)$. Assume for $0 \le k \le 0.8$ and $0 \le u \le 20$. Find $\text{cd}^{-1}(0, 0.5)$.

12.3 Write a *Mathematica* program that plots the Jacobi functions $\text{sn}(yK(k), k)$, $\text{cn}(yK(k), k)$, and $\text{cd}(yK(k), k)$. Assume for $0 \le k \le 0.8$ and $0 \le y \le 4$. Determine the periods of the Jacobi functions.

12.4 Write a *Mathematica* program that plots the modular constant, $q(k)$. Compare the accuracy of the modular constant computed using complete elliptic integrals and series representation. Find $q(0)$, $q(1/\sqrt{2})$, and $q(1)$.

12.5 Write a *Mathematica* program that plots the Chebyshev polynomials $C(8, x)$, $C(4, C(2, x))$, $C(2, C(4, x))$, and $C(2, C(2, C(2, x)))$. Plot the Chebyshev polynomials $C(n, x)$, $\cos(n\cos^{-1}(x))$, and $\cosh(n\cosh^{-1}(x))$ for $0 \le x \le 1.2$ and $1 \le n \le 5$.

12.6 Write a *Mathematica* program that plots the function $\mathcal{R}(n, k, p, x) = \text{cd}(n \frac{K(p)}{K(k)} \text{cd}^{-1}(x, k), p)$, for $-1 \leq x \leq 1$, $k = 0.8$, $p = 0.1$, $k = -0.8$, and $n \in \{1, 2, 3, 4, 5\}$. Find p which satisfies $\mathcal{R}(1, k, p, x) = \frac{1}{p}$ and $\mathcal{R}(2, k, p, x) = \frac{1}{p}$, assuming $k = 0.95$.

12.7 Write a *Mathematica* program that plots the elliptic rational functions $\mathcal{R}(n, \xi, x)$, in terms of Jacobi elliptic functions for $0 \leq x \leq 1.2$ and $1 \leq n \leq 5$. Plot $\mathcal{R}(10, 1.1, x)$ and $\mathcal{R}(2, \mathcal{R}(5, 1.1, 1.1), x)$ for $0 \leq x \leq 1$, $1 \leq x \leq 1.1$, and $1.1 \leq x \leq 10$.

12.8 Write a *Mathematica* program that computes the zeros, the poles, and the discrimination factor of the elliptic transfer function. Compute all poles and zeros of the 18th-order transfer function. Assume the ripple factor $\epsilon = 0.1$ and the selectivity factor $\xi = 1.01$.

CHAPTER 13

ELLIPTIC RATIONAL FUNCTION

Design of *elliptic-function filters* is based on the elliptic rational function. A *rational function* is defined as a ratio of two polynomials. The *elliptic rational function* can be derived from the Jacobi elliptic functions as shown in the previous chapter. However, in some cases, the elliptic rational function can be derived without knowledge of the Jacobi functions. In this chapter we introduce the elliptic rational function as a natural generalization of the Chebyshev polynomial and we bypass mathematical theory of special functions required in the previous chapter.

First, we repeat the basic properties of the Chebyshev polynomial. Next, we show the similarities between the Chebyshev polynomial and the elliptic rational function. Some properties are illustrated graphically, and we omit strict mathematical proofs for some relations.

In this chapter, we prefer to give our reader an intuitive feel of the basic properties of the elliptic rational function. Our goal is to build the knowledge of the elliptic rational function using simple algebraic manipulations, even without mentioning the Jacobi elliptic functions.

13.1 CHEBYSHEV POLYNOMIALS

The simplest way to generate the Chebyshev polynomial $C(n, x)$ is to use the recurrence relation

$$\begin{aligned} C(0, x) &= 1 \\ C(1, x) &= x \\ C(n, x) &= 2x\, C(n-1, x) - C(n-2, x) \end{aligned} \tag{13.1}$$

where n represents the order and x stands for the independent variable. For example, let us generate the Chebyshev polynomials up to the fifth order:

$$\begin{aligned} C(0,x) &= 1 \\ C(1,x) &= x \\ C(2,x) &= 2x^2 - 1 \\ C(3,x) &= 4x^3 - 3x \\ C(4,x) &= 8x^4 - 8x^2 + 1 \\ C(5,x) &= 16x^5 - 20x^3 + 5x \\ \cdots &= \cdots \end{aligned} \tag{13.2}$$

One of the most important properties of the Chebyshev polynomial is [110, p. 196]

$$C(mp, x) = C(m, C(p, x)) \tag{13.3}$$

which we call the *nesting property*. A function $F(n, x)$ is said to have the nesting property if $F(m, F(p, x)) = F(mp, x)$. In other words, if we replace the independent variable x, in an mth-order function $F(m, x)$, with the function of an order p, we obtain the same function, but with the order equal to the product of orders $m \cdot p$.

The nesting property is useful for generating higher-order Chebyshev polynomials from the lower-order polynomials. For example, we can generate the eighth-order polynomial $C(8, x)$ from the second-order polynomial $C(2, x)$ by applying Eq. (13.3):

$$C(8, x) = C(2, C(4, x)) = C(2, C(2, C(2, x))) \tag{13.4}$$

Chebyshev polynomials are plotted in Fig. 13.1 for $|x| \leq 1$. All curves take the unit value for $x = 1$.

By increasing the polynomial order, n, we increase the slope at $x = 1$ as shown in Fig. 13.2.

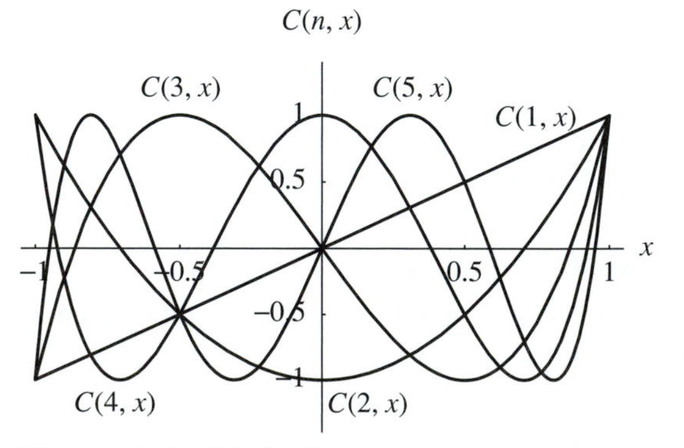

Figure 13.1 Chebyshev polynomials $C(n, x)$, $|x| \leq 1$.

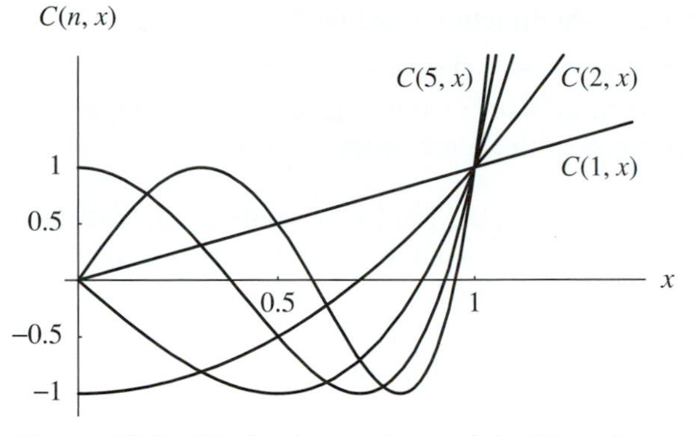

Figure 13.2 Chebyshev polynomials $C(n, x)$, $0 \leq x < 1.2$.

For $x > 1$, Chebyshev polynomials are monotonically increasing (Fig. 13.3).

From Figs. 13.2 and 13.3 we conclude that the slope of the Chebyshev polynomial at $x = 1$ increases when the zeros tend to $x = 1$. The Chebyshev polynomials have the highest slope of all polynomials, at $x = 1$ (the maximal value of the magnitude of the polynomial for $x \leq 1$ must be 1). Further increasing the slope at $x = 1$ is possible only by introducing poles for $x > 1$; that is, instead of polynomials, we have to use rational functions.

13.2 SECOND-ORDER ELLIPTIC RATIONAL FUNCTION

Suppose that we want to construct an nth-order rational function $R(n, x)$ that has to meet the following constraints:

1. Over the unit interval $-1 \leq x \leq 1$ the function should be such that $R^2(n, x) \leq 1$.
2. At the end points $x = \pm 1$ the squared function must have the unit value $R^2(n, \pm 1) = 1$.
3. For a given n the function is either even or odd; that is, $R^2(n, -x) = R^2(n, x)$.

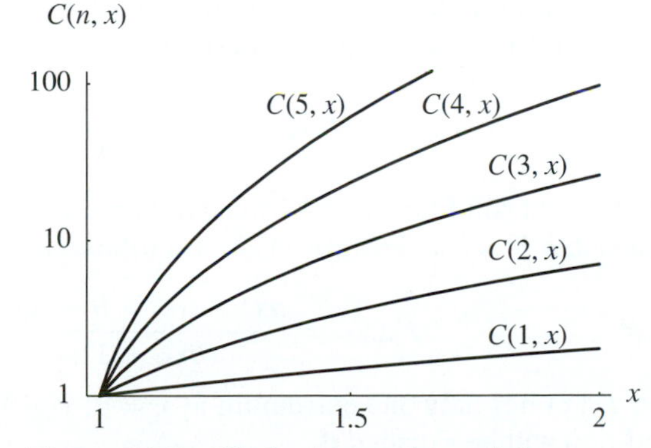

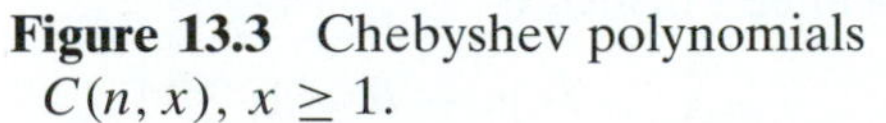

Figure 13.3 Chebyshev polynomials $C(n, x)$, $x \geq 1$.

4. For $x > 1$ the function must be $R^2(n, x) > 1$.
5. The slope at $x = 1$ should be the largest possible.
6. The function $R(n, x)$ must have a larger slope at $x = 1$ than the Chebyshev polynomial of the same order, $C(n, x)$.

It can be shown that (1) the Chebyshev polynomial satisfies the first five constraints and (2) the Chebyshev polynomial has the steepest slope at $x = \pm 1$ compared to any other polynomial of the same order, under the listed constraints.

A first-order rational function $R(1, x) = (ax + c)/(bx + d)$ satisfying our conditions degenerates to the polynomial

$$R(1, x) = x \tag{13.5}$$

It has one real zero at $x = 0$. Formally, we can say that $R(1, x)$ has one pole in infinity, because $\lim_{x\to\infty} R(1, x) = \infty$.

Consider a general second-order rational function:

$$R(2, x) = \frac{ax^2 + cx + e}{bx^2 + dx + f} \tag{13.6}$$

There exist three candidate functions that are either even or odd:

$$\begin{aligned} F_1(x) &= \frac{cx}{bx^2 + f} \\ F_2(x) &= \frac{ax^2 + e}{dx} \\ F_3(x) &= \frac{ax^2 + e}{bx^2 + f} \end{aligned} \tag{13.7}$$

The first function, $F_1(x)$, vanishes for $x \to \pm\infty$ and fails to meet the fourth condition. The second function, $F_2(x)$, discontinues at $x = 0$, for $e \neq 0$, or it degenerates to the first-order function for $e = 0$, thus failing to satisfy the first condition.

In the third function, $F_3(x)$, the coefficient f must not be zero, because we want to avoid the discontinuity at $x = 0$. Without lack of generality we can assume $f = 1$, which is equivalent to normalizing a, b, and e by f:

$$F_3(x) = \frac{ax^2 + e}{bx^2 + 1} \tag{13.8}$$

According to the first condition, bx^2+1 must not have zeros over the interval $-1 \le x \le 1$, which means that $b > -1$. From $F_3(1) = 1$ it follows $e = -a+b+1 = -(a-b-1)$:

$$F_3(x) = \frac{ax^2 - (a - b - 1)}{bx^2 + 1} \tag{13.9}$$

The function $F_3(x)$ has only one extremum at $x = 0$, $F_3(0) = -(a - b - 1)$; therefore, the first condition will be satisfied if

$$|a - b - 1| \le 1 \tag{13.10}$$

The slope of $F_3(x)$ at $x = 1$ is

$$\left.\frac{\mathrm{d}F_3(x)}{\mathrm{d}x}\right|_{x=1} = 2\frac{a-b}{1+b} \tag{13.11}$$

and its modulus reaches the maximum, $4/(1+b)$, for

$$a - b = 2 \tag{13.12}$$

yielding

$$F_3(x) = \frac{ax^2 - 1}{(a-2)x^2 + 1} \tag{13.13}$$

From the fourth condition we have $|a| > |a-2|$, and we obtain $a > 1$. The sixth condition is met for $4/(1+b) > 4$, and we find $b < 0$ and

$$1 < a < 2 \tag{13.14}$$

It is convenient to express the coefficient a as $a = 1 + t$:

$$R(2, x) = \frac{(t+1)x^2 - 1}{(t-1)x^2 + 1}, \quad 0 < t < 1 \tag{13.15}$$

The zeros of $R(2, x)$ are

$$\begin{aligned} x_1 &= \frac{1}{\sqrt{t+1}} \\ x_2 &= -x_1 \end{aligned} \tag{13.16}$$

and the poles are

$$\begin{aligned} x_{p,1} &= \frac{1}{\sqrt{1-t}} \\ x_{p,2} &= -x_{p,1} \end{aligned} \tag{13.17}$$

For $0 < t < 1$ the zero x_1 remains within the interval $1/\sqrt{2} < x_1 < 1$, while the pole $x_{p,1}$ lies in $1 < x_{p,1} < +\infty$. For $t = 1$ the function $R(2, x)$ becomes the Chebyshev second-order polynomial $C(2, x)$. For $t = 0$ the rational function degenerates to a constant $R(2, x) = -1$.

The slope at $x = 1$ can be made arbitrarily large by choosing t closer to zero, $t \to 0$. In that case the positive zero and pole tend to unity, as shown in Figs. 13.4 and 13.5. The above analysis shows that there exist a rational function $R(2, x)$ which is steeper at $x = 1$ than the Chebyshev polynomial $C(2, x)$.

The limit of $|R(2, x)|$ for $x \to +\infty$ is

$$\lim_{x \to +\infty} |R(2, x)| = -\frac{t+1}{t-1} \tag{13.18}$$

Let us determine the positive value of the independent variable x for which $R(2, x)$ has the same value as the above limit, and denote this value by ξ:

$$R(2, \xi) = \lim_{x \to +\infty} |R(2, x)| \tag{13.19}$$

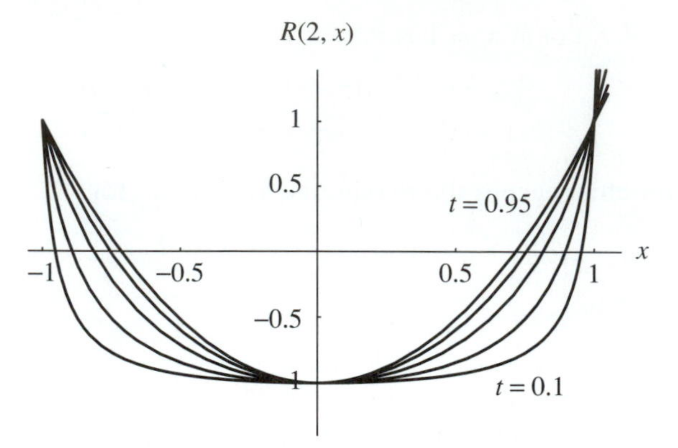

Figure 13.4 Rational function $R(2, x)$, $x \leq 1$, for $t = 0.1, 0.25, 0.5, 0.75, 0.95$.

We find

$$\xi = \frac{1}{1 - t^2} \tag{13.20}$$

Obviously ξ can take any value from the interval $1 < \xi < \infty$.

Alternatively, the parameter t is uniquely defined by ξ:

$$t = \sqrt{1 - \frac{1}{\xi^2}} \tag{13.21}$$

The rational function $R(2, x)$ can be expressed in terms of ξ as follows:

$$R(2, x) = \frac{\left(1 + \sqrt{1 - \frac{1}{\xi^2}}\right) x^2 - 1}{\left(-1 + \sqrt{1 - \frac{1}{\xi^2}}\right) x^2 + 1}, \qquad \xi > 1 \tag{13.22}$$

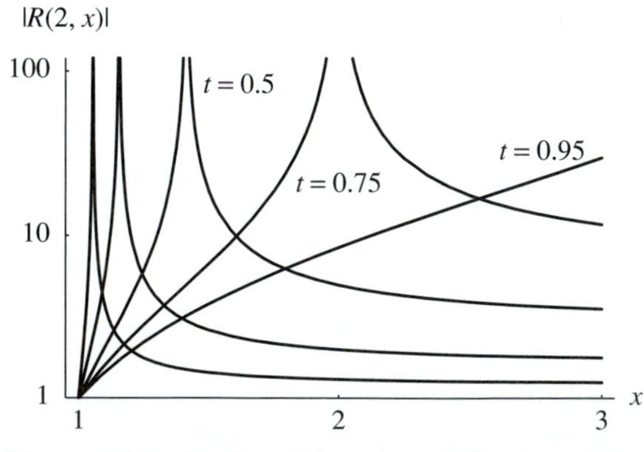

Figure 13.5 Rational function $R(2, x)$, $x \geq 1$, for $t = 0.1, 0.25, 0.5, 0.75, 0.95$.

The behavior of $R(2, x)$ is summarized in Eq. (13.23):

$$\begin{array}{rlllll} & |R(2,x)| & = 1 & \text{for} & x & = 0 \\ & |R(2,x)| & \le 1 & \text{for} & 0 \le x & \le 1 \\ 1 < & |R(2,x)| & < R(2,\xi) & \text{for} & 1 < x & < \xi \\ R(2,\xi) \le & |R(2,x)| & & \text{for} & \xi \le x & \\ R(2,\xi) = & |R(2,x)| & & \text{for} & x & \to \infty \end{array} \tag{13.23}$$

From Eqs. (13.22) and (13.23) we conclude that ξ is a very important parameter of $R(n, x)$. We call ξ the *selectivity factor*, and it can be interpreted as a measure of the slope at $x = 1$. Also, it marks the interval $\xi < x < \infty$ over which the rational function $|R(2, x)|$ is no less that $R(2, \xi)$.

The fact that $R(2, x)$ depends on ξ as well as on x gives us an idea to represent R as function of three arguments:

$$R(2, \xi, x) = \frac{\left(1 + \sqrt{1 - \frac{1}{\xi^2}}\right) x^2 - 1}{\left(-1 + \sqrt{1 - \frac{1}{\xi^2}}\right) x^2 + 1}, \qquad \xi > 1 \tag{13.24}$$

Formally, we can rewrite the first-order rational function Eq. (13.5) as follows:

$$R(1, \xi, x) = x \tag{13.25}$$

and notice that $R(1, \xi, \xi) = \xi$.

The functions $R(1, \xi, x)$ and $R(2, \xi, x)$ are known as the *elliptic rational functions* and are traditionally generated by using the Jacobi elliptic functions.

From now on, we shall call $R(n, \xi, x)$ the elliptic rational function of the order n, the selectivity factor ξ, and the independent variable x.

13.3 ELLIPTIC RATIONAL FUNCTIONS OF ORDER 2^i

The fourth-order Chebyshev polynomial can be generated by using the nesting property:

$$C(4, x) = C(2, C(2, x)) \tag{13.26}$$

This means that by substituting the independent variable X by another second-order Chebyshev polynomial $X = C(2, x)$ in the second-order Chebyshev polynomial, $C(2, X)$, we derive $C(4, x)$

$$C(4, x) = C(2, X), \qquad X = C(2, x) \tag{13.27}$$

The same procedure we can apply to rational functions. In the elliptic rational function $R(2, \Xi, X)$, we substitute the independent variable X by another elliptic rational function $R(2, \xi, x)$. The resulting function is again a rational function but of the fourth order. We denote this new function by $\mathcal{R}(4, \Xi, \xi, x)$:

$$\mathcal{R}(4, \Xi, \xi, x) = R(2, \Xi, X), \qquad X = R(2, \xi, x) \tag{13.28}$$

which can be rewritten as

$$\mathcal{R}(4, \Xi, \xi, x) = R(2, \Xi, R(2, \xi, x)) \tag{13.29}$$

We use a different symbol for the new function, $\mathcal{R}$, instead of R, because $\mathcal{R}$ is a function of the four arguments, $(4, \Xi, \xi, x)$, while the elliptic rational function R is a function of the three arguments, $(4, \xi, x)$.

Our target is to derive $R(4, \xi, x)$. Therefore, we have to calculate the argument Ξ in terms of ξ.

We have defined the selectivity factor ξ as the smallest value of positive x which marks the interval $\xi \leq x < \infty$ over which the rational function $R(2, \xi, x)$ is no less than $R(2, \xi, \xi)$. Also, the selectivity factor, ξ, of the rational function $R(4, \xi, x)$ marks the interval $\xi \leq x < \infty$ over which $R(4, \xi, x)$ is no less than $R(4, \xi, \xi)$. Therefore, we can calculate Ξ for the smallest x that satisfies the inequality

$$\mathcal{R}(4, \Xi, \xi, x) \geq \mathcal{R}(4, \Xi, \xi, \xi) \qquad \text{for} \quad x \geq \xi \tag{13.30}$$

If we start from the condition that is valid for the second-order function $|R(2, \xi, x)|_{x \to \infty} = R(2, \xi, \xi)$, we obtain

$$|\mathcal{R}(4, \Xi, \xi, x)|_{x \to \infty} = \mathcal{R}(4, \Xi, \xi, \xi) \tag{13.31}$$

in which Ξ disappears, and we cannot determine Ξ.

Starting from

$$R(2, \xi, x) = \frac{(1+t)\; x^2 - 1}{(-1+t)\; x^2 + 1}, \qquad \text{for} \quad t = \sqrt{1 - \frac{1}{\xi^2}} \tag{13.32}$$

$$R(2, \Xi, X) = \frac{(1+T)\; X^2 - 1}{(-1+T)\; X^2 + 1}, \qquad \text{for} \quad T = \sqrt{1 - \frac{1}{\Xi^2}} \tag{13.33}$$

we find

$$\mathcal{R}(4, \Xi, \xi, x) = \frac{\left(4t + T(1+t)^2\right) x^4 + (-4t - 2T(1+t))\, x^2 + T}{\left(-4t + T(1+t)^2\right) x^4 + (4t - 2T(1+t))\, x^2 + T} \tag{13.34}$$

and

$$|\mathcal{R}(4, \Xi, \xi, x)|_{x \to \infty} = \frac{4t + T(1+t)^2}{-4t + T(1+t)^2} \tag{13.35}$$

For $X = \Xi$, Eq. (13.33) becomes

$$R(2, \Xi, \Xi) = \frac{(1+T)\left(1 + T - T^2\right)}{1 - 2T + T^3} \tag{13.36}$$

Solving the equation $|\mathcal{R}(4, \Xi, \xi, x)|_{x \to \infty} = R(2, \Xi, \Xi)$, we calculate T:

$$T = \sqrt{\frac{2t}{1+t}} \tag{13.37}$$

Since $t = \sqrt{1 - 1/\xi^2}$ and $T = \sqrt{1 - 1/\Xi^2}$ we can express Ξ in terms of ξ:

$$\Xi = R(2, \xi, \xi) \tag{13.38}$$

Finally, we conclude that

$$\mathcal{R}(4, R(2, \xi, \xi), \xi, x) = R(4, \xi, x) \tag{13.39}$$

or in the nesting form [Eq. (13.29)]

$$\mathcal{R}(4, R(2, \xi, \xi), \xi, x) = R(4, \xi, x) = R\left(2, R\left(2, \xi, \xi\right), R\left(2, \xi, x\right)\right) \tag{13.40}$$

has the minimal value $R(4, \xi, \xi)$:

$$R(4, \xi, \xi) = R(2, \Xi, \Xi) = R\left(2, R\left(2, \xi, \xi\right), R\left(2, \xi, \xi\right)\right) \tag{13.41}$$

for $x \geq \xi$.

Examining the properties of $R(4, \xi, x)$

$$\begin{array}{ccccc}
0 \leq x \leq 1 & \rightarrow & 0 \leq X \leq 1 & \rightarrow & |R(4, \xi, x)| \leq 1 \\
1 < x < \xi & \rightarrow & 1 < X < \Xi & \rightarrow & 1 < |R(4, \xi, x)| < R(4, \xi, \xi) \\
\xi \leq x & \rightarrow & \Xi \leq X & \rightarrow & R(4, \xi, \xi) \leq |R(4, \xi, x)| \\
\text{for} \quad X = |R(2, \xi, x)| & & & & \text{for} \quad |R(4, \xi, x)| = |R(2, \Xi, X)| \\
\text{and} \quad \Xi = R(2, \xi, \xi) & & & & \text{and} \quad R(2, \Xi, \Xi)| = R(4, \xi, \xi)
\end{array} \tag{13.42}$$

we find that this function satisfies the six constraints from the previous section.

The function $R(2, \Xi, X)$, for $|x| \leq 1$, is shown in Fig. 13.6a. The intermediate variable $X = R(2, \xi, x)$, shown in Fig. 13.6b, causes the two initial oscillations ($1 > X > -1$, $-1 < X < 1$) to be repeated twice in the resultant function $R(4, \xi, x)$, given in Fig. 13.6d. The corresponding points in all figures are denoted by the same letters. So, for instance, the zeros of the $R(2, \Xi, X)$, denoted by A and B in Fig. 13.6a, are projected to the $R(2, \xi, x)$ curve in Fig. 13.6b, whose vertical coordinates are zeros of the resultant function $R(4, \xi, x)$. In order to fix these points on the curve $R(4, \xi, x)$, the $R(2, \xi, x)$ is redrawn in Fig. 13.6c but with x on the abscissa. The $R(2, \xi, x)$, presented in Fig. 13.6c, point out that all zeros and extremes (a, b, c) of $R(2, \xi, x)$ determine the position of the extremes of the resultant function $R(4, \xi, x)$.

The function $1/R(4, \xi, x)$ is plotted for $|x| > 1$ in Fig. 13.7. The reciprocals $1/R$ and $1/x$ are introduced in order to avoid infinite values. As a consequence, the edge $x = \pm 1$ remains at $1/x = \pm 1$ while the infinite values of x are brought to the origin. The functions $R(2, \xi, x)$ and $R(2, \Xi, X)$ are monotonic over the interval $1 < x < \xi$, implying that the nested function $R(2, R(2, \xi, \xi), R(2, \xi, x)) = R(4, \xi, x)$ has to be monotonic, too.

Notice that the curves in Figs. 13.6 and 13.7 have the same shape. The reciprocal of $R(2, \xi, x)$ oscillates twice between the limiting values $\pm 1/R(2, \xi, \xi)$, while, within the limits $\pm 1/L = \pm 1/R(4, \xi, \xi)$, $1/R(2, \Xi, X)$ varies two times and $1/R(4, \xi, x)$ varies four times.

Let us now express $R(4, \xi, x)$ in terms of x and ξ. According to Eq. (13.32), the expression $R(2, R(2, \xi, \xi), R(2, \xi, x))$ simplifies to

$$R(4, \xi, x) = \frac{(1+t)\left(1+\sqrt{t}\right)^2 x^4 - 2(1+t)(1+\sqrt{t})x^2 + 1}{(1+t)\left(1-\sqrt{t}\right)^2 x^4 - 2(1+t)(1-\sqrt{t})x^2 + 1} \tag{13.43}$$

where

$$t = \sqrt{1 - \frac{1}{\xi^2}} \tag{13.44}$$

From the analysis of the rational functions $R(n, \xi, x)$ for $n = 1$, $n = 2$, and $n = 4$, we find that

- $R(n, \xi, x)$ oscillates between the values ± 1 in the interval $-1 \le x \le 1$ and
- $1/R(n, \xi, x)$ oscillates between the values $\pm 1/R(n, \xi, \xi)$ for $-1/\xi \le 1/x \le 1/\xi$,

which is the property of the elliptic rational function [23]. This means that a quite general analysis, which starts from the six constrains that a rational function has to meet, yields

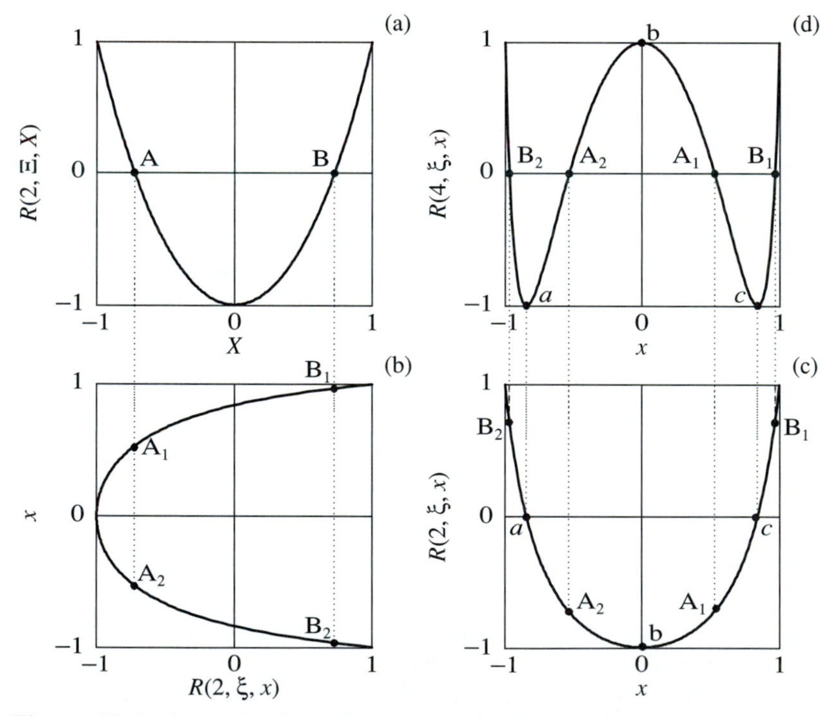

Figure 13.6 Construction of rational elliptic function $R(4, \xi, x) = R(2, \Xi, R(2, \xi, x))$ for $|x| \le 1$.

rational functions that have the equiripple property in the interval $-1 \le x \le 1$, and $1/R(n, \xi, x)$ has the equiripple property in the interval $-1/\xi \le 1/x \le 1/\xi$. In addition, it could be shown [23] that

$$R(n, \xi, x) = \frac{R(n, \xi, \xi)}{R\left(n, \xi, \dfrac{\xi}{x}\right)} \tag{13.45}$$

which means that the equiripple property in the interval $-1 \le x \le 1$ causes the equiripple property in another interval, $-1/\xi \le 1/x \le 1/\xi$.

Equation (13.45) shows that the poles and zeros of $R(n, \xi, x)$ are closely related by

$$x_{\text{zero}} = \frac{\xi}{x_{\text{pole}}}$$

$$R(n, \xi, x_{\text{zero}}) = 0$$

$$\frac{1}{R(n, \xi, x_{\text{pole}})} = 0$$

In the case of the second-order rational function the nesting property holds:

$$\mathcal{R}(2, R(1, \xi, \xi), \xi, x) = R(2, \xi, x) = R\left(2, R\left(1, \xi, \xi\right), R\left(1, \xi, x\right)\right) \tag{13.46}$$

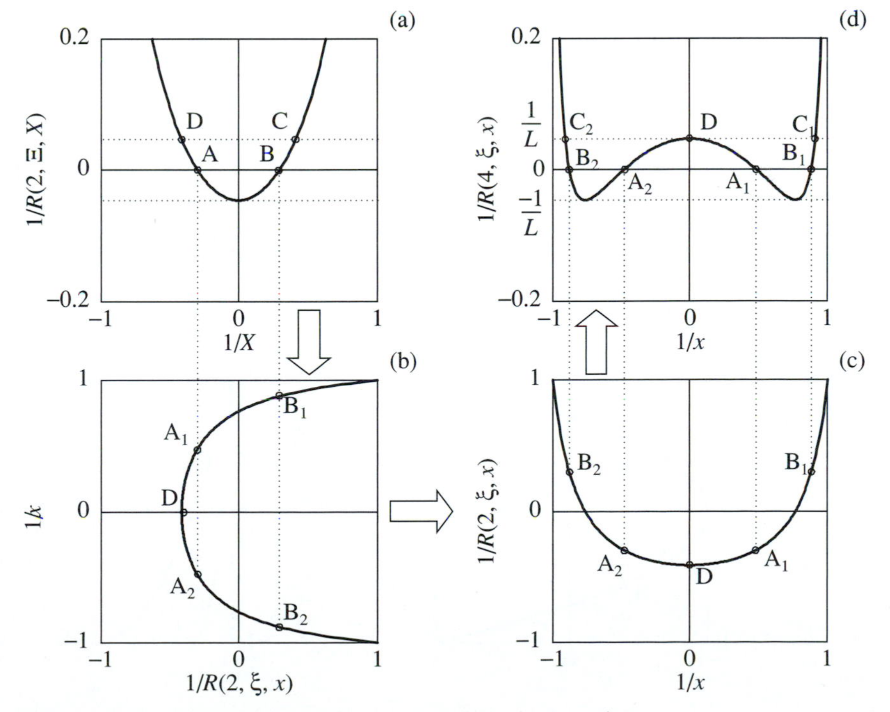

Figure 13.7 Construction of rational elliptic function $R(4, \xi, x) = R(2, \Xi, R(2, \xi, x))$ for $|x| > 1$.

because $R(1,\xi,x) = x$ and $R(1,\xi,\xi) = \xi$.

Similarly, we generate the eighth-order elliptic rational function by nesting

$$\mathcal{R}(8,\, R(4,\xi,\xi),\xi,x) = R(8,\xi,x) = R\left(4,\, R\left(2,\xi,\xi\right),\, R\left(2,\xi,x\right)\right) \tag{13.47}$$

or

$$R(8,\xi,x) = R(4,\Xi,X) \qquad \text{for} \quad \begin{cases} X = R(2,\xi,x) \\ \Xi = R(2,\xi,\xi) \end{cases} \tag{13.48}$$

as shown in Fig. 13.8.

The function $R(4,\Xi,X)$, for $|x| \le 1$, is presented in Fig. 13.8a. The intermediate variable $X = R(2,\xi,x)$, shown in Fig. 13.8b, causes four initial oscillations ($1 > X > -1$, $-1 < X < 1$, $1 > X > -1$, $-1 < X < 1$) to be repeated twice in the resultant function $R(8,\xi,x)$, given in Fig. 13.8d. The corresponding points in all figures are denoted by the same letters. So, for instance, the zeros of the $R(4,\Xi,X)$, denoted by A, B, C, and D in Fig. 13.8a, are projected to the $R(2,\xi,x)$ curve in Fig. 13.8b, whose vertical coordinates are zeros of the resultant function $R(8,\xi,x)$. In order to fix these points on the curve $R(8,\xi,x)$, the $R(2,\xi,x)$ is redrawn in Fig. 13.8c but with x on the abscissa. The function $R(4,\xi,x)$, presented in Fig. 13.8c, point out that all zeros and extremes (a,b,c,d,e,f,g) of $R(4,\xi,x)$ determine the position of the extremes of the resultant function $R(8,\xi,x)$.

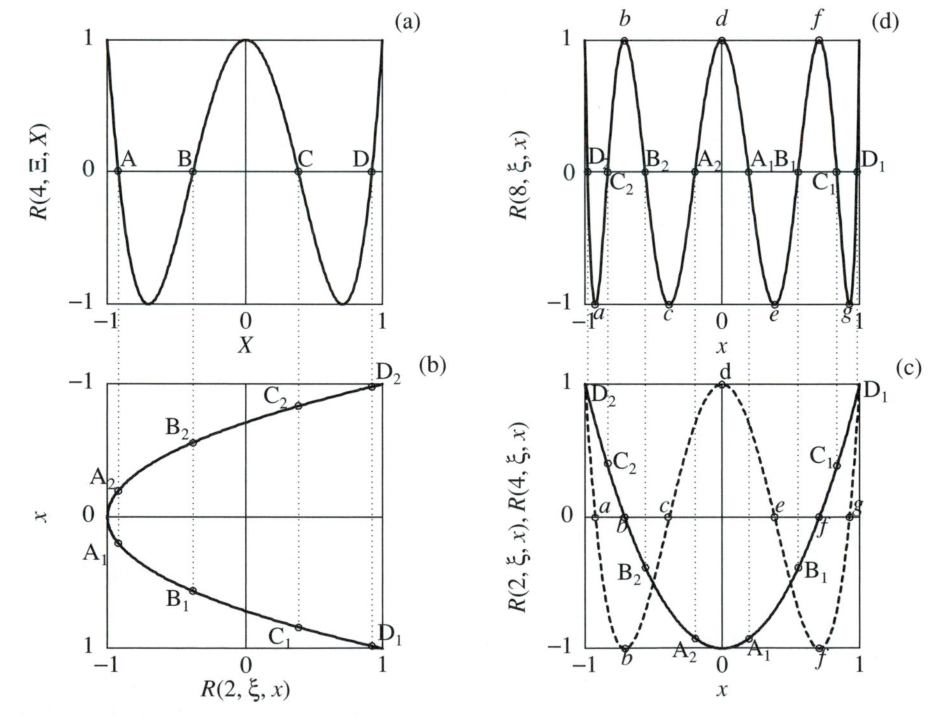

Figure 13.8 Construction of rational elliptic function $R(8,\xi,x) = R(4,\Xi,R(2,\xi,x))$ for $|x| < 1$.

The function $1/R(4, \xi, x)$ is plotted for $|x| > 1$ in Fig. 13.9. The reciprocals $1/R$ and $1/x$ are introduced in order to avoid infinite values. As a consequence, the edge $x = \pm 1$ remains at $1/x = \pm 1$ while the infinite values of x are brought to the origin. The functions $R(2, \xi, x)$ and $R(4, \Xi, X)$ are monotonic over the interval $1 < x < \xi$, implying that the nested function $R(4, R(2, \xi, \xi), R(2, \xi, x)) = R(8, \xi, x)$ has to be monotonic, too.

Notice that the curves in Figs. 13.8 and 13.9 have the same shape. The reciprocal of $R(2, \xi, x)$ oscillates twice between the limiting values $\pm 1/R(2, \xi, \xi)$, while, within the limits $\pm 1/L = \pm 1/R(4, \xi, \xi)$, $1/R(4, \Xi, X)$ varies four times and $1/R(8, \xi, x)$ varies eight times.

After some algebraic manipulation the closed-form expression for $R(8, \xi, x)$ becomes

$$R(8, \xi, x) = \frac{a_8 x^8 + a_6 x^6 + a_4 x^4 + a_2 x^2 + 1}{b_8 x^8 + b_6 x^6 + b_4 x^4 + b_2 x^2 + 1} \tag{13.49}$$

$$a_8 = (1+t)^2 \left(1+\sqrt{t}\right)^2 \left(\left(1+\sqrt{t}\right)^2 + 2\sqrt{2(1+t)\sqrt{t}}\right)$$

$$a_6 = -2(1+t)\left(1+\sqrt{t}\right)^2 \left(2(1+t)\left(1+\sqrt{t}\right) + (3+t)\sqrt{2(1+t)\sqrt{t}}\right)$$

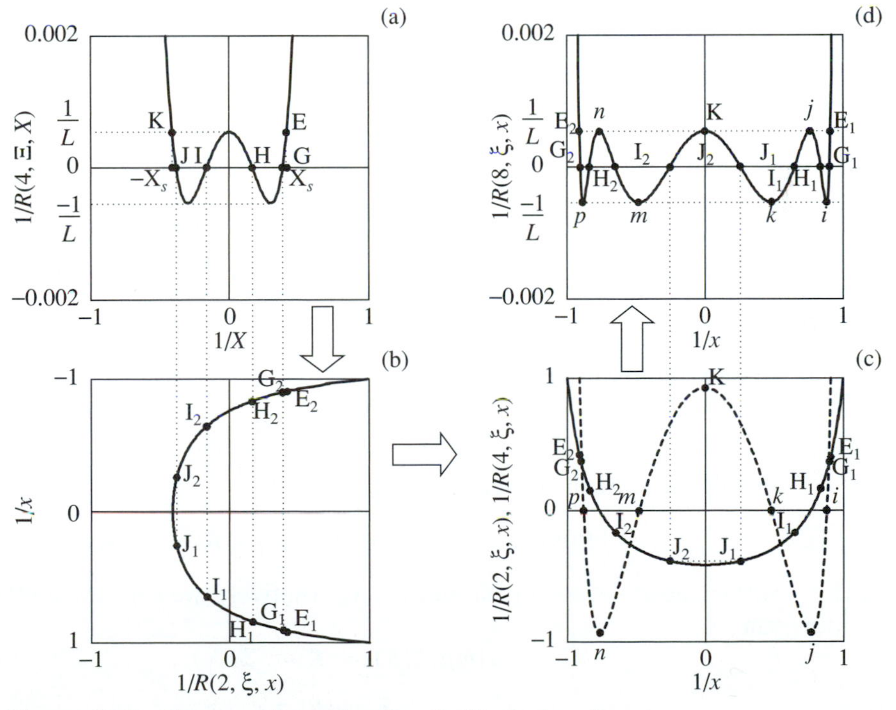

Figure 13.9 Construction of rational elliptic function $R(8, \xi, x) = R(4, \Xi, R(2, \xi, x))$ for $|x| > 1$.

$$a_4 = 2(3+2t)\left(1+\sqrt{t}\right)^2\left(1+t+\sqrt{2(1+t)\sqrt{t}}\right)$$

$$a_2 = -2\left(1+\sqrt{t}\right)\left(2(1+t)+\left(1+\sqrt{t}\right)\sqrt{2(1+t)\sqrt{t}}\right)$$

$$b_8 = (1+t)^2\left(1+\sqrt{t}\right)^2\left(\left(1+\sqrt{t}\right)^2-2\sqrt{2(1+t)\sqrt{t}}\right)$$

$$b_6 = -2(1+t)\left(1+\sqrt{t}\right)^2\left(2(1+t)\left(1+\sqrt{t}\right)-(3+t)\sqrt{2(1+t)\sqrt{t}}\right)$$

$$b_4 = 2(3+2t)\left(1+\sqrt{t}\right)^2\left(1+t-\sqrt{2(1+t)\sqrt{t}}\right)$$

$$b_2 = -2\left(1+\sqrt{t}\right)\left(2(1+t)-\left(1+\sqrt{t}\right)\sqrt{2(1+t)\sqrt{t}}\right)$$

$$t = \sqrt{1-\frac{1}{\xi^2}}$$

It is worth to point out a logical fact that can be useful as a test for the correctness of the $R(n, \xi, x)$ expressions. Namely, by letting $\xi \to \infty$ the elliptic rational function $R(n, \xi, x)$ degrades into the Chebyshev polynomial $C(n, x)$. As examples, the elliptic rational function of the second, fourth, and eighth order are considered.

For infinite ξ the elliptic rational functions $R(2, \xi, x)$, $R(4, \xi, x)$, and $R(8, \xi, x)$ become the Chebyshev polynomials

$$R(2, \xi, x)|_{\xi\to\infty} = C(2, x) = 2x^2 - 1 \tag{13.50}$$

$$R(4, \xi, x)|_{\xi\to\infty} = C(4, x) = 8x^4 - 8x^2 + 1 \tag{13.51}$$

$$R(8, \xi, x)|_{\xi\to\infty} = C(8, x) = 128x^8 - 256x^6 + 160x^4 - 32x^2 + 1 \tag{13.52}$$

and, for an arbitrary order,

$$C(n, x) = R(n, \xi, x)|_{\xi\to\infty} \tag{13.53}$$

Finally, we state the general nesting property:

$$R(mp, \xi, x) = R(m, R(p, \xi, \xi), R(p, \xi, x))$$

for arbitrary positive integers m and p. Alternatively, we can rewrite this relation in the form

$$\begin{aligned} R(mp, \xi, x) &= R(m, \Xi, X) \\ \Xi &= R(p, \xi, \xi) \\ X &= R(p, \xi, x)) \end{aligned} \tag{13.54}$$

If we derive the analytic formulas for $R(n, \xi, x)$, for prime orders 1, 3, 5, 7, 11, and so on, we can exploit the nesting property to generate a closed-form expression for $R(n, \xi, x)$ of an arbitrary order.

13.4 THIRD-ORDER ELLIPTIC RATIONAL FUNCTIONS

The third-order elliptic rational function can be expressed as

$$R(3, \xi, x) = x \frac{(x^2 - x_z^2)}{(x^2 - x_p^2)} \frac{(1 - x_p^2)}{(1 - x_z^2)} \tag{13.55}$$

where the pole, x_p, is a function of ξ [54],

$$x_p^2 = \frac{2\xi^2\sqrt{G}}{\sqrt{8\xi^2(\xi^2 + 1) + 12G\xi^2 - G^3} - \sqrt{G^3}} \tag{13.56}$$

and an auxiliary parameter G,

$$G = \sqrt{4\xi^2 + \left(4\xi^2(\xi^2 - 1)\right)^{\frac{2}{3}}} \tag{13.57}$$

while the zero, x_z, can be found as a function of x_p,

$$x_z^2 = x_p^2(3 - 2x_p^2) + 2x_p\sqrt{(x_p^2 - 1)^3} \tag{13.58}$$

The squared third-order elliptic rational function is presented in Fig. 13.10.

In order to prove Eqs. (13.56) and (13.57), we start with Eq. (13.55) and the condition

$$R^2(3, \xi, \pm x_e) = 1 \tag{13.59}$$

that is,

$$\frac{x_e^2(x_e^2 - x_z^2)^2(1 - x_p^2)^2}{(x_e^2 - x_p^2)^2(1 - x_z^2)^2} = 1 \tag{13.60}$$

and for $x = \pm x_e$ the squared rational function has the local maximum.

Also, we use the fact that $R^2(3, \xi, \pm 1) = 1$:

$$R^2(3, \xi, x) - 1 = \frac{(x^2 - x_e^2)^2(x^2 - 1)}{x_e^4} \tag{13.61}$$

Combining Eqs. (13.60) and (13.61) for $x = x_e$ we find

$$x_e^4 + (x_z^2 - 3x_p^2)x_e^2 + x_z^2 x_p^2 = 0 \tag{13.62}$$

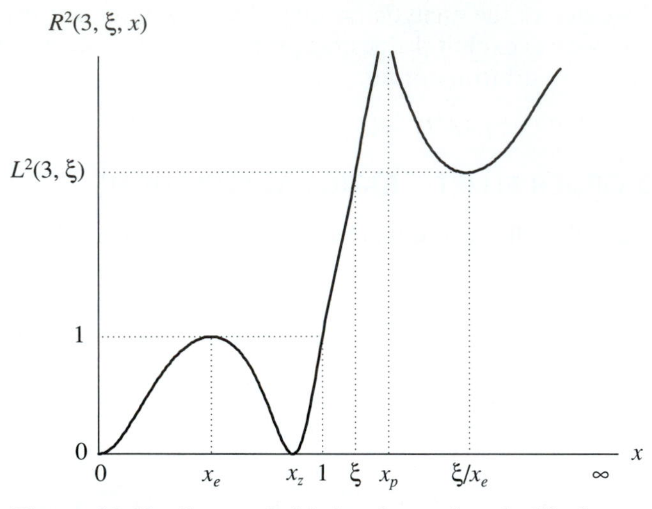

Figure 13.10 Squared third-order rational elliptic function $R^2(3, \xi, x)$.

We find x_e^2 from the above equation as

$$x_e^2 = x_{e1}^2 + x_{e2}^2 \tag{13.63}$$

with

$$x_{e1}^2 = \frac{3x_p^2 - x_z^2}{2}, \quad x_{e2}^2 = -\frac{\sqrt{9x_p^4 - 10x_p^2x_z^2 + x_z^4}}{2} \tag{13.64}$$

From Eq. (13.59) , one obtains

$$\frac{(1 - x_z^2)^2}{(1 - x_p^2)^2}(x_e^2 - x_p^2)^2 = x_e^2(x_e^2 - x_z^2)^2 \tag{13.65}$$

To solve the above equation, it is necessary to collect the x_{e1}^2 terms and even powers of x_{e2}^2 on one side and the odd powers of x_{e2}^2 on the other side of the equation. By squaring the equation, we avoid the square roots and put the equation in the form

$$\left(x_z^4 + (4x_p^4 - 6x_p^2)x_z^2 - 3x_p^4 + 4x_p^2\right)(x_p^2 - x_z^2)^2\left(x_z^2(x_p^2 + 1)^2 - 3x_p^4 - 2x_p^2 + 1\right)^2 = 0 \tag{13.66}$$

The last two terms in Eq. (13.66) are not eligible as a solution since $x_z < 1$. Therefore, only the first term, the equation

$$x_z^4 + (4x_p^4 - 6x_p^2)x_z^2 - 3x_p^4 + 4x_p^2 = 0 \tag{13.67}$$

is acceptable, which immediately gives Eq. (13.58).

Since the selectivity factor ξ can be related to x_p and x_z as

$$\xi = x_z x_p \tag{13.68}$$

the pole, x_p, from Eq. (13.56), is found from the fourth-order equation

$$x_p^8 - \frac{4}{3}(\xi^2 + 1)x_p^6 + 2\xi^2 x_p^4 - \frac{\xi^4}{3} = 0 \tag{13.69}$$

It should be mentioned that we can find an alternative closed-form expression for the third-order elliptic rational function

$$R(3, x, \xi) = (1 + r)^2\, x\, \frac{x^2 - \dfrac{1 + 2r}{(1 + r)^2}}{-(1 - r^2)x^2 + 1}, \qquad 0 < r < 1$$

$$r = \frac{1}{2}\left(-\frac{1 - 2t}{\sqrt{1 + b + b^2}} - 1 + \sqrt{2 + b + 2\sqrt{1 + b + b^2}}\right)$$

$$b = \sqrt[3]{2t\,(1 - 2t)}$$

$$t = \sqrt{1 - \frac{1}{\xi^2}}$$

For a given r the selectivity factor is determined from

$$(1 - r)(1 + r)^3\xi^2 - (1 + 2r) = 0$$

as

$$\xi = \sqrt{\frac{1 + 2r}{(1 - r)(1 + r)^3}}$$

13.5 ELLIPTIC RATIONAL FUNCTIONS OF ORDER $2^i\,3^j$

We have derived explicit analytic expressions for $R(2, \xi, x)$ and $R(3, \xi, x)$ and our goal is to generate functions of the order $n = 2^i 3^j$, $(i, j = 0, 1, 2, 3, \ldots)$. For instance, $R(6, \xi, x)$ can be generated by using the following relations

$$R(6, \xi, x) = R(2, \Xi, X) \qquad \text{for} \quad \begin{cases} X = R(3, \xi, x) \\ \Xi = R(3, \xi, \xi) \end{cases} \tag{13.70}$$

as shown in Fig. 13.11.

The same elliptic rational function is obtained by substituting $R(2, ., .)$ into $R(3, ., .)$:

$$R(6, \xi, x) = R(3, \Xi, X) \qquad \text{for} \quad \begin{cases} X = R(2, \xi, x) \\ \Xi = R(2, \xi, \xi) \end{cases} \tag{13.71}$$

as shown in Fig. 13.12.

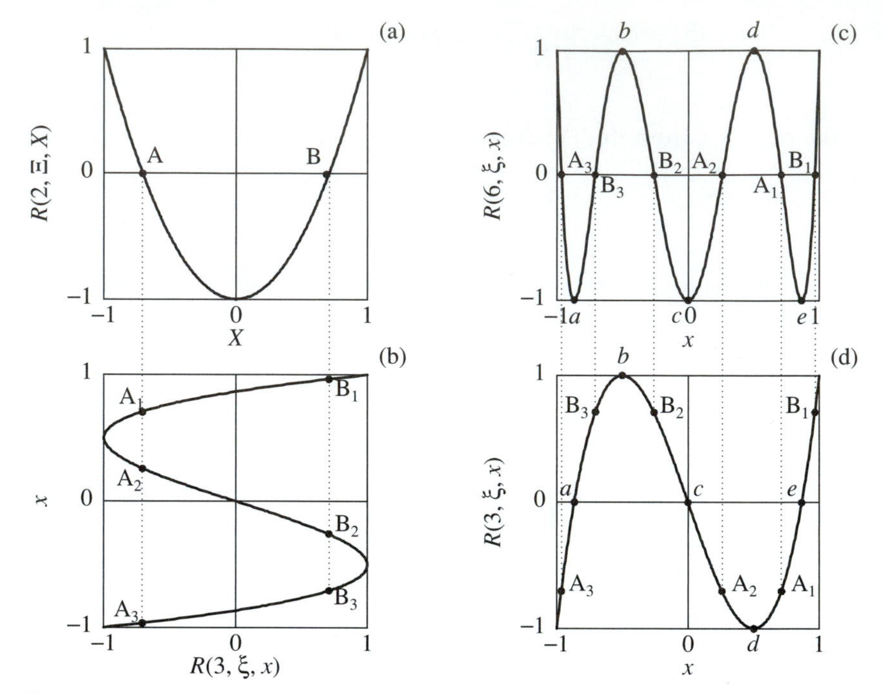

Figure 13.11 Construction of rational elliptic function $R(6, \xi, x) = R(2, \Xi, R(3, \xi, x))$ for $|x| < 1$.

The function $R(2, \Xi, X)$, for $|x| \leq 1$, is presented in Fig. 13.11a. The intermediate variable $X = R(3, \xi, x)$, shown in Fig. 13.11b, causes the two initial oscillations ($1 > X > -1$, $-1 < X < 1$) to be repeated three times in the resultant function $R(6, \xi, x)$, given in Fig. 13.11c. The corresponding points in all figures are denoted by the same letters. So, for instance, the zeros of the $R(2, \Xi, X)$, denoted by A and B in Fig. 13.11a, are projected to the $R(3, \xi, x)$ curve in Fig. 13.11b, whose vertical coordinates are zeros of the resultant function $R(6, \xi, x)$. In order to fix these points on the curve $R(6, \xi, x)$, the $R(3, \xi, x)$ is redrawn in Fig. 13.11d but with x on the abscissa. The $R(3, \xi, x)$, presented in Fig. 13.11d, point out that all zeros and extremes (a, b, c, d, e) of $R(3, \xi, x)$ determine the position of the extremes of the resultant function $R(6, \xi, x)$.

The function $1/R(6, \xi, x)$ is plotted for $|x| > 1$ in Fig. 13.13. The functions $R(2, \xi, x)$ and $R(3, \Xi, X)$ are monotonic over the interval $1 < x < \xi$, implying that the nested function $R(3, R(2, \xi, \xi), R(2, \xi, x)) = R(6, \xi, x)$ has to be monotonic, too.

Notice that the curves in Figs. 13.11 and 13.13 have the same shape. The reciprocal of $R(2, \xi, x)$ oscillates twice between the limiting values $\pm 1/R(2, \xi, \xi)$, while, within the limits $\pm 1/L = \pm 1/R(6, \xi, \xi)$, $1/R(3, \Xi, X)$ varies three times and $1/R(6, \xi, x)$ varies six times. Note that $1/R(3, \Xi, X)$ has smaller range of x than does $1/R(6, \xi, x)$, in which it is equiripple. In Fig. 13.12, $R(3, \Xi, X)$ has the same range of x as does $R(6, \xi, x)$, in which it is equiripple, $|x| \leq 1$.

After some algebraic manipulation the closed-form expression for $R(6, \xi, x)$ becomes

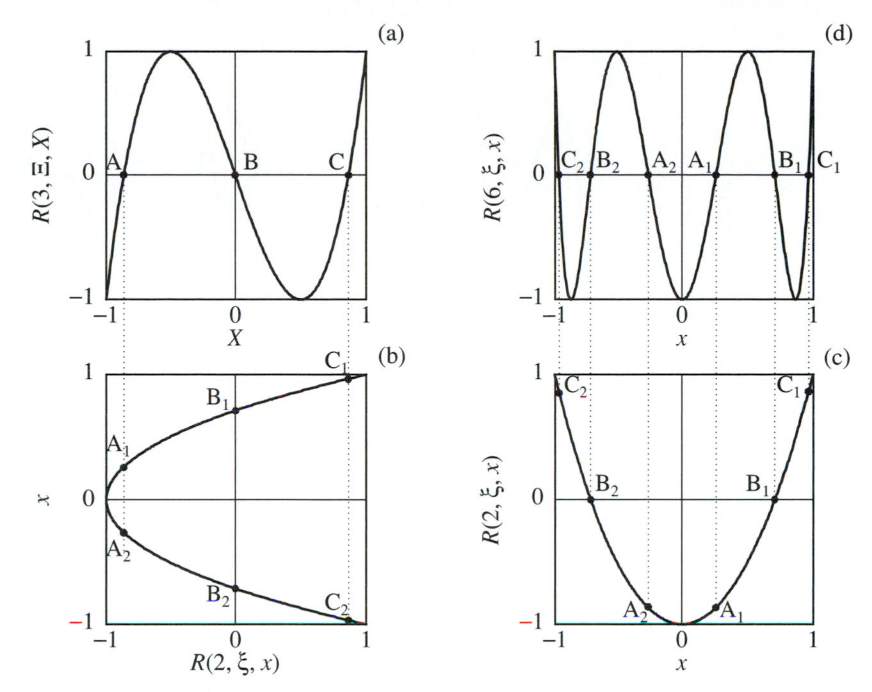

Figure 13.12 Construction of rational elliptic function $R(6, \xi, x) = R(3, \Xi, R(2, \xi, x))$ for $|x| \leq 1$.

$$R(6, \xi, x) = \frac{a_6 x^6 - a_4 x^4 + a_2 x^2 - 1}{b_6 x^6 + b_4 x^4 + b_2 x^2 + 1}$$

$$a_6 = (1 + r)^4 \left(1 + \sqrt{\frac{r\,(2 + r)^3}{(1 + 2r)^3}} \right)$$

$$a_4 = (1 + r)^2 \left(2 + (1 + r)^2 + 2\sqrt{\frac{r\,(2 + r)^3}{1 + 2r}} \right)$$

$$a_2 = 1 + 2(1 + r)^2 + (2 + r)\sqrt{r\,(2 + r)\,(1 + 2r)}$$

$$b_6 = (1 + r)^4 \left(-1 + \sqrt{\frac{r\,(2 + r)^3}{(1 + 2r)^3}} \right)$$

$$b_4 = (1 + r)^2 \left(2 + (1 + r)^2 - 2\sqrt{\frac{r\,(2 + r)^3}{1 + 2r}} \right)$$

$$b_2 = -1 - 2(1 + r)^2 + (2 + r)\sqrt{r\,(2 + r)\,(1 + 2r)}$$

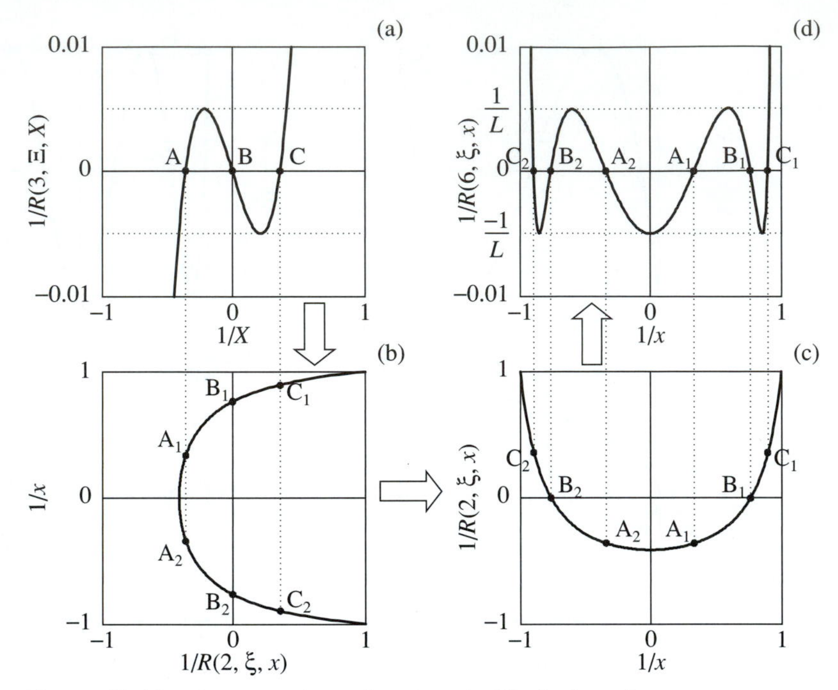

Figure 13.13 Construction of rational elliptic function $R(6, \xi, x) = R(3, \Xi, R(2, \xi, x))$ for $|x| > 1$.

where

$$b = \sqrt[3]{4\frac{\xi^2 - 1}{\xi^4}}$$

$$r = \frac{1}{2}\left(\frac{1 - \dfrac{2}{\xi^2}}{\sqrt{1 + b + b^2}} - 1 + \sqrt{2 + b + 2\sqrt{1 + b + b^2}}\right)$$

$$\frac{a_6}{b_6} = \frac{\left(1 + \sqrt{\dfrac{r\,(2 + r)^3}{(1 + 2r)^3}}\right)^2}{-1 + \dfrac{r\,(2 + r)^3}{(1 + 2r)^3}} = (1 + 2r)^3\,\frac{\left(1 + \sqrt{\dfrac{r\,(2 + r)^3}{(1 + 2r)^3}}\right)^2}{-(1 + 2r)^3 + r\,(2 + r)^3}$$

and

$$R(6, \xi, \xi) = \frac{\sqrt{r\,(2 + r)^3} + \sqrt{(1 + 2r)^3}}{\sqrt{r\,(2 + r)^3} - \sqrt{(1 + 2r)^3}}$$

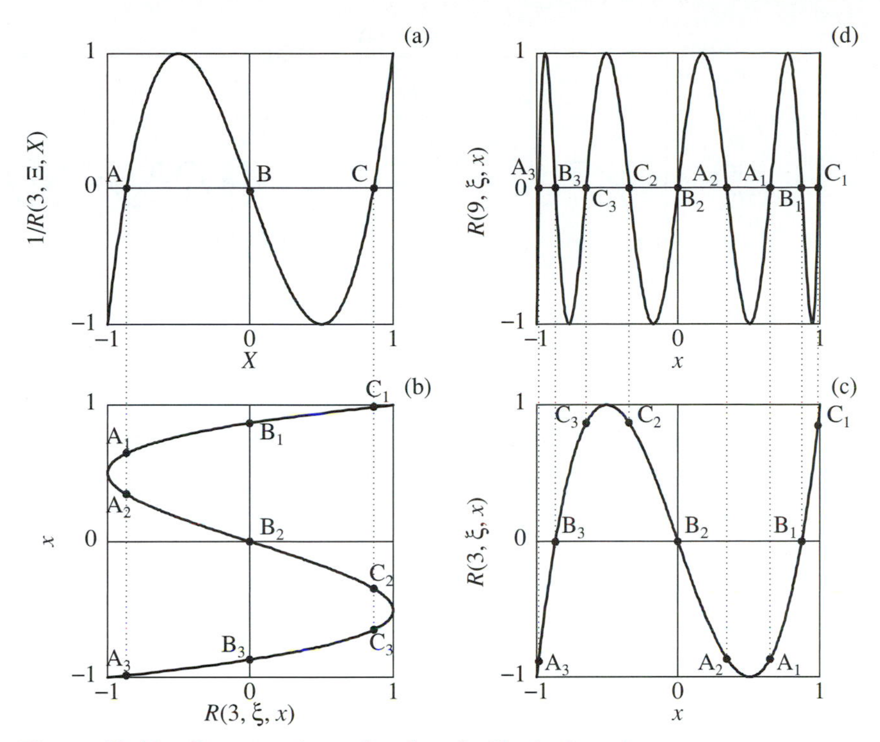

Figure 13.14 Construction of rational elliptic function $R(9, \xi, x) = R(3, \Xi, R(3, \xi, x))$ for $|x| \leq 1$.

It is worth to point out that the edge value, for $x = 1$, of all elliptic rational functions is always 1:

$$\begin{aligned} R(n, \xi, 1) &= 1 \\ R(n, \Xi, 1) &= 1 \end{aligned} \tag{13.72}$$

$R(9, \xi, x)$ can be generated by using only $R(3, ., .)$:

$$R(9, \xi, x) = R(3, \Xi, X) \qquad \text{for} \qquad \begin{cases} X = R(3, \xi, x)) \\ \Xi = R(3, \xi, \xi)) \end{cases} \tag{13.73}$$

as shown in Fig. 13.14. From $R(3, \Xi, X) = 0$, the zero A is transformed in three different zeros of $R(9, \xi, x)$, A_1, A_2 and A_3; one zero is negative and two others are positive. The zero B is transformed into three different zeros of $R(9, \xi, x)$: B_1, B_2, and B_3; one zero is positive and two others are negative. The difference between this zeros is in sign only: $x(A_1) = -x(B_3)$, $x(A_2) = -x(B_2)$, $x(A_3) = -x(B_1)$. The zero of $R(3, \Xi, X)$ at origin is transformed into three zeros: one at the origin, while two others are of the same magnitude but with different sign; those zeros are, also, the zeros of $R(3, \xi, x)$.

The equiripple property of $1/R(9, \xi, x)$, for $x > \xi$, is presented in Fig. 13.15.

The function $1/R(3, \xi, x)$ oscillates three times between the limiting values $\pm 1/R(3, \xi, \xi)$ for $|x| > \xi$, while $1/R(3, \Xi, X)$ vary three times between the limits

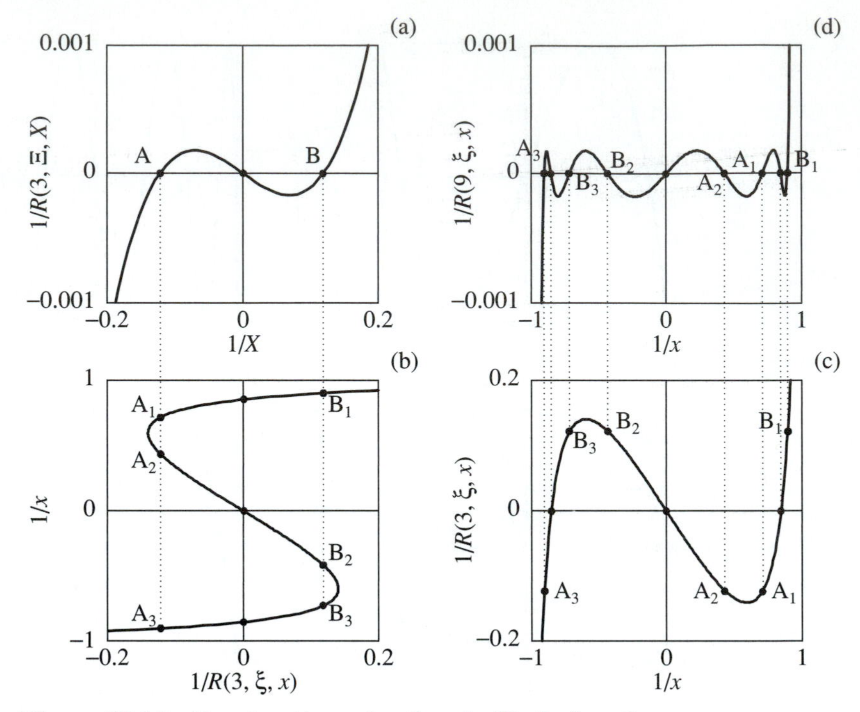

Figure 13.15 Construction of rational elliptic function $R(9, \xi, x) = R(3, \Xi, R(3, \xi, x))$ for $|x| > 1$.

$\pm 1/L = \pm 1/R(9, \xi, \xi)$ for $|x| > R(3, \xi, \xi)$. The resultant function, $1/R(9, \xi, x)$, oscillates nine times between the limiting values $\pm 1/L$ for $|x| > 1/\xi$. $1/R(3, \Xi, X)$ has a smaller range of x than does $1/R(3, \xi, x)$, in which it is equiripple. In Fig. 13.14, $R(3, \Xi, X)$ has the same range of x as $R(3, \xi, x)$ in which it is equiripple, $|x| \leq 1$.

13.6 FIFTH-ORDER ELLIPTIC RATIONAL FUNCTIONS

A straightforward approach to derive $R(5, \xi, x)$ in terms of ξ is not particularly attractive because two very nonlinear simultaneous equations arise [27, 31, 108]. However, $R(5, \xi, x)$ can be expressed in terms of auxiliary quantities z_1 and z_2 in much simpler form:

$$\begin{aligned} R(5, \xi, x) &= \frac{b_5 x^5 + b_3 x^3 + b_1 x}{b_4 x^4 + b_2 x^2 + 1} \\ &= x \frac{(x^2 - x_{z1}^2)(x^2 - x_{z2}^2)(1 - x_{p1}^2)(1 - x_{p2}^2)}{(x^2 - x_{p1}^2)(x^2 - x_{p2}^2)(1 - x_{z1}^2)(1 - x_{z2}^2)} \end{aligned}$$

$$\begin{aligned} b_5 &= (1 + z_1)^2 (1 + z_2)^2 \\ b_3 &= -\big(1 - z_1^2 z_2^2 + (1 + z_1)^2 (1 + z_2)^2 + 2(z_1 + z_2)\big) \\ b_1 &= (1 + 2(z_1 + z_2)) \end{aligned}$$

$$\begin{aligned} b_4 &= (1 - z_1^2)(1 - z_2^2) \\ b_2 &= -(2 - z_1^2 - z_2^2) \end{aligned} \tag{13.74}$$

where

$$\begin{aligned} z_1 &= \sqrt{1 - \frac{x_1^2}{\xi^2}} = \sqrt{1 - \frac{1}{x_{p,1}^2}}, \qquad 0 < z_1 \le 1 \\ z_2 &= \sqrt{1 - \frac{x_2^2}{\xi^2}} = \sqrt{1 - \frac{1}{x_{p,2}^2}}, \qquad 0 < z_2 \le 1 \end{aligned} \tag{13.75}$$

z_1 and z_2 are functions of ξ only. However, z_1 and z_2 are related by

$$z_1^3 - 2z_1z_2 + z_1^2z_2 + z_1z_2^2 - 2z_1^2z_2^2 + z_2^3 = 0 \tag{13.76}$$

From Eq. (13.76) we can calculate z_1 in terms of z_2, or z_2 in terms of z_1.

In addition, we can derive an equation which shows the relationships between ξ and z_1 or z_2:

$$\begin{aligned} &(-1 + 2z_i + 4z_i^2) + \xi^2(1 - z_i)(1 + z_i)^3(3 - 12z_i + 12z_i^2 - 8z_i^3) \\ &\quad - \xi^4(1 - z_i)^2(1 + z_i)^4(3 - 12z_i + 15z_i^2 - 10z_i^3) + \xi^6(1 - z_i)^7(1 + z_i)^5 = 0, \\ &i = 1, 2 \end{aligned} \tag{13.77}$$

or

$$\begin{aligned} &(\xi^2 - 1)^3 - 2(\xi^2 - 1)^3 z_i - 4(\xi^2 - 1)^3 z_i^2 + 10\xi^2(\xi^2 - 1)^2 z_i^3 \\ &\quad + 5\xi^2(\xi^2 - 1)^2 z_i^4 - 4\xi^2(\xi^2 - 1)(5\xi^2 - 3) z_i^5 - 4\xi^2(\xi^2 - 1) z_i^6 \\ &\quad + 4\xi^2(\xi^2 - 1)(5\xi^2 - 2) z_i^7 - 5\xi^4(\xi^2 - 1) z_i^8 - 10\xi^4(\xi^2 - 1) z_i^9 \\ &\quad + 4\xi^6 z_i^{10} + 2\xi^6 z_i^{11} - \xi^6 z_i^{12} = 0, \\ &i = 1, 2 \end{aligned} \tag{13.78}$$

It is interesting to notice that the substitution of ξ^2 by $1 - \xi^2$ and of z_i by $1/z_i$ in Eqs. (13.77) and (13.78) gives the same Eqs. (13.77) and (13.78).

If ξ takes the edge values $\xi = 1$ and $\xi = \infty$, then

$$\begin{aligned} z_1 &= z_2 = 0 \qquad \text{for} \quad \xi = 1 \\ z_1 &= z_2 = 1 \qquad \text{for} \quad \xi = \infty \end{aligned} \tag{13.79}$$

When $\xi \to \infty$ the fifth-order elliptic rational function transforms into the fifth-order Chebyshev polynomial

$$R(5, \xi, x)\,|_{\xi\to\infty} = \frac{16x^5 - 20x^3 + 5x}{1} = C(5, x) \tag{13.80}$$

As an example, we consider $\xi = \sqrt{2}$; the roots of Eq. (13.78) are

$$
\begin{aligned}
z_i \in \{&0.2787682579 \pm j0.2429341359,\\
&-0.8968022467 \pm j0.2429341359,\\
&0.6360098248 \pm j0.8930756889,\\
&-0.6360098248 \pm j0.1069243111,\\
&0.73098799703435,\\
&0.88704599171554,\\
&-1.1711250356,\\
&2.7891590243\}
\end{aligned}
$$

where $j = \sqrt{-1}$. Since $0 < z_i \leq 1$ and assuming that $z_1 \leq z_2$, we find the two roots $z_1 = 0.73098799703435$ and $z_2 = 0.88704599171554$.

Next, we calculate the roots of Eq. (13.76).

Substituting $z_1 = 0.73098799703435$ into Eq. (13.76), we find the roots of Eq. (13.76):

$$z_2 \in \{-0.992853163,\ \ 0.88704599171554,\ \ 0.44350607794511\}$$

Only the second root is acceptable because we assume that $0 \leq z_1 \leq z_2 \leq 1$. The first root is not acceptable because it is negative, while the third has a smaller value than z_1.

Substituting $z_2 = 0.88704599171554$ into Eq. (13.76), we find the roots of Eq. (13.76):

$$z_1 \in \{-0.999574195,\ \ 0.95524138933686,\ \ 0.73098799703436\}$$

Only the third root is acceptable for z_1 because we assume that $0 \leq z_1 \leq z_2 \leq 1$. The first root is not acceptable because it is negative, while the second has larger value than z_2.

If we substitute z_1 and z_2 into Eq. (13.77), we find

$$\xi^2 \in \{-685.0008,\ 2,\ 1.1974896\} \qquad \text{for} \quad z_1 = 0.73098799703435$$

$$\xi^2 \in \{-81325.357,\ 4.295084,\ 2\} \qquad \text{for} \quad z_2 = 0.88704599171554$$

The negative roots are discarded because $\xi > 1$. If we use $z_i = z_1$ and $z_1 < z_2$, then the larger root is the solution. On the contrary, if we use $z_i = z_2$ and $z_1 < z_2$, the smaller root is the solution.

By using the above analysis we can solve the third order equation and find ξ as a function of z_i

$$\xi = \sqrt{\frac{3 - 12z_1 + 15z_1^2 - 10z_1^3 + \sqrt{32z_1^5(3 - 4z_1 + 3z_1^2)}\cos\frac{\phi}{3}}{3(1 - z_1)^5(1 + z_1)}}$$

$$\phi = \cos^{-1}\left(\frac{-(45 - 100z_1 + 142z_1^2 - 100z_1^3 + 45z_1^4)\sqrt{2z_1(3 - 4z_1 + 3z_1^2)}}{4(1 + z_1)^2(3 - 4z_1 + 3z_1^2)^2}\right) \tag{13.81}$$

and

$$\xi = \sqrt{\frac{3 - 12z_2 + 15z_2^2 - 10z_2^3 + \sqrt{32z_2^5(3 - 4z_2 + 3z_2^2)}\cos\frac{\phi + 4\pi}{3}}{3(1 - z_2)^5(1 + z_2)}}$$

$$\phi = \cos^{-1}\left(\frac{-(45 - 100z_2 + 142z_2^2 - 100z_2^3 + 45z_2^4)\sqrt{2z_2(3 - 4z_2 + 3z_2^2)}}{4(1 + z_2)^2(3 - 4z_2 + 3z_2^2)^2}\right) \tag{13.82}$$

We can express the fifth-order elliptic rational function in terms of z_1. For a given z_1 we calculate ξ from Eq. (13.81), and determine z_2 as

$$z_2 = \frac{z_1^2 + (z_1^2 - 1)(1 - \xi^2 + \xi^2 z_1^2)}{z_1^2 - (z_1^2 - 1)(1 - \xi^2 + \xi^2 z_1^2)} \tag{13.83}$$

Finally, substituting z_2 into Eq. (13.74), we find $R(5, \xi, x)$ as a function of z_1 and x.

13.7 MODIFIED ELLIPTIC RATIONAL FUNCTIONS

So far, we have considered higher-order elliptic rational functions, generated by using the nesting property $R(n, \xi, x) = R(m, \Xi, X)$, where $X = R(p, \xi, x)$, $\Xi = R(p, \xi, \xi)$, and $n = mp$. Let us examine what happens if the condition $\Xi = R(p, \xi, \xi)$ is not satisfied—that is, what is the behavior of the general rational function $\mathcal{R}(n, \Xi, \xi, x) = R(m, \Xi, R(p, \xi, x))$? In this analysis, for a given order n, we have two independent parameters, Ξ and ξ. In Figs. 13.16–13.18, the magnitude of an eighth-order rational functions is plotted for different Ξ.

If $\Xi < R(2, \xi, \xi)$, the minimal values of magnitude for $x > \xi$ are smaller and one real pole disappears.

If $\Xi > R(2, \xi, \xi)$, the minimal values of the magnitude are increased in the range about ξ, $x > \xi$, but asymptotic value for $x \to \infty$ is smaller.

For all three cases the magnitude function in range $x \leq 1$ is practically unchanged; zeros are slightly moved, but magnitude is $|R| \leq 1$ as shown in Fig. 13.19.

The same elliptic rational function $R(8, \xi, x) = R(4, R(2, \xi, \xi), R(2, \xi, x))$ can be constructed as $R(8, \xi, x) = R(2, \Xi, R(4, \xi, x))$. In Figs. 13.20 and 13.21, the magnitudes of rational functions are shown for $\Xi \neq R(4, \xi, \xi)$.

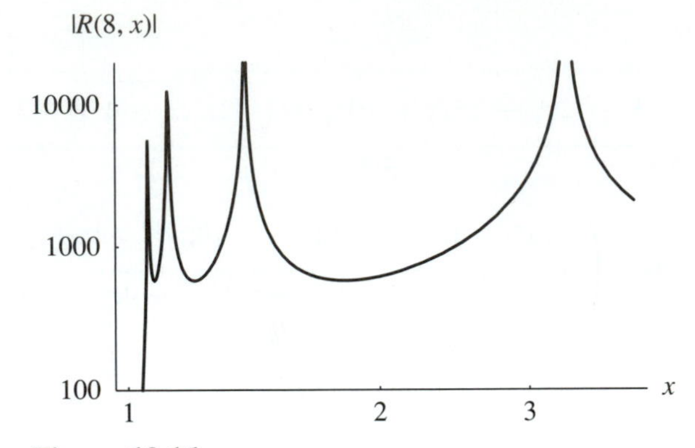

Figure 13.16
$R(8, x) = R(8, \xi, x) = R(4, \Xi, R(2, \xi, x))$ for $\Xi = R(2, \xi, \xi)$.

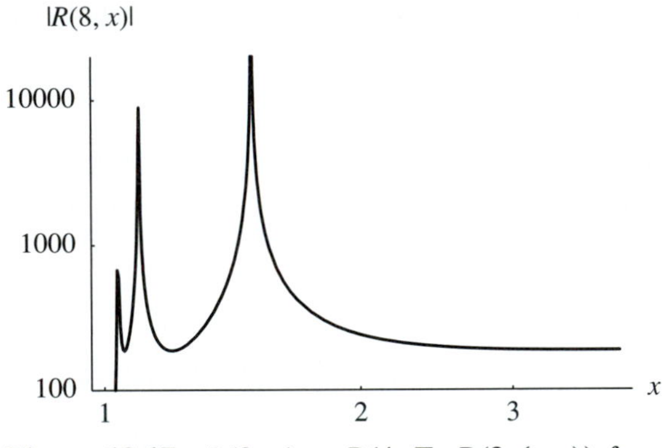

Figure 13.17 $R(8, x) = R(4, \Xi, R(2, \xi, x))$ for $\Xi < R(2, \xi, \xi)$.

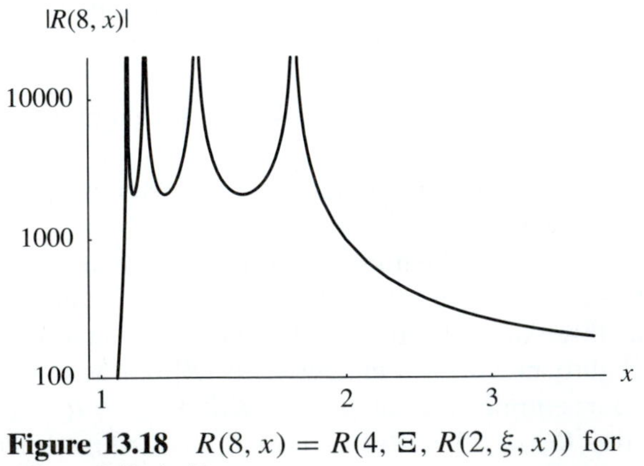

Figure 13.18 $R(8, x) = R(4, \Xi, R(2, \xi, x))$ for $\Xi > R(2, \xi, \xi)$.

Generally, if $\Xi \neq R(4, \xi, \xi)$, the magnitude of the resultant rational function can have a quite different behavior for $x > \xi$ depending on the value of Ξ. We can use this property in constructing rational functions with the equiripple characteristic for $x \leq 1$ and nonuniform requirements for $x > \xi$.

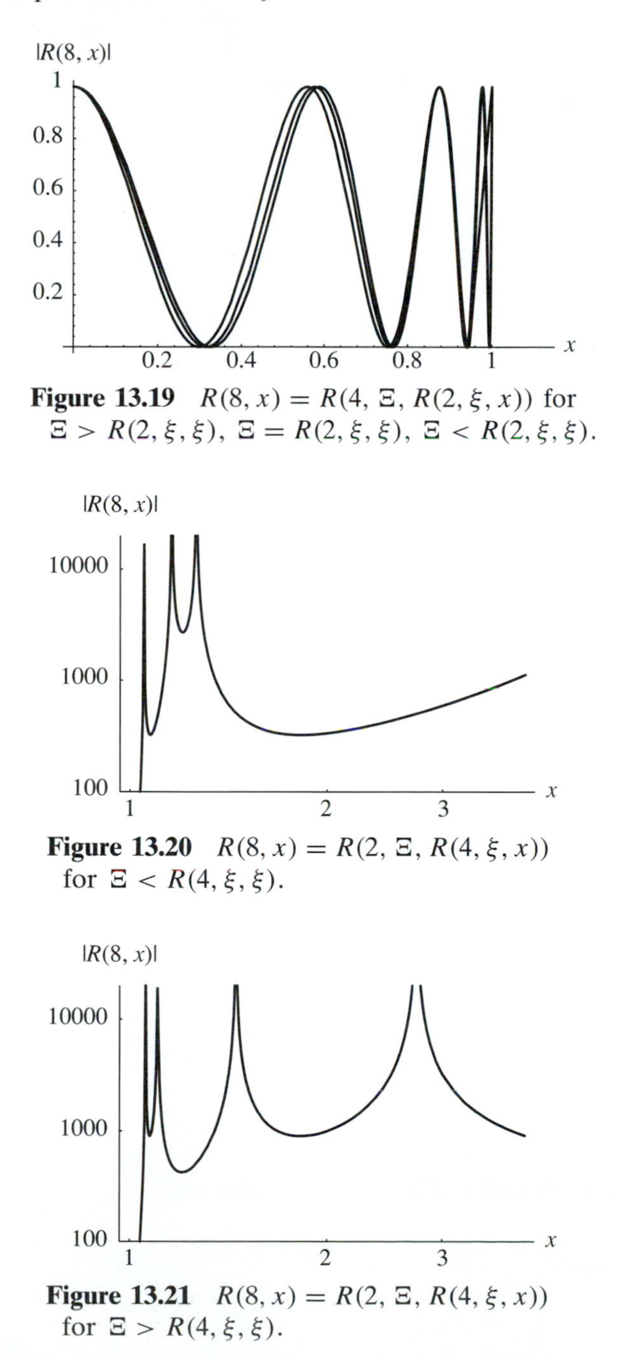

Figure 13.19 $R(8, x) = R(4, \Xi, R(2, \xi, x))$ for $\Xi > R(2, \xi, \xi)$, $\Xi = R(2, \xi, \xi)$, $\Xi < R(2, \xi, \xi)$.

Figure 13.20 $R(8, x) = R(2, \Xi, R(4, \xi, x))$ for $\Xi < R(4, \xi, \xi)$.

Figure 13.21 $R(8, x) = R(2, \Xi, R(4, \xi, x))$ for $\Xi > R(4, \xi, \xi)$.

13.8 POLES, ZEROS, AND SELECTIVITY FACTOR

The elliptic rational function can be expressed in the zero-pole form:

$$R(n,\xi,x) = r_0 \frac{\prod_{i=1}^{n/2}(x^2 - x_i^2)}{\prod_{i=1}^{n/2}(x^2 - x_{p,i}^2)}$$

$$r_0 = \frac{\prod_{i=1}^{n/2}(1 - x_{p,i}^2)}{\prod_{i=1}^{n/2}(1 - x_i^2)} \qquad n \text{ even} \tag{13.84}$$

$$R(n,\xi,x) = r_0\, x \frac{\prod_{i=1}^{(n-1)/2}(x^2 - x_i^2)}{\prod_{i=1}^{(n-1)/2}(x^2 - x_{p,i}^2)}$$

$$r_0 = \frac{\prod_{i=1}^{(n-1)/2}(1 - x_{p,i}^2)}{\prod_{i=1}^{(n-1)/2}(1 - x_i^2)} \qquad n \text{ odd} \tag{13.85}$$

The coefficient r_0 provides

$$R(n,\xi,1) = 1 \tag{13.86}$$

All poles and zeros appear in pairs

$$\begin{aligned} x_i &= -x_{n+1-i} \\ x_{p,i} &= -x_{p,n+1-i} \\ i &= 1, 2, \ldots, n \end{aligned} \tag{13.87}$$

except for odd-order elliptic rational functions

$$\begin{aligned} x_{(n+1)/2} &= 0 \\ x_{p,(n+1)/2} &= \infty \qquad n \text{ odd} \end{aligned} \tag{13.88}$$

From

$$R(n, \xi, x) = \frac{R(n, \xi, \xi)}{R\left(n, \xi, \frac{\xi}{x}\right)} \tag{13.89}$$

we can show that each pole has its corresponding zero related by

$$x_i x_{p,i} = \xi, \quad i = 1, \ldots, n \tag{13.90}$$

13.9 EXACT FORMULAS FOR ZEROS

In this section we find the zeros of the nth-order elliptic rational function for $n = 2^i 3^j$ $(i, j = 0, 1, 2, \ldots)$ as explicit functions of ξ. By using the nesting property we compute the zeros of a higher-order function in terms of the zeros of the lower-order functions.

We have to solve the equation

$$R(mv, \xi, x_i) = R\left(m, \Xi, R(v, \xi, x_i)\right) = R(m, \Xi, X_i) = 0 \tag{13.91}$$

where X_i are the zeros of the elliptic rational function $R(m, \Xi, X)$ with $\Xi = R(v, \xi, \xi)$. The zeros x_i can be calculated from the equation

$$R(v, \xi, x_i) = X_i \tag{13.92}$$

Since we have derived the very basic elliptic rational functions for $n = 2$ and $n = 3$ as explicit functions of the selectivity factor ξ [52, 111], with the help of the nesting properties, we determine the zeros as shown in Table 13.1. After finding the zeros x_i we compute the poles $x_{p,i}$ from Eq. (13.90)

$$x_{p,i} = \frac{\xi}{x_i} \tag{13.93}$$

and the elliptic rational function from Eqs. (13.84) and (13.85).

Table 13.1 Zeros of elliptic rational functions, x_i $(i = 1, \ldots, n/2)$, $x_{n+1-i} = -x_i$, in terms of the selectivity factor ξ

$n = 2$

$$t = \sqrt{1 - \frac{1}{\xi^2}}$$

$$x_{(2)1} = \xi\sqrt{1-t}, \quad x_{(2)2} = -x_{(2)1}$$

$n = 3$

$$G = \sqrt{4\xi^2 + \left(4\xi^2\left(\xi^2 - 1\right)\right)^{\frac{2}{3}}}$$

$$x_{(3)2} = 0$$

$$x_{(3)1} = \sqrt{\frac{\sqrt{8\xi^2(\xi^2+1) + 12G\xi^2 - G^3} - \sqrt{G^3}}{2\sqrt{G}}}, \quad x_{(3)3} = -x_{(3)1}$$

$n = 4$

$$t = \sqrt{1 - \frac{1}{\xi^2}}$$

$$x_{(4)1} = \xi\sqrt{(1-\sqrt{t})\left(1 + t - \sqrt{t^2+t}\right)}, \quad x_{(4)4} = -x_{(4)1}$$

$$x_{(4)2} = \xi\sqrt{(1-\sqrt{t})\left(1 + t + \sqrt{t^2+t}\right)}, \quad x_{(4)3} = -x_{(4)2}$$

$n = 6$

$$t = \sqrt{1 - \frac{1}{\xi^2}}$$

$$L_2 = \xi^2(1+t)^2$$

$$G = \sqrt{4L_2^2 + \left(4L_2^2\left(L_2^2 - 1\right)\right)^{\frac{2}{3}}}$$

$$X_{(3)1} = \sqrt{\frac{\sqrt{8L_2^2(L_2^2+1) + 12GL_2^2 - G^3} - \sqrt{G^3}}{2\sqrt{G}}}$$

$$x_{(6)1} = \sqrt{\frac{1 - X_{(3)1}}{1 + t - (1-t)X_{(3)1}}}, \quad x_{(6)6} = -x_{(6)1}$$

$$x_{(6)2} = \xi\sqrt{1-t}, \quad x_{(6)5} = -x_{(6)2}$$

$$x_{(6)3} = \sqrt{\frac{1 + X_{(3)1}}{1 + t + (1-t)X_{(3)1}}}, \quad x_{(6)4} = -x_{(6)3}$$

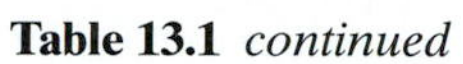

Table 13.1 *continued*

$n = 8$

$$t = \sqrt{1 - \frac{1}{\xi^2}}$$

$$t_2 = \sqrt{1 - \frac{1}{\xi^4(1+t)^4}}$$

$$X_{(4)1} = \xi^2(1+t)^2\sqrt{(1-\sqrt{t_2})\left(1 + t_2 - \sqrt{t_2^2 + t_2}\right)}$$

$$X_{(4)2} = \xi^2(1+t)^2\sqrt{(1-\sqrt{t_2})\left(1 + t_2 + \sqrt{t_2^2 + t_2}\right)}$$

$$x_{(8)1} = \sqrt{\frac{1 - X_{(4)1}}{1 + t - (1-t)X_{(4)1}}}, \qquad x_{(8)8} = -x_{(8)1}$$

$$x_{(8)4} = \sqrt{\frac{1 + X_{(4)1}}{1 + t + (1-t)X_{(4)1}}}, \qquad x_{(8)5} = -x_{(8)4}$$

$$x_{(8)2} = \sqrt{\frac{1 - X_{(4)2}}{1 + t - (1-t)X_{(4)2}}}, \qquad x_{(8)7} = -x_{(8)2}$$

$$x_{(8)3} = \sqrt{\frac{1 + X_{(4)2}}{1 + t + (1-t)X_{(4)2}}}, \qquad x_{(8)6} = -x_{(8)3}$$

$n = 9$

$$G = \sqrt{4\xi^2 + \left(4\xi^2\left(\xi^2 - 1\right)\right)^{\frac{2}{3}}}$$

$$X_1^2 = \frac{2\xi^2\sqrt{G}}{\sqrt{8\xi^2(\xi^2+1) + 12G\xi^2 - G^3} - \sqrt{G^3}}$$

$$L_3 = R(3(\xi, \xi) = \xi^3\left(\frac{1 - X_1^2}{\xi^2 - X_1^2}\right)^2$$

$$G_3 = \sqrt{4L_3^2 + \left(4L_3^2\left(L_3^2 - 1\right)\right)^{\frac{2}{3}}}$$

$$X_{(3)1} = \sqrt{\frac{\sqrt{8L_3^2(L_3^2+1) + 12G_3L_3^2 - G_3^3} - \sqrt{G_3^3}}{2\sqrt{G_3}}}$$

$$a_1 = -\frac{X_{(3)1}(X_1^2 - \xi^2)}{6X_1^2(1 - X_1^2)}$$

$$a_2 = -\frac{\xi^2}{6X_1^2}$$

Table 13.1 *continued*

$$a_3 = \frac{X_{(3)1}(X_1^2 - \xi^2)}{6(1 - X_1^2)}$$

$$\phi = \cos^{-1}\left((6a_1a_2 - 3a_3 - 8a_1^3)\left(4a_1^2 - 2a_2\right)^{-\frac{3}{2}} \right)$$

$$y = \sqrt{16a_1^2 - 8a_2}$$

$$x_{(9)1} = -2a_1 + y\cos\frac{\phi}{3}, \qquad x_{(9)9} = -x_{(9)1}$$

$$x_{(9)2} = 2a_1 - y\cos\frac{\phi + 2\pi}{3}, \qquad x_{(9)8} = -x_{(9)2}$$

$$x_{(9)3} = 2a_1 - y\cos\frac{\phi + 4\pi}{3}, \qquad x_{(9)7} = -x_{(9)3}$$

$$x_{(9)4} = \frac{\xi}{X_1}, \qquad x_{(9)6} = -x_{(9)4}$$

$$x_{(9)5} = 0$$

$n = 12$

$$t = \sqrt{1 - \frac{1}{\xi^2}}$$

$$t_2 = \sqrt{1 - \frac{1}{\xi^4(1+t)^4}}$$

$$L_4 = \xi^4(1+t)^2(1+\sqrt{t})^4$$

$$G = \sqrt{4L_4^2 + \left(4L_4^2\left(L_4^2 - 1\right)\right)^{\frac{2}{3}}}$$

$$X_{(3)1} = \sqrt{\frac{\sqrt{8L_4^2(L_4^2+1) + 12GL_4^2 - G^3} - \sqrt{G^3}}{2\sqrt{G}}}$$

$$X_{(6)1} = \sqrt{\frac{1 - X_{(3)1}}{1 + t_2 - (1 - t_2)X_{(3)1}}}$$

$$X_{(6)2} = \sqrt{\frac{1 + X_{(3)1}}{1 + t_2 + (1 - t_2)X_{(3)1}}}$$

$$x_{(12)1} = \sqrt{\frac{1 - X_{(6)1}}{1 + t - (1 - t)X_{(6)1}}}, \qquad x_{(12)2} = \sqrt{\frac{1 + X_{(6)1}}{1 + t + (1 - t)X_{(6)1}}}$$

$$x_{(12)3} = \sqrt{\frac{1 - X_{(6)2}}{1 + t - (1 - t)X_{(6)2}}}, \qquad x_{(12)4} = \sqrt{\frac{1 + X_{(6)2}}{1 + t + (1 - t)X_{(6)2}}}$$

Table 13.1 *continued*

$$x_{(12)5} = \xi\sqrt{(1-\sqrt{t}\,)\left(1+t-\sqrt{t^2+t}\,\right)}, \qquad x_{(12)8} = -x_{(12)5}$$

$$x_{(12)6} = \xi\sqrt{(1-\sqrt{t}\,)\left(1+t+\sqrt{t^2+t}\,\right)}, \qquad x_{(12)7} = -x_{(12)6}$$

$$x_{(12)12} = -x_{(12)1}, \qquad x_{(12)11} = -x_{(12)2}, \qquad x_{(12)10} = -x_{(12)3}, \qquad x_{(12)9} = -x_{(12)4}$$

$n = 16$

$$t = \sqrt{1-\frac{1}{\xi^2}}$$

$$t_2 = \sqrt{1-\frac{1}{\xi^4(1+t)^4}}$$

$$L_4 = \xi^4(1+t)^2(1+\sqrt{t})^4$$

$$t_3 = \sqrt{1-\frac{1}{L_4^2}}$$

$$X_{(4)1} = L_4\sqrt{(1-\sqrt{t_3})\left(1+t_3-\sqrt{t_3^2+t_3}\right)}$$

$$X_{(4)2} = L_4\sqrt{(1-\sqrt{t_3})\left(1+t_3+\sqrt{t_3^2+t_3}\right)}$$

$$X_{(8)1} = \sqrt{\frac{1-X_{(4)1}}{1+t_2-(1-t_2)X_{(4)1}}}, \qquad X_{(8)2} = \sqrt{\frac{1+X_{(4)1}}{1+t_2+(1-t_2)X_{(4)1}}}$$

$$X_{(8)3} = \sqrt{\frac{1-X_{(4)2}}{1+t_2-(1-t_2)X_{(4)2}}}, \qquad X_{(8)4} = \sqrt{\frac{1+X_{(4)2}}{1+t_2+(1-t_2)X_{(4)2}}}$$

$$x_{(16)1} = \sqrt{\frac{1-X_{(8)1}}{1+t-(1-t)X_{(8)1}}}, \qquad x_{(16)2} = \sqrt{\frac{1+X_{(8)1}}{1+t+(1-t)X_{(8)1}}}$$

$$x_{(16)3} = \sqrt{\frac{1-X_{(8)2}}{1+t-(1-t)X_{(8)2}}}, \qquad x_{(16)4} = \sqrt{\frac{1+X_{(8)2}}{1+t+(1-t)X_{(8)2}}}$$

$$x_{(16)5} = \sqrt{\frac{1-X_{(8)3}}{1+t+(1-t)X_{(8)3}}}, \qquad x_{(16)6} = \sqrt{\frac{1+X_{(8)3}}{1+t+(1-t)X_{(8)3}}}$$

$$x_{(16)7} = \sqrt{\frac{1-X_{(8)4}}{1+t+(1-t)X_{(8)4}}}, \qquad x_{(16)8} = \sqrt{\frac{1+X_{(8)4}}{1+t+(1-t)X_{(8)4}}}$$

$$x_{(16)16} = -x_{(16)1}, \qquad x_{(16)15} = -x_{(16)2}, \qquad x_{(16)14} = -x_{(16)3}, \qquad x_{(16)13} = -x_{(16)4}$$

$$x_{(16)12} = -x_{(16)5}, \qquad x_{(16)11} = -x_{(16)6}, \qquad x_{(16)10} = -x_{(16)7}, \qquad x_{(16)9} = -x_{(16)8}$$

Table 13.1 *continued*

$$n = 18$$

$$t = \sqrt{1 - \frac{1}{\xi^2}}$$

$$L_2 = \xi^2(1+t)^2$$

$$G = \sqrt{4L_2^2 + \left(4L_2^2\left(L_2^2 - 1\right)\right)^{\frac{2}{3}}}$$

$$X_1^2 = \frac{2L_2^2\sqrt{G}}{\sqrt{8L_2^2(L_2^2+1) + 12GL_2^2 - G^3} - \sqrt{G^3}}$$

$$L_3 = L_2^3\left(\frac{1 - X_1^2}{L_2^2 - X_1^2}\right)^2$$

$$G_3 = \sqrt{4L_3^2 + \left(4L_3^2\left(L_3^2 - 1\right)\right)^{\frac{2}{3}}}$$

$$X_{(3)1} = \sqrt{\frac{\sqrt{8L_3^2(L_3^2+1) + 12G_3L_3^2 - G_3^3} - \sqrt{G_3^3}}{2\sqrt{G_3}}}$$

$$a_1 = -\frac{X_{(3)1}(X_1^2 - L_2^2)}{6X_1^2(1 - X_1^2)}$$

$$a_2 = -\frac{L_2^2}{6X_1^2}$$

$$a_3 = \frac{X_{(3)1}(X_1^2 - L_2^2)}{6(1 - X_1^2)}$$

$$\phi = \cos^{-1}\left(\left(6a_1a_2 - 3a_3 - 8a_1^3\right)\left(4a_1^2 - 2a_2\right)^{-\frac{3}{2}}\right)$$

$$y = \sqrt{16a_1^2 - 8a_2}$$

$$X_{(9)1} = -2a_1 + y\cos\frac{\phi}{3}$$

$$X_{(9)2} = 2a_1 - y\cos\frac{\phi + 2\pi}{3}$$

$$X_{(9)3} = 2a_1 - y\cos\frac{\phi + 4\pi}{3}$$

Table 13.1 *continued*

$$X_{(9)4} = \frac{L_2}{X_i}$$

$$x_{(18)1} = \sqrt{\frac{1 - X_{(9)1}}{1 + t - (1-t)X_{(9)1}}}, \qquad x_{(18)2} = \sqrt{\frac{1 + X_{(9)1}}{1 + t + (1-t)X_{(9)1}}}$$

$$x_{(18)3} = \sqrt{\frac{1 - X_{(9)2}}{1 + t - (1-t)X_{(9)2}}}, \qquad x_{(18)4} = \sqrt{\frac{1 + X_{(9)2}}{1 + t + (1-t)X_{(9)2}}}$$

$$x_{(18)5} = \sqrt{\frac{1 - X_{(9)3}}{1 - t - (1-t)X_{(9)3}}}, \qquad x_{(18)6} = \sqrt{\frac{1 + X_{(9)3}}{1 + t + (1-t)X_{(9)3}}}$$

$$x_{(18)7} = \sqrt{\frac{1 - X_{(9)4}}{1 + t - (1-t)X_{(9)4}}}, \qquad x_{(18)8} = \sqrt{\frac{1 + X_{(9)4}}{1 + t + (1-t)X_{(9)4}}}$$

$$x_{(18)9} = \xi\sqrt{1-t}, \qquad x_{(18)10} = -x_{(18)9}$$

$$x_{(18)18} = -x_{(18)1}, \qquad x_{(18)17} = -x_{(18)2}, \qquad x_{(18)16} = -x_{(18)3}, \qquad x_{(18)15} = -x_{(18)4}$$

$$x_{(18)14} = -x_{(18)5}, \qquad x_{(18)13} = -x_{(18)6}, \qquad x_{(18)12} = -x_{(18)7}, \qquad x_{(18)11} = -x_{(18)8}$$

It is interesting to notice that the extremes of $R(2n, \xi, x)$ are at the same time zeros and extremes of $R(n, \xi, x)$. Equating Eq. (13.91 with ± 1, we obtain

$$R(2n, \xi, x_e) = R(2, \Xi, R(n, \xi, x_e)) = R(2, \Xi, X_e) = \pm 1 \tag{13.94}$$

with

$$X_e = R(n, \xi, x_e) \tag{13.95}$$

The extremes are

$$X_{e1} = -1 = R(n, \xi, x_e) \tag{13.96}$$

$$X_{e2} = 0 = R(n, \xi, x_e) \tag{13.97}$$

$$X_{e3} = 1 = R(n, \xi, x_e) \tag{13.98}$$

From Eqs. (13.96), (13.97), and (13.98) it follows that x_e are the minima, maxima, and zeros of the function $R(n, \xi, x)$.

13.10 APPROXIMATE DETERMINATION OF ZEROS

In Section 13.9 we have presented the exact formulas for the zeros of some elliptic rational functions. The underlining method has been based on the exact formulas for the second-order and third-order elliptic rational function and the nesting property of higher-order elliptic rational functions $n = 2^i 3^j$. The nesting property could be used, also, in the case of other functions if we could find explicit closed-form expressions for orders that are prime numbers. At this moment, we do not know analytic expressions for the zeros for $n = 5, 7, 11, 13, 17, \ldots$; therefore, we resort to approximate relations based on the Darlington approach [84, 112].

Once we calculate the zeros of the fifth-order elliptic rational function, we can use the nesting property to calculate the zeros for $n = 10$ and $n = 15$; however, for $n = 25$ the nesting method is not computationally superior in comparison to the Darlington approach.

The approximate algorithm for $n \neq 2^i 3^j$ follows:

$$x_{(n)i} = 2\sqrt{\xi\sqrt{q}}\,\frac{\sum\limits_{m=0}^{\infty}(-1)^m q^{m(m+1)}\sin\dfrac{(2m+1)(n+2i-1)\pi}{2n}}{1+2\sum\limits_{m=1}^{\infty}(-1)^m q^{m^2}\cos\dfrac{m(n+2i-1)\pi}{n}}, \qquad i = 1, 2, \ldots n \tag{13.99}$$

where

$$q(k) = \begin{cases} q_0 = \dfrac{1}{2}\dfrac{1-\sqrt[4]{1-k^2}}{1+\sqrt[4]{1-k^2}} \\ q = q_0 + 2q_0^5 + 15q_0^9 + 150q_0^{13} \\ \quad + 1707q_0^{17} + 20{,}910q_0^{21} \\ \quad + 268{,}616q_0^{25} + 3{,}567{,}400q_0^{29} \\ \quad + 48{,}555{,}069q_0^{33} + 673{,}458{,}874q_0^{37}, & k \le \dfrac{1}{\sqrt{2}} \\ q_p = \dfrac{1}{2}\dfrac{1-\sqrt{k}}{1+\sqrt{k}} \\ q' = q_p + 2q_p^5 + 15q_p^9 + 150q_p^{13} \\ \quad + 1707q_p^{17} + 20'\,910'q_p^{21} \\ \quad + 268{,}616q_p^{25} + 3{,}567{,}400q_p^{29} \\ \quad + 48{,}555{,}069q_p^{33} + 673{,}458{,}874q_p^{37}, & \dfrac{1}{\sqrt{2}} < k < 1 \\ q = e^{\pi^2/\ln q'} \end{cases} \tag{13.100}$$

and

$$k = \frac{1}{\xi} \tag{13.101}$$

Note that the symbol $x_{(n)i}$ denotes the ith zero of the nth-order elliptic rational function for the selectivity factor ξ.

The expansion of the sum from Eq. (13.99) yields

$$x_{(n)i} = 2\sqrt{\xi\sqrt{q}}\,\frac{\begin{aligned}&\sin\frac{(n+2i-1)\pi}{2n} - q^2\sin\frac{3(n+2i-1)\pi}{2n} + q^6\sin\frac{5(n+2i-1)\pi}{2n}\\ &\quad - q^{12}\sin\frac{7(n+2i-1)\pi}{2n} + q^{20}\sin\frac{9(n+2i-1)\pi}{2n}\\ &\quad - q^{30}\sin\frac{11(n+2i-1)\pi}{2n} + q^{42}\sin\frac{13(n+2i-1)\pi}{2n} + \cdots\end{aligned}}{\begin{aligned}&1 - 2q\cos\frac{(n+2i-1)\pi}{n} + 2q^4\cos\frac{2(n+2i-1)\pi}{n} - 2q^9\cos\frac{3(n+2i-1)\pi}{n}\\ &\quad + 2q^{16}\cos\frac{4(n+2i-1)\pi}{n} - 2q^{25}\cos\frac{5(n+2i-1)\pi}{n}\\ &\quad + 2q^{36}\cos\frac{6(n+2i-1)\pi}{n} - 2q^{49}\cos\frac{7(n+2i-1)\pi}{n} + \cdots\end{aligned}}$$

Particularly, for $n = 5$, the above expression simplifies to

$$n = 5$$

$$x_{1(5)} = \sqrt{2\xi\sqrt{q}}\,\frac{\sqrt{5+\sqrt{5}}\,(1 - q^{20} - q^{30}) + \sqrt{5-\sqrt{5}}\,(q^2 - q^{12} - q^{42})}{2 - 4q^{25} + \left(\sqrt{5}+1\right)(q - q^{16} - q^{36}) + \left(\sqrt{51}\right)(q^4 - q^9 - q^{49})}$$

$$x_{2(5)} = \sqrt{2\xi\sqrt{q}}\,\frac{\sqrt{5-\sqrt{5}}\,(1 - q^{20} - q^{30}) - \sqrt{5+\sqrt{5}}\,(q^2 - q^{12} - q^{42})}{2 - 4q^{25} - \left(\sqrt{5}-1\right)(q - q^{16} - q^{36}) - \left(\sqrt{5}+1\right)(q^4 - q^9 - q^{49})}$$

$$x_{3(5)} = 0$$

Similarly, the zeros of the seventh-order function are calculated from

$$n = 7$$

$$x_{1(7)} = 2\sqrt{\xi\sqrt{q}}\,\frac{(1-q^{42})\sin\frac{3\pi}{7} + (q^2-q^{30})\sin\frac{2\pi}{7} + (q^6-q^{20})\sin\frac{\pi}{7}}{1-2q^{49}+2(q-q^{36})\cos\frac{\pi}{7} + 2(q^4-q^{25})\cos\frac{2\pi}{7} + 2(q^9-q^{16})\cos\frac{3\pi}{7}}$$

$$x_{2(7)} = 2\sqrt{\xi\sqrt{q}}\,\frac{(1-q^{42})\sin\frac{2\pi}{7} - (q^2-q^{30})\sin\frac{\pi}{7} - (q^6-q^{20})\sin\frac{3\pi}{7}}{1-2q^{49}+2(q-q^{36})\cos\frac{3\pi}{7} - 2(q^4-q^{25})\cos\frac{\pi}{7} - 2(q^9-q^{16})\cos\frac{2\pi}{7}}$$

$$x_{3(7)} = 2\sqrt{\xi\sqrt{q}}\,\frac{(1-q^{42})\sin\frac{\pi}{7} - (q^2-q^{30})\sin\frac{3\pi}{7} + (q^6-q^{20})\sin\frac{2\pi}{7}}{1-2q^{49}-2(q-q^{36})\cos\frac{2\pi}{7} - 2(q^4-q^{25})\cos\frac{3\pi}{7} + 2(q^9-q^{16})\cos\frac{\pi}{7}}$$

$$x_{4(7)} = 0$$

It should be observed that there exist zeros whose exact values are known. For instance, for odd n, we exactly know $x_{(n+1)/2} = 0$. Next, it can be shown that

$$x_{(10)3} = x_{(14)4} = \frac{1}{\sqrt{1+\sqrt{1-1/\xi^2}}}$$

$$x_{(15)3} = x_{(3)1}$$

which can be used to improve computational efficiency, or as a reference value for testing the accuracy of the numerical procedure.

13.11 DISCRIMINATION FACTOR

We define the *discrimination factor* as the value of the elliptic rational function at $x = \xi$, and we denote it by $L(n, \xi)$:

$$L(n, \xi) = R(n, \xi, \xi) \tag{13.102}$$

For a known $R(n, \xi, x)$ we can directly compute $L(n, \xi)$; however, we shall derive the discrimination factor in an alternative way, in a more compact form.

The discrimination factor of the first-order elliptic rational function is

$$L(1, \xi) = \xi \tag{13.103}$$

The second-order discrimination factor $L(2, \xi)$ is given by

$$L(2, \xi) = 2\xi^2 - 1 + 2\xi\sqrt{\xi^2 - 1} \tag{13.104}$$

or

$$L(2, \xi) = \left(\xi + \sqrt{\xi^2 - 1}\right)^2 \tag{13.105}$$

Formally, we can express $L(2, \xi)$ from Eq. (13.105) in the nested form in terms of a lower-order discrimination factor $L(1, \xi)$:

$$L\left(2, L(1, \xi)\right) = \left(L(1, \xi) + \sqrt{L(1, \xi)^2 - 1}\right)^2 \tag{13.106}$$

Similarly, after some simplification, the discrimination factor of the fourth-order elliptic rational function is found to be [53]

$$L(4, \xi) = \left(\sqrt{\xi} + \sqrt[4]{\xi^2 - 1}\right)^4 \left(\xi + \sqrt{\xi^2 - 1}\right)^2 \tag{13.107}$$

Alternatively, the same result follows from the nested expression

$$L(4, \xi) = L(2, L(2, \xi)) \tag{13.108}$$

or

$$L(4, \xi) = \left(L(2, \xi) + \sqrt{L(2, \xi)^2 - 1}\right)^2 \tag{13.109}$$

Generally, the nesting property of the discrimination factor can be formulated as

$$L(uv, \xi) = L(u, L(v, \xi)) \tag{13.110}$$

For example, Eq. (13.108) combined with Eqs. (13.105) and (13.107) allows us to determine higher-order discrimination factors: $L(8, \xi)$, $L(16, \xi)$, $L(32, \xi)$, and so on. For instance, $L(8, \xi)$ can be computed in several ways:

$$L(8, \xi) = L(2, L(4, \xi)) = L\left(2, L(2, L(2, \xi))\right) = L(4, L(2, \xi)) \tag{13.111}$$

which gives, after some labor,

$$L(8, \xi) = \left(\left(\sqrt{\xi + \sqrt{\xi^2 - 1}} + \sqrt[4]{4\xi}\,\sqrt[8]{\xi^2 - 1}\right)\left(\sqrt{\xi} + \sqrt[4]{\xi^2 - 1}\right)\left(\xi + \sqrt{\xi^2 - 1}\right)\right)^4 \tag{13.112}$$

Since the explicit expressions for poles and zeros of the third-order elliptic rational function as a function of ξ are already known, the possibility to exactly calculate $L(n, \xi)$ is extended to all functions of the order $n = 2^i 3^j \, (i, j = 0, 1, 2, 3, \ldots)$.

The discrimination factor $L(3, \xi)$ is obtained from

$$L(3, \xi) = \xi^3 \left(\frac{1 - x_z^2}{\xi^2 - x_z^2}\right)^2 \tag{13.113}$$

where

$$x_z^2 = \frac{2\xi^2\sqrt{G}}{\sqrt{8\xi^2(\xi^2 + 1) + 12G\xi^2 - G^3} - \sqrt{G^3}} \tag{13.114}$$

$$G = \sqrt{4\xi^2 + \left(4\xi^2(\xi^2 - 1)\right)^{2/3}} \tag{13.115}$$

The discrimination factor of the order $n \neq 2^i 3^j$ can be approximated by

$$L(n, \xi) \approx L_{approx}(n, \xi) = \sqrt{L(n+1, \xi)L(n-1, \xi)} \qquad (13.116)$$

which is motivated by a relation that is valid for the Chebyshev polynomial [110, p. 196]

$$C^2(n, x) = C(n+1, x)C(n-1, x) + 1 - x^2 \qquad (13.117)$$

or approximately

$$C(n, x) \approx C_{approx}(n, x) = \sqrt{C(n+1, x)C(n-1, x)}, \qquad x \geq 1 \qquad (13.118)$$

From the filter design viewpoint the error caused by neglecting the term $(1 - x^2)$ in Eq. (13.117) can be defined by

$$\begin{aligned} E(n, \epsilon, x) &= 10 \log_{10}(1 + \epsilon^2 C^2_{approx}(n, x)) - 10 \log_{10}(1 + \epsilon^2 C^2(n, x)) \\ &= 10 \log_{10}\left(1 + \epsilon^2 \frac{x^2 - 1}{1 + \epsilon^2 C^2(n, x)}\right) \end{aligned} \qquad (13.119)$$

where we assume that ϵ (known as the *ripple factor*) takes a value from the interval $0 < \epsilon < 1$.

The error for the polynomials $C(5, x)$ and $C(7, x)$ is shown in Fig. 13.22. For $x > 1$ the error decreases rapidly because the neglected term becomes insignificant; that is, $x^2 - 1 << C^2(n, x)$.

For the sake of comparison, Fig. 13.22 also contains the error curves of the two lower-order discrimination factors $L(5, \xi)$ and $L(7, \xi)$ for which we have to use the approximate relation, Eq. (13.116).

The similarity of the L and C curves, as well as the magnitude of encountered errors, justifies the use of the approximate relation, Eq. (13.116). Even maximal errors found in Fig. 13.22 are small for most practical applications. These errors, drawn as solid lines in Fig. 13.22 and enlarged in Fig. 13.23, are negligible in the usual domain of the selectivity factor ξ.

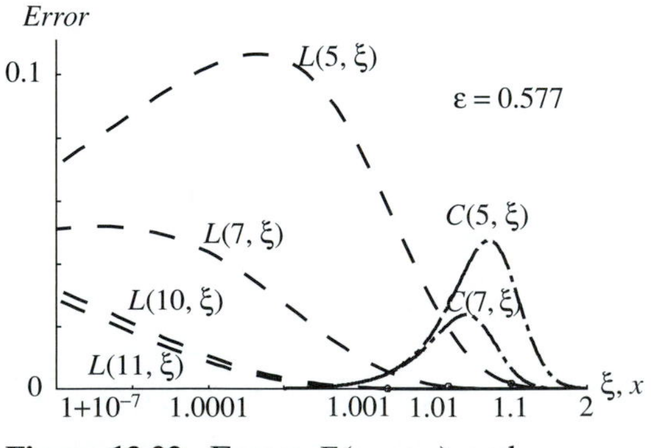

Figure 13.22 Errors $E(n, \epsilon, x)$ and $(10 \log_{10}(1 + \epsilon^2 L^2_{approx}(n, \xi)) - 10 \log_{10}(1 + \epsilon^2 L^2(n, \xi)))$.

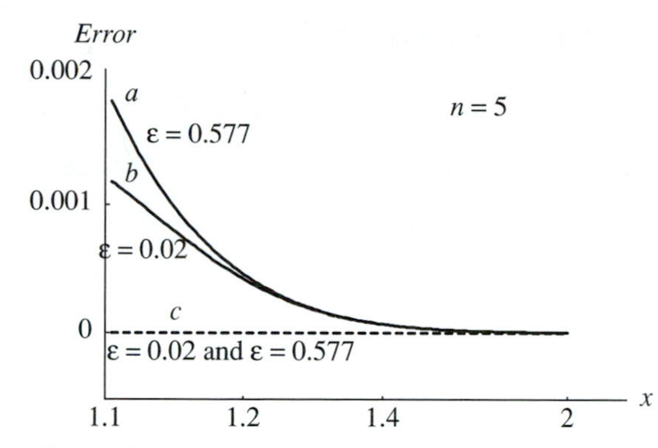

Figure 13.23 Error ($10\ \log_{10}(1 + \epsilon^2 L^2_{approx}(n, \xi)) - 10\ \log_{10}(1 + \epsilon^2 L^2(5, \xi))$) a) $\epsilon = 0.577$ and Eq. (13.116), b) $\epsilon = 0.02$ and Eq. (13.116), c) $\epsilon = 0.577$, $\epsilon = 0.02$ and Eqs. (13.120) and (13.130).

The curves in Fig. 13.23, drawn for the two extreme ripple factors, $\epsilon = 0.577$ and $\epsilon = 0.02$, show that the errors increase with ϵ, which is again in full agreement with Eq. (13.119).

If the exact expressions for $L(n+1, \xi)$ and $L(n-1, \xi)$ are not available, the more general relation can be applied:

$$L^{r-p}(n, \xi) \approx L^{n-p}(r, \xi)\ L^{r-n}(p, \xi) \tag{13.120}$$

where $p < n < r$.

Theorem 13.1 If the nth-order discrimination factor $L(n, x)$ satisfies the condition

$$L^2(n, \xi) = L(n + 1, \xi)L(n - 1, \xi) \tag{13.121}$$

then

$$L^{r-p}(n, \xi) = L^{n-p}(r, \xi)\ L^{r-n}(p, \xi), \qquad r > n > p \tag{13.122}$$

Proof: Throughout this proof we shall use the short notation $L_n = L(n, \xi)$. The second argument ξ is to be understood in every case.

If we replace n by $n + 1$ in Eq. (13.121) we obtain

$$L_{n+2} = \frac{L^2_{n+1}}{L_n} = \frac{\left(\dfrac{L^2_n}{L_{n-1}}\right)^2}{L_n} = \frac{L^3_n}{L^2_{n-1}} \tag{13.123}$$

From Eq. (13.123) with n replaced by $n+1$ we have

$$L_{n+3} = \frac{L_{n+1}^3}{L_n^2} = \frac{\left(\frac{L_n^2}{L_{n-1}}\right)^3}{L_n^2} = \frac{L_n^4}{L_{n-1}^3} \tag{13.124}$$

Repeating this process m times, we obtain

$$L_{n+m+1} = \frac{L_{n+1}^{m+1}}{L_n^m} = \frac{\left(\frac{L_n^2}{L_{n-1}}\right)^{m+1}}{L_n^m} = \frac{L_n^{m+2}}{L_{n-1}^{m+1}}, \qquad m > 0 \tag{13.125}$$

Denoting $n+m+1$ by r, we obtain

$$L_r = \frac{L_n^{r-n+1}}{L_{n-1}^{r-n}}, \qquad r > n+1 \tag{13.126}$$

The proof for L_p is similar:

$$L_p = \frac{L_n^{n+1-p}}{L_{n+1}^{n-p}}, \qquad p < n-1 \tag{13.127}$$

Substituting Eqs. (13.126) and (13.127) into the right-hand side of Eq. (13.122), we find

$$\begin{aligned} L_r^{n-p} L_p^{r-n} &= \frac{L_n^{(r-n+1)(n-p)}}{L_{n-1}^{(r-n)(n-p)}} \frac{L_n^{(n+1-p)(r-n)}}{L_{n+1}^{(n-p)(r-n)}} \\ &= \frac{L_n^{n-p+r-n+2(r-n)(n-p)}}{(L_{n-1}L_{n+1})^{(r-n)(n-p)}} = \frac{L_n^{r-p} L_n^{2(r-n)(n-p)}}{\left(L_n^2\right)^{(r-n)(n-p)}} \\ &= L_n^{r-p} = L_n^{r-p} \end{aligned} \tag{13.128}$$

which concludes the proof. ■

The fact that higher-order functions for a given ξ have smaller errors implies a possibility of a further reduction of the error by calculating $L(2n, \xi)$ instead of $L(n, \xi)$. Required $L(n, \xi)$ is then obtained by means of Eqs. (13.104) and (13.110) as

$$L(2n, \xi) = \left(L(n, \xi) + \sqrt{L(n, \xi)^2 - 1}\right)^2 \tag{13.129}$$

from which

$$L(n, \xi) = \frac{L(2n, \xi) + 1}{2\sqrt{L(2n, \xi)}} \tag{13.130}$$

Since Eq. (13.130) is exact, the error of $L(n, \xi)$ now conforms to that of $L(2n, \xi)$, which is more precisely determined.

To illustrate the above statement we shall consider an example of the fifth-order elliptic rational function with $\xi = 1/\sin 65^o = 1.1033779$ and $\epsilon = 1/\sqrt{3} = 0.57735027$.

The approximate expression, Eq. (13.116), yields an error less than 2×10^{-3}. The approximate procedure suggested in reference [84] gives a similar error.

We obtain $L(5, \xi)$ more precisely by the following procedure:

1. Determine $L(3, \xi) = 7.273070086$ from the exact equations (13.114)–(13.113); next, $L(9, \xi) = L(3, L(3, \xi))$, where in Eqs. (13.114)–(13.113) ξ is replaced by $L(3, \xi)$, $L(9, \xi) = L(3, 7.273070086) = 6068.31197$;
2. Compute $L(12, \xi) = L(4, L(3, \xi)) = L(4, 7.273070086) = 17570.2197$ from Eq. (13.106);
3. Find $L(10, \xi)$ from Eq. (13.120), $L(10, \xi) \approx \left(L(9, \xi)^2 L(12, \xi)\right)^{1/3}$;
4. Calculate $L(5, \xi)$ from Eq. (13.130).

This procedure provides a considerably smaller error 2×10^{-8}.

Notice that the discrimination factor can be calculated from the known zeros x_i of the elliptic rational function, along with the selectivity factor ξ:

$$L(n, \xi) = \frac{1}{\xi^n} \frac{\prod_{i=1}^{\frac{n}{2}} (\xi^2 - x_i^2)^2}{\prod_{i=1}^{\frac{n}{2}} (1 - x_i^2)^2}, \qquad n \text{ even} \tag{13.131}$$

$$L(n, \xi) = \frac{1}{\xi^{n-2}} \frac{\prod_{i=1}^{\frac{n-1}{2}} (\xi^2 - x_i^2)^2}{\prod_{i=1}^{\frac{n-1}{2}} (1 - x_i^2)^2}, \qquad n \text{ odd} \tag{13.132}$$

13.12 SELECTIVITY FACTOR VERSUS DISCRIMINATION FACTOR

Until now we have considered (and defined) the discrimination factor L as a function of the order n and the selectivity factor ξ. We have used the short notation $L = L(n, \xi)$ because we assumed that n and ξ were known. A question arises as to whether we can find the inverse function of $L(n, \xi)$, for a given n, and compute ξ in terms of L. We shall use $\mathcal{L}(n, \xi)$ to denote the function that maps n and ξ into L; however, we shall introduce the symbol $\mathcal{L}^{-1}(n, L)$ to stand for mapping of n and L into ξ

$$\begin{aligned} L &= \mathcal{L}(n, \xi) \\ \xi &= \mathcal{L}^{-1}(n, L) \end{aligned} \tag{13.133}$$

The first-order discrimination factor is

$$L = \mathcal{L}(1, \xi) = \xi \tag{13.134}$$

The corresponding selectivity factor is

$$\xi = \mathcal{L}^{-1}(1, L) = L \tag{13.135}$$

The second-order discrimination factor is

$$L = \mathcal{L}(2, \xi) = \left(\xi + \sqrt{\xi^2 - 1}\right)^2 \tag{13.136}$$

and the corresponding selectivity factor is

$$\xi = \mathcal{L}^{-1}(2, L) = \frac{L + 1}{2\sqrt{L}} \tag{13.137}$$

The third-order discrimination factor can be derived to be

$$L = \mathcal{L}(3, \xi) = \sqrt{\frac{(1 + 2z)^3}{(1 - z)^3(1 + z)}} \tag{13.138}$$

where z is one root of the fourth-order polynomial

$$\xi^2(1 - z^2)(1 + z)^2 - 1 + 2z = 0, \qquad 0 \leq z \leq 1 \tag{13.139}$$

The corresponding selectivity factor is

$$\xi = \mathcal{L}^{-1}(3, L) = \sqrt{\frac{1 + 2z}{(1 - z)(1 + z)^3}} \tag{13.140}$$

where z is a root of the fourth-order polynomial

$$L^2(1 - z)^2(1 - z^2) - (1 + 2z)^3 = 0, \qquad 0 \leq z \leq 1 \tag{13.141}$$

Substituting Eq. (13.138) into Eq. (13.140), we can prove that $\mathcal{L}^{-1}(3, L)$ is the inverse function of $\mathcal{L}(3, \xi)$:

$$\xi = \mathcal{L}^{-1}(3, L) = \mathcal{L}^{-1}(3, \mathcal{L}(3, \xi)) \tag{13.142}$$

We derive the fifth-order discrimination factor

$$L = \mathcal{L}(5, \xi) = \sqrt[4]{\frac{(1 + z_1)(1 + z_2)\,(1 + 2(z_1 + z_2))^5}{(1 - z_1)^5(1 - z_2)^5}} \tag{13.143}$$

where, for a known ξ, the auxiliary variables z_1 and z_2 can be found as roots of the 12th-order polynomial ($0 \le z_1 \le z_2 \le 1$)

$$\begin{aligned}(\xi^2-1)^3 &- 2(\xi^2-1)^3 z + 4(\xi^2-1)^3 z^2 + 10\xi^2(\xi^2-1)^2 z^3 \\ &+ 5\xi^2(\xi^2-1)^2 z^4 - 4\xi^2(\xi^2-1)(5\xi^2-3) z^5 - 4\xi^2(\xi^2-1) z^6 \\ &+ 4\xi^2(\xi^2-1)(5\xi^2-2) z^7 - 5\xi^4(\xi^2-1) z^8 - 10\xi^4(\xi^2-1) z^9 \\ &+ 4\xi^6 z^{10} + 2\xi^6 z^{11} - \xi^6 z^{12} = 0, \qquad 0 \le z_1 \le z_2 \le 1\end{aligned} \tag{13.144}$$

The fifth-order selectivity factor is determined by

$$\xi = \mathcal{L}^{-1}(5, L) = \sqrt[4]{\frac{1 + 2(z_1 + z_2)}{(1 - z_1)(1 - z_2)(1 + z_1)^3(1 + z_2)^3}} \tag{13.145}$$

where z_1 and z_2 are computed from the system of equations

$$\begin{aligned} z_1^3 - 2z_1z_2 + z_1^2z_2 + z_1z_2^2 - 2z_1^2z_2^2 + z_2^3 &= 0 \\ L^4(1-z_1)^5(1-z_2)^5 - (1+z_1)(1+z_2)\,(1+2(z_1+z_2))^5 &= 0 \end{aligned} \tag{13.146}$$

By using the nesting property, the $2n$th-order discrimination factor can be calculated from the known nth-order discrimination factor, and vice versa:

$$\begin{aligned} L = \mathcal{L}(2n, \xi) &= \left(\mathcal{L}(n, \xi) + \sqrt{\mathcal{L}(n, \xi)^2 - 1}\right)^2 \\ \mathcal{L}(n, \xi) &= \frac{\mathcal{L}(2n, \xi) + 1}{2\sqrt{\mathcal{L}(2n, \xi)}} \end{aligned} \tag{13.147}$$

Alternatively, the selectivity factor of twice higher or lower order can be calculated from

$$\begin{aligned} \xi = \mathcal{L}^{-1}(2n, L) &= \frac{\mathcal{L}^{-1}(n, L) + 1}{2\sqrt{\mathcal{L}^{-1}(n, L)}} \\ \mathcal{L}^{-1}(n, L) &= \left(\mathcal{L}^{-1}(2n, L) + \sqrt{\left(\mathcal{L}^{-1}(2n, L)\right)^2 - 1}\right)^2 \end{aligned} \tag{13.148}$$

From the symmetry of Eqs. (13.147) and (13.148), we postulate the fundamental relation

$$\mathcal{L}^{-1}(n, x) = \mathcal{L}\left(\frac{1}{n}, x\right), \qquad x > 1 \tag{13.149}$$

where x stands for ξ or L. This means that the selectivity factor of the nth order can be calculated as the discrimination factor of the order $1/n$.

The discrimination factor calculated as a function of order n (n = 1/16, 1/8, 1/4, 1/2, 1, 2, 4, 8, 16) for $\xi = 10$ is plotted in Fig. 13.24. The logarithmic scale is selected for n as well as for $\mathcal{L}(n, \xi)$.

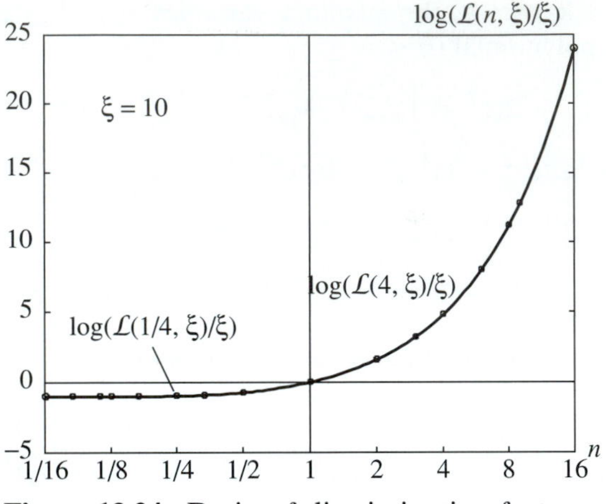

Figure 13.24 Ratio of discrimination factor and selectivity factor, $\mathcal{L}(n, \xi)/\xi$, for $\xi = 10$.

For $n > 1$, the discrimination factor is an increasing function of n:

$$\mathcal{L}(1, \xi) < \mathcal{L}(2, \xi) < \mathcal{L}(4, \xi) < \mathcal{L}(8, \xi) \tag{13.150}$$

For a known discrimination factor L, we can calculate the corresponding selectivity factor ξ. Figure 13.25 shows the selectivity factor as a function of n, ($n = 1/16, 1/8, 1/4, 1/2, 1, 2, 4, 8, 16$), for $L = 10$.

The selectivity factor is a decreasing function of n:

$$\mathcal{L}^{-1}(1, L) > \mathcal{L}^{-1}(2, L) > \mathcal{L}^{-1}(4, L) > \mathcal{L}^{-1}(8, L) \tag{13.151}$$

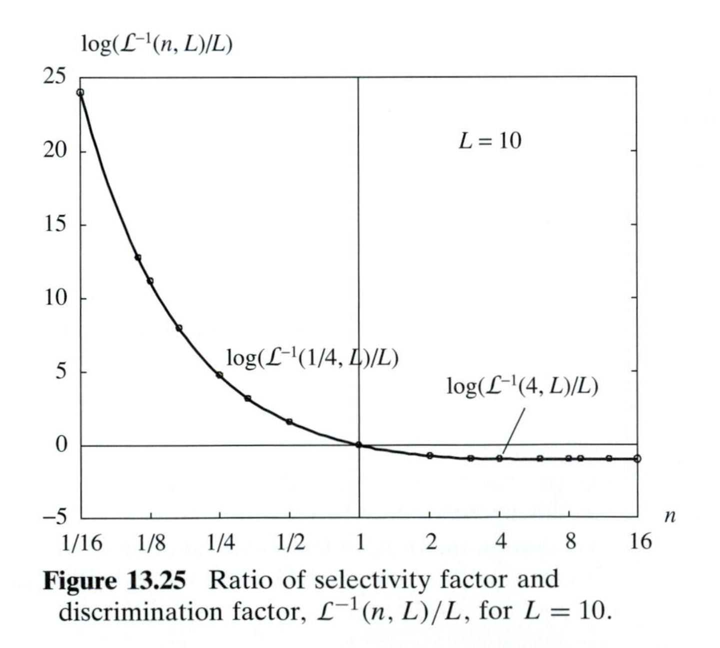

Figure 13.25 Ratio of selectivity factor and discrimination factor, $\mathcal{L}^{-1}(n, L)/L$, for $L = 10$.

From Figs. 13.24 and 13.25 we find that the curves of the selectivity factor and the discrimination factor are symmetric with respect to the vertical axis for the case $L = 10$ and $\xi = 10$. The joint plot of the discrimination factor and the selectivity factor is shown in Fig. 13.26.

Some of the computed values of ξ and L for $n = 2^i 3^j$, derived by means of Eqs. (13.138) and (13.140), are also indicated on the curves in Figs. 13.24 and 13.25. The discrimination factor for $\xi = 2$, $\xi = 10$, and $\xi = 100$ is shown in Fig. 13.27.

Generally, by using the nesting property, we can calculate the discrimination factor for a noninteger order n/m,

$$L = \mathcal{L}\left(\frac{n}{m}, \xi\right) = \mathcal{L}\left(n, \mathcal{L}\left(\frac{1}{m}, \xi\right)\right) = \mathcal{L}(n, \mathcal{L}^{-1}(m, \xi)) \tag{13.152}$$

or, alternatively, the selectivity factor for the noninteger order n/m,

$$\xi = \mathcal{L}^{-1}\left(\frac{n}{m}, L\right) = \mathcal{L}^{-1}\left(n, \mathcal{L}^{-1}\left(\frac{1}{m}, L\right)\right) = \mathcal{L}^{-1}(n, \mathcal{L}(m, L)) \tag{13.153}$$

We exploit Eqs. (13.152) and (13.153) for approximate calculation when $n \neq 2^i 3^j$. For instance, let us determine ξ for given $L = 20$ and $n = 5$. We use the approximate relation, Eq. (13.122):

$$\mathcal{L}^{r-p}(n, \xi) \approx \mathcal{L}^{n-p}(r, \xi)\mathcal{L}^{r-n}(p, \xi), \qquad r > n > p, \quad \xi >> 1 \tag{13.154}$$

The intermediate results of the calculation are summarized below. The value $L = 20$ is rather low for practice [84].

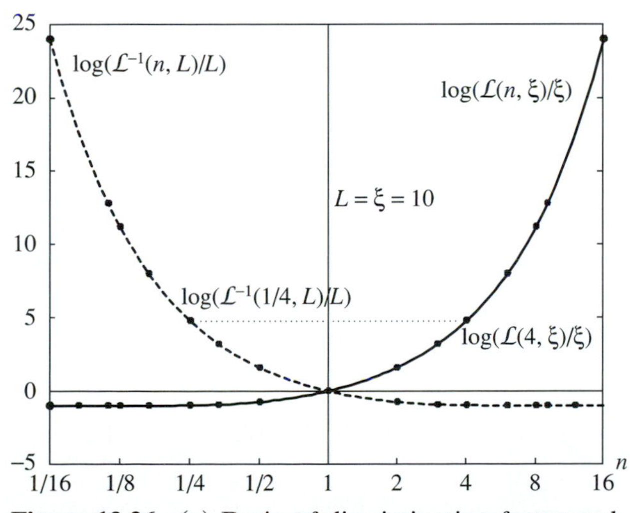

Figure 13.26 (a) Ratio of discrimination factor and selectivity factor, $\mathcal{L}(n, \xi)/\xi$, for $\xi = 10$ (solid line). (b) Ratio of selectivity factor and discrimination factor, $\mathcal{L}^{-1}(n, L)/L$, for $L = 10$ (dashed line).

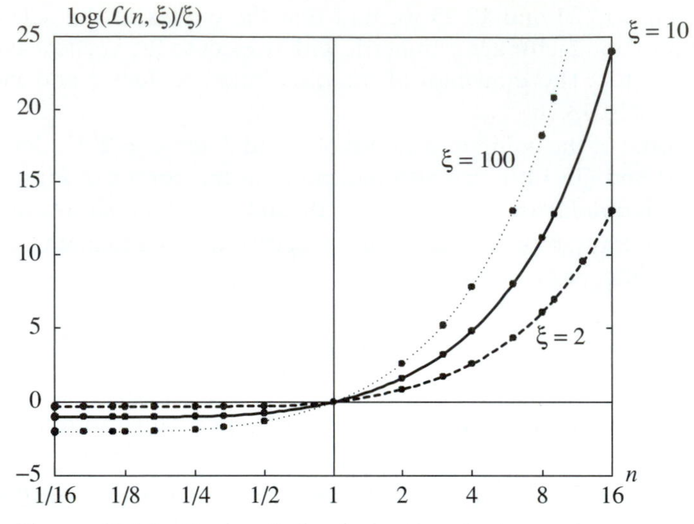

Figure 13.27 Ratio of discrimination factor and selectivity factor $\mathcal{L}(n, \xi)/\xi$, for $\xi = 2$, $\xi = 10$, and $\xi = 100$.

$$L_3 = \mathcal{L}(3, L) = 127759.9812089$$

$$L_4 = \mathcal{L}(4, L) = 10214406.00000$$

$$\mathcal{L}\left(\frac{16}{5}, L\right) = 306872.8697020, \qquad \mathcal{L}\left(\frac{16}{5}, L\right) \approx L_4^{\frac{16}{5}-3} \cdot L_3^{4-\frac{16}{5}}$$

$$\mathcal{L}^{-1}\left(\frac{5}{16}, L\right) = 306872.8697020, \qquad \mathcal{L}^{-1}(n, x) = \mathcal{L}\left(\frac{1}{n}, x\right)$$

$$\mathcal{L}^{-1}\left(\frac{5}{8}, L\right) = 276.9814387758, \qquad \mathcal{L}^{-1}\left(2\frac{5}{16}, L\right)$$

$$\mathcal{L}^{-1}\left(\frac{5}{4}, L\right) = 8.351422769894, \qquad \mathcal{L}^{-1}\left(2\frac{5}{8}, L\right)$$

$$\mathcal{L}^{-1}\left(\frac{5}{2}, L\right) = 1.617958810217, \qquad \mathcal{L}^{-1}\left(2\frac{5}{4}, L\right)$$

$$\mathcal{L}^{-1}(5, L) = 1.029079870071, \qquad \mathcal{L}^{-1}\left(2\frac{5}{2}, L\right)$$

The above procedure can be compared with another procedure presented in reference 84. The error of the method reported in reference 84 is about 7×10^{-5}. However, our algorithm gives the smaller error which depends on the computational complexity. For instance, $\mathcal{L}^{-1}(5/8, L) = \sqrt[5]{(\mathcal{L}(2, L))^3\,(\mathcal{L}(1, L))^2}$ gives an error of 6×10^{-6}, while the error of $\mathcal{L}^{-1}(5/16, L) = \sqrt[5]{(L(3, L))^4 \, L(4, L)}$ is 1×10^{-13}. A more involved formula $\mathcal{L}^{-1}(5/32, L) = \sqrt[5]{(\mathcal{L}(6, L))^4 \, \mathcal{L}(8, L)}$ reduces the error to only 3×10^{-25}. Assuming that all computations are performed in *Mathematica* with much higher numerical precision, practically, all errors are caused by the approximate relation Eq. (13.154).

13.13 TRANSFER FUNCTION POLES

The elliptic rational function is an even or odd function with real coefficients. Therefore, $R^2(n, \xi, x)$ is an even function and can be expressed in terms of x^2, as a ratio of two even polynomials with real coefficients:

$$R^2(n, \xi, x) = \frac{N(x^2)}{D(x^2)} \tag{13.155}$$

From now on, let us consider an nth-order filter whose normalized squared magnitude response can be written in the form

$$|\mathcal{H}(s)|^2 = \frac{1}{1 + \epsilon^2 R^2(n, \xi, x)} = \frac{D(x^2)}{D(x^2) + \epsilon^2 N(x^2)}, \qquad s^2 = -x^2 \tag{13.156}$$

where ϵ denotes the ripple factor ($\epsilon > 0$), and s represents the normalized complex frequency. The transfer function poles are the left half-plane roots of the equation

$$D(-s^2) + \epsilon^2 N(-s^2) = 0, \qquad s^2 = -x^2 \tag{13.157}$$

in which it is assumed that x^2 is replaced by $-s^2$ in the polynomials $N(x^2)$ and $D(x^2)$. Also, the poles of the transfer function are the left half-plane roots of the equation

$$1 + \epsilon^2 R^2(n, \xi, x) = 0, \qquad x = -js \tag{13.158}$$

where $j = \sqrt{-1}$.

We denote by s_i the values of s that satisfy Eq. (13.158), which can be rewritten as follows:

$$R^2(n, \xi, -js_i) = -\frac{1}{\epsilon^2} \tag{13.159}$$

Since ϵ^2 is a positive number, $R^2(n, \xi, -js_i)$ must have a negative value; that is, $R(n, \xi, -js_i)$ must be a complex (purely imaginary) number:

$$R(n, \xi, -js_i) = \pm j\frac{1}{\epsilon} \tag{13.160}$$

The poles s_i ($i = 1, 2, \ldots, n$) of the normalized transfer function $\mathcal{H}(s)$ are

$$s_i = \frac{-\zeta\sqrt{1-\zeta^2}\sqrt{1-x_i^2}\sqrt{1-\dfrac{x_i^2}{\xi^2}} + jx_i\sqrt{1-\left(1-\dfrac{1}{\xi^2}\right)\zeta^2}}{1-\zeta^2\left(1-\dfrac{x_i^2}{\xi^2}\right)} \tag{13.161}$$

where x_i are the zeros of the elliptic rational function $R(n, \xi, x)$, while ζ denotes a function of n, ξ and ϵ, $\zeta\ (n, \xi, \epsilon)$, and it can be expressed by the Jacobi sine function.

13.13.1 First-Order ζ-Function

The first-order elliptic rational function $R(1,\xi,x) = x$ has only one zero: $x_1 = 0$. The equation $1 + \epsilon^2 R^2(1,\xi,-js) = 0$ yields the transfer function pole:

$$s_1 = -\frac{1}{\epsilon} \tag{13.162}$$

Alternatively, from Eq. (13.161) with $x_1 = 0$ we obtain

$$s_1 = -\frac{\zeta\sqrt{1-\zeta^2}}{1-\zeta^2} \tag{13.163}$$

By equating Eqs. (13.162) and (13.163)

$$-\frac{1}{\epsilon} = -\frac{\zeta\sqrt{1-\zeta^2}}{1-\zeta^2} \tag{13.164}$$

we determine ζ in terms of ϵ:

$$\zeta = \zeta\,(1,\xi,\epsilon) = \frac{1}{\sqrt{1+\epsilon^2}} \tag{13.165}$$

13.13.2 Second-Order ζ-Function

In the simplest case of a second-order function, which we are now going to analyze, zeros x_i in Eq. (13.161) are

$$\begin{aligned} x_1 &= \xi\sqrt{1-t} \\ x_2 &= -\xi\sqrt{1-t}, \qquad t = \sqrt{1-\frac{1}{\xi^2}} \end{aligned} \tag{13.166}$$

The transfer function poles from Eq. (13.161) are found to be

$$\begin{aligned} s_1 &= \frac{-t\zeta\sqrt{1-\zeta^2} + j\sqrt{1-t^2\zeta^2}}{(1-t\zeta^2)\sqrt{1+t}} \\ s_2 &= \frac{-t\zeta\sqrt{1-\zeta^2} - j\sqrt{1-t^2\zeta^2}}{(1-t\zeta^2)\sqrt{1+t}} \end{aligned} \tag{13.167}$$

The squared magnitude of the poles is

$$|s_1|^2 = |s_2|^2 = \frac{1+t\zeta^2}{(1+t)(1-t\zeta^2)} \tag{13.168}$$

On the other hand, the poles of the second-order elliptic rational function in terms of t and ϵ have been already determined in [52, 111]

$$|s_1|^2 = |s_2|^2 = \frac{\sqrt{1+\epsilon^2}}{\sqrt{(1-t)^2 + \epsilon^2(1+t)^2}} \tag{13.169}$$

Equating the right-hand sides of Eqs. (13.168) and (13.169), we find ζ in terms of ϵ and ξ:

$$\zeta = \zeta(2, \xi, \epsilon) = \frac{2}{(1+t)\sqrt{1+\epsilon^2} + \sqrt{(1-t)^2 + \epsilon^2(1+t)^2}}, \qquad t = \sqrt{1 - \frac{1}{\xi^2}} \tag{13.170}$$

The function $\zeta(2, \xi, \epsilon)$ is shown in Fig. 13.28.

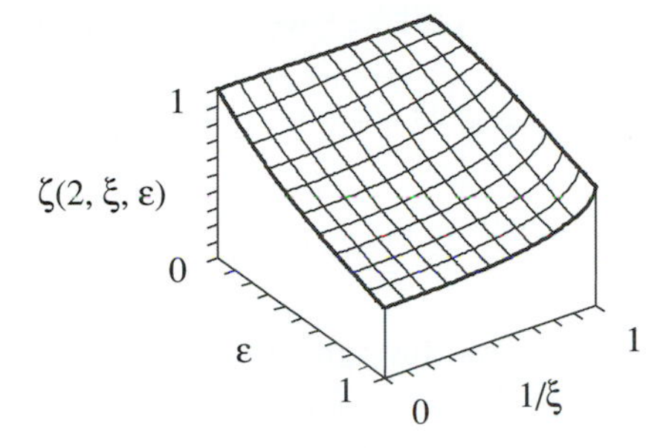

Figure 13.28 $\zeta(2, \xi, \epsilon)$ for $\xi > 1$ and $0 < \epsilon < 1$.

13.13.3 Third-Order ζ-Function

A derivation of the third-order ζ-function is more involved [52, 111]. However, we propose a straightforward procedure that starts from the transfer function real pole σ_1, which can be found as follows:

$$s_1 = \sigma_1 = -\sqrt{\frac{a_1}{3a_0} - \sqrt[3]{A} - \sqrt[3]{B}} \tag{13.171}$$

where

$$A = -q + \sqrt{q^2 + p^3} \tag{13.172}$$

$$B = -q - \sqrt{q^2 + p^3} \tag{13.173}$$

$$p = \frac{3a_0a_2 - a_1^2}{9a_0^2} \tag{13.174}$$

$$q = \frac{27a_0^2a_3 - 9a_0a_1a_2 + 2a_1^3}{54a_0^3} \tag{13.175}$$

$$a_0 = \epsilon^2\left(1 - \frac{\xi^2}{x_z^2}\right)^2 \tag{13.176}$$

$$a_1 = \left(1 - x_z^2\right)^2 - 2\epsilon^2x_z^2\left(1 - \frac{\xi^2}{x_z^2}\right)^2 \tag{13.177}$$

$$a_2 = -2\frac{\xi^2}{x_z^2}\left(1 - x_z^2\right)^2 + \epsilon^2x_z^4\left(1 - \frac{\xi^2}{x_z^2}\right)^2 \tag{13.178}$$

$$a_3 = \frac{\xi^4}{x_z^4}(1 - x_z^2)^2 \tag{13.179}$$

$$x_z = \frac{\xi}{Z} \tag{13.180}$$

$$Z^2 = \frac{2\xi^2\sqrt{G}}{\sqrt{8\xi^2(\xi^2+1) + 12G\xi^2 - G^3} - \sqrt{G^3}} \tag{13.181}$$

$$G = \sqrt{4\xi^2 + \left(4\xi^2(\xi^2 - 1)\right)^{2/3}} \tag{13.182}$$

In order to determine the numerical value of the real pole s_1, it is necessary to proceed in the opposite direction, from Eq. (13.182) toward Eq. (13.171). The expression for pole s_1 obtained by this procedure should be equated to the expression derived from Eq. (13.161) for $x_1 = 0$:

$$s_i = \frac{-\zeta}{\sqrt{1 - \zeta^2}} \tag{13.183}$$

which yields

$$\zeta = \zeta(3, \xi, \epsilon) = \frac{1}{\sqrt{1 + \dfrac{1}{\sigma_1^2}}} \tag{13.184}$$

The function $\zeta(3, \xi, \epsilon)$ is shown in Fig. 13.29.

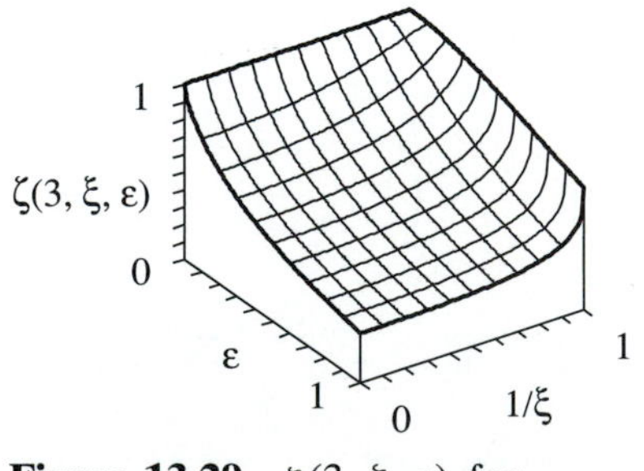

Figure 13.29 $\zeta(3, \xi, \epsilon)$ for $\xi > 1$ and $0 < \epsilon < 1$.

13.13.4 Higher-Order ζ-Function

We use the closed-form expressions for $\zeta(2, \xi, \epsilon)$ and $\zeta(3, \xi, \epsilon)$ to find $\zeta(n, \xi, \epsilon)$ analytically for $n = 2^l 3^m$, $(l, m = 0, 1, 2, 3, \ldots)$ [52, 113]. We rely on the relation

$$\zeta = \zeta(n, \xi, \epsilon_1) = \zeta(u, \xi, \epsilon_v) = \zeta(1, \xi, \epsilon_n) \tag{13.185}$$

and

$$L_n = \mathcal{L}(n, \xi), \qquad L_u = \mathcal{L}(u, \xi), \quad L_1 = \mathcal{L}(1, \xi) \tag{13.186}$$

We do not know explicit formulas for $\zeta(n, \xi, \epsilon)$ when $n > 3$, so we want to exploit the relation $\zeta = \zeta(1, L_1, \epsilon_n)$. The quantity ϵ_n has to be determined [52]:

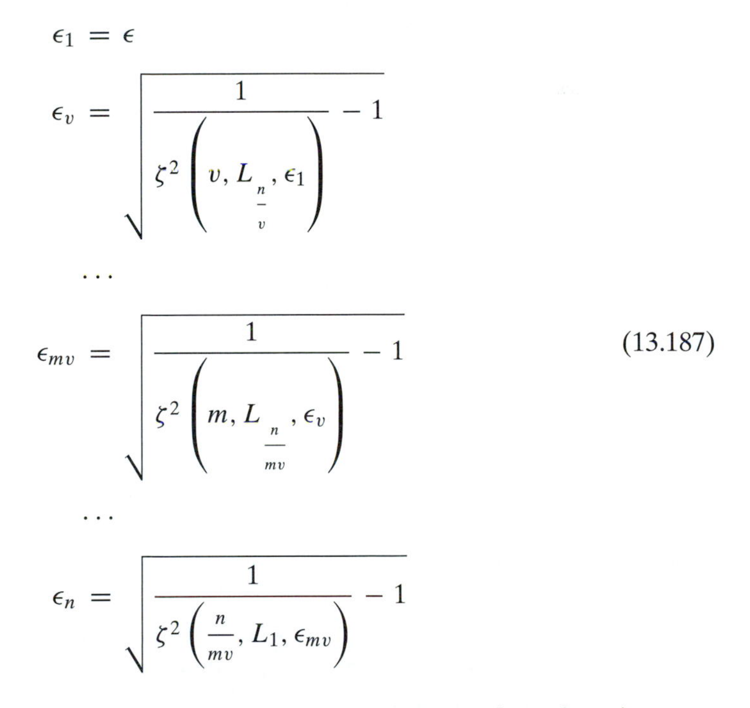

$$\begin{aligned}
\epsilon_1 &= \epsilon \\
\epsilon_v &= \sqrt{\frac{1}{\zeta^2\left(v, L_{\frac{n}{v}}, \epsilon_1\right)} - 1} \\
&\ldots \\
\epsilon_{mv} &= \sqrt{\frac{1}{\zeta^2\left(m, L_{\frac{n}{mv}}, \epsilon_v\right)} - 1} \\
&\ldots \\
\epsilon_n &= \sqrt{\frac{1}{\zeta^2\left(\frac{n}{mv}, L_1, \epsilon_{mv}\right)} - 1}
\end{aligned} \tag{13.187}$$

Our procedure assumes that the order n can be represented as a product of two integers, $n = uv$.

Let us compute $\zeta(n, \xi, \epsilon)$ for $n = 18$, $\xi = 1.01$, and $\epsilon = 0.1$. First, we calculate the discrimination factors:

$$\begin{aligned}
L_1 &= \xi = 1.01 \\
L_3 &= \mathcal{L}(3, L_1) = 2.3958806103 \\
L_9 &= \mathcal{L}(3, L_3) = 191.11276848
\end{aligned}$$

Next, we find intermediate variables:

$$\epsilon_1 = \epsilon = 0.1$$

$$\epsilon_2 = \sqrt{\frac{1}{\zeta^2\left(2, L_{\frac{18}{2}}, \epsilon_1\right)} - 1} = \sqrt{\frac{1}{\zeta^2(2, L_9, \epsilon_1)} - 1} = 0.4700859474$$

$$\epsilon_6 = \epsilon_{2\cdot 3} = \sqrt{\frac{1}{\zeta^2\left(3, L_{\frac{18}{2\cdot 3}}, \epsilon_2\right)} - 1} = \sqrt{\frac{1}{\zeta^2(3, L_3, \epsilon_2)} - 1} = 1.8075200187$$

$$\epsilon_{18} = \epsilon_{6\cdot 3} = \sqrt{\frac{1}{\zeta^2\left(3, L_{\frac{18}{6\cdot 3}}, \epsilon_6\right)} - 1} = \sqrt{\frac{1}{\zeta^2(3, L_1, \epsilon_6)} - 1} = 2.686794$$

Finally, we obtain the required quantity:

$$\zeta = \zeta(1, L_1, \epsilon_{18}) = \frac{1}{\sqrt{1 + \epsilon_{18}^2}} = 0.3488141990$$

The corresponding (1) zeros of the elliptic rational function, (2) poles of the elliptic function, and (3) poles of the normalized transfer function are summarized in Table 13.2.

Table 13.2 Zeros and poles of elliptic rational function and transfer function poles for $n = 18$, $\xi = 1.01$, and $\epsilon = 0.1$

x_i	ξ / x_i	s_i
0.18462741750413	5.4704767778	$-0.357971765868310 \pm j\,0.20898361957$
0.50861382205562	1.9857895248	$-0.267460821713425 \pm j\,0.55874731405$
0.73380564837877	1.3763862437	$-0.161901564167992 \pm j\,0.77759509348$
0.86617972088486	1.1660397671	$-0.086814745023169 \pm j\,0.89391042583$
0.93643354624049	1.0785602503	$-0.043717742641114 \pm j\,0.95156219211$
0.97165432916900	1.0394643133	$-0.021285619955791 \pm j\,0.97936525680$
0.98870006363558	1.0215433751	$-0.010062311866523 \pm j\,0.99255413850$
0.99657968566646	1.0134663736	$-0.004403361641965 \pm j\,0.99859176210$
0.99965250272743	1.0103510942	$-0.001233419840232 \pm j\,1.00093610545$

The above procedure is valid generally for arbitrary values of ξ and ϵ which can be given as symbols, too [51].

In the general case we represent the order as $n = 2^l 3^m u$, where u stands for an integer which cannot be further divided by 2 or 3. The next step is finding $\zeta(u, L_u, \epsilon_v)$. The arguments L_u and ϵ_v are computed from Eqs. (13.186) and (13.187). The poles of the normalized transfer function

$$|\mathcal{H}(s)|^2 = \frac{1}{1 + \epsilon_v^2 R^2(u, \xi, x)} = \frac{D(x^2)}{D(x^2) + \epsilon_v^2 N(x^2)}, \qquad s^2 = -x^2 \tag{13.188}$$

are roots of the equation

$$D(-s^2) + \epsilon_v^2 N(-s^2) = 0, \qquad s^2 = -x^2 \tag{13.189}$$

We need only the real pole which determines ζ:

$$\zeta(u, L_u, \epsilon_v) = \frac{1}{\sqrt{1 + \dfrac{1}{\sigma^2_{(u+1)/2}}}} \tag{13.190}$$

The function $\zeta(4, \xi, \epsilon)$ is shown in Fig. 13.30, and $\zeta(n, \xi, \epsilon)$ for $n = 2, 3, 4$ is presented in Fig. 13.31.

To show the advantages of the proposed method, we have compared in Table 13.3 the number of floating point operations (FLOPS) necessary for the computation of the transfer function poles by MATLAB [1] with the number of FLOPS required by our closed-form method. Obviously, our method is computationally more efficient because its requires fewer FLOPS. Also, the method that we propose requires the same number of FLOPS irrespective of the value of parameters ξ or ϵ.

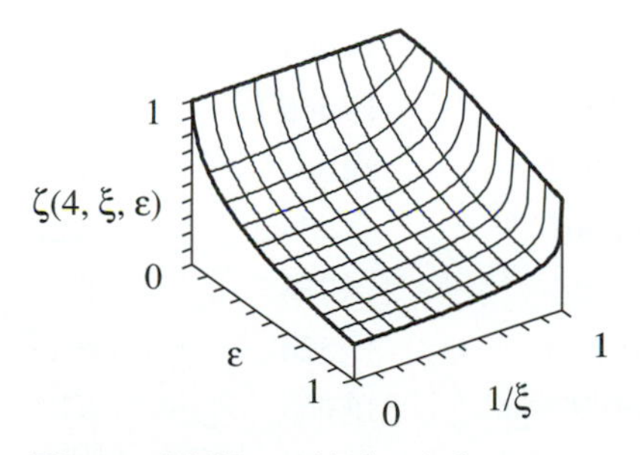

Figure 13.30 $\zeta(4, \xi, \epsilon)$ for $\xi > 1$ and $0 < \epsilon < 1$.

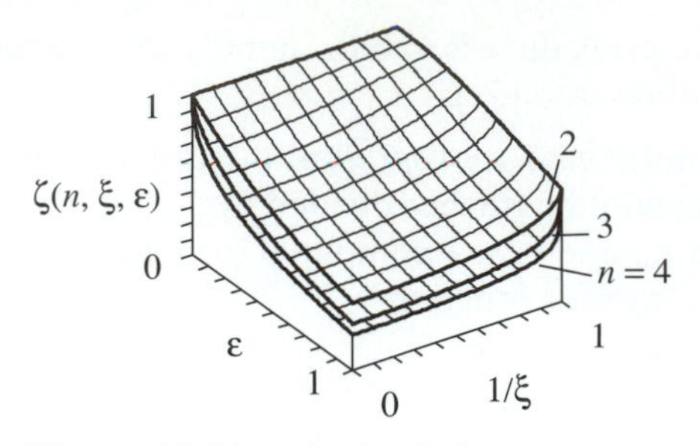

Figure 13.31 $\zeta(n, \xi, \epsilon)$ for $n = 2, 3, 4$.

13.13.5 Alternative Method for Computing Transfer Function Poles

Alternatively to the proposed procedure, the transfer function poles could be determined directly from Eq. (13.158) and the nesting property of the elliptic rational function.

First, we solve the equation

$$R(u, L(v, \xi), js) = \pm \frac{j}{\epsilon} \quad \Rightarrow \quad s_{(u)a}, \qquad a = 1, \ldots, u \tag{13.191}$$

which simplifies to finding the roots, $s_{(u)a}$, of a uth-order polynomial in s.

Second, we solve u equations:

$$R(v, \xi, js) = js_{(u)a} \quad \Rightarrow \quad s_{(uv)b}, \qquad b = 1, \ldots, uv \tag{13.192}$$

which simplifies to finding the roots, $s_{(uv)b}$, of vth-order polynomials in s. The root finding procedure, Eq. (13.192), is repeated until $v = 1$.

Table 13.3 Number of FLOPS for $A_p = 0.2$ dB

order	FLOPS_M $A_s = 30$ dB	FLOPS_M $A_s = 60$ dB	FLOPS_A $A_s = 30$ dB	FLOPS_A $A_s = 60$ dB	$\frac{\text{FLOPS}_M}{\text{FLOPS}_A}$	
2	13 458	16 231	83	83	162	196
3	12 246	14 064	235	235	52	59
4	10 513	13 468	164	164	64	82
6	11 962	12 064	356	356	33	33
8	14 467	14 188	302	302	47	46
9	14 669	14 238	577	577	25	24
12	15 680	15 324	626	626	25	24
16	16 717	15 665	538	538	31	29
18	19 134	17 404	855	855	22	20

As an illustration, for $n = 18$, $\xi = 1.01$, and $\epsilon = 0.1$, the alternative method yields the results summarized in Table 13.4.

Table 13.4 Computation of transfer function poles for $n = 18$, $\xi = 1.01$, and $\epsilon = 0.1$

$\dfrac{-1 - 1.9999863103 s_{(2)i}^2}{1 + 0.0000136896886044 s_{(2)i}^2} = j10$	Eq. (13.191)
$s_{(2)1} = -1.504098821 + j1.662124824$ $s_{(2)2} = -1.504098821 - j1.662124824$	
$j\,\dfrac{-2.8614777965 s_{(3)i} - 3.727752693 s_{(3)i}^3}{1 + 0.1337251033 s_{(3)i}^2} = j s_{(2)1}$	Eq. (13.192)
$s_{(3)1} = -0.3540085936 - j0.825899117$ $s_{(3)2} = -0.1191974638 + j1.082281576$ $s_{(3)3} = 0.5271623648 - j0.316007610$	
$j\,\dfrac{-1.46976287742 s_{(9)i} - 1.52493216767 s_{(9)i}^3}{1 + 0.94483070975 s_{(9)i}^2} = j s_{(3)1}$ $j\,\dfrac{-1.46976287742 s_{(9)i} - 1.52493216767 s_{(9)i}^3}{1 + 0.94483070975 s_{(9)i}^2} = j s_{(3)2}$ $j\,\dfrac{-1.46976287742 s_{(9)i} - 1.52493216767 s_{(9)i}^3}{1 + 0.94483070975 s_{(9)i}^2} = j s_{(3)3}$	Eq. (13.192)
$s_i = -\lvert \mathrm{Real}(s_{(9)i})\rvert \pm j \lvert \mathrm{Imag}(s_{(9)i})\rvert$	see Table 13.2

The alternative method exposed in this section is general and it does not make use of Eq. (13.161) and ζ-functions. However, it assumes that we have a numeric procedure for finding roots of a polynomial with complex coefficients. Therefore, it might be computationally intensive.

13.14 TRANSFER FUNCTION WITH MINIMAL Q-FACTORS

Q-factor of a transfer function pole s_i is defined as

$$Q_i = -\frac{|s_i|}{2\,\mathrm{Re}\,s_i} \tag{13.193}$$

In fact, the Q-factor is a function of the argument of the pole s_i,

$$Q_i = -\frac{1}{2\,\cos(\arg(s_i))} \tag{13.194}$$

and it does not depend on the pole magnitude. The minimal value of the Q-factor is 1/2, and it occurs when the pole is real. The Q-factor takes the infinite value for poles on the imaginary axis. We assume that the real part of a transfer function pole is always negative, implying that the pole Q-factor is always positive.

In order to find the minimal Q-factor of transfer function poles, we consider, without lack of generality, the odd-order elliptic rational function

$$R(n, \xi, x) = r_0\, x \frac{\prod_{i=1}^{\frac{n-1}{2}} (x^2 - x_i^2)}{\prod_{i=1}^{\frac{n-1}{2}} \left(x^2 - \frac{\xi^2}{x_i^2}\right)} \tag{13.195}$$

$$r_0 = \frac{\prod_{i=1}^{\frac{n-1}{2}} \left(1 - \frac{\xi^2}{x_i^2}\right)}{\prod_{i=1}^{\frac{n-1}{2}} (1 - x_i^2)} \qquad n \text{ odd}$$

The poles of the normalized transfer function $\mathcal{H}(s)$ are the left half-plane roots of the denominator of $|\mathcal{H}(s)|^2 = 1/\left(1 + \epsilon^2 R^2(n, \xi, js)\right)$:

$$|\mathcal{H}(s)|^2 = \frac{\prod_{i=1}^{\frac{n-1}{2}} (1 - x_i^2)^2 \left(-s^2 - \frac{\xi^2}{x_i^2}\right)^2}{\prod_{i=1}^{\frac{n-1}{2}} (1 - x_i^2)^2 \left(-s^2 - \frac{\xi^2}{x_i^2}\right)^2 - \epsilon^2 s^2 \prod_{i=1}^{\frac{n-1}{2}} \left(1 - \frac{\xi^2}{x_i^2}\right)^2 (-s^2 - x_i^2)^2} \tag{13.196}$$

that is, we have to solve the equation

$$\prod_{i=1}^{\frac{n-1}{2}} (1 - x_i^2)^2 \left(-s^2 - \frac{\xi^2}{x_i^2}\right)^2 - \epsilon^2 s^2 \prod_{i=1}^{\frac{n-1}{2}} \left(1 - \frac{\xi^2}{x_i^2}\right)^2 (-s^2 - x_i^2)^2 = 0 \tag{13.197}$$

The zeros x_i and the poles ξ/x_i are functions of the selectivity factor ξ and the order n. The transfer function poles, s_i, depend on ξ, n, and ϵ:

$$s_i = \frac{-\zeta\sqrt{1-\zeta^2}\sqrt{1-x_i^2}\sqrt{1-\dfrac{x_i^2}{\xi^2}} + jx_i\sqrt{1-\left(1-\dfrac{1}{\xi^2}\right)\zeta^2}}{1-\zeta^2\left(1-\dfrac{x_i^2}{\xi^2}\right)} \tag{13.198}$$

where ζ is a function of ξ, n, and ϵ.

From Eq. (13.197) one can see that the poles s_i are purely imaginary for $\epsilon \to \infty$, and they take values $s_i = 0$ and $s_i = \pm jx_i$:

$$s^2 \prod_{i=1}^{\frac{n-1}{2}} \left(-s^2 - x_i^2\right)^2 = 0$$

For $\epsilon = 0$ the poles become $s_i = \pm j\xi/x_i$, which follows from

$$\prod_{i=1}^{\frac{n-1}{2}} \left(-s^2 - \frac{\xi^2}{x_i^2}\right)^2 = 0$$

In both cases the Q-factors of the poles are infinite. Between these two extremes the poles occur in complex conjugate pairs [47, 114] as shown in Fig. 13.32.

Let us find ϵ which corresponds to the minimal value of the Q-factor of a transfer function pole s_i. We have to solve the equation

$$\frac{\mathrm{d}Q_i}{\mathrm{d}\epsilon} = \frac{\mathrm{d}}{\mathrm{d}\epsilon}\left(-\frac{|s_i|}{2\,\mathrm{Re}\, s_i}\right) = 0$$

which simplifies to

$$\frac{\mathrm{Im}\, s_i}{\mathrm{Re}\, s_i} = \frac{\mathrm{Im}\,\dfrac{\mathrm{d}s_i}{\mathrm{d}\epsilon}}{\mathrm{Re}\,\dfrac{\mathrm{d}s_i}{\mathrm{d}\epsilon}} \tag{13.199}$$

Since ζ is a real monotonic function in ϵ, we have

$$\frac{\mathrm{d}s_i}{\mathrm{d}\epsilon} = \frac{\mathrm{d}s_i}{\mathrm{d}\zeta}\frac{\mathrm{d}\zeta}{\mathrm{d}\epsilon}$$

yielding

$$\frac{\mathrm{Im}\, s_i}{\mathrm{Re}\, s_i} = \frac{\mathrm{Im}\,\dfrac{\mathrm{d}s_i}{\mathrm{d}\zeta}}{\mathrm{Re}\,\dfrac{\mathrm{d}s_i}{\mathrm{d}\zeta}} \tag{13.200}$$

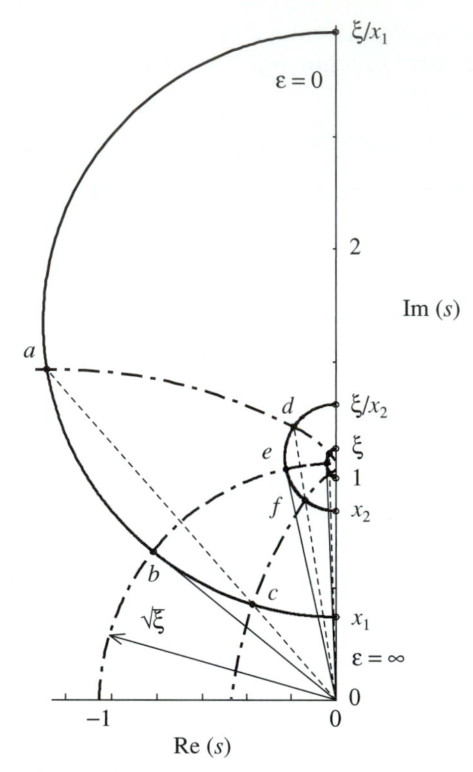

Figure 13.32 Root loci of the sixth-order transfer function poles.

Substituting Eq. (13.198) into Eq. (13.200) and simplifying, we find

$$x_i\,\xi^4\sqrt{1-x_i^2}\sqrt{1-\frac{x_i^2}{\xi^2}}\sqrt{1-\zeta^2+\frac{\zeta^2}{\xi^2}}\left(\zeta^4-\xi^2(1-\zeta^2)^2\right)=0 \tag{13.201}$$

Because $0 < x_i^2 < 1$ and $\xi > 1$, from Eq. (13.201) we find

$$\zeta^2-\xi(1-\zeta^2)=0 \tag{13.202}$$

which determines $\zeta = \zeta_{minQ}$ for which Q-factors are minimal:

$$\zeta_{minQ}=\sqrt{\frac{\xi}{1+\xi}} \tag{13.203}$$

It is important to notice that ζ_{minQ} depends only on the selectivity factor ξ. This means that for $\zeta = \zeta_{minQ}$ all poles simultaneously have their minimal Q-factor.

Substituting ζ from Eq. (13.203) into Eq. (13.198) yields a simple expression for transfer function poles

$$s_i = \sqrt{\xi}\frac{\sqrt{1-x_i^2}\sqrt{\xi^2-x_i^2}+jx_i(\xi+1)}{\xi+x_i^2} \tag{13.204}$$

from which

$$|s_i| = \sqrt{\xi} \tag{13.205}$$

The minimal Q-factor of poles is found in closed form:

$$Q_{min,i} = \frac{\xi+x_i^2}{2\sqrt{1-x_i^2}\sqrt{\xi^2-x_i^2}} \tag{13.206}$$

From Eqs. (13.205) and (13.206) the transfer function simplifies to

$$\mathcal{H}_{minQ}(s) = \frac{\prod_{i=1}^{\frac{n}{2}} x_i^2}{\sqrt{1+\epsilon^2}\sqrt{\xi^n}} \prod_{i=1}^{\frac{n}{2}} \frac{s^2+\dfrac{\xi^2}{x_i^2}}{s^2+\dfrac{\sqrt{\xi}}{Q_{min,i}}s+\xi}, \qquad n \text{ even} \tag{13.207}$$

$$\mathcal{H}_{minQ}(s) = \frac{\prod_{i=1}^{\frac{n-1}{2}} x_i^2}{\sqrt{\xi^{n-2}}} \frac{1}{s+\sqrt{\xi}} \prod_{i=1}^{\frac{n-1}{2}} \frac{s^2+\dfrac{\xi^2}{x_i^2}}{s^2+\dfrac{\sqrt{\xi}}{Q_{min,i}}s+\xi}, \qquad n \text{ odd} \tag{13.208}$$

Equations (13.207) and (13.208) lend itself readily to the filter implementation because all poles $|s_i|$ are equal. According to Eq. (13.206), $Q_{min,i}$ is determined by the selectivity factor ξ and the zeros x_i $(i = 1, 2, \ldots, n)$ only. For $n = 2^l 3^m (l, m = 1, 2, \ldots)$, the zeros x_i can be found analytically, implying that the minimal Q-factors can also be found in closed form.

In any case, it will be useful to remind the reader that Eqs. (13.207) and (13.208) allow a straightforward computation of transfer function poles, in contrast to the usual practice of the cumbersome extraction of the poles from the squared magnitude function from the left half-plane.

Simplifying $|\mathcal{H}_{minQ}(j\xi)|^2/|\mathcal{H}_{minQ}(j1)|^2$, we find

$$\frac{|\mathcal{H}_{minQ}(j\xi)|^2}{|\mathcal{H}_{minQ}(j1)|^2} = \frac{1}{L(n,\xi)} \tag{13.209}$$

On the other hand, from the definition of $\mathcal{H}(s)$, we know

$$\frac{|\mathcal{H}(j\xi)|^2}{|\mathcal{H}(j1)|^2} = \frac{1+\epsilon^2}{1+\epsilon^2 L^2(n,\xi)} \tag{13.210}$$

From Eqs. (13.209) and (13.210) we derive the ripple factor for which the Q-factors are minimal:

$$\epsilon = \epsilon_{minQ} = \frac{1}{\sqrt{L(n,\xi)}} \tag{13.211}$$

Equation (13.211) relates directly the ripple factor required for Q_{min} to the value of the discrimination factor $L(n,\xi)$. Since the above relation does not depend on x_i all Q's are simultaneously minimal.

As an illustration of the above considerations, root loci of the sixth-order elliptic function are presented in Fig. 13.32. All curves start at the zeros of the elliptic rational function, at jx_i, for $\epsilon = \infty$, and terminate at the poles $j\xi/x_i$, for $\epsilon = 0$. For given ξ and an arbitrary order n, all poles of $\mathcal{H}_{minQ}(s)$ lie on a circle of the radius $\sqrt{\xi}$.

PROBLEMS

13.1 Sketch the first five Chebyshev polynomials, $C(n,x)$, for $-1 \le x \le 1$, $1 \le x \le 1.1$, and $1 \le x \le 2$. Next, sketch $C(8,x)$, $C(2,C(4,x))$, and $C(2,C(2,C(2,x)))$. What do you observe?

13.2 Find the minimal and maximal value of the zero and the pole of the second-order elliptic rational function. Find the zero of the second-order Chebyshev polynomial.

13.3 Sketch the fourth-order rational function, for $t = \frac{1}{2}$,

$$\mathcal{R}(4,T,t,x) = \frac{\left(4t + T(1+t)^2\right)x^4 + (-4t - 2T(1+t))\,x^2 + T}{\left(-4t + T(1+t)^2\right)x^4 + (4t - 2T(1+t))\,x^2 + T}$$

Find T which maximizes the minimum of $\mathcal{R}(4,T,t,x)$ over the interval $x > \dfrac{1}{1-t^2}$.

MATLAB EXERCISES

13.1 Write a MATLAB program that computes the first five Chebyshev polynomials, $C(n,x)$, for $-1 \le x \le 1$, $1 \le x \le 1.1$, and $1 \le x \le 2$. Plot $C(8,x)$, $C(2,C(4,x))$, and $C(2,C(2,C(2,x)))$.

13.2 Write a MATLAB program that computes zeros of the second-order elliptic rational functions. Find the range of the zeros.

13.3 Write a MATLAB program that plots the fourth-order rational function $\mathcal{R}(4,T,t,x)$ given in Problem 13.3. Find T to maximize the minimum of $\mathcal{R}(4,T,t,x)$ for $x > \dfrac{1}{1-t^2}$. Assume $t = \dfrac{1}{2}$.

13.4 Write a MATLAB program that plots the elliptic rational function of the order $n = 3$, $n = 6$, and $n = 9$. Assume $\xi = 1.01$, $\xi = 1.1$, $\xi = 1.4$.

13.5 Write a MATLAB program that plots the elliptic rational function of the order $n = 5$, $n = 10$, and $n = 15$. Assume $\xi = 1.01$, $\xi = 1.1$, $\xi = 1.4$. Use the nesting property.

13.6 Write a MATLAB program that computes zeros of the elliptic rational function of the order $n = 8$, $n = 16$, and $n = 18$. Assume $\xi = 1.1$.

13.7 Write a MATLAB program that computes the zeros of elliptic rational function of the order $n = 13$, $n = 14$, and $n = 17$. Assume $\xi = 1.1$.

13.8 Write a MATLAB program that computes the selectivity factor in terms of discrimination factor for $n = 12$, $n = 15$, and $n = 16$. Assume $\xi = 1.1$.

13.9 Write a MATLAB program that computes the discrimination factor in terms of selectivity factor for $n = 12$, $n = 15$, and $n = 16$. Assume $\xi = 1.1$.

13.10 Write a MATLAB program that computes the ζ-function in terms of selectivity and ripple factor for $n = 12$, $n = 16$, and $n = 18$. Assume $\xi = 1.1$, $\epsilon = 0.1$.

13.11 Write a MATLAB program that computes the ζ-function in terms of selectivity and ripple factor for $n = 5$, $n = 10$, and $n = 15$. Assume $\xi = 1.1$, $\epsilon = 0.1$.

13.12 Write a MATLAB program that computes the transfer function poles for $n = 10$, $n = 16$, and $n = 18$. Assume $\xi = 1.01$, $\epsilon = 0.1$.

13.13 Write a MATLAB program that computes the selectivity factor for a given maximal Q-factor for $n = 6$, $n = 8$, and $n = 9$. Assume $\epsilon = 0.1$.

MATHEMATICA EXERCISES

13.1 Write a *Mathematica* program that computes the first five Chebyshev polynomials, $C(n, x)$, for $-1 \leq x \leq 1$, $1 \leq x \leq 1.1$, and $1 \leq x \leq 2$. Evaluate and expand $C(8, x)$, $C(2, C(4, x))$, and $C(2, C(2, C(2, x)))$.

13.2 Write a *Mathematica* program that computes poles and zeros of the second-order elliptic rational functions. Find the range of the zeros.

13.3 Write a *Mathematica* program that plots the fourth-order rational function $\mathcal{R}(4, T, t, x)$ given in Problem 13.3. Find T to maximize the minimum of $\mathcal{R}(4, T, t, x)$ for $x > \dfrac{1}{1 - t^2}$. Assume $t = \dfrac{1}{2}$.

13.4 Write a *Mathematica* program that plots the elliptic rational function of the order $n = 3$, $n = 6$, and $n = 9$. Assume $\xi = 1.01$, $\xi = 1.1$, $\xi = 1.4$.

13.5 Write a *Mathematica* program that plots the elliptic rational function of the order $n = 5$, $n = 10$, and $n = 15$. Assume $\xi = 1.01$, $\xi = 1.1$, $\xi = 1.4$.

13.6 Write a *Mathematica* program that computes the zeros of elliptic rational function of the order $n = 8$, $n = 16$, and $n = 18$. Assume $\xi = 1.1$.

13.7 Write a *Mathematica* program that computes the zeros of elliptic rational function of the order $n = 13$, $n = 14$, and $n = 17$. Assume $\xi = 1.1$.

13.8 Write a *Mathematica* program that computes the selectivity factor in terms of discrimination factor for $n = 12$, $n = 15$, and $n = 16$. Assume $\xi = 1.1$.

13.9 Write a *Mathematica* program that computes the discrimination factor in terms of selectivity factor for $n = 12$, $n = 15$, and $n = 16$. Assume $\xi = 1.1$.

13.10 Write a *Mathematica* program that computes the ζ-function in terms of selectivity and ripple factor for $n = 12$, $n = 16$, and $n = 18$. Assume $\xi = 1.1$, $\epsilon = 0.1$.

13.11 Write a *Mathematica* program that computes the ζ-function in terms of selectivity and ripple factor for $n = 5$, $n = 10$, and $n = 15$. Assume $\xi = 1.1$, $\epsilon = 0.1$.

13.12 Write a *Mathematica* program that computes the transfer function poles for $n = 10$, $n = 16$, and $n = 18$. Assume $\xi = 1.01$, $\epsilon = 0.1$.

13.13 Write a *Mathematica* program that computes the selectivity factor for a given maximal Q-factor for $n = 6$, $n = 8$, and $n = 9$. Assume $\epsilon = 0.1$.

APPENDIX A

EXAMPLE *MATHEMATICA* NOTEBOOKS

Filter Design

For Signal Processing

Using MATLAB and *Mathematica*

Miroslav D. Lutovac, Ph. D.

Dejan V. Tošić, Ph. D.

Brian L. Evans, Ph. D.

A.1 Analysis by Transform Method of Analog LTI Circuits

Miroslav D. Lutovac, Dejan V. Tosic and Brian L. Evans
lutovac@iritel.bg.ac.yu tosic@galeb.etf.bg.ac.yu bevans@ece.utexas.edu

■ A.1.1 References

1. G. S. Moschytz and P. Horn,
Active Filter Design Handbook
For Use with Programmable Pocket Calculators and Mini Computers.
John Wiley and Sons, Ltd., New York, 1981, pp. 38–39

2. R. P. Sallen and E. L. Key,
"A practical method of designing RC active filters,"
IRE Trans. Circuit Theory, CT–2, Mar. 1955, pp. 74–85.

3. D.V. Tosic, M.D. Lutovac, B.L. Evans, I.M. Markoski,
"A tool for symbolic analysis and design of analog active filters,"
5th Int. Workshop, Symbolic Methods, Applications, Circuit Design,
SMACD'98, Kaiserslautern, Germany, pp. 71–74, Oct. 1998.

4. D. V. Tosic, M. D. Lutovac,
"Symbolic computation of impulse, step and sine responses of linear time-invariant systems,"
Proc. 10th Int. Symp. Theor. Electrical Engineering,
ISTET99, Magdeburg, Germany, Sep. 1999, pp. 653–657.

5. D. V. Tosic,
"SALECAS - a package for symbolic analysis of linear circuits and systems,"
4th Int. Workshop, Symbolic Methods, Applications, Circuit Design,
SMACD, Leuven, Belgium, pp. 227–230, Oct. 1996.

6. M. D. Lutovac, D. V. Tosic and B. L. Evans,
"Advanced Filter Design for Signal Processing using MATLAB and Mathematica,"
http://iritel.iritel.bg.ac.yu/~lutovac/www/afdhome.htm
http://galeb.etf.bg.ac.yu/~tosic/afdhome.htm
http://www.ece.utexas.edu/~bevans/

■ A.1.2 Initialization

```
SetDirectory[HomeDirectory[]];
<<Calculus`LaplaceTransform`
<<afd\math\m\clearall.m
<<afd\math\m\drawafil.m
<<afd\math\m\drawacc.m
```

A.1.3 Schematic

```
DrawLPSK[0, 0, 1/2, 1.25, 8];
```

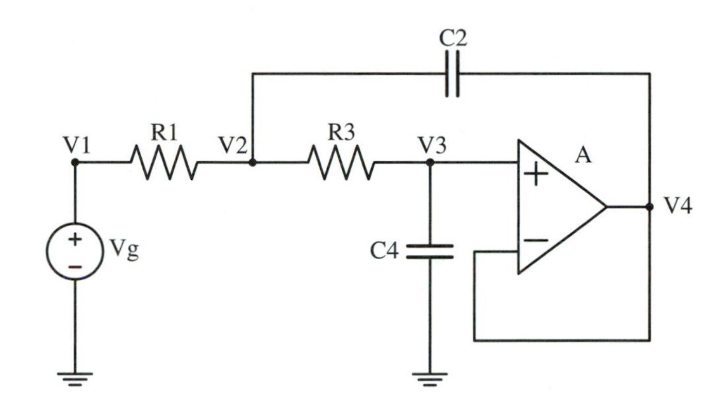

A.1.4 Circuit Analysis

Reduced Modified Nodal Analysis (RMNA)

```
CircuitEquations = {
 V1 == Vg
,(V2-V1)/R1 + (V2-V3)/R3 + (V2-V4)*(s*C2) == 0
,(V3-V2)/R3 + V3*(s*C4) == 0
,V3 == V4
};
NodeVoltages = {V1,V2,V3,V4};
CircuitResponse = Together[Flatten[
 Solve[CircuitEquations,NodeVoltages]
]];
```

A.1.5 Response

```
V1s = Simplify[V1 /. CircuitResponse];
Collect[Numerator[%],s]/Collect[Denominator[%],s]
```

```
Vg
```

```
V2s = Simplify[V2 /. CircuitResponse];
Collect[Numerator[%],Vg]/Collect[Denominator[%],s]
```

```
           (1 + C4 R3 s) Vg
-------------------------------------
                                    2
1 + (C4 R1 + C4 R3) s + C2 C4 R1 R3 s
```

```
V3s = Simplify[V3 /. CircuitResponse];
Collect[Numerator[%],s]/Collect[Denominator[%],s]

                  Vg
-------------------------------------
                                         2
1 + (C4 R1 + C4 R3) s + C2 C4 R1 R3 s

V4s = Simplify[V4 /. CircuitResponse];
Collect[Numerator[%],s]/Collect[Denominator[%],s]

                  Vg
-------------------------------------
                                         2
1 + (C4 R1 + C4 R3) s + C2 C4 R1 R3 s
```

A.1.6 Voltage Transfer Function

```
H = V4s/Vg;
Collect[Numerator[%],s]/Collect[Denominator[%],s]

                  1
-------------------------------------
                                         2
1 + (C4 R1 + C4 R3) s + C2 C4 R1 R3 s

M = Abs[H] /. {C2->4*C, C4->C, R1->R, R3->R} ./ s -> I*2*Pi*f

                  1
Abs[---------------------------------]
                              2 2   2 2
    1 + 4 * C f Pi R - 16 C F Pi R

Plot[ {M ./ {C->10^(-8), R->10^4}}, {f,0,4000}
, PlotRange -> All
, AxesLabel -> {"f (Hz)", "M(f)"}
, AxesOrigin -> {0,0}
, PlotLabel -> "Magnitude response"];
```

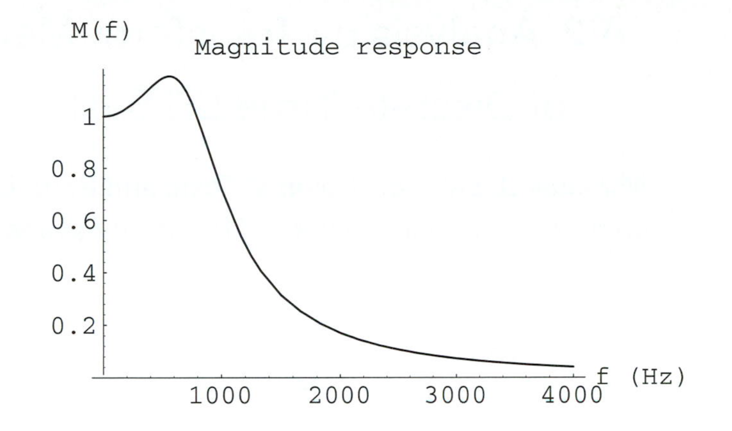

A.1.7 Step Response

```
v4t = InverseLaplaceTransform[
H/s //. {C2->4*C, C4->C, R1->R, R3->R}
, s, t];
v4t //Together //Simplify
```

```
        Sqrt[3] t            Sqrt[3] t
    Cos[---------]       Sin[---------]
          4 C R                4 C R
1 - -------------- - ------------------
        t/(4 C R)                t/(4 C R)
       E              Sqrt[3] E
```

```
Plot[v4t /. {C->10^(-8), R->10^4}
, {t,0,4*10^(-3)}
, PlotRange -> All
, AxesLabel -> {"t (s)", "v4"}
, Ticks -> {{0,0.002,0.004},{0,0.5,1}}
, PlotLabel -> "step response"];
```

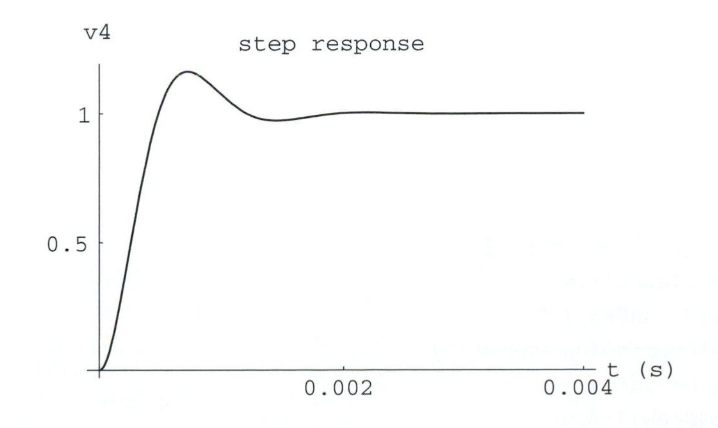

A.2 Analysis by Transform Method of Discrete-Time LTI Systems

Miroslav D. Lutovac, Dejan V. Tosic and Brian L. Evans
lutovac@iritel.bg.ac.yu tosic@galeb.etf.bg.ac.yu bevans@ece.utexas.edu

■ A.2.1 References

1. Alan Oppenheim, Ronald Schafer, "Digital Signal Processing," Prentice-Hall, Englewood Cliffs, New Jersey, 1975.

2.Sanjit Mitra, James Kaiser, "Handbook for Digital Signal Processing," John Wiley, New York, 1993, pp. 127–128.

3. D. V. Tosic, M. D. Lutovac,
"Symbolic computation of impulse, step and sine responses of linear time-invariant systems,"
Proc. 10th Int. Symp. Theor. Electrical Engineering,
ISTET99, Magdeburg, Germany, Sep. 1999, pp. 653–657.

4. John M. Novak and Brian L. Evans,
"Mathematica, Signals and Systems,"
Georgia Tech Research Corp., Atlanta, Georgia, 1995.

5. D. V. Tosic, M. D. Lutovac, and I. M. Markoski,
"Symbolic derivation of transfer functions of discrete-time systems,"
ISTET'97, Palermo, Italia, 9–11 June 1997, pp. 311–314.

6. D. V. Tosic,
"SALECAS - a package for symbolic analysis of linear circuits and systems,"
SMACD, Leuven, Belgium, pp. 227–230, Oct. 1996.

7. M. D. Lutovac, D. V. Tosic and B. L. Evans,
"Advanced Filter Design for Signal Processing using MATLAB and Mathematica,"
http://iritel.iritel.bg.ac.yu/~lutovac/www/afdhome.htm
http://galeb.etf.bg.ac.yu/~tosic/afdhome.htm
http://www.ece.utexas.edu/~bevans/

■ A.2.2 Initialization

```
SetDirectory[HomeDirectory[]];
<<afd\math\m\clearall.m
AppendTo[$Path, "APPS"];"
Needs["SignalProcessing`Master`"]
<<afd\math\m\drawdfil.m
<<afd\math\m\drawiirf.m
```

A.2.3 Block Diagram

```
DrawTDF2[0,0,1,1/0.8,10];
```

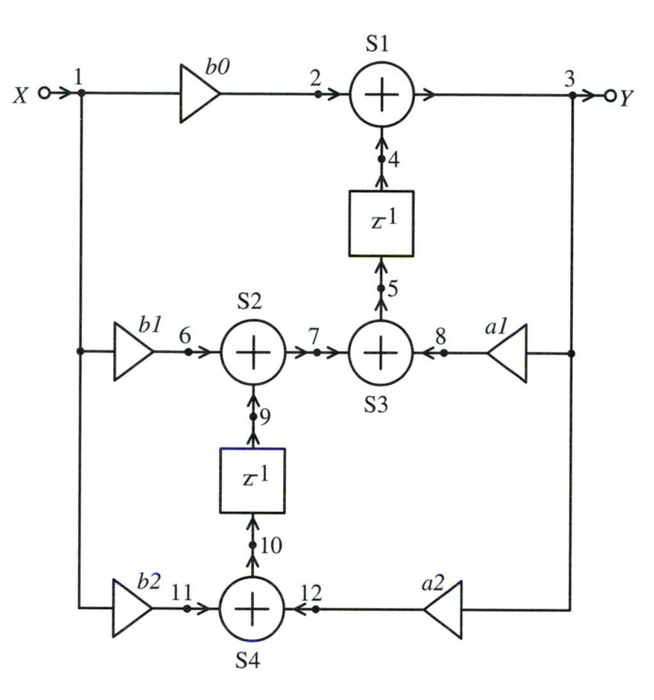

A.2.4 Analysis (v=1/z)

```
ElementEquations = {
  Y1 == X
, Y2 == b0*Y1 + Xb0
, Y3 == Y2 + Y4
, Y4 == v*Y5
, Y5 == Y7 + Y8
, Y6 == b1*Y1 + Xb1
, Y7 == Y6 + Y9
, Y8 == a1*Y3 + Xa1
, Y9 ==  v*Y10
, Y10 == Y11 + Y12
, Y11 == b2*Y1 + Xb2
, Y12 == a2*Y3 + Xa2
};
```

```
NodeSignals = {Y1,Y2,Y3,Y4,Y5,Y6,Y7,Y8,Y9,Y10,Y11,Y12};
Response = Flatten[Solve[ElementEquations,NodeSignals]];
Y = Together[Y3/.Response]
```

```
                          2               2                        2
-(b0 X) - b1 v X - b2 v   X - v Xa1 - v   Xa2 - Xb0 - v Xb1 - v   Xb2
---------------------------------------------------------------------
                                              2
                           -1 + a1 v + a2 v
```

■ A.2.5 Transfer Function and Noise Transfer Functions

```
Hv   = Y /. {X -> 1, Xa1 -> 0, Xa2 -> 0, Xb0 -> 0, Xb1 -> 0, Xb2 -> 0};
Ha1v = Y /. {X -> 0, Xa1 -> 1, Xa2 -> 0, Xb0 -> 0, Xb1 -> 0, Xb2 -> 0};
Ha2v = Y /. {X -> 0, Xa1 -> 0, Xa2 -> 1, Xb0 -> 0, Xb1 -> 0, Xb2 -> 0};
Hb0v = Y /. {X -> 0, Xa1 -> 0, Xa2 -> 0, Xb0 -> 1, Xb1 -> 0, Xb2 -> 0};
Hb1v = Y /. {X -> 0, Xa1 -> 0, Xa2 -> 0, Xb0 -> 0, Xb1 -> 1, Xb2 -> 0};
Hb2v = Y /. {X -> 0, Xa1 -> 0, Xa2 -> 0, Xb0 -> 0, Xb1 -> 0, Xb2 -> 1};
  H = Collect[-Numerator[Hv],v] / (-Collect[Denominator[Hv],v]);
Ha1 = Collect[-Numerator[Ha1v],v]/(-Collect[Denominator[Ha1v],v]);
Ha2 = Collect[-Numerator[Ha2v],v]/(-Collect[Denominator[Ha2v],v]);
Hb0 = Collect[-Numerator[Hb0v],v]/(-Collect[Denominator[Hb0v],v]);
Hb1 = Collect[-Numerator[Hb1v],v]/(-Collect[Denominator[Hb1v],v]);
Hb2 = Collect[-Numerator[Hb2v],v]/(-Collect[Denominator[Hb2v],v]);
v2invz = {v->z^"-1", v^2->z^("-2")};
Print["H(z) = ", H /. v2invz ]
Print["Ha1(z) = ", Ha1 /. v2invz ]
Print["Ha2(z) = ", Ha2 /. v2invz ]
Print["Hb0(z) = ", Hb0 /. v2invz ]
Print["Hb1(z) = ", Hb1 /. v2invz ]
Print["Hb2(z) = ", Hb2 /. v2invz ]
```

```
                  -1        -2
       b0 + b1 z    + b2 z
H(z) = --------------------
                -1        -2
       1 - a1 z    - a2 z
                   -1
                  z
Ha1(z) = -------------------
                  -1        -2
         1 - a1 z    - a2 z
                   -2
                  z
Ha2(z) = -------------------
                  -1        -2
         1 - a1 z    - a2 z
```

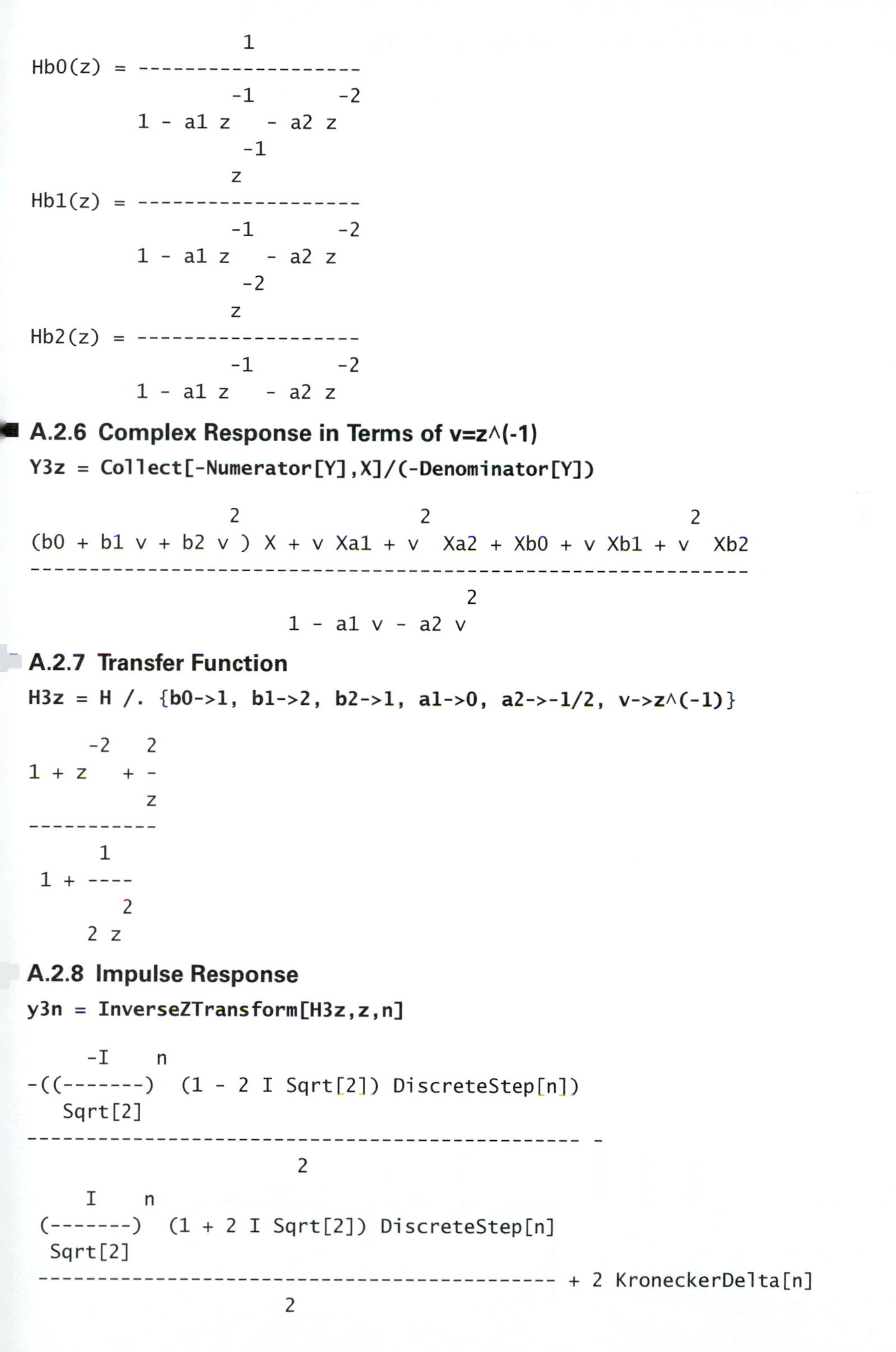

```
                 1
Hb0(z) = ------------------
                -1       -2
         1 - a1 z   - a2 z
                 -1
                z
Hb1(z) = ------------------
                -1       -2
         1 - a1 z   - a2 z
                 -2
                z
Hb2(z) = ------------------
                -1       -2
         1 - a1 z   - a2 z
```

A.2.6 Complex Response in Terms of v=z^(-1)

```
Y3z = Collect[-Numerator[Y],X]/(-Denominator[Y])
```

```
                   2                  2                           2
(b0 + b1 v + b2 v ) X + v Xa1 + v  Xa2 + Xb0 + v Xb1 + v  Xb2
---------------------------------------------------------------
                                          2
                         1 - a1 v - a2 v
```

A.2.7 Transfer Function

```
H3z = H /. {b0->1, b1->2, b2->1, a1->0, a2->-1/2, v->z^(-1)}
```

```
     -2   2
1 + z   + -
          z
-----------
      1
 1 + ----
        2
     2 z
```

A.2.8 Impulse Response

```
y3n = InverseZTransform[H3z,z,n]
```

```
     -I     n
-((-------)  (1 - 2 I Sqrt[2]) DiscreteStep[n])
   Sqrt[2]
----------------------------------------------- -
                       2

     I     n
 (-------)  (1 + 2 I Sqrt[2]) DiscreteStep[n]
  Sqrt[2]
--------------------------------------------- + 2 KroneckerDelta[n]
                      2
```

```
y3nDS = ComplexExpand[Coefficient[y3n,DiscreteStep[n]]]

      n Pi
  Cos[----]
       2            1/2 - n/2     n Pi
-(---------) + 2 2              Sin[----]
     n/2                             2
    2

y3nKD = Coefficient[y3n,KroneckerDelta[n]]

2

h3n = y3nDS*DiscreteStep[n] + y3nKD*KroneckerDelta[n]

                                                   n Pi
                                              Cos[----]
                                                   2            1/2 - n/2     n Pi
2 KroneckerDelta[n] + DiscreteStep[n] (-(---------) + 2 2              Sin[----])
                                                 n/2                             2
                                                2

DiscreteSignalPlot[h3n
,{n,0,10}
,AxesLabel -> {"n","y3(n)"}
,PlotLabel -> "impulse response"
];
```

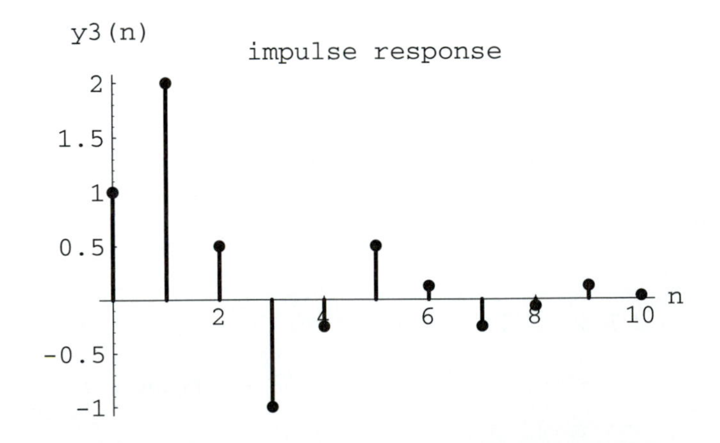

```
Table[{n,h3n}, {n,0,10}]
% //TableForm //N

                         1                    1         1        1           1
{{0, 1}, {1, 2}, {2, -}, {3, -1}, {4, -(-)}, {5, -}, {6, -}, {7, -(-)},
                         2                    4         2        8           4

      1         1         1
{8, -(--)}, {9, -}, {10, --}}
      16        8         32

0      1.
1.     2.
2.     0.5
3.     -1.
4.     -0.25
5.     0.5
6.     0.125
7.     -0.25
8.     -0.0625
9.     0.125
10.    0.03125
```

A.3 Switched Capacitor Filter - Mode 3a Analysis and Design

Miroslav D. Lutovac, Dejan V. Tosic and Brian L. Evans
lutovac@iritel.bg.ac.yu tosic@galeb.etf.bg.ac.yu bevans@ece.utexas.edu

■ A.3.1 References

1. Linear Technology Corporation
Monolithic Filter Handbook, 1990.

2. M. D. Lutovac, D. Novakovic, I. Markoski,
"Selective SC-filters with low passive sensitivity,"
Electronics Letters vol. 33, no. 8, pp. 674–675, Apr. 1997.

3. M.D.Lutovac, D. V. Tosic, D.Novakovic,
"Programmable low-pass/high-pass SC-filters,"
Proc.9th Conf. MELECON '98, Tel-Aviv, Israel May 1998, pp.673–677

4. D.V. Tosic, M.D. Lutovac, B.L. Evans, I.M. Markoski,
"A tool for symbolic analysis and design of analog active filters,"
5th Int. Workshop, Symbolic Methods, Applications, Circuit Design,
SMACD'98, Kaiserslautern, Germany, pp. 71–74, Oct. 1998.

5. M. D. Lutovac, D. V. Tosic and B. L. Evans,
"Advanced Filter Design for Signal Processing using MATLAB and Mathematica,"
http://iritel.iritel.bg.ac.yu/~lutovac/www/afdhome.htm
http://galeb.etf.bg.ac.yu/~tosic/afdhome.htm
http://www.ece.utexas.edu/~bevans/

■ A.3.2 Initialization

```
SetDirectory[HomeDirectory[]];
<<afd\math\m\clearall.m
<<afd\math\m\drawafil.m
<<afd\math\m\drawasc.m
```

■ A.3.3 Schematic

```
DrawMode3a[0,0,1/2,1/0.8,8];
```

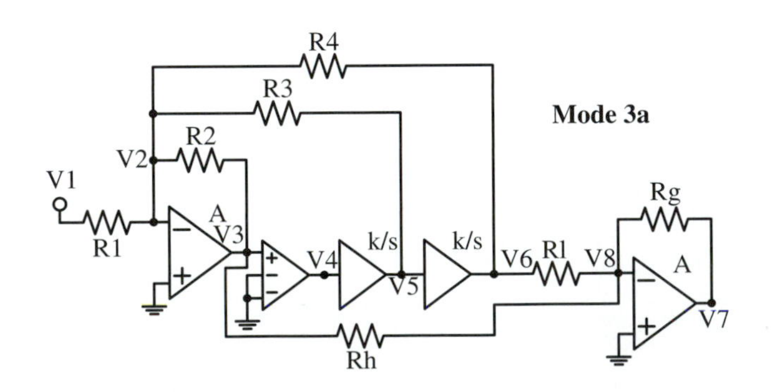

A.3.4 Circuit Analysis

Reduced Modified Nodal Analysis

```
CircuitEquations = {
  V1 == Vg
, (V2-V1)/R1 + (V2-V3)/R2 + (V2-V5)/R3 + (V2-V6)/R4 == 0
, V3 == -A*V2
, V4 == V3
, V5 == V4*k/s
, V6 == V5*k/s
, (V8-V6)/Rl + (V8-V3)/Rh + (V8-V7)/Rg == 0
, V7 == -A*V8
};
NodeVoltages = {V1,V2,V3,V4,V5,V6,V7,V8};
CircuitResponse = Together[Flatten[
 Solve[CircuitEquations,NodeVoltages]
]];
Print["V1 = ", V1 /. CircuitResponse]
Print["V7 = ", V7 /. CircuitResponse]
```

```
V1 = Vg
            2                      2                      2
V7 = -((A   R2 R3 R4 s (-(k   Rg Rh) - Rg Rl  s ) Vg) /
    ((Rg Rh + Rg Rl + Rh Rl + A Rh Rl)
                                                              3
      ((R1 R2 R3 + R1 R2 R4 + R1 R3 R4 + R2 R3 R4) s   -
                 2               2
        A (-(k   R1 R2 R3 s) + s   (-(k R1 R2 R4) - R1 R3 R4 s)))))
```

A.3.5 Voltage Transfer Function

```
H = V7/V1 /. CircuitResponse //Together //Simplify;
H3a = Limit[H, A->Infinity];
Print["H(s) = ",
 Factor[Collect[Numerator[H3a],s]]/Factor[Collect[Denominator[H3a],s]]
]
```

```
                                   2            2
                Rg R2 R3 R4 (k   Rh + Rl  s )
H(s) = ----------------------------------------
                     2                               2
       Rh Rl R1 (k   R2 R3 + k R2 R4 s + R3 R4 s )
```

A.3.6 Definitions and Procedures

```
PoleQpole[H_,s_] := Module[{den,fp,Qp},
  den = Denominator[H];
  fp = Sqrt[Coefficient[den,s,0]/Coefficient[den,s,2]]/(2*Pi);
  Qp = (Coefficient[den,s,2]/Coefficient[den,s,1])*(2*Pi*fp);
  Simplify[{fp, Qp}]];
```

```
ZeroQzero[H_,s_] := Module[{fz,num,Qz0},
  num = Numerator[H];
  Qz0 = (Coefficient[num,s,2]/Coefficient[num,s,1]);
  fz = Sqrt[Coefficient[num,s,0]/Coefficient[num,s,2]]/(2*Pi);
  Simplify[{fz, Qz0*fz}]];
Sensitivity[F_,x_] := (x/F)*D[F,x];
GSP[F_,A_] := Limit[A*Sensitivity[F,A],A->Infinity]//Simplify;
PrintLabeledList[expressions_List,labels_List] := Map[
 Print[#[[1]]," = ",#[[2]]]&
,Transpose[{labels,expressions}]
];
```

■ A.3.7 Poles, Zeros, Q-Factors

```
{fp,Qp} = Simplify[PoleQpole[H,s]];
PrintLabeledList[{fp,Qp},{"fp","Qp"}];
```

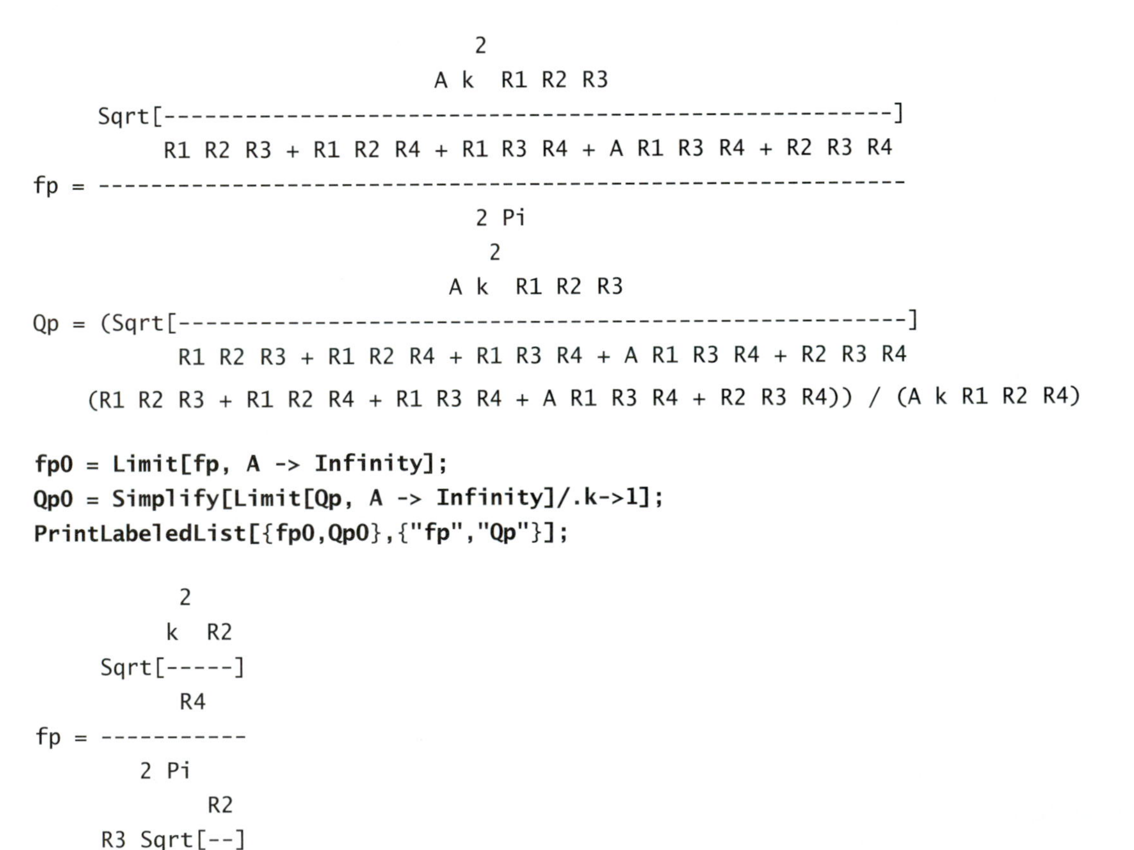

```
                                   2
                               A k   R1 R2 R3
     Sqrt[-------------------------------------------------------]
          R1 R2 R3 + R1 R2 R4 + R1 R3 R4 + A R1 R3 R4 + R2 R3 R4
fp = -------------------------------------------------------------
                               2 Pi
                                2
                            A k   R1 R2 R3
Qp = (Sqrt[-------------------------------------------------------]
            R1 R2 R3 + R1 R2 R4 + R1 R3 R4 + A R1 R3 R4 + R2 R3 R4
    (R1 R2 R3 + R1 R2 R4 + R1 R3 R4 + A R1 R3 R4 + R2 R3 R4)) / (A k R1 R2 R4)
```

```
fp0 = Limit[fp, A -> Infinity];
Qp0 = Simplify[Limit[Qp, A -> Infinity]/.k->1];
PrintLabeledList[{fp0,Qp0},{"fp","Qp"}];
```

```
           2
          k   R2
     Sqrt[-----]
           R4
fp = -----------
        2 Pi
              R2
     R3 Sqrt[--]
              R4
Qp = -----------
         R2
```

```
{fz,Qz} = Simplify[ZeroQzero[H,s]];
PrintLabeledList[{fz,Qz},{"fz","Qz"}];

                                     1
Power::infy: Infinite expression - encountered.
                                     0
           2
          k   Rh
     Sqrt[-----]
           Rl
fz = -----------
        2 Pi
Qz = ComplexInfinity
```

A.3.8 Gain-Sensitivity Product (GSP)

```
GSPfp = GSP[fp,A];
GSPQp = GSP[Qp,A];
PrintLabeledList[{GSPfp,GSPQp},{"GSPfp","GSPQp"}];

        1    R2     R2     R2
GSPfp = - + ---- + ---- + ----
        2   2 R1   2 R3   2 R4
          1     R2     R2     R2
GSPQp = -(-) - ---- - ---- - ----
          2    2 R1   2 R3   2 R4
```

A.3.9 Design

Find Element Values

```
DesignMode3a[K_,Qp_,wp_,wz_,fclk_,P_:100,R1_:R1nom,R2_:R2nom,Rh_:Rhnom]:=
 Module[{R3,R4,Rl,Rg},
 R4 = R2*(2*Pi*fclk/(P*wp))^2;
 R3 = Qp*Sqrt[R2*R4];
 Rl = Rh*(2*Pi*fclk/(P*wz))^2;
 Rg = K*Rh*R1/R2;
{R1,R2,R3,R4,Rl,Rh,Rg}
];
{R1,R2,R3,R4,Rl,Rh,Rg} = Together[
                        DesignMode3a[K,Q,W,Z,Fc,P,R1n,R2n,Rhn]];
PrintLabeledList[{R1,R2,R3,R4,Rl,Rh,Rg}
               ,{"R1","R2","R3","R4","Rl","Rh","Rg"}];

R1 = R1n
R2 = R2n
                 2    2     2
               Fc   Pi   R2n
R3 = 2 Q Sqrt[------------]
                   2   2
                 P   W
```

```
          2    2
     4 Fc   Pi   R2n
R4 = -------------
           2  2
          P  W
          2    2
     4 Fc   Pi   Rhn
Rl = -------------
           2  2
          P  Z
Rh = Rhn
     K Rhn R1n
Rg = ---------
        R2n
```

■ **Test**

```
H3atest  = Simplify[ExpandAll[Together[Limit[H , A -> Infinity]] /.
 {Sqrt[x_^2*y_^2/z_^2] -> x*y/z, Sqrt[x_^2*y_^2*p_^2/z_^2] -> x*y*p/z
, Sqrt[x_^2*y_^2*p_^2/(z_^2*n_^2)] -> x*y*p/(z*n)} ]];
num = Numerator[H3atest];
den = Denominator[H3atest];
numlist = CoefficientList[num,s];
denlist = CoefficientList[den,s];
K3at =(numlist[[3]]/denlist[[3]]);
H3at =  K3at *
Simplify[num/numlist[[3]]]/Simplify[den/denlist[[3]]] /. k-> 2*Pi*Fc/P
```

```
     2    2
 K (s  + Z )
-------------
 2   s W    2
s  + --- + W
      Q
```

Design Examples

```
values = {K ->1, Q -> 1.0349, W -> 2*Pi*1710.9457, Z -> 2*Pi*5129.3034
   , Fc -> 256.*10^3, P -> 100, R12 ->23.16*10^3
   , R22 ->10.*10^3, Rh2 ->238.6*10^3} //N;
h1 = H  /. k-> 2*Pi*Fc/P /.values //N;
H3atest = Limit[h1, A->Infinity];
H3atest =H3atest /. Sqrt[x_^2] -> x;
num = Numerator[H3atest];
den = Denominator[H3atest];
numlist = CoefficientList[num,s];
denlist = CoefficientList[den,s];
K3at =(numlist[[3]]/denlist[[3]]);
H3at =  K3at * Simplify[num/numlist[[3]]]/Simplify[den/denlist[[3]]]
```

```
                  9         2
   1. (1.03867 10   + 1. s )
-------------------------------
          8                   2
1.15567 10   + 10387.7 s + 1. s
```

```
Rexample1 = N[{R1,R2,R3,R4,Rl,Rh,Rg}/. values] /. Sqrt[R2n^2]->R2n;
PrintLabeledList[Rexample1,{"R1","R2","R3","R4","Rl","Rh","Rg"}];
```

```
R1 = R1n
R2 = R2n
R3 = 1.54847 R2n
R4 = 2.23876 R2n
Rl = 0.249094 Rhn
Rh = Rhn
     1. Rhn R1n
Rg = ----------
        R2n
```

```
Plot[{Abs[H3at] /. s -> I*2*Pi*f}
, {f, 100, 10000}
,PlotRange -> All
,Ticks -> {{0,2000,5000,10000},{0,2,4,6,8,10}}
, AxesLabel -> {"f (Hz)","M(f)"}
];
```

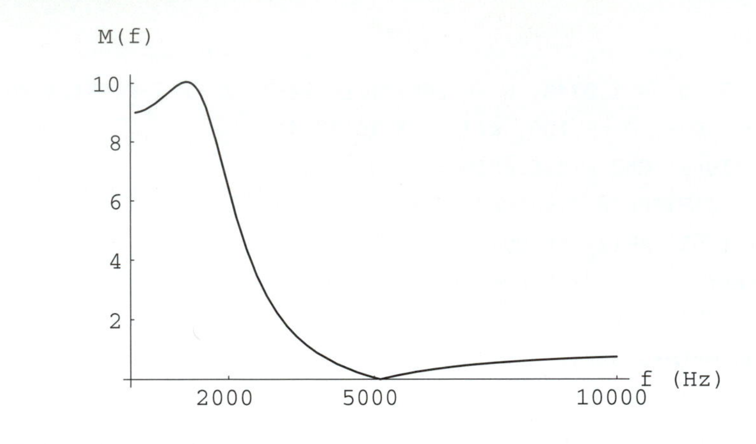

A.3.10 Optimization

Find R2/R1 for Low Gain-Sensitivity Product

```
sf = (Simplify[N[Together[GSPfp /. values]]
     /. Sqrt[x_^2] -> x/. Sqrt[x_^2*y_^2] -> x*y ]
     /. Sqrt[R2n^2] -> R2n)

          0.5 R2n
1.04624 + -------
           R1n
```

Remark

We have to choose R1 > R2 to minimize GSP.

A.4 OTA-C General Biquad Analysis and Design

Miroslav D. Lutovac, Dejan V. Tosic and Brian L. Evans
lutovac@iritel.bg.ac.yu tosic@galeb.etf.bg.ac.yu bevans@ece.utexas.edu

A.4.1 References

1. Wai-Kai-Chen, Ed.,
"The Circuits and Filters Handbook," p. 2479,
"CRC Press," "Boca Raton, Florida," "1995."

2. D.V. Tosic, M.D. Lutovac, B.L. Evans, I.M. Markoski,
"A tool for symbolic analysis and design of analog active filters,"
5th Int. Workshop, Symbolic Methods, Applications, Circuit Design,
SMACD'98, Kaiserslautern, Germany, pp. 71–74, Oct. 1998.

3. M. D. Lutovac, D. V. Tosic and B. L. Evans,
"Advanced Filter Design for Signal Processing using MATLAB and Mathematica,"
http://iritel.iritel.bg.ac.yu/~lutovac/www/afdhome.htm
http://galeb.etf.bg.ac.yu/~tosic/afdhome.htm
http://www.ece.utexas.edu/~bevans/

A.4.2 Initialization

```
SetDirectory[HomeDirectory[]];
<<afd\math\m\clearall.m
<<afd\math\m\drawafil.m
<<afd\math\m\drawaota.m
```

A.4.3 Schematic

```
DrawOTAb[0, 0, 1/2, 1/08, 8]
```

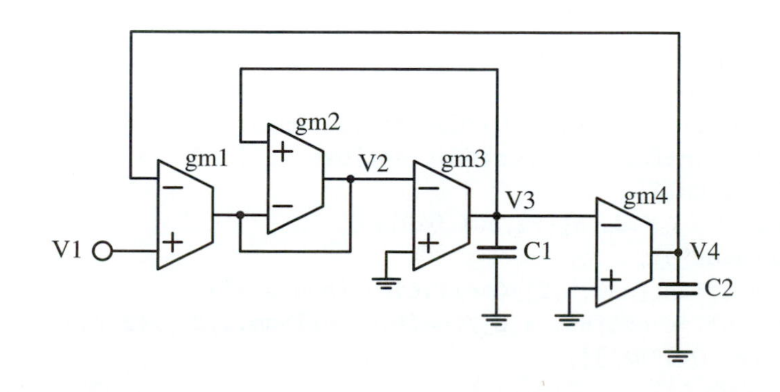

A.4.4 Circuit Analysis

Reduced Modified Nodal Analysis

```
CircuitEquations = {
  V1 == Vg
, -(V1-V4)*gm1 - (V3-V2)*gm2 == 0
, -(-V2)*gm3 + V3/(1/(s*C1)) == 0
, -(-V3)*gm4 + V4/(1/(s*C2)) == 0
};

NodeVoltages = {V1,V2,V3,V4};
CircuitResponse = Together[Flatten[
 Solve[CircuitEquations,NodeVoltages]
]];
```

A.4.5 Voltage Transfer Function

```
H = V4/V1 /. CircuitResponse //Together //Simplify;
Print["H(s) = ", H]
Hhp = V2/V1 /. CircuitResponse //Together //Simplify;
Print["Hhp(s) = ", Hhp]
Hbp = V3/V1 /. CircuitResponse //Together //Simplify;
Print["Hbp(s) = ", Hbp]
```

```
                              gm1 gm3 gm4
H(s) = -----------------------------------------
                                                 2
       gm1 gm3 gm4 + C2 gm2 gm3 s + C1 C2 gm2 s
                                     2
                              C1 C2 gm1 s
Hhp(s) = -----------------------------------------
                                                   2
         gm1 gm3 gm4 + C2 gm2 gm3 s + C1 C2 gm2 s
                                C2 gm1 gm3 s
Hbp(s) = -(-----------------------------------------)
                                                     2
           gm1 gm3 gm4 + C2 gm2 gm3 s + C1 C2 gm2 s
```

A.4.6 Definitions and Procedures

```
PoleQpole[H_,s_] := Module[{den,fp,Qp},
  den = Denominator[H];
  fp = Sqrt[Coefficient[den,s,0]/Coefficient[den,s,2]]/(2*Pi);
  Qp = (Coefficient[den,s,2]/Coefficient[den,s,1])*(2*Pi*fp);
  Simplify[{fp, Qp}]];
ZeroQzero[H_,s_] := Module[{fz,num,Qz0},
  num = Numerator[H];
  Qz0 = (Coefficient[num,s,2]/Coefficient[num,s,1]);
  fz = Sqrt[Coefficient[num,s,0]/Coefficient[num,s,2]]/(2*Pi);
  Simplify[{fz, Qz0*fz}]];
```

```
PrintLabeledList[expressions_List,labels_List] := Map[
 Print[#[[1]]," = ",#[[2]]]&
,Transpose[{labels,expressions}]
];
```

A.4.7 Poles, Zeros, Q-Factors

```
{fp,Qp} = Simplify[PoleQpole[H,s]];
Klp = H /. s-> 0
PrintLabeledList[{fp,Qp},{"fp","Qp"}];
```

```
1
          gm1 gm3 gm4
     Sqrt[-----------]
           C1 C2 gm2
fp = -----------------
           2 Pi
             gm1 gm3 gm4
     C1 Sqrt[-----------]
              C1 C2 gm2
Qp = --------------------
             gm3
```

A.4.8 Design

Find Element Values

```
DesignOTA1[Qp_,wp_,C1_,C2_,gm1_,gm2_] :=
 Module[{gm3,gm4},
 gm3 = C1*wp/Qp;
 gm4 = C2*wp*Qp*gm2/gm1;
{C1,C2,gm1,gm2,gm3,gm4}
];
{C1,C2,gm1,gm2,gm3,gm4} = Together[DesignOTA1[Q,W,c1,c2,g1,g2]];
PrintLabeledList[{C1,C2,gm1,gm2,gm3,gm4},{"C1","C2","gm1","gm2","gm3","gm4"}];
```

```
C1 = c1
C2 = c2
gm1 = g1
gm2 = g2
      c1 W
gm3 = ----
       Q
      c2 g2 Q W
gm4 = ---------
         g1
```

A.4.9 Test

```
Simplify[H]

          2
       Q W
-----------------
   2            2
Q s  + s W + Q W
```

Design Examples

```
values = {Q -> 4., W -> N[2*Pi*10^6] ,
  c1 -> 10.*10^(-12), c2 ->10.*10^(-12), g1 -> g , g2 -> g} //N;
h1 = Together[H  /. values] /. 1. -> 1;
Print["gm3 = ",10^3*gm3 /. values, " mS "]
Print["gm4 = ",10^3*gm4 /. values, " mS "]
h = (Numerator[h1]/g)/ (Simplify[Denominator[h1]/g])

gm3 = 0.015708 mS
gm4 = 0.251327 mS
                          13
          3.94784 10
--------------------------------
          13             6      2
3.94784 10   + 1.5708 10  s + s

PrintLabeledList[N[{Q,W/(2*Pi)} /. values]
,{"Qp","fp (Hz)"}];
Print["------------------------"]
Rexample1 = N[{C1*10^(12),C2*10^(12),gm1,gm2,gm3*10^(3),gm4*10^(3)}/. values] ;
PrintLabeledList[Rexample1
                ,{"C1 (pF)","C2 (pF)","gm1","gm2","gm3 (mS)","gm4 (mS)"}];

Qp = 4.
                 6
fp (Hz) = 1. 10
------------------------
C1 (pF) = 10.
C2 (pF) = 10.
gm1 = g
gm2 = g
gm3 (mS) = 0.015708
gm4 (mS) = 0.251327

Plot[{Abs[h] /. s -> I*2*Pi*f*10^6}
, {f, 0.01, 4}
,PlotRange -> All
, AxesLabel -> {"f (MHz)","M(f)"}
];
```

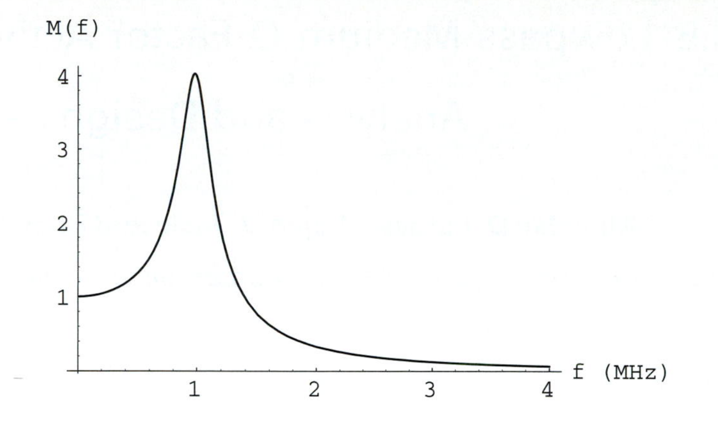
M(f)
4
3
2
1
1
2
3
4
f (MHz)

A.5 Lowpass-Medium-Q-Factor Active RC Filter Analysis and Design

Miroslav D. Lutovac, Dejan V. Tosic and Brian L. Evans

lutovac@iritel.bg.ac.yu tosic@galeb.etf.bg.ac.yu bevans@ece.utexas.edu

A.5.1 References

1. G. S. Moschytz and P. Horn,
Active Filter Design Handbook
For Use with Programmable Pocket Calculators and Mini Computers
John Wiley and Sons, Ltd., New York, 1981, pp. 38–39.

2. R. P. Sallen and E. L. Key,
"A practical method of designing RC active filters,"
IRE Trans. Circuit Theory, CT-2, Mar. 1955, pp. 74–85.

3. D.V. Tosic, M.D. Lutovac, B.L. Evans, I.M. Markoski,
"A tool for symbolic analysis and design of analog active filters,"
5th Int. Workshop, Symbolic Methods, Applications, Circuit Design,
SMACD'98, Kaiserslautern, Germany, pp. 71–74, Oct. 1998.

4. M. D. Lutovac, D. V. Tosic and B. L. Evans,
"Advanced Filter Design for Signal Processing using MATLAB and Mathematica,"
http://iritel.iritel.bg.ac.yu/~lutovac/www/afdhome.htm
http://galeb.etf.bg.ac.yu/~tosic/afdhome.htm
http://www.ece.utexas.edu/~bevans/

A.5.2 Initialization

```
SetDirectory[HomeDirectory[]];
<<afd\math\m\clearall.m
<<afd\math\m\drawafil.m
<<afd\math\m\drawarc.m
```

A.5.3 Schematic

```
DrawLPMQ[0,0,1/2,1/0.8];
```

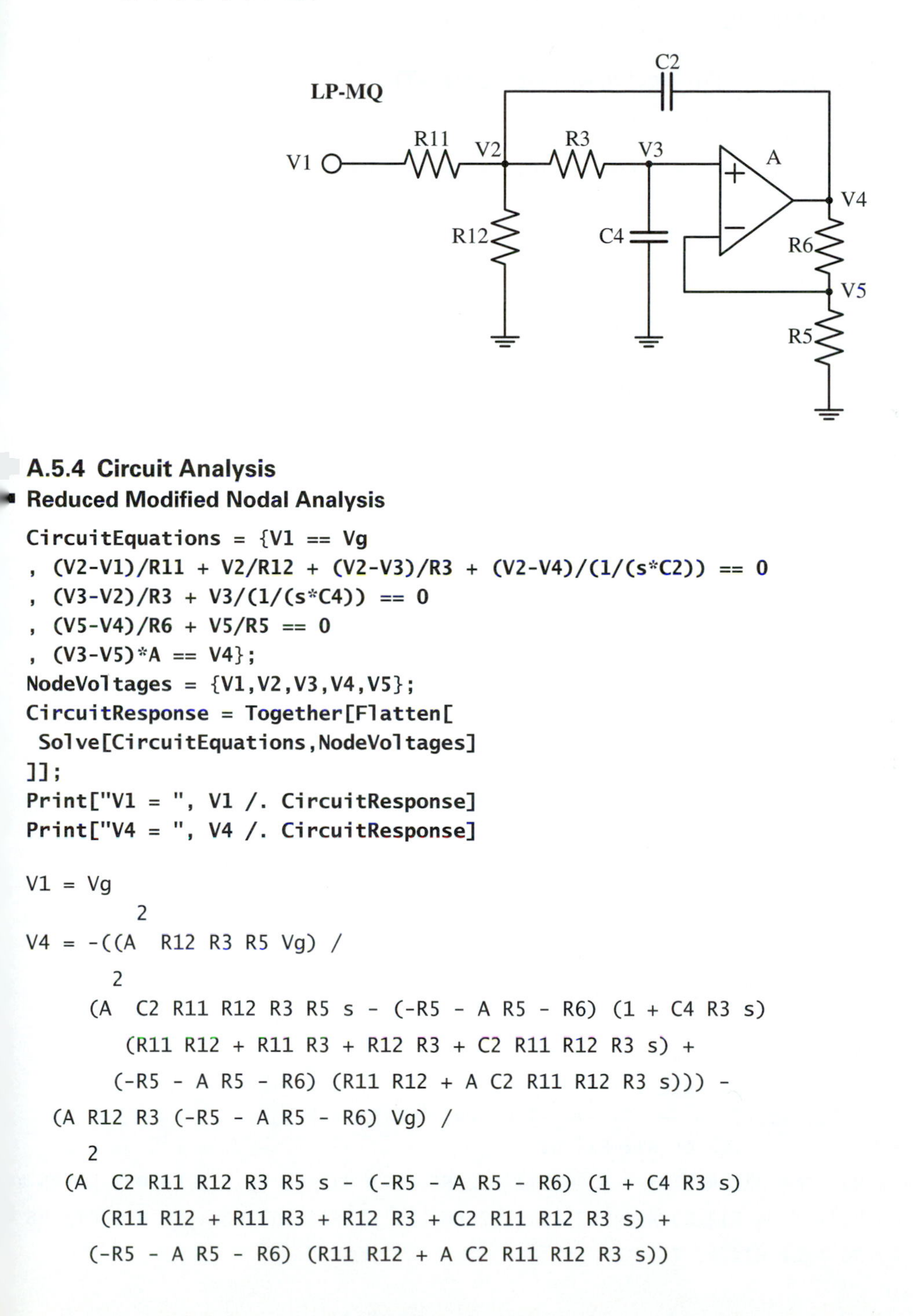

A.5.4 Circuit Analysis

Reduced Modified Nodal Analysis

```
CircuitEquations = {V1 == Vg
, (V2-V1)/R11 + V2/R12 + (V2-V3)/R3 + (V2-V4)/(1/(s*C2)) == 0
, (V3-V2)/R3 + V3/(1/(s*C4)) == 0
, (V5-V4)/R6 + V5/R5 == 0
, (V3-V5)*A == V4};
NodeVoltages = {V1,V2,V3,V4,V5};
CircuitResponse = Together[Flatten[
 Solve[CircuitEquations,NodeVoltages]
]];
Print["V1 = ", V1 /. CircuitResponse]
Print["V4 = ", V4 /. CircuitResponse]
```

```
V1 = Vg
           2
V4 = -((A   R12 R3 R5 Vg) /
         2
      (A   C2 R11 R12 R3 R5 s - (-R5 - A R5 - R6) (1 + C4 R3 s)
         (R11 R12 + R11 R3 + R12 R3 + C2 R11 R12 R3 s) +
        (-R5 - A R5 - R6) (R11 R12 + A C2 R11 R12 R3 s))) -
   (A R12 R3 (-R5 - A R5 - R6) Vg) /
       2
    (A   C2 R11 R12 R3 R5 s - (-R5 - A R5 - R6) (1 + C4 R3 s)
       (R11 R12 + R11 R3 + R12 R3 + C2 R11 R12 R3 s) +
      (-R5 - A R5 - R6) (R11 R12 + A C2 R11 R12 R3 s))
```

A.5.5 Voltage Transfer Function

```
H = V4/V1 /. CircuitResponse //Together //Simplify;
Ha = Limit[H, A->Infinity];
Print["H(s) = ",
 Collect[Numerator[Ha],s]/Collect[Denominator[Ha],s]]

H(s) = (R12 (R5 + R6)) /
  (R11 R5 + R12 R5 + (C4 R11 R12 R5 + C4 R11 R3 R5 + C4 R12 R3 R5 - C2 R11 R12 R6) s
                              2
    C2 C4 R11 R12 R3 R5 s )
```

A.5.6 Definitions and Procedures

```
PoleQpole[H_,s_] := Module[{den,fp,Qp},
  den = Denominator[H];
  fp = Sqrt[Coefficient[den,s,0]/Coefficient[den,s,2]]/(2*Pi);
  Qp = (Coefficient[den,s,2]/Coefficient[den,s,1])*(2*Pi*fp);
  Simplify[{fp, Qp}]];
ZeroQzero[H_,s_] := Module[{fz,num,Qz0},
  num = Numerator[H];
  Qz0 = (Coefficient[num,s,2]/Coefficient[num,s,1]);
  fz = Sqrt[Coefficient[num,s,0]/Coefficient[num,s,2]]/(2*Pi);
  Simplify[{fz, Qz0*fz}]];
Sensitivity[F_,x_] := (x/F)*D[F,x];
GSP[F_,A_] := Limit[A*Sensitivity[F,A],A->Infinity]//Simplify;
PrintLabeledList[expressions_List,labels_List] := Map[
 Print[#[[1]]," = ",#[[2]]]&
,Transpose[{labels,expressions}]
];
```

A.5.7 Poles, Zeros, Q-Factors

```
{fp,Qp} = Simplify[PoleQpole[H,s]];
PrintLabeledList[{fp,Qp},{"fp","Qp"}];

             R11 + R12
      Sqrt[----------------]
           C2 C4 R11 R12 R3
fp = ----------------------
              2 Pi
                                 R11 + R12
Qp = (C2 C4 R11 R12 Sqrt[----------------] R3 (R5 + A R5 + R6)) /
                         C2 C4 R11 R12 R3
  (C2 R11 R12 R5 + C4 R11 R12 R5 + A C4 R11 R12 R5 + C4 R11 R3 R5 + A C4 R11 R3 R5 +
    C4 R12 R3 R5 + A C4 R12 R3 R5 + C2 R11 R12 R6 - A C2 R11 R12 R6 + C4 R11 R12 R6
    C4 R11 R3 R6 + C4 R12 R3 R6)
```

A.5.8 Gain-Sensitivity Product (GSP)

```
GSPfp = GSP[fp,A];
GSPQp = GSP[Qp,A];
PrintLabeledList[{GSPfp,GSPQp},{"GSPfp","GSPQp"}];
```

```
GSPfp = 0

                                                     2
                         C2 R11 R12 (R5 + R6)
GSPQp = ---------------------------------------------------------------------
        R5 (C4 R11 R12 R5 + C4 R11 R3 R5 + C4 R12 R3 R5 - C2 R11 R12 R6)
```

A.5.9 Design

Find Element Values

```
DesignLPMQ[K_,Qp_,wp_,P_,C2x_,C4x_,R5x_] :=  Module[
{C2,C4,R1,R11,R12,R3,R5,R6,Ko},
 C2 = C2x;
 C4 = C4x;
 R1 = 1/(wp*Sqrt[C2x*C4x*P]);
 R3 = P*R1;
 R5 = R5x;
 R6 = R5*((1+P)*C4/C2-Sqrt[P*C4/C2]/Qp);
 Ko = 1+R6/R5;
 R11 = R1*Ko/K;
 R12 = R1*Ko/(Ko-K);
{R11,R12,C2,R3,C4,R5,R6}
];
{R11,R12,C2,R3,C4,R5,R6} = Together[DesignLPMQ[K,Q,W,P,C2x,C4x,R5x]];
```

A.5.10 Design Example

```
values = {K -> 1., Q -> 7.5, W -> 2*Pi*2500., P -> 1.5333078
 , C2x -> 33.*10^(-9), C4x -> 10.*10^(-9), R5x -> 6800.} //N;
PrintLabeledList[N[{K,Q,W/(2*Pi),P} /. values]
,{"K","Qp","fp (Hz)","P"} ];
Print["----------------"]
PrintLabeledList[Together[{R11,R12,C2*10^9,R3,C4*10^9,R5,R6} /. values],
{"R11 (ohm)","R12 (ohm)","C2 (nF)","R3 (ohm)","C4 (nF)","R5 (ohm)","R6 (ohm)"}];
```

```
K = 1.
Qp = 7.5
fp (Hz) = 2500.
P = 1.53331
----------------
R11 (ohm) = 4745.54
R12 (ohm) = 7011.9
C2 (nF) = 33.
R3 (ohm) = 4339.48
C4 (nF) = 10.
R5 (ohm) = 6800.
R6 (ohm) = 4602.13
```

A.5.11 Optimization

Find P for Low Gain-Sensitivity Product

```
values = {K -> 1.476288, Q -> 7.5, W -> 2*Pi*2500.
 , C2x -> 100.*10^(-9), C4x -> 15.*10^(-9), R5x -> 10000.} //N;
gspQp = Together[Simplify[GSPQp] /. values]
```

```
                                          2
0.435711 (7.66667 - 0.344265 Sqrt[P] + 1. P)
--------------------------------------------
                  Sqrt[P]
```

```
P1 = 1.;
P2 = 4.;
Plot[gspQp /. values
, {P, P1, P2}
, AxesLabel -> {"P","GSP Qp"}
];
```

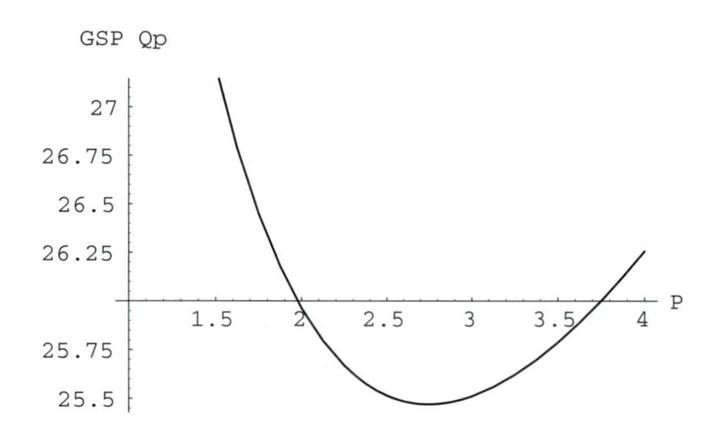


```
{GSPmin,Pset} = FindMinimum[gspQp,{P,P1,P2}]
```

```
{25.4701, {P -> 2.74571}}
```

```
PrintLabeledList[N[{K,Q,W/(2*Pi),P} /. values /.Pset]
,{"K","Qp","fp (Hz)","P"} ];
Print["----------------"]
PrintLabeledList[Together[{R11,R12,C2*10^9,R3,C4*10^9,R5,R6} /. values /.Pset],
{"R11 (ohm)","R12 (ohm)","C2 (nF)","R3 (ohm)","C4 (nF)","R5 (ohm)","R6 (ohm)"}];
```

```
K = 1.47629
Qp = 7.5
fp (Hz) = 2500.
P = 2.74571
----------------
R11 (ohm) = 991.99
```

```
                    9
R12 (ohm) = 1.81368 10
C2 (nF) = 100.
R3 (ohm) = 2723.72
C4 (nF) = 15.
R5 (ohm) = 10000.
R6 (ohm) = 4762.89
```

```
Hhpmq = Together[Ha  /.Pset/. values] ;
num = Numerator[Hhpmq];
den = Denominator[Hhpmq];
numlist = CoefficientList[num,s];
denlist = CoefficientList[den,s];
Hhpmq = (1/denlist[[3]]) * num/(den/denlist[[3]])//Factor
```

```
                 8
        3.64259 10
----------------------------
        8                  2
2.4674 10  + 2094.4 s + 1. s
```

```
Plot[{Abs[Hhpmq] /. s -> N[I*2*Pi*f]}
, {f, 100, 10000}
, PlotRange -> All
, Ticks -> {{0,2000,5000,10000},{0,2,4,6,8,10}}
, AxesLabel -> {"f (Hz)","M(f)"} ];
```


Remark

We choose P=2.7457 as a good choice because we obtain small GSP and 1/R12=0.

A.6 General Purpose High-Q-Factor RC Filter Analysis and Design

Miroslav D. Lutovac, Dejan V. Tosic and Brian L. Evans
lutovac@iritel.bg.ac.yu tosic@galeb.etf.bg.ac.yu bevans@ece.utexas.edu

A.6.1 References

1. G. S. Moschytz and P. Horn,
Active Filter Design Handbook
For Use with Programmable Pocket Calculators and Mini Computers
John Wiley and Sons, Ltd., New York, 1981 pp. 64–65.

2. W. J. Kerwin, L. P. Huelsman, and R. W. Newcomb,
State-variable synthesis for insensitive integrated circuit transfer functions,
IEEE J. Solid-State Circuits, SC–2, pp. 87–92, September, 1967.

3. D.V. Tosic, M.D. Lutovac, B.L. Evans, I.M. Markoski,
"A tool for symbolic analysis and design of analog active filters,"
5th Int. Workshop, Symbolic Methods, Applications, Circuit Design,
SMACD'98, Kaiserslautern, Germany, pp. 71–74, Oct. 1998.

4. M. D. Lutovac, D. V. Tosic and B. L. Evans,
"Advanced Filter Design for Signal Processing using MATLAB and Mathematica,"
http://iritel.iritel.bg.ac.yu/~lutovac/www/afdhome.htm
http://galeb.etf.bg.ac.yu/~tosic/afdhome.htm
http://www.ece.utexas.edu/~bevans/

A.6.2 Initialization

```
SetDirectory[HomeDirectory[]];
<<afd\math\m\clearall.m
<<afd\math\m\drawafil.m
<<afd\math\m\drawarc.m
```

A.6.3 Schematic

```
DrawGPHQ[0,0,1/2,1/0.8];
```

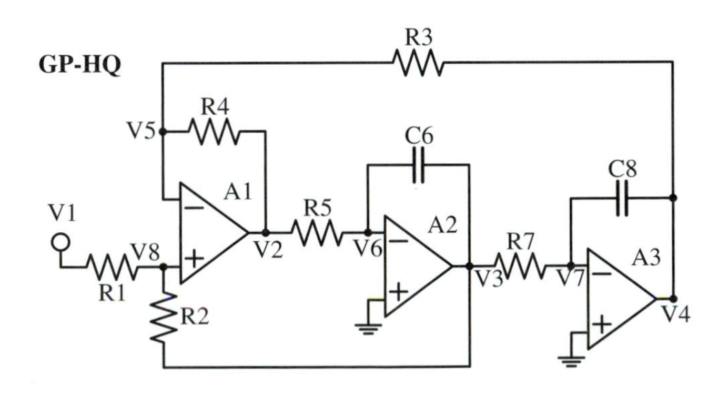

A.6.4 Circuit Analysis

Reduced Modified Nodal Analysis

```
CircuitEquations = {V1 == Vg
, V2 == A*(V8 - V5)
, V3 == -A*V6
, V4 == -A*V7
, (V5-V2)/R4 + (V5-V4)/R3 == 0
, (V6-V2)/R5 + (V6-V3)/(1/(s*C6)) ==0
, (V7-V3)/R7 + (V7-V4)/(1/(s*C8)) ==0
, (V8-V1)/R1 + (V8-V3)/R2 ==0};
NodeVoltages = {V1,V2,V3,V4,V5,V6,V7,V8};
CircuitResponse = Together[Flatten[
 Solve[CircuitEquations,NodeVoltages]
]];
```

A.6.5 Voltage Transfer Function

```
H = V4/V1 /. CircuitResponse //Together //Simplify;
Ha = Limit[H, A->Infinity];
Print["H(s) = ",
 Collect[Numerator[Ha],s]/Collect[Denominator[Ha],s]]
```

```
H(s) = (R2 (R3 + R4)) /
  (R1 R4 + R2 R4 + (C8 R1 R3 R7 + C8 R1 R4 R7) s +
                                                2
    (C6 C8 R1 R3 R5 R7 + C6 C8 R2 R3 R5 R7) s )
```

```
Klp = Factor[Ha /. s->0]
```

```
R2 (R3 + R4)
------------
(R1 + R2) R4
```

```
Hh = V2/V1 /. CircuitResponse //Together //Simplify;
Hha = Limit[Hh, A->Infinity];
        Print["Hh(s) = ",
 Collect[Numerator[Hha],s]/Collect[Denominator[Hha],s]]
```

```
                                      2
Hh(s) = (C6 C8 R2 (R3 + R4) R5 R7 s ) /
  (R1 R4 + R2 R4 + (C8 R1 R3 R7 + C8 R1 R4 R7) s +
                                                2
    (C6 C8 R1 R3 R5 R7 + C6 C8 R2 R3 R5 R7) s )
```

```
Khp = Factor[Limit[Hha , s->Infinity]]
```

```
R2 (R3 + R4)
------------
(R1 + R2) R3
```

```
Hb = V3/V1 /. CircuitResponse //Together //Simplify;
Hba = Limit[Hb, A->Infinity];
Print["Hb(s) = ",
 Collect[Numerator[Hba],s]/Collect[Denominator[Hba],s]]
```

```
Hb(s) = -((C8 R2 (R3 + R4) R7 s) /
    (R1 R4 + R2 R4 + (C8 R1 R3 R7 + C8 R1 R4 R7) s +
                                                2
      (C6 C8 R1 R3 R5 R7 + C6 C8 R2 R3 R5 R7) s ))
```

```
num = Numerator[Hba];
den = Denominator[Hba];
numlist = CoefficientList[num,s];
denlist = CoefficientList[den,s];
Hbp = Factor[numlist[[2]]/denlist[[2]]]
```

```
  R2
-(--)
  R1
```

A.6.6 Definitions and Procedures

```
PoleQpole[H_,s_] := Module[{den,fp,Qp},
  den = Denominator[H];
  fp = Sqrt[Coefficient[den,s,0]/Coefficient[den,s,2]]/(2*Pi);
  Qp = (Coefficient[den,s,2]/Coefficient[den,s,1])*(2*Pi*fp);
  Simplify[{fp, Qp}]];
ZeroQzero[H_,s_] := Module[{fz,num,Qz0},
  num = Numerator[H];
  Qz0 = (Coefficient[num,s,2]/Coefficient[num,s,1]);
  fz = Sqrt[Coefficient[num,s,0]/Coefficient[num,s,2]]/(2*Pi);
  Simplify[{fz, Qz0*fz}]];
Sensitivity[F_,x_] := (x/F)*D[F,x];
GSP[F_,A_] := Limit[A*Sensitivity[F,A],A->Infinity]//Simplify;
GSPepsA[F_,epsA_] := -(1/epsA)*Sensitivity[F,epsA]//Together;
PrintLabeledList[expressions_List,labels_List] := Map[
 Print[#[[1]]," = ",#[[2]]]&
,Transpose[{labels,expressions}]
];
```

A.6.7 Poles, Zeros, Q-Factors

```
{fp,Qp} = Simplify[PoleQpole[H,s]];
fp0 = Together[Limit[fp,A-> Infinity]];
Qp0 = Limit[Qp,A-> Infinity];
PrintLabeledList[{fp0,Qp0},{"fp","Qp"}];
```

```
                R4
     Sqrt[--------------]
          C6 C8 R3 R5 R7
fp = --------------------
            2 Pi
                                    R4
     C6 (R1 + R2) R3 R5 Sqrt[--------------]
                              C6 C8 R3 R5 R7
Qp = ---------------------------------------
                R1 (R3 + R4)
```

A.6.8 Gain-Sensitivity Product (GSP)

```
fpepsA = Together[fp /. A->1/e];
QpepsA = Together[Qp /. A->1/e];
GSPQp = Simplify[GSPepsA[QpepsA,e] /. e->0];
GSPfp = Simplify[GSPepsA[fpepsA,e] /. e->0];
PrintLabeledList[{GSPfp,GSPQp},{"GSPfp","GSPQp"}];
```

```
                  2                               2         2
        -(R1 R3 ) + 2 R1 R3 R4 + 3 R2 R3 R4 + R1 R4  + R2 R4
GSPfp = ------------------------------------------------------
                         2 R3 (R1 R4 + R2 R4)
                  2   2                     2                2   2               2   3
GSPQp = (2 C6 R1  R3  R4 R5 + 4 C6 R1 R2 R3  R4 R5 + 2 C6 R2  R3  R4 R5 - C8 R1  R3  R7 -
 
         2   2                         2                2   2               2         2
     C8 R1  R3  R4 R7 + 3 C8 R1 R2 R3  R4 R7 + 2 C8 R2  R3  R4 R7 - 3 C8 R1  R3 R4  R7 -
 
                      2            2   3                     3
     2 C8 R1 R2 R3 R4  R7 - C8 R1  R4  R7 - C8 R1 R2 R4  R7) /
 
   (2 R3 (R1 R4 + R2 R4) (C8 R1 R3 R7 + C8 R1 R4 R7))
```

A.6.9 Design

Find Element Values

```
DesignGPHQ[Qp_,wp_,Cx_,Rx_] :=
 Module[{C6,C8,R1,R2,R3,R4,R5,R7,Ro},
 C6 = Cx;
 C8 = Cx;
 Ro = 1/(wp*Cx);
 R1 = Rx;
 R3 = Rx;
 R5 = Rx;
 R7 = Rx;
 R4 = Rx*(Rx/Ro)^2;
 R2 = Rx*(Qp*(1+R4/Rx)/Sqrt[R4/Rx]-1);
 {R1,R2,R3,R4,R5,C6,R7,C8}];
 {R1,R2,R3,R4,R5,C6,R7,C8} = DesignGPHQ[Q,Wp,Co,Rd];
```

Design Examples

```
values = {Q -> 6., Wp -> 2*Pi*1500.
, Co -> 68.*10^(-9), Rd -> 1800.} //N;
gspQpfp = Together[Abs[GSPQp/2]+Abs[Q*GSPfp] /. values];
```

```
PrintLabeledList[N[{Q,Wp/(2*Pi),gspQpfp} /. values]
,{"Qp","fp (Hz)","GSP"} ];
Print["------------------"]
PrintLabeledList[Together[{R1,R2,R3,R4,R5,C6*10^9,R7,C8*10^9} /. values]
,{"R1 (ohm)","R2 (ohm)","R3 (ohm)","R4 (ohm)","R5 (ohm)","C6 (nF)"
,"R7 (ohm)","C8 (nF)"}];

Qp = 6.
fp (Hz) = 1500.
GSP = 17.1412
------------------
R1 (ohm) = 1800.
R2 (ohm) = 20020.9
R3 (ohm) = 1800.
R4 (ohm) = 2395.4
R5 (ohm) = 1800.
C6 (nF) = 68.
R7 (ohm) = 1800.
C8 (nF) = 68.

values = {Q -> 25., Wp -> 2*Pi*1500.
, Co -> 22.*10^(-9), Rd -> 4700.} //N;
gspQpfp = Together[Abs[GSPQp/2]+Abs[Q*GSPfp] /. values];
PrintLabeledList[N[{Q,Wp/(2*Pi),gspQpfp} /. values]
,{"Qp","fp (Hz)","GSP"} ];
Print["------------------"]
PrintLabeledList[Together[{R1,R2,R3,R4,R5,C6*10^9,R7,C8*10^9} /. values]
,{"R1 (ohm)","R2 (ohm)","R3 (ohm)","R4 (ohm)","R5 (ohm)","C6 (nF)"
,"R7 (ohm)","C8 (nF)"}];

Qp = 25.
fp (Hz) = 1500.
GSP = 74.014
------------------
R1 (ohm) = 4700.
R2 (ohm) = 230378.
R3 (ohm) = 4700.
R4 (ohm) = 4463.56
R5 (ohm) = 4700.
C6 (nF) = 22.
R7 (ohm) = 4700.
C8 (nF) = 22.
```

■ A.6.10 Optimization

■ Find Rd for Low Gain-Sensitivity Product

```
values = {Q -> 25., Wp -> 2*Pi*1500.
, Co -> 22.*10^(-9)} //N;
gspQpfp1 = Together[Abs[GSPQp/2.] + Abs[Q*GSPfp] /. values] //Simplify;
```

```
r1 = 2000;
r2 = 7000;
Plot[{gspQpfp1}
, {Rd, r1, r2}
, AxesLabel -> {"Rd (ohm)","GSP"}
];
```

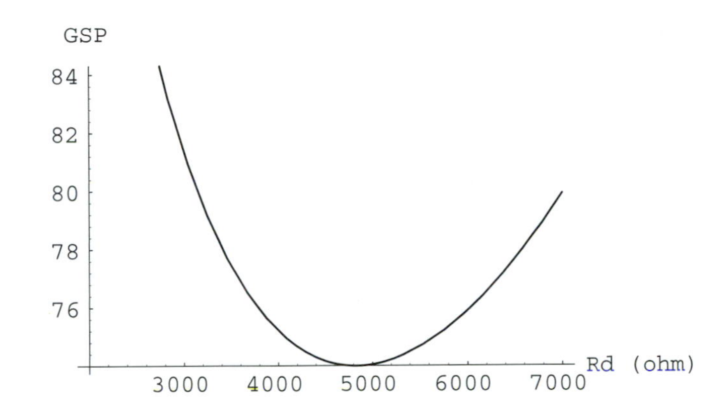


```
{Gspmin,Rset} = FindMinimum[gspQpfp1, {Rd, r1, r2}]

{73.99, {Rd -> 4822.22}}

values = {Q -> 25., Wp -> 2*Pi*1500.
, Co -> 22.*10^(-9)} //N;
gspQpfp = Together[Abs[GSPQp/2]+Abs[Q*GSPfp] /. values /. Rset];
PrintLabeledList[N[{Q,Wp/(2*Pi),gspQpfp} /. values/. Rset]
,{"Qp","fp (Hz)","GSP"} ];
Print["------------------"]
PrintLabeledList[Together[{R1,R2,R3,R4,R5,C6*10^9,R7,C8*10^9} /. values/.Rset]
,{"R1 (ohm)","R2 (ohm)","R3 (ohm)","R4 (ohm)","R5 (ohm)","C6 (nF)"
,"R7 (ohm)","C8 (nF)"}];

Qp = 25.
fp (Hz) = 1500.
GSP = 73.99
------------------
R1 (ohm) = 4822.22
R2 (ohm) = 236289.
R3 (ohm) = 4822.22
R4 (ohm) = 4820.9
R5 (ohm) = 4822.22
C6 (nF) = 22.
R7 (ohm) = 4822.22
C8 (nF) = 22.
```

```
Hlp = Ha /. values /. Rset//Together;
num = Numerator[Hlp];
den = Denominator[Hlp];
numlist = CoefficientList[num,s];
denlist = CoefficientList[den,s];
Hlp = 1/denlist[[3]] * num/(den/denlist[[3]])//Factor

                     8
          1.74124 10
-------------------------------
          7                   2
8.88264 10  + 376.991 s + 1. s

Hhp = Hha /. values /. Rset//Together;
num = Numerator[Hhp];
den = Denominator[Hhp];
numlist = CoefficientList[num,s];
denlist = CoefficientList[den,s];
Hhp = 1/denlist[[3]] * num/(den/denlist[[3]])//Factor

                    2
          1.95973 s
-------------------------------
          7                   2
8.88264 10  + 376.991 s + 1. s

Hbp = Hba /. values /. Rset//Together;
num = Numerator[Hbp];
den = Denominator[Hbp];
numlist = CoefficientList[num,s];
denlist = CoefficientList[den,s];
Hbp = 1/denlist[[3]] * num/(den/denlist[[3]])//Factor

          -18472.6 s
-------------------------------
          7                   2
8.88264 10  + 376.991 s + 1. s

AHlp = 20*Log[10,Abs[Hlp] /. s -> N[I*2*Pi*f]];
AHhp = 20*Log[10,Abs[Hhp] /. s -> N[I*2*Pi*f]];
AHbp = 20*Log[10,Abs[Hbp] /. s -> N[I*2*Pi*f]];
Plot[{AHlp, AHhp, AHbp}
, {f, 100, 4000}
,PlotRange -> {-30,40}
,PlotStyle -> {Dashing[{}],
               Dashing[{.02}],
               Dashing[{.04}]}
, AxesLabel -> {"f (Hz)","20*log M(f)"}
];
```

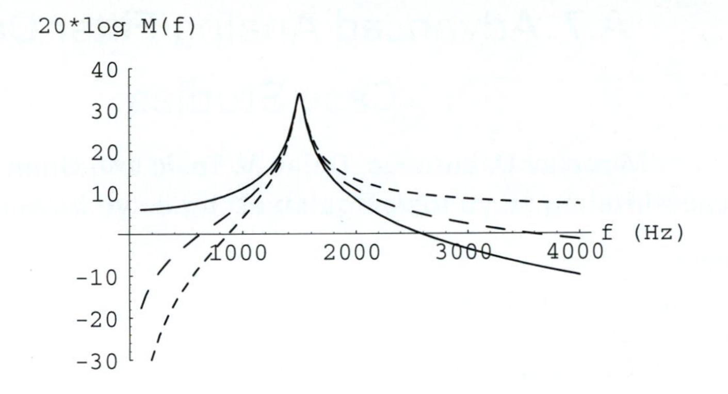
20*log M(f)
40
30
20
10
-10
-20
-30
1000
2000
3000
4000
f (Hz)

A.7 Advanced Analog Filter Design Case Studies

Miroslav D. Lutovac, Dejan V. Tosic and Brian L. Evans
lutovac@iritel.bg.ac.yu tosic@galeb.etf.bg.ac.yu bevans@ece.utexas.edu

■ A.7.1 References

1. D.V. Tosic, M.D.Lutovac, B.L.Evans,
"Advanced filter design,"
Proc. IEEE Asilomar Conf. Signal, Systems, Computer,
Nov. 1997, pp.710–715.

2. M. D. Lutovac, D. V. Tosic and B. L. Evans,
"Advanced Filter Design for Signal Processing using MATLAB and Mathematica,"
http://iritel.iritel.bg.ac.yu/~lutovac/www/afdhome.htm
http://galeb.etf.bg.ac.yu/~tosic/afdhome.htm
http://www.ece.utexas.edu/~bevans/

■ A.7.2 Initialization

```
SetDirectory[HomeDirectory[]];
<<afd\math\m\clearall.m
<<graphics`graphics`
```

■ A.7.3 Notation

a – selectivity factor
ap – maximum passband loss, dB, of realized filter
Ap – maximum passband loss, dB, in specification
as – minimum stopband loss, dB, of realized filter
As – minimum stopband loss, dB, in specification
A2a(n,Ap,As) – minimum selectivity factor from attenuation spec
A2K(A) – characteristic function in terms of attenuation in dB
e – ripple factor
fp – passband edge [Hz] of realized filter
Fp – passband edge [Hz] in specification
fs – stopband edge [Hz] of realized filter
Fs – stopband edge [Hz] in specification
Ke(n,a,e,x) – characteristic function
L(n,a) – discrimination factor
n – order
nbut(Fp,Fs,Ap,As) minimum Butterworth order from specification
ncheb(Fp,Fs,Ap,As) minimum Chebyshev order from specification
nellip(Fp,Fs,Ap,As) minimum order from specification
q(k) – modular constant
R(n,a,x) – elliptic rational function
S(n,a,e) – list of transfer function poles
S(n,a,e,i) – transfer function pole

SA – attenuation specification
SK – characterictic function specification
X(n,a) – list of zeros of elliptic rational function
X(n,a,i) – zero of elliptic rational function
Z(n,a,e) – zeta function

A.7.4 Definitions and Procedures

```
X[n_Integer,a_,i_Integer] :=  -JacobiCD[
 (2*i-1)*EllipticK[1/a^2]/n, 1/a^2
];
X[n_Integer,a_,i_Integer] := 0 /; And[i==(n+1)/2,OddQ[n]];
X[n_Integer, a_] :=  X[n,a,#]& /@ Range[n];
L[n_Integer,a_] := Block[{i,r},
 If[EvenQ[n],
  r = (1/a^n)*Product[(a^2 - X[n,a,i]^2)^2, {i,n/2}]/
            Product[(1   - X[n,a,i]^2)^2, {i,n/2}];,
  r = (1/a^(n-2))*Product[(a^2 - X[n,a,i]^2)^2, {i,(n-1)/2}]/
                Product[(1   - X[n,a,i]^2)^2, {i,(n-1)/2}];
 ];
 r
];
R[n_Integer,a_,x_] := Block[{i,r,r0},
 If[EvenQ[n],
  r = Product[x^2 -      X[n,a,i]^2, {i,n/2}]/
      Product[x^2 - a^2/X[n,a,i]^2, {i,n/2}];
  r0 = Product[1 -      X[n,a,i]^2, {i,n/2}]/
       Product[1 - a^2/X[n,a,i]^2, {i,n/2}];,
  r = x*Product[x^2 -      X[n,a,i]^2, {i,(n-1)/2}]/
        Product[x^2 - a^2/X[n,a,i]^2, {i,(n-1)/2}];
  r0 = Product[1 -      X[n,a,i]^2, {i,(n-1)/2}]/
       Product[1 - a^2/X[n,a,i]^2, {i,(n-1)/2}];
 ];
 r/r0
];
Z[n_Integer,a_,e_] :=  JacobiSN[
 InverseJacobiSN[1/Sqrt[1+e^2],1-1/(L[n,a])^2]*
 EllipticK[1-1/a^2]/EllipticK[1-1/(L[n,a])^2],1-1/a^2
];
S[n_Integer, a_, e_, i_Integer] := Block[
{den,num,numim,numre,x,z},
 x = X[n,a,i];
 z = Z[n,a,e];
 numre = -z*Sqrt[1 - z^2]*Sqrt[1 - x^2]*Sqrt[1 - x^2/a^2];
 numim = x*Sqrt[1 - (1-1/a^2)*z^2];
 num = numre + I*numim;
```

```
  den = 1 - (1 - x^2/a^2)*z^2;
  num/den
 ];
 S[n_Integer, a_, e_] :=  S[n,a,e,#]& /@ Range[n];
 A2K[A_] := Sqrt[1 - 10^(-A/10)]/10^(-A/20);
 Ke[n_Integer, a_, e_, x_] := e*Abs[R[n,a,x]];
```

A.7.5 Specification

```
SA = {3000., 3225., 0.2, 40.};
{Fp, Fs, Ap, As} = SA;
Kp = A2K[Ap];
Ks = A2K[As];
SK = {Fp, Fs, Kp, Ks}

{3000., 3225., 0.217091, 99.995}
```

A.7.6 Minimum Order

```
nellip[Fp_,Fs_,Ap_,As_] := Block[
 {num, den,
  k = Fp/Fs,
  L = Sqrt[(-1 + 10^(As/10))/(-1 + 10^(Ap/10))]},
 num = EllipticK[1-1/L^2]/EllipticK[1/L^2];
 den = EllipticK[1-k^2]/EllipticK[k^2];
 Ceiling[num/den//N]
];
ncheb[Fp_,Fs_,Ap_,As_] := Block[
 {L = Sqrt[(-1 + 10^(As/10))/(-1 + 10^(Ap/10))],
  aspec = Fs/Fp},
 Ceiling[ArcCosh[L]/ArcCosh[aspec]//N]
];
nbut[Fp_,Fs_,Ap_,As_] := Block[
 {L = Sqrt[(-1 + 10^(As/10))/(-1 + 10^(Ap/10))],
  aspec = Fs/Fp},
 Ceiling[Log[10,L]/Log[10,aspec]//N]
];
{nellip[Fp,Fs,Ap,As], ncheb[Fp,Fs,Ap,As], nbut[Fp,Fs,Ap,As]}
nmin = nellip[Fp,Fs,Ap,As];
nmax = 2*nmin;
nlist = Range[nmin,nmax]

{8, 18, 85}
{8, 9, 10, 11, 12, 13, 14, 15, 16}
```

■ A.7.7 Range of Selectivity Factor and Ripple Factor

```
q[k_] := Block[{c,e,r,s,t},
 If[k<=1/Sqrt[2.0],
   t = (1/2)*(1 - (1-k^2)^(1/4))/(1 + (1-k^2)^(1/4));,
   t = (1/2)*(1 - Sqrt[k])/(1 + Sqrt[k]);
 ];
 e = {1,5, 9, 13,  17,   21,    25,     29,      33,       37};
 c = {1,2,15,150,1707,20910,268616,3567400,48555069,673458874};
 s = Sum[c[[i]]*(t^e[[i]]),{i,Length[e]}];
 If[k<=1/Sqrt[2.0],
   r = s;,
   r = Exp[Pi^2/Log[s]];
 ];
 N[r]
];
A2a[n_,Ap_,As_] := Block[
 {m, num, den, terms=9, L, qL},
 L = Sqrt[(-1 + 10^(As/10))/(-1 + 10^(Ap/10))];
 qL = q[1/L]^(1/n);
 num = 1 + 2*Sum[(-1)^m*(qL)^(m^2), {m,1,terms}];
 den = 1 + 2*Sum[(qL)^(m^2), {m,1,terms}];
 1/Sqrt[1 - (num/den)^4]
];
amin8  = A2a[nmin,Ap,As];
amax8  = a/. FindRoot[R[nmin,a,Fs/Fp]==Ks/Kp, {a,Fs/Fp,1.1}];
amin9  = A2a[9,Ap,As];
amax9  = a/. FindRoot[R[9,a,Fs/Fp]==Ks/Kp, {a,amax8,1.1}];
amin10 = A2a[10,Ap,As];
amax10 = a/. FindRoot[R[10,a,Fs/Fp]==Ks/Kp, {a,amax9,1.2}];
amin11 = A2a[11,Ap,As];
amax11 = a/. FindRoot[R[11,a,Fs/Fp]==Ks/Kp, {a,amax10,1.2}];
amin12 = A2a[12,Ap,As];
amax12 = a/. FindRoot[R[12,a,Fs/Fp]==Ks/Kp, {a,amax11,1.2}];
amin13 = A2a[13,Ap,As];
amax13 = a/. FindRoot[R[13,a,Fs/Fp]==Ks/Kp, {a,amax12,1.3}];
amin14 = A2a[14,Ap,As];
amax14 = a/. FindRoot[R[14,a,Fs/Fp]==Ks/Kp, {a,amax13,1.4}];
amin15 = A2a[15,Ap,As];
amax15 = a/. FindRoot[R[15,a,Fs/Fp]==Ks/Kp, {a,amax14,1.6}];
amin16 = A2a[16,Ap,As];
amax16 = a/. FindRoot[R[16,a,Fs/Fp]==Ks/Kp, {a,amax15,1.9}];
aminlist = {amin8,amin9,amin10,amin11,amin12,amin13,amin14,amin15,amin16};
amaxlist = {amax8,amax9,amax10,amax11,amax12,amax13,amax14,amax15,amax16};
eminlist = Table[Ks/L[n,amaxlist[[n-8+1]]], {n,nmin,nmax}];
{emin8,emin9,emin10,emin11,emin12,emin13,emin14,emin15,emin16} = eminlist;
emax = Kp;
emaxlist = Table[Kp,{nmax-nmin+1}];
```

```
TableForm[Transpose[{nlist,aminlist,amaxlist,eminlist,emaxlist}]
, TableHeadings->{{},{"n","amin","amax","emin","emax"}}]
```

n	amin	amax	emin	emax
8	1.04285	1.08323	0.0757872	0.217091
9	1.022	1.09807	0.0184689	0.217091
10	1.01135	1.12013	0.00368669	0.217091
11	1.00587	1.15172	0.000578651	0.217091
12	1.00304	1.19663	0.0000676012	0.217091
13	1.00158	1.26158	5.39161 10^{-6}	0.217091
14	1.00082	1.35951	2.53319 10^{-7}	0.217091
15	1.00042	1.51914	5.28479 10^{-9}	0.217091
16	1.00022	1.82219	2.51893 10^{-11}	0.217091

■ A.7.8 Range of Edge Frequencies

```
fpminlist = Fs / amaxlist;
{fpmin8,fpmin9,fpmin10,fpmin11,fpmin12,fpmin13,
 fpmin14,fpmin15,fpmin16} = fpminlist;
fpmaxlist = Fs / aminlist;
{fpmax8,fpmax9,fpmax10,fpmax11,fpmax12,fpmax13,
 fpmax14,fpmax15,fpmax16} = fpmaxlist;
fsminlist = Fp * aminlist;
{fsmin8,fsmin9,fsmin10,fsmin11,fsmin12,fsmin13,
 fsmin14,fsmin15,fsmin16} = fsminlist;
fsmaxlist = Fp * amaxlist;
{fsmax8,fsmax9,fsmax10,fsmax11,fsmax12,fsmax13,
 fsmax14,fsmax15,fsmax16} = fsmaxlist;
TableForm[Transpose[{nlist,fpminlist,fpmaxlist,fsminlist,fsmaxlist}]
, TableHeadings->{{},{"n","fpmin","fpmax","fsmin","fsmax (Hz)"}}]
```

n	fpmin	fpmax	fsmin	fsmax (Hz)
8	2977.2	3092.49	3128.55	3249.7
9	2936.97	3155.57	3066.01	3294.21
10	2879.14	3188.79	3034.06	3360.38
11	2800.16	3206.17	3017.62	3455.16
12	2695.06	3215.22	3009.13	3589.89
13	2556.32	3219.92	3004.73	3784.74
14	2372.18	3222.36	3002.45	4078.53
15	2122.91	3223.63	3001.27	4557.42
16	1769.85	3224.29	3000.66	5466.56

■ A.7.9 Design D1

```
Plot[Evaluate[{Kp
              , Ke[ 8, Fs/Fp, emax, f/Fp]
              , Ke[ 9, Fs/Fp, emax, f/Fp]
              , Ke[13, Fs/Fp, emax, f/Fp]
             }]
, {f,0,Fp}
, AxesLabel -> {"f (Hz)", "K"}
, PlotStyle -> {Dashing[{0.04}],
               Dashing[{}],
               Dashing[{0.02}],
               Dashing[{0.01}]
              }
, Ticks -> {{1000, 2000, 3000}, {0, 0.1, 0.2}}
, PlotRange -> All];
```

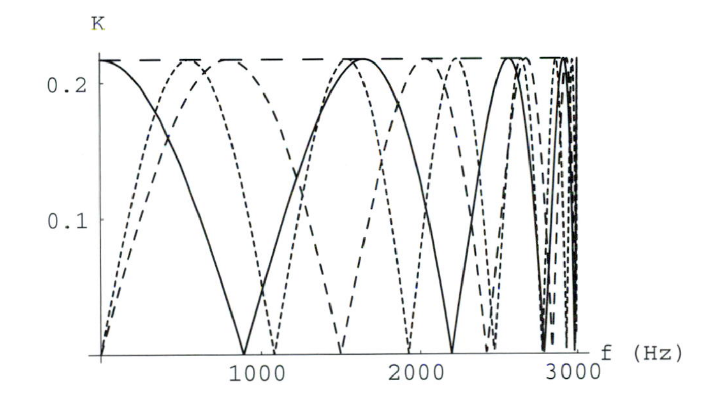


```
LogPlot[Evaluate[{Ks
              , Ke[ 8, Fs/Fp, emax, f/Fp]
              , Ke[ 9, Fs/Fp, emax, f/Fp]
              , Ke[13, Fs/Fp, emax, f/Fp]
             }]
, {f,Fp,2*Fs}
, AxesLabel -> {"f (Hz)", "K"}
, PlotStyle -> {Dashing[{0.04}],
               Dashing[{}],
               Dashing[{0.02}],
               Dashing[{0.01}]
              }
, Ticks -> {{3000, 4000, 5000, 6000}, Automatic}
, PlotRange -> {1,10^6}];
```

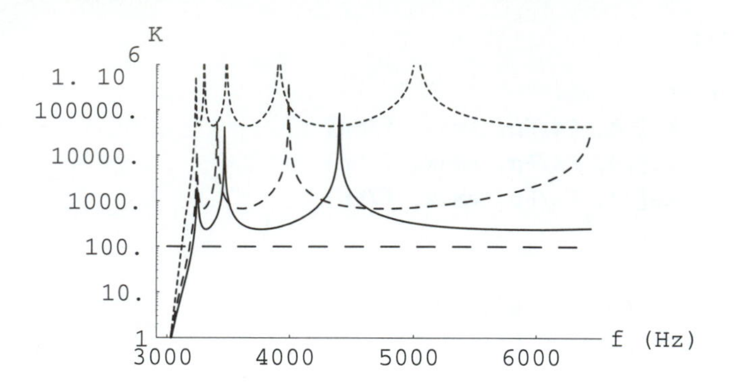

■ A.7.10 Design D2

```
Plot[Evaluate[{Kp
              , Ke[ 8, Fs/Fp, Ks/L[ 8,Fs/Fp], f/Fp]
              , Ke[ 9, Fs/Fp, Ks/L[ 9,Fs/Fp], f/Fp]
              , Ke[13, Fs/Fp, Ks/L[13,Fs/Fp], f/Fp]
             }]
, {f,0,Fp}
, AxesLabel -> {"f (Hz)", "K"}
, PlotStyle -> {Dashing[{0.04}],
                Dashing[{}],
                Dashing[{0.02}],
                Dashing[{0.01}]
               }
, Ticks -> {{1000, 2000, 3000}, {0, 0.1, 0.2}}
, PlotRange -> All];
```

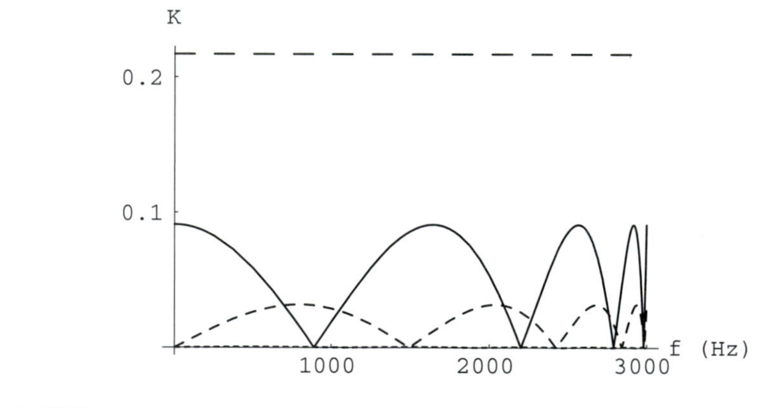


```
LogPlot[Evaluate[{Ks
              , Ke[ 8, Fs/Fp, Ks/L[ 8,Fs/Fp], f/Fp]
              , Ke[ 9, Fs/Fp, Ks/L[ 9,Fs/Fp], f/Fp]
              , Ke[13, Fs/Fp, Ks/L[13,Fs/Fp], f/Fp]
             }]
, {f,Fp,2*Fs}
```

```
, AxesLabel -> {"f (Hz)", "K"}
, PlotStyle -> {Dashing[{0.04}],
                Dashing[{}],
                Dashing[{0.02}],
                Dashing[{0.01}]
               }
, Ticks -> {{3000, 4000, 5000, 6000}, Automatic}
, PlotRange -> {1,10^4}];
```

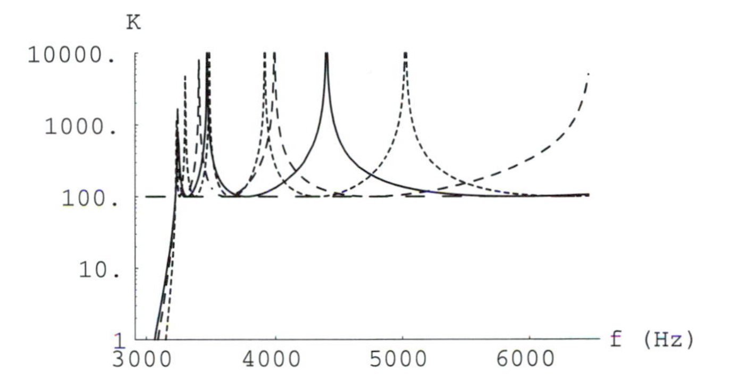

A.7.11 Design D3a

```
Plot[Evaluate[{Kp
             , Ke[ 8, amin8,  emax, f/Fp]
             , Ke[ 9, amin9,  emax, f/Fp]
             , Ke[13, amin13, emax, f/Fp]
            }]
, {f,0,Fp}
, AxesLabel -> {"f (Hz)", "K"}
, PlotStyle -> {Dashing[{0.04}],
                Dashing[{}],
                Dashing[{0.02}],
                Dashing[{0.01}]
               }
, Ticks -> {{1000, 2000, 3000}, {0, 0.1, 0.2}}
, PlotRange -> All];
```

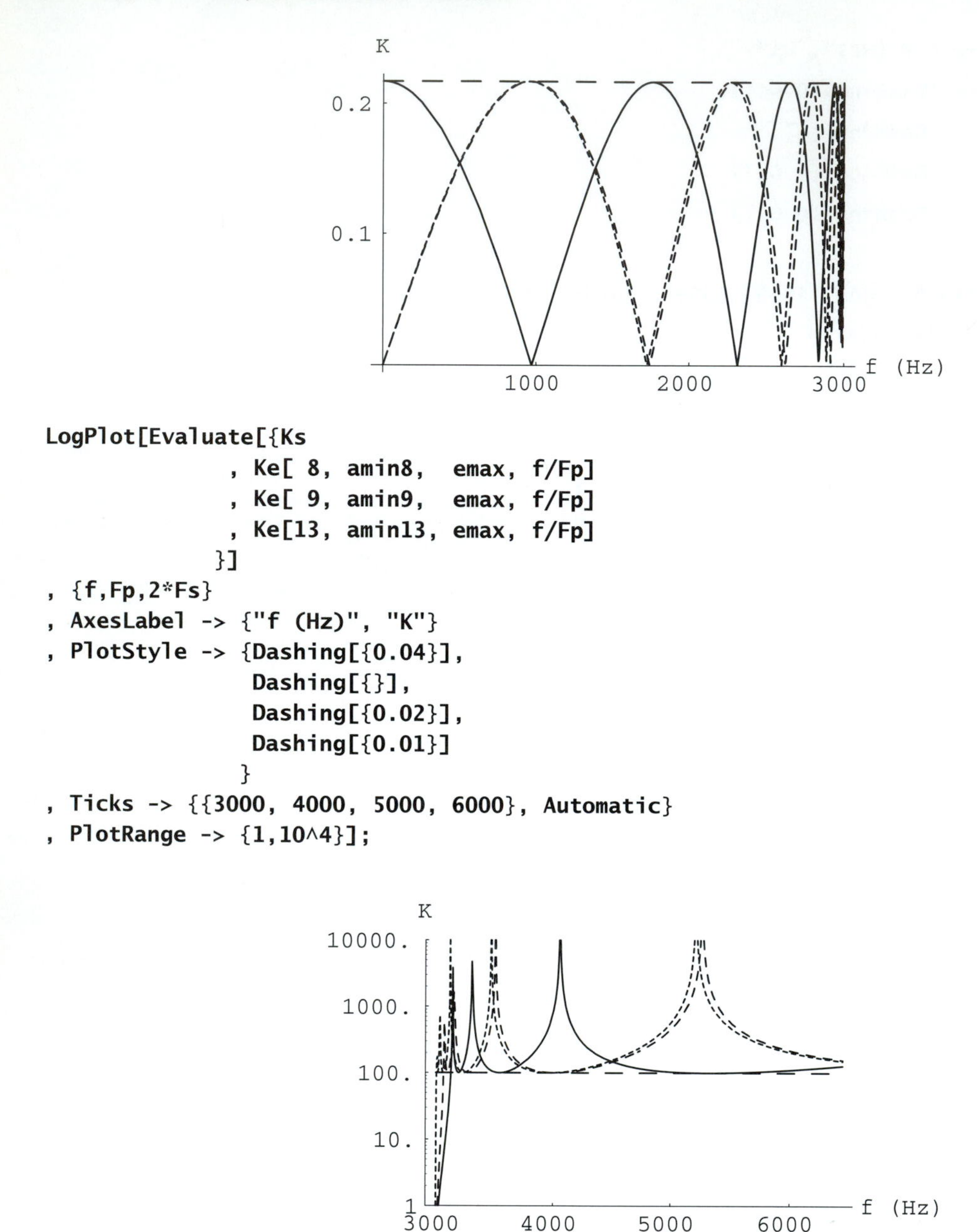


```
LogPlot[Evaluate[{Ks
              , Ke[ 8, amin8,  emax, f/Fp]
              , Ke[ 9, amin9,  emax, f/Fp]
              , Ke[13, amin13, emax, f/Fp]
             }]
, {f,Fp,2*Fs}
, AxesLabel -> {"f (Hz)", "K"}
, PlotStyle -> {Dashing[{0.04}],
               Dashing[{}],
               Dashing[{0.02}],
               Dashing[{0.01}]
              }
, Ticks -> {{3000, 4000, 5000, 6000}, Automatic}
, PlotRange -> {1,10^4}];
```



```
LogPlot[Evaluate[{Kp,Ks
              , Ke[ 8, amin8,  emax, f/Fp]
              , Ke[ 9, amin9,  emax, f/Fp]
              , Ke[13, amin13, emax, f/Fp]
             }]
, {f,Fp,Fs}
, AxesLabel -> {"f (Hz)", "K"}
```

```
, PlotStyle -> {Dashing[{0.04}],
               Dashing[{0.04}],
               Dashing[{}],
               Dashing[{0.02}],
               Dashing[{0.01}]
              }
, Ticks -> {{3000, 3100, 3200}, {0.1, 1, 10, 100}}
, PlotRange -> {0.1,200}];
```

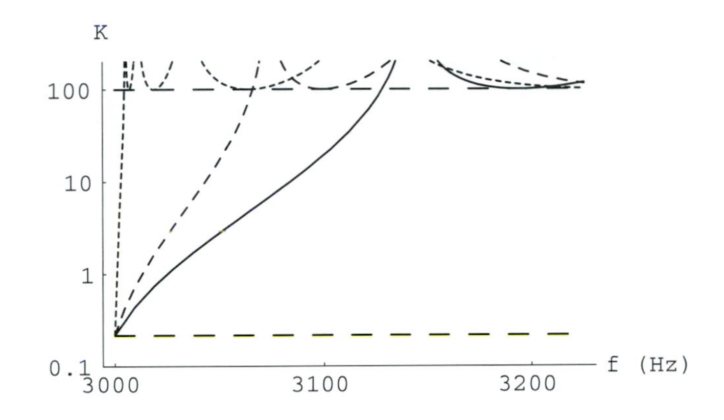

■ A.7.12 Design D3b

```
Plot[Evaluate[{Kp
             , Ke[ 8, amin8,  emax, f/fpmax8]
             , Ke[ 9, amin9,  emax, f/fpmax9]
             , Ke[13, amin13, emax, f/fpmax13]
            }]
, {f,0,Fp}
, AxesLabel -> {"f (Hz)", "K"}
, PlotStyle -> {Dashing[{0.04}],
               Dashing[{}],
               Dashing[{0.02}],
               Dashing[{0.01}]
              }
, Ticks -> {{1000, 2000, 3000}, {0, 0.1, 0.2}}
, PlotRange -> All];
```

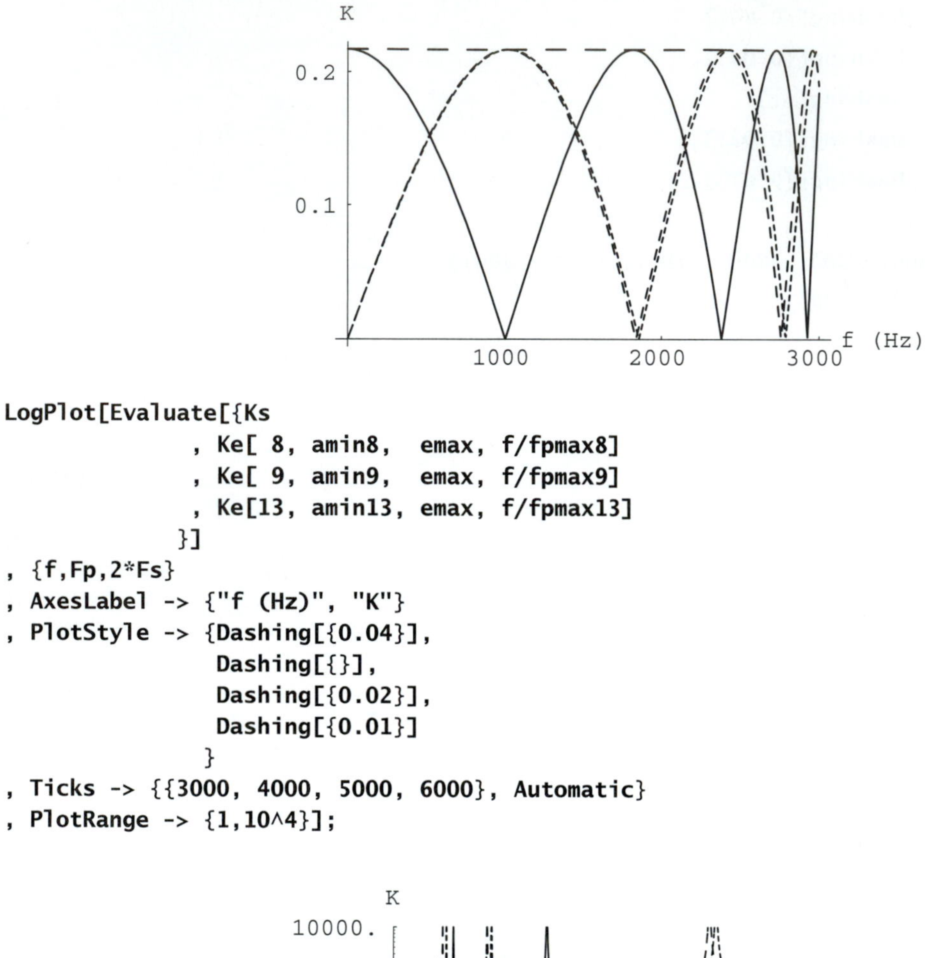


```
LogPlot[Evaluate[{Ks
              , Ke[ 8, amin8,  emax, f/fpmax8]
              , Ke[ 9, amin9,  emax, f/fpmax9]
              , Ke[13, amin13, emax, f/fpmax13]
             }]
, {f,Fp,2*Fs}
, AxesLabel -> {"f (Hz)", "K"}
, PlotStyle -> {Dashing[{0.04}],
               Dashing[{}],
               Dashing[{0.02}],
               Dashing[{0.01}]
              }
, Ticks -> {{3000, 4000, 5000, 6000}, Automatic}
, PlotRange -> {1,10^4}];
```

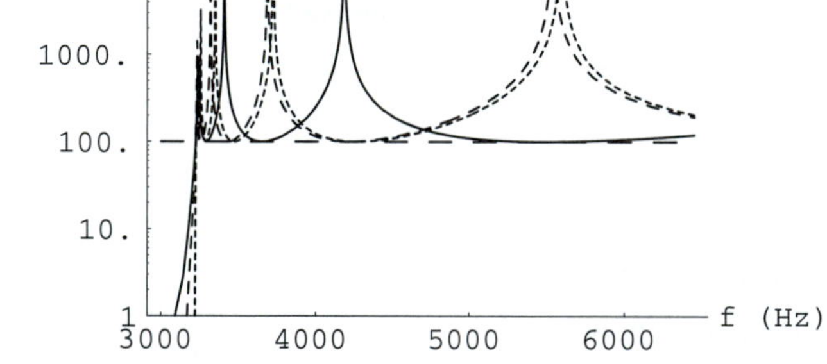


```
LogPlot[Evaluate[{Kp,Ks
              , Ke[ 8, amin8,  emax, f/fpmax8]
              , Ke[ 9, amin9,  emax, f/fpmax9]
              , Ke[13, amin13, emax, f/fpmax13]
             }]
, {f,Fp,Fs}
, AxesLabel -> {"f (Hz)", "K"}
```

```
, PlotStyle -> {Dashing[{0.04}],
                Dashing[{0.04}],
                Dashing[{}],
                Dashing[{0.02}],
                Dashing[{0.01}]
               }
, Ticks -> {{3000, 3100, 3200}, {0.1, 1, 10, 100}}
, PlotRange -> {0.1,200}];
```

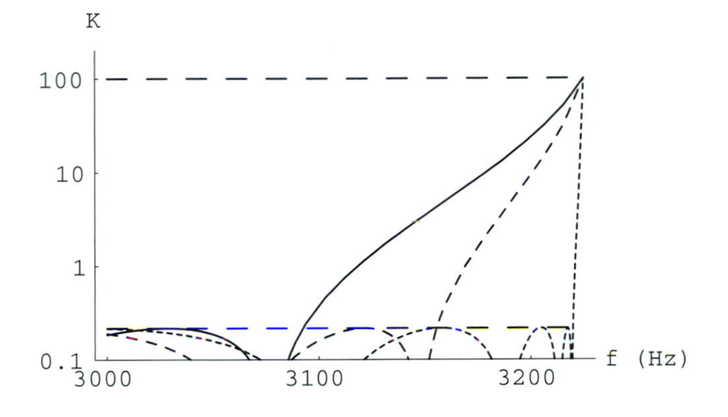

A.7.13 Design D4a

```
Plot[Evaluate[{Kp
              , Ke[ 8, amax8,  emax, f/Fp]
              , Ke[ 9, amax9,  emax, f/Fp]
              , Ke[13, amax13, emax, f/Fp]
             }]
, {f,0,Fp}
, AxesLabel -> {"f (Hz)", "K"}
, PlotStyle -> {Dashing[{0.04}],
                Dashing[{}],
                Dashing[{0.02}],
                Dashing[{0.01}]
               }
, Ticks -> {{1000, 2000, 3000}, {0, 0.1, 0.2}}
, PlotRange -> All];
```

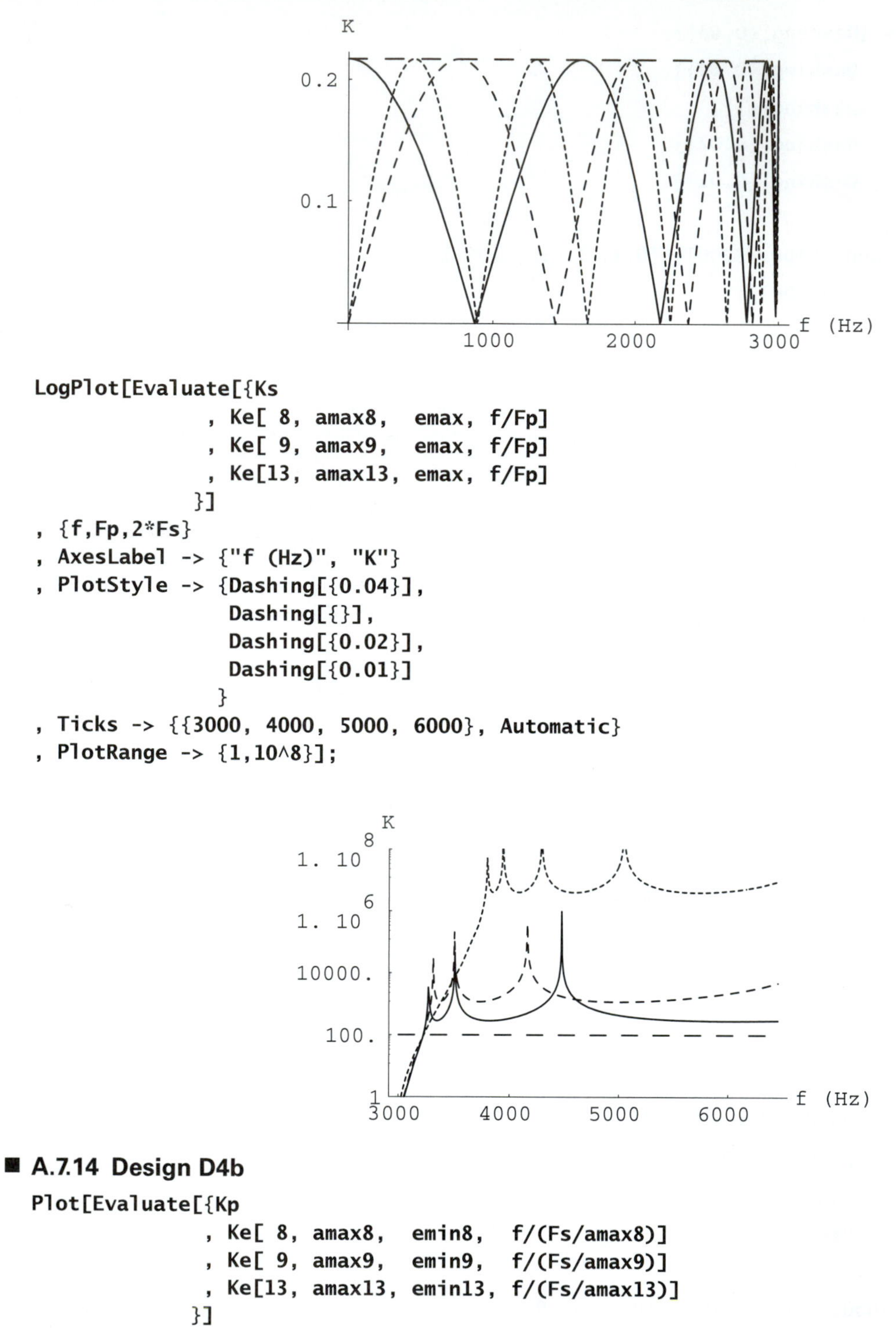


```
LogPlot[Evaluate[{Ks
              , Ke[ 8, amax8,  emax, f/Fp]
              , Ke[ 9, amax9,  emax, f/Fp]
              , Ke[13, amax13, emax, f/Fp]
             }]
, {f,Fp,2*Fs}
, AxesLabel -> {"f (Hz)", "K"}
, PlotStyle -> {Dashing[{0.04}],
                Dashing[{}],
                Dashing[{0.02}],
                Dashing[{0.01}]
               }
, Ticks -> {{3000, 4000, 5000, 6000}, Automatic}
, PlotRange -> {1,10^8}];
```


■ A.7.14 Design D4b

```
Plot[Evaluate[{Kp
              , Ke[ 8, amax8,  emin8,  f/(Fs/amax8)]
              , Ke[ 9, amax9,  emin9,  f/(Fs/amax9)]
              , Ke[13, amax13, emin13, f/(Fs/amax13)]
             }]
, {f,0,Fp}
```

```
, AxesLabel -> {"f (Hz)", "K"}
, PlotStyle -> {Dashing[{0.04}],
               Dashing[{}],
               Dashing[{0.02}],
               Dashing[{0.01}]
              }
, Ticks -> {{1000, 2000, 3000}, {0, 0.1, 0.2}}
, PlotRange -> All];
```

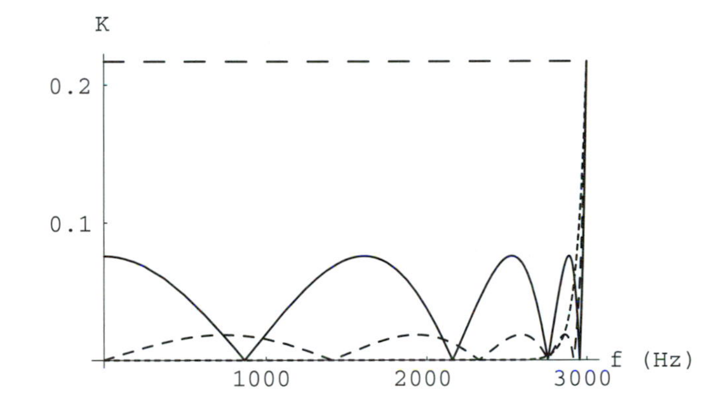


```
LogPlot[Evaluate[{Ks
             , Ke[ 8, amax8,  emin8,  f/(Fs/amax8)]
             , Ke[ 9, amax9,  emin9,  f/(Fs/amax9)]
             , Ke[13, amax13, emin13, f/(Fs/amax13)]
            }]
, {f,Fp,2*Fs}
, AxesLabel -> {"f (Hz)", "K"}
, PlotStyle -> {Dashing[{0.04}],
               Dashing[{}],
               Dashing[{0.02}],
               Dashing[{0.01}]
              }
, Ticks -> {{3000, 4000, 5000, 6000}, {1,10,100}}
, PlotRange -> {1,10^3}];
```

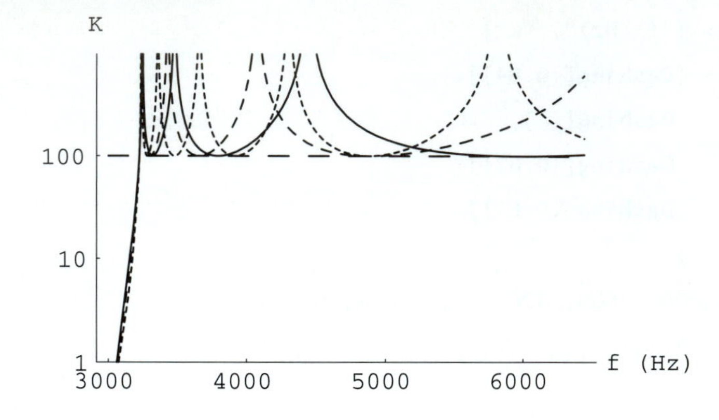
K
100
10
1
3000
4000
5000
6000
f (Hz)

A.8 Classical Digital Filters

Transpose Direct Form II

IIR Second-Order Realization

Miroslav D. Lutovac, Dejan V. Tosic and Brian L. Evans

lutovac@iritel.bg.ac.yu tosic@galeb.etf.bg.ac.yu bevans@ece.utexas.edu

A.8.1 References

1. Alan Oppenheim, Ronald Schafer, "Digital Signal Processing,"

Prentice-Hall, Englewood Cliffs, New Jersey, 1975.

2. Sanjit Mitra, James Kaiser, "Handbook for Digital Signal Processing,"

John Wiley, New York, 1993, pp. 127–128.

3. M. D. Lutovac, D. V. Tosic and B. L. Evans,

"Advanced Filter Design for Signal Processing using MATLAB and Mathematica,"

http://iritel.iritel.bg.ac.yu/~lutovac/www/afdhome.htm

http://galeb.etf.bg.ac.yu/~tosic/afdhome.htm

http://www.ece.utexas.edu/~bevans/

A.8.2 Initialization

```
SetDirectory[HomeDirectory[]];
<<afd\math\m\clearall.m
<<afd\math\m\drawdfil.m
<<afd\math\m\drawiirf.m
```

■ A.8.3 Block Diagram

```
DrawTDF2[0,0,1,1/0.8,10];
```

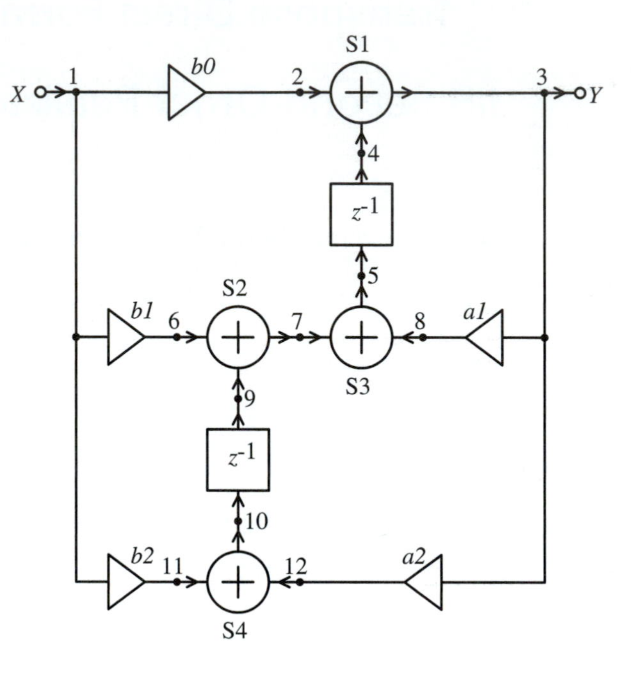

■ A.8.4 Analysis (v=1/z)

```
ElementEquations = {
  Y1 == X
, Y2 == b0*Y1 + Xb0
, Y3 == Y2 + Y4
, Y4 == v*Y5
, Y5 == Y7 + Y8
, Y6 == b1*Y1 + Xb1
, Y7 == Y6 + Y9
, Y8 == a1*Y3 + Xa1
, Y9 ==  v*Y10
, Y10 == Y11 + Y12
, Y11 == b2*Y1 + Xb2
, Y12 == a2*Y3 + Xa2
};
NodeSignals = {Y1,Y2,Y3,Y4,Y5,Y6,Y7,Y8,Y9,Y10,Y11,Y12};
Response = Flatten[Solve[ElementEquations,NodeSignals]];
Y = Together[Y3/.Response];
Hv = Y /. {X -> 1, Xa1 -> 0, Xa2 -> 0, Xb0 -> 0, Xb1 -> 0, Xb2 -> 0};
Hz = Hv /. v -> 1/z //Together;
Print["H(v) = ", Hv]
Print["H(z) = ", Hz]
```

```
                        2
        -b0 - b1 v - b2 v
H(v) = ------------------
                       2
        -1 + a1 v + a2 v
                        2
        -b2 - b1 z - b0 z
H(z) = ------------------
                      2
          a2 + a1 z - z
```

A.8.5 Transfer Function and Noise Transfer Functions

```
Ha1v = Y /. {X -> 0, Xa1 -> 1, Xa2 -> 0, Xb0 -> 0, Xb1 -> 0, Xb2 -> 0};
Ha2v = Y /. {X -> 0, Xa1 -> 0, Xa2 -> 1, Xb0 -> 0, Xb1 -> 0, Xb2 -> 0};
Hb0v = Y /. {X -> 0, Xa1 -> 0, Xa2 -> 0, Xb0 -> 1, Xb1 -> 0, Xb2 -> 0};
Hb1v = Y /. {X -> 0, Xa1 -> 0, Xa2 -> 0, Xb0 -> 0, Xb1 -> 1, Xb2 -> 0};
Hb2v = Y /. {X -> 0, Xa1 -> 0, Xa2 -> 0, Xb0 -> 0, Xb1 -> 0, Xb2 -> 1};
H = Collect[Numerator[Hv],v]/Collect[Denominator[Hv],v];
Ha1 = Collect[Numerator[Ha1v],v]/Collect[Denominator[Ha1v],v];
Ha2 = Collect[Numerator[Ha2v],v]/Collect[Denominator[Ha2v],v];
Hb0 = Collect[Numerator[Hb0v],v]/Collect[Denominator[Hb0v],v];
Hb1 = Collect[Numerator[Hb1v],v]/Collect[Denominator[Hb1v],v];
Hb2 = Collect[Numerator[Hb2v],v]/Collect[Denominator[Hb2v],v];
v2invz = {v->z^"-1", v^2->z^("-2"), v^3->z^("-3"), v^4->z^("-4")};
Print["H(z) = ", H /. v2invz ]
Print["Ha1(z) = ", Ha1 /. v2invz ]
Print["Ha2(z) = ", Ha2 /. v2invz ]
Print["Hb0(z) = ", Hb0 /. v2invz ]
Print["Hb1(z) = ", Hb1 /. v2invz ]
Print["Hb2(z) = ", Hb2 /. v2invz ]
```

```
                   -1        -2
        -b0 - b1 z    - b2 z
H(z) = ---------------------
                 -1        -2
        -1 + a1 z    + a2 z
                       -1
                      z
Ha1(z) = -(--------------------)
                       -1        -2
            -1 + a1 z    + a2 z
                       -2
                      z
Ha2(z) = -(--------------------)
                       -1        -2
            -1 + a1 z    + a2 z
```

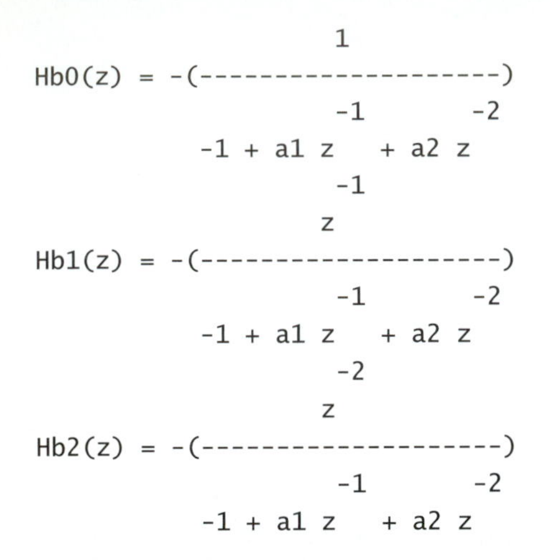

```
                    1
Hb0(z) = -(--------------------)
                      -1        -2
          -1 + a1 z    + a2 z
                      -1
                    z
Hb1(z) = -(--------------------)
                      -1        -2
          -1 + a1 z    + a2 z
                      -2
                    z
Hb2(z) = -(--------------------)
                      -1        -2
          -1 + a1 z    + a2 z
```

■ A.8.6 VQNR
Variance of Quantization Noise Due to Rounding the Output of the Multiplier

```
VQNR[H_,z_Symbol,a_Symbol,b_Symbol
    ,r_Symbol,theta_Symbol] := Module[
  {ax, bx, d0, d1, d2, denH2, H0, H0inv, H2, numH2
  ,res0, res1, res2, rtheta2ab, sumres, var=Infinity
  ,z1, z2},
  H0 = Together[H] /. {a -> ax, b -> bx};
  H0inv = Together[H0 /. z->1/z];
  H2 = Together[Cancel[z^2*H0]];
  numH2 = Collect[Numerator[H2],z];
  denH2 = Collect[Denominator[H2],z];
  If[Exponent[denH2,z] == 2
  ,{d0,d1,d2} = CoefficientList[denH2,z];
   {z1,z2} = Flatten[Solve[denH2==0,z]];
   res0 = D[H2*H0inv,{z,2}]/2 /. z->0;
   res1 = numH2*H0inv/(d2*(z-(z/.z2))*z^3) /. z1;
   res2 = numH2*H0inv/(d2*(z-(z/.z1))*z^3) /. z2;
   sumres = Simplify[res0 + res1 + res2];
   rtheta2ab = Solve[{d0 == d2*(r^2),
                      d1 == d2*(-2*r*Cos[theta])}
               ,{ax,bx}] //Flatten;
   var = Together[sumres /. rtheta2ab];
  ,Print["Error in denominator!   ", denH2];
  ];
 var]
Haz = Together[Ha2 /. v -> 1/z]
vara = Simplify[VQNR[Haz,z,a1,a2,r,theta], Trig -> True]
```

```
        1
-(--------------)
               2
  a2 + a1 z - z

                    2
               -1 - r
--------------------------------------
       2         4      2
(-1 + r ) (1 + r  - 2 r  Cos[2 theta])
```

■ A.8.7 Frequency Response (z=Exp[I*2*Pi*f])

```
values = {a1 -> 0.8, a2 -> -.64, b0 -> 1/10, b1 -> 2/10, b2 -> 1/10};
Hf = Hz /. z->Exp[I*2*Pi*f] /. values //N;
Plot[Evaluate[Abs[Hf]]
, {f, 0, 1/2}
, AxesLabel -> {"f", "M(f)"}
];
```

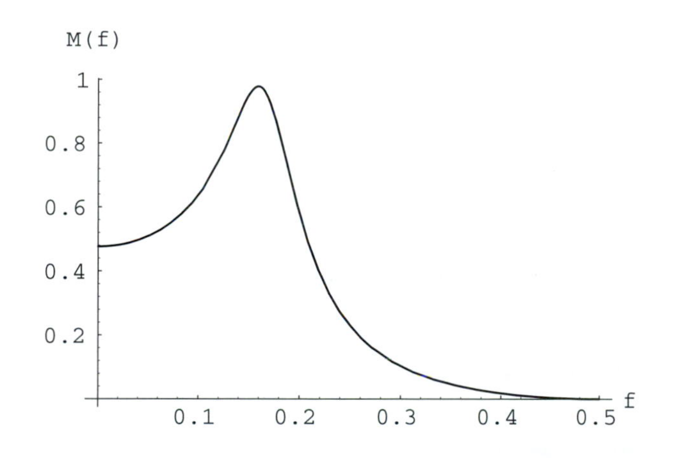


```
Plot[(Evaluate[Arg[Hf]])/Degree
, {f, 0, 1/2}
, AxesLabel -> {"f", "phase (deg)"}
, AxesOrigin -> {0, -180}
, PlotRange -> All
];
```

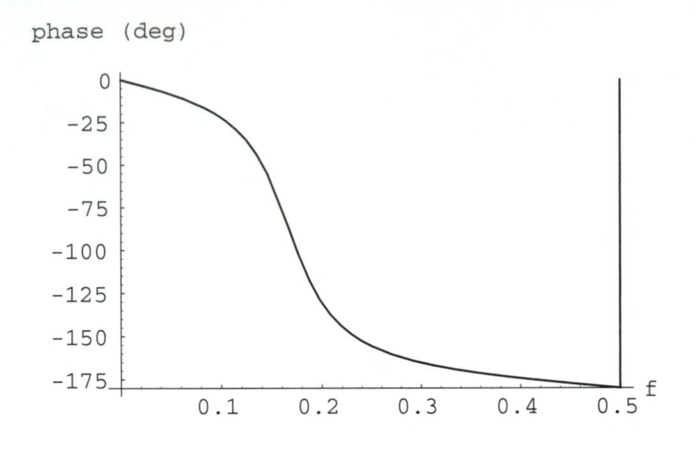

A.8.8 Dynamics Magnitude Response of Partial Transfer Functions

```
excitation = {X -> 1, Xa1 -> 0, Xa2 -> 0, Xb0 -> 0, Xb1 -> 0, Xb2 -> 0};
values = {a1 -> 0.8, a2 -> -.64, b0 -> 1/10, b1 -> 2/10, b2 -> 1/10};
H7 = Together[Y7/.Response] /. excitation;
H7z = H7 /. v -> 1/z //Together;
H7f = H7z /. z->Exp[I*2*Pi*f] /. values //N;
H5 = Together[Y5/.Response] /. excitation;
H5z = H5 /. v -> 1/z //Together;
H5f = H5z /. z->Exp[I*2*Pi*f] /. values //N;
H10 = Together[Y10/.Response] /. excitation;
H10z = H10 /. v -> 1/z //Together;
H10f = H10z /. z->Exp[I*2*Pi*f] /. values //N;
Plot[Evaluate[
  {Abs[H5f], Abs[H7f], Abs[H10f]}]
, {f, 0, 1/2}
, AxesLabel -> {"f", "M5, M7, M10"}
, PlotRange -> All
, PlotStyle -> {Dashing[{}]
             , Dashing[{.02}]
             , Dashing[{.04}]
}
];
```

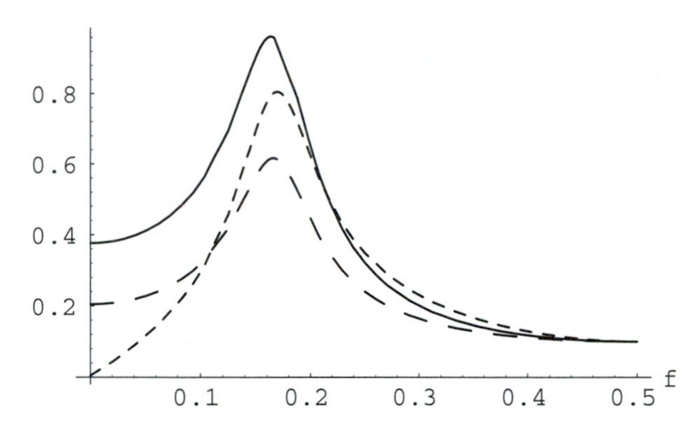

A.8.9 Relative Output Noise as a Function of Pole Position Due to All 5 Multipliers

```
valuesb = {b0 -> 1/10, b1 -> 2/10, b2 -> 1/10};
Plot3D[5*(vara /. valuesb)
, {theta, 0, Pi}
, {r, 0, 0.5}
, AxesLabel -> {"thetha", "r", "noise"}
, PlotRange -> All];
```

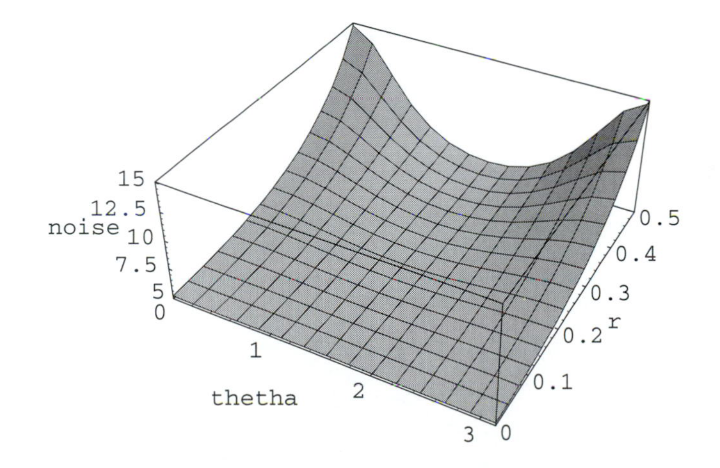

Remark

In this example all multipliers have equal noise variance, i.e. sum of variances is equal to 6*vara, where vara represents the noise variance due to the multiplier a2.

```
valuesb = {b0 -> 1/10, b1 -> 2/10, b2 -> 1/10};
Plot3D[5*(vara /. valuesb)
, {theta, Pi/2, Pi}
, {r, 0, .8}
```

```
, AxesLabel -> {"thetha", "r", "noise"}
, PlotRange -> All
(*, Ticks -> {{Pi/2,Pi},{0,1},{1,2,3,4}}*)
];
```

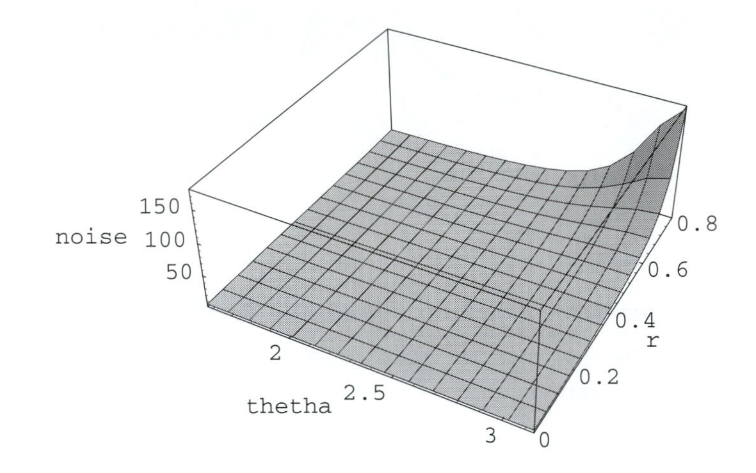

A.8.10 Filtering and Quantization

Filtering without Quantization

```
IIRTDF2[a1_, a2_, b0_, b1_, b2_, X_, d1_:0, d2_:0] := Module[
{vY5, vY10, Y, Y1, Y2, Y3, Y4, Y5, Y6, Y7, Y8, Y9, Y10, Y11,Y12},
 vY5 = d1;
 vY10 = d2;
 Y1  = X;
 Y4  = vY5;
 Y9  = vY10;
 Y2  = b0*Y1;
 Y6  = b1*Y1;
 Y11 = b2*Y1;
 Y3  = Y2 + Y4;
 Y8  = a1*Y3;
 Y12 = a2*Y3;
 Y7 = Y6 + Y9;
 Y5 = Y7 + Y8;
 Y10 = Y11 + Y12;
 Y = Y3;
{Y,Y5,Y10}];
```

Filtering with Quantization

```
Quantize[x_,b_:8,q_:0] := Module[{},
 If[q == 0
 , y = Round[x*2^b]/2^b;
 , y = Floor[2^nm*x]/2^nm ];
y];
```

```
QIIRTDF2[a1_, a2_, b0_, b1_, b2_, X_, d1_:0, d2_:0, bits_:8, mode_:0] := Module[
{vY5, vY10, Y, Y1, Y2, Y3, Y4, Y5, Y6, Y7, Y8, Y9, Y10, Y11,Y12},
 vY5 = d1;
 vY10 = d2;
 Y1  = X;
 Y4  = vY5;
 Y9  = vY10;
 Y2  = Quantize[b0*Y1, bits, mode];
 Y6  = Quantize[b1*Y1, bits, mode];
 Y11 = Quantize[b2*Y1, bits, mode];
 Y3  = Y2 + Y4;
 Y8  = Quantize[a1*Y3, bits, mode];
 Y12 = Quantize[a2*Y3, bits, mode];
 Y7 = Y6 + Y9;
 Y5 = Y7 + Y8;
 Y10 = Y11 + Y12;
 Y = Y3;
{Y,Y5,Y10}];
```

- **Filtering**

```
filterQIIRTDF2[a1_, a2_, b0_, b1_, b2_, Xdata_List, d1_:0, d2_:0
               , bits_Integer:8, mode_Integer:0] := Module[
{n, Qd1, Qd2, QYdata, QYn, QY3, Xn},
Qd1 = d1;
Qd2 = d2;
QYdata = {};
Do[Xn = Xdata[[n]];
   QY3 = QIIRTDF2[a1, a2, b0, b1, b2, Xn, Qd1, Qd2, bits, mode];
   {QYn, Qd1, Qd2} = QY3;
   AppendTo[QYdata,QYn];
,{n,1,Length[Xdata]}];
{QYdata,Qd1,Qd2}];
```

- **Example of Filtering**

```
values = {a1 -> 0.8, a2 -> -.64, b0 -> 0.1, b1 -> 0.2, b2 -> 0.1};
Hv /. values
{aD1,aD2,bD0,bD1,bD2} = {a1,a2,b0,b1,b2} /. values
Ndata = 100;                  (* number of samples *)
Xdata = Table[0,{Ndata}]; Xdata[[1]] = 1;
{QYdata,Qd1,Qd2} = filterQIIRTDF2[aD1,aD2,bD0,bD1,bD2, Xdata, 0, 0, 8, 0]//N;
{Ydata,Qd1,Qd2} = filterQIIRTDF2[aD1,aD2,bD0,bD1,bD2, Xdata, 0, 0, 12, 0]//N;
```

```
                      2
-0.1 - 0.2 v - 0.1 v
---------------------
                     2
-1 + 0.8 v - 0.64 v
{0.8, -0.64, 0.1, 0.2, 0.1}
```

- **Impulse Response**

```
ListPlot[QYdata
 , PlotJoined->True
 , PlotRange -> All
 , AxesOrigin -> {.1,-.1}
 , AxesLabel -> {"samples", "impulse response"}
];
```

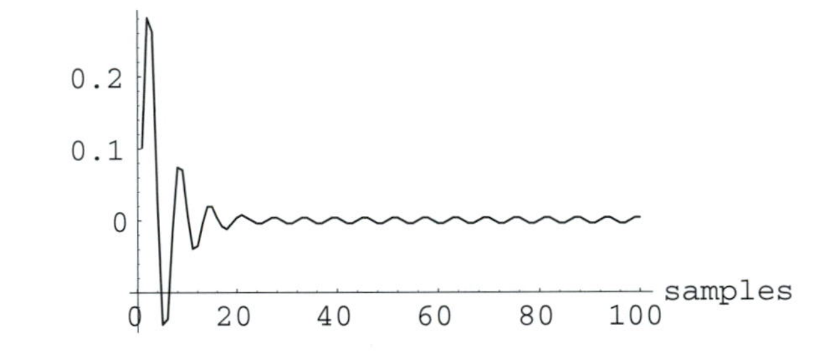

- **Spectrum of Impulse Response**
  Define Electrical Engineering FFT

```
EEfft[data_List] := InverseFourier[data]*Sqrt[Length[data]];
EEifft[data_List] := Fourier[data]/Sqrt[Length[data]];
spectrum  = EEfft[Ydata] //Chop;
qspectrum = EEfft[N[QYdata]] //Chop;
```

A.8.11 Verify Frequency Response by Spectrum of Impulse Response

- **12 Bits Filtering Quantization**

```
pfs = Table[{(k-1)/Length[spectrum], Abs[spectrum[[k]]]}
            , {k,1,Length[spectrum]/2+1}];
ListPlot[pfs
, PlotJoined -> True
, PlotRange -> All
, AxesLabel -> {"f", "M(f) 12 bits"}
];
```

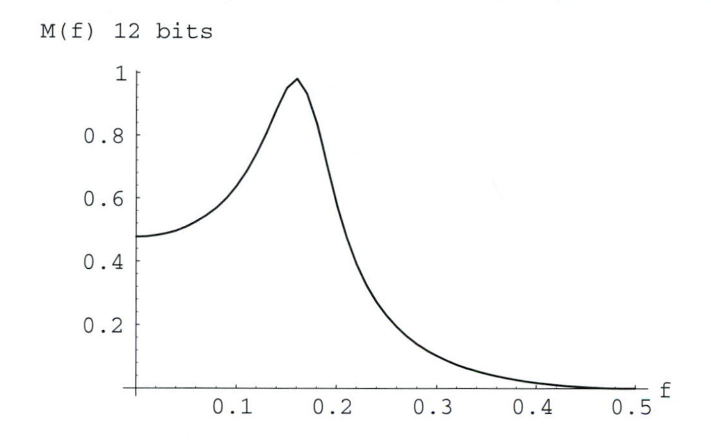


```
pfsdB = Table[{(ind-1)/Length[spectrum], 20*Log[10,Abs[spectrum[[ind]]]]}
              , {ind,1,Length[spectrum]/2+1}];
ListPlot[pfsdB
, PlotJoined -> True
, PlotRange -> All
, AxesLabel -> {"f", "M(f) (dB) 12 bits"}
];
```

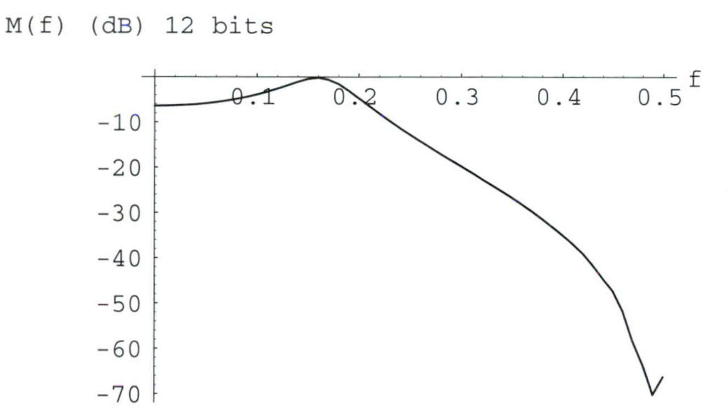

8 Bits Filtering Quantization

```
pfqsdB = Table[{(ind-1)/Length[qspectrum], 20*Log[10,Abs[qspectrum[[ind]]]]}
              , {ind,1,Length[qspectrum]/2+1}];
ListPlot[pfqsdB
, PlotJoined -> True
, PlotRange -> All
, AxesLabel -> {"f", "M(f) (dB) 8 bits"}
];
```

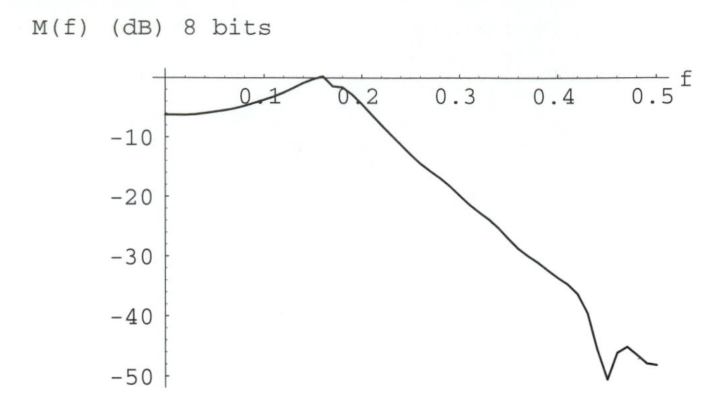


```
pfp = Table[{(ind-1)/Length[spectrum], Arg[spectrum[[ind]]]/Degree}
            , {ind,1,Length[qspectrum]/2+1}];
ListPlot[pfp
, PlotJoined -> True
, PlotRange -> {-180,0}
, AxesOrigin -> {0,-180}
, AxesLabel -> {"f", "phase (deg) 8 bits"}
];
```

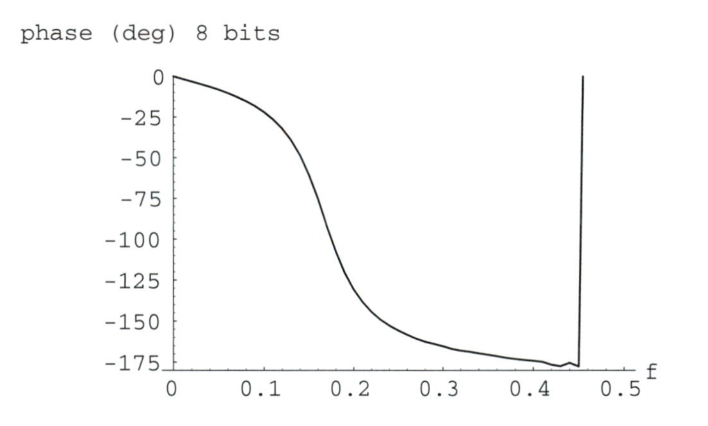

■ Error in Magnitude Response Due to Quantization

```
deltapfs = Table[{(k-1)/Length[spectrum]
        , 20*Log[10,Abs[qspectrum[[k]]]] - 20*Log[10,Abs[spectrum[[k]]]]}
          , {k,1,Length[spectrum]/2+1}];
ListPlot[deltapfs
, PlotJoined -> True
, PlotRange -> {-1,1}
, AxesLabel -> {"f", "error in M(f) (dB)"}
];
```

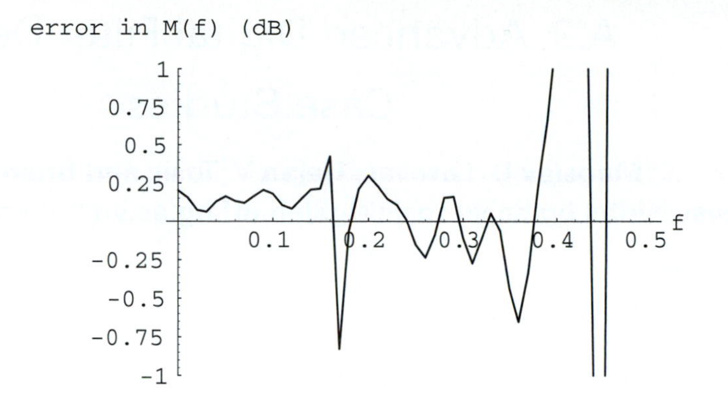
error in M(f) (dB)
1
0.75
0.5
0.25
-0.25
-0.5
-0.75
-1
0.1
0.2
0.3
0.4
0.5
f

A.9 Advanced Digital Filter Design Case Studies

Miroslav D. Lutovac, Dejan V. Tosic and Brian L. Evans
lutovac@iritel.bg.ac.yu tosic@galeb.etf.bg.ac.yu bevans@ece.utexas.edu

■ A.9.1 References

1. D.V. Tosic, M.D.Lutovac, B.L.Evans,
"Advanced filter design,"
Proc. IEEE Asilomar Conf. Signal, Systems, Computer,
Nov. 1997, pp.710–715.

2. D.V.Tosic, M.D.Lutovac, B.L.Evans,
"Advanced digital filter design,"
Proc. European Conf. Circuit Theory Design ECCTD'99, Stresa, Italy,
Sep.1999,vol.2, pp.1323–1326.

3. M. D. Lutovac, D. V. Tosic and B. L. Evans,
"Advanced Filter Design for Signal Processing using MATLAB and Mathematica,"
http://iritel.iritel.bg.ac.yu/~lutovac/www/afdhome.htm
http://galeb.etf.bg.ac.yu/~tosic/afdhome.htm
http://www.ece.utexas.edu/~bevans/

■ A.9.2 Initialization

```
SetDirectory[HomeDirectory[]];
<<afd\math\m\clearall.m
<<afd\math\m\dfdellip.m
<<graphics'graphics'
```

■ A.9.3 Notation

```
!!afd\math\m\dfdnotat.m
```

a – selectivity factor
AdB – attenuation (dB) in terms of digital frequency f
ap – maximum passband attenuation, (dB), of realized filter
Ap – maximum passband attenuation, (dB), in specification
as – minimum stopband attenuation, (dB), of realized filter
As – minimum stopband attenuation, (dB), in specification
A2a(n,Ap,As) – minimum selectivity factor from attenuation spec
A2K(A) – characteristic function in terms of attenuation in dB
ba – coefficients of the second-order section
e – ripple factor
f – digital frequency (0<f<0.5)
fp – passband edge (0<fp<0.5) of realized filter
Fp – passband edge (0<Fp<0.5) in specification
fs – stopband edge (0<fs<0.5) of realized filter
Fs – stopband edge (0<Fs<0.5) in specification

hz – transfer function in z
Hz(n,a,e,Fp,z) – transfer function in z
Ke(n,a,e,x) – characteristic function
L(n,a) – discrimination factor
n – order
nbut(Fp,Fs,Ap,As) – minimum Butterworth order from specification
ncheb(Fp,Fs,Ap,As) – minimum Chebyshev order from specification
nellip(Fp,Fs,Ap,As) – minimum elliptic order from specification
nminQ – minimum order of elliptic minimal Q-factor design
q(k) – modular constant
Qfactor – Qfactor of the second-order section
rtan(f1,f2) – tan(pi*f1)/tan(pi*f2)
R(n,a,x) – elliptic rational function
S(n,a,e) – list of transfer function poles
S(n,a,e,i) – transfer function pole
SA – attenuation-limit specification
SK – characterictic-function-limit specification
xtan – tan(pi*Fs)/tan(pi*Fp)
X(n,a) – list of zeros of elliptic rational function
X(n,a,i) – zero of elliptic rational function
z – complex variable
Z(n,a,e) – zeta function
Zbl(Spole,Fp) – bilinear transformation

A.9.4 Definitions and Procedures

```
!!afd\math\m\dfdellip.m
(* DFDELLIP.M
   7:06PM  9/11/98
*)

A2a[n_,Ap_,As_] := Module[
 {m, num, den, terms=9, L, qL},
 L = Sqrt[(-1 + 10^(As/10))/(-1 + 10^(Ap/10))];
 qL = q[1/L]^(1/n);
 num = 1 + 2*Sum[(-1)^m*(qL)^(m^2), {m,1,terms}];
 den = 1 + 2*Sum[(qL)^(m^2), {m,1,terms}];
 1/Sqrt[1 - (num/den)^4]
];

A2K[A_] := Sqrt[1 - 10^(-A/10)]/10^(-A/20);

AdB[hz_,z_,f_] :=  -20*Log[10,Abs[hz /. z-> N[E^(I*2*Pi*f)] ]];

ba[n_,a_,e_,Fp_,z_] := Module[{i,g,t},
   {-2*Re[Zbl[S[n,a,e,1],Fp]], Abs[Zbl[S[n,a,e,1],Fp]]^2}
];

DGDelay[A_,B_,z_,f_] := Module[
```

```
{Ai,Bi,gA,gAd,gAn,nA,tAd,tAn,gB,gBd,gBn,nB,tBd,tBn},
   Ai = CoefficientList[A,z];
   Bi = CoefficientList[B,z];
   nA = Length[Ai];
   tAn = Join[{Sum[(i-1)*Ai[[i]]^2,{i,1,nA}]},
      Table[Cos[(k-1) w]*Sum[(2*i-k-1)*Ai[[i]]*Ai[[i-k+1]],{i,k,nA}],{k,2,nA}]];
   tAd = Join[{Sum[Ai[[i]]^2,{i,1,nA}]},
      Table[2*Cos[(k-1) w]*Sum[Ai[[i]]*Ai[[i-k+1]],{i,k,nA}],{k,2,nA}]];
  gAn = Sum[tAn[[i]],{i,1,nA}];
  gAd = Sum[tAd[[i]],{i,1,nA}];
  gA = Simplify[gAn/gAd];
  nB = Length[Bi];
  tBn = Join[{Sum[(i-1)*Bi[[i]]^2,{i,1,nB}]},
      Table[Cos[(k-1) w]*Sum[(2*i-k-1)*Bi[[i]]*Bi[[i-k+1]],{i,k,nB}],{k,2,nB}]];
  tBd = Join[{Sum[Bi[[i]]^2,{i,1,nB}]},
      Table[2*Cos[(k-1) w]*Sum[Bi[[i]]*Bi[[i-k+1]],{i,k,nB}],{k,2,nB}]];
  gBn = Sum[tBn[[i]],{i,1,nB}];
  gBd = Sum[tBd[[i]],{i,1,nB}];
  gB = Simplify[gBn/gBd];
 gB - gA +nA -nB /. z-> E^(I*2*N[Pi]*f) /. w-> 2*N[Pi]*f
];

Hz[n_,a_,e_,Fp_,z_] := Module[{i,g,t},
 If[EvenQ[n],
   g = Product[(2-2*Re[Zbl[I*a/X[n,a,i],Fp]])/
               (1-2*Re[Zbl[S[n,a,e,i],Fp]]
                +Abs[Zbl[S[n,a,e,i],Fp]]^2)
         ,{i,1,n/2}]*Sqrt[1+e^2];
   t = (1/g)*Product[(z^2-2*Re[Zbl[I*a/X[n,a,i],Fp]]*z+1)/
       (z^2-2*Re[Zbl[S[n,a,e,i],Fp]]*z
        +Abs[Zbl[S[n,a,e,i],Fp]]^2)
         ,{i,1,n/2}];,
   g = 2*Product[(1-2*Re[Zbl[I*a/X[n,a,i],Fp]]+1)/
       (1-2*Re[Zbl[S[n,a,e,i],Fp]]
         +Abs[Zbl[S[n,a,e,i],Fp]]^2)
         ,{i,1,(n-1)/2}]/(1-Zbl[S[n,a,e,(n+1)/2],Fp]);
   t = (1/g)*(z+1)*Product[(z^2-2*Re[Zbl[I*a/X[n,a,i],Fp]]*z+1)/
       (z^2-2*Re[Zbl[S[n,a,e,i],Fp]]*z
        +Abs[Zbl[S[n,a,e,i],Fp]]^2)
         ,{i,1,(n-1)/2}]/(z-Zbl[S[n,a,e,(n+1)/2],Fp]);
 ];
 t
];

Ke[n_Integer, a_, e_, x_] := e*Abs[R[n,a,x]];
```

```
L[n_Integer,a_] := Module[{i,r},
 If[EvenQ[n],
  r = (1/a^n)*Product[(a^2 - X[n,a,i]^2)^2, {i,n/2}]/
              Product[(1   - X[n,a,i]^2)^2, {i,n/2}];,
  r = (1/a^(n-2))*Product[(a^2 - X[n,a,i]^2)^2, {i,(n-1)/2}]/
                  Product[(1   - X[n,a,i]^2)^2, {i,(n-1)/2}];
 ];
 r
];

nbut[Fp_,Fs_,Ap_,As_] := Module[
 {L = Sqrt[(-1 + 10^(As/10))/(-1 + 10^(Ap/10))],
  aspec = Tan[Pi*Fs]/Tan[Pi*Fp]},
 Ceiling[Log[10,L]/Log[10,aspec]//N]
];

ncheb[Fp_,Fs_,Ap_,As_] := Module[
 {L = Sqrt[(-1 + 10^(As/10))/(-1 + 10^(Ap/10))],
  aspec = Tan[Pi*Fs]/Tan[Pi*Fp]},
 Ceiling[ArcCosh[L]/ArcCosh[aspec]//N]
];

nellip[Fp_,Fs_,Ap_,As_] := Module[
 {num, den,
  k = Tan[Pi*Fp]/Tan[Pi*Fs],
  L = Sqrt[(-1 + 10^(As/10))/(-1 + 10^(Ap/10))]},
 num = EllipticK[1-1/L^2]/EllipticK[1/L^2];
 den = EllipticK[1-k^2]/EllipticK[k^2];
 Ceiling[num/den//N]
];

nminQ[Fp_, Fs_, Ap_, As_] := Block[{ai, i, Kp, Ks, x, x1, x2},
 i = nellip[Fp,Fs,Ap,As];
 Kp = A2K[Ap];
 Ks = A2K[As];
 x1 = Tan[Pi*Fs]/Tan[Pi*Fp];
 x2 = x1^2;
 ai = x /. FindRoot[Sqrt[L[i, x]] == Ks, {x,x1,x2}];
 While[Ke[i, ai, 1/Ks, Fp/(Fs/ai)] > Kp,
  i += 1;
  x2 = ai;
  ai = x /. FindRoot[Sqrt[L[i, x]] == Ks, {x,x1,x2}];
 ];
 i
];
```

```
plotstyle012345 :=  PlotStyle -> {Dashing[{}],
                                  Dashing[{0.01}],
                                  Dashing[{0.02}],
                                  Dashing[{0.03}],
                                  Dashing[{0.04}],
                                  Dashing[{0.05}]}

plotPTS[MdB_,f_,spec_List] := Block[
{Fpass=spec[[1]], Fstop=spec[[2]],
 Apass=spec[[3]], Astop=spec[[4]], c1, c2, g1 ,g2, g3, M},
  M = MdB;
  c1 = PlotStyle -> {RGBColor[1,0,0],RGBColor[1,0,0],RGBColor[0,0,1]};
  c2 = PlotStyle -> {RGBColor[0,1,0],RGBColor[0,1,0],RGBColor[0,0,1]};
 g1 = Plot[{Apass,0,M}, {f,0,Fpass}, Evaluate[c1]
      , PlotRange -> All
      , Ticks -> {{0, Fp},{0,Ap/2,Ap}}
      , AxesOrigin->{-0.001,-0.01}];
 g2 = Plot[{Apass,Astop,M}, {f,Fpass,Fstop}, Evaluate[c2]
      , PlotRange -> All
      , Ticks -> {{Fp,Fs},{Ap, As/2, As}}
      , AxesOrigin->{Fpass-0.001,0}];
 g3 = Plot[{Astop,Astop,M}, {f,Fstop,0.5}, Evaluate[c1]
      , PlotRange -> {Astop,Astop+41}
      , Ticks -> {{Fs,0.5},{As, As+20, As+40}}
      , AxesOrigin->{Fstop-0.001,Astop-2}];
 Show[GraphicsArray[{g1,g2,g3}
]];
];

q[k_] := Module[{c,e,r,s,t},
 If[k<=1/Sqrt[2.0],
   t = (1/2)*(1 - (1-k^2)^(1/4))/(1 + (1-k^2)^(1/4));,
   t = (1/2)*(1 - Sqrt[k])/(1 + Sqrt[k]);
 ];
 e = {1,5, 9, 13,  17,   21,    25,     29,      33,       37};
 c = {1,2,15,150,1707,20910,268616,3567400,48555069,673458874};
 s = Sum[c[[i]]*(t^e[[i]]),{i,Length[e]}];
 If[k<=1/Sqrt[2.0],
   r = s;,
   r = Exp[Pi^2/Log[s]];
 ];
 N[r]
];

Qfactor[a_,b_] := Sqrt[(1+a+b)*(1+a-b)]/(2*(1-a));

rtan[fnum_,fden_,prec_:16] := N[Tan[Pi*fnum]/Tan[Pi*fden],prec];
```

```
R[n_Integer,a_,x_] := Module[{i,r,r0},
 If[EvenQ[n],
  r = Product[x^2 -      X[n,a,i]^2, {i,n/2}]/
      Product[x^2 - a^2/X[n,a,i]^2, {i,n/2}];
  r0 = Product[1 -      X[n,a,i]^2, {i,n/2}]/
       Product[1 - a^2/X[n,a,i]^2, {i,n/2}];,
  r = x*Product[x^2 -      X[n,a,i]^2, {i,(n-1)/2}]/
        Product[x^2 - a^2/X[n,a,i]^2, {i,(n-1)/2}];
  r0 = Product[1 -      X[n,a,i]^2, {i,(n-1)/2}]/
       Product[1 - a^2/X[n,a,i]^2, {i,(n-1)/2}];
 ];
 r/r0
];
```

```
S[n_Integer, a_, e_] :=  S[n,a,e,#]& /@ Range[n];
```

```
S[n_Integer, a_, e_, i_Integer] := Module[
{den,num,numim,numre,x,z},
 x = X[n,a,i];
 z = Z[n,a,e];
 numre = -z*Sqrt[1 - z^2]*Sqrt[1 - x^2]*Sqrt[1 - x^2/a^2];
 numim = x*Sqrt[1 - (1-1/a^2)*z^2];
 num = numre + I*numim;
 den = 1 - (1 - x^2/a^2)*z^2;
 num/den
];
```

```
X[n_Integer, a_] :=  X[n,a,#]& /@ Range[n];
```

```
X[n_Integer,a_,i_Integer] :=  -JacobiCD[
 (2*i-1)*EllipticK[1/a^2]/n, 1/a^2
];
X[n_Integer,a_,i_Integer] := 0 /; And[i==(n+1)/2,OddQ[n]];
```

```
Z[n_Integer,a_,e_] :=  JacobiSN[
 InverseJacobiSN[1/Sqrt[1+e^2],1-1/(L[n,a])^2]*
 EllipticK[1-1/a^2]/EllipticK[1-1/(L[n,a])^2],1-1/a^2
];
```

```
Zbl[sp_,Fp_] := (1+sp*(Tan[Pi*Fp]))/(1-sp*(Tan[Pi*Fp]));
```

A.9.5 Specification

```
SA = {0.2, 0.212, 0.2, 40.};
{Fp, Fs, Ap, As} = SA;
Kp = A2K[Ap];
Ks = A2K[As];
SK = {Fp, Fs, Kp, Ks}
```

```
{0.2, 0.212, 0.217091, 99.995}
```

A.9.6 Minimum Order

```
{nellip[Fp,Fs,Ap,As], ncheb[Fp,Fs,Ap,As], nbut[Fp,Fs,Ap,As]}
nmin = nellip[Fp,Fs,Ap,As];
nmax = 2*nmin;
nlist = Range[nmin,nmax]
```

```
{8, 18, 79}
{8, 9, 10, 11, 12, 13, 14, 15, 16}
```

A.9.7 Range of Selectivity Factor and Ripple Factor

```
xtan = rtan[Fs,Fp];
amin8  = A2a[nmin,Ap,As];
amax8  = a/. FindRoot[R[nmin,a,xtan]==Ks/Kp, {a,xtan,1.095}];
amin9  = A2a[9,Ap,As];
amax9  = a/. FindRoot[R[ 9,a,xtan]==Ks/Kp, {a,amax8,1.1}];
amin10 = A2a[10,Ap,As];
amax10 = a/. FindRoot[R[10,a,xtan]==Ks/Kp, {a,amax9,1.2}];
amin11 = A2a[11,Ap,As];
amax11 = a/. FindRoot[R[11,a,xtan]==Ks/Kp, {a,amax10,1.2}];
amin12 = A2a[12,Ap,As];
amax12 = a/. FindRoot[R[12,a,xtan]==Ks/Kp, {a,amax11,1.2}];
amin13 = A2a[13,Ap,As];
amax13 = a/. FindRoot[R[13,a,xtan]==Ks/Kp, {a,amax12,1.3}];
amin14 = A2a[14,Ap,As];
amax14 = a/. FindRoot[R[14,a,xtan]==Ks/Kp, {a,amax13,1.4}];
amin15 = A2a[15,Ap,As];
amax15 = a/. FindRoot[R[15,a,xtan]==Ks/Kp, {a,amax14,1.6}];
amin16 = A2a[16,Ap,As];
amax16 = a/. FindRoot[R[16,a,xtan]==Ks/Kp, {a,amax15,1.9}];
aminlist = {amin8,amin9,amin10,amin11,amin12,amin13,amin14,amin15,amin16};
amaxlist = {amax8,amax9,amax10,amax11,amax12,amax13,amax14,amax15,amax16};
eminlist = Table[Ks/L[n,amaxlist[[n-8+1]]], {n,nmin,nmax}];
{emin8,emin9,emin10,emin11,emin12,emin13,emin14,emin15,emin16} = eminlist;
emax = Kp;
emaxlist = Table[Kp,{nmax-nmin+1}];
TableForm[Transpose[{nlist,aminlist,amaxlist
                    ,ScientificForm/@eminlist,emaxlist}]
, TableHeadings->{{},{"n","amin","amax","emin","emax"}}]
```

n	amin	amax	emin	emax
8	1.04285	1.09245	$6.25782\ 10^{-2}$	0.217091
9	1.022	1.11016	$1.43176\ 10^{-2}$	0.217091
10	1.01135	1.13668	$2.62202\ 10^{-3}$	0.217091
11	1.00587	1.17518	$3.65611\ 10^{-4}$	0.217091

12	1.00304	1.23116	3.61027 10^{-5}	0.217091
13	1.00158	1.31502	2.23759 10^{-6}	0.217091
14	1.00082	1.44884	6.96526 10^{-8}	0.217091
15	1.00042	1.69029	6.68381 10^{-10}	0.217091
16	1.00022	2.26906	4.39641 10^{-13}	0.217091

A.9.8 Range of Edge Frequencies

```
fpminlist = ArcTan[Tan[Pi*Fs] / amaxlist]/Pi //N;
{fpmin8,fpmin9,fpmin10,fpmin11,fpmin12,fpmin13,
 fpmin14,fpmin15,fpmin16} = fpminlist;
fpmaxlist = ArcTan[Tan[Pi*Fs] / aminlist]/Pi //N;
{fpmax8,fpmax9,fpmax10,fpmax11,fpmax12,fpmax13,
 fpmax14,fpmax15,fpmax16} = fpmaxlist;
fsminlist = ArcTan[Tan[Pi*Fp] * aminlist]/Pi //N;
{fsmin8,fsmin9,fsmin10,fsmin11,fsmin12,fsmin13,
 fsmin14,fsmin15,fsmin16} = fsminlist;
fsmaxlist = ArcTan[Tan[Pi*Fp] * amaxlist]/Pi //N;
{fsmax8,fsmax9,fsmax10,fsmax11,fsmax12,fsmax13,
 fsmax14,fsmax15,fsmax16} = fsmaxlist;
TableForm[Transpose[{nlist,fpminlist,fpmaxlist,fsminlist,fsmaxlist}]
, TableHeadings->{{},{"n","fpmin","fpmax","fsmin","fsmax"}}]
```

n	fpmin	fpmax	fsmin	fsmax
8	0.198485	0.205546	0.20639	0.213552
9	0.196065	0.208643	0.203305	0.216049
10	0.192535	0.210256	0.201712	0.21973
11	0.187606	0.211095	0.200887	0.224951
12	0.180824	0.21153	0.20046	0.23229
13	0.171447	0.211756	0.200239	0.242744
14	0.158187	0.211874	0.200124	0.258162
15	0.138517	0.211934	0.200064	0.282469
16	0.106119	0.211966	0.200033	0.326442

■ A.9.9 Transfer Functions and Plots of Attenuation in dB Versus Frequency

■ A.9.10 Design D1

```
hz8D1 = Hz[8, rtan[Fs,Fp], emax, Fp, z] //N;
AdB8D1 = AdB[hz8D1,z,f];
{bln8,aln8} = ba[8, rtan[Fs,Fp], emax, Fp, z] //N;
qln8 = Qfactor[aln8,bln8];
plotPTS[AdB8D1,f,{0.2,0.212,0.2,40}]
```

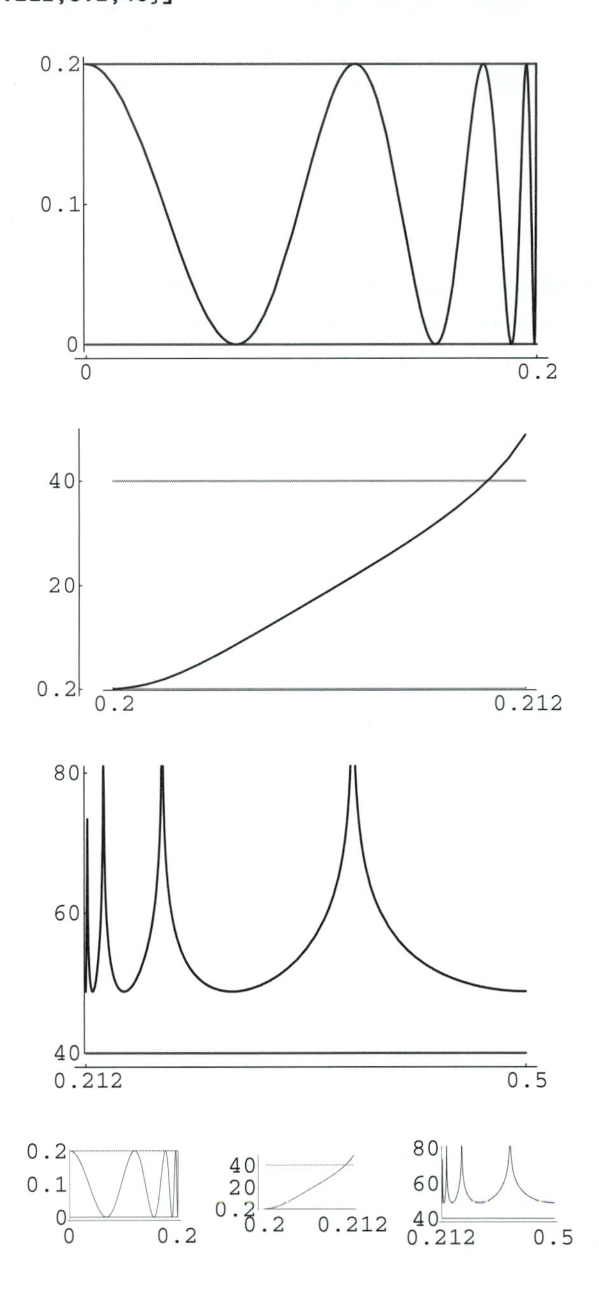

■ A.9.11 Design D2

```
hz8D2 = Hz[8, rtan[Fs,Fp], Ks/L[8,rtan[Fs,Fp]], Fp, z] //N;
AdB8D2 = AdB[hz8D2,z,f];
{b2n8,a2n8} = ba[8, rtan[Fs,Fp], Ks/L[8,rtan[Fs,Fp]], Fp, z] //N;
q2n8 = Qfactor[a2n8,b2n8];
plotPTS[AdB8D2,f,{0.2,0.212,0.2,40}]
```

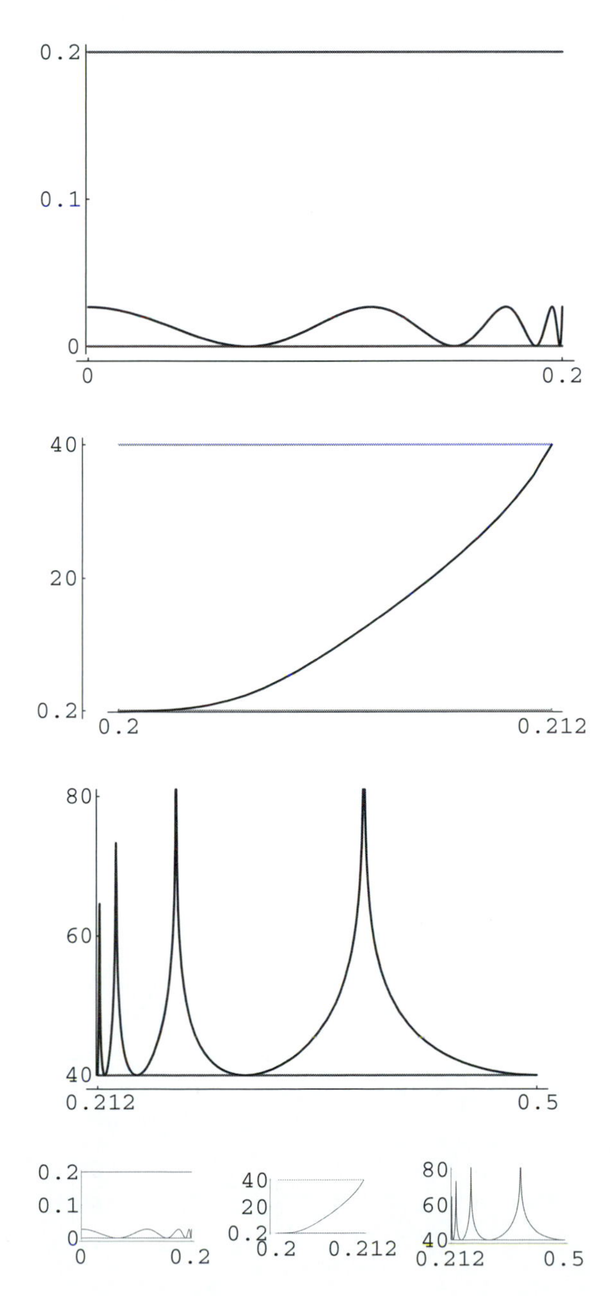

■ A.9.12 Design D3a

```
hz8D3a = Hz[8, amin8, emax, Fp, z] //N;
AdB8D3a = AdB[hz8D3a,z,f];
{b3an8,a3an8} = ba[8, amin8, emax, Fp, z] //N;
q3an8 = Qfactor[a3an8,b3an8];
plotPTS[AdB8D3a,f,{0.2,0.212,0.2,40}]
```

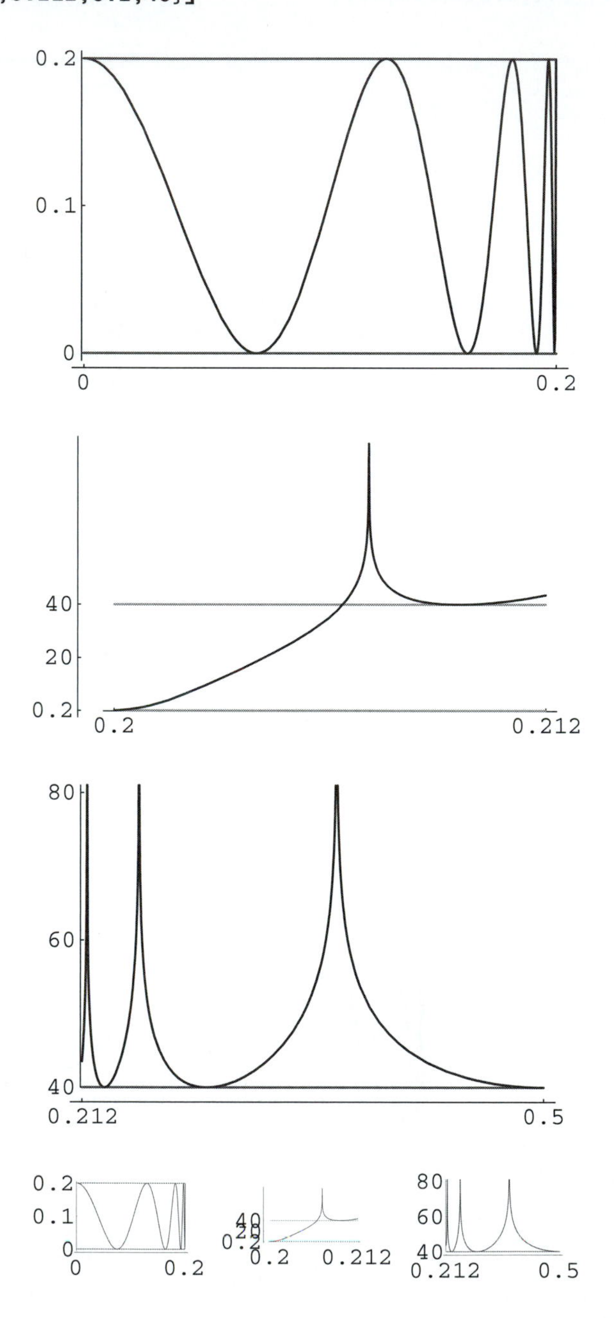

A.9.13 Design D3b

```
hz8D3b = Hz[8, amin8, emax, fpmax8, z] //N;
AdB8D3b = AdB[hz8D3b,z,f];
{b3bn8,a3bn8} = ba[8, amin8, emax, fpmax8, z] //N;
q3bn8 = Qfactor[a3bn8,b3bn8];
plotPTS[AdB8D3b,f,{0.2,0.212,0.2,40}]
```

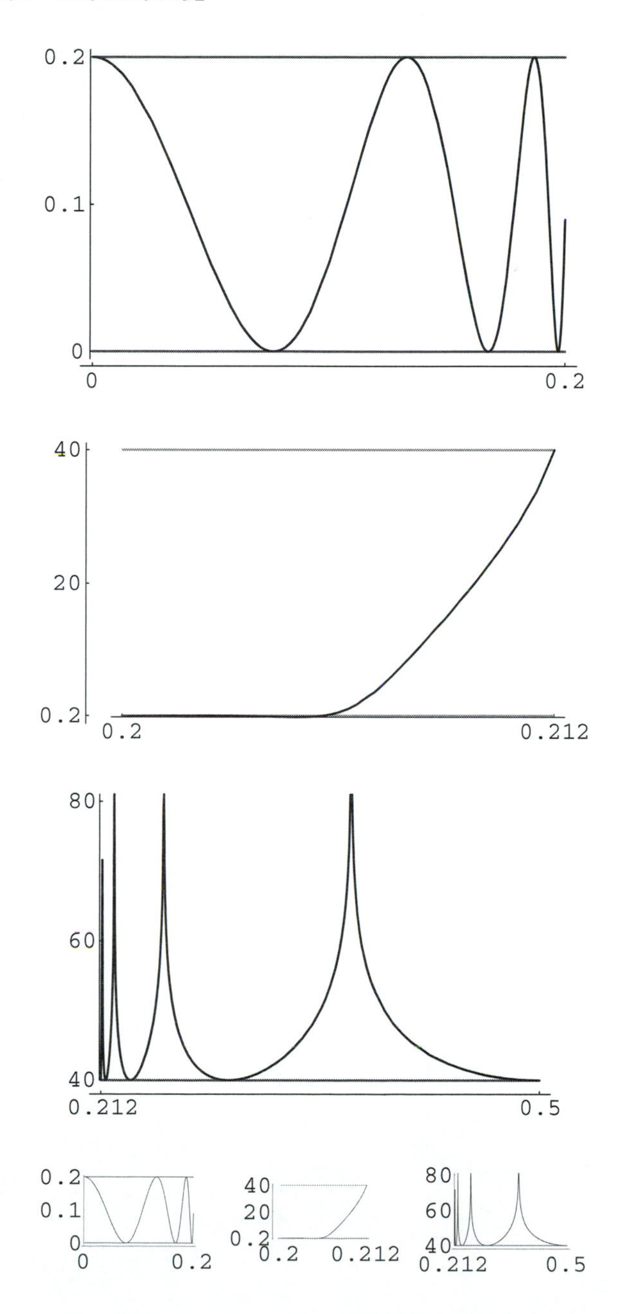

■ A.9.14 Design D4a

```
hz8D4a = Hz[8, amax8, emax, Fp, z] //N;
AdB8D4a = AdB[hz8D4a,z,f];
{b4an8,a4an8} = ba[8, amax8, emax, Fp, z] //N;
q4an8 = Qfactor[a4an8,b4an8];
plotPTS[AdB8D4a,f,{0.2,0.212,0.2,40}]
```

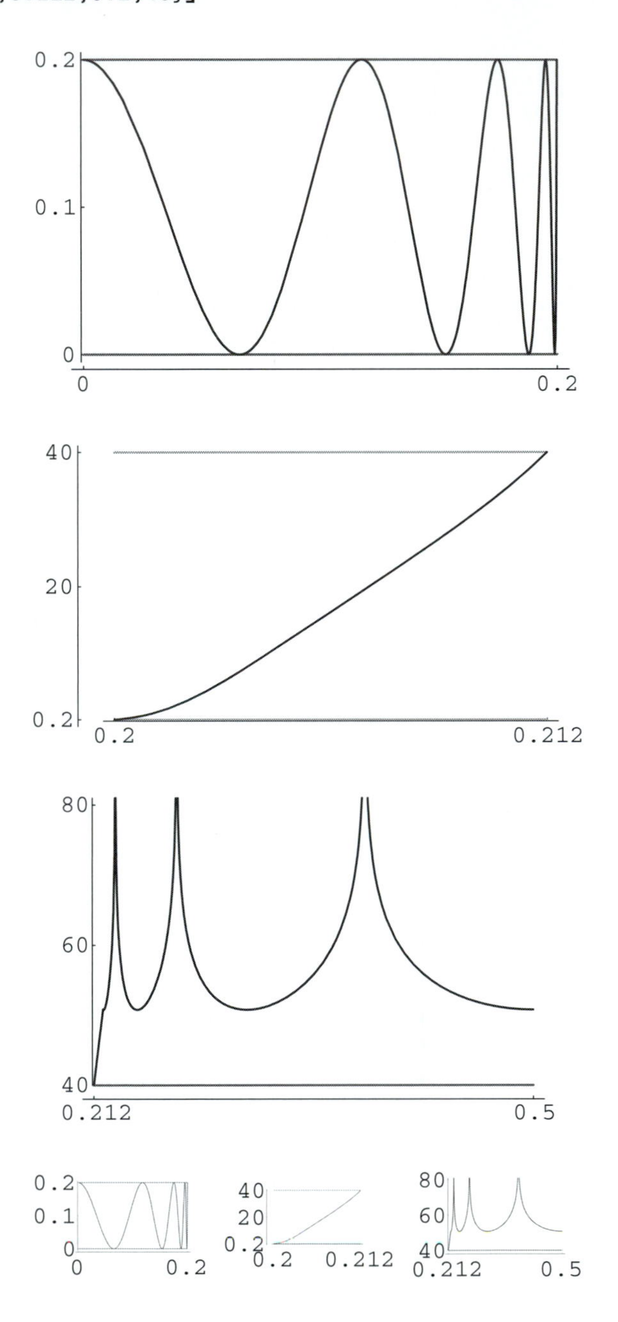

A.9.15 Design D4b

```
hz8D4b = Hz[8, amax8, emin8, fpmin8, z] //N;
AdB8D4b = AdB[hz8D4b,z,f];
{b4bn8,a4bn8} = ba[8, amax8, emin8, fpmin8, z] //N;
q4bn8 = Qfactor[a4bn8,b4bn8];
plotPTS[AdB8D4b,f,{0.2,0.212,0.2,40}]
```

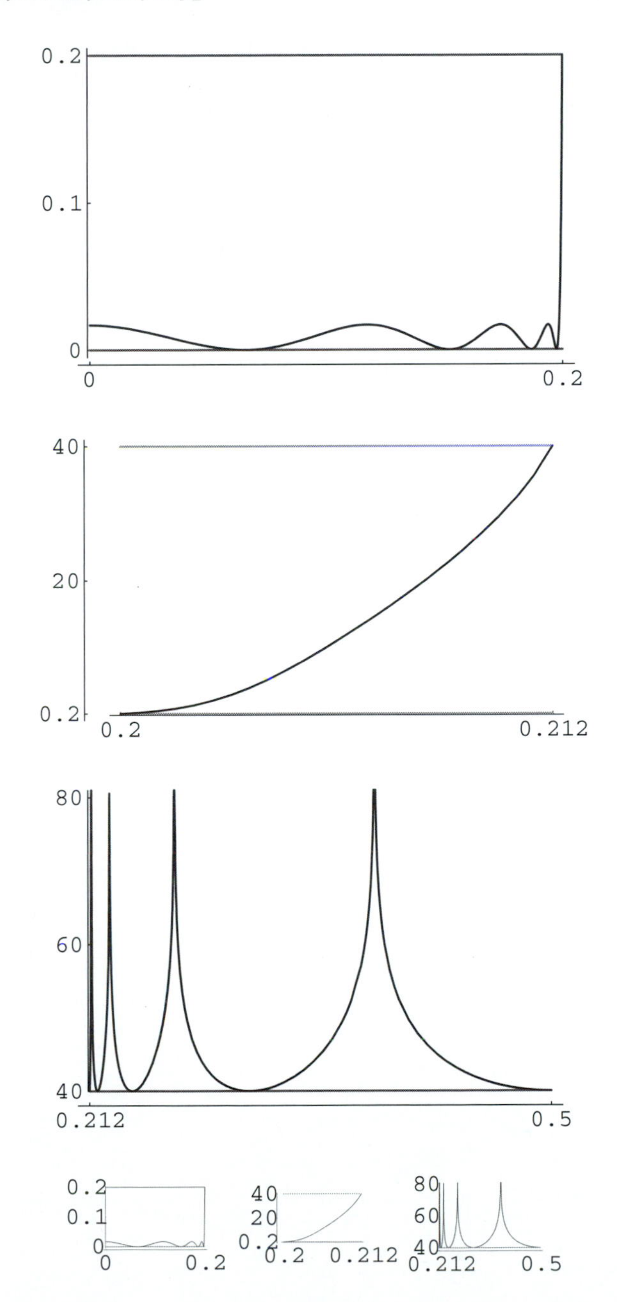

■ A.9.16 Coefficients a and b, Q-Factor, and Sensitivity of the Second-Order Section with the Maximal Q-Factor

```
alist = {a1n8, a2n8, a3an8, a3bn8, a4an8, a4bn8};
blist = {b1n8, b2n8, b3an8, b3bn8, b4an8, b4bn8};
qlist = {q1n8, q2n8, q3an8, q3bn8, q4an8, q4bn8};
slist = 1/(1-alist)
```

```
{30.4061, 23.9766, 44.7132, 44.2529, 28.3548, 21.5643}
```

```
 TableForm[Transpose[{blist,alist,qlist,slist,slist^3}]
, TableHeadings->{{"D1","D2","D3a","D3b","D4a","D4b"}
   , {"b","a","Q-factor","1/(1-a)","1/(1-a)^3"}}]
```

	b	a	Q-factor	1/(1-a)	1/(1-a)^3
D1	-0.593746	0.967112	28.5112	30.4061	28111.3
D2	-0.565538	0.958293	22.4763	23.9766	13783.6
D3a	-0.600922	0.977635	42.1227	44.7132	89393.7
D3b	-0.534747	0.977403	42.1227	44.2529	86661.3
D4a	-0.592153	0.964733	26.5596	28.3548	22797.2
D4b	-0.572418	0.953627	20.1398	21.5643	10027.8

■ Remark

The expressions 1/(1-a) and 1/(1-a)^3 are used to estimate the magnitude response sensitivity to the transfer function coefficients.

A.10 Jacobi Elliptic Functions

Miroslav D. Lutovac, Dejan V. Tosic and Brian L. Evans
lutovac@iritel.bg.ac.yu tosic@galeb.etf.bg.ac.yu bevans@ece.utexas.edu

A.10.1 References

1. M.D.Lutovac, D.V. Tosic, I.M.Markoski,
"Symbolic computation of elliptic rational functions,"
5th Int. Workshop, Symbolic Methods, Applications, Circuit Design,
SMACD'98, Kaiserslautern, Germany,
Oct. 1998, pp.177–180.

2. M. Abramowitz and I. Stegun,
"Handbook of Mathematical Functions,"
Dover, New York, 1972.

3. M. D. Lutovac, D. V. Tosic and B. L. Evans,
"Advanced Filter Design for Signal Processing using MATLAB and Mathematica,"
http://iritel.iritel.bg.ac.yu/~lutovac/www/afdhome.htm
http://galeb.etf.bg.ac.yu/~tosic/afdhome.htm
http://www.ece.utexas.edu/~bevans/

A.10.2 Initialization

```
SetDirectory[HomeDirectory[]];
<<afd\math\m\clearall.m
```

A.10.3 Notation

cd[u,k] – Jacobi sine shifted
cn[u,k] – Jacobi cosine
k – modulus
K[k] – complete elliptic integral of the first kind
Kp[k] – complementary complete elliptic integral of the first kind
sn[u,k] – Jacobi sine

A.10.4 Definitions

```
sn[u_,k_] := JacobiSN[u,k^2];
cn[u_,k_] := JacobiCN[u,k^2];
cd[u_,k_] := JacobiCD[u,k^2];
dn[u_,k_] := JacobiDN[u,k^2];
invsn[v_,k_] := InverseJacobiSN[v,k^2];
K[k_] := EllipticK[k^2];
Kp[k_] := K[Sqrt[1-k^2]];
```

A.10.5 Basic properties

```
{sn[0,k], sn[K[k],k], cn[0,k], cn[K[k],k]}
```

```
{0, 1, 1, 0}
```

```
a = 1.1;
b = 0.9;
{sn[I*a,b],
 I*sn[a,Sqrt[1-b^2]]/cn[a,Sqrt[1-b^2]]}
{cn[I*a,b],
 1/cn[a,Sqrt[1-b^2]]}
{dn[I*a,b],
 dn[a,Sqrt[1-b^2]]/cn[a,Sqrt[1-b^2]]}
```

```
{1.81411 I, 1.81411 I}
{2.07147, 2.07147}
{1.9146, 1.9146}
```

```
c = sn[a,b]
d = invsn[c,b]
a - d
```

```
0.819407
1.1
            -16
6.66134 10
```

A.10.6 Plots

```
Plot3D[sn[u,k]
, {u,0,20},{k,0,0.999}
, AxesLabel -> {"u", "k", "sn"}
, PlotLabel -> "Jacobi sine"
, PlotPoints -> 40
];
```

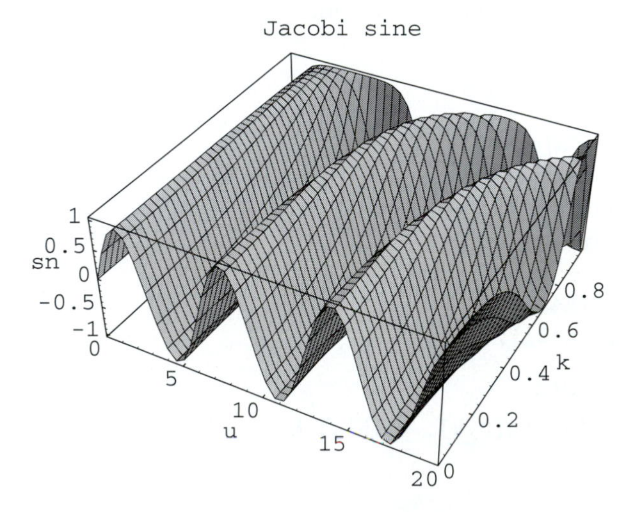

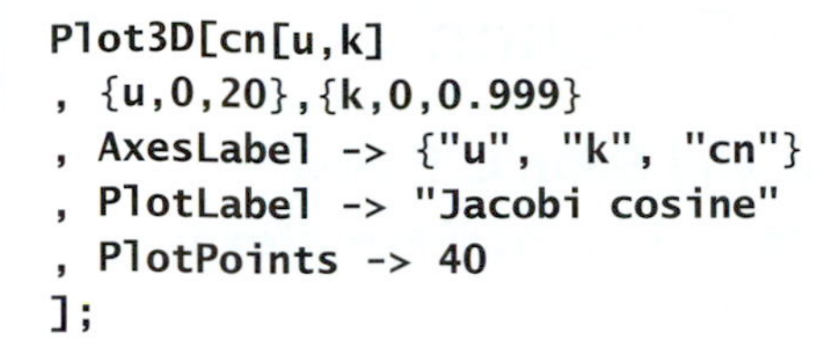

```
Plot3D[cn[u,k]
, {u,0,20},{k,0,0.999}
, AxesLabel -> {"u", "k", "cn"}
, PlotLabel -> "Jacobi cosine"
, PlotPoints -> 40
];
```

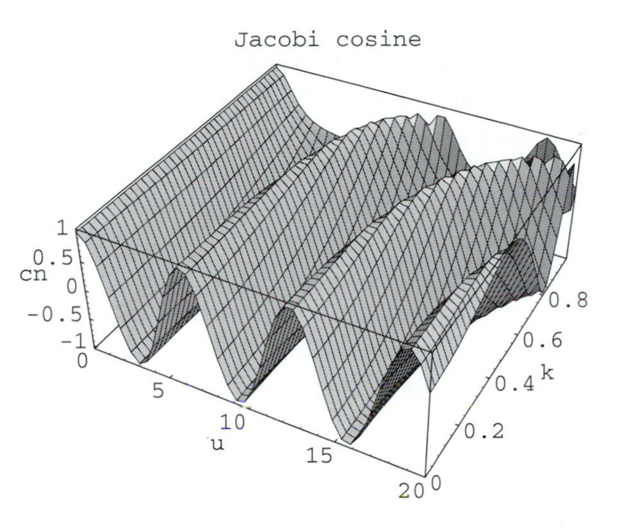


```
Plot3D[cd[u,k]
, {u,0,20},{k,0,0.999}
, AxesLabel -> {"u", "k", "cd"}
, PlotLabel -> "Jacobi shifted sine"
, PlotPoints -> 40
];
```

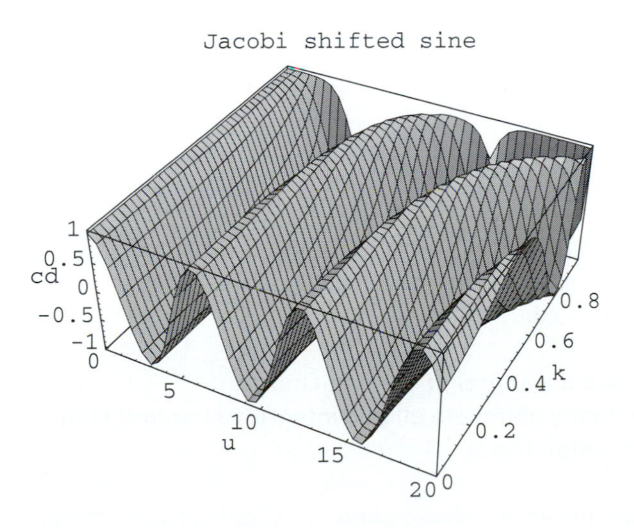

A.11 Elliptic Rational Function

Miroslav D. Lutovac, Dejan V. Tosic and Brian L. Evans
lutovac@iritel.bg.ac.yu tosic@galeb.etf.bg.ac.yu bevans@ece.utexas.edu

A.11.1 References

1. M. D. Lutovac, D.M.Rabrenovic,
"A simplified design of some Cauer filters without jacobian elliptic functions,"
IEEE Trans. Circ. Systems: Part II, vol. 39, no. 9, pp. 666–671, Sept. 1992.

2. M. D. Lutovac, D.M.Rabrenovic,
"Algebraic design of some lower-order elliptic filters,"
Electronics Letters, vol. 29, no. 2, pp. 192–193, Jan. 1993.

3. D.M.Rabrenovic, M. D. Lutovac,
"Minimum stopband attenuation of the Cauer filters without elliptic functions and integrals,"
IEEE Trans. Circ. Systems: Part I, vol. 40, no. 9, pp. 618–621, Sept. 1993.

4. M. D. Lutovac, D.M.Rabrenovic,
"Exact determination of the natural modes of some Cauer filters by means of a standard analytical procedure,"
IEE Proc. Circuits Devices Syst., vol. 143, no. 3, pp. 134–138, June 1996.

5. M. D. Lutovac, D. V. Tosic and B. L. Evans,
"Advanced Filter Design for Signal Processing using MATLAB and Mathematica,"
http://iritel.iritel.bg.ac.yu/~lutovac/www/afdhome.htm
http://galeb.etf.bg.ac.yu/~tosic/afdhome.htm
http://www.ece.utexas.edu/~bevans/

A.11.2 Initialization

```
SetDirectory[HomeDirectory[]];
<<afd\math\m\clearall.m
<<afd\math\m\jeldf.m
<<graphics`graphics`
```

```
Discrimination Factor L(n,1/k)
n = 1,2,3,4,6,8,9,12,16,18
```

A.11.3 Notation

cd[u,k] – Jacobi sine shifted
invcd[v,k] – inverse of cd
k – modulus
K[k] – complete elliptic integral of the first kind
Kp[k] – complementary complete elliptic integral of the first kind
L[n,1/k] – discrimination factor
n – order, n>0
R[n,k,p,x] – function from which we generate elliptic rational function

A.11.4 Definitions

```
cd[u_,k_] := JacobiCD[u,k^2];
invcd[v_,k_] := InverseJacobiCD[v,k^2];
K[k_] := EllipticK[k^2];
Kp[k_] := K[Sqrt[1-k^2]];
R[n_,k_,p_,x_] := cd[n*K[p]*invcd[x,k]/K[k],p]
x[k_,u_,v_] := cd[u+I*v,k];
Ruv[n_Integer,k_,u_,v_] := cd[n*(u+I*v),1/L[n,1/k]];
```

A.11.5 Example

```
{R[2,0.8,0.1,-1.0], R[2,0.8,0.1,0], R[2,0.8,0.1,1.0]}

{1., -1., 1.}
```

Remark

We know that invcd(0,k)=K(k), so, we add a rule

```
{invcd[0,0.9], K[0.9]}
invcd[0.0,0.9]

{2.28055, 2.28055}
                                               1
Power::infy: Infinite expression -- encountered.
                                               0.
Infinity::indet: Indeterminate expression 0. ComplexInfinity encountered.
2.28054913842277 - 2 EllipticLog[{Indeterminate, ComplexInfinity}, {3.62, 0.0361}] -
 2 (1.654616667522527 I Round[0.302185
       Im[0. - 2 EllipticLog[{Indeterminate, ComplexInfinity}, {3.62, 0.0361}]]] +
    2.28054913842277 Round[0.219245 Re[0. -
         2 EllipticLog[{Indeterminate, ComplexInfinity}, {3.62, 0.0361}]]])

invcd[0.0,k_] := K[k];
{invcd[0,0.9], invcd[0.0,0.9]}

{2.28054913842277, 2.28055}
```

A.11.6 Plots

```
Plot[{R[0, 0.8, 0.1, x],
      R[1, 0.8, 0.1, x],
      R[2, 0.8, 0.1, x],
      R[3, 0.8, 0.1, x],
      R[4, 0.8, 0.1, x],
      R[5, 0.8, 0.1, x]}
, {x,-1,1}
, AxesLabel -> {"x", "R[n,k,p,x]"}];
```

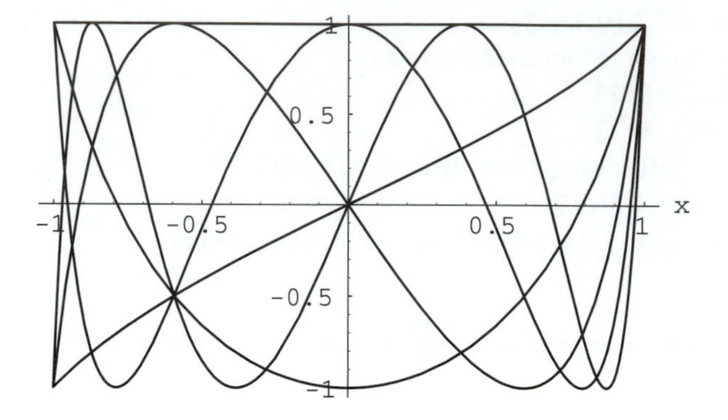

A.11.7 Can we Construct a Rational Function from R(n,k,p,x) by Varying k and p?

For n=1

R(1,k,p,x) = cd(K(p)*invcd(x,k)/K(k),p)

and simplyfies to R(1,k,p,x)=x for k=p.

R(1,k,p,0) = 0

R(1,k,p,1) = 1

R(1,k,p,1/k) = 1/k = 1/p

A.11.8 Find p Which Satisfies R(n,k,p,1/k)=1/p for n=1

```
n1 = 1
k1 = N[95/100,24]
Plot[R[n1,k1,p,1/k1] - 1/p
, {p,0.9,0.99}
, AxesLabel -> {"p", "R(1,k,p,1/k)-1/p"}];
```

```
1
0.95
```

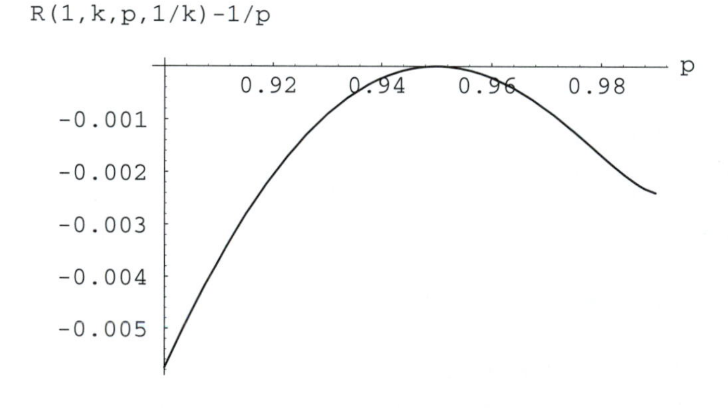

```
sol = FindRoot[
 R[n1,k1,p,1/k1] == 1/p
, {p,0.949,0.951}
, AccuracyGoal -> 16
, MaxIterations -> 64
];
p1 = N[p/.sol,16]
p1exact = k1
```

```
                                         -17
0.950000006790337 + 5.259681766043407 10     I
0.95
```

A.11.9 Find p Which Satisfies R(n,k,p,1/k)=1/p for n=2

```
n2 = 2
k2 = N[95/100,24]
Plot[R[n2,k2,p,1/k2] - 1/p
, {p,0.2,0.8}
, AxesLabel -> {"p", "R(2,k,p,1/k)-1/p"}];
```

```
2
0.95
```

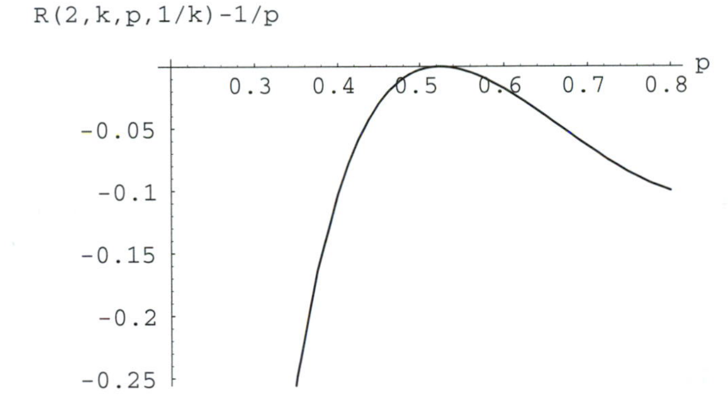


```
sol = FindRoot[
 R[n2,k2,p,1/k2] == 1/p
, {p,0.52,0.53}
, AccuracyGoal -> 16
, MaxIterations -> 64
];
p2 = N[p/.sol,16]
p2exact = 1/(1/k2+Sqrt[1/k2^2-1])^2
```

```
0.5240999634235638
0.524099944775800752814475
```

```
Plot[R[n2,k2,p2,x]
, {x,0,1/k2}
, AxesLabel -> {"x", "R(2,k,p,x)"}
, PlotLabel -> "R(n,k,p,1/k)=1/p"
, GridLines -> {{1},{-1,1}}
];
```

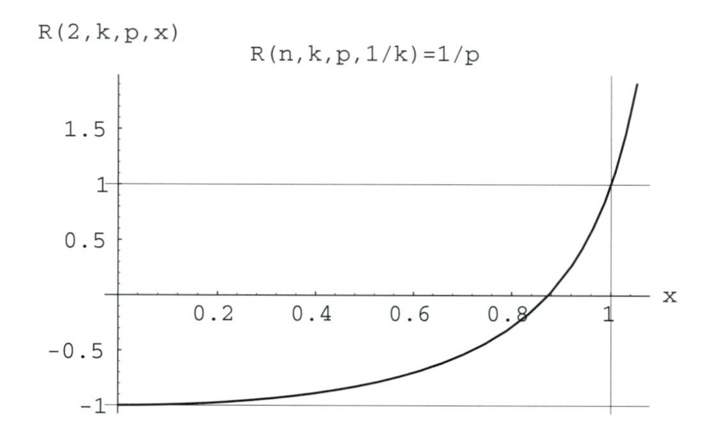

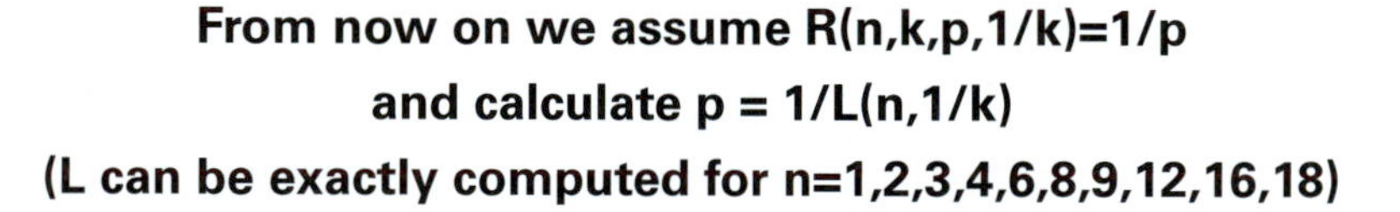

From now on we assume R(n,k,p,1/k)=1/p
and calculate p = 1/L(n,1/k)
(L can be exactly computed for n=1,2,3,4,6,8,9,12,16,18)

■ A.11.10 Parametric Plot of x in Terms of Complex Parametric Variable w=u+I*v

```
n1 = 3;
k1 = 0.7;
K1 = K[k1];
ParametricPlot3D[
     {{t, 0,      x[k1, t*K1, 0]},
      {4, t-3,    x[k1, 4*K1, (t-3)*K1]},
      {t, 1,   Re[x[k1, t*K1, K1]]}
     }
, {t,3,4}
, Ticks -> {{3,4},{0,1},{0,1,2,3,4}}
, AspectRatio -> 0.6
, ViewPoint -> {10,-30,20}
, AxesLabel -> {"u/K", "v/K'", "x"}
];
```

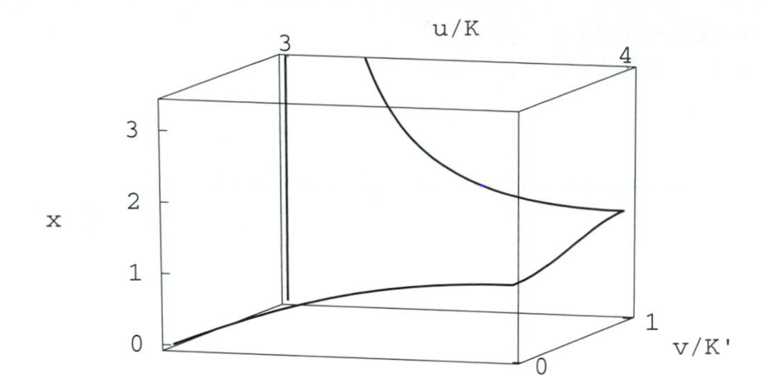

■ A.11.11 Parametric Plot of R in Terms of Complex Parametric Variable w=u+I*v

```
nl = 3;
kl = 0.7;
Knl = K[1/L[nl,1/kl]];
r0 = 1/20;
ParametricPlot3D[
     {{t, 0,   r0*Abs[Ruv[nl, kl, t*Knl, 0]]},
      {4, t-3, r0*Abs[Ruv[nl, kl, 4*Knl, (t-3)*Knl]]},
      {t, 1,   r0*Abs[Ruv[nl, kl, t*Knl, Knl]]}
     }
, {t,3,4}
, Ticks -> {{3,4},{0,1},{1,3,5}}
, AspectRatio -> 0.5
, ViewPoint -> {10,-25,20}
, AxesLabel -> {"u/K", "v/K'", "|Ruv|/20"}
];
```

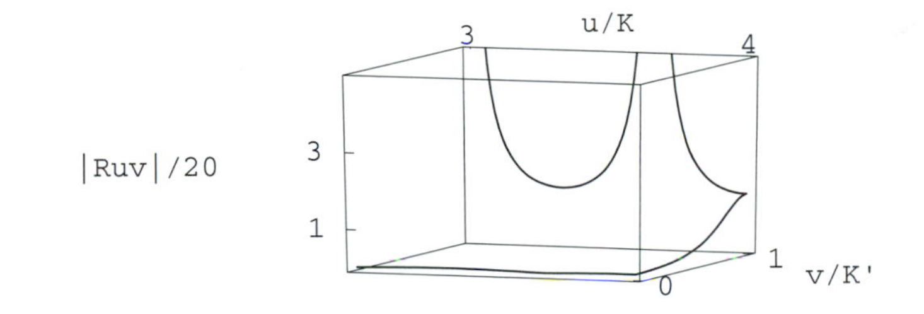


```
nl = 3;
kl = 0.7;
Knl = K[1/L[nl,1/kl]];
ParametricPlot3D[
     {{t, 0,        Abs[Ruv[nl, kl, t*Knl, 0]]},
      {4, (t-3)/2, Abs[Ruv[nl, kl, 4*Knl, (t-3)*Knl/2]]}
     }
, {t,3,4}
, Ticks -> {{3,4},{0,0.5},{1,3,5}}
, AspectRatio -> 0.5
```

```
, ViewPoint -> {10,-25,20}
, AxesLabel -> {"u/K", "v/K'", "|Ruv|"}
];
```

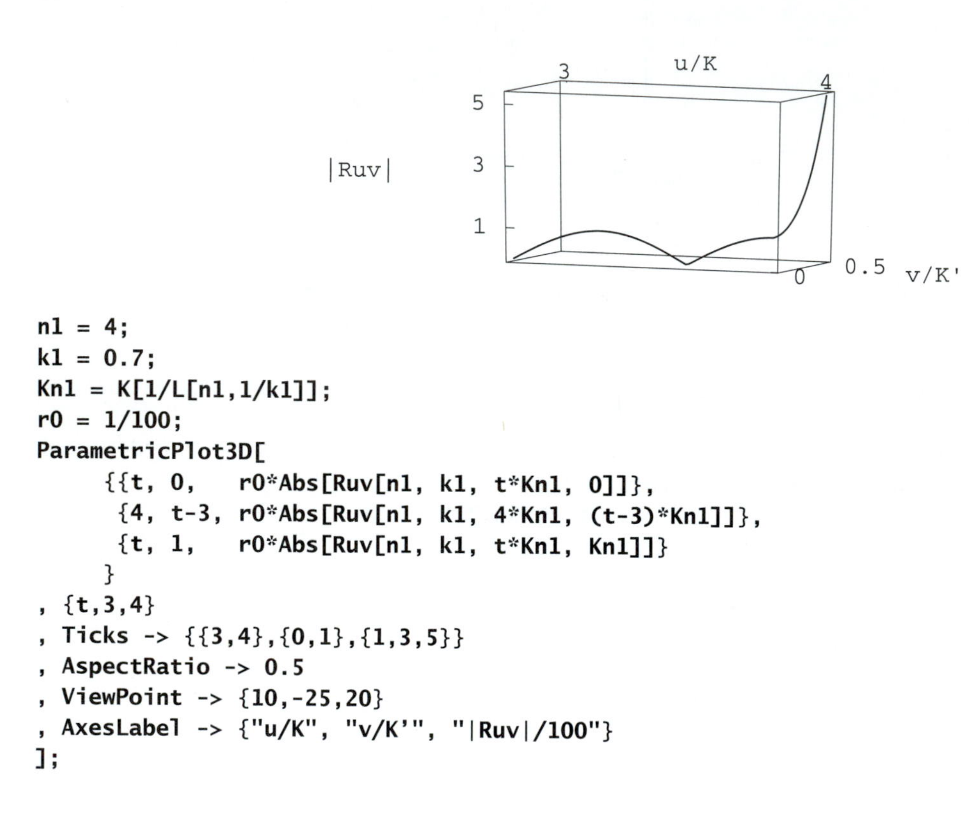


```
n1 = 4;
k1 = 0.7;
Kn1 = K[1/L[n1,1/k1]];
r0 = 1/100;
ParametricPlot3D[
     {{t, 0,   r0*Abs[Ruv[n1, k1, t*Kn1, 0]]},
      {4, t-3, r0*Abs[Ruv[n1, k1, 4*Kn1, (t-3)*Kn1]]},
      {t, 1,   r0*Abs[Ruv[n1, k1, t*Kn1, Kn1]]}
     }
, {t,3,4}
, Ticks -> {{3,4},{0,1},{1,3,5}}
, AspectRatio -> 0.5
, ViewPoint -> {10,-25,20}
, AxesLabel -> {"u/K", "v/K'", "|Ruv|/100"}
];
```

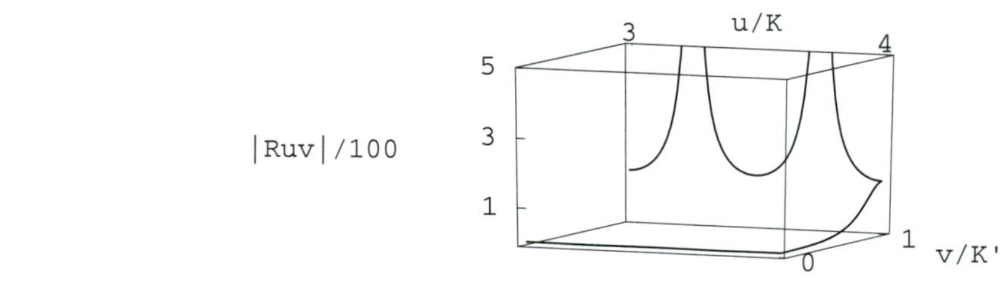


```
n1 = 4;
k1 = 0.7;
Kn1 = K[1/L[n1,1/k1]];
ParametricPlot3D[
     {{t, 0,          Abs[Ruv[n1, k1, t*Kn1, 0]]},
      {4, (t-3)/2.5, Abs[Ruv[n1, k1, 4*Kn1, (t-3)*Kn1/2.5]]}
     }
, {t,3,4}
, Ticks -> {{3,4},{0,0.4},{1,3,5}}
, AspectRatio -> 0.5
```

```
, ViewPoint -> {10,-25,20}
, AxesLabel -> {"u/K", "v/K'", "|Ruv|"}
];
```

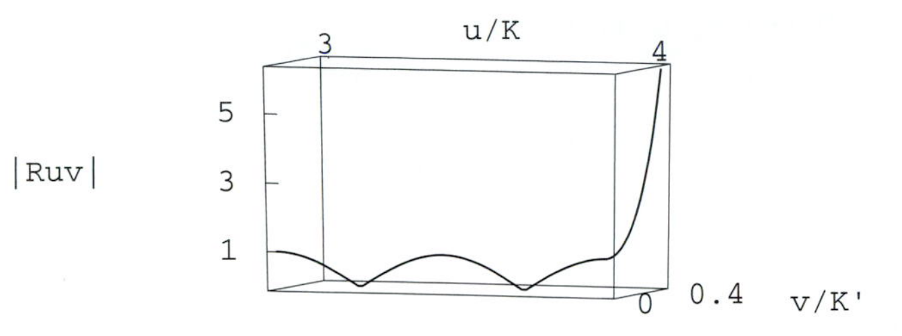

■ A.11.12 Elliptic Rational Function in Terms of Jacobi Elliptic Functions: a=1/k

```
Rjel[0, a_, x_] := 1
Rjel[1, a_, x_] := x
Rjel[2, a_, x_] := (( 1+Sqrt[1-1/a^2])*x^2-1)/
                   ((-1+Sqrt[1-1/a^2])*x^2+1) ;
Rjel[3,a_,x_] := Block[
{k=1/a,z1,z2,b1,b2,b3},
 z1 = JacobiDN[2*EllipticK[k^2]/3,k^2];
 b3 = (1 + z1)^2;
 b1 = -1 - 2*z1;
 b2 = -1 + z1^2;
(b3*x^3 + b1*x)/(b2*x^2 + 1)
];
Rjel[4, a_, x_] := Block[{t},
 t = Sqrt[1 - 1/a^2];
((1+t)*(1+Sqrt[t])^2*x^4-2(1+t)*(1+Sqrt[t])*x^2+1)/
((1+t)*(1-Sqrt[t])^2*x^4-2(1+t)*(1-Sqrt[t])*x^2+1)];
Rjel[5,a_,x_] := Block[
{k=1/a,z1,z2,b1,b2,b3,b4,b5},
 z1 = JacobiDN[2*EllipticK[k^2]/5,k^2];
 z2 = JacobiDN[4*EllipticK[k^2]/5,k^2];
 b5 = (1+z1)^2*(1+z2)^2;
 b3 = -( 1-z1^2*z2^2+(1+z1)^2*(1+z2)^2+2*(z1+z2));
 b1 = ( 1+2*(z1+z2));
 b4 = (1-z1^2)*(1-z2^2);
 b2 = -(2-z1^2-z2^2);
(b5*x^5 + b3*x^3 + b1*x)/(b4*x^4 + b2*x^2 + 1)
];
a1 = 1.1;
Plot[{Rjel[1, a1, x],
      Rjel[2, a1, x],
      Rjel[3, a1, x],
      Rjel[4, a1, x],
      Rjel[5, a1, x]}
```

```
, {x,-1,1}
, AxesLabel -> {"x", "Rjel(n,a,x)"}
, PlotStyle -> {Dashing[{0.04}],
                Dashing[{0.03}],
                Dashing[{0.02}],
                Dashing[{0.01}],
                Dashing[{}]}
];
```

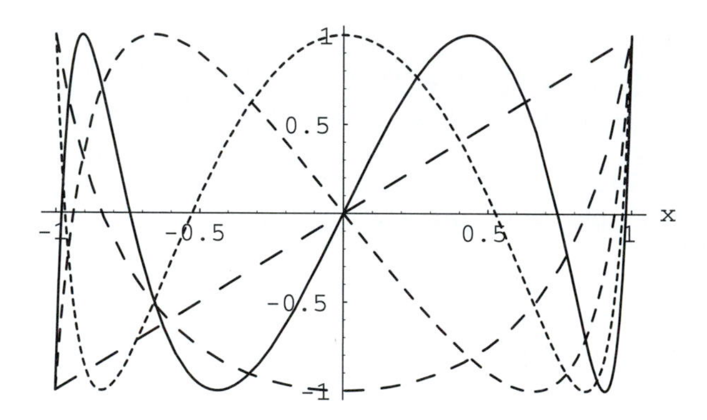


```
a1 = 1.1;
Plot[{Rjel[1, a1, x],
      Rjel[2, a1, x],
      Rjel[3, a1, x],
      Rjel[4, a1, x],
      Rjel[5, a1, x]}
, {x,1,a1}
, AxesLabel -> {"x", "Rjel(n,a,x)"}
, PlotStyle -> {Dashing[{0.04}],
                Dashing[{0.03}],
                Dashing[{0.02}],
                Dashing[{0.01}],
                Dashing[{}]}
];
```

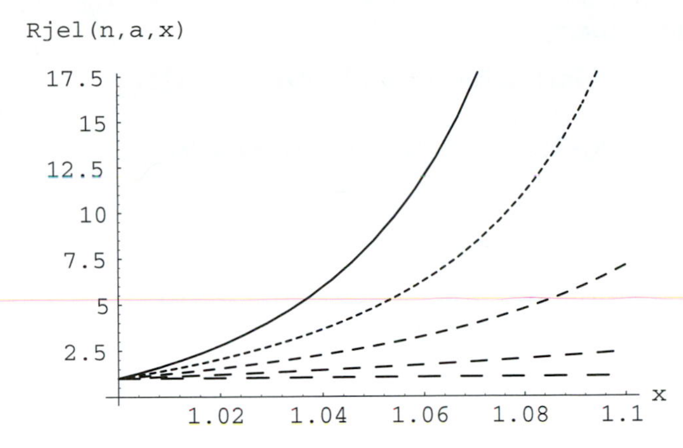


```
al = 1.1;
LogPlot[{Abs[Rjel[1, al, x]],
         Abs[Rjel[2, al, x]],
         Abs[Rjel[3, al, x]],
         Abs[Rjel[4, al, x]],
         Abs[Rjel[5, al, x]]}
, {x,al,2*al}
, AxesLabel -> {"x", "Rjel(n,a,x)"}
, PlotRange -> {1,1000}
, Ticks -> {{al,1.5,2},{1,10,100}}
, PlotStyle -> {Dashing[{0.04}],
                Dashing[{0.03}],
                Dashing[{0.02}],
                Dashing[{0.01}],
                Dashing[{}]}
];
```

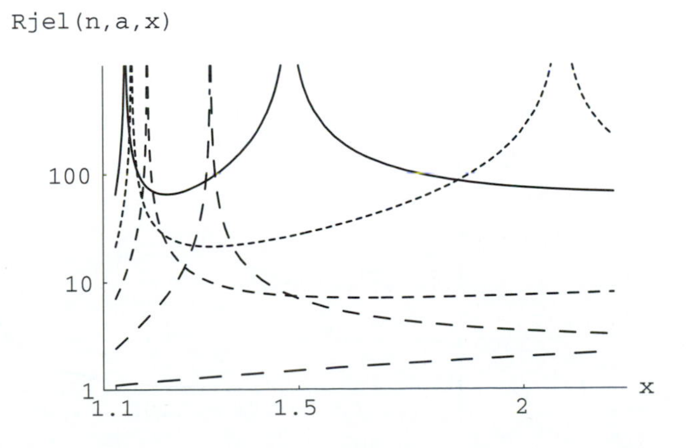

■ A.11.13 Nesting Property

```
Rjel[10, a_, x_] := Rjel[2, Rjel[5,a,a], Rjel[5,a,x]];
a1 = 1.1;
{Rjel[10,a1,-1.0], Rjel[10,a1,0.0], Rjel[10,a1,1.0]}
{1., -1., 1.}
a1 = 1.1;
Plot[Evaluate[Rjel[10, a1, x]]
, {x,-1,1}
, AxesLabel -> {"x", "Rjel(10,a,x)"}
];
```

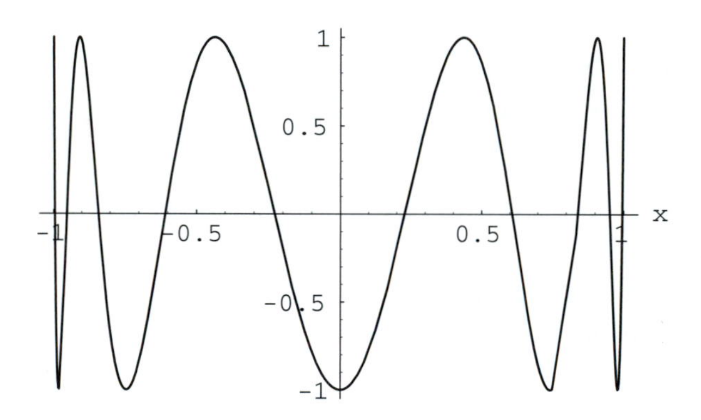


```
a1 = 1.1;
Plot[Evaluate[Rjel[10, a1, x]]
, {x,1,a1}
, AxesLabel -> {"x", "Rjel(10,a,x)"}
];
```

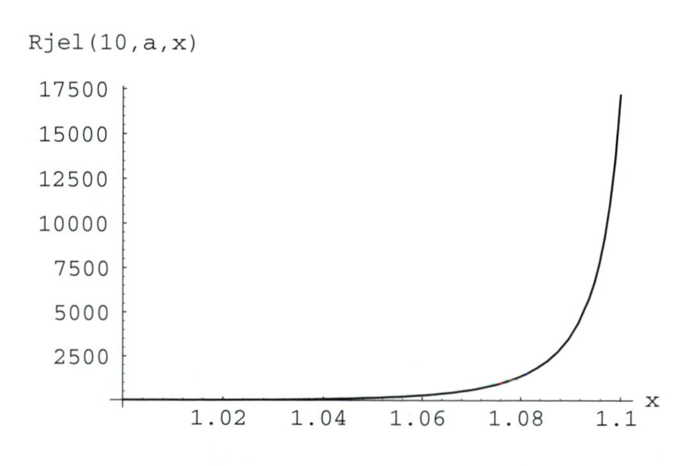

```
a1 = 1.1;
LogLogPlot[Evaluate[Abs[Rjel[10, a1, x]]]
, {x,1,10*a1}
, AxesLabel -> {"x", "Rjel(10,a,x)"}
, PlotRange -> {10^4,1.1*10^5}
, Ticks -> {{1,2,3,5,10}, {10^4, 10^5}}
];
```

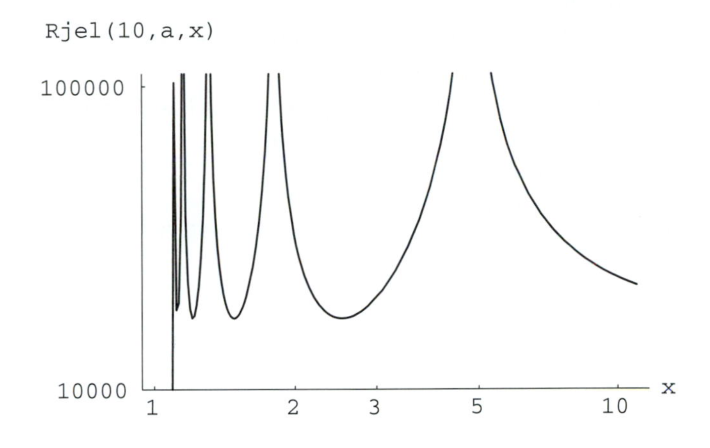


```
Rjel[10,a1,x] //Together
```

$$\frac{211.208 - 5361.41\,x^{2} + 28064.4\,x^{4} - 57774.2\,x^{6} + 52056.\,x^{8} - 17198.7\,x^{10}}{-211.208 + 528.325\,x^{2} - 484.596\,x^{4} + 194.543\,x^{6} - 30.7152\,x^{8} + 1.\,x^{10}}$$

A.12 Transfer Function with Minimal Q-Factors

Miroslav D. Lutovac, Dejan V. Tosic and Brian L. Evans
lutovac@iritel.bg.ac.yu tosic@galeb.etf.bg.ac.yu bevans@ece.utexas.edu

A.12.1 References

1. D.M.Rabrenovic, M. D. Lutovac,
"Elliptic filters with minimal Q-factors,"
Electronics Letters, vol. 30, no. 3, pp. 206–207, Feb. 1994.

2. M. D. Lutovac, D. V. Tosic and B. L. Evans,
"Advanced Filter Design for Signal Processing using MATLAB and Mathematica,"
http://iritel.iritel.bg.ac.yu/~lutovac/www/afdhome.htm
http://galeb.etf.bg.ac.yu/~tosic/afdhome.htm
http://www.ece.utexas.edu/~bevans/

A.12.2 Initialization

```
SetDirectory[HomeDirectory[]];
<<afd\math\m\clearall.m
<<afd\math\m\erfdf.m
<<afd\math\m\erfez.m
<<afd\math\m\erfzh.m
<<afd\math\m\erf235.m
<<graphics`graphics`
```

```
Discrimination Factor L(n,a)
n = 1,2,3,4,6,8,9,12,16,18
Exact Formulas for Zeros: n=2^i 3^j
X(n,a)={list-of-zeros}, X(n,a,i)
n = 1,2,3,4,6,8,9,12,16,18
Zeta Function Z(n,a,e)
Elliptic rational functions: n=2^i 3^j
Relliptic(n,a,x)
n = 1,2,3,4,5,6,8,9,10,12,15,16,18
```

A.12.3 Notation

a – selectivity factor
e – ripple factor
n – order
L(n,a) – discrimination factor
H - normalized lowpass transfer function
HminQ - minQ normalized lowpass transfer function
Q(s) – Q-factor of a pole s
QminQ(n,a) – list of pole Q-factors of minQ transfer function
QminQ(n,a,i) – pole Q-factor of minQ transfer function
Relliptic(n,a,x) – elliptic rational function
s - normalized compex frequency
S(n,a,e) – list of transfer function poles

S(n,a,e,i) – transfer function pole
SminQ(n,a,e) – list of minQ transfer function poles
SminQ(n,a,e,i) – minQ transfer function pole
X(n,a) – list of zeros of elliptic rational function
X(n,a,i) – zero of elliptic rational function
w - normalized angular frequency
Z(n,a,e) – zeta function

A.12.4 Definition and Procedures

```
Q[s_] := - Abs[s]/(2*Re[s]);

QminQ[n_Integer,a_,i_Integer] :=
 (a + X[n,a,i]^2)/
 (2 Sqrt[1- X[n,a,i]^2]*Sqrt[a^2 - X[n,a,i]^2]);

QminQ[n_Integer,a_] :=
 (a + X[n,a]^2)/
 (2 Sqrt[1- X[n,a]^2]*Sqrt[a^2 - X[n,a]^2]);

SminQ[n_Integer,a_,i_Integer] :=  Block[
{den,num,numim,numre,x,z},
 numre = -Sqrt[1- X[n,a,i]^2]*Sqrt[a^2- X[n,a,i]^2];
 numim = X[n,a,i]*(a + 1);
 num = numre + I*numim;
 den = a + X[n,a,i]^2;
 Sqrt[a]*num/den
];

SminQ[n_Integer,a_] :=  Block[
{den,num,numim,numre,x,z},
 numre = -Sqrt[1- X[n,a]^2]*Sqrt[a^2- X[n,a]^2];
 numim = X[n,a]*(a + 1);
 num = numre + I*numim;
 den = a + X[n,a]^2;
 Sqrt[a]*num/den
];

S[n_Integer, a_, e_, i_Integer] := Block[
{den,num,numim,numre,x,z},
 x = X[n,a,i];
 z = Z[n,a,e];
 numre = -z*Sqrt[1 - z^2]*Sqrt[1 - x^2]*Sqrt[1 - x^2/a^2];
 numim = x*Sqrt[1 - (1-1/a^2)*z^2];
 num = numre + I*numim;
 den = 1 - (1 - x^2/a^2)*z^2;
 num/den
];

S[n_Integer, a_, e_] :=  S[n,a,e,#]& /@ Range[n]
```

A.12.5 Example

```
n1 = 18;
a1 = 1.1;
e1 = 0.1;
i1 = 1;

S[n1,a1,e1,i1]
SminQ[n1,a1,i1]

Q[S[n1,a1,e1,i1]]
Q[SminQ[n1,a1,i1]]
QminQ[n1,a1,i1]

QminQ[n1,a1]
ListPlot[%
,PlotStyle->{AbsolutePointSize[4]}
,AxesLabel -> {"i", "QminQ(n,a,i)"}
,Ticks->{{1,3,6,9,12,15,18},{10,20,30,40}}
,PlotLabel -> "n = 18, a = 1.1"
];
```

```
-0.0053858 - 1.00386 I
-0.0124234 - 1.04874 I
93.1967
42.2111
42.2111
{42.2111, 13.0034, 6.71209, 3.88702, 2.34521, 1.45084, 0.930164, 0.642843, 0.515239,
 0.515239, 0.642843, 0.930164, 1.45084, 2.34521, 3.88702, 6.71209, 13.0034, 42.2111}
```

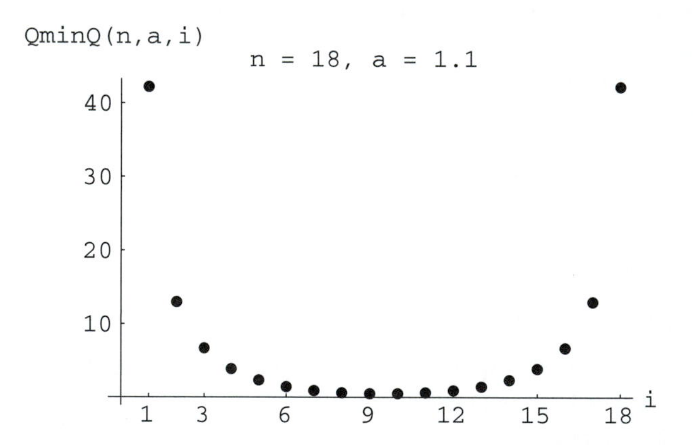


```
Plot[QminQ[18,a,1]
,{a,1.1,1.5}
,AxesLabel -> {"a", "QminQ(18,a,1)"}
];
```

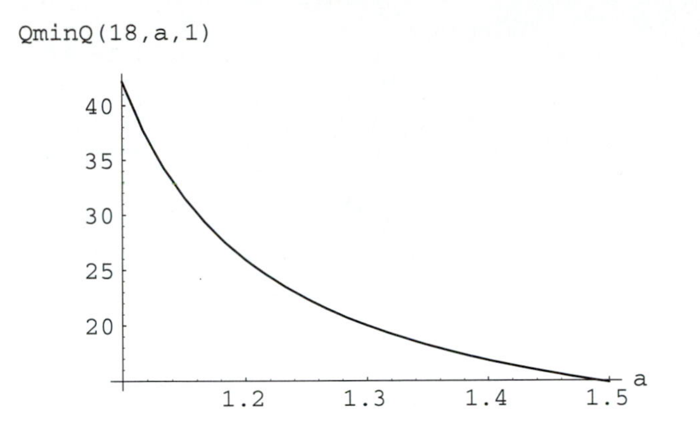

A.12.6 Find Selectivity Factor for Given Maximal Q-Factor

```
FindRoot[QminQ[18,a,1]==20, {a,1.1,1.5}]
```

```
{a -> 1.29948}
```

```
Plot[{QminQ[18,a,1],
      QminQ[16,a,1],
      QminQ[12,a,1],
      QminQ[9, a,1],
      QminQ[8, a,1],
      QminQ[6, a,1],
      QminQ[4, a,1],
      20,
      QminQ[3,a,1],
      QminQ[2,a,1]}
 ,{a,1.01,1.3}
 ,PlotRange -> {{0.99,1.3},{10,45}}
 ,PlotStyle -> {Dashing[{.04}],
                Dashing[{}],
                Dashing[{.02}]}
 ,AxesLabel -> {"a", "QminQ(n,a,1)"}
 ,Ticks->{{1.0,1.1,1.2,1.3},{10,20,30,40}}
 ,PlotLabel -> "n= 2,3,4,6,8,9,12,16,18"
 ];
```

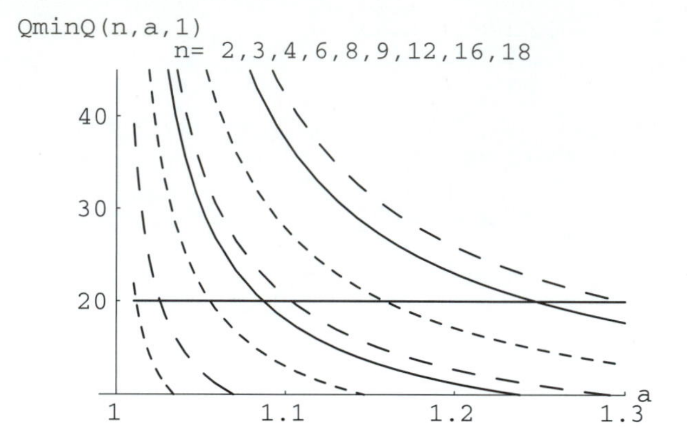


```
nminQ = {18,16,12,9,8,6,4,3,2};
aminQ = a /.
{FindRoot[QminQ[18,a,1]==20, {a,1.1,  1.3}],
 FindRoot[QminQ[16,a,1]==20, {a,1.1,  1.3}],
 FindRoot[QminQ[12,a,1]==20, {a,1.1,  1.3}],
 FindRoot[QminQ[9, a,1]==20, {a,1.1,  1.3}],
 FindRoot[QminQ[8, a,1]==20, {a,1.05, 1.1}],
 FindRoot[QminQ[6, a,1]==20, {a,1.04, 1.1}],
 FindRoot[QminQ[4, a,1]==20, {a,1.03, 1.1}],
 FindRoot[QminQ[3, a,1]==20, {a,1.02, 1.1}],
 FindRoot[QminQ[2, a,1]==20, {a,1.001,1.1}]
};
t = Transpose[{nminQ,aminQ}]
ListPlot[t
, PlotJoined -> True
, PlotRange -> {1,1.3}
, AxesLabel -> {"n", "a"}
, Ticks->{{2,3,4,6,8,9,12,16,18},{1,1.1,1.2,1.3}}
, PlotLabel -> "maximal Q-factor = 20"
];
```

```
{{18, 1.29948}, {16, 1.24774}, {12, 1.15906}, {9, 1.10338}, {8, 1.08654}, {6, 1.05495},
 {4, 1.02538}, {3, 1.01146}, {2, 1.00125}}
```

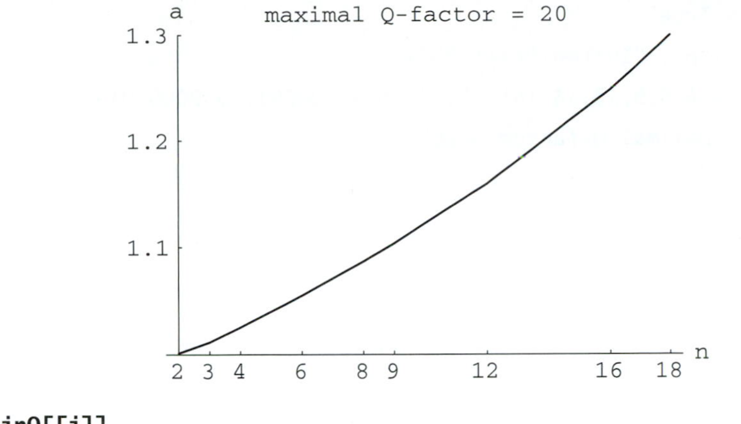


```
tL = Table[{nminQ[[i]],
           10*Log[10,1+L[nminQ[[i]],aminQ[[i]]]]}
     ,{i,1,9}
     ]
ListPlot[tL
, PlotJoined -> True
, PlotRange -> {0,110}
, AxesLabel -> {"n", "10*Log10(1+L(n,a))"}
, Ticks->{{2,3,4,6,8,9,12,16,18},{0,5,20,40,60,80,100}}
, PlotLabel -> "maximal Q-factor = 20"
];
```

```
{{18, 106.763}, {16, 89.5348}, {12, 58.386}, {9, 37.8317}, {8, 31.5114}, {6, 19.7043},
 {4, 9.39632}, {3, 5.4385}, {2, 3.23299}}
```

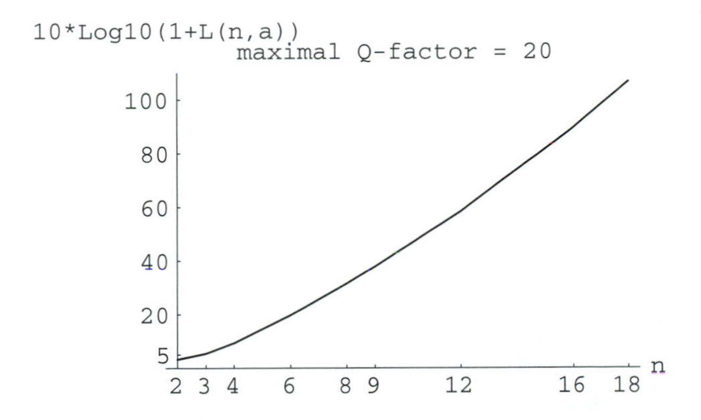

■ A.12.7 Ripple Factor e_minQ = 1 / Sqrt[L(n, a)]

```
te = Table[{nminQ[[i]],
           10*Log[10,1+1/L[nminQ[[i]],aminQ[[i]]]]}
     ,{i,1,9}
     ]
LogListPlot[te
```

```
, PlotJoined -> True
, AxesLabel -> {"n", "10*Log10(1+e^2)"}
, Ticks->{{2,3,4,6,8,9,12,16,18},{1,.1,.001,.00001,.000000001}}
, PlotLabel -> "maximal Q-factor = 20"
];
```

```
                  -11                    -9                    -6
{{18, 9.15089 10   }, {16, 4.83395 10  }, {12, 6.29779 10  }, {9, 0.00071557},
 {8, 0.0030676}, {6, 0.0467403}, {4, 0.530139}, {3, 1.46215}, {2, 2.79847}}
```

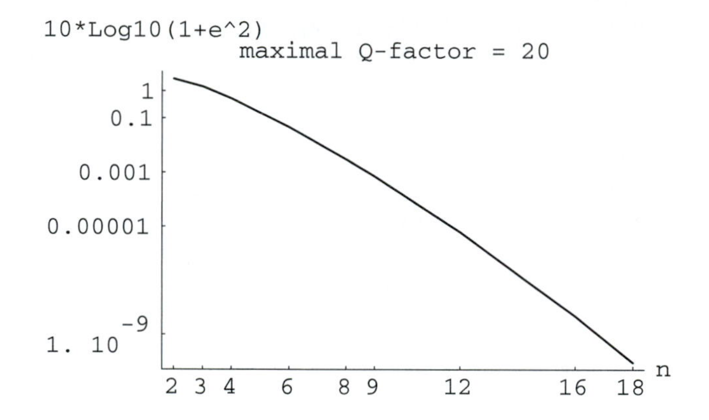

A.12.8 Pole Q-Factor in Terms of Ripple Factor

```
a1 = 1.1;
Plot[{Q[S[4,a1,e,1]],
      Q[S[3,a1,e,1]],
      Q[S[2,a1,e,1]]}
,{e,0.01,1}
,PlotRange -> {0,16}
,PlotStyle -> {Dashing[{.04}],
               Dashing[{}],
               Dashing[{.02}]}
,AxesLabel -> {"e", "Q(S(n,a,e,1))"}
,Ticks->{{0,0.2,0.4,0.6,0.8,1},{0,5,10,15}}
,PlotLabel -> "n= 2,3,4,  a=1.1"
];
```

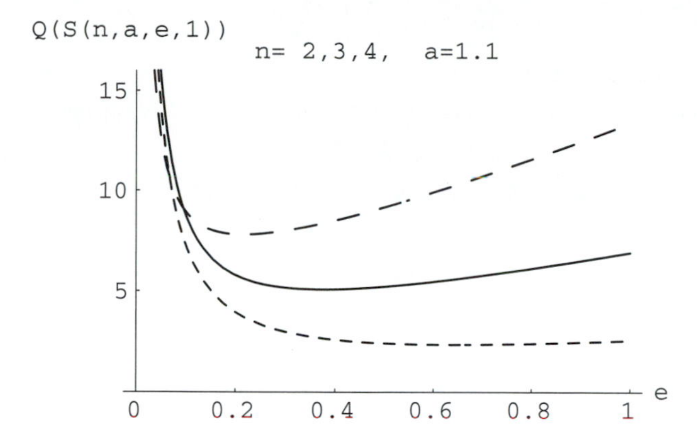

A.12.9 Minimal Q-Factor Transfer Function

```
HminQ[n_,a_,s_] := Block[{e,i,K,t},
 e = 1/Sqrt[L[n,a]];
 If[EvenQ[n],
   K = Product[X[n,a,i]^2,{i,1,n/2}]/(Sqrt[1+e^2]*Sqrt[a^n]);
   t = K*Product[(s^2+(a^2/X[n,a,i]^2))/(s^2+(Sqrt[a]/QminQ[n,a,i])*s+a)
         ,{i,1,n/2}];,
   K = Product[X[n,a,i]^2,{i,1,(n-1)/2}]/((s+Sqrt[a])*Sqrt[a^(n-2)]);
   t = K*Product[(s^2+(a^2/X[n,a,i]^2))/(s^2+(Sqrt[a]/QminQ[n,a,i])*s+a)
         ,{i,1,(n-1)/2}];
 ];
 t
];
a1 = 1.1;
H1 = HminQ[1,a1,s]
H2 = HminQ[2,a1,s]
H3 = HminQ[3,a1,s]
H4 = HminQ[4,a1,s]
H6 = HminQ[6,a1,s]
H8 = HminQ[8,a1,s];
H9 = HminQ[9,a1,s];
```

```
  1.04881
-----------
1.04881 + s
                          2
0.540094 (1.71408 + s )
-----------------------
                          2
 1.1 + 0.447214 s + s
```

```
                          2
      0.841917 (1.37031 + s )
-------------------------------------

                                  2
(1.04881 + s) (1.1 + 0.206892 s + s )
                           2              2
   0.210643 (1.29093 + s ) (4.34993 + s )
----------------------------------------------

                    2                     2
(1.1 + 0.134615 s + s ) (1.1 + 1.24828 s + s )
                              2              2              2
       0.0705794 (1.24336 + s ) (1.71408 + s ) (8.82646 + s )
----------------------------------------------------------------

                     2                      2                     2
(1.1 + 0.0806565 s + s ) (1.1 + 0.447214 s + s ) (1.1 + 1.63152 s + s )
```

```
H6w = H6 /. s->I*w;
```

■ A.12.10 Magnitude Response in Terms of Angular Frequency

```
Plot[Abs[H6w]
,{w,0,1}
,AxesLabel -> {"w", "|H6(jw)|"}
];
Plot[Abs[H6w]
,{w,1,a1}
,AxesLabel -> {"w", "|H6(jw)|"}
];
```

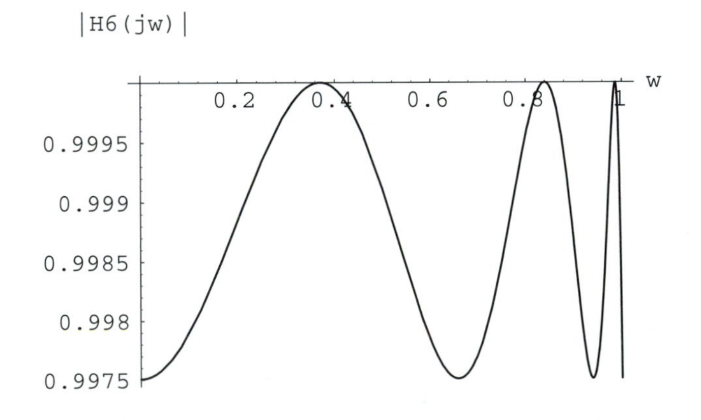

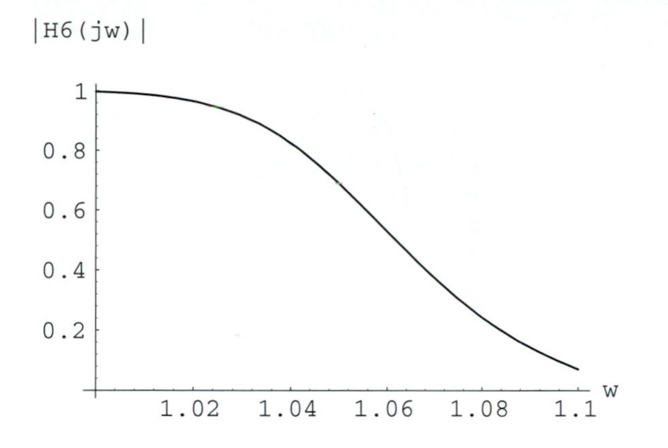


```
Plot[Abs[H6w]
,{w,a1,5*a1/4}
,AxesLabel -> {"w", "|H6(jw)|"}
];
Plot[Abs[H6w]
,{w,a1,3*a1}
,AxesLabel -> {"w", "|H6(jw)|"}
,PlotPoints->75
];
```

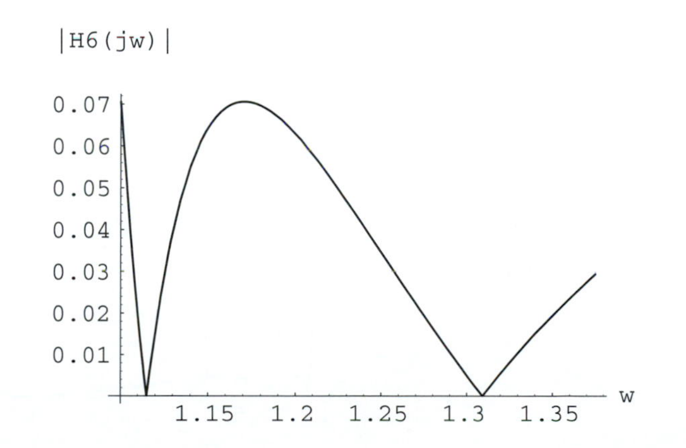

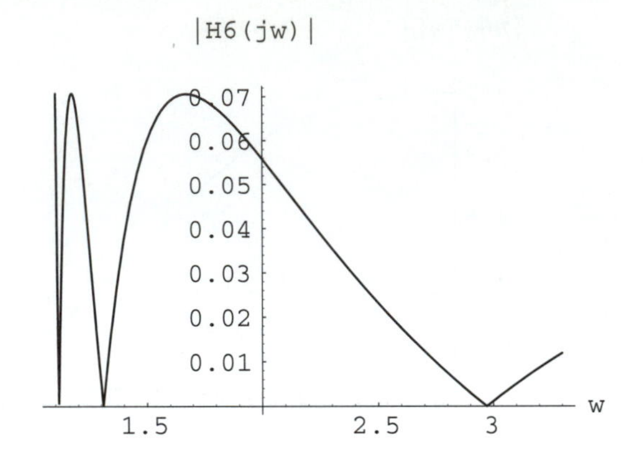


```
Plot[Abs[H6w]
,{w,0,3*al}
,AxesLabel -> {"w", "|H6(jw)|"}
];
Plot[Abs[H6w]
,{w,0,12*al}
,AxesLabel -> {"w", "|H6(jw)|"}
];
```

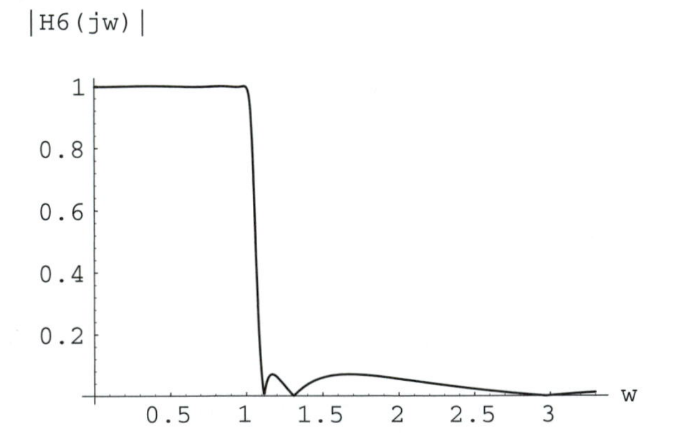

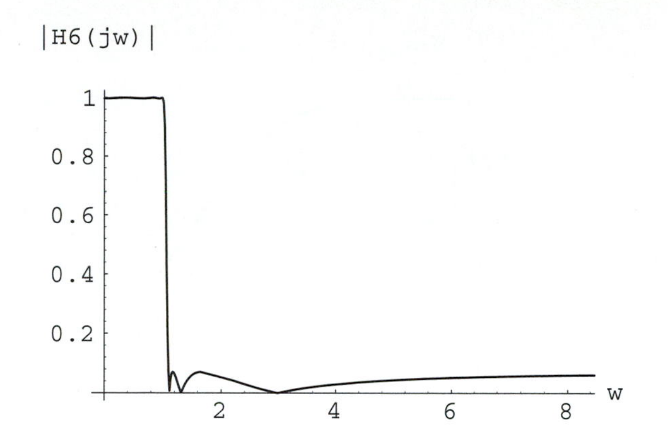


```
H9w = H9 /. s->I*w;
```

```
Plot[Abs[H9w]
,{w,0,1}
,AxesLabel -> {"w", "|H9(jw)|"}
];
```

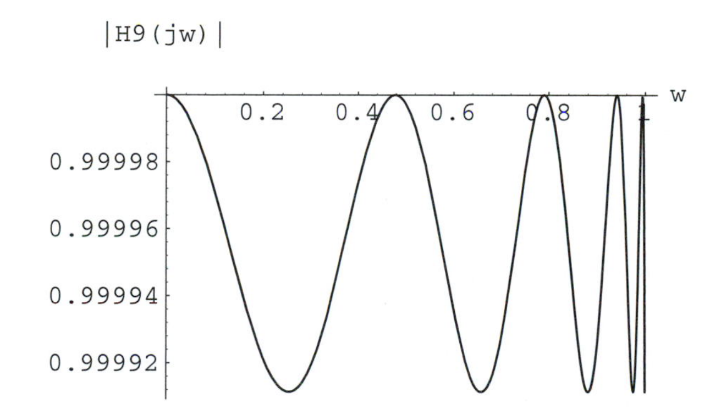


```
Plot[Abs[H9w]
,{w,1,a1}
,AxesLabel -> {"w", "|H9(jw)|"}
];
```

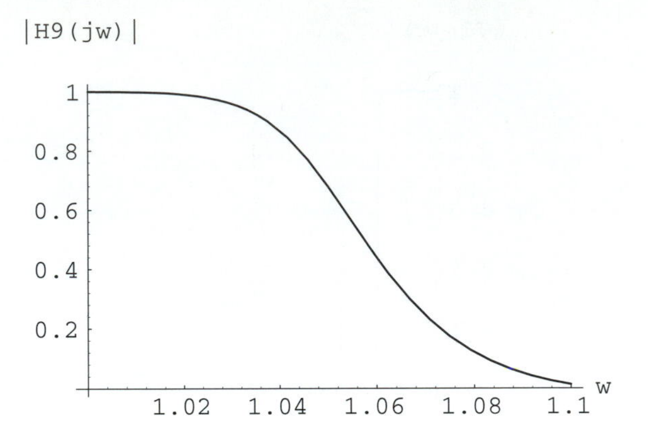


```
Plot[Abs[H9w]
,{w,a1,5*a1}
,AxesLabel -> {"w", "|H9(jw)|"}
];
```

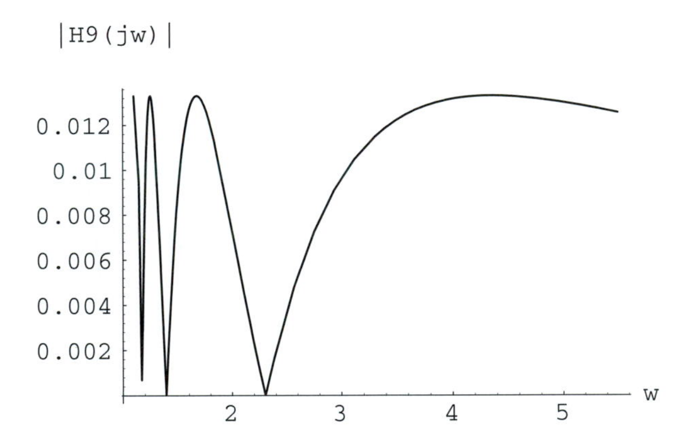


```
Plot[Abs[H9w]
,{w,0,5*a1}
,AxesLabel -> {"w", "|H9(jw)|"}
];
```

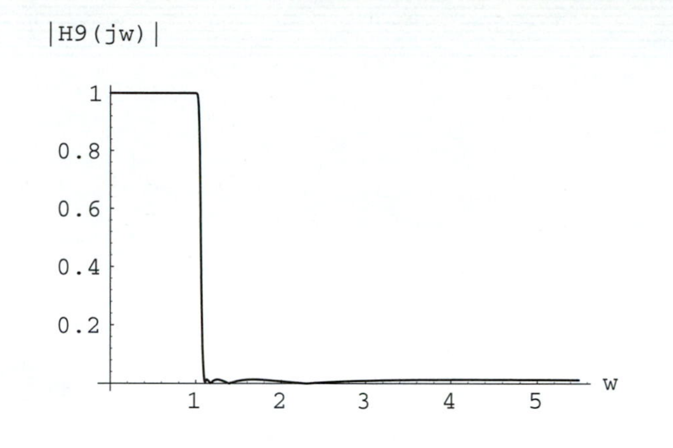
|H9(jw)|
1
0.8
0.6
0.4
0.2
1
2
3
4
5
w

REFERENCES

[1] *MATLAB Signal Processing Toolbox*, The MathWorks, Natick, MA, 1991.

[2] S. Wolfram, *Mathematica: A System for Doing Mathematics by Computer*, Addison-Wesley, Reading, MA, 1991.

[3] Charles A. Desoer and Ernest S. Kuh, *Basic Circuit Theory*, McGraw-Hill, New York, NY, 1969.

[4] M. A. Aizerman, *Theory of automatic control*, Addison-Wesley, Reading, MA, 1963.

[5] M. J. Lighthill, *Introduction to Fourier analysis and generalized functions*, Cambridge University Press, London, UK, 1958.

[6] Laurent Schwartz, *Theorie des distributions*, Hermann and Cie, Paris, France, 1957.

[7] John M. Novak and Brian L. Evans, *Mathematica, Signals and Systems*, Georgia Tech Research Corp., Atlanta, Georgia, 1995.

[8] Benjamin J. Leon, *Lumped Systems*, Holt, Rinehart and Winston, Inc., New York, NY, 1968.

[9] Paul Procter et al., *Longman dictionary of contemporary English*, Longman Group Limited, London, Great Britain, 1978.

[10] Alan V. Oppenheim and Alan S. Willsky, *Signals and Systems*, Prentice Hall, Englewood Cliffs, NJ, 1983.

[11] Sanjit K. Mitra and James F. Kaiser, *Handbook for digital signal processing*, John Wiley and Sons, New York, NY, 1993.

[12] Chi-Tsong Chen, *Introduction to Linear System Theory*, Holt, Rinehart and Winston, Inc., New York, NY, 1970.

[13] Lotfi A. Zadeh and E. Polak, *System Theory*, McGraw-Hill, New York, NY, 1969.

[14] H. Baher, *Analog and Digital Signal processing*, John Wiley and Sons, New York, NY, 1990.

[15] G. S. Moschytz and P. Horn, *Active filter design handbook*, John Wiley, New York, 1981.

[16] Ed. Wai-Kai-Chen, *The Circuits and Filters Handbook*, CRC Press, Boca Raton, Florida, 1995.

[17] Alexander D. Poularikas and Samuel Seely, *Signals and systems*, Prindle, Weber and Schmidt, Boston, MA, 1985.

[18] Robert D. Strum and Donald E. Kirk, *Contemporary Linear Systems Using MATLAB*, PWS Publishing Company, Boston, MA, 1994.

[19] D. V. Tošić, M. F. Hribšek, and B. D. Reljin, "Generation and design of new continuous-time second order gain equalizers using program SALEC," *International Journal of Electronics and Communications*, vol. 50, no. 3, pp. 226–229, May 1996.

[20] D. V. Tošić, B. D. Kovačević, and B. D. Reljin, "Symbolic analysis of linear dynamic systems," *Control and Computers*, vol. 24, no. 2, pp. 54–59, 1996.

[21] D. V. Tošić and M. D. Lutovac, "Symbolic computation of impulse, step and sine response of linear time-invariant systems," in *Proc. 10th Int. Symp. Theoretical Electrical Engineering, ISTET'99*, Magdeburg, Germany, Sept. 1999, pp. 653–657.

[22] Dejan V. Tošić, *A contribution to algorithms in computer-aided symbolic analysis of linear electric circuits and systems*, Doctoral dissertation, University of Belgrade, School of Electrical Engineering, Belgrade, Serbia, Yugoslavia, 14. December 1996.

[23] R. W. Daniels, *Approximation Methods for Electronic Filter Design*, McGraw–Hill, New York, 1974.

[24] G. Daryanani, *Principles of Active Network Synthesis and Design*, John Wiley, New York, 1976.

[25] M. S. Ghausi and K. R. Laker, *Modern filter design, active RC and switch capacitor*, Prentice–Hall, London, 1981.

[26] K. Geher, *Theory of Network Tolerances*, Akademiai Kiado, Budapest, 1971.

[27] D. Humpherys, *The analysis, design and synthesis of electrical filters*, Prentice–Hall, New York, 1970.

[28] G. S. Moschytz, *Linear Integrated Networks–Fundamentals*, Van Nostrand Reinhold Company, New York, 1974.

[29] G. S. Moschytz, *Linear Integrated Networks–Design*, Van Nostrand Reinhold Company, New York, 1975.

[30] R. Saal, *Handbuch zum filterenwurf*, AEG–Telefunken, Backnang, 1979.

[31] A.S. Sedra and P.O. Brackett, *Filter Theory and Design: Active and Passive*, Pitman, London, 1978.

[32] J.K. Skwirzynski, *Design Theory and Data for Electrical Filters*, Van Nostrand, London, 1965.

[33] K. Su, *Time Domain Synthesis of Linear Networks*, Prentice–Hall, New Jersey, 1971.

[34] J. Vlach, *Computerized Approximation and Synthesis of Linear Networks*, John Wiley, New York, NY, 1969.

[35] R. P. Sallen and E. L. Key, "A practical method of designing RC active filters," *IRE Trans. Circuit Theory*, vol. CT–2, no. 0, pp. 74–85, Mar. 1955.

[36] C. E. Cohn, "Note on the simulation of higher-order linear systems with single operational amplifier," *Proc. IEEE*, p. 874, July 1964.

[37] T. Deliyannis, "RC active allpass sections," *Electronics Letters*, vol. 5, no. 3, pp. 59–60, Feb. 1969.

[38] T. Deliyannis, "High-Q factor circuit with reduced sensitivity," *Electronics Letters*, vol. 4, no. 26, pp. 577–579, Dec. 1968.

[39] J. J. Friend, C. A. Harris, and D. Hilberman, "STAR:an active biquadratic filter section," *IEEE Trans. Circuits Systems*, vol. CAS–22, no. 2, pp. 115–121, Feb. 1975.

[40] W. B. Mikhael and B. B. Bhattacharyya, "A practical design for insensitive RC-active filters," *IEEE Trans. Circuits Systems*, vol. CAS–22, no. 5, pp. 407–415, May 1975.

[41] N. Fliege, "A new class of second-order RC-active filters with two operational amplifiers," *Nachrichtentechn Zeitung*, vol. 26, no. 6, pp. 279–282, 1973.

[42] W. J. Kerwin, L. P. Huelsman, and R. W. Newcomb, "State-variable synthesis for insensitive integrated circuit transfer functions," *IEEE J. Solid-State Circuits*, vol. SC–2, no. 3, pp. 87–92, Sept. 1967.

[43] National Semiconductor, *Introducing the MF10: A Versatile Monolithic Active Filter Building Block*, Santa Clara, CA, 1982.

[44] Linear Technology Corporation, *Monolithic Filter Handbook*, Milpitas, CA, 1990.

[45] MAXIM, *Analog Design Guide Services, Book 1*, Pangbourne Reading, 1992.

[46] M. D. Lutovac, D. Novaković, and I. Markoski, "Selective SC-filters with low passive sensitivity," *Electronics Letters*, vol. 33, no. 8, pp. 674–675, Apr. 1997.

[47] D. M. Rabrenović and M. D. Lutovac, "Elliptic filters with minimal Q-factors," *Electronics Letters*, vol. 30, no. 3, pp. 206–207, Feb. 1994.

[48] C. Toumazou, F.J. Lidgey, and D.G.Haigh, *Analogue IC Design: the Current-mode Approach*, Peter Petegrinus, London, United Kingdom, 1990.

[49] M. Biey and A. Premoli, "Low–pole–Q transfer functions for the design of selective active filters: a review," in *Proc. of 5th European Conf. Circuit Theory and Design*, Den Haag, Sept. 1981, pp. 272–279.

[50] T. W. Parks and C. S. Burrus, *Digital Filter Design*, John Wiley and Sons, New York, NY, 1987.

[51] M. D. Lutovac, D. V. Tošić, and B. L. Evans, "Algorithm for symbolic design of elliptic filters," in *4th International conference Symbolic Methods, Application, Circuit Design SMACD96*, Leuven, Belgium, Oct. 1996.

[52] M. D. Lutovac and D. M. Rabrenović, "Exact determination of the natural modes of some Cauer filters by means of a standard analytical procedure," *IEE Proc. Circuits Devices Syst.*, vol. 143, no. 3, pp. 134–138, June 1996.

[53] D. M. Rabrenović and M. D. Lutovac, "Minimum stopband attenuation of the Cauer filters without elliptic functions and integrals," *IEEE Trans. Circ. Systems: Part I*, vol. 40, no. 9, pp. 618–621, Sept. 1993.

[54] D. M. Rabrenović and M. D. Lutovac, "A Chebyshev rational function with low Q factors," *Int. J. Circuit Theory and Appl.*, vol. 19, no. 3, pp. 229–240, June 1991.

[55] M. D. Lutovac and D. Novaković, "Efficient low–sensitive selective SC–filters," in *Third IEEE Conf. Electronics, Circuits, Systems ICECS96*, Rodos, Greece, Oct. 1996, pp. 211–214.

[56] M. D. Lutovac, D. V. Tošić, and B. L. Evans, "Advanced filter design," in *IEEE Asilomar Conf. on Signal, Systems and Computer*, Asilomar, Nov. 1997, pp. 710–715.

[57] M. D. Lutovac, D. V. Tošić, and B. L. Evans, "Symbolic design and synthesis of digital IIR and analog filters," *Rev. Roum. Sci Techn. - Electrotechn. et Energ., Romania*, vol. 42, no. 2, pp. 229–233, Apr. 1997.

[58] D. V. Tošić, M. D. Lutovac, B. L. Evans, and I. M. Markoski, "A tool for symbolic analysis and design of analog active filters," in *Fifth International Workshop on Symbolic Methods and Applications in Circuit Design, SMACD'98*, Kaiserslautern, Germany, Oct. 1998, pp. 71–74.

[59] A. S. Sedra and L. Brown, "A refined classification of single amplifier filters," *Int. J. Circuit Theory and Appl.*, vol. 7, pp. 127–137, 1979.

[60] A. S. Sedra and J.L. Espinoza, "Sensitivity and frequency limitations of biquadratic active filters," *IEEE Trans., CAS*, vol. 22, pp. 122–130, 1975.

[61] T. Saramaki and K.-P. Estola, "Design of linear-phase partly digital anti-aliasing filters," in *Proc. IEEE Int. Conf. Acoust., Speech, and Signal Processing*, Tampa, FL, 1985, pp. 907–980.

[62] S. Lawson and T. Wicks, "Improved design of digital filters satisfying a combined loss and delay specification," *IEE Proceedings G: Circuits, Devices and Systems*, vol. 140, pp. 223–229, June 1993.

[63] S. Wright, "Convergence of SQP-like methods for constrained optimization," *SIAM Journal on Control and Optimization*, vol. 27, pp. 13–26, Jan. 1989.

[64] K. Schittkowski, "NLPQL: A Fortran subroutine solving constrained nonlinear programming problems," *Annals of Operations Research*, vol. 5, no. 1–4, pp. 485–500, 1986.

[65] S. Wolfram, *The Mathematica Book*, Champaign, IL: Wolfram Media Inc., 1996.

[66] Matlab 5, *User's Guide*, Natick, MA: The MathWorks Inc., 1997.

[67] A. Grace, *Optimization Toolbox*, Natick, MA: The MathWorks, Inc., 1992.

[68] N. Damera-Venkata and B. L. Evans, "An automatic framework for multicritera optimization of analog filter designs," *IEEE Trans. Circuits Syst.–II*, vol. 46, no. 8, pp. 981–990, Aug. 1999.

[69] D. V. Tošić, M. D. Lutovac, and I. M. Markoski, "Symbolic derivation of transfer functions of discrete-time systems," in *Proc. 9th Int. Symp. Theoretical Electrical Engineering ISTET97*, Palermo, Italia, June 1997, pp. 311–314.

[70] Sanjit K. Mitra and James F. Kaiser, *Handbook for Digital Signal Processing*, John Wiley and Sons, New York, NY, 1993.

[71] L. Gazsi, "Explicit formulas for lattice wave digital filters," *IEEE Trans. Circuits Syst.*, vol. CAS–32, pp. 68–88, Jan. 1985.

[72] A. Fettweis, H. Levin, and A. Sedlemeyer, "Wave digital lattice filters," *Int. J. Circuit Theory Appl.*, vol. 2, no. 2, pp. 203–211, June 1974.

[73] A. Fettweis, "Digital circuits and systems," *IEEE Trans. Circuits Syst.*, vol. CAS–31, no. 1, pp. 321–348, Jan. 1974.

[74] A. Fettweis, "Digital filter structures related to classical networks," *Arch. Elek. ubertragung*, vol. 25, no. 1, pp. 79–89, Feb. 1971.

[75] A. N. Willson and H. J. Orchard, "Insights into digital filters made as the sum of two allpass functions," *IEEE Trans. Circuits Syst.–I*, vol. 42, no. 3, pp. 129–137, Mar. 1995.

[76] M. D. Lutovac, D. V. Tošić, and I. M. Markoski, "Symbolic computation of elliptic rational functions," in *Fifth International Workshop on Symbolic Methods and Applications in Circuit Design, SMACD'98*, Kaiserslautern, Germany, Oct. 1998, pp. 177–180.

[77] Lj. Milić and M. D. Lutovac, "Multiplierless elliptic IIR filters with minimal coefficient–quantization error," in *3rd Internat. Conf. TELSIKS*, Niš, Yugoslavia, Oct. 1997, pp. 199–202.

[78] L. D. Milić and M. D. Lutovac, "Reducing the number of multipliers in the parallel realization of half-band elliptic IIR filters," *IEEE Trans. Signal Processing*, vol. 44, no. 10, pp. 2619–2623, Oct. 1996.

[79] Lj. Milić and M. Lutovac, "Design of multiplierless elliptic IIR filters with a small quantization error," *IEEE Trans. Signal Processing*, vol. 47, no. 2, pp. 469–479, Feb. 1999.

[80] M. D. Lutovac and L. D. Milić, "Short wordlength selective IIR filters," in *Proc. of European Conference on Circuit Theory and Design, ECCTD'99*, Stresa, Italy, Aug. 1999, pp. 277–280.

[81] D. V. Tošić, M. D. Lutovac, and B. L. Evans, "Advanced digital IIR filter design," in *Proc. of European Conference on Circuit Theory and Design, ECCTD'99*, Stresa, Italy, Aug. 1999, pp. 1323–1326.

[82] R. Crochiere and R. Lawrence, *Multirate Digital Signal Processing*, Prentice-Hall, New Jersey, 1983.

[83] S. Powell and M. Chau, "A technique for realizing linear phase IIR filters," *IEEE Trans. Signal Process.*, vol. 39, no. 11, pp. 2425–2435, Nov. 1991.

[84] H. J. Orchard, "Adjusting the parameters in elliptic–function filters," *IEEE Trans. on CAS*, vol. 37, pp. 631–633, May 1990.

[85] M. D. Lutovac and Lj. D. Milić, "Elliptic half-band IIR filters," *Facta Universitatis, Yugoslavia*, vol. 9, no. 1, pp. 43–59, June 1996.

[86] M. D. Lutovac and Lj. D. Milić, "Lattice wave digital filters with a reduced number of multipliers," *Yugoslav IEEE MTT Chapter Inform.*, vol. 2, no. 3, pp. 29–39, June 1996.

[87] Lj. D. Milić, M. D. Lutovac, and D. M. Rabrenović, "Facilities in design and implementation of digital Butterworth and elliptic filters," in *Proc. 12th European Conference Circuit Theory Design, ECCTD95*, Istanbul, Aug. 1995, pp. 549–552.

[88] P. A. Regalia, "Special filter design," in S.K.Mitra and J.F.Kaiser, *Handbook of Digital Signal Processing*, John Wiley, 1993, pp. 907–980.

[89] C. M. Rader, "A simple method for sampling in-phase and quadrature components," *IEEE Trans. Aerospace Electronic systems*, vol. 20, no. 6, pp. 821–824, Nov. 1984.

[90] A. G. Dempster and M. D. Macleod, "Multiplier blocs and complexity of IIR structures," *Electronics Letters*, vol. 30, no. 22, pp. 1841–1843, Oct. 1994.

[91] A. G. Dempster, *Digital filter design for low-complexity implementation*, Cambrige University Engineering Department, PhD thesis, 1995.

[92] M. D. Lutovac and L. D. Milić, "Design of computationally efficient elliptic IIR filters with a reduced number of shift-and-add operations in multipliers," *IEEE Trans. Signal Processing*, vol. 45, no. 10, pp. 2422–2430, Oct. 1997.

[93] L. Milić and M. D. Lutovac, "Low-complexity lattice wave digital filters," in *Proc. 13th European Conf. Circuit Theory Design, ECCTD97*, Budapest, Hungary, Aug. 1997, pp. 1141–1146.

[94] Lj. Milić and M. D. Lutovac, "Design of multiplierless IIR filters," in *IEEE Int. Conf. Acoustics, Speech, Signal Processing, ICASSP97*, Munich, Germany, Apr. 1997, pp. 2201–2204.

[95] R. Ansari and B. Liu, "A class of low-noise computationally efficient recursive digital filters with applications to sampling rate alterations," *IEEE Trans. Acoust., Speech, Signal Process.*, vol. ASSP–33, no. 1, pp. 90–97, Feb. 1985.

[96] R. Boite, H. Dubois, and H. Leich, "Optimisation of digital filters in the discrete space of coefficients," *Electronics Letters*, vol. 10, no. 10, pp. 179–180, May 1974.

[97] S. J. Varoufakis and A. N. Venetsanopoulos, "Fast optimum design of IIR digital filters with finite precision coefficients," *Int. J. Electronics*, vol. 57, no. 2, pp. 207–216, 1984.

[98] Z. Jing and A. T. Fam, "A new scheme for designing IIR filters with finite wordlength coefficients," *IEEE Trans. Acoust., Speech, Signal Process.*, vol. ASSP–34, no. 5, pp. 1335–1336, Oct. 1986.

[99] A. G. Dempster and M. D. Macleod, "Variable statistical wordlength in digital filters," *IEE Proc. - Vision, Image, Signal Processing*, vol. 143, no. 1, pp. 62–66, Feb. 1996.

[100] A. G. Dempster and M. D. Macleod, "Variable wordlength in filters with least-squares error criterion," *Electronics Letters*, vol. 33, no. 3, pp. 201–202, Jan. 1997.

[101] P. P. Vaidyanathan, S. K. Mitra, and Y. Neuvo, "A new approach to the realization of low-sensitivity IIR digital filters," *IEEE Trans. Acoust., Speech, Signal Process.*, vol. ASSP–34, no. 2, pp. 350–361, Apr. 1986.

[102] A. G. Dempster and M. D. Macleod, "Constant integer multiplication using minimum adders," *IEE Proc. - Circuits, Devices, Systems*, vol. 141, no. 5, pp. 407–413, Oct. 1994.

[103] J. J. Kormylo and V. K. Jain, "Two-pass recursive digital filter with zero phase shift," *IEEE Trans. Acoust., Speech, Signal Processing*, vol. ASSP–30, pp. 384–387, Oct. 1974.

[104] A. N. Willson and H. J. Orchard, "An improvement to the Powell and Chau linear phase IIR filters," *IEEE Trans. Signal Processing*, vol. SP–42, no. 10, pp. 2842–2848, Oct. 1994.

[105] B. Djokić, M. Popović, and M. Lutovac, "A new improvement to the Powell and Chau linear phase IIR filters," *IEEE Trans. Signal Processing*, vol. 46, no. 6, pp. 1685–1688, June 1998.

[106] S. Nordebo and Z. Zang, "Semi-infinite linear programming: A unified approach to digital filter design with time- and frequency-domain specifications," *IEEE Trans. Circuits Syst.–II*, vol. 46, no. 6, pp. 765–775, June 1999.

[107] P. Gill, W. Murray, and M. Wright, *Numerical Linear Algebra and Optimization*, Addison-Wesley, 1990.

[108] Andreas Antoniou, *Digital Filters: Analysis and Design*, McGraw-Hill, New York, NY, 1979.

[109] M. Abramowitz and I. Stegun, *Handbook of Mathematical Functions*, Dover, New York, 1972.

[110] W. W. Bell, *Special functions for scientists and engineers*, Van Nostrand Reinhold Company, London, UK, 1968.

[111] M. D. Lutovac and D. M. Rabrenović, "Algebraic design of some lower-order elliptic filters," *Electronics Letters*, vol. 29, no. 2, pp. 192–193, Jan. 1993.

[112] A. J. Grosman, "Synthesis of tschebycheff parameter symmetrical filters," *Proc. IRE*, pp. 454–473, Apr. 1957.

[113] M. D. Lutovac and D. M. Rabrenović, "Determination of the natural modes of some Cauer filters without elliptic functions," in *Proc. 12th European Conference Circuit Theory Design, ECCTD95*, Istanbul, Aug. 1995, pp. 611–614.

[114] M. D. Lutovac and D. M. Rabrenović, "Comments on the effects of lower Q values on the filters having equal ripples in the passband," *IEEE Trans. Circ. Systems: Part I*, vol. 40, no. 4, pp. 294–295, Apr. 1993.

INDEX

C

D

E

F

G

H

I

J

K

L

S

T

U

V

W

Z